CROP PHYSIOLOGY

APPLICATIONS FOR GENETIC IMPROVEMENT AND AGRONOMY

SECOND EDITION

CROP PHYSIOLOGY

APPLICATIONS FOR GENETIC IMPROVEMENT AND AGRONOMY

SECOND EDITION

Edited by

VICTOR O. SADRAS
SARDI, Adelaide, Australia

DANIEL F. CALDERINI
Universidad Austral de Chile, Valdivia, Chile

AMSTERDAM • BOSTON • HEIDELBERG • LONDON • NEW YORK • OXFORD • PARIS
SAN DIEGO • SAN FRANCISCO • SINGAPORE • SYDNEY • TOKYO

Academic Press is an Imprint of Elsevier

Academic Press is an imprint of Elsevier
32 Jamestown Road, London NW1 7BY, UK
225 Wyman Street, Waltham, MA 02451, USA
525 B Street, Suite 1800, San Diego, CA 92101-4495, USA

Cover Photo. Widespread use of plastic mulch in spring maize crops in the Loess Plateau (see Chapter 3). This photo was previously published in Zhang, S., Sadras, V.O., Chen, X., Zhang, F., 2014. Water use efficiency of dryland maize in the Loess Plateau of China in response to crop management. Field Crops Research 163, 55–63.

British Library Cataloguing-in-Publication Data
A catalogue record for this book is available from the British Library

Library of Congress Cataloging-in-Publication Data
A catalog record for this book is available from the Library of Congress

ISBN : 978-0-12-4171046

For information on all Academic Press publications
visit our website at http://store.elsevier.com

Typeset by Thomson Digital
visit our website at www.thomsondigital.com

Printed and bound in United States of America
Transferred to Digital Printing, 2014

Contents

List of contributors

L. Gabriela Abeledo Department of Crop Production, University of Buenos Aires, Buenos Aires, Argentina

Luis Aguirrezábal Unidad Integrada Balcarce (Facultad de Ciencias Agrarias, Universidad Nacional de Mar del Plata – Instituto Nacional de Tecnología Agropecuaria, Estación Experimental Balcarce, Balcarce, Argentina

Fernando H. Andrade INTA-Universidad de Mar del Plata, CONICET, Argentina

Maria L. Appendino University of Buenos Aires, Buenos Aires, Argentina

Senthold Asseng University of Florida, Gainesville, USA

Delfina Barabaschi Consiglio per la Ricerca e la sperimentazione in Agricoltura – Genomics Research Centre, Fiorenzuola d'Arda, Italy

Lucas Borrás CONICET & University of Rosario, Rosario, Santa Fe, Argentina

Grazia M. Borrelli Consiglio per la Ricerca e la sperimentazione in Agricoltura – Cereal Research Centre, Foggia, Italy

Helen Bramley University of Western Australia, Crawley, Australia

Timothy J. Brodribb University of Tasmania, Australia

Daniel F. Calderini Universidad Austral de Chile, Instituto de Produccion y Sanidad Vegetal, Valdivia, Chile

Kenneth G. Cassman University of Nebraska, Department of Agronomy and Horticulture, USA

Sebastián Castro Universidad de Mar del Plata, Argentina

Luigi Cattivelli Consiglio per la Ricerca e la sperimentazione in Agricoltura – Genomics Research Centre, Fiorenzuola d'Arda, Italy

Karine Chenu The University of Queensland, Queensland Alliance for Agriculture and Food Innovation (QAAFI), Toowoomba, Queensland, Australia

Ignacio Ciampitti Kansas State University, Department of Agronomy, USA

C. Mariano Cossani CIMMYT (International Maize and Wheat Improvement Center), Mexico, DF, Mexico

Pasquale De Vita Consiglio per la Ricerca e la sperimentazione in Agricoltura – Cereal Research Centre, Foggia, Italy

Philippe Debaeke INRA, UMR1248 Agroécologie, Innovations, Territoires, Castanet-Tolosan; INPT, Université Toulouse, UMR1248 Agroécologie, Innovations, Territoires, Toulouse, France

Aixing Deng Institute of Crop Sciences, Chinese Academy of Agricultural Sciences, Beijing, China

R. Ford Denison Department of Ecology, Evolution and Behavior, University of Minnesota, USA

John Dimes Queensland Alliance for Agriculture and Food Innovation (QAAFI), The University of Queensland, Toowoomba, QLD, Australia

Jean-Louis Durand Institut National de la Recherche Agronomique (INRA), Lusignan, France

María Mercedes Echarte Unidad Integrada Balcarce (Facultad de Ciencias Agrarias, Universidad Nacional de Mar del Plata – Instituto Nacional de Tecnología Agropecuaria, Estación Experimental Balcarce, Balcarce, Argentina

M.J. Foulkes University of Nottingham, School of Biosciences, Nottingham, UK

François Gastal Institut National de la Recherche Agronomique (INRA), Lusignan, France

Patricio Grassini University of Nebraska, Department of Agronomy and Horticulture, USA

Kaija Hakala MTT Agrifood Research Finland, Plant Production, Jokioinen, Finland

Zhonghu He Institute of Crop Sciences, Chinese Academy of Agricultural Sciences, Beijing; International Maize and Wheat Improvement Center (CIMMYT) China Office, Beijing, China

Meisha-Marika Holloway-Phillips Australian National University, Australia

Natalia Izquierdo Unidad Integrada Balcarce (Facultad de Ciencias Agrarias, Universidad Nacional de Mar del Plata – Instituto Nacional de Tecnología Agropecuaria, Estación Experimental Balcarce, Balcarce, Argentina

Hannu Känkänen MTT Agrifood Research Finland, Plant Production, Jokioinen, Finland

Adriana G. Kantolic University of Buenos Aires, Buenos Aires, Argentina

Gilles Lemaire Institut National de la Recherche Agronomique (INRA), Lusignan, France

Alberto León Advanta Seeds, Argentina

X. Carolina Lizana Universidad Austral de Chile, Instituto de Produccion y Sanidad Vegetal, Valdivia, Chile

Gaëtan Louarn Institut National de la Recherche Agronomique (INRA), Lusignan, France

Delphine Luquet CIRAD, Department of Biological Systems, UMR1334 Genetic Improvement and Adaptation of Mediterranean and Tropical Plants, Montpellier, France

Gustavo A. Maddonni IFEVA-CONICET & University of Buenos Aires, Buenos Aries, Argentina

Pierre Martre INRA, UMR1095 Genetics, Diversity and Ecophysiology of Cereals, Clermont-Ferrand; Blaise Pascal University, UMR1095 Genetics, Diversity and Ecophysiology of Cereals, Aubière, France

Anna M. Mastrangelo Consiglio per la Ricerca e la sperimentazione in Agricoltura – Cereal Research Centre, Foggia, Italy

Mario Mera Instituto de Investigaciones Agropecuarias (INIA); Universidad de La Frontera, Temuco, Chile

Daniel J. Miralles University of Buenos Aires, Buenos Aires, Argentina; IFEVA-CONICET, Buenos Aires, Argentina.

Luigi Orrù Consiglio per la Ricerca e la sperimentazione in Agricoltura – Genomics Research Centre, Fiorenzuola d'Arda, Italy

Maria E. Otegui IFEVA-CONICET & University of Buenos Aires, Buenos Aries, Argentina

Helen Ougham IBERS, Edward Llwyd Building, Aberystwyth University, Ceredigion, UK

Mohammed-Mahmoud Ould-Sidi Memmah INRA, UR1115, Plantes et Systèmes de Culture Horticoles, Avignon, France

Pirjo Peltonen-Sainio MTT Agrifood Research Finland, Plant Production, Jokioinen, Finland

Gustavo Pereyra-Irujo Unidad Integrada Balcarce (Facultad de Ciencias Agrarias, Universidad Nacional de Mar del Plata – Instituto Nacional de Tecnología Agropecuaria, Estación Experimental Balcarce, Balcarce, Argentina

Ana C. Pontaroli INTA-Universidad de Mar del Plata, CONICET, Argentina

Andries Potgieter Queensland Alliance for Agriculture and Food Innovation (QAAFI), The University of Queensland, Toowoomba, QLD, Australia

Bénédicte Quilot-Turion INRA, UR1052 Genetics and Improvement of Fruit and Vegetables, Avignon, France

Ari Rajala MTT Agrifood Research Finland, Plant Production, Jokioinen, Finland

M.P. Reynolds CIMMYT Wheat Program, Mexico

Daniel Rodriguez Queensland Alliance for Agriculture and Food Innovation (QAAFI), The University of Queensland, Toowoomba, QLD, Australia

Victor O. Sadras South Australian Research & Development Institute, Waite Campus, Adelaide, Australia

Rodrigo G. Sala Monsanto, Argentina

Roxana Savin Department of Crop and Forest Sciences & AGROTECNIO (Center for Research in Agrotechnology), University of Lleida, Lleida, Spain

Gustavo A. Slafer Department of Crop and Forest Sciences & AGROTECNIO (Center for Research in Agrotechnology), University of Lleida, Lleida, Spain; ICREA (Catalonian Institution for Research and Advanced Studies)

Zhenwei Song Institute of Crop Sciences, Chinese Academy of Agricultural Sciences, Beijing, China

James E. Specht University of Nebraska, Department of Agronomy and Horticulture, USA

Howard Thomas IBERS, Edward Llwyd Building, Aberystwyth University, Ceredigion, UK

Matthijs Tollenaar Monsanto Company, USA

Gabriela Tranquilli Instituto de Recursos Biológicos, INTA, Argentina

Enli Wang CSIRO, Australia

Weijian Zhang Institute of Crop Sciences, Chinese Academy of Agricultural Sciences, Beijing, China

Chengyan Zheng Institute of Crop Sciences, Chinese Academy of Agricultural Sciences, Beijing, China

Yan Zhu Nanjing Agricultural University, China

Preface

In the most current and rigorous statistical analysis of yield trends, Grassini et al. (2013) found 'widespread deceleration in the relative rate of increase of average yields of the major cereal crops during the 1990–2010 period in countries with greatest production of these crops, and strong evidence of yield plateaus or an abrupt drop in rate of yield gain in 44% of the cases which, together, account for 31% of total global rice, wheat and maize production'. This highlights the urgency to increase crop yield which, in turn, requires allocation of research and development resources where significant returns are more likely in the next two decades. The time lag from innovation to adoption adds to this urgency (Hall and Richards, 2013). Whereas we cannot predict the discovery of the unknown (Osmond, 1995), it is also evident, as expressed by Tony Fischer, that we cannot afford expensive distractions.

In scales from molecular to field, we can ask three types of questions (Sadras and Richards, 2014). First, we can ask: *how does it work?* This elicits proximal, physiological answers that require scaling down from the level where the question is asked; for example, to understand crop flowering patterns we need to consider the major photoperiod, earliness *per se* and vernalization genes and the epigenetic modulation of gene expression. Next, we can ask *why?* In our example, why do cereals have a particular set of vernalization or photoperiod responsive genes in the first place? This elicits evolutionary explanations (e.g. Rhone et al., 2010). Finally, we can ask: *so what?* This refers to agronomic relevance; following on our example, a certain genetic makeup, combined with environmental and agronomic information, allows the modeling of flowering time to inform agronomic and breeding applications. The *so what* question requires scaling up to the crop level where yield is defined, hence the importance in understanding crop-level processes.

In common with the first edition, the objective of this book is to provide a contemporary appreciation of crop physiology as a mature scientific discipline. We want to show that crop physiology is relevant to agriculture. Progress in agriculture, however, depends directly on progress in agronomy, plant breeding, and their interaction. Hence crop physiology can contribute to agriculture only to the extent that it is meaningfully engaged with breeding and agronomy; this is the theme of this book. Eight chapters are updated versions of the first edition and twelve chapters are new, collectively presenting original viewpoints on the carbon, nitrogen and water economy of crops, plant hydraulics, biological nitrogen fixation, and senescence; crop phenological development, quantitative environmental characterization, ideotypes and phenotyping; yield potential and grain quality; the role of biotechnology; the impact of climate change; and case studies of cropping systems of China, Northern Europe, Africa and South and North America.

V.O. Sadras, Adelaide
D.F. Calderni, Valdivia

References

Grassini, P., Eskridge, K.M., Cassman, K.G., 2013. Distinguishing between yield advances and yield plateaus in historical crop production trends. Nat Commun. 4, Article 2918.

Hall, A.J., Richards, R.A., 2013. Prognosis for genetic improvement of yield potential and water-limited yield of major grain crops. Field Crops Res. 143, 18–33.

Osmond, C., 1995. Quintessential inefficiencies of plant bioenergetics: Tales of two cultures. Aust. J. Plant Physiol. 22, 123–129.

Rhone, B., Vitalis, R., Goldringer, I., Bonnin, I., 2010. Evolution of flowering time in experimental wheat populations: A comprehensive approach to detect genetic signatures of natural selection. Evolution 64, 2110–2125.

Sadras, V.O., Richards, R.A., 2014. Improvement of crop yield in dry environments: benchmarks, levels of organisation and the role of nitrogen. J. Exp. Bot. 65, 1981–1995.

Acknowledgments

We are grateful to the fundamental contribution of all the authors; those who updated their chapters through a detailed review of new information and those who wrote new chapters for the 2nd edition. We wish to express our gratitude to all of them for their willingness, knowledge and time.

We thank our host organizations, the South Australian Research and Development Institute and Universidad Austral de Chile, for their support to this project.

We thank Elsevier for inviting us to produce the 2nd edition of *Crop Physiology. Application for Genetic Improvement and Agronomy,* especially to Nancy Maragioglio (Senior Acquisitions Editor of Academic Press) for her encouragement and support, and Carrie Bolger and Caroline Johnson for their professional advice.

Throughout the book, key concepts developed by the authors are supported with material previously published in several journals; we thank the publishers who permitted reproduction of their material.

To Ana and Magda, for their love, support and patience to the editors.

CHAPTER

1

Crop physiology: applications for breeding and agronomy

Victor O. Sadras[1], *Daniel F. Calderini*[2]

[1]South Australian Research & Development Institute, Waite Campus, Adelaide, Australia
[2]Universidad Austral de Chile, Instituto de Produccion y Sanidad Vegetal, Chile

organization levels cut Nature at its joints
W.C. Wimsatt

1 INTRODUCTION

The economic, social and environmental challenges of global agriculture have been dissected thoroughly in the context of the urgent need to increase food production (Fischer et al., 2014). Three complementary factors contribute directly to larger agricultural output: superior varieties, better agronomic practices for growing individual crops, and enhanced spatial and temporal arrangement of the components of the farming system. Breeding, agronomy and farming systems are therefore the three disciplines with direct impact on agricultural output. The synergy between breeding and agronomy has been widely recognized (Evans, 2005; Fischer, 2009). Synergies between breeding, agronomy and farming systems are also important. One example is the combination of no-till technology, glyphosate-resistant soybean, imidazoline-resistant sunflower and Interfield® herbicide that allow for soybean–sunflower intercropping in high-tech production systems (Calviño and Monzon, 2009). Cotton- and maize-based cropping systems had to be redesigned to allow for the deployment of transgenic crops expressing Bt-proteins for the control of lepidopteran pests (Carroll et al., 2012; Downes and Mahon, 2012; Head and Greenplate, 2012; Siegfried and Hellmich, 2012).

Disciplines including crop physiology, genetics, entomology, pathology, edaphology, hydrology, climate science and so forth are one step removed from, and contribute to, agricultural production only to the extent that they engage with breeding, agronomy and farming system research. Whereas curiosity-driven research remains a legitimate motivation for science, engagement with the proximate drivers of agricultural production is essential for relevance. Heinze (2012) highlighted how 'scientific creativity springs from the fundamental tension between originality and scientific relevance'.

Crop Physiology. DOI: 10.1016/B978-0-12-417104-6.00001-7

It could be argued that any good piece of science will be relevant, sooner or later; the history of science abounds on apparently irrelevant curiosities that supported transforming technologies many years after their discovery (Hellemans and Bunch, 1988). Owing to the pressing challenges of agricultural production on a global scale, however, a definition of relevance must account for the short- to mid-term challenges. In the design of more productive cropping systems, we need a sharp focus and allocation of resources where larger impacts in the next two decades are more likely; in the terms of Tony Fischer, we cannot afford 'expensive distractions'. The dual motivation of scientific insight and agricultural relevance is at the core of this book; to ensure relevance, insights from crop physiology, genetics, modeling and other disciplines are thus linked with breeding, agronomy and/or farming systems.

2 LEVELS OF ORGANIZATION AND SCALABILITY

Biological sciences are the foundation of contemporary agriculture. Following Wimsatt (1994), compositional levels of organization can be defined as hierarchical divisions of biological entities 'organized by part-whole relations in which wholes at one level function as parts at the next (and all higher) levels'; furthermore, levels 'are constituted by families of entities usually of comparable size and dynamical properties, which characteristically interact primarily with one another'. Notwithstanding cross-level interactions, this perspective emphasizes that molecules interact mostly with molecules (Rietman et al., 2011) and plants with plants (Harper, 1977). This partially accounts for the compartmentalized theories and methods associated with each level (Sadras and Richards, 2014). Rates of key processes decline from low to high levels, partially because it takes longer for casual effects to propagate larger distances (Wimsatt, 1994). For agriculture, relevant levels of organization include molecule, organelle, cell, tissue, organ, individual, population and community; ecosystem, biome and biosphere are particularly relevant to other aspects of agriculture such as biodiversity (Fig. 1.1).

Levels of organization have therefore distinct components, sizes, rates, interactions, dynamics, methods and theories (Sadras and Richards, 2014). Consequently, a biotechnologist working in crop improvement is in many regards closer to a biotechnologist working in medicine than to the crop physiologist, whereas a crop physiologist is closer to an ecologist than to the plant biotechnologist; organization levels cut Nature at its joints (Wimsatt, 1976).

In the hierarchy of biological organization from molecular to global (Fig. 1.1), we ask questions at a given level (e.g. how do plants absorb water from soil?), find answers at lower levels (e.g. root and leaf morphology and physiology down to the molecular structure of aquaporins) and seek for relevance at higher levels (e.g. implications for crop water management). The crop level is therefore a reference for relevance at lower organization levels (plant to molecular) and a platform for explanations at higher organization levels (paddock to global).

Sadras and Richards (2014) proposed an operational definition in the context of crop improvement: a trait scales up if it remains agronomically relevant at higher levels, and eventually at the population level where yield is defined. They classified crop traits in three categories: those that generally scale up (e.g. herbicide resistance), those that do not (e.g. grain yield), and traits that might scale up provided they are considered in an integrated manner with scientifically sound scaling assumptions, appropriate growing conditions and screening techniques (e.g. stay-green). Traits for which scaling depends on higher-level interactions are challenging. The contrast between traits that scale up from molecular to crop level and traits that do not, including many aspects of

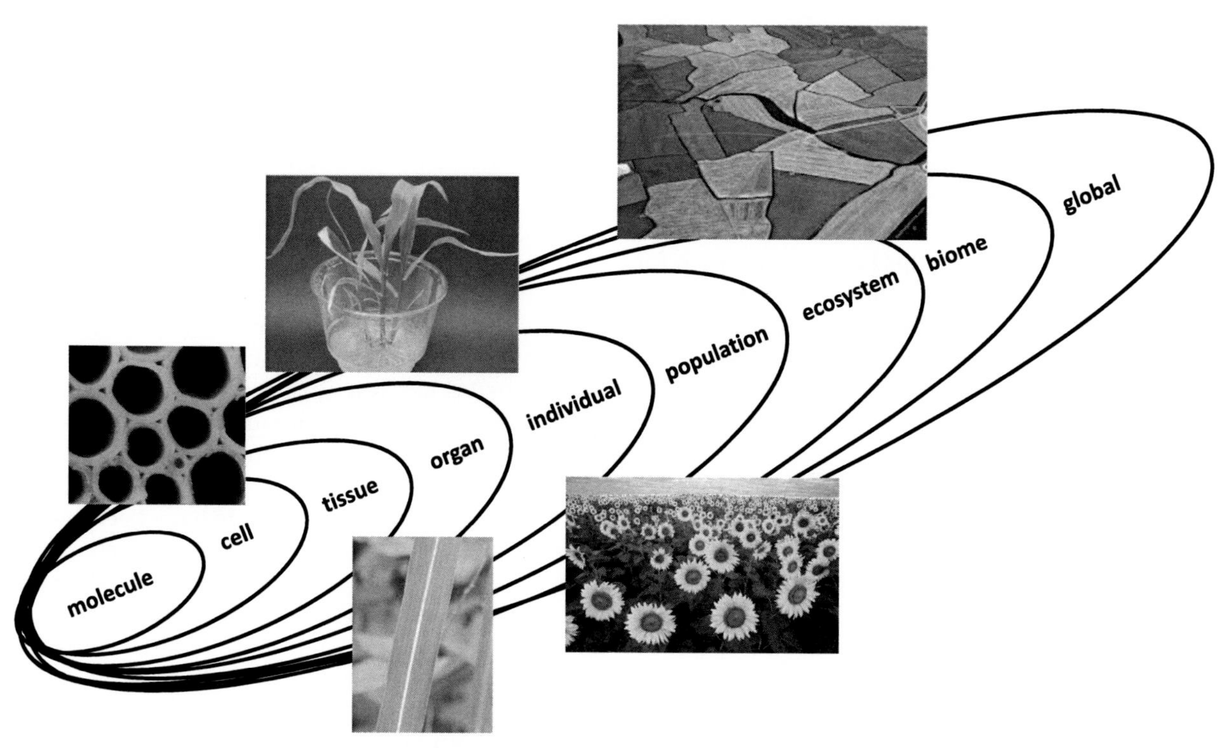

FIG. 1.1 Nested hierarchy of biological scales relevant to agriculture. Source: Sadras and Richards (2014).

crop morphology, photosynthesis, growth and yield (Fig. 1.2) partially accounts for the success of biotechnology in crop protection and its limited impact on yield potential and crop adaptation to abiotic stress (Sadras and Richards, 2014). In Chapter 10 of this book, Thomas and Ougham nicely illustrate the problem of unconditional extrapolation; considering that the net carbon assimilation of each leaf of *Lolium temulentum* could contribute to make about 3.7 more leaves, in the absence of scale, ontogeny and environmental constraints, 'it would require only about 33 leaves to achieve the estimated net primary productivity of the entire biosphere!'

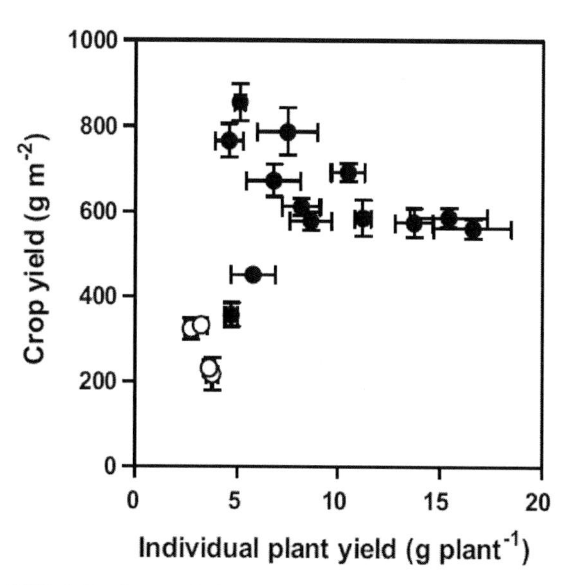

FIG. 1.2 Crop and plant yield are unrelated. Source: Pedró et al. (2012).

We can also look at scalability from an epistemological perspective confronting reduction and emergence. Le Boutillier (2013) thus asked: 'can the properties of the emergent whole be reduced to the properties of its parts?' A positive answer means the traits are scalable, and a negative answer means the system cannot be explained by reduction to its components. From both practical (Denison, 2012) and epistemological perspectives (Wimsatt, 1976; Le Boutillier, 2013), reduction and emergence are not mutually exclusive. Hence, this book makes room for both approaches, for example in the discussion of the role of biotechnology in agriculture from molecular (Chapter 18) and crop perspectives (Chapter 19).

3 BOOK STRUCTURE AND THEMES

The book is organized in three parts. Part 1 focuses on farming systems, Part 2 on aspects of the nitrogen, water and carbon economies of plants and crops, and Part 3 deals with breeding and agronomic applications.

3.1 Part 1: Farming Systems

The aim of Part 1 is to provide a framework for relevance, as questions at lower levels of organization became agronomically meaningful at crop and farming system levels. Six sharply contrasting cropping systems are presented: high-yield maize–soybean systems in the USA (Chapter 2); cereal-based single-, double- and triple-cropping systems in China (Chapter 3); high-latitude, short-season cropping systems of northern Europe centered in cereals and rapeseed (Chapter 4); low-input maize-based systems in eastern and southern Africa (Chapter 5); high yield potential systems in southern Chile based on cereals and potato (Chapter 6); and nitrogen and water constrained, cereal-based systems of Mediterranean environments (Chapter 7). Against the backdrop of diverse environments and technologies, these cropping systems share questions, challenges and opportunities.

Chapters 2, 3 and 4 compare the role of agronomy and breeding underlying the long-term trends in grain production. The USA's share of global grain production is 38% for maize and 35% for soybean. Of these, more than 85% are produced in the north-central region known as the 'Corn Belt', the focus of Chapter 2. Continuous maize and 2-year maize–soybean rotation are dominant. China is the largest consumer, producer and importer of grain in the world, and agricultural research and development has played a key role in the progress of grain production over the last decades. Maize, wheat and rice are the core elements in production systems; about 1178 wheat varieties, 4081 rice varieties and 3941 maize hybrids have been released since 2001. The Nordic European countries have 9.1 Mha of arable land, including 2 Mha above 60°N in Finland, the northernmost agricultural country. High latitude agriculture is exceptional as it has to deal with harsh winters, long days, rapidly increasing but generally cool mean temperatures, risk of night frosts in early and late growing season, early summer drought and risk of abundant precipitation close to harvest. For all three cropping systems, rates of genetic gain and underlying changes in phenotype are discussed alongside the main changes in agronomic practices including water and nitrogen management. Chapter 6 highlights the favorable conditions of southern Chile to temperate crops including wheat, oat, barley, potato, seed-rape and grain legumes supported by mild temperature and high photothermal quotient. Trends in acreage and yield have been contrasting among crops, especially recently, and a large yield gap between entrepreneurial and small farmers is remarkable in potato cropping systems. Despite the reduction in acreage of some of these crops, opportunities arise to increase productivity in cropping systems based on annual crops.

Water and nitrogen are a common theme, and comparisons highlight that the blanket notion that cropping systems should shift to low nitrogen supply is unjustified, as the management of water and nitrogen has to be solved locally. Part 1 shows cases where nitrogen supply is excessive and needs to be reduced, but also systems where nitrogen supply is well balanced or deficient. Leading farmers in the USA are close to reaching the attainable yield of current varieties, with relatively high water-use efficiency and near-zero nitrogen balance. The risk of excess nitrogen, with undesirable environmental consequences is a concern in China, and shapes agronomic practices in Finland. Cropping systems of eastern and southern Africa and Mediterranean environments share chronic shortage of rainfall and unfertile soils, where increasing nitrogen supply is essential to lift yield. Scatter plots of yield against water use or rainfall show widespread and often large yield gaps relative to the locally determined attainable water-use efficiency in both Africa and Mediterranean environments. In both cases, shortage of nitrogen is a common major cause of these gaps. The proximal cause is therefore the same (shortage of nitrogen), but the ultimate cause is fundamentally different. In Africa, aspects of infrastructure, access to markets, yield stability and food security inhibit the adoption of higher fertilizer rates. In countries of southern Europe and particularly in Australia, shortage of nitrogen is more related to uncertain rainfall and associated risk.

Climate change, realized and projected, is another thread. Whereas the negative effects of warming feature profusely in the literature, there are important benefits at both high and intermediate latitudes that can be captured with alternative combinations of varieties and management practices, as highlighted for cropping systems of the USA (Chapter 2), China (Chapter 3) and northern Europe (Chapter 4). In the US Corn Belt, maize and soybean sowing time had advanced at rate of about 4 days per decade during the last 30 years. This is associated with climate change (i.e. warmer spring) and technological developments including new cultivars with improved germination and seedling establishment at low temperature, seed protection against pathogens and multirow seeders that shorten the sowing period. In Finland, annual temperature has increased by 0.93°C between 1909 and 2008, and at 0.03°C per year between 1959 and 2008. Between 1965 and 2007, the advancement of the start of sowing was 0.6–1.7 days per decade for cereals, 2.5 days per decade for sugar beet and 3.4 days per decade for potato. Currently, the growing season at Jokioinen (approx 60°N) in Finland is 169 days, and is projected to increase to 181 days by 2025, 196 days by 2055 and 219 days by 2085, in comparison to the current season of 225 days in Denmark (approx. 55°N). In China, experiments showed warming may promote plant development and growth, thus improving yield of late rice and winter wheat, but may be detrimental to early and mid rice as heat stress affects grain set and filling. In common with the USA and Finland, the cropping seasons in China have been adjusted during the past decades. In the winter wheat–maize cropping system, wheat sowing and maize harvest have been delayed simultaneously, resulting in a large increase in annual yield. Similarly, wheat sowing and rice harvesting have been delayed in the wheat–rice cropping system. In response to warming, breeders, agronomists and growers are actively working to reduce negative effects and capture the many opportunities for higher production and expanded agricultural boundaries at higher latitudes.

Conservation agriculture, involving maintenance of a vegetative cover or mulch on the soil surface, minimal soil disturbance and diversified crop rotations, is another thread connecting the diverse farming systems in Part 1. Short- and long-term effects of these practices are discussed in systems with different degrees of adoption. In the US Corn Belt, conservation tillage has been widely adopted over the last few decades. Conservation tillage, combined with other measures, reduces run-off and erosion risk, particularly on sloping land. In maize crops, conservation practices combined with center-pivot systems to replace less efficient gravity irrigation, increasing efficiency in the manufacture of key agricultural inputs and higher yields with a stabilized input of nitrogen, have reduced greenhouse gas emissions per unit grain produced. Chapter 2 further discusses the benefits and trade-offs of the 2-year maize–soybean rotation, which represents ≈65% of the Corn Belt, in relation to continuous maize representing the remaining 35%. In Africa, a number of factors are identified that hinder adoption of conservation practices, including the need to increase nitrogen fertilization associated with nitrogen immobilization in the transition to reduced tillage and retention of stubble with high C:N ratio, weed control, the opportunity costs of low volumes of stubble, all compounded with the constraints imposed by limited access to finance and markets. Chapter 5 further analyzes the drivers of cropping systems in eastern and southern Africa, where maize is intercropped, predominantly with legumes, but also with vegetable crops. In China, substantial efforts have been made on the research and demonstration of conservation agriculture over the last five decades, leading to the current adoption of these practices in ≈6.4 Mha. In addition to the conventional plant materials used for ground cover, local growers have access to subsidized plastic film mulching which is spreading widely in maize crops, as illustrated on the cover of the book. On its own or combined with stubble, plastic mulch helps to harvest water into the soil root zone, reduces soil evaporation and significantly increases water-use efficiency. Owing to the diversity of soil and climate, a variety of cropping systems combine one or more of the major cereals – rice, wheat and maize – with other crops in single-, double- or triple-cropping systems (Fig. 1.3). In northern Europe, most fields sown to annual crops were traditionally autumn plowed. The EU environmental program stimulated the adoption of reduced tillage to decrease erosion and nutrient loss. No-tillage has been practiced on increasing areas in Finland until 2007, when 30% of the winter cereal, 10% of spring cereal and 17% of oilseed rape area was direct drilled. Afterwards the no-till area has stagnated at 13% of total agricultural land. Chapter 4 further explores contradictory evidence on the role of reduced tillage on soil carbon storage and greenhouse gas emissions.

Chapter 7 deals with cereal crops in Mediterranean environments, and challenges the paradigms on terminal drought, the superior stress adaptation of barley in relation to wheat and the role of nitrogen fertilization in water-stressed crops. Its focus on crop growth in response to water and nitrogen bridges Parts 1 and 2.

3.2 Part 2: Resources – carbon, nitrogen and water

The aim of Part 2 is to present new views on the carbon, nitrogen and water economies of crops. Chapters 8, 9 and 10 have a strong focus on nitrogen and carbon, and Chapter 11 deals with the links between water and carbon assimilation.

Agriculture has been a historical source of ideas and methods for ecology and evolutionary biology. In his opening editorial article for the journal *Agricultural Ecosystems*, John Harper (1974) observes that 'Agriculture and forestry were both concerned with the practical management of ecosystems before the concept was ever introduced into ecology'. He further states '…it is salutary to remember that most of the major advances in ecology, both theoretical and practical, have stemmed from the study of problems in agriculture, forestry and fisheries'. Evans (1984) examines Darwin's use of the analogy between selection under domestication and that under nature, not only for the genesis of his theory of evolution but also for its development and presentation. In a letter to J.D. Hooker, Darwin writes: '…All my notions about how species change are derived from long-continued study of the works of agriculturists and horticulturists…'.

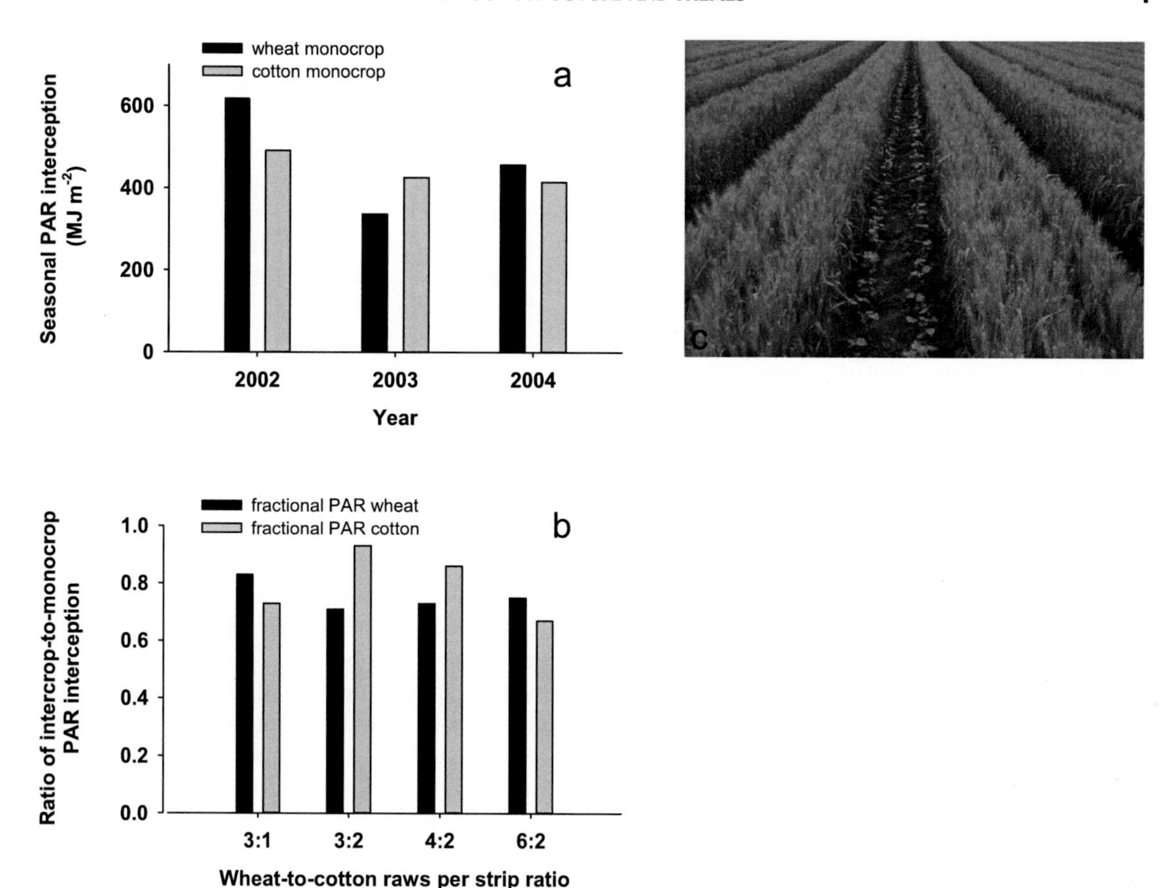

FIG. 1.3 Relay cotton–wheat intercropping in northern China. (a) Approximate seasonal interception of photosynthetically active radiation (PAR) in monocrops. (b) Ratio of PAR interception between intercrop components and their respective monocrops. (c) The 3:2 arrangement in June, shortly before wheat harvest. Land equivalent ratios were 1.39 for all arrangements except 6:2 which had a ratio = 1.28. Capture of radiation (a, b) accounted for these improvements in production, with no effect of radiation-use efficiency. Adapted from Zhang et al. (2007, 2008).

Reciprocally, insights from ecology and evolution are relevant to agriculture. Biologically, the rice plants in a crop form a population, which is ruled by the principles of plant population ecology (Harper, 1977). Interactions between neighboring plants involve competition for resources as well as non-resource signals (Aphalo and Ballaré, 1995). An ecological perspective inspired research showing that ectopic expression of the Arabidopsis phytochrome B gene, a photoreceptor involved in detecting red to far-red light ratio associated with plant density, can increase the yield of transgenic potato in comparison to wild types (Boccalandro et al., 2003). Natural selection favors, whereas selection for yield in crops reduces, the competitiveness of individual plants (Donald, 1981; Evans, 1993; Denison, 2012).This has implications for phenotyping protocols, as many yield-related traits do not scale from plant to crop (Fig. 1.2). Natural selection experiments, where wheat populations were grown in a transect from southern to northern France for 12 generations, revealed genes that had strong (VERNALIZATION-1, Flowering

Locus T and CONSTANS) or weak (Ppd-1, LUMINI DEPENDENS and GIGANTEA) selection signatures (Rhone et al., 2010). Transitions in individuality with evolution (Maynard Smith and Szathmáry, 1995) bring a novel focus on intra-plant allocation of resources (Sadras and Denison, 2009). The theory of Smith and Fretwell (1974) relating energy expenditure in individual offspring and the fitness of parents has only recently been incorporated in the conceptual model of grain yield thus providing a new angle on the trade-off between seed size and number in crops (Sadras, 2007; Gambín and Borrás, 2010; Sadras and Slafer, 2012; Slafer et al., 2014). Furthermore, it has been argued that an evolutionary focus might help to break epistemological barriers between scientists working at different levels of organization (Sadras and Richards, 2014).

Chapter 8 is informed by ecological perspectives to scale from organ to plant and from plant to crop and community of mixed plant species. Combining these perspectives with the principles of capture and efficiency in the use of resources (Monteith et al., 1994), Chapter 8 revises the drivers of crop nitrogen uptake, emphasizing its co-regulation by soil nitrogen availability and plant growth as an emerging property of the whole plant–soil system. The need to account for the allometric relationships between nitrogen uptake and crop growth is summarized in nitrogen dilution curves, and derived nitrogen nutrition index as a method to quantify unequivocally the nitrogen status of crops. Where allometric relationships are ignored, assessment of cropping practices and comparisons of genotypes includes trivial factors (e.g. higher nitrogen uptake in crops with larger biomass) which are confounded with more interesting traits (e.g. high nitrogen uptake at the same biomass).

Ecological and evolutionary ideas inform Chapters 9, 10 and 11. In Chapter 9, the evolutionary-trade-off hypothesis is used to analyze potential avenues to improve nitrogen-fixation efficiency (gN/gC) of legume crops and forages. This hypothesis states that past natural selection is unlikely to have missed simple trade-off-free improvements, where 'simple' is defined as arising frequently via mutation, and 'trade-off-free' means increasing individual fitness across a wide range of past environments. The potential of less effective 'free rider' strains of rhizobia that divert most available resources from nitrogen fixation to their own reproduction is confronted with the notion of 'host sanctions', whereby legume responses to less-beneficial nodules reduce their reproduction. The chapter concludes that increasing nitrogen-fixation efficiency may be possible via more complex genetic changes previously untested by natural selection or by accepting trade-offs rejected by natural selection. Further enhancing host sanctions could lead to legumes that selectively enrich soils with only the most-beneficial local rhizobia.

Chapter 10 examines the genetic and environmental control of senescence. Low irradiance, low red:far red ratio, elevated ultraviolet radiation, waterlogging, drought, nutrient deficiency, extreme temperatures, high atmospheric CO_2 concentration can all lead to increased expression of senescence-associated genes. Senescence is considered at different levels of organization including organelle, cell, tissue, organ, plant and crop. Chloroplasts redifferentiate into gerontoplasts in senescing leaves, and into chromoplasts in ripening fruit. At the cellular level, senescence is part of the program that specifies cell fate; thus, differential senescence in tissues and organs results in complex anatomies and morphologies that change and adapt over time. Evolution has shaped variations on the senescence program to give rise to a diversity of structures within the angiosperm life cycle. At the organ level, the chapter discusses the senescence of leaves, associated with catabolism of chlorophyll and proteins, and change in organelle structure and function under the control of hormones and reactive oxygen species. It also outlines the role of senescence in root turnover, the lifespan of nodules in legumes and reproductive structures including differentiation of floral parts and gametes, embryogenesis, seed maturation and fruit ripening. At the plant and crop levels, senescence is defined as the point of transition from the carbon capture phase of development to the period of remobilization of nitrogen and other nutrients, hence the focus on source–sink relations and the recycling of resources from obsolete to new developing structures. This stage of resource remobilization 'could offer a feast to potential pathogens', thus explaining the upregulation of defense pathways during senescence and the dual role of some hormones, such as salicylic acid and jasmonic acid, in senescence and defense.

Chapter 11 defines the plant's water transport system as including the anatomical attributes and network assembly of the xylem which determine the physical limits to water flow, and the size and architecture of the root system which dictates the accessibility of soil water. A vascular-centric approach to plant water use is developed, emphasizing opportunities and ramifications for biomass improvement in both water-stressed and unstressed conditions. Midday depression of photosynthesis is thus analyzed in terms of hydraulic constraints. Inter-specific variation in xylem conductivity of woody species is associated with variation in maximum photosynthetic rate and growth, hence the question on the intra-specific variation of traits related to hydraulic conductivity in crop species. In leaves, the transpiration stream flows through two pathways: the non-living xylem 'plumbing' and the living cells, thus replacing water evaporating from mesophyll tissue near the stomata. The extra-vascular pathway constitutes a major hydraulic resistance, and its length is largely determined by the density of minor veins. In addition, the conductivity of extra-vascular pathway is actively regulated by protein water channels (aquaporins) in the plasma membrane. Both vein density and aquaporins are therefore potential targets to manipulate hydraulic conductivity. Xylem cavitation is initiated when hydraulic tension in the xylem exceeds the ability of the xylem membranes to prevent air from being sucked into the water column. Once air bubbles are sucked into xylem conduits they block the passage of water, rendering these cavitated conduits non-functional. Xylem vulnerability to cavitation is quantified with P50, the water potential at which 50% of the xylem capacity to transport water has been lost due to cavitation and the hydraulic safety margin defined as the difference in water potential between the point of stomatal closure and P50. Preliminary data from grasses grown in controlled environments suggest negative safety margins, meaning that stomatal closure occurs after the onset of cavitation, thus exposing plants to significant diurnal xylem cavitation. Chapter 11 thus presents a vascular-centric approach to plant water use that is largely unexplored in crop species. This perspective is important for a wider and deeper understanding of the links between the water and carbon economies of plants. Application in crop improvement, however, requires filling many gaps including matters of scale, e.g. how do short-term, leaf-level traits scale to seasonal crop growth (section 2 in this chapter; Sadras and Richards, 2014) and phenotyping methods, as current techniques to measure vascular traits in crop species are incipient and time consuming.

As a consequence of the ecological and evolutionary perspectives in Part 2, trade-offs feature prominently. Some of trade-offs considered are those between:

- photosynthesis rate per unit leaf area and leaf area expansion; capture of radiation and radiation-use efficiency; radiation-use efficiency and nitrogen-use efficiency; nitrogen uptake per unit soil nitrogen and biomass per unit nitrogen uptake; and between root traits including denser root systems, efficiency of nitrate and ammonium transport system, and quantity and quality of carbon exudates (Chapter 8)
- individual fitness and the collective performance of crop communities; fitness

of rhizobia and fitness of their legume hosts; accumulation of energy-rich polyhydroxybutyrate in nodules and nitrogen fixation; competitiveness and nitrogen-fixation efficiency of rhizobia (Chapter 9)

- survival and reproduction; root longevity and nitrogen content (Chapter 10)
- leaf photosynthesis and transpiration; stomatal sensitivity to temperature and humidity and leaf-level water-use efficiency (associated with ratio of vein density to stomatal density); vein density and direct (lignin) and indirect (area of photosynthetic tissue) carbon cost; root growth and reproduction; aquaporin expression and root size; hydraulic conductivity under water stress and transport of water under high tension (associated with cavitation) (Chapter 11).

Making these trade-offs explicit is an insurance against oversimplistic attempts to manipulate individual crop traits in isolation using agronomic practices, breeding or both; these are the themes of Part 3.

3.3 Part 3: Physiological applications in breeding and agronomy

As discussed in the Introduction, crop yield can only be improved with better varieties and practices. The aim of Part 3 is therefore to illustrate applications of physiological principles and tools in breeding and agronomy.

Chapter 12 deals with crop phenological development, the single most important trait for crop adaptation. This statement is justified by the existence of critical periods for grain-yield determination, when crops are particularly vulnerable to stress. Hence, combining sowing date and variety choice is the cornerstone decision to minimize the likelihood of critical periods to be exposed to extreme stress. The chapter uses soybean and wheat as model crops, outlines the models integrating genetic and environmental regulation of development, and updates the discussion of phenology manipulation to improve grain yield.

Following on the notion of critical periods in Chapter 12, Chapter 13 reviews the methods to quantify environmental conditions in a context of plant breeding. Whereas the environment is the major source of yield variation, environmental characterization is often superficial, and it is not rare to find environments defined nominally in terms of location and season. This chapter thus explores different facets of environment characterization, including how to describe the target population of environments (TPE; i.e. conditions to which future-release cultivars are likely to be exposed), the extent to which multi-environment breeding trials match the target population of environments, how to identify relevant environment classes where genotypes are expected to perform similarly and the role of managed-environment trials combining, for example, irrigation and rainout shelters to generate specific patterns of water supply. Modeling, it is proposed, is a cost-effective means to explore the complex genotype × environment × management (G × E × M) interactions and, in particular, to assess the potential value of traits and alleles depending on 'genetic backgrounds' (G × G and trait × trait interactions), environments (e.g. current and future climates) and management practices.

After the in-depth consideration of environmental quantification, Chapters 14 and 15 focus on agronomically important aspects of the phenotype, particularly yield, as the outcome of genetic, environmental, management and interaction effects. Chapter 15 shows how new ecophysiological models scale up individual traits to whole-plant and crop phenotypes. The integration of genetic controls in ecophysiological models has allowed analysis of the genetic control of phenotypic plasticity across wide ranges of environments, and the G × E × M space is now explored using efficient algorithms to find ideotypes optimizing many antagonist criteria. Lack of quantitative relationships between genes and model parameters remains a challenge. Shifting from a primarily modeling perspective in Chapter 14, Chapter 15 presents a compelling case study in maize, where conventional physiological approaches are boosted to reach breeding relevance. Whereas molecular genetics continues to develop at astonishing rates (Chapter 18), methods to characterize the phenotype lag in terms of quantity, i.e. the numbers of plots that need to be measured, and qualitatively, as high-throughput techniques are largely unsuitable to dissect the relevant traits. Against the backdrop of this genotype–phenotype gap, Chapter 15 reviews the physiological model of yield and its dissection in minor traits, outlines a field-based approach for phenotyping traits critical for yield determination at the crop level, and explores the genetic controls of these traits. Central to this proposition is the expanded capability to measure the physiological components of yield in large, genetically suitable populations (e.g. families of recombinant inbred lines).

Crop yield potential is the maximum yield that can be achieved by a particular cultivar in an environment to which it is adapted when pests and diseases are effectively controlled and nutrients and water are non-limiting (Evans and Fischer, 1999). Chapter 16 outlines the rationale for increasing yield potential, including the slowing trends in genetic yield gain for major cereal crops, the narrowing of the exploitable gap in some cropping systems (Chapter 2), and the straightforward delivery of technologies embedded in crop seeds, in comparison to agronomic practices. The chapter focuses on the three main cereals – wheat, rice and maize – and emphasizes the progress in physiology, genetics and breeding for yield potential in the last five years; five key areas are highlighted. First, advances have been made in the genetic engineering of photosynthesis though realization is likely to be measured in decades rather than years. Second, previous propositions about the importance of fruiting efficiency (grains per gram ear/panicle at anthesis) in raising grain number per unit area have been reinforced and associations with genetic gains in grains per unit are demonstrated, particularly in wheat and maize. Strategies to optimize phenological pattern to maximize fruiting efficiency have been proposed in wheat (Chapter 12). Third, the need to increase grain weight and avoid effects on grain number is emphasized as a priority for breeding. There have been advances in understanding the role of expansins in regulating potential grain size. Recent work indicates expansion of the pericarp may be controlled by the rheological properties of the cell wall through the action of expansins. The potential for combining high grain number and grain weight has been demonstrated in some high yield potential environments (Bustos et al., 2013). Fourth, the benefits of trait-based breeding to combine physiological traits for yield potential have been demonstrated in wheat by recent work at CIMMYT. Selecting for physiological traits offers promise for increasing the probability of achieving cumulative gene action. Fifth, high-throughput field-based phenotyping techniques have been developed for application in plant breeding not only for canopy traits (e.g. hyperspectral radiometry, thermal imagery from drones) but also for root traits (e.g. root DNA concentration assay in soil).

Seed quality traits, the focus of Chapter 17, are increasingly important. The term 'quality' hides different meanings, which are clarified with relevant definitions and illustrated using two model crops, wheat and sunflower. Wheat grain proteins and sunflower fatty acids, tocopherols and phytosterols are examined in detail. Similarly to the analysis of the key developmental window for grain number determination (Chapters 12 and 15), Chapter 17 identifies developmental windows relevant to the concentration of grain protein in wheat and oil in sunflower. Two aspects of climate change, dimming and warming (Chapter 20) would challenge oil concentration due to the sensitivity of this trait to radiation during the window 250–450°Cd after flowering. On the other hand, oleic acid would increase under these conditions. The impact of temperature and other environmental factors on oil concentration and composition is extensively reviewed stressing the biochemical processes and genotype-dependent responses. Environmentally-driven changes in grain protein composition are shown to be associated with the altered expression of genes encoding storage proteins in response to signals indicating the availability of nitrogen and sulfur. Quality simulation models and their empirical or functional bases integrating physiological and environmental variables are described for wheat (*SiriusQuality*) and sunflower (Pereyra-Irujo and Aguirrezábal's model). These models complement modeling primarily aiming at growth- and yield-related traits (Chapters 13, 14, 20). The improvement of quantitative analysis of genotype by environment interactions supported by physiological understanding and models would in turn facilitate the design of quality traits. This novel area is developing fast to assist breeding in the future.

Chapters 18 and 19 deal with biotechnology. Chapter 18 updates molecular science and technology in a context of plant breeding. Recent molecular marker technologies are analyzed and examples of their use, some of them in animals, are presented. Lower prices and novel statistical methods allow the integration between phenotypic and genotypic knowledge thus bridging both approaches. Chapter 18 emphasizes that the relevance of molecular selection in plant breeding is largely dependent on the species of interest, on the trait under selection and on the cost:benefit ratio. To capture fully the potential of molecular breeding described in this chapter, these technologies need to be combined with the most advanced methods for environmental characterization (Chapter 13) and phenotyping (Chapters 14 and 15). Chapter 19 demonstrates the benefits from these linkages. It presents clear examples where physiological insight helps the interpretation of complex quantitative trait loci (QTL) × environment interactions where sunflower grain oil concentration and phenological development are confounded. On the other hand, the chapter highlights the usefulness of QTL information for understanding crop trait regulation. The benefits of multidisciplinary work are a common theme of many chapters (12, 13, 14, 16 and 18).

Chapter 20 deals with climate change, its consequences for crop development, growth and yield, and explores agronomic and breeding adaptations. It relies primarily on modeling to link crops and climate. The present knowledge and gaps about climate change are reviewed considering uncertainties of general circulation models (GCMs) and downscaling GCM scenarios to farm conditions. A recent analysis shows, however, that the uncertainty associated with crop models is larger than the uncertainty associated with GCMs. Despite the generally assumed negative impact of warming on crops, this chapter illustrates past trends where beneficial responses were found for cropping systems like that of North China Plain (Chapter 3) and simulated results showing a positive effect of higher CO_2 concentration on wheat. Differences in crop response to climate change depending on soil type are illustrated and the need for complementing simulation with field experiments is stressed. Adaptation to climate change is analyzed considering management, weather forecasting and breeding.

Crop production is the unifying theme across all the three book parts. Water and nitrogen are also prominent in linking aspects of cropping systems in Part 1, economy of resources in Part 2, and applications in Part 3. Climate change and modeling are common features of many chapters. All chapters in Part 1 use, in different degrees, simulation models to investigate yield gaps and agronomic practices. Chapter 13 applies models for quantitative environmental characterization. Chapter 14 outlines modeling approaches aimed at crop improvement, with emphasis on the links between phenotype and genotype. Chapter 17 applies models with emphasis on grain quality. Chapter 20 uses models as the main tool to investigate likely consequences and adaptations to climate change. Comparison of models in these chapters highlights common elements, significant differences depending on the intended applications, the central role of crop physiology providing models' building blocks, and the need for continued effort to update models for research, development and extension in agriculture.

References

Aphalo, P.J., Ballaré, C.L., 1995. On the importance of information-acquiring systems in plant-plant interactions. Funct. Ecol. 9, 5–14.

Boccalandro, H.E., Ploschuk, E.L., Yanovsky, M.J., Sanchez, R.A., Gatz, C., Casal, J.J., 2003. Increased phytochrome B alleviates density effects on tuber yield of field potato crops. Plant Physiol. 133, 1539–1546.

Bustos, D.V., Hasan, A.K., Reynolds, M.P., Calderini, D.F., 2013. Combining high grain number and weight through a DH-population to improve grain yield potential of wheat in high-yielding environments. Field Crops Res. 145, 106–115.

Calviño, P.A., Monzon, J.P., 2009. Farming systems of Argentina: yield constraints and risk management. In: Sadras, V.O., Calderini, D.F. (Eds.), Crop physiology: applications for genetic improvement and agronomy. Academic Press, San Diego, pp. 55–70.

Carroll, M.W., Head, G., Caprio, M., 2012. When and where a seed mix refuge makes sense for managing insect resistance to Bt plants. Crop Protect. 38, 74–79.

Cassman, K.G., 2007. Climate change, biofuels, and global food security. Environ. Res. Lett. 2, 011002.

Connor, D.J., Mínguez, M.I., 2012. Evolution not revolution of farming systems will best feed and green the world. Glob. Food Secur. 1, 106–113.

Denison, R.F., 2012. Darwinian agriculture: how understanding evolution can improve agriculture. Princeton University Press, Princeton, NJ.

Donald, C.M., 1981. Competitive plants, communal plants, and yield in wheat crops. In: Evans, L.T., Peacock, W.J. (Eds.), Wheat science – today and tomorrow. Cambridge University Press, Cambridge, pp. 223–247.

Downes, S., Mahon, R., 2012. Evolution, ecology and management of resistance in Helicoverpa spp. to Bt cotton in Australia. J. Invert. Pathol. 110, 281–286.

Evans, L.T., 1984. Darwin's use of the analogy between artificial and natural selection. J. Hist. Biol. 17, 113–140.

Evans, L.T., 1993. Crop evolution, adaptation and yield. Cambridge University Press, Cambridge.

Evans, L.T., 2005. The changing context for agricultural science. J. Agric. Sci. 143, 7–10.

Evans, L.T., Fischer, R.A., 1999. Yield potential: its definition, measurement and significance. Crop Sci. 39, 1544–1551.

Fischer, R.A., 2009. Farming systems of Australia: exploiting the synergy between genetic improvement and agronomy. In: Sadras, V.O., Calderini, D.F. (Eds.), Crop physiology: applications for genetic improvement and agronomy. Academic Press, San Diego, pp. 23–54.

Fischer, R.A., Edmeades, G.O., 2010. Breeding and cereal yield progress. Crop Sci. 50, S85–S98.

Frow, E., Ingram, D., Powell, W., Steer, D., Vogel, J., Yearley, S., 2009. The politics of plants. Food Secur. 1, 17–23.

Fischer, R.A., Byerlee, D., Edmeades, G.O., 2014. Crop yields and global food security. Will yield increase continue to feed the world? ACIAR, Canberra.

Gambín, B.L., Borrás, L., 2010. Resource distribution and the trade-off between seed number and seed weight: a comparison across crop species. Ann. Appl. Biol. 156, 91–102.

Grassini, P., Cassman, K.G., 2012. High-yield maize with large net energy yield and small global warming intensity. Proc. Natl. Acad. Sci. USA 109, 1074–1079.

Harper, J.L., 1974. Agricultural ecosystems. Agro-Ecosystems 1, 1–6.

Harper, J.L., 1977. The population biology of plants. Academic Press, London.

Head, G.P., Greenplate, J., 2012. The design and implementation of insect resistance management programs for Bt crops. GM Crops Food 3, 144–153.

Heinze, T., 2012. What are creative accomplishments in science? Conceptual considerations using examples from science history and bibliometric findings. Kolner Zeitsch. Soziol. Sozialpsychol. 64, 583–599.

Hellemans, A., Bunch, B., 1988. The timetables of science. Simon and Schuster, New York.

Le Boutillier, S., 2013. Emergence and reduction. J. Theor. Social Behav. 43, 205–225.

Lobell, D.B., Cassman, K.G., Field, C.B., 2009. Crop yield gaps: their importance, magnitudes, and causes. Annu. Rev. Environ. Resour. 34, 179–204.

Maynard Smith, J., Szathmáry, E., 1995. The major transitions in evolution. W.H. Freeman & Co Ltd, Oxford.

Miller, F.P., 2008. After 10,000 years of agriculture, whither agronomy? (reprinted from Agron. J., 100, 22-34, 2008). Agron. J. 100, S40–S52.

Monteith, J.L., Scott, R.K., Unsworth, M.H., 1994. Resource capture by crops, Proc. 52nd Easter School, Univ of Nottingham, School of Agriculture. Nottingham University Press, Nottingham, pp. 1-15.

Pedró, A., Savin, R., Slafer, G.A., 2012. Crop productivity as related to single-plant traits at key phenological stages in durum wheat. Field Crops Res. 138, 42–51.

Pinstrup-Andersen, P., 2009. Food security: definition and measurement. Food Secur. 1, 5–7.

Pretty, J., 2008. Agricultural sustainability: concepts, principles and evidence. Philosoph. Transact. R. Soc. B Biol. Sci. 363, 447–465.

Rhone, B., Vitalis, R., Goldringer, I., Bonnin, I., 2010. Evolution of flowering time in experimental wheat populations: A comprehensive approach to detect genetic signatures of natural selection. Evolution 64, 2110–2125.

Rietman, E.A., Karp, R.L., Tuszynski, J.A., 2011. Review and application of group theory to molecular systems biology. Theor. Biol. Med. Model. 8, article number 21.

Rockstrom, J., Falkenmark, M., Lannerstad, M., Karlberg, L., 2012. The planetary water drama: dual task of feeding humanity and curbing climate change. Geophys. Res. Lett. 39, L15401.

Sadras, V.O., 2007. Evolutionary aspects of the trade-off between seed size and number in crops. Field Crops Res. 100, 125–138.

Sadras, V.O., Denison, R.F., 2009. Do plant parts compete for resources? An evolutionary perspective. New Phytol. 183, 565–574.

Sadras, V.O., Richards, R.A., 2014. Improvement of crop yield in dry environments: benchmarks, levels of organisation and the role of nitrogen. J. Exp. Bot. 65, 1981-1995

Sadras, V.O., Slafer, G.A., 2012. Environmental modulation of yield components in cereals: Heritabilities reveal a hierarchy of phenotypic plasticities. Field Crops Res. 127, 215–224.

Siegfried, B.D., Hellmich, R.L., 2012. Understanding successful resistance management: the European corn borer and Bt corn in the United States. GM Crops Food 3, 184–193.

Slafer, G., Savin, R., Sadras, V.O., 2014. Coarse and fine regulation of wheat yield components in response to genotype and environment. Field Crops Res. 157, 71–83.

Smith, C.C., Fretwell, S.D., 1974. The optimal balance between size and number of offspring. Am. Natural. 108, 499–506.

Tilman, D., Cassman, K.G., Pamela, A., Matson, P.A., Naylor, R., Polasky, S., 2002. Agricultural sustainability and intensive production practices. Nat. Biotechnol. 418, 671–677.

Wimsatt, W.C., 1976. Reductionism, levels of organization and the mind-body problem. In: Globus, G.G., Maxwell, G., Savodnik, I. (Eds.), Consciousness and the brain. Pleum, New York, pp. 199–267.

Wimsatt, W.C., 1994. The ontology of complex systems: levels of organization, perspectives and causal thickets. Can. J. Philosoph. (Suppl. 20), 207–274.

Zhang, L., van der Werf, W., Bastiaans, L., Zhang, S., Li, B., Spiertz, J.H.J., 2008. Light interception and utilization in relay intercrops of wheat and cotton. Field Crops Res. 107, 29–42.

Zhang, L., van der Werf, W., Zhang, S., Li, B., Spiertz, J.H.J., 2007. Growth, yield and quality of wheat and cotton in relay strip intercropping systems. Field Crops Res. 103, 178–188.

Zhou, G.H., Zhang, W.G., Xu, X.L., 2012. China's meat industry revolution: Challenges and opportunities for the future. Meat Sci. 92, 188–196.

PART I

FARMING SYSTEMS

CHAPTER

2

High-yield maize–soybean cropping systems in the US Corn Belt

Patricio Grassini[1], *James E. Specht*[1], *Matthijs Tollenaar*[2], *Ignacio Ciampitti*[3], *Kenneth G. Cassman*[1]

[1]**University of Nebraska, Department of Agronomy and Horticulture, USA**
[2]**Monsanto Company, USA**
[3]**Kansas State University, Department of Agronomy, USA**

1 INTRODUCTION

The USA accounts for 38 and 35% of global maize and soybean production, producing a respective 320 and 84 Mt of these crops annually (FAOSTAT & USDA-NASS, 2007–2011). More than 85% of those totals are produced in the north-central region known as the 'Corn Belt', where continuous maize (≈35%) and 2-year maize–soybean rotation (≈65%) are the dominant cropping systems (Fig. 2.1). The western edge of the Corn Belt includes the eastern Great Plains states of North Dakota, South Dakota, Nebraska, and Kansas, where irrigated agriculture accounts for more than 50% of the total maize and soybean production, though only 37% of the total area sown with these crops is irrigated (Fig. 2.1). Of total maize and soybean production in the Corn Belt, a respective 14 and 7% is irrigated. The present chapter describes the climate, soil, and management practices of high-yield maize–soybean cropping systems in the Corn Belt. Major drivers for higher yields and resource-use efficiency are evaluated and opportunities for further improvement are discussed.

2 CROPPING SYSTEMS

2.1 Weather and soils

Annual patterns of solar radiation, temperature, rainfall, and crop evapotranspiration (ET_C) for three representative locations along a west–east transect across the Corn Belt are presented in Figure 2.2. The climate is mid-continental and temperate with cold winters. Rainfall distribution follows a monsoonal pattern with 70, 67, and 55%

Crop Physiology. DOI: 10.1016/B978-0-12-417104-6.00002-9

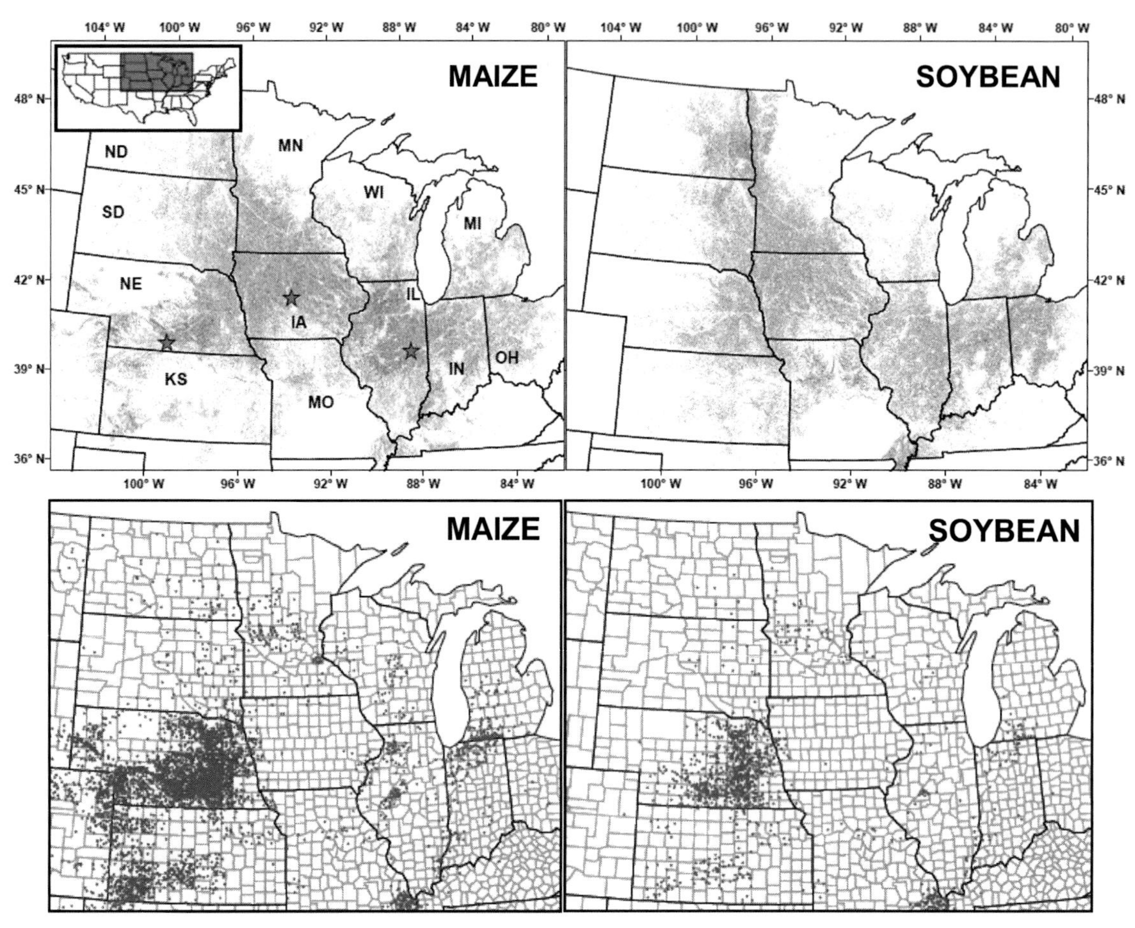

FIG. 2.1 Map of the Corn Belt showing harvested maize and soybean area (top panels) and irrigated area (bottom panels); one dot represents 810 ha. Total harvested area (2007–2011 average) was 28 Mha for maize and 24 Mha for soybean. State boundaries are shown (IA: Iowa, IL: Illinois; IN: Indiana; KS: Kansas; MI: Michigan; MN: Minnesota; MO: Missouri; NE: Nebraska; ND: North Dakota; OH: Ohio; SD: South Dakota; WI: Wisconsin). Stars in top left panel indicate location of weather stations shown in Fig. 2.2. *Source: USDA-NASS.*

of annual total rainfall concentrated in the summer-crop growing season in Nebraska, Iowa, and Illinois, respectively. Key stages for yield determination coincide with the peak in solar radiation, temperature, and crop water requirements: silking stage in maize and pod development in soybean typically occur during mid-July, with grain filling occurring in both crops during August.

Annual rainfall decreases almost linearly from east to west, with a parallel increase in the evaporative demand (Fig. 2.2). Crop water deficit, estimated as the difference between total cumulative ET_C and rain from sowing to physiological maturity, is 32 mm in Champaign (Illinois), 50 mm in Ames (Iowa), and 253 mm in Holdrege (Nebraska). It is thus clear that probability and severity of water stress during the growing season of summer crops increases along the east–west transect.

In this chapter, yield potential is taken as a benchmark to estimate yield gaps (section 4), that is, the difference between average on-farm

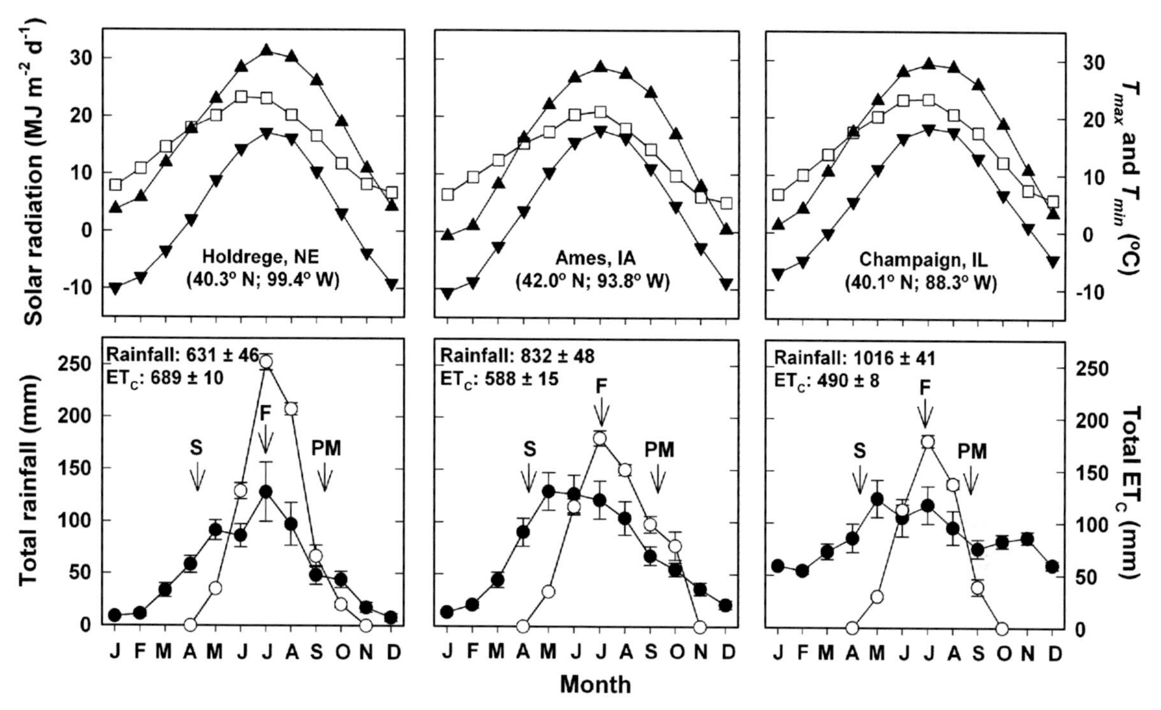

FIG. 2.2 Monthly average incoming solar radiation (□), maximum (Tmax [▲]) and minimum temperature (Tmin [▼]), total rainfall (●), and total crop evapotranspiration without water limitation (ET_C [○]) based on long-term (21-year) weather data in Holdrege (Nebraska), Ames (Iowa), and Champaign (Illinois); Fig. 2.1 maps these locations. ET_C was simulated for maize using Hybrid-Maize model based on location-specific typical sowing date and hybrid maturity. Arrows in bottom panels indicate average dates of sowing (S), silking (F), and physiological maturity (PM) for maize. Except for an average 15-d later sowing, soybean pod development (stages R3-R4) and PM (R7) generally correspond to maize F and PM. Error bars are two standard errors of the mean.

yield and the yield potential defined by solar radiation, temperature, and genotype, whereas water-limited yield potential is also defined by water availability in rain-fed systems (Lobell et al., 2009; Van Ittersum et al., 2013). Simulations of water-limited maize yield potential using a crop model (Hybrid-Maize, Yang et al., 2004) and a 21-year climate series indicated that rain-fed yield exceeded 75% of potential yield achievable without water limitation in 90% of years at Illinois sites and 66% in Iowa, with an inter-annual variability of 20% (Fig. 2.3). In contrast, rain-fed yield in Holdrege (Nebraska) exceeded 75% of the potential yield without water limitation in just 5 (24%) years of the 21 simulated years, with higher inter-annual variability (CV = 42%). On average, simulated rain-fed yield, expressed as a percentage of potential yield achievable without water limitation, was 90% in Champaign (Illinois), 81% in Ames (Iowa), and 61% in Holdrege (Nebraska). Not surprisingly, most irrigated agriculture is located in the western region (Fig. 2.1), where groundwater and surface water resources are available to supplement the low, erratic, and ill-distributed rainfall in most years. Irrigation in the western region can thus ensure yield and stability that are comparable to those achieved in the more rainfall-favorable central and eastern regions of the Corn Belt.

Soils are generally deep, fertile, rich in soil organic carbon (SOC), and have large water-holding capacity. By sowing time of maize and

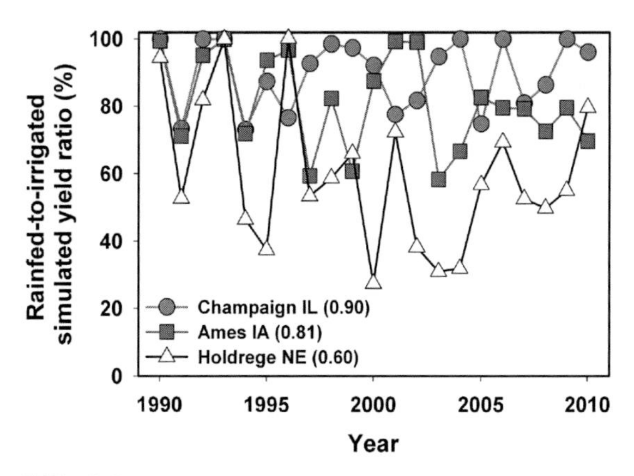

FIG. 2.3 Simulated rain-fed maize yield potential, expressed as a percentage of the simulated irrigated yield potential, at Holdrege (Nebraska), Ames (Iowa), and Champaign (Illinois) from 1990 to 2010. Average (21-year) irrigated yield potential was 16.0, 15.6, and 15.0 t ha^{-1}, respectively. Simulations were based on actual weather, soil properties, and typical management practices at each location. Average (21-year) rain-fed-to-irrigated yield ratios are indicated in parenthesis for each location.

soybean, the soil profile is typically filled to field capacity by cumulative rainfall in the fallow period between fall harvest and spring sowing. Suitable soils for annual crop production typically belong to the Alfisols and Mollisols orders. Alfisols were formed in (pre-settlement) forested lands. Dominant Alfisols suborders are Aqualfs and Udalfs, which are mostly located in the central and eastern regions of the Corn Belt. In contrast, Mollisols were formed in grassland prairies and have higher SOC content and a thicker A-horizon than Alfisols. Dominant Mollisols suborders are Udolls (humid central and eastern regions) and Ustolls (sub-humid to arid western regions).

Long-term studies have indicated no change or small net loss in SOC over time in maize-based systems (Baker and Griffis, 2005; Verma et al., 2005; Blanco-Canqui and Lal, 2008). Soils are prone to erosion when crops are grown on sloping land without conservation tillage and other measures to reduce runoff. In wet years, transient early-season waterlogging is likely in soils with moderate-to-low infiltration rates in the central and eastern regions of the Corn Belt, where subsurface tile drainage is currently used on about one-third of total cropland to mitigate this problem (Sugg, 2007).

2.2 Crop management

Individual fields are typically >50 ha and crop production is highly mechanized, resulting in systems where the average labor input is only 4–6 h ha^{-1} per crop. Management practices are adjusted according to thermal and water regimens. The length of the crop growing season is technically defined (on a calendar date basis) by the probability of a late spring frost occurring on or after crop emergence and by the probability of an early frost before physiological maturity. Average sowing date for maize ranges from mid-April in southern locations to mid-May in northern locations, whereas soybean sowing begins 12–15 days after the start of maize sowing (Fig. 2.4). Similarly, recommended maize hybrid maturity (range: 1200 to 1600 growing degree days; Tb = 10°C) and soybean maturity group (range: 00 to IV) decreases from south to north in concordance with the long-term probability of frost occurrence during the final phases of grain filling. When water supply is not limiting, delays in sowing reduce yield with linear rates about 17 to 42 kg ha^{-1} d^{-1} after May 1 in soybean (Bastidas et al., 2008; Villamil et al., 2012) and 33 to 72 kg ha^{-1} d^{-1} after the end of April in maize (Nafziger, 1994; Grassini et al., 2011a; Roekel and Coulter, 2011). Over the past three decades, crop producers have persistently shifted their maize and soybean sowing times to ever-earlier calendar dates at a rate of about 0.4 d year^{-1} (Fig. 2.4). This shift may be attributable to climate change (i.e. warmer early springs), genetics (better early-season germination and seedling cold tolerance in new cultivars), and agronomic technologies (multirow planters that lessen the number of total sowing days, improved seed treatments that mitigate pathogen losses) (Sacks and Kucharik, 2011;

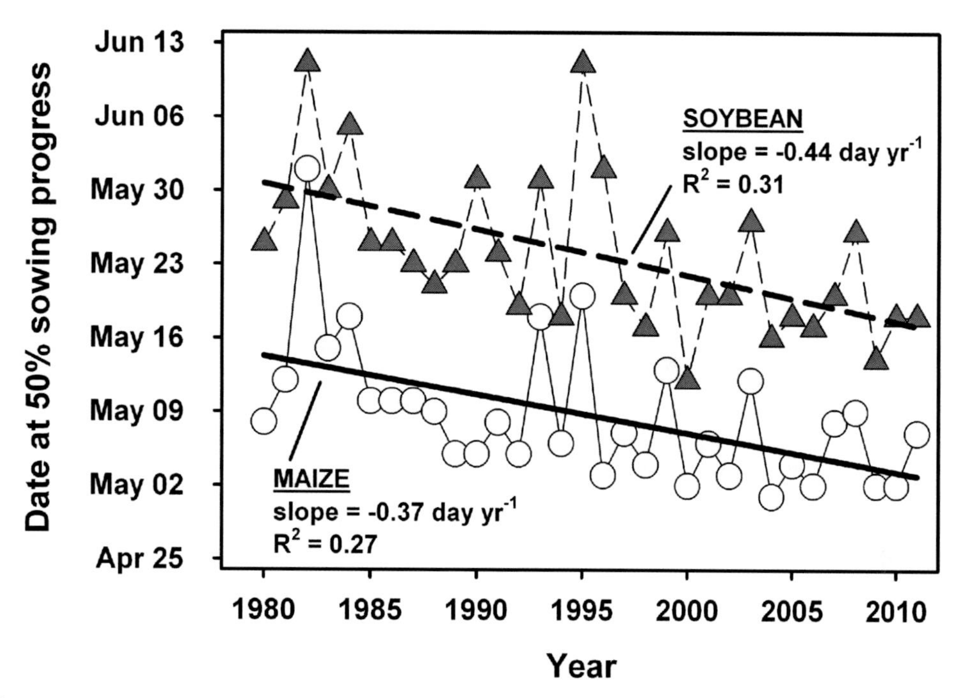

FIG. 2.4 Time-trend in the date at which 50% of sowing progress was achieved for maize and soybean total sown area between 1980 and 2011 in Nebraska. *Source: USDA-NASS.*

Tollenaar and Lee, 2011). Similarly, a combination of environmental, largely warming, and technological innovations including new varieties, are shifting the spatial and temporal boundaries of cropping systems in China (Chapter 3) and northern Europe (Chapter 4).

In the regions of the Corn Belt where maize crops are not severely limited by water, such as the eastern and central areas or in the western irrigated area, average plant density is about 7.5–8.5 plants m^{-2}. Although higher yields can be achieved with higher plant densities, there are economic trade-offs because the yield benefit from additional plants does not compensate for the greater seed cost (Grassini et al., 2011a; Roekel and Coulter, 2011). In contrast, recommended plant density for rain-fed maize in the western Corn Belt decreases following the east–west gradient in water deficit from about 8 in central Iowa to 3 plants m^{-2} in eastern Colorado (Fig. 2.5). On-farm average soybean plant density is typically 35–45 plants m^{-2}, which is higher than the required plant density for maximum yields (25–35 plants m^{-2}) due to lower seed cost compared with hybrid maize, varying little across regions or water regimens (De Bruin and Pedersen, 2009).

During the past decades, there has been a shift from continuous maize under conventional tillage to a 2-year maize–soybean rotation under reduced tillage (defined as any tillage method that leaves ≥30% of soil surface covered by plant residues). Of total area sown with maize and soybean, a respective 52% and 75% is under reduced till (Horowitz et al., 2010). Aside from economic and marketing factors, there are agronomic incentives for adoption of the reduced-till maize–soybean rotation system as opposed to conventional-till continuous maize. First, there is a yield advantage of maize grown in rotation with soybean compared with continuous maize, in both irrigated and rain-fed production systems. This yield advantage ranges from +3

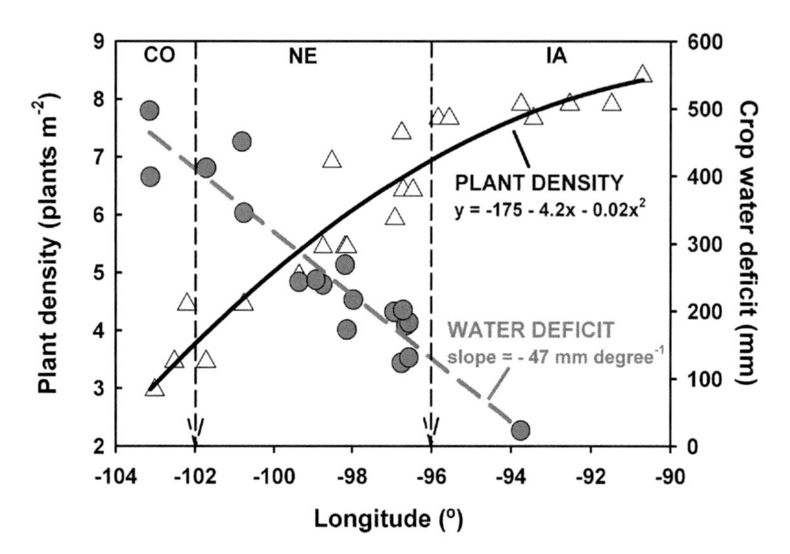

FIG. 2.5 Actual rain-fed maize plant density (solid line) and crop water deficit (in-season ET_C minus rain, dashed line) along a west–east transect in the western Corn Belt. Vertical lines with arrows indicate approximate borders of the states (CO: Colorado; IA: Iowa, NE: Nebraska). *Adapted from Grassini et al. (2009).*

to +15% in both experimental plots (Crookston et al., 1991; Porter et al., 1997; Vyn et al., 2006) and producer fields (Grassini et al., 2011a). This is generally true for cereal–legume crop rotations, and has been attributed to lower disease pressure and improved weed management with rotation, more favorable seedbed for maize sown into soybean residue, than into maize residue (i.e. second-year maize) in reduced-tilled fields, and less N immobilization with soybean residue. Second, soil residue cover with reduced- or no-till minimizes soil erosion, increases precipitation-storage efficiency during the non-growing season, and reduces surface runoff and soil evaporation which, altogether, improve crop water availability and reduce irrigation water requirements (Nielsen et al., 2005; Klocke et al., 2009; Grassini et al., 2011b).

3 PRODUCTIVITY AND RESOURCE-USE EFFICIENCY

3.1 Resource requirements for high yields

Biophysical limits to crop resource-use efficiency are defined by fundamental physiological processes including photosynthesis, respiration, transpiration, nutrient uptake, and dry matter partitioning. Relationships between crop productivity and resource use follow linear (solar radiation and evapotranspiration) or curvilinear functions (nitrogen, N) (Fig. 2.6a, b, c; solid lines).

The physiological efficiencies, calculated from the first derivative of these relationships, are constant (radiation- and water-use efficiency) or decrease (N-use efficiency) with increasing absorbed resource. Although data scattering along these functions reflects variation in physiological efficiencies due to variable weather across environments, variation in the temporal patterns of resource availability and absorption during the growing season, imbalances with other nutrients, grain biomass composition, and incidence of other limiting factors, there is still an upper limit for the efficiencies (Fig. 2.6b, c; dashed lines). The relationships shown in Figure 2.6 can be used to estimate the amounts of absorbed radiation, N, and water that are required to achieve a certain yield. For example, a maize crop that produces about 13 t ha^{-1} grain (harvest index of 0.50 and standard grain moisture content of 0.155 kg H_2O kg^{-1} grain), grown without the interference of weeds, pathogens, and insect pests, absorbs 5.7 TJ of photosynthetic-active solar radiation, 200 kg of N, and transpires 400 mm

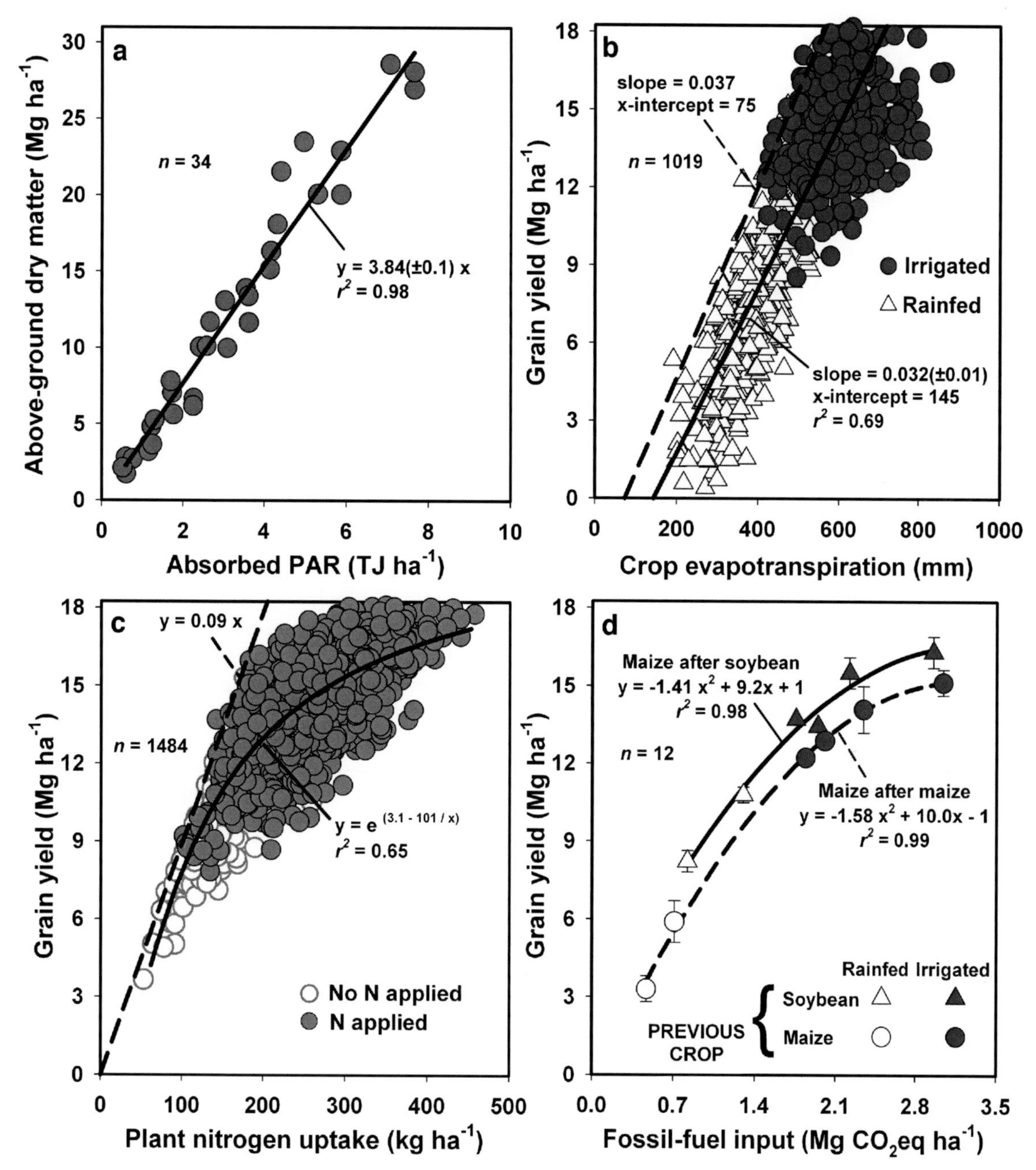

FIG. 2.6 Relationships between maize (a) above-ground dry matter and absorbed photosynthetic-active radiation (PAR), and between grain yield and (b) nitrogen (N) uptake, (c) ET_C, and (d) fossil-fuel inputs. Data points correspond to actual measurements (a, c, Lindquist et al., 2005, Setiyono et al., 2010a) or crop-model simulations (b, Grassini et al., 2009). Averages for maize-based systems are shown in (d) including rain-fed maize in western Nebraska (Drew Lyon, personal communication), rain-fed and irrigated maize in central (Grassini & Cassman, 2012) and eastern Nebraska (Verma et al., 2005, Adviento-Borbe et al., 2007), and rain-fed maize in Iowa (Connor et al., 2011). Boundary functions (dashed lines) were fitted (b, c). Fossil-fuel inputs (d) were expressed as carbon-dioxide equivalents (CO_2eq) and separate regressions were fitted for continuous maize and soybean–maize systems.

of water. A corollary of this analysis is that relatively high amounts of resources must be absorbed by the crop to achieve high yields. Light harvesting can be maximized through earlier sowing and proper selection of cultivar maturity, plant density, and stand spatial arrangement. When seasonal rainfall and indigenous soil N are not enough to satisfy water and N requirements to build and maintain a green canopy, the extra resource requirement has to be provided through supplemental irrigation and fertilizer (Fig. 2.6b, c). For example, assuming that indigenous N provides 130 kg N ha^{-1} (which can be retrieved from grain yield measured in unfertilized field strips, so-called 'omission plots'), the remaining 70 kg N ha^{-1} must be provided by applied N. If average recovery efficiency of applied fertilizer is 40% (Cassman et al., 2002), then an N fertilizer rate of 175 kg N ha^{-1} must be applied to meet crop N demand. This estimated N rate compares very well with actual on-farm fertilizer N rate (183 kg N ha^{-1}) reported for irrigated maize in Nebraska that achieves average yields of ≈13 t ha^{-1} (Grassini et al., 2011a). Likewise, irrigation is fundamental to ensure high yields with low inter-annual variation where growing-season rainfall and stored available soil water at sowing are not sufficient to satisfy total crop water requirements. For example, average maize and soybean yields in Nebraska were 66 and 40% higher, respectively, and three and four times less variable with irrigation than under rain-fed conditions (Fig. 2.7).

Table 2.1 shows an inventory of applied inputs in rain-fed maize and soybean production in Iowa. While maize receives relatively large N, P, and K inputs from fertilizer or livestock manure, soybean largely relies on mineralization of soil organic matter (≈50%) and N symbiotic fixation (≈50%) to meet N requirements and residual P and K fertilizer from previous maize crop (Salvagiotti et al., 2009). Lime is applied to offset soil acidification caused by application of N fertilizer and N fixation. Herbicide use is higher than fungicide and insecticide inputs as a result of the cold winter, which interrupts the cycle of most insect pests and diseases, and the widespread adoption of transgenic herbicide- and insect-resistance crop varieties. In fact, about 85% of maize fields and 90% of soybean fields are sown

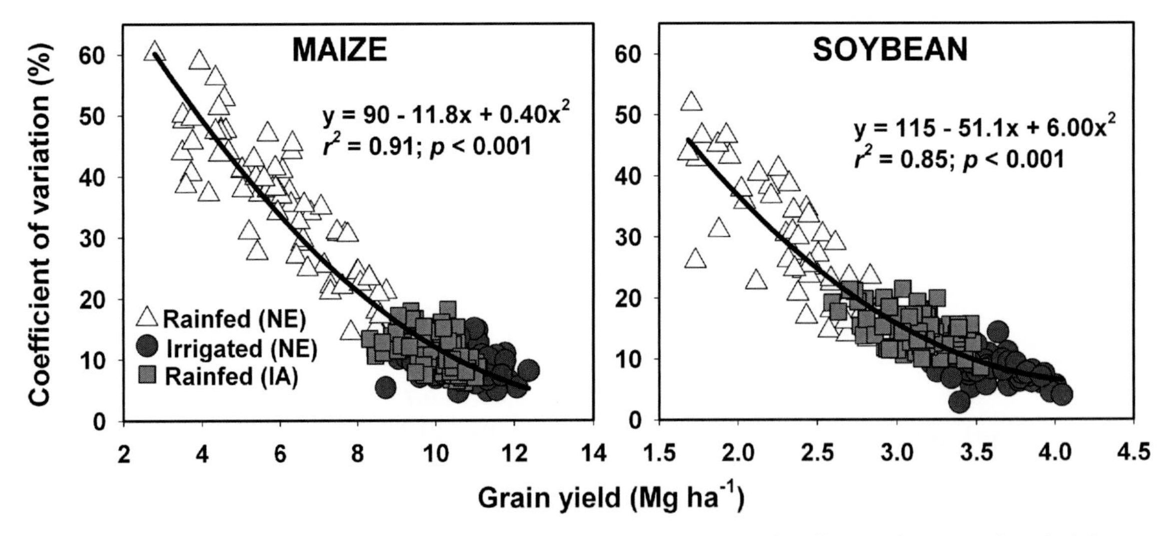

FIG. 2.7 County-average (2000–2009) maize and soybean yield and inter-annual coefficient of variation for rain-fed crops in Iowa and irrigated and rain-fed crops in Nebraska. Data from counties with harvested area <1500 ha were excluded. *Source: USDA-NASS.*

TABLE 2.1 Average fertilizer, N manure, and pesticides rates per treated hectare in rain-fed maize and soybean in Iowa

Applied inputs	Maize	Soybean
Average fertilizer rates		
Nitrogen (kg N ha^{-1})	158 (95%)	16 (7%)
Phosphorous (kg P ha^{-1})	32 (72%)	26 (12%)
Potassium (kg K ha^{-1})	75 (68%)	79 (20%)
N from manure (kg N ha^{-1})*	281 (17%)	145 (2%)
Pesticide application (kg a.i. ha^{-1})		
Insecticide	NA	0.2 (9%)
Herbicide	2.2 (97%)	1.6 (98%)
Fungicide	0.1 (10%)	0.1 (2%)

Percentage of treated hectares is indicated in parenthesis.
** N manure was calculated assuming 50% of dry matter in applied manure and 0.018 kg N kg^{-1} dry matter manure. a.i.: active ingredient. NA: Data not available.*
Source: USDA-NASS (2005).

with genotypes possessing transgenic traits (USDA-ERS, 2005–2011). Most fungicide and insecticide is applied as seed treatment but foliar applications are becoming increasingly popular, even in the absence of pathogens or insect pests (Villamil et al., 2012).

High-yield cropping systems require fossil-fuel inputs to substitute human and animal labor and maximize capture and conversion of solar radiation into crop biomass. Fossil-fuel cost associated with agricultural production can be quantified as the sum of greenhouse-gas (GHG) emissions derived from manufacturing, packaging, and transportation of applied inputs including fertilizers, manure, pesticides, lime, seed, machinery, and fossil fuel used for field operations, irrigation water pumping, and grain drying. In rain-fed maize, about 75% of total energy inputs are accounted for by N fertilizer, grain drying and field operations, while fuel use for irrigation pumping accounts for more than 40% of total energy inputs in irrigated maize (Connor et al., 2011; Grassini & Cassman, 2012). Figure 2.6d presents the relationship between grain yield and fossil-fuel inputs for maize systems of diverse degrees of intensification in the western Corn Belt, from low-input low-yield rain-fed maize in western Nebraska to high-input, high-yield maize under irrigation. The relationship is fairly linear up to relatively high levels of intensification that correspond to grain yields of ≈13 t ha^{-1}. Above this yield, the relationship becomes quadratic as actual yield approaches yield potential. Interestingly, the fossil-fuel requirement to achieve the same yield is higher in maize monoculture than in maize–soybean rotation (Fig. 2.6d).

3.2 Time trends in yields and input-use efficiency

Rates of increase in average maize and soybean yields in the Corn Belt have been markedly linear between 1964 and 2010, despite evidence of an incipient yield plateau for irrigated maize in Nebraska in recent years (Fig. 2.8). Annual rate of yield gain (i.e. slope of regression) decreased and inter-annual yield variation (r^2) increased from favorable to less favorable environments for crop production (Nebraska irrigated > Iowa rain-fed > Nebraska rain-fed). Simulations of maize water-limited yields indicate that rain-fed crops can potentially achieve yields similar to those under irrigated conditions in years with adequate water supply (Fig. 2.3). However, actual yield trends show that the average difference between rain-fed and irrigated crops in Nebraska during the 1964–2009 interval was, on average, 3.4 t ha^{-1} for maize and 0.7 t ha^{-1} for soybean and, even in the best years for rain-fed crops, this yield difference was never smaller than 1.3 for maize and 0.3 t ha^{-1} for soybean (Fig. 2.8, left panels). It seems like allocation of the best land to irrigated agriculture, lower applied inputs, later sowing, and lower plant density (in maize) under rain-fed conditions impose a limit to rain-fed crop yields, even in years when growing-season rainfall can support yield above the long-term trend.

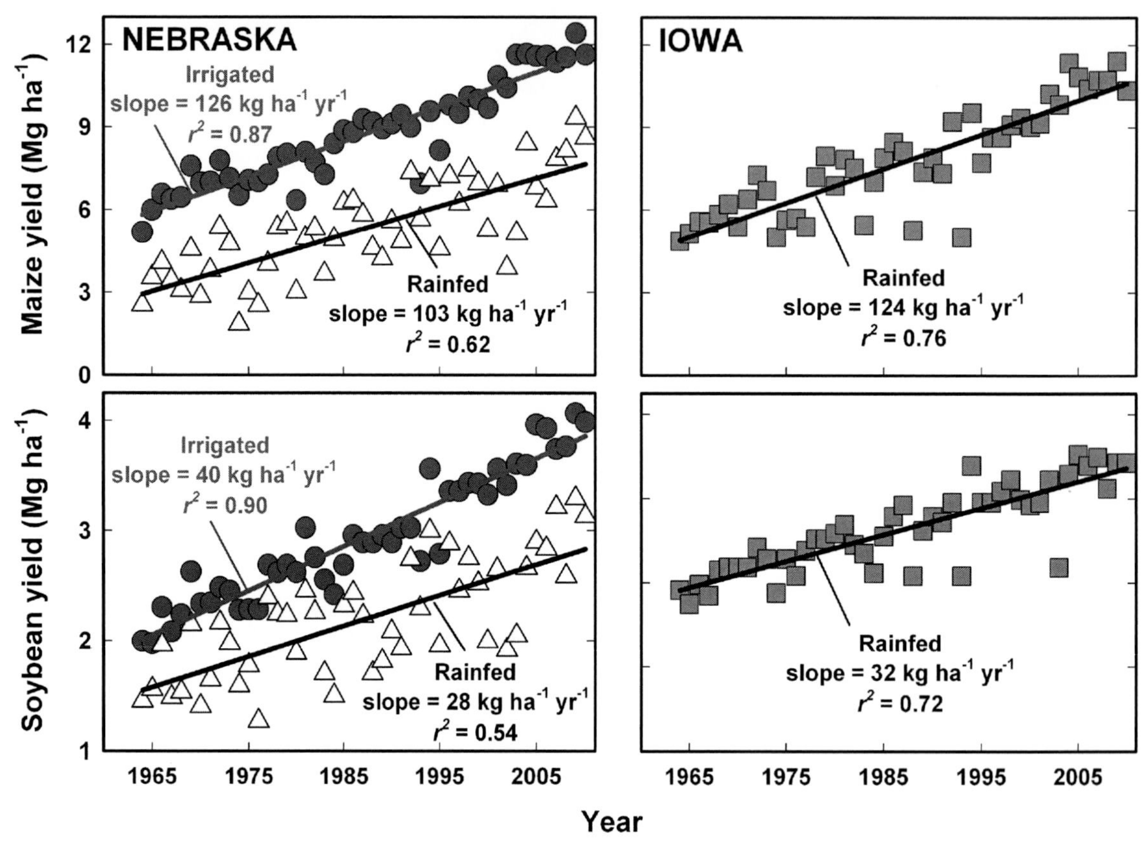

FIG. 2.8 Trends in maize and soybean yield in Nebraska (left panels) and Iowa (right panels). Separate trends are shown for rain-fed (triangles) and irrigated crops (circles) in Nebraska. *Source: USDA-NASS.*

Partial productivity factors, estimated as the ratio between grain yield to in-season rainfall or applied N fertilizer, can be used as proxies to assess trends in input-use efficiency. Increases in yield per unit rainfall and yield per unit N fertilizer (after 1975) have accompanied the increases in yield (Fig. 2.9). Average (10-year) maize yield, yield per unit rainfall, and yield per unit fertilizer in Iowa in the 2000s were 70, 50, and 40% higher than in the 1970s. Given the lack of any detectable long-term decline in in-season rainfall and applied N fertilizer (after the first decade), long-term improvement in efficiency is attributed to increases in crop yield.

Nutrient status of Corn Belt soils is determined by the balance between long-term trends in annual applications of fertilizer or livestock manure and nutrient amounts annually removed with grain. Fertilizer N rates used by maize farmers since 1965 have exceeded nutrient removal, but that excess has declined in recent years (Fig. 2.9). The steady improvement in yield per unit N fertilizer in maize since the late 1970s coincides with a decline in the apparent N fertilizer surplus (difference between applied N fertilizer and N removed with grain), which has narrowed from 50–70 (1976–1985 average) to 10–20 kg N ha^{-1} (2001–2010 average) (Fig. 2.9, insets). The declining N surplus implies that yield-scaled nitrous oxide emissions (kg N_2O kg^{-1} grain) have also decreased over time (van Groeningen et al., 2010; Venterea et al., 2011).

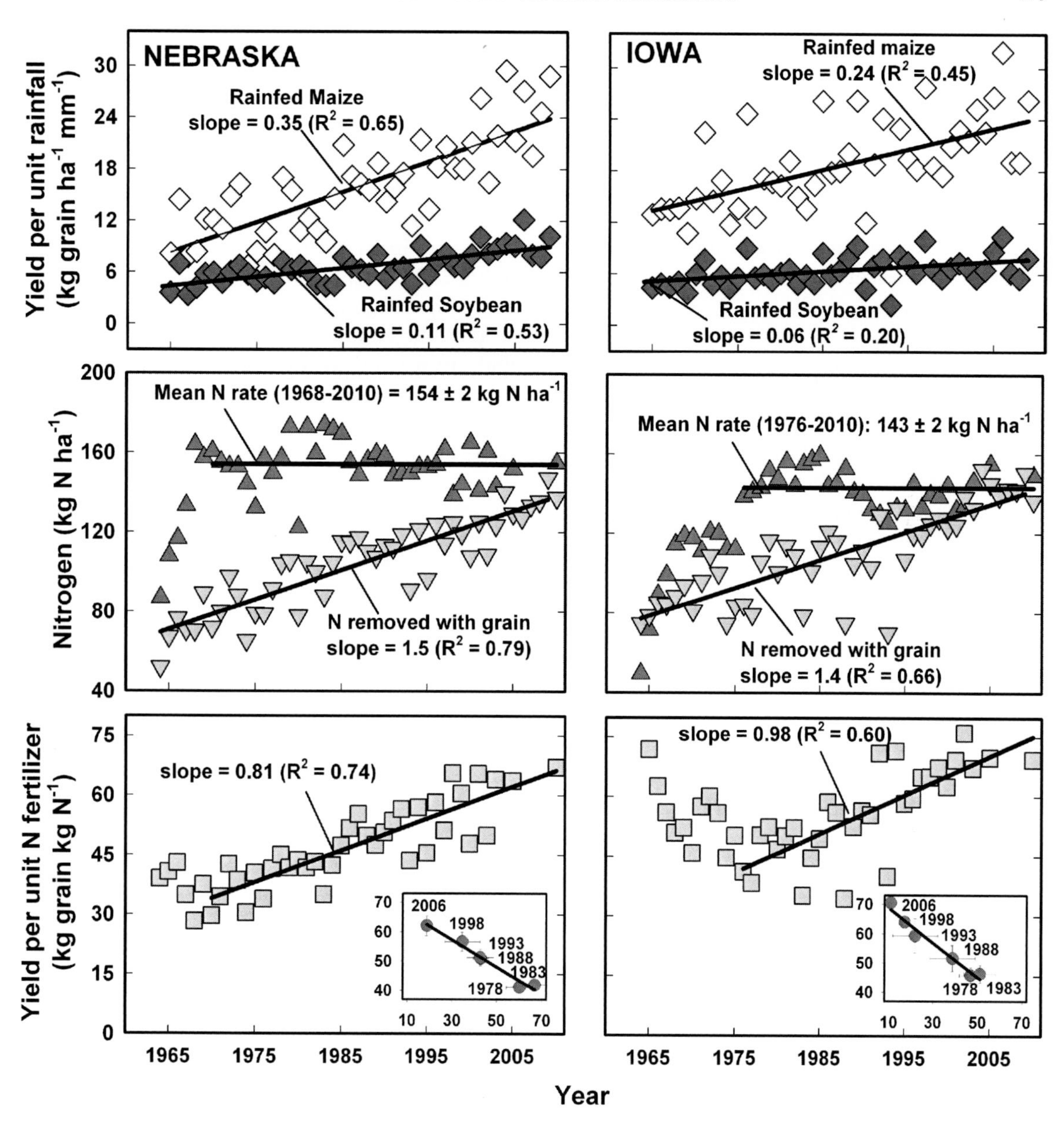

FIG. 2.9 Trends in yield per unit rainfall for rain-fed maize and soybean (upper panels), applied N fertilizer and grain N removal in maize (middle panels), and yield per unit N fertilizer for maize (bottom panels) in Nebraska and Iowa. Rainfall is total between May and August, measured at 4–5 weather stations located in major producing regions within each state (average: 340 mm in Nebraska and 433 mm in Iowa). Grain N removal was calculated based on average maize yields in Nebraska and Iowa and grain N concentration reported by Duvick et al. (2004) for hybrids released during the time interval. Applied N manure averaged 8 kg N ha^{-1} in Nebraska and 42 kg N ha^{-1} in Iowa during the last decade. Insets: yield per unit N fertilizer against N-fertilizer surplus (N fertilizer minus N grain removal, in kg N ha^{-1}). Values represent 5-year averages (± standard error of the mean), centered on the reported years, except for 2006 which is a 4-year (2001, 2003, 2005, and 2010) average. *Source: USDA-NASS.*

In fact, during recent decades, energy-use efficiency increased and yield-scaled GHG emissions decreased as a result of increasing yields without increases in applied N fertilizer rates, widespread adoption of conservation tillage practices, center-pivot systems to replace less efficient gravity irrigation, and increasing efficiency in the manufacture of various agricultural inputs (Shapouri et al., 2003; Connor et al., 2011; Grassini & Cassman, 2012).

3.3 Drivers for higher yields and efficiencies

Improvement in on-farm crop yield and efficiencies arises from the adoption by producers of (1) genetic (G) technology in the form of newly released maize hybrids and soybean cultivars that have greater genetic yield potential and (2) agronomic (A) technology in the form of crop management practices that enhance the yield of the production environment. Transgenes that increased yield of the production environment (i.e. by reducing the impact of insects or weeds) are considered here as agronomic technologies. Aside from the separate impact of the foregoing technologies, (3) synergistic G × A interaction is also important, given that the greater genetic yield potential available in modern vs obsolete cultivars is coupled with the greater yield of modern vs dated agronomic practices. For comparison, other chapters in this book outline breeding and agronomic progress in cereal-based systems of China (Chapter 3) and northern Europe (Chapter 4).

Unlike genetic improvement, which is more or less continuous (i.e. new cultivars are released almost every year), agronomic yield improvement is more episodic. Moreover, if there is a steep learning curve, producer adoption of a new management practice will be cautiously measured. As a result, agronomic yield improvement still tends to be nominally gradual – not immediate. Moreover, the degree to which a given agronomic practice can improve system productivity is typically finite – for example, advancing the calendar date of sowing will ultimately be constrained by an increasing probability of a seedling-killing spring freeze that will accompany any sowing date advance. Yet, even if intervallic, agronomic improvement in the yield of modern era production environments is critical. Indeed, producers with highly fertile fields and favorable weather become yield contest winners because they can bring to bear their crop management skills to create the conditions that are needed to capture and thus realize the high genetic yield potential of current cultivars.

A question of interest is how much of the annual on-farm yield improvement can be attributed to an increase in genetic crop yield potential. The answer to that question is complex and controversial because it is fraught with methodological difficulties. A typical approach to determine the annual rate of genetic gain in yield potential is to evaluate the yields of an array of widely used commercial cultivars released over a given timeframe and grown simultaneously in current production environments. The assumption is that the ratio between the rate of genetic yield gain (i.e. slope of the linear regression between yield and year of cultivar release) and on-farm yield gain provides an estimate of the relative contribution of genetics to on-farm yield improvement; the remaining portion of the on-farm yield gain is attributable to agronomic technologies and to the G × A interaction. This way of distinguishing the relative contributions of genetics and agronomy to on-farm yield gain is subjected to multiple sources of error, including differences between old and new cultivars in growth duration, grain composition (carbohydrate, oil, and protein), susceptibility to pathogens and insect pests, and acclimation to changing climate. Also, site-management conditions of the trials in which the genetic yield gain is estimated are generally better than those for the average crop producer, resulting in an overestimation of the genetic gain because new cultivars exhibited greater yield response to improvements in the production environment compared with old ones. The following section discusses the genetic and agronomic technologies that have contributed to on-farm maize and soybean yield gain in the Corn Belt and highlights the synergies between the two. The relative contribution of each to the overall yield gain, however, remains uncertain.

3.3.1 *Maize*

GENETIC DRIVERS

Genetic improvement in maize has been extensively documented in a set of material called the ERA hybrids, which consisted of successful, widely grown hybrids developed and released by Pioneer Hi-Bred International from the 1930s to the 2000s (Duvick, 2005 and references cited therein). Hybrids in these trials were grown at 3.0, 5.4, and 7.9 plants m^{-2} and for each hybrid the highest average yield was taken for the regression of yield against year of hybrid release. Yields of the hybrids from the 1930s to the 2000s grown in ERA-hybrid trials increased linearly at a rate of 77 kg ha^{-1} $year^{-1}$ (Duvick et al., 2004). Genetic improvement has been the result of selection for yield *per se*, although molecular technologies have been increasingly used since the early 1990s as an aid in the selection, supported by a nearly fourfold increase in the number of breeders and annual inflation-adjusted investment in maize breeding from the 1970s to 1990s (Duvick and Cassman, 1999). Although maize breeders did generally not select for specific traits, a number of traits have changed in association with genetic improvement. Newer hybrids are more tolerant to high plant density and exhibit enhanced tolerance to specific abiotic stresses such as drought (Dwyer et al., 1992; Nissanka et al., 1997), low N (McCullough et al., 1994a,b; Echarte et al., 2008; Ciampitti and Vyn, 2012), and low air temperature (Dwyer and Tollenaar, 1989; Ying et al., 2000, 2002). The mechanisms responsible for enhanced stress tolerance in newer hybrids are unknown, but they appear to be associated with greater canopy-level photosynthesis and kernel setting.

Seasonal canopy photosynthesis has increased due to greater light interception and higher canopy photosynthesis. Light interception has increased because of higher maximum green leaf area index (LAI) of newer hybrids, which resulted from increases in plant density, and delayed leaf area senescence during the grain-filling period (Duvick et al., 2004). Aspects of the 'functional stay green' syndrome are discussed in Chapters 8 and 10. Increase in canopy photosynthesis can be attributed to a better light distribution within the canopy due to higher leaf angle and, especially, to the capacity of green leaves to maintain high photosynthetic rates during grain filling (Tollenaar et al., 2000; Ying et al., 2000; Echarte et al., 2008) (Fig. 2.10).

Duvick et al. (2004) reported that grain yield and kernel weight of ERA hybrids during the 1934–2001 period increased linearly at a rate of 1.45 and 0.22% $year^{-1}$, respectively, indicating that 85% of the yield improvement was associated with kernel number and the other 15% with kernel weight. Kernel set, defined as the number of fertilized florets that continue to accumulate dry matter during the grain-filling period, depends on the supply of assimilates from current photosynthesis during a 3-week period bracketed by silking (Tollenaar et al., 1992; Otegui and Bonhomme, 1998; Andrade et al., 1999). Differences in rate of dry matter accumulation during this period in old vs new hybrids are not large (Tollenaar et al., 1994; Echarte et al., 2000). Hence, increased kernel number per unit area in new vs old hybrids has resulted from increases in kernel set per unit of dry matter accumulation as is indicated by (1) a lower threshold of plant growth rate for kernel set and associated reduction in the number of barren plants at high plant density and (2) greater kernel set over the whole range of plant growth rates above the threshold (Echarte and Andrade, 2003; Echarte et al., 2004; Echarte and Tollenaar, 2006).

Other changes that have occurred are: shorter anthesis–silking interval, slightly longer duration of the grain-filling period, reduction

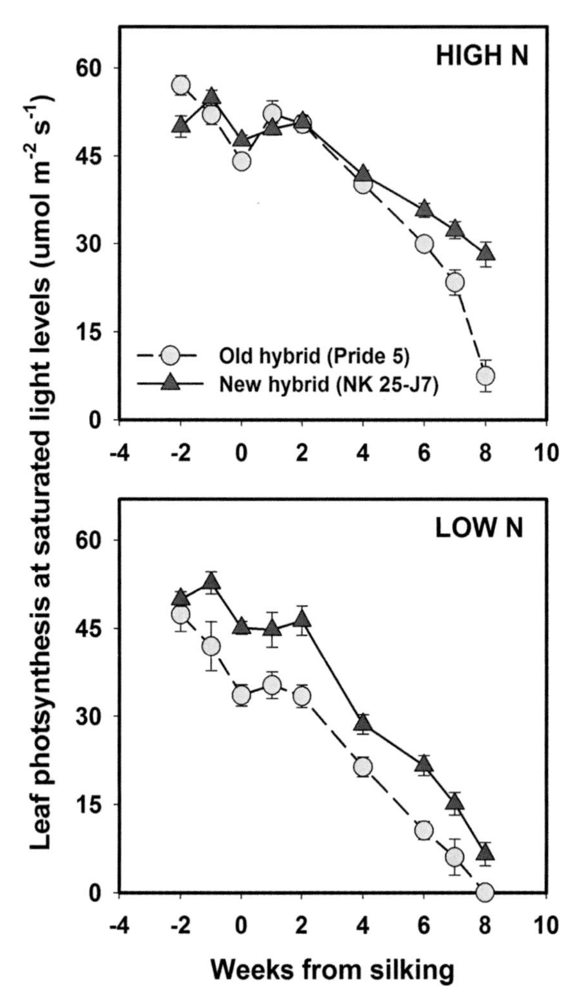

FIG. 2.10 Leaf photosynthesis at saturated light levels as a function of weeks from silking in old (Pride 5) and new (NK 25-J7) maize hybrids exposed to high (upper panel) and low nitrogen (N) conditions (bottom panel). Vertical bars represent two standard errors of the mean. *Adapted from Echarte et al. (2008).*

in tassel dry weight and branch number, and decline in grain N concentration with concurrent increase in grain starch (Cavalieri and Smith, 1985; Russell, 1985; Duvick, 1997; Duvick and Cassman, 1999; Duvick et al., 2004). Modeling showed that deeper rooting can allow access to additional water from deeper soil layers and increase grain yield (Hammer et al., 2009), but experimental evidence is required to confirm whether or not this was a contributor to past genetic gain. In contrast, a number of traits associated with potential yield per plant have not changed, such as the number of leaves, leaf area per plant, plant height, potential leaf photosynthesis, and plant N uptake (Russell, 1985; Tollenaar et al., 1994, 2000; Duvick, 1997; Duvick et al., 2004; Tollenaar and Lee, 2002; Ying et al., 2000, 2002; Lee and Tollenaar, 2007; Echarte et al., 2008; Ciampitti and Vyn, 2012). Contribution of heterosis to yield increases, expressed as a percentage of the average of the parental genotypes, has not changed over time, nor has it declined (Duvick, 2005). In contrast to the steep increase in harvest index detected in Argentinean hybrids released from 1965 to 1997 (Luque et al., 2006), harvest index has not changed in North American hybrids, although there is some evidence of higher harvest index in recent years (Duvick, 2005; Tollenaar and Lee, 2002, 2011).

AGRONOMIC DRIVERS

Use of N fertilizer and higher plant densities were the most important agronomic technologies associated with maize yield gain in the Corn Belt. Plant density increased from 3 plants m^{-2} in the 1930s to 8 plants m^{-2} in the 2000s. Higher N fertilizer rates have contributed to the yield gain, especially during the 1960–1975 period (Fig. 2.9). Since 1975, average N rates leveled off at ≈150 kg ha^{-1}, while yield per unit of N fertilizer increased steadily due to higher yields and better N fertilizer management. Improvements in N management included significant reductions in fall applied N fertilizer with a shift to applications in spring or at sowing, greater use of split N-fertilizer applications during the growing season rather than a single large N application, and development and extension of N-fertilizer recommendations that give N 'credits' for manure, legume rotations, and residual soil nitrate as determined by soil testing (Cassman et al., 2002). Each of these practices helped to better match the amount and timing of applied N to crop-N demand and the N supply from indigenous resources. The expansion of maize irrigated area in the Corn Belt, from 0.9 Mha in 1970 to 3.2 Mha in 2010, increased yields in regions where rainfall cannot fully meet crop water requirements by eliminating water deficits and amplifying the benefits of N fertilizer and higher plant densities.

Better weed and pest control is another factor that has contributed to increased on-farm yield. Better weed control was achieved by increasing use of herbicides, starting with 2,4-D in the 1940s and atrazine that was first used commercially in the 1960s. The use of broad-spectrum herbicides such as glyphosate in herbicide-tolerant maize (GT) expanded rapidly in the late 1990s and it has simplified and improved weed control. Owing to widespread use of glyphosate in herbicide-resistant maize and soybean, weed resistance is becoming a substantial problem (Mortensen et al., 2012). Duvick (2005) reported improved insect and disease resistance in ERA-hybrid studies using hybrids from the 1930s to the 1990s, but it was during the 1990s when European corn borer (*Ostrinia nubilalis*, ECB) and corn rootworm (*Diabrotica spp.*, RCW) resistance was introduced into commercial maize hybrids using molecular technologies. Edgerton et al. (2012) compared isogenic hybrids, each with and without the ECB and RCW Bt traits (2400 pairs), as well as non-isogenic pairs (15000 pairs) during a 5-year period (2005–2009) in the Corn Belt. Mean trait effects were similar for the two groups, with a statistically significant average trait effect in the non-isogenic comparisons of 0.35 t ha^{-1} (3.5% of average yield). Shi et al. (2013) compared yield among conventional and transgenic (ECB-, RCW-, and GT-resistant) hybrids based on field experiments conducted in agricultural research stations and producer fields in Wisconsin that included a total of 4748 hybrids tested in the past 21 years, of which 2653 were conventional hybrids and 2095 were transgenic. These authors reported that, despite the lack of strong, consistent positive yield differences between conventional and transgenic hybrids (except for the outperforming ECB-resistant hybrids), the transgenes reduced production risk by reducing the variance and increasing skewness compared with conventional hybrids (i.e. yield distribution is less skewed to the left indicating a lower exposure to losses and downside risk).

Other agronomic changes that are unrelated to the use of greater production inputs or transgenes also explained part of the maize yield gain in the Corn Belt. Maize sowing date has occurred earlier during the last three decades (Fig. 2.4). Earlier maize sowing can increase yield because of increased duration of the growing season and, consequently, an increase in the total absorbed solar radiation. Kucharik (2008) showed that 19 to 53% of state-level yield increase during the 1979–2005 period can be attributed to earlier sowing in 6 out of the 12 states in the northern and western Corn Belt. Also, based on total maize and soybean harvested area data reported for the three major maize producer states, it can be inferred that the proportion of maize grown in a 2-year maize–soybean rotation has increased from 47% (1968–1972 average) to 65% (2007–2011 average). Given the yield advantage when maize is grown in rotation with soybean compared with monoculture, part of the observed increase in on-farm maize yield is attributable to the increasing proportion of maize grown in a 2-year maize–soybean rotation.

Aside from advancing the sowing to earlier spring and rotating maize with soybean, there are some other agronomic practices that might have enhanced the yield of the production environment, but whose impact has not yet been experimentally documented. Producers have increasingly adopted no-till or reduced-till prior to sowing. An undisturbed crop residue can increase capture of both pre- and in-season precipitation in the soil profile that buffer dry spells between rainfall events. Such stored water is also useful in irrigated production, since it can reduce or delay the need for a water application (Grassini et al., 2011b). One would expect some yield enhancement arising from the gradual increase in the cropland area under conservation tillage in the Corn Belt. The benefits and drawbacks of these conservation practices are further discussed for contrasting cropping systems of China in Chapter 3, and Africa in Chapter 5.

G × A INTERACTION

Effects of genetics and agronomic technologies on maize yield improvement are frequently not additive. Indeed, yield improvement would not have been nearly as large if either genetics or management had occurred in isolation from each other. The relationship between yield and plant density was the trait most affected by G × A interaction. Genetic yield improvement of US maize hybrids has been associated with an increase in the optimum plant density for grain yield (Tollenaar and Lee, 2011). Plant densities were increased markedly in the 1960s to take advantage of the greater N fertilizer rates (Fig. 2.9). Although some hybrids yielded more at higher plant density, other hybrids exhibited similar or even lower yields when grown at high density. Producers preferred those hybrids that produced higher yield at high density and breeders selected new hybrids that can take advantage of even higher plant densities. An unintended consequence of this process was selection for hybrids more tolerant to plant-to-plant resource competition, reduced root or stalk lodging, and an increase in leaf angle which allows greater light penetration into the canopy at the higher leaf area index that results from higher plant densities (Duvick and Cassman, 1999; Duvick et al., 2004). The upper limit for the increase in maize canopy photosynthesis associated with an increase in leaf angle from 30 to 60° has been estimated to range from 15 to 30%, which translates to an ≈15% increase in total dry matter (Tollenaar and Dwyer, 1998).

Crop N uptake is greater in new (1991–2011) than in old hybrids (1940–1990) due to greater tolerance of modern hybrids to high plant

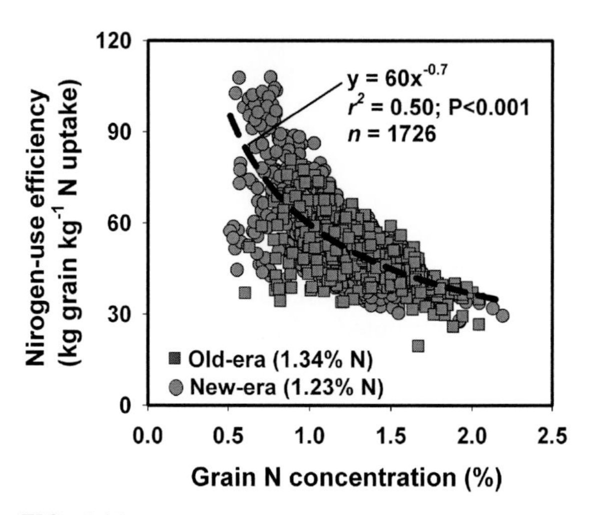

FIG. 2.11 Relationship between nitrogen-use efficiency and grain nitrogen (N) percentage for the old- (blue squares; n = 356) and new-era North American maize hybrids (red circles; n = 1370). Average grain N concentration is shown for each group. *Adapted from Ciampitti and Vyn (2012).*

densities with no change in N uptake per plant (Ciampitti and Vyn, 2012). Differences in N-use efficiency (kg grain yield per unit N uptake) between new and old hybrids, and also among hybrids within the old- and new-era groups, are mostly associated with variation in grain N content (Fig. 2.11). Comparison of Figure 2.11 in this chapter with Figure 8.16a in Chapter 8 highlights the robustness of this relationship across crop species and environments.

The traits that will likely drive future maize yield improvement are the same that have driven genetic gain in the past: abiotic- and biotic-stress tolerance, increased functional stay-green, and increased kernel set per unit area (Tollenaar and Lee, 2011). G and A factors are to a large extent interchangeable and, consequently, yield improvement will result from the combination of both. For instance, abiotic-stress tolerance can be improved by genetic means and/or improved site-specific management through selection of the best combination of hybrid, plant density, pest control, and nutrient inputs for management zones within a field.

3.3.2 *Soybean*

GENETIC DRIVERS

Soybean breeders seek to increase the genetic yield potential by releasing new cultivars that have demonstrably higher yield than currently available cultivars of the same maturity group (MG). Rincker et al. (2014) recently reported the results of a 2-year multilocation set of ≈27 site-year trials in which 60 MG II, or 59 MG III, or 49 MG IV soybean cultivars were evaluated. These cultivars had been developed by public or proprietary breeders and released to US producers during an eight-decade timeframe (≈1925–≈2005). Using this database, Specht et al. (2014) reported a 12 kg ha^{-1} $year^{-1}$ rate of genetic yield improvement prior to 1969 for the three MG, followed by faster 30 kg ha^{-1} yr^{-1} rate thereafter. Specht et al. (2014) attempted to address that issue by assuming that the genetic yield potential bred into modern cultivars should be more readily evident if it could be expressed in a high-yield production system, such as irrigated production sites with fertile silt loam soils with high water-holding capacity in Nebraska. A post-1969 linear rate of genetic improvement of 28 kg ha^{-1} $year^{-1}$ was estimated for the MG II and III historic cultivar sets that were grown in 2010–11 in an irrigated site (Rincker et al., 2014). In a near-concordant timeframe of 1971 to 2012, on-farm yield improvement in irrigated soybean production systems rose at a rate of 42 kg ha^{-1} $year^{-1}$. Simple comparison of the genetic (28 kg ha^{-1} $year^{-1}$) and the on-farm rate (42 kg ha^{-1} $year^{-1}$) suggests that breeding could account for two-thirds (28/42) of the total yield improvement, at least at a high-yield site that provides ample opportunity for the expression of genetic yield potential in an eight-decade array of soybean cultivar releases. As was discussed previously, this estimate of the contribution of genetic improvement to on-farm yield gain is likely to be lower if changes in cultivar maturity, grain composition, and site-management differences between experimental trials vs average producer production environment are taken into account. For example, Rincker et al. (2014) found that the R8 maturity date, which breeders use to assess relative maturity differences among cultivars, had advanced by 0.09 d $year^{-1}$ (i.e. 0.9 d per decade; totaling 7 days over the eight-decade timeframe). There is a highly positive correlation between yield and pre-fall-frost maturity, so an intense breeder selection focus on ever-greater cultivar yield may have led to concurrent indirect selection for an imperceptibly small advance in R8 cultivar maturity each year. In hindsight, this cumulative advance in cultivar maturity might have been facilitated by warmer fall temperatures in the selection environments. Another unintended consequence of breeder selection for ever-greater yield is a concordant decline in seed protein content by 0.02% $year^{-1}$(Rincker et al., 2014). Despite physiological attributions to the contrary, the generation of soybean seed storage proteins and their deposition in the seed actually requires more energy than seed oil (Chung et al., 2003). Given the long-known highly negative correlation of seed protein content with seed yield, in the eight-decade timeframe, breeders have released high-yielding cultivars whose yield advantage over obsolete cultivars may have come (to some degree at least) at the expense of seed protein content.

Other traits altered during eight decades of breeder selection for yield include a reduction in plant height and reduction in lodging potential which may have been achieved by the former (Rincker et al., 2014). However, there was no significant change over time in soybean seed mass suggesting that yield improvement was arising from ever-greater seed numbers produced per ha by modern cultivars. Rowntree et al. (2014) recently observed that breeding efforts over the 85-year period had improved total dry matter at physiological maturity, harvest index, and the length of the seed-fill period, and the latter is consistent with an observed concordant reduction in the length of the vegetative period. New varieties also exhibit greater N uptake during the seed-filling period due to greater contribution from N symbiotic fixation, which did not compensate for maintaining seed N concentration (Specht et al., 1999).

AGRONOMIC DRIVERS

Several agronomic practices likely contributed to higher yield in soybean production environments. The impact of irrigation and reduced-till adoption has been discussed for maize, and these factors have also contributed to the soybean yield gain. Sowing soybeans early to take advantage of the now warmer springs is one such practice that Nebraska producers have readily embraced (Fig. 2.4). Producers have also halved their row spacing from 0.76 to 0.39 m, but have moved away from narrower 0.19 m rows, probably because drill-planters do not work well in heavy residue, no-till systems. Soybean yield can theoretically be enhanced by 17 to 42 kg ha^{-1} for every day that the sowing date can be moved closer to May 1 for MG II and III cultivars grown in the Corn Belt [see Bastidas et al. (2008) for Nebraska data, and references therein for other Corn Belt states]. Using a soybean simulation model (SoySim; Setiyono et al., 2010b), Specht et al. (2014) estimated the yield impact if Nebraska producers had not advanced irrigated soybean sowing date to the degree shown in Figure 2.4. They used 1983 to 2012 actual weather data at two sites to compare the simulated yield impact of (1) using the date of 50% sowing progress for each of the 30 years (as shown in Fig. 2.4) and (2) using 1983 baseline 50% sowing progress date (May 30) that was kept constant for each of the subsequent 30 years. These authors documented that the sowing date advance of 0.44 d $year^{-1}$ depicted in Figure 2.4 may have accounted for one-eighth to one-fifth of the observed on-farm irrigated yield gain of 54 kg ha^{-1} $year^{-1}$ that concurrently occurred in the same 1983 to 2012 timeframe. More agronomic yield enhancement could be had if the 50% sowing progress date were to be moved even closer to May 1.

A soybean crop following maize in a 2-year maize–soybean rotation outyields soybean in monoculture (Fox et al., 2013 and references cited therein). A monoculture soybean system is rarely practiced nowadays in the Corn Belt, so any soybean yield enhancement arising from a shift from continuous soybean to a maize–soybean rotation would have been relevant in the past, but not now. It is noteworthy, however, that when a soybean crop follows two or more successive maize crops, the yield of the soybean crop is even greater than that expected when soybean follows one prior year of maize (Bonin et al., 2014). Crookston et al. (1991) found that, relative to monoculture soybean, annually rotated soybean (with maize) yielded 8% more and first-year soybean (after 5 years of maize) yielded even more (17%). The practice of two or more maize crops before a soybean crop is infrequent, but can occur when continuous maize production in a particular field must be interrupted with soybean to mitigate the build-up of some maize-specific disease or pest, or when soybean price at sowing time is substantively more favorable than concurrent maize price. The latter scenario could induce continuous maize producers to take advantage of that soybean price advantage and couple it with the substantive soybean yield bump that occurs.

Almost all of the soybean area in the Corn Belt is sown with transgenic cultivars and the vast majority of those cultivars possess a transgene that provides tolerance to the glyphosate. The use of glyphosate-resistant cultivars has not only provided near-total weed control (though glyphosate-resistant weeds are now increasingly appearing), but it has also accelerated producer adoption of reduced- or no-till practices as well as the narrowing of row spacing.

G × A INTERACTION

Rincker et al. (2014) noted that the estimated rate of genetic improvement was significantly greater when the historic cultivar sets were evaluated in highly productive environments vs less productive environments. This is a particular case of a widespread phenomenon, as illustrated in Chapter 16. For MGs II, III, and IV, the rates estimated in high vs low yield systems were a respective 24 vs 10 kg ha^{-1} $year^{-1}$, 23 vs 17 kg ha^{-1} $year^{-1}$, and 24 vs 18 kg ha^{-1} $year^{-1}$. Several recent publications have documented synergistic G × A interactions in which the yield difference between modern and obsolete cultivars widens when a dated agronomic practice is updated or replaced. For example, Rowntree et al. (2013) evaluated the MG II and MG III historic cultivar sets in additional experiments to assess the impact of a 30-d difference in sowing date and noted that genetic yield gain was greater in the early sown MG III tests (23 vs 20 kg ha^{-1} $year^{-1}$).

In another study done on the historic set of MG II and III cultivars, Suhre et al. (2014) examined the interactions between year of release and plant density. They sowed the historic set at a high (45 plants m^{-2}) and a low density (15 plants m^{-2}) and observed a significant interaction between the year of cultivar release and plant density, with the newer cultivars taking greater advantage of the high density than older ones. At high density, the yield increase was largely the result of increased yield on the main stems of plants, whereas at the low density, the yield increases over generations were the result of increased yield on both the main stems and branches. Finally, Wilson et al. (2014) evaluated the interaction between N supply and year of release and reported that new MG III cultivars had greater yield response to N-fertilizer addition compared with old ones. In contrast, yield of MG II cultivars did not increase with N-fertilizer addition irrespective of the year of cultivar release.

Unlike maize, soybean is a C_3 photosynthetic species in which photorespiration lessens the degree of potential dry matter accumulation. Thus, the rise in atmospheric CO_2 from 315 ppmv in 1958 to 394 ppmv in 2013 would be expected to incrementally improve on-farm soybean yields over time. Specht et al. (1999) estimated that continuously rising CO_2 levels could have contributed as much as 3–5 kg ha^{-1} $year^{-1}$ to the on-farm rate of annual yield improvement. Estimates of 0.7–5.4 kg ha^{-1} $year^{-1}$ were reported by Specht et al. (2014), who simulated the impact of a 2 ppmv annual rise in CO_2 from 1983 to 2011 on irrigated soybean yield at two locations in Nebraska. Global warming, on the other hand, has contributed to the increase in soybean production area in the northern Great Plains states of South and North Dakota and in the adjacent Canadian provinces (Fig. 2.1). Chapters 3 and 4 further document the benefits of global warming for agriculture at high latitudes.

4 CHALLENGES TO HIGHER YIELDS AND EFFICIENCIES

Yield potential can be taken as a benchmark to estimate yield gaps, that is, the difference between average on-farm yield and the yield potential defined by solar radiation, temperature, and genotype and also by water availability in rain-fed crop systems (Lobell et al., 2009; Van Ittersum et al., 2013). This definition of yield potential reflects an upper biophysical limit to what might be ultimately attainable for crop yields on a given field; hence, the magnitude of the yield gap estimates the degree of yield improvement that could still be captured on that farm with adjustments in crop management. Based on site-specific weather and current management practices (sowing date, hybrid maturity, and plant density) collected from irrigated maize fields over three years, Grassini et al. (2011a) estimated that the average yield gap for irrigated maize was 11% of the simulated yield potential of 14.7 t ha^{-1} (Fig. 2.12).These authors also noted that average yield potential could be increased to 17.0 t ha^{-1}, for the same subset of field-year cases, by simply increasing plant density and hybrid maturity from those currently used by producers. This upper yield potential compares well with the 18.2 t ha^{-1} yield-potential estimated by Duvick and Cassman (1999) based on contest-winning irrigated yields. Using this maximum yield potential of 17.0 t ha^{-1}as a benchmark, the current yield gap is 23% (Fig. 2.12), which is similar to the estimated yield gap for US irrigated maize reported by Van Wart et al. (2013). However, there are trade-offs associated with management practices that seek to achieve maximum yields such as increasing seed costs, lodging, and potential difficulties in using later hybrid maturity with regard to fall frost risk, additional grain drying costs, and problems with crop harvest due to inclement later fall weather (Grassini et al., 2011a; Roekel and Coulter, 2011; Novacek et al., 2013). Also, much higher (and probably not economically and environmentally sound) N additions are needed to satisfy the large amount of extra crop N uptake required to generate an additional yield unit because the law of diminishing returns prevails at high yields (Fig. 2.6c). Hence, yield potential based on current practices may be a more realistic benchmark for a population of crop producers who seek to maximize net return and reduce risk. Together with the observation of small yield gaps, average irrigated maize yield in high-yielding districts in Nebraska has not increased since 2001, which may indicate that highly productive areas in the Corn Belt are already approaching an upper biophysical limit for crop productivity (Grassini et al., 2011a). In contrast, no incipient yield plateau is evident in rain-fed maize and soybean. Specht et al. (1999) and Sinclair and Rufty (2012) reported that a yield potential of ≈6 t ha^{-1} can be used as

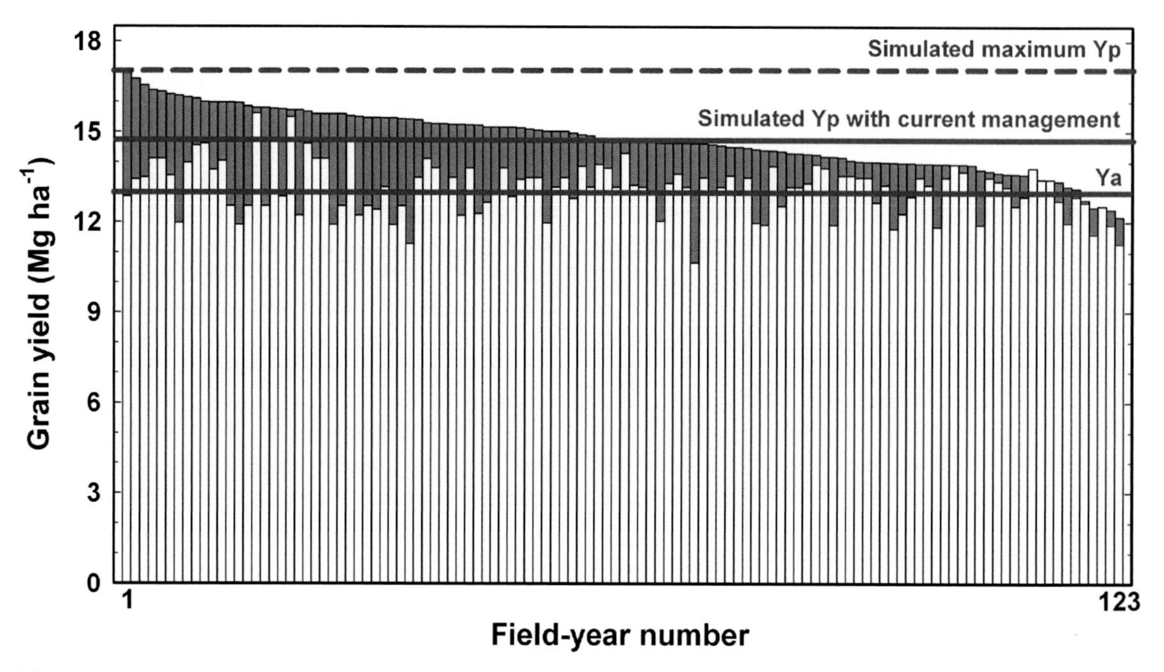

FIG. 2.12 Simulated irrigated maize yield potential (Yp) based on site-specific weather and management collected from 123 producer fields in central Nebraska. Each bar corresponds to an individual field-year case. Fields were sorted from highest to lowest Yp. Yellow and red portions of the bars indicate actual yield (Ya) and yield gap (Yg), respectively. Horizontal solid lines indicate average Yp and Ya (14.7 and 13.0 t ha^{-1}, CV = 7%). Horizontal dashed line indicates average simulated Yp based on the combination of sowing date, hybrid maturity, and plant density that gives highest long-term Yp (17.0 t ha^{-1}). *Adapted from van Ittersum et al. (2013).*

a 'functional' upper limit for on-farm soybean yields. Using this benchmark, and based on actual yield average of 3.8 t ha^{-1} between 2001 and 2010, the yield gap of irrigated soybean in Nebraska is about 37% of the potential. However, on-farm soybean yields ca.6 t/ha might be achieved but only under the best possible genotype × location × year × management interaction across a large geographic area. Based on producer data collected from six regions in Nebraska during eight years (2004-2011), Grassini et al. (2014) estimated the average yield gap for irrigated soybean using the 95th percentile of the yield distribution in each region-year as a proxy to yield potential. These authors found that average yield gap ranged from 12 to 20% of estimated yield potential across the six regions, with the latter ranging from 4.5 to 5.0 t/ha.

Seed companies are now employing diverse strategies for improving crop tolerance to drought. Non-transgenic so-called 'drought-tolerant' hybrids are already available commercially and transgenic versions were commercially available in 2013. Yield gains of 12–55% under drought conditions were claimed by a private seed company for transgenic drought-tolerant vs non-transgenic isogenic lines (Nelson et al., 2007). Also, yield gains of 9% under drought and 2% under favorable conditions were claimed by another seed company for non-transgenic drought-tolerant vs conventional hybrids (Pioneer, 2013). However, Roth et al. (2013) reported no yield differences between commercially available non-transgenic drought-tolerant vs conventional hybrids in thoroughly designed, independent trials over two growing seasons (one of which was a drought year). Specht et al. (2001) indicated that selection for drought tolerance and yield responsiveness to water are mutually exclusive if selection is solely targeted for yield stability in harsh environments. This can be illustrated with the comparison of 'yield responsive' rain-fed maize and 'drought-tolerant' sorghum yields in the western Corn Belt (Fig. 2.13, left). On average, sorghum yielded 17% more than maize in low-rainfall county-year cases when maize yielded <3.5 t ha^{-1}. In absolute terms, however, there is

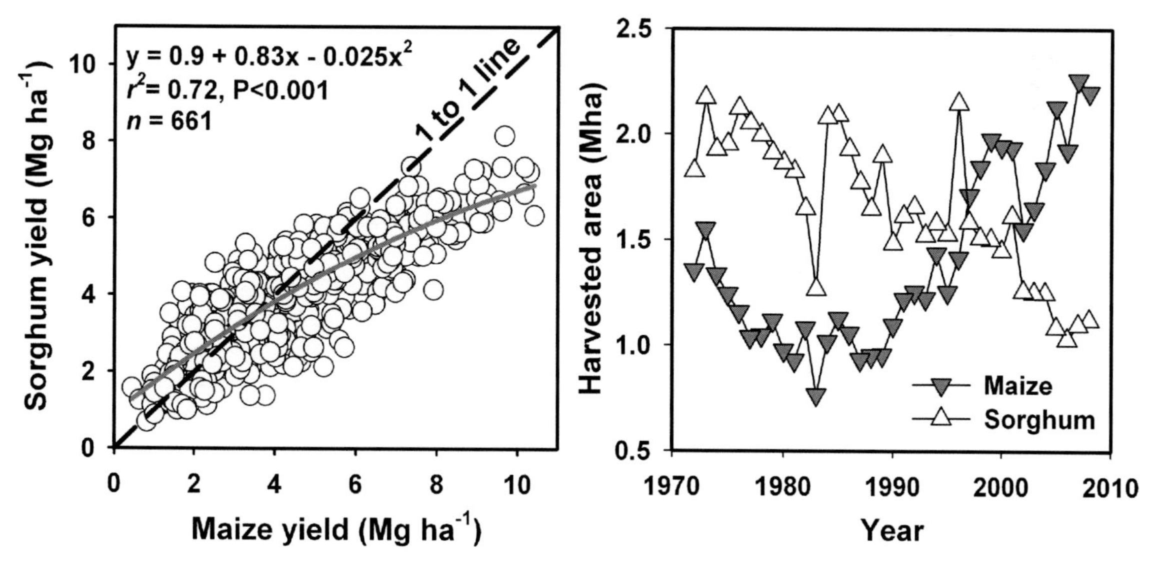

FIG. 2.13 Relationship between county-year rain-fed sorghum and maize yield (2000–2009) averages in Nebraska and Kansas (left panel) and trends in total (Nebraska and Kansas) rain-fed harvested area for both crops (right panel). *Source: USDA-NASS.*

a substantive yield advantage of maize in wetter county-year cases, which more than offsets any yield disadvantage it suffers in drought years as reflected by the steady expansion in maize area and parallel decline in sorghum area during the last three decades (Fig. 2.13, right). Further field evaluation of the so-called 'drought-tolerant' cultivars under varying water-stress conditions (intensity, duration, and timing) is required to understand which specific scenarios make these traits effective at increasing water-limited on-farm yield while maintaining yield responsiveness in years with higher water supply (Messina et al., 2011, Cooper et al., 2014).

Ample opportunities exist for increasing on-farm input-use efficiency by improving the temporal and spatial congruence among applied inputs, indigenous sources, and crop requirements. For example, a previous review on N-use efficiency in maize systems in the Corn Belt revealed that, on average, only 40% of the applied N fertilizer is absorbed by the crop (Cassman et al., 2002). A similar fertilizer N recovery (44%) was documented for modern hybrids (Ciampitti and Vyn, 2012). However, data from well-managed crops showed an average recovery efficiency of 64% in fields that receive optimum N fertilizer rates to achieve maximum profit (Dobermann et al., 2011; Wortmann et al., 2011). Available options to improve fertilizer N recovery include: in-season applications instead of pre-sowing fertilization; use of split N-fertilizer applications instead of a single large application; within-field site-specific N management and adjustment of N-fertilizer recommendations according to residual soil N; previous crop and reasonable yield goals calculated based on field-specific historical yields and/or long-term simulated yields (Cassman et al., 2002; Setiyono et al., 2010a; Dobermann et al., 2011).It is noteworthy that improvements in N-recovery efficiency will also help reduce N surplus and mitigate N_2O emissions and N leaching (Broadbent and Carlton, 1978; Venterea et al., 2011).

Similarly, there appears to be significant scope to improve the efficiency in use of irrigation water because applied irrigation amounts typically exceed crop water requirements for high yields. For example, Irmak et al. (2012) reported that similar maize yields can be produced with 30% less irrigation water by optimizing irrigation scheduling based on crop phenology and soil water status. Grassini et al. (2011b) found that about 50% of the irrigated maize fields in central Nebraska received irrigation amounts in excess to crop water requirements to achieve maximum yields and reported that 30% of current water use for maize irrigation can be saved, with little yield penalty, by replacing existing surface irrigation systems by center pivot and by adjusting irrigation management according to real-time crop water requirements.

5 CONCLUDING REMARKS

Yield and input-use efficiency have increased steadily during the last 40 years as a result of improvements in agronomic management and crop genetics. While both genetic and agronomic improvement contributed to the on-farm yield improvement, the synergistic interaction of relentlessly recurrent genetic improvement with intermittently phased periods of agronomic improvement has also made a significant contribution, such that all three ensured the persistent rise in maize and soybean yields in the Corn Belt. However, future increases may be difficult to achieve as on-farm yields approach yield potential, particularly in highly productive irrigated areas where the upward trend in on-farm maize yield has slowed in the last decade, which may be indicative of an incipient yield plateau. Still, some of the yield gap between on-farm yield and simulated yield potential might be captured by fine-tuning crop management in a manner that increases yield, while simultaneously reducing the resource input amount or cost. Indeed, substantive opportunities exist for increased input-use efficiency by scheduling just-in-time irrigation events of the minimum amount needed, and by optimizing management of N fertilizer to be temporally and spatially effective.

References

Adviento-Borbe, M.A.A., Haddix, M.L., Binder, D.L., Walters, D.T., Dobermann, A., 2007. Soil greenhouse gas fluxes and global warming potential in four high-yielding maize systems. Glob. Change Biol. 13, 1972–1988.

Andrade, F.H., Vega, C.R., Uhart, S.A., Cirilo, A.G., Cantarero, M., Valentinuz, O., 1999. Kernel number determination in maize. Crop Sci. 39, 453–459.

Baker, J.M., Griffis, T.J., 2005. Examining strategies to improve the carbon balance of corn/soybean agriculture using eddy covariance and mass balance techniques. Agric. For. Meteorol. 128, 163–177.

Bastidas, A.M., Setiyono, T.D., Dobermann, A., Cassman, K.G., Elmore, R.W., Graef, G.L., 2008. Soybean sowing date: the vegetative, reproductive, and agronomicimpacts. Crop Sci. 48, 727–740.

Blanco-Canqui, H., Lal, R., 2008. No-tillage and soil-profile carbon sequestration: an on-farm assessment. Soil Sci. Soc. Am. J. 72, 693–701.

Broadbent, F.E., Carlton, A.B., 1978. Field trials with isotopically labeled nitrogen fertilizer. In: Nielsen, D.R. (Ed.), Nitrogen in the environment. Academic Press, San Diego, CA.

Cassman, K.G., Dobermann, A., Walters, D.T., 2002. Agroecosystems, nitrogen-use-efficiency, and nitrogen management. Ambio 31, 132–140.

Cavalieri, A.J., Smith, O.S., 1985. Grain filling and field drying of a set of maize hybrids released from 1930 to 1982. Crop Sci. 25, 856–860.

Chung, J., Babka, H.L., Graef, G.L., et al., 2003. The seed protein, oil, and yield QTL on soybean linkage group I. Crop Sci. 43, 1053–1067.

Ciampitti, I.A., Vyn, T., 2012. Physiological perspectives of changes over time in maize yield dependency on nitrogen uptake and associated nitrogen efficiencies: a review. Field Crops Res. 133, 48–67.

Connor, D.J., Loomis, R.S., Cassman, K.G., 2011. Crop Ecology. Productivity and management in agricultural systems. Cambridge University Press, Cambridge.

Cooper, M., Gho, C., Leafgren, R., Tang, T., Messina, C., 2014. Breeding drought-tolerant maize hybrids for the US corn-belt: discovery to product. J.Exp. Bot. doi: 10.1093/jxb/eru064

Crookston, R.K., Kurle, J.E., Copeland, P.J., Ford, J.H., Lueschen, W.E., 1991. Rotational cropping sequence affects yield of corn and soybean. Agron. J. 83, 108–113.

De Bruin, J.L., Pedersen, P., 2009. New and old soybean cultivar responses to plant density and intercepted light. Crop Sci. 49, 2225–2232.

Dobermann, A., Wortmann, C.S., Ferguson, R.B., et al., 2011. Nitrogen response and economics for irrigated corn in Nebraska. Agron. J. 103, 67–75.

Duvick, D.N., 1997. What is yield? In: Developing drought, low N-tolerant maize, Edmeades, G.O., Bänziger, M., Mickelson, H.R., Peña-Valdivia, C.B., (eds), Proc. Symp., 25–29 March 1996, CIMMYT, El Batan, Mexico. CIMMYT, Mexico, D.F.

Duvick, D.N., 2005. The contribution of breeding to yield advances in maize (Zea mays L.). Adv. Agron. 86, 83–145.

Duvick, D.N., Cassman, K.G., 1999. Post-green revolution trends in yield potential of temperate maize in the North-Central United Sates. Crop Sci. 39, 1622–1630.

Duvick, D.N., Smith, J.S.C., Cooper, M., 2004. Long-term selection in a commercial hybrid maize breeding program. Plant Breed. Rev. 24, 109–151.

Dwyer, L.M., Tollenaar, M., 1989. Genetic improvement in photosynthetic response of hybrid maize cultivars, 1959 to 1988. Can. J. Plant Sci. 69, 81–91.

Dwyer, L.M., Stewart, D.W., Tollenaar, M., 1992. Analysis of maize leaf photosynthesis under drought stress. Can. J. Plant Sci. 72, 477–481.

Echarte, L., Andrade, F.H., 2003. Harvest index stability of Argentinean maize hybrids released between 1965 and 1993. Field Crops Res. 82, 1–12.

Echarte, L., Tollenaar, M., 2006. Kernel set in maize hybrids and their inbred lines exposed to stress. Crop Sci. 46, 870–878.

Echarte, L., Luque, S., Andrade, F.H., et al., 2000. Response of maize kernel number to plant density in Argentinean hybrids released between 1965and 1993. Field Crops Res. 68, 1–8.

Echarte, L., Andrade, F.H., Vega, C.R.C., Tollenaar, M., 2004. Kernel number determination in Argentinean maize hybrids released between 1965 and 1993. Crop Sci. 44, 1654–1661.

Echarte, L., Rothstein, S., Tollenaar, M., 2008. The response of leaf photosynthesis and dry matter accumulation to nitrogen supply in an older and a newer maize hybrid. Crop Sci. 48, 656–665.

Edgerton, M.D., Fridgen, J., Anderson, Jr., J.R., et al., 2012. Transgenic insect resistance traits increase corn yield and yield stability. Nat. Biotech. 30, 493–496.

Food and Agriculture Organization of the United Nations (FAO). FAOSTAT Database–Agricultural Production [online WWW]. Available URL: http://faostat.fao.org/.

Fox, C.M., Cary, T.R., Colgrove, A.L., et al., 2013. Estimating soybean genetic gain for yield in the northern USA – Influencing of cropping history. Crop Sci. 53, 2473–2482.

Grassini, P., Cassman, K.G., 2012. High-yield maize with large net energy yield and small global warming intensity. Proc. Natl. Acad. Sci. USA 109, 1074–1079.

Grassini, P., Yang, H., Cassman, K.G., 2009. Limits to maize productivity in Western Corn-Belt: a simulation analysis for fully-irrigated and rainfed conditions. Agric. For. Meteorol. 149, 1254–1265.

Grassini, P., Thorburn, J., Burr, C., Cassman, K.G., 2011a. High-yield irrigated maize systems in the Western U.S. Corn-Belt. I. On-farm yield, yield-potential, and impact of agronomic practices. Field Crops Res. 120, 142–150.

Grassini, P., Torrion, J.A., Cassman, K.G., Yang, H.S., Specht, J.E., 2014. Drivers of spatial and temporal variation in soybean yield and irrigation requirements. Field Crops Res. 163, 32–46.

Grassini, P., Yang, H., Irmak, S., Thorburn, J., Burr, C., Cassman, K.G., 2011b. High-yield irrigated maize systems in the Western U.S. Corn-Belt. II. Irrigation management and crop water productivity. Field Crops Res. 120, 133–141.

Hammer, G.L., Dong, Z., McLean, G., et al., 2009. Can changes in canopy and/or root system architecture explain historical maize yield trends in the U.S. Corn Belt? Crop Sci. 49, 299–312.

Horowitz, J., Ebel, R., Ueda, K., 2010. No-till farming is a growing practice. USDA-ERS Economic Information Bulletin Number 70.

Irmak, S., Burgert, M.J., Yang, H., et al., 2012. Large-scale implementation of soil moisture-based irrigation strategies for increasing maize water productivity. Trans. ASABE 55, 881–894.

Klocke, N.L., Currie, R.S., Aiken, R.M., 2009. Soil water evaporation and crop residues. Trans. ASABE 52, 103–110.

Kucharik, C.J., 2008. Contribution of planting date trends to increased maize yields in the central United States. Agron. J. 100, 328–336.

Lee, E.A., Tollenaar, M., 2007. Physiological basis of successful breeding strategies for maize grain yield. Crop Sci. 47, 202–215.

Lindquist, J.L., Arkebauer, T.J., Walters, D.T., Cassman, K.G., Dobermann, A., 2005. Maize radiation use efficiency under optimal growth conditions. Agron. J. 97, 72–78.

Lobell, D.B., Cassman, K.G., Field, C.B., 2009. Crop yield gaps: their importance, magnitudes,and causes. Ann. Rev. Environ. Resour. 34, 179–204.

Luque, S.F., Cirilo, A.G., Otegui, M.E., 2006. Genetic gains in grain yield and related physiological attributes in Argentine maize hybrids. Field Crops Res. 96, 383–397.

McCullough, D.E., Girardin, Ph., Mihajlovic, M., Aguilera, A., Tollenaar, M., 1994a. Influence of N supply on development and dry matter accumulation of an old and new maize hybrid. Can. J. Plant Sci. 74, 471–477.

McCullough, D.E., Aguilera, A., Tollenaar, M., 1994b. N uptake, N partitioning, and photosynthetic N-use efficiency of an old and a new maize hybrid. Can. J. Plant Sci. 74, 479–484.

Messina, C.D., Podlich, D., Dong, Z., Samples, M., Cooper, M., 2011. Yield-trait performance landscapes: from theory to application in breeding maize for drought tolerance. J. Exp. Bot. 62, 855–868.

Mortensen, D.A., Egan, J.F., Maxwell, B.D., Ryan, M.R., Smith, R.G., 2012. Navigating a critical juncture for sustainable weed management. Bioscience 62, 75–84.

Nafziger, E.D., 1994. Corn planting date and plant population. J. Prod. Agric. 7, 59–62.

Nelson, D.E., Repetti, P.P., Adams, T.R., et al., 2007. Plant nuclear factor Y (NF-Y) B subunits confer drought tolerance and lead to improved corn yields on water-limited acres. Proc. Natl. Acad. Sci. USA 104, 16450–16455.

Nielsen, D.C., Unger, P.W., Miller, P.R., 2005. Efficient water use in dry land cropping systems in the Great Plains. Agron. J. 97, 364–372.

Nissanka, S.P., Dixon, M.A., Tollenaar, M., 1997. Canopy gas exchange response to moisture stress in old and new maize hybrid. Crop Sci. 37, 172–181.

Novacek, M.J., Mason, S.C., Galusha, T.D., Yaseen, M., 2013. Twin rows minimally impact irrigated maize yield, morphology, and lodging. Agron. J. 105, 268–276.

Otegui, M.E., Bonhomme, R., 1998. Grain yield components in maize. I. Ear growth and kernel set. Field Crops Res. 56, 247–256.

Pioneer, 2013. Optimum® AQUA max TM products from DuPont Pioneer. DuPont Pioneer. www.pioneer.com/CMRoot/Pioneer/US/products/seed_trait_technology/see_the_difference/AQUAmax_Product_Offerings.pdf

Porter, P.M., Lauer, J.G., Lueschen, W.E., et al., 1997. Environment affects the corn and soybean rotation effect. Agron. J. 89, 441–448.

Rincker, K., Fox, C., Cary, T., et al., 2014. Genetic improvement of soybean in North American maturity groups II, III, and IV. Crop Sci. (in press).

Roekel, R.J.V., Coulter, J.A., 2011. Agronomic responses of corn to planting date and plant density. Agron. J. 103, 1414–1422.

Roth, J.A., Ciampitti, I.A., Vyn, T.J., 2013. Physiological evaluations of recent drought-tolerant maize hybrids at varying stress levels. Agron. J. 105, 1129–1141.

Rowntree, S.C., Suhre, J.J., Wilson, E.W., et al., 2013. Genetic gain x management interactions in soybean: I. Planting date. Crop Sci. 53, 1128–1138.

Rowntree, S., Suhre, J., Wilson, E., et al., 2014. Physiological and phenological responses of historical soybean cultivar releases to earlier planting. Crop Sci. 54, 804–816.

Russell, W.A., 1985. Evaluations for plant, ear, and grain traits of maize cultivars representing seven eras of breeding. Maydica 25, 85–96.

Sacks, W.J., Kucharik, C.J., 2011. Crop management and phenology trends in the U.S. Corn Belt: impacts on yields, evapotranspiration and energy balance. Agric. For. Meteorol. 151, 882–894.

Salvagiotti, F., Specht, J.E., Cassman, K.G., Walters, D.T., Weiss, A., Dobermann, A., 2009. Growth and nitrogen fixation in high-yielding soybean: impact of nitrogen fertilization. Agron. J. 101, 958–970.

Setiyono, T.D., Walters, D.T., Cassman, K.G., Witt, C., Dobermann, A., 2010a. Estimating maize nutrient uptake requirements. Field Crops Res. 118, 158–168.

Setiyono, T.D., Cassman, K.G., Specht, J.E., et al., 2010b. Simulation of soybean growth and yield in near-optimal growth conditions. Field Crops Res. 119, 161–174.

Shapouri, H., Duffield, J.A., Wang, M., 2003. The energy balance of corn ethanol revisited. Trans. ASAE 46, 959–968.

Shi, G., Chavas, J.P., Lauer, J., 2013. Commercialized transgenic traits, maize productivity and yield risk. Nat. Biotech. 31, 111–114.

Sinclair, T.R., Rufty, T.W., 2012. Nitrogen and water resources commonly limit crop yield increases, not necessarily plant genetics. Glob. Food Secur. 1, 94–98.

Specht, J.E., Hume, D.J., Kumudini, S.V., 1999. Soybean yield potential – A genetic and physiological perspective. Crop Sci. 39, 1560–1570.

Specht, J.E., Chase, K., Macrander, M., et al., 2001. Soybean response to water: a QTL analysis of drought tolerance. Crop Sci. 41, 493–509.

Specht, J.E., Diers, B.W., Nelson, R.L., Toledo, J.F., Torrion, J.A., Grassini, P., 2014. Soybean (Glycine max (L.) Merr.). In: Smith, J.S.C., Carver, B., Diers, B.W., Specht, J.E. (Eds.), Yield gains in major US field crops: contributing factors and future prospects. ASA-CSSA, Madison, WI.

Sugg, Z., 2007. Assessing U.S. farm drainage: can GIS lead to better estimates of subsurface drainage extent? World Resources Institute, Washington D.C [online WWW]. Available URL:http://pdf.wri.org/assessing_farm_drainage.pdf.

Suhre, J.J., Weidenbenner, N.H., Rowntree, S.C., et al., 2014. Genetic gain x management interactions in soybean: III. Seeding rate. Crop Sci. (in press).

Tollenaar, M., Lee, E.A., 2002. Yield potential yield, yield stability and stress tolerance in maize. Field Crops Res. 75, 161–170.

Tollenaar, M., Lee, E.A., 2011. Strategies for enhancing grain yield in maize. Plant Breed. Rev. 34, 37–81.

Tollenaar, M., Dwyer, L.M., Stewart, D.W., 1992. Ear and kernel formation in maize hybrids representing three decades of grain yield improvement in Ontario. Crop Sci. 32, 432–438.

Tollenaar, M., McCullough, D.E., Dwyer, L.M., 1994. Physiological basis of the genetic improvement of corn. In: Slafer, G.A. (Ed.), Genetic improvement of field crops. Marcel Dekker Inc, New York.

Tollenaar, M., Dwyer, L.M., 1998. Physiology of maize. In: Smith, D.L., Hamel, C. (Eds.), Crop yield, physiology and processes. Springer Verlag, New York.

Tollenaar, M., Ying, J., Duvick, D.N., 2000. Genetic gain in corn hybrids from the northern and central Corn Belt. In: 55th Ann. Corn Sorghum Ind. Res. Conf. American Seed Trade Association, Chicago, IL.

USDA-Economic Research Service (ERS), 2005. Crop production practices (online WWW). Available URL: http://www.ers.usda.gov/Data/ARMS/.

USDA-National Agricultural Statistics Service (NASS). Crops U.S. state and county databases. Washington DC [online WWW]. Available URL: http://www.nass.usda.gov/index.asp.

van Groenigen, J.W., Velthof, G., Oenema, O., van Groenigen, K.J., van Kessel, C., 2010. Towards an agronomic assessment of N2O emissions: a case study for arable crops. Eur. J. Soil Sci. 61, 903–913.

van Ittersum, M.K., Cassman, K.G., Grassini, P., Wolf, J., Tittonell, P., Hochman, Z., 2013. Yield gap analysis with local to global relevance –a review. Field Crops Res. 143, 4–17.

van Wart, J., Kersebaum, K.C., Peng, S., Milner, M., Cassman, K.G., 2013. Estimating crop yield potential at regional to national scale. Field Crops Res. 143, 34–43.

Venterea, R.T., Maharjan, B., Dolan, M.S., 2011. Fertilizer source and tillage effects on yield-scaled nitrous oxide emissions in a corn cropping system. J. Environ. Qual. 40, 1521–1531.

Verma, S.B., Dobermann, A., Cassman, K.G., et al., 2005. Annual carbon dioxide exchange in irrigated and rainfed maize-based agroecosystems. Agri. For. Meteorol. 131, 77–96.

Villamil, M.B., Davis, V.M., Nafziger, E.D., 2012. Estimating factor contributions to soybean yield from farm field data. Agron. J. 104, 881–887.

Vyn, T.J., 2006. Meeting the ethanol demand: consequences and compromises associated with more corn on corn in Indiana. Purdue University Extension Publication ID-336. [online WWW]. Available URL: http://www.agry.purdue.edu/staffbio/Vyn_ID-336.pdf

Wilson, E.W., Rowntree, S.C., Suhre, J.J., et al., 2014. Genetic gain x management interactions in soybean: II. Nitrogen utilization. Crop Sci. 54, 340–348.

Wortmann, C.S., Tarkalson, D.D., Shapiro, C.A., et al., 2011. Nitrogen use efficiency in irrigated corn for three cropping systems in Nebraska. Agron. J. 103, 76–84.

Yang, H.S., Dobermann, A., Cassman, K.G., et al., 2004. Hybrid-maize: a maize simulation model that combines two crop modelling approaches. Field Crops Res. 87, 131–154.

Ying, J., Lee, E.A., Tollenaar, M., 2000. Response of maize leaf photosynthesis to low temperature during the grain-filling period. Field Crops Res. 68, 87–96.

Ying, J., Lee, E.A., Tollenaar, M., 2002. Response of maize leaf photosynthesis during the grain-filling period of maize to duration of cold exposure, acclimation, and incident PPFD. Crop Sci. 42, 1164–1172.

CHAPTER

3

Farming systems in China: Innovations for sustainable crop production

Weijian Zhang[1], *Chengyan Zheng*[1], *Zhenwei Song*[1], *Aixing Deng*[1], *Zhonghu He*[1,2]

[1]**Institute of Crop Sciences, Chinese Academy of Agricultural Sciences, Beijing, China**
[2]**International Maize and Wheat Improvement Center (CIMMYT) China Office, Beijing, China**

1 INTRODUCTION

China is the world's largest grain producer and consumer. Chinese grain production was more than 530 Mt in 2012; however, China also became the world's largest cereal importer in 2013. Increasing crop production is therefore important for both Chinese and world food security. Although China has successfully increased grain production since 2003, the resource-use efficiency remains low. Resource shortage and environmental pollution are the big challenges to sustainable grain production in China over the next decades, whereas projected climate change represents both challenges and new opportunities. Therefore, it is critically important to evaluate the experiences of Chinese farming-system innovations for sustainable crop production.

This chapter focuses on the progress of grain crop production through genetic and agronomic improvements in China. Section 2 summarizes the abiotic conditions and limitations to main crop production. Lack of information precludes the analysis of historical changes of biotic factors. Section 3 summarizes the basic features of major grain crop production and farming systems. We discuss the contributions of genetic improvement and agronomic innovation to yield increases (section 4) and the resource-use efficiency (section 5) over past decades. Section 6 outlines responses of major grain crops to future climate patterns and outlines adaptive practices.

2 THE ABIOTIC ENVIRONMENTS FOR CROP PRODUCTION

2.1 Climatic conditions and historical changes

There are large spatial (Fig. 3.1) and temporal (Fig. 3.2) variations in air temperature and

Crop Physiology. DOI: 10.1016/B978-0-12-417104-6.00003-0

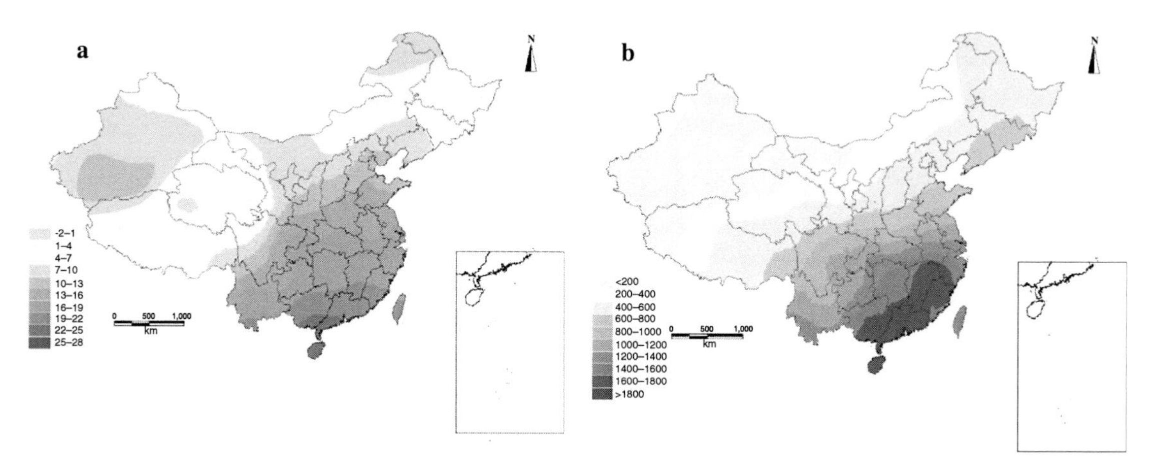

FIG. 3.1 Spatial differences in (a) annual mean air temperature (°C) and (b) the annual precipitation (mm) in China. Data are the mean values during 1990–2010 period. Please see color plate at the back of the book.

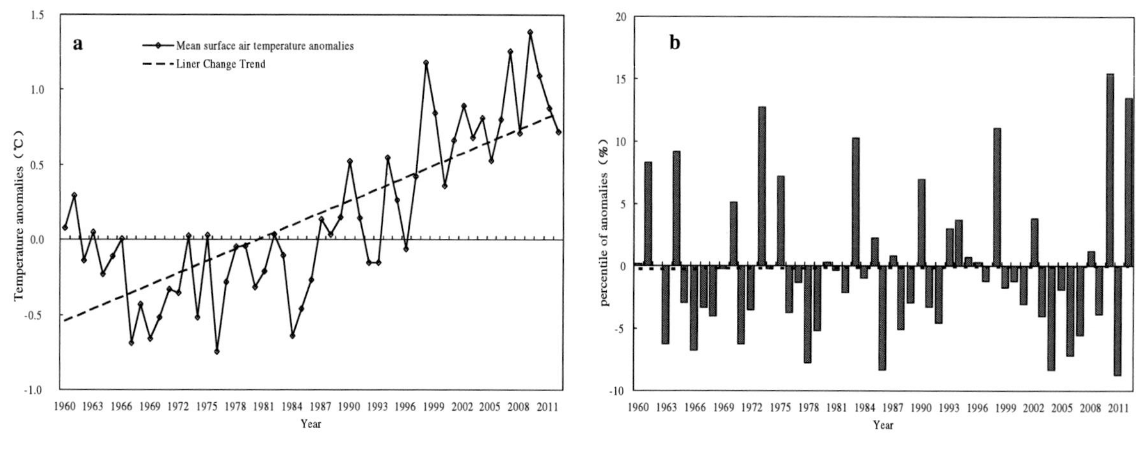

FIG. 3.2 (a) Mean surface air temperature anomalies and (b) percentile of annual precipitation anomalies in China (relative to 1971–2000).

precipitation. The annual mean air temperature is below 10°C in the north and west, and above 20°C in the south (Fig. 3.1a). Thus, only one crop can be sown annually in the north and west, and two or more crops can be sown in the south. Annual precipitation ranges from more than 1500 mm in the south to less than 500 mm in the north and the west. Thus, the limitations to crop production are low temperature and precipitation in the north and the west, and heat stress and short sunshine duration in the south and the east. Extreme weather events, such as chilling injury during early rice seedling and winter wheat booting, heat stress during wheat grain filling and during rice and maize flowering, frequently cause large reductions in crop production.

During the past 100 years, particularly over the last decades, the Earth has experienced significant warming. The mean air temperature has increased about 0.74°C since 100 years ago, and it is expected to rise about 2.0–5.4°C (depending on region and CO_2 increase scenarios) by 2100 (IPCC, 2007; Chapter 20). Similar warming trends were observed in China (Fig. 3.2a). The mean annual air temperature in the past 50 years has increased by 1.33°C, and no significant change in annual precipitation has been recorded since the early 20th century (Fig. 3.2b). However, rainfall seasonality has changed; the summer rainfall showed an increasing trend, while significant reductions occurred in spring and autumn. Extremely high temperature is common, and the frequencies of drought and flooding have increased. Some modeling studies indicate that climate change might reduce crop production in China (Chapter 20), thus potentially compromising food security. Section 6, however, shows experimental and modeling evidence for a more diverse range of responses, including significant shifts in the boundaries of cropping opportunities.

2.2 Soil conditions, land use and historical changes

China is one of the world's largest countries in both area and population; it covers 9.6 million km^2 with a large diversity in landscape and land use. The altitude increases from less than 500 m in the southern hilly region to higher than 1500 m in the northern Loess Plateau region, and from less than 50 m in the eastern plain to higher than 5000 m in the western mountain area.

The land area suitable for agriculture only accounts for 17.3% of the total. In 2012, the farmland area was 121.7 million ha, making up less than 7% of the world's total acreage. The farmland area per capita is about 0.09 ha, 43% of the world average farmland size per capita. Farmland soil quality varies among the cropping regions, and only a fourth of this represents high quality soil. The plow layers of major farmlands are less than 15 cm in depth with diverse physical and chemical limitations for root growth. According to the Chinese second soil survey, finished in the 1980s, soil organic matter (SOM) below 2.0% accounts for more than 65% of total farmland area. Moreover, serious acidification affects soils in the south and salinization in the northern areas.

Long-term cropping intensification plays an important role in farmland soil quality. Intensified maize cropping has caused a large reduction in the concentration of SOM in northeast China, the largest maize-growing region. The main reasons for the reduction of SOM concentration are long-term inorganic fertilizer application without any organic amendment and long-term conventional intensive tillage. In the north and south, however, the SOM concentration showed an increasing trend during the last decades, and higher SOM concentration was found in paddy fields than in the dry land (Wang et al., 2010; Huang et al., 2013). The main reasons for the increase in SOM concentration in north China are wheat straw retention and less tillage during the maize growing season; however, this has led to the increased occurrence of head scab for wheat. The higher SOM concentration in paddy fields can be attributed to wheat or rice straw retention and the anaerobic soil condition during rice growing seasons (Huang et al., 2013). Long-term fertilizer application with intensive conventional tillage has degraded soil aggregate structure, though soluble fertilizers combined with organic matter amendments (i.e. crop straw and manure) can enhance soil physical structure (Chen et al., 2010; Huang et al., 2010). Soil acidification is a serious issue. A recent study showed that over-use of nitrogen fertilizer, especially urea, caused significant soil acidification in major Chinese cropping lands (Guo et al., 2010). Due to mining, industrial and urban activities, heavy metal pollution is becoming a severe problem in farmland soil, especially in the southeast paddy fields where heavy metal concentration (e.g. Pb and Cd) of grain exceeds the state standard in rice crops around the mining areas. To ensure environmental health and food safety, the Chinese government is allocating resources to research and extend sustainable cropping techniques.

3 FARMING SYSTEM DIVERSITY AND SPATIAL DISTRIBUTION

3.1 Major grain crops

With only 7% of the world's arable land, China has produced enough food for 22% of the world's population. In 2012, China produced 589.6Mt of grain. As shown in Table 3.1, maize, rice and wheat are the three major grain crops and the production of these three crops accounts for 23.8, 28.4 and 17.9% of the world respectively. Owing to the increased demand from livestock and industrial development, maize has become the largest crop, increased from around 20 million ha to 35 million ha. Major grain cropping areas are located in the northeast, north, east and south (Fig. 3.3). The major rice cropping areas are located in the south and the east, while maize cropping areas are mainly found in the north and northeast. More than 80% of wheat is sown in autumn in the north plain, which is also the major cropping area of maize.

3.2 Grain-based cropping systems

There is large diversity in Chinese cropping systems including single, double and triple cropping (Fig. 3.4). The major reasons for multiple cropping systems are to increase grain production and to maximize resource-use efficiency, which is facilitated by family labor (Zhou et al., 2007). China has a small farmland area per capita with the largest population in the world; hence the role of the multiple cropping systems to ensure food security is crucially important.

In the northeast and the west, only one crop (maize, rice or soybean) can be sown annually due to temperature and precipitation limitations. Crops are commonly sown in May and harvested from September to October. Between the Great Wall and the northern Huaihe River, the double-cropping system is dominant with two dryland crops annually, such as wheat–maize, wheat–soybean, or wheat–peanut. Wheat is sown in October and harvested in June, and the other crops are sown in June and harvested

TABLE 3.1 Sown area, yield and production of the main crops in China in 2012

Crop	Area (million ha)	Yield (t ha^{-1})	Production (Mt)	Area ratio to total cropping area (%)	Production ratio to the total of the world (%)
Maize	35.0	5.9	205.6	21.4	23.3
Rice	30.1	6.8	204.2	18.4	28.4
Wheat	24.3	5.0	121.0	14.9	18.0
Rapeseed	7.4	1.9	14.0	4.5	21.5
Soybean	7.2	1.8	13.0	4.4	5.4
Potato	5.5	3.4	18.6	3.4	5.1
Peanut	4.6	3.6	16.7	2.8	40.5
Millet	0.7	2.4	1.8	0.4	6.0
Barley	0.5	3.3	1.6	0.3	1.2
Sorghum	0.6	4.1	2.6	0.4	4.5

Data are from China Statistical Year book (2012) and FAO (2013). The yield of potato is converted into that of grain at the ratio 5:1, i.e. 5kg of fresh tuber was equivalent to 1kg of grain.

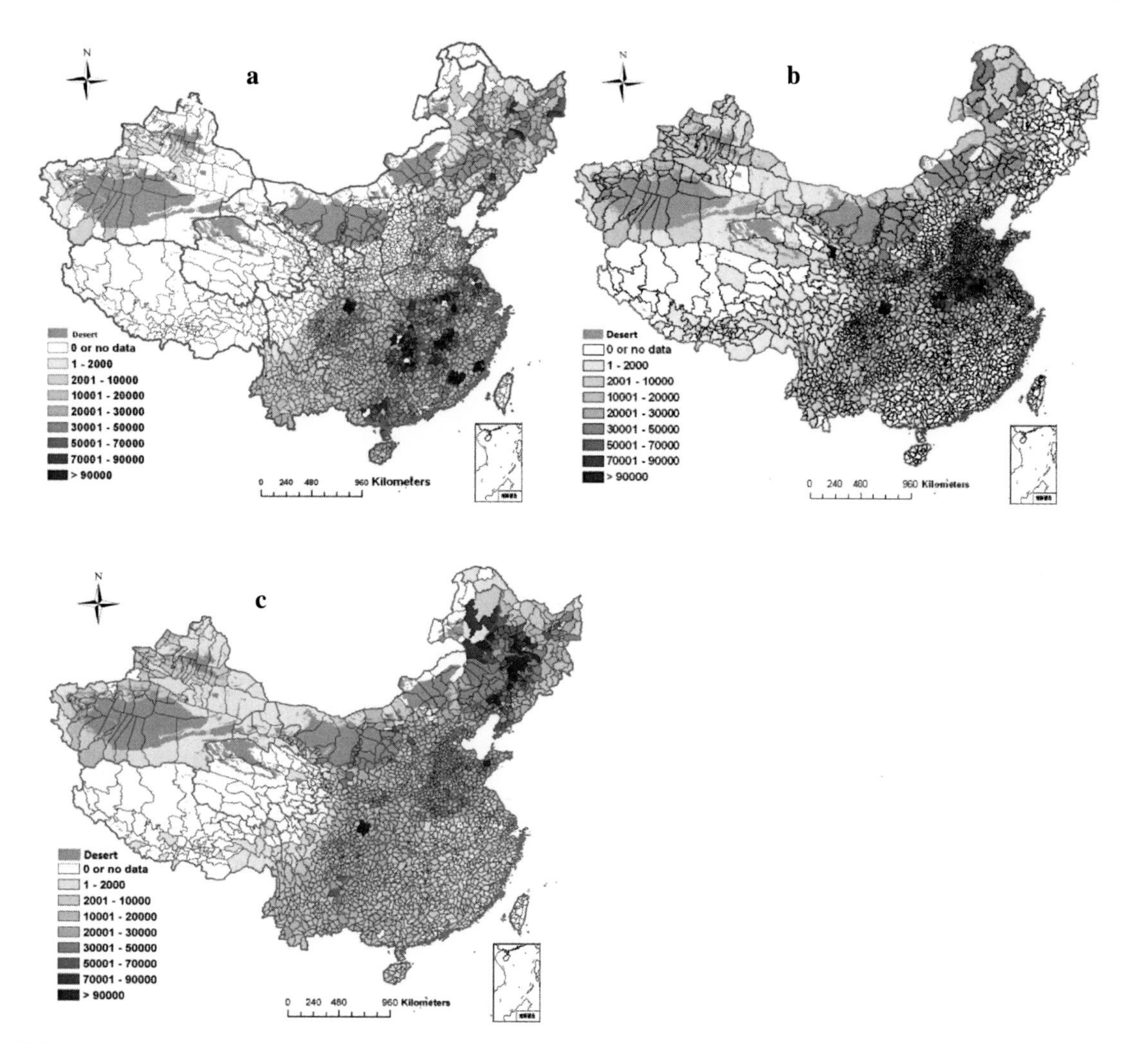

FIG. 3.3 Spatial layouts of sown areas (ha) of (a) rice, (b) maize and (c) wheat in China (2012). Please see color plate at the back of the book.

in October. From the southern Huaihe River to the Yangtze River, the annual double cropping changes to rice–wheat, rice–rapeseed, and rice–vegetables (e.g. cabbage, faba bean, cauliflower, carrot, garlic, potato, etc.). Rice is commonly sown in May, transplanted in June and harvested in October, and the other crops are sown in October and harvested in June. In the south, an annual double-rice cropping system is dominant.

Rice is sown early or late in the cropping season. Early rice is sown in March to April, transplanted in May and harvested in July, while late rice is sown in June, transplanted in July and harvested in October. During winter, vegetables (e.g. cabbage, faba bean, cauliflower, carrot, garlic, potato, etc.), green manure and rapeseed are planted in the paddy fields. Multiple cropping areas decreased recently due to the shortage of rural labor and the development of the

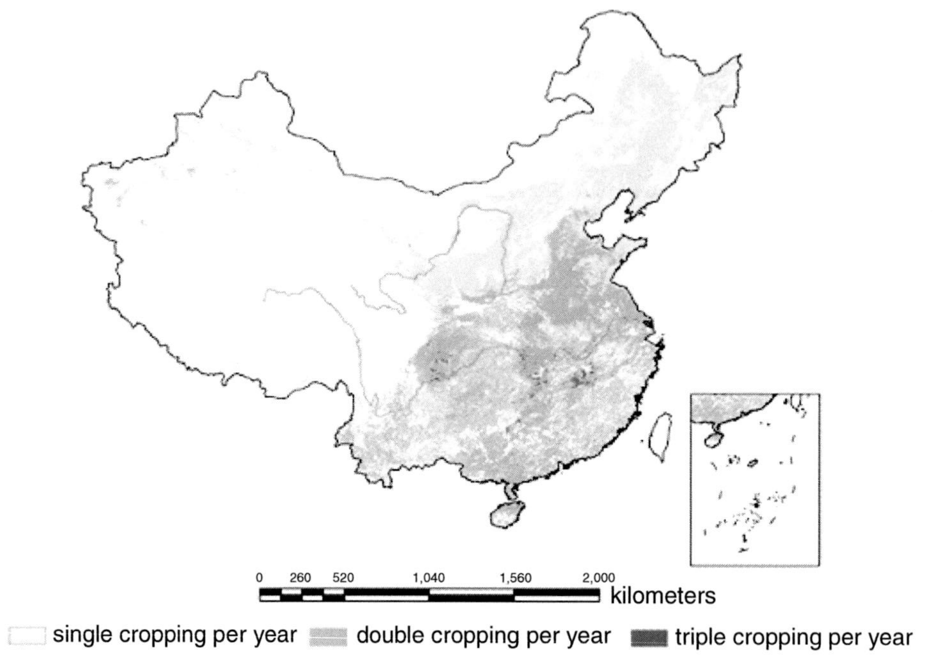

FIG. 3.4 Spatial layout of cropping systems in China, (Liu et al., 2013). Please see color plate at the back of the book.

rural economy. As a consequence, the double rice cropping system is being replaced by single rice with winter crops in southern China. Wheat–maize rotation or a relay cropping system is being replaced by spring maize per season in the northwest (Gansu and Shaanxi), or in the northern China Plain such as northern Hebei, Tianjin, and Beijing.

Although the cropping system is being simplified due to shortage of rural labor and mechanical planting, multiple cropping is still dominant. In the southwest, intercropping, relay cropping and mixed cropping are still widely applied for the dryland due to the combination of sufficient farm labor and farmland shortage. For example, in the dryland area of Sichuan province, maize is commonly sown together with soybean or sweet potato; and pea is sown in the field before the maize harvest and after soybean harvest. Before the wheat harvest, rice is sown in the wheat field with a direct-seeding method; and wheat is sown in paddy fields before rice harvesting in the wheat–rice cropping system. Recently, a novel relay strip inter-cropping system of wheat–maize/soybean has been developing in the southwest which outperforms economically the traditional wheat–maize cropping system by 30% or more (Yong et al., 2012). Before the maize harvest, vegetable crops or wheat are sown. Even in the paddy fields, farmers often grow soybean and vegetable crops on the paddy ridges during the rice-growing season. These integrated cropping systems improve the farmers' income and food production stability.

4 YIELD ENHANCEMENT VIA GENETIC IMPROVEMENT AND AGRONOMIC INNOVATION

4.1 Trends in grain production

Over the last 60 years, there has been a remarkable growth in crop production in China (Figs 3.5 and 3.6). Production of rice, wheat and maize has increased from 48.6Mt to 204.2Mt,

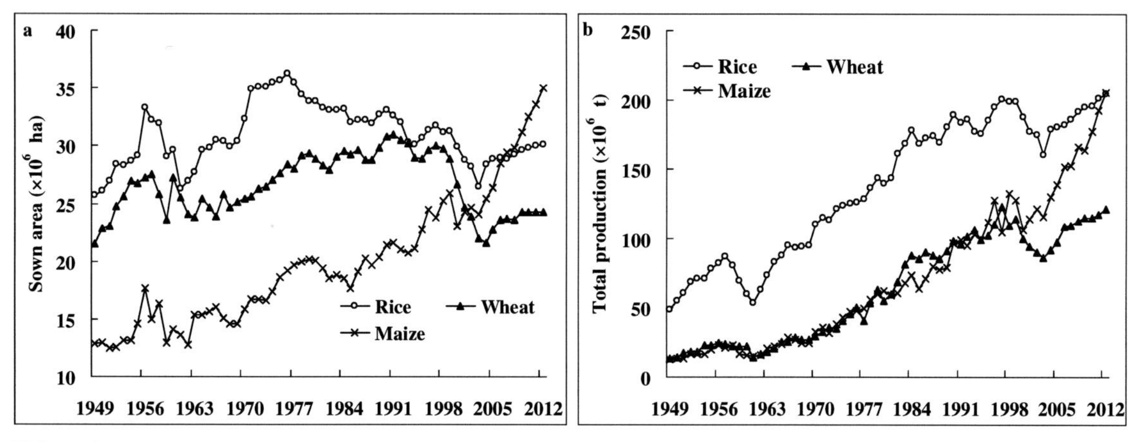

FIG. 3.5 (a) Sown area and (b) total production of rice, wheat and maize during the period 1949–2012 in China. Data from National Bureau of Statistics of China, 1949–2012.

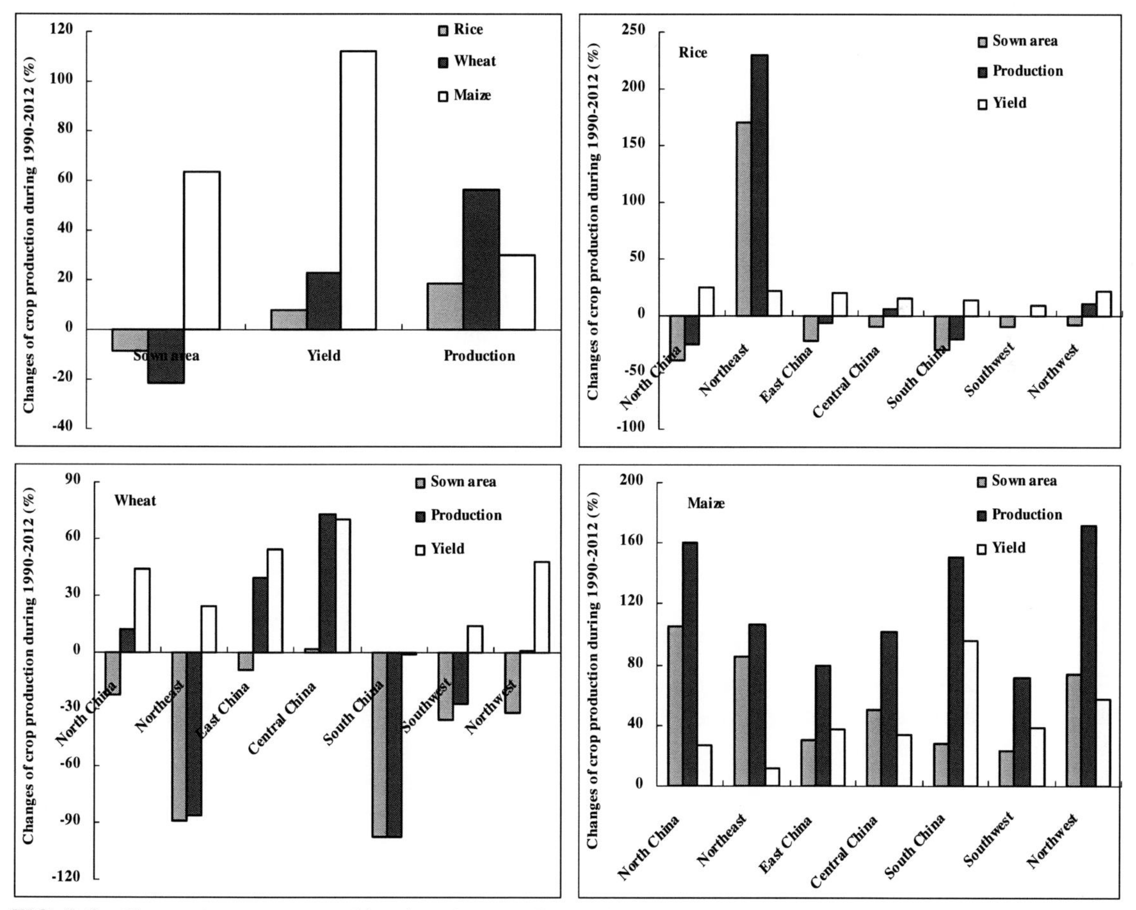

FIG. 3.6 Changes in sown area, yield and production of rice, wheat and maize between 1990 and 2012. Data from National Bureau of Statistics of China, 1990–2012.

13.8 Mt to 121.0 Mt, and 12.4 Mt to 205.6 Mt between 1949 and 2012, accounting for 66.2% of the total country cereal production in 1949 and 90% in 2012, respectively.

During the last 20 years, significant changes were observed in sown area and production of the main grain crops. From 1990 to 2012, total rice and wheat acreage decreased by 8.9% and 21.1%; in contrast, the maize area increased by 63.7% (Fig. 3.6). He et al. (2013) also showed that a significant reduction in the soybean area has allowed expansion of the maize area in both the Yellow and Huai River Valleys and northeastern China from 2000 to 2012. However, grain yields increased by 7.9% for rice, 23.2% for wheat and 112.4% for maize, and the corresponding increases in total production were 18.4%, 56.1%, and 29.7%, respectively (Fig. 3.6). Compared to the sown area, yield increase contributed much more to increased production. The main contribution to the increase in grain production was maize in the north and northwest, where the sown area increased by 105% and 73.3% and production by 160.5% and 171.7%, respectively. Rice acreage increased by 171% and production by 230% in the northeast, and Heilongjiang province became the last rice producer in China. The increase in wheat production was mainly in central (72.7%) and eastern (39.2%) regions (Fig. 3.6). Since farmland and natural resources are in short supply, further increase in grain production will mainly depend on the increase in yield rather than extension of the sown area in China as in most of the other world cropping systems.

4.2 Contribution of genetic improvement

Grain production depends on the yield potential (Chapter 16) and stress tolerance, hence genetic improvement plays an important role in increasing productivity. He et al. (2013) have summarized the genetic improvement of yield potential for major cereals. Conventional breeding has played a dominant role in developing new varieties and hybrids, although molecular marker-assisted selection has also been used (He et al., 2013). Improvement of wheat for the irrigated wheat/maize rotation system in the Yellow and Huai River Valleys since the 1960s has produced varieties characterized by reduced plant height, increased harvest index, and significantly increased yield potential, largely due to the utilization of semi-dwarf genes and the 1B/1R translocation (Zhou et al., 2007). Recent studies indicate that wheat yield in Henan and Shandong, China's largest wheat-producing provinces, has increased significantly since 1990. In Henan, the average annual genetic gain in grain yield between 1981 and 2008 was 0.60%, or 51.3 kg ha^{-1} $year^{-1}$, and it can largely be attributed to increased thousand kernel weight and harvest index (Zheng et al., 2011). These authors also found that grain yield was closely and positively associated with stomatal conductance and transpiration rate at 30 days post-anthesis. In Shandong, the genetic gain in grain yield between 1969 and 2006 was 0.85%, or 62 kg ha^{-1} $year^{-1}$, mainly due to increased kernels m^{-2} and biomass as a result of improved photosynthetic efficiency at and after heading (Xiao et al., 2012). Plant height was maintained and harvest index increased, with greater photosynthetic efficiency after anthesis, indicating there is potential for a physiological approach to yield improvement.

Historical gains of maize yield recorded in our experiment (Fig. 3.7) were 17.9 g $plant^{-1}$ $decade^{-1}$ and 936 kg ha^{-1} $decade^{-1}$ over the period 1970–2010 in northeast China, which were similar to previous reports (Xie et al., 2009; Ci et al., 2012). Yield improvements were significant from 1950 to 1970 and from 1990 to 2000 (Ci et al., 2013). From 1970 to 2000, the gain in maize yield averaged 94.7 kg ha^{-1} $year^{-1}$, 53% of which was attributed to breeding. New hybrids had increased tolerance to multiple stresses, including lodging resistance, which allowed plant densities to increase from 60000 to 75000 plants ha^{-1} (Ci et al., 2011). A comparison of Chinese and US hybrids grown at different planting densities from 1964 to 2001 indicated that US hybrids showed

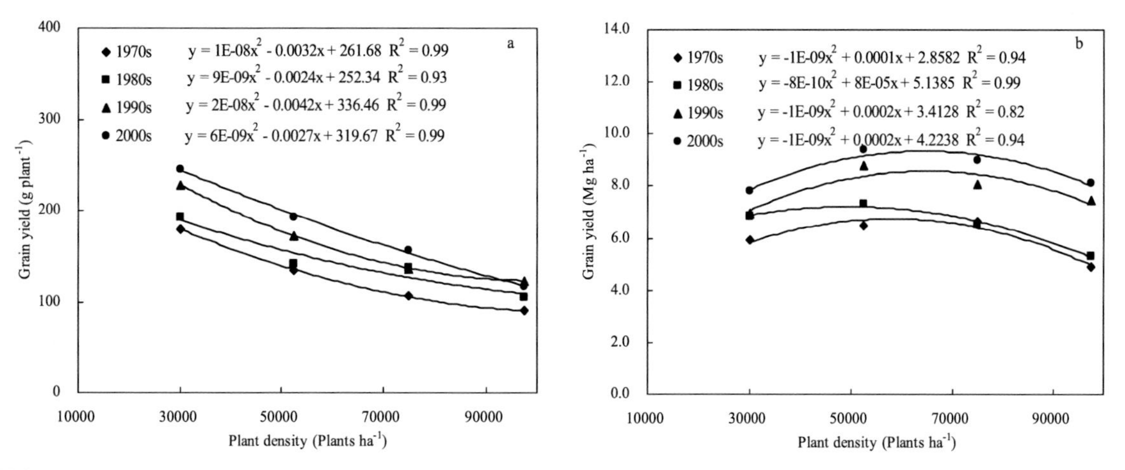

FIG. 3.7 Responses of grain yields (a) per plant and (b) per unit area to plant density for maize hybrids released from different eras over the 1970s–2000s.

the highest rate of gain (81 kg ha^{-1} year^{-1}) at the highest planting density (67550 plants ha^{-1}), whereas the highest rate of gain for Chinese hybrids was 62 kg ha^{-1} year^{-1} at a medium planting density (52500 plants ha^{-1}). Pedigree and molecular marker data showed that US and Chinese hybrids were based on very different germplasm, with decades-old US germplasm contributing to recently developed and widely used Chinese hybrids (Li et al., 2011). For comparison, Chapter 2 outlines breeding and agronomic progress in maize-based systems of the USA, including a detailed account for the role of sowing density to accommodate hybrids and environments.

The significant yield improvement achieved for the cereal crops in favorable environments in China is in agreement with reports from other countries, in the sense that the yield improvements came from increasing biomass through increased photosynthesis (Fischer and Edmeades, 2010). This trend will probably continue in the future, but probably at a lower rate of increase than in the past. Conventional breeding will continue to play a significant role in improving yield potential, with likely larger contributions of biotechnology particularly for pyramiding genes for disease resistance and processing quality; Chapters 18 and 19 update the role of biotechnology on crop improvement. While detailed strategies for improving rice, maize, and wheat yields may be different, there are also many similarities, such as combining high yield potential with broad adaptation, increasing spikelet or kernel number per unit area, improving lodging resistance under high planting densities suitable for mechanized harvesting, and improving resistance/tolerance to abiotic and biotic stresses. To achieve such goals, a combined approach is required, including continued selection of elite parents for crossing, alternative selection in different environments, and multilocation testing at advanced stages. Increasing the number of advanced lines screened in multiple environments will greatly assist in identifying elite genotypes. For all three cereals, international germplasm exchange will continue to play a key role in further yield improvement in China (He et al., 2013).

4.3 Contribution of agronomic innovation

A multiple and proper cultivation system is essential for farmland productivity, food security and efficient use of agricultural resources. Along with the great development of science and technology and the progress of the society economy, Chinese cropping techniques have achieved much progress during the past decades, resulting in a large contribution to the increase of grain production. Here we outline three innovations in tillage, cropping season optimization, and rice production technologies.

4.3.1 *Soil tillage*

Traditional tillage has caused significant soil erosion and degradation, resulting in large decreases in SOM and fragile soil physical structure (Tang et al., 2004). To decrease the cost and to avoid the negative impacts of conventional tillage, reduced/no tillage has been encouraged (Xie et al., 2008; Li et al., 2009). However, no tillage could have long-term impacts on weed populations and soil compaction. Thus, some novel tillage practices have been developed to counteract these effects. In the northeast, less tillage combined with sub-soiling tillage was widely adopted in maize cropping. In the north and center, a rotation of conventional tillage for winter wheat and no tillage for summer maize or soybean is becoming dominant. For the annual double rice cropping system, a rotation of conventional tillage with less tillage is widely used.

4.3.2 *Cropping season optimization*

Field experiments showed that climate change benefits late rice and winter wheat production by promoting plant development and growth, but is detrimental to early and mid rice because heat stress affects floret fertilization and grain filling (Dong et al., 2011; Tian et al., 2012; Chen et al., 2013). To avoid the negative impacts of warming, the cropping seasons have been adjusted greatly over the past decades. In the winter wheat–maize cropping system, wheat sowing and maize harvest have been delayed simultaneously, resulting in a large increase in annual yield. Similarly, wheat sowing and rice harvesting have been delayed in the wheat–rice cropping system. Due to higher air temperature, the sowing of rice, maize and soybean has been significantly advanced with unchanged harvest dates, consequently resulting in a longer growth period in the northeast.

4.3.3 *Rice cropping technique innovation*

During the last decades, Chinese rice cropping techniques have been innovated in three aspects (Fig. 3.8). First, alternate dry-wetting seedling-nursery has progressively replaced continuous flooded seedling-nursery. This technique can save water and enhance rice seedling quality.

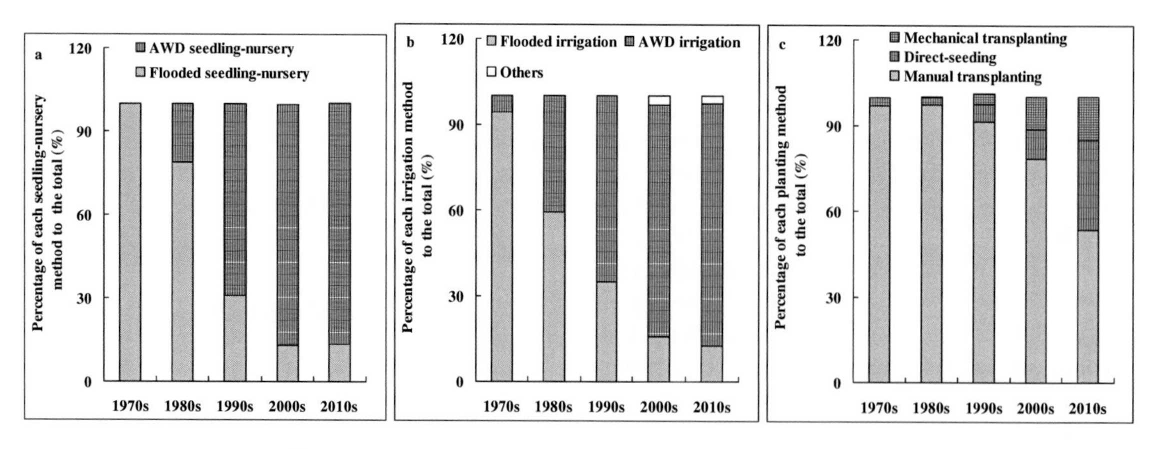

FIG. 3.8 Rice cropping technique innovations: (a) seedling nursery, (b) irrigation and (c) planting method in China since 1970s.

Secondly, alternate dry-wetting (AWD) irrigation has replaced the continuous flooded irrigation in major rice cropping areas. AWD can increase O_2 concentration in paddy soil to enhance rice root activity after anthesis, consequently benefiting grain filling and yield. One more innovation is the planting method. Traditional manual transplanting is being replaced by mechanical transplanting and dry direct seeding because of the shortage of rural labor. Mechanical rice planting can significantly decrease rice production costs and increase the labor efficiency.

5 ATTEMPTS TO IMPROVE RESOURCE-USE EFFICIENCY

5.1 Conservation agriculture for high water-use efficiency

Conservation agriculture (CA) is an approach developed to manage farmland for sustainable crop production, while simultaneously preserving soil and water resources (Erenstein, 2011). Generally, conservation agriculture relies on three major principles: maintenance of a permanent vegetative cover or mulch on the soil surface; minimal soil disturbance (no/reduced tillage); and diversified crop rotation (FAO, 2013). Given the effects of conservation agriculture on soil, water and economic viability, this management has been widely recommended and adopted with mixed success; Chapter 5 discusses the short- and long-term implications of CA in Africa, and the trade-offs involved in its adoption.

Since the 1970s, large efforts have been made on research and demonstration of conservation agriculture in China, where it has been applied in 6.4×10^6 ha by 2012 with a projected increase to 11×10^6 ha by 2015 (GOV, 2009). Conservation agriculture practices in China include no tillage, reduced tillage, straw application, plastic film mulching, water-saving irrigation, and limited irrigation (Zhang et al., 2013a).

Crop straw and plastic film mulching have been extended because material is easily accessible and low cost. In the north China Plain and Loess Plateau, for example, many efforts have been paid to increasing water-use efficiency (WUE, yield per unit evapotranspiration) with an acceptable crop yield (Zhang et al., 2013a). However, the range of WUE was very large for the cropping practices, suggesting opportunities for maintaining or increasing WUE with high yield in the region. Mulching with crop residues can improve water-use efficiency by 10–20% through reduced soil evaporation and increased plant transpiration (Table 3.2). Mulching with crop residues during the summer fallow can increase soil water retention. Wang et al. (2004) demonstrated straw mulching significantly

TABLE 3.2 The effects of mulching on soil evaporation, grain yield and water-use efficiency of wheat and maize

Crop	Treatment	Evapotranspiration (mm)	Soil evaporation (mm)	Grain yield ($g\ m^{-2}$)	WUE ($kg\ m^{-3}$)
Winter wheat	Mulching	367	75	714	1.94
	No mulching	390	117	669	1.72
Maize	Mulching	386	86	712	1.84
	No mulching	431	129	666	1.55
Summer fallow	Mulching	----	39	----	----
	No mulching	----	107	----	----

---- *Data not available. Source:* Deng et al., 2006.

increased the harvesting of rainwater and yield. Plastic film mulched furrow-ridge cropping (PMF) is a recent modification of maize cropping in the semiarid Loess Plateau. Recent studies indicated that soil evaporation significantly decreased in PMF after the jointing stage of spring maize as compared with uncovered and flat sowing (CK). The grain yield and water-use efficiency in the PMF treatment was increased by 333.1% and 290.6%, respectively. In addition, the harvest index of maize in PMF treatment was 132.5% higher than CK (Wang et al., 2011, 2013). Plastic film mulched furrow-ridge can significantly enhance maize grain yields and water-use efficiency by reducing soil evaporation and improving precipitation water movement from ridges to furrows and increasing soil temperature. Based on a 20-year experiment, Zhang et al. (2013a) also found the maximum wheat yield and WUE in the fields with plastic film and crop straw mulching. Consequently, mulching has been widely extended on the semi-arid Loess Plateau for maize production.

Although conservation agriculture can benefit crop production (Wang et al., 2012b), it can also reduce crop production through undesirable effects on soil physiochemical and biological conditions (Deubel et al., 2011; Chapter 5). The key limiting factor to the application of conservation agriculture in China is the uncertainty of the long-term effects on soil and crop yield. To quantify the impacts of conservation agriculture practices (i.e. NT: no/reduced-tillage only, CTSR: conventional tillage with straw retention, NTSR: NT with straw retention) on crop yield we conducted a meta-analysis to compare with conventional tillage without straw retention (CT).

Conservation agriculture significantly increased crop yield by 4.6% compared to the CT (Fig. 3.9), though there were large differences among the practices (Qb, R and $P = 0.0346$). The yield gains of CTSR and NTSR were 4.9 and 6.3%, respectively, while there was no significant effect in the NT compared to the CT. Meanwhile, the longer duration of conservation agriculture, the higher the increase in crop yield (Qb, R and $P = 0.004$, Fig. 3.9).

The effects of conservation agriculture on crop yield decreased with increasing annual precipitation (Qb, R and $P = 0.042$, Fig. 3.10). Significant positive effects occurred where annual precipitation is below 600 mm, whereas

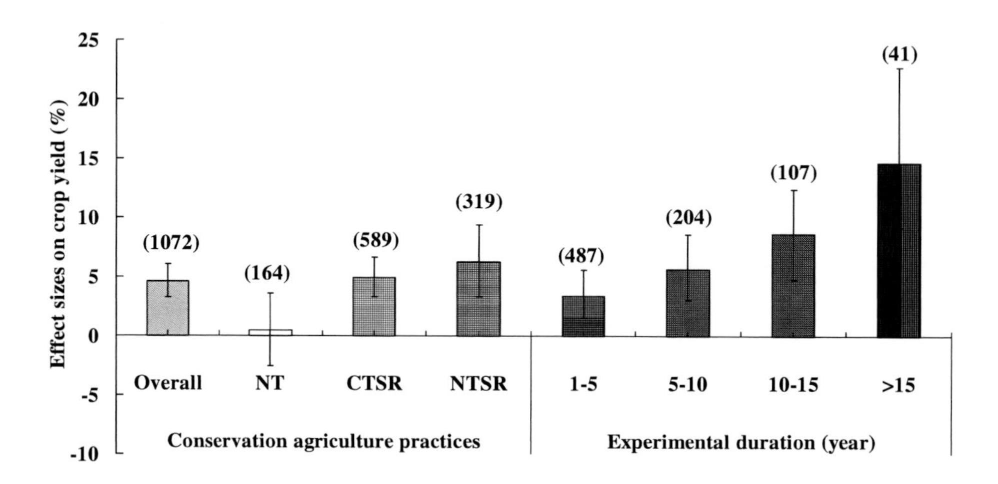

FIG. 3.9 Differences in the effect sizes among the conservation agriculture practices and among the experimental durations. Numbers between brackets are the observations.

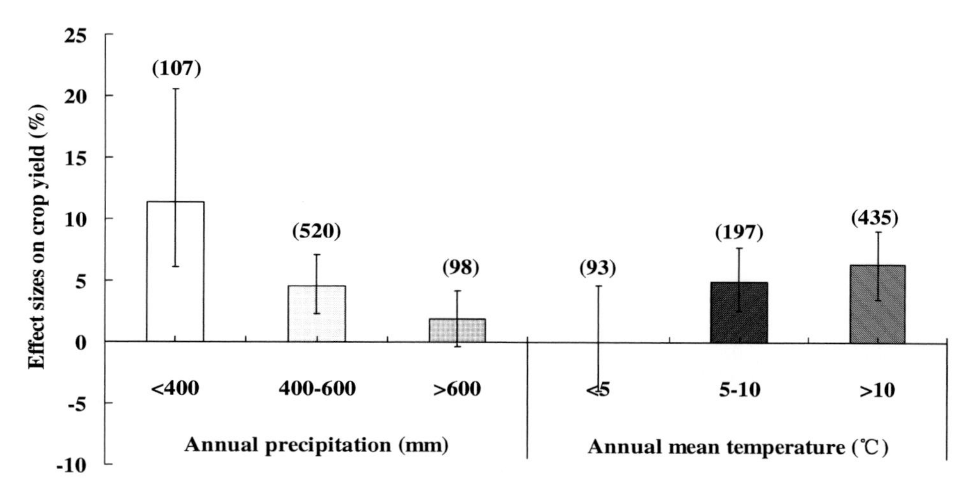

FIG. 3.10 Differences in the effect sizes of conservation agriculture practices in relation to precipitation and air temperature. Numbers between brackets are the observations.

no clear effects were found when precipitation was above 600 mm. The higher the mean annual temperature, the higher yield gains under conservation agriculture, although the differences were not significant between the temperature ranges (Qb, R and $P = 0.1042$, Fig. 3.10). The highest enhancing effects on crop yield occurred when mean annual temperature was higher than 10°C, whereas the effect was not significant when mean annual temperature was lower than 5°C.

5.2 Innovations for improving nitrogen fertilizer-use efficiency

Nitrogen (N) is the most important mineral nutrient for cereal production, and an adequate supply is essential for high yields (Chapter 8). Global consumption of synthetic N has increased from 11.6 Tg in 1961 to 104 Tg in 2006 (FAO, 2013). Furthermore, the annual total global N use is projected to grow to approximately 112 Tg in 2020, and up to 171 Mt in 2050, assuming no change in N-use efficiency. China has become one of the world's largest producers and consumers of N, P (P_2O_5) and K (K_2O) fertilizer; its use increased from 8.3 to 24.0 Tg, 2.2 to 8.3 Tg, and 0.3 to 6.2 Tg between 1979 and 2012, respectively (Fig. 3.11).

Excessive N application causes environmental pollution, increases the cost, reduces nitrogen-use efficiency and may reduce grain yield. The N-use efficiency (NUE: grain yield per unit N applied, kg kg^{-1} N), decreased by about 24%, from 43 kg kg^{-1} N in 1980 to 32 kg kg^{-1} N in 2005. China's national average N application rate is higher than the world's average, hence the average NUE in China is lower than the estimated global average of 44 kg kg^{-1} (Wang et al., 2011). Huang et al. (2010) indicated that the N-uptake efficiency (% fertilizer N recovered in above-ground crop biomass) in China is 30–35% for the three main crops. Improving N-uptake efficiency in the areas where it is lower than 30–50% could save 2.8–6.6 Tg N $year^{-1}$ (Huang et al., 2010). To improve synchrony between N supply and crop demand, large efforts have been made in research and demonstration. To increase crop yield and reduce fertilizer application, a national soil testing and fertilization recommendation (STFR) project has been conducted since 2006. STFR resolved the conflicts between crop fertilizer demand and soil fertilizer supply, increased crop yields and farmers' incomes, and improved the

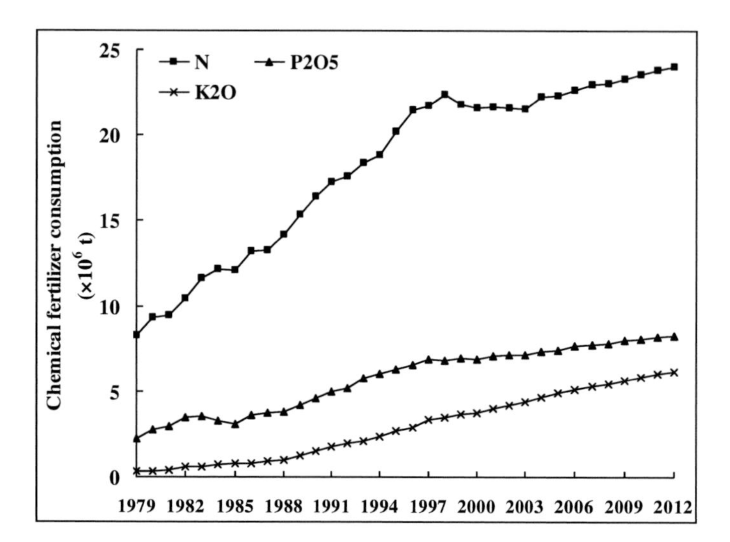

FIG. 3.11 Fertilizer consumption in China between 1979 and 2012.

quality of agricultural products. Chinese farmland under STFR has reached 8×10^7 ha in 2012. Compared to traditional fertilization, STFR increased rice, wheat, and maize yields by 3.7, 3.8 and 5.9%, respectively, and reduced fertilization to 700 Mt by 2011.

Site-specific N management (SSNM), where fertilization is based on the crop N demand, was developed to increase fertilizer-uptake efficiency of irrigated rice. During the seedling and booting stages, leaf N status measured with a chlorophyll meter (SPAD) is a good indicator of crop N demand to adjust predetermined N topdressings at seedling and booting. With this approach, the timing and number of N applications are fixed while the rate of N topdressing changes across the seasons and locations. Averaged across 107 farmers, SSNM produced 5% higher grain yield than the farmers' practice, and the N rate was reduced from 195 to 133 kg ha^{-1}, or 32% (Table 3.3). Consequently, SSNM almost doubled NUE and had a 55% higher partial factor productivity of applied N than the farmers' fertilization practice. Xue et al. (2013) developed

TABLE 3.3 Grain yield, total N rate, yield response to N application, agronomic N-use efficiency, and partial factor productivity of applied N (PFP) of farmers' fertilizer practice and site-specific N management (SSNM)

Parameters	Farmers' practice	SSNM	Difference (%)
Grain yield (t ha^{-1})	7.1^b	7.5^a	5.0
N rate (kg ha^{-1})	195.0^a	133.0^b	38.0
N response (t ha^{-1})	1.4^b	1.8^a	25.0
Agronomic NUE (kg kg^{-1})	7.1^b	13.4^a	61.0
PFP (kg kg^{-1})	36.3^b	56.2^a	43.0

Source: Peng et al., (2010). Within a row, means followed by different letters are significantly different at the 0.05 probability level according to the least significant difference (LSD) test.

Data were from on-farm demonstrations conducted by 107 farmers from six provinces in China between 2003 and 2007. Average grain yield of zero-N control was 5.7 t ha^{-1} across the 107 farmers.

an improved rice management system combining both SSNM and alternate wetting and moderate drying irrigation technologies (AWMD). Compared with local practice, this system increased rice yield, NUE and irrigation water-use efficiency (grain yield over amount of irrigation water) by 14.4, 64.1 and 36.4%, respectively.

An in-season N management strategy (INM) has been developed for the intensive wheat–maize system in the north China Plain. According to this strategy, the total amount of N fertilizer is divided between two or three applications during the growing season, with the optimal N rate for each application being determined from soil nitrate-N tests in the root zone and a target N value for the corresponding growth period of the crop. In this way, the effect of N mineralization, immobilization and N losses on plant available N during the previous growth period can be included in the result of the next soil nitrate-N analysis thus affecting the next N fertilization recommendation. When employing the INM in north China, experiments demonstrated that N fertilizer was reduced by 66% compared with regular practices without sacrificing crop yield (Cui et al., 2010).

6 CROPPING RESPONSES AND ADAPTATIONS TO WARMING

6.1 Crop phenology responses

Temperature is the key factor controlling crop development and growth (Chapter 12). Even a moderate increase in air temperature can affect crop phenology and growth duration significantly (Tao et al., 2006; Lobell et al., 2012). The impacts of warming on crop phenology in China have been investigated in the field (Table 3.4) and modeling (Chapter 20) experiments. For example, an air temperature increase of around 1.5°C carried out in the Free Air Temperature Increase (FATI) in the Yangtze Delta Plain significantly advanced crop phenophases of wheat and rice (Dong et al., 2011; Tian et al., 2012). Wheat time to maturity was shortened by 10 days ($P < 0.05$); this was mainly associated with shorter pre-anthesis phase, while the length of post-anthesis was slightly prolonged (Tian et al., 2012). The length of rice pre-heading phase was shortened by 3.3, 1.7 and 2.0 days in the all-day warming, daytime warming and night-time warming plots compared to the

TABLE 3.4 Responses of crop phenology to warming under field conditions in China

		Length of phenophase (d)			
Crop	**Treatment**	**Pre-anthesis**	**Post-anthesis**	**Entire period**	**Data source**
Winter wheat	Unwarmed	146	53	198	Tian et al., 2012
	All-day warming	134	53	187	
Middle rice	Unwarmed	72	49	120	Dong et al., 2011
	All-day warming	68	50	119	
	Daytime warming	70	48	118	
	Night-time warming	70	49	118	
Maize	Unwarmed	82	71	152	Qian et al., 2012
	Night-time warming	86	70	155	

The magnitudes of warming were 1.5°C and 1.1°C in daily mean temperatures in the experiments of Tian et al. (2012) and Qian et al. (2012), respectively. In the experiment of Dong et al. (2011), the air temperature elevations were 2.0°C in the daily mean temperature for the all-day warming treatment, 1.1°C in the daytime mean temperature for daytime warming treatment and 1.8°C in the night-time mean temperature for the night-time warming treatment.

no-warmed control, while the post-heading phase stayed almost unchanged (Dong et al., 2011). Another experiment conducted in northeast China, showed night-time warming before anthesis advanced the spring maize phenophases (Qian et al., 2012).

Similar responses of the crop growth period to warming were found in historical data analyses. Wang et al. (2012a) reported that the increase in temperature shortened the growth duration of winter wheat mainly by shortening the period from sowing to jointing at two northern sites. Tao et al. (2013) indicated that single rice transplanting, heading and maturity dates were generally advanced, but the heading and maturity dates of single rice in the middle and lower reaches of the Yangtze River and the northeast China Plain were delayed between 1981 and 2009. In general, warming shortened the crop pre-anthesis phase, while the post-anthesis phase was prolonged or not changed. Chapter 20 provides further evidence and advances an explanation for this pattern.

6.2 Crop yield responses

The effects of warming on crop production may depend on the crop type, the warming extent and the local background temperature (Lobell et al., 2011; Chapter 20). Theoretically, warming can shorten the length of the crop growth period, likely resulting in reduced biomass production and N uptake and accumulation. Warming may also aggravate high temperature stress potentially decreasing grain number and weight. Because multiple cropping systems dominate in China, the short growing period and the post-anthesis high temperature are the major constraints on crop production. Thus, warming may aggravate the constraints on grain production. On the other hand, temperature increase may directly reduce frost/chilling and indirectly reduce heat injury due to warming-led earlier anthesis. Since frost/chilling before flowering and high temperature stress after flowering are common in winter crops, warming may enhance yield and grain quality. For example, based on historical data analysis and the assumption of similar rainfall, Xiao et al. (2008) predicted that warming might increase wheat yield by 3.1% at low altitude and 4.0% at high altitude by 2030. Sommer et al. (2013) predicted that an increase in air temperature may mostly benefit wheat production in central Asia. Obviously, there are still major uncertainties about Chinese crop production under future climate that need further modeling and experimental work.

Many experiments have been conducted to quantify warming impacts on wheat yield and quality during the past decades. However, many of them were conducted under artificial environments (e.g. greenhouse or open top chambers), rather than an agroecosystem scale under field conditions. Meanwhile, existing experiments in China mainly focused on high temperature impacts on starch and protein depositions in grain during post-anthesis, and only a few studies have been conducted across an entire growing cycle. The impacts of predicted warming on wheat yield and grain quality may have been overestimated. Recently, some warming experiments were conducted under field conditions in China (Tian et al., 2012; Dong et al., 2011). Tian et al. (2012) found that anticipated warming may facilitate winter wheat production in eastern China (Table 3.5). An increase in air mean temperature of 1.5°C enhanced wheat grain yield by 16.3% ($P < 0.05$) mainly because of the warming-led increases in green leaf area and grain size. The area of flag leaf and total green leaves at anthesis and grain size were 36.0, 19.2 and 5.9% higher in the warmed plots than the unwarmed control ($P < 0.05$). Similarly, Hou et al. (2012) found that warming by 1.6°C also increased wheat yield in Yucheng, the center of Chinese winter wheat cropping, where there was no soil moisture limitation. Fang et al. (2013) reported that warming by 2°C decreased wheat yield under a large moveable rain shelter in Gucheng, the northern edge of Chinese

TABLE 3.5 Responses of crop productivity to climate warming under field conditions in China

Crop	Treatment	Effective panicles (plant m^{-2})	Grain number per panicle (grain $spike^{-1}$)	1000-Grain weight (g)	Above-ground biomass (g m^{-2})	Grain yield (g m^{-2})	Data source
Winter wheat	Unwarmed	545.7	38.9	42.4	1546.4	645.2	Tian et al., 2012
	All-day warming	558.7	40.4	44.9	1666.6	750.4	
Single rice	Unwarmed	248.6	153.6	25.8	2091.7	709.4	Dong et al., 2011
	All-day warming	260.1	146.6	24.3	1902.5	702.6	
	Daytime warming	265.1	147.0	24.9	1864.2	664.9	
	Night-time warming	253.5	148.5	24.7	2021.8	665.6	
Maize	Unwarmed	5.6	621.5	372.0	1265.0	2460.0	Qian et al., 2012
	Night-time warming	5.6	617.5	348.0	1160.0	2275.0	

winter wheat cropping, where the irrigation was controlled at 100 mm without any precipitation. With an addition of 20 mm, however, warming increased wheat yield significantly. Together, previous field experiments showed that predicted climatic warming will benefit winter wheat production in east China if there is no limitation in precipitation or irrigation. Similarly, cereal yield is likely to increase in milder climates projected for northern Europe, unless increasing water stress cancels the benefits of warming (Chapter 4). For the middle rice at the same site, warming slightly decreased the above-ground biomass by an average of 9.1, 10.3 and 3.3%, and the grain yield by an average of 0.9, 6.4 and 6.1% in the all-day warming, daytime warming and night-time warming plots compared to the ambient control, respectively (Table 3.5). Warming tended to decrease rice photosynthesis and stimulate night-time respiration. For maize in the northeast, night-time warming (less than 1.5°C) before anthesis increased maize above-ground biomass and grain yield by 8.2 and 9.3% (Table 3.5). Maize green leaf area and three-ear-leaves (i.e. the ear leaf, and the leaves immediately below and above the ear) area under night-time warming were 13.5 and 14.6% larger than for the unwarmed control.

6.3 Adaptations of cropping systems to warming in northeast China

Although some reports showed negative effects of climate warming in crop yield, some studies showed that crop production might benefit from warming if suitable adaptations are implemented (Lobell et al., 2008), especially in the high latitude areas (Chen et al., 2012; Chapter 4). Therefore, many cropping practices have been recommended as adaptations to warming, such as adjusting sowing date and cropping pattern, adopting heat-tolerant, higher-yielding varieties and improving crop management (Chen et al., 2012; Chapter 20).

In northeast China, where mean air temperature has increased 1.0°C over the last 20 years, present cropping boundaries can be theoretically extended northward about 80 km with a prolonged growing period of 10 days compared to the 1970s. For example, based on the records of the rice cropping area of Heilongjiang province in 1970 and 2006, the actual changes of the spatial distributions of rice sowing area were analyzed with GIS (Fig. 3.12). The total area of rice cropping increased by 17.1 times in this period, with a greatest increment occurring in the east of the province. Meanwhile, the cropping region has also enlarged over the past years. In 1970, the main cropping region was located south of 46°N, especially around the line of 45°N and extended northward of 47°N, mainly around the line of 46°N during the period 1970–2006. The rice sowing areas north of 46°N increased from 3.6×10^4 ha in 1970 to 130×10^4 ha in 2006. Rice production north of 46°N increased from 11×10^4 t in 1970 to 851×10^4 t in 2006. The gravity center, a point that can maintain a balance of force in all directions in a region space (Griffith, 1984), of rice cropping region shifted from E128°52′, N45°37′in 1970 to E129°53′, N46°29′ in 2006. The actual cropping center moved about 80 km northward from 1970 to 2006.

The growth durations of newly approved varieties of rice, maize and soybean have been prolonged by 14.0, 7.0 and 2.7 days since the 1950s, respectively (Fig. 3.13). Significant differences in the growth durations were found among different provinces. The increase in the growth duration of rice over the years of 1950–2008 was 9.1 days for Heilongjiang, 13.4 days for Jilin and 19.7 days for Liaoning province (Fig. 3.13). Similar increases were found for the growth duration of maize and soybean (Fig. 3.13). The rice growth duration showed the greatest increment among the three crops during the variety improving process. These changes highlight how crop breeding has contributed to crop adaptation to warming.

The adjustment of sowing and harvest dates since the 1990s prolonged the actual growing period of both rice (6 d) and maize (4 d) (Chen et al., 2012). Since the 1980s, improved varieties

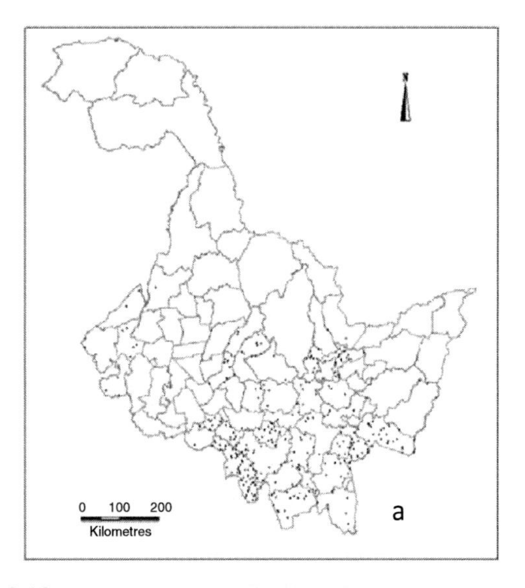

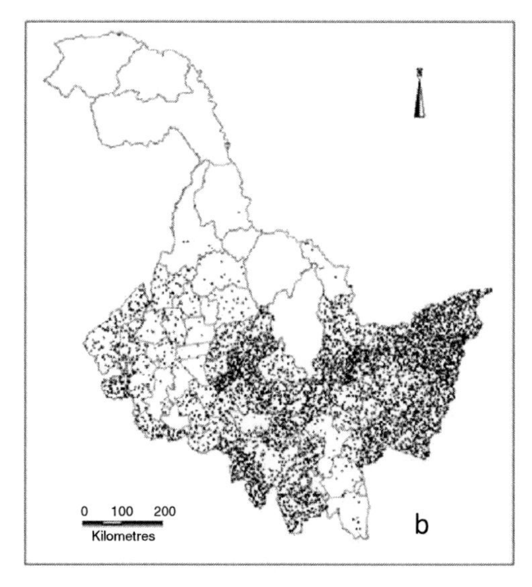

FIG. 3.12 Actual spatial distributions of rice cropping area in (a) 1970 and (b) 2006 in Heilongjiang province in northeast China. One dot means 300 ha. *Source: Chen et al. (2012).*

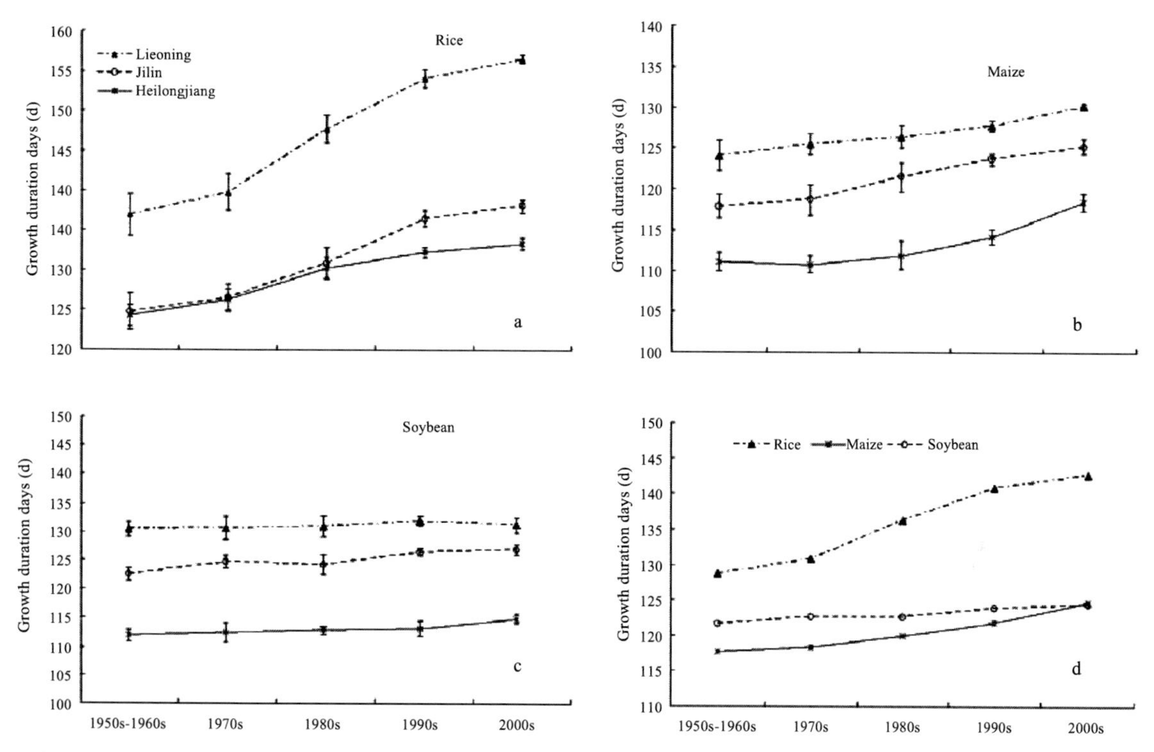

FIG. 3.13 Crop growth duration in northeast China for varieties released since 1950s. (a) Rice, (b) maize, (c) soybean and (d) their average. *Source: Chen et al. (2012).*

were able to compensate for the negative impact of climate in the north China Plain. The variety changes of wheat and maize maintained the length of the pre-flowering period against the shortening effect of warming and extended the length of the grain-filling period (Liu et al., 2010). Historical increase in air temperature before the over-wintering stage enabled late sowing of wheat and late harvesting of maize, leading to an overall 4–6% increase in annual production of the wheat–maize system. Mechanical sowing and less tillage also shortened the time for field preparation, which facilitated the later harvest of summer maize. Zhang et al. (2013b) found that a major, temperature-induced change in the rice growth duration was underway in China and that using a short-duration cultivar has been accelerating the process for late rice.

Climate change has shifted the cropping boundary of different cropping systems in China (Yang et al., 2010). Compared to the cropping boundary in the period from the 1950s to 1980, the northern limits of double cropping were displaced in Shaanxi, Shanxi, Hebei, Beijing and Liaoning provinces from 1981 to 2007. The northern limits of the three-crop system were displaced in Hunan, Hubei, Anhui, Jiangsu, and Zhejiang provinces. The northern limits of winter wheat moved northwards and westwards to different degrees in Liaoning, Hebei, Shanxi, Shaanxi, Inner Mongolia, Ningxia, Gansu and Qinghai provinces, compared with the period 1950–1980. The northern limits of double-rice cropping in Zhejiang, Anhui, Hubei, and Hunan provinces moved northwards. The stable-yield northern limits of rain-fed winter wheat–summer maize moved south-eastwards in most regions, mainly in response to decreased rainfall during recent years. During the past 50 years, warming caused the northwards movement of the northern limits of the cropping system, and the northern limits of winter wheat and double rice.

7 CONCLUDING REMARKS

China has increased grain production continuously for ten years, but there are concerns with the secondary effects of intensive cropping on farmland soil and water. To reduce the negative impacts of intensive cropping on natural resources and the environment, the Chinese government plans to decrease the grain self-sufficiency rate. Simultaneously, increasing efforts are allocated to research and application of cropping techniques to achieve the multiple goals of food security, environmental safety and climate change adaptation. Integrated innovations in crop breeding and agronomy, informed by crop physiology, will be the key to achieve these goals.

References

Chen, C.Q., Qian, C.R., Deng, A.X., Zhang, W.J., 2012. Progressive and active adaptations of cropping system to climate change in Northeast China. Eur. J. Agron. 38, 94–103.

Chen, J., Tian, Y.L., Zhang, X., et al.,2013. Nighttime warming will increase winter wheat yield through improving plant development and grain growth in north. China J. Plant Growth Regul.doi: 10.1007/s00344-013-9390-0.

Chen, Y., Zhang, X.D., He, H.B., et al.,2010. Carbon and nitrogen pools in different aggregates of a Chinese Mollisol as influenced by long-term fertilization. J. Soil Sediments 10, 1018–1026.

Ci, X.K., Li, M.H., Li, X.L., et al.,2011. Genetic contribution to advanced yield for maize hybrids released from 1970 to 2000 in China. Crop Sci. 51, 13–20.

Ci, X.K., Li, M.S., Xu, J.S., Lu, Z.Y., Bai, P.F., Ru, G.L., 2012. Trends of grain yield and plant traits in Chinese maize cultivars from the 1950s to the 2000s. Euphytica 185, 395–406.

Ci, X.K., Zhang, D.G., Li, X.H., et al.,2013. Trends in ear traits of Chinese maize cultivars from the 1950s to the 2000s. Agron. J. 105, 20–27.

Cui, Z., Zhang, F., Chen, X., Dou, Z., Li, J., 2010. In-season nitrogen management strategy for winter wheat: Maximizing yields, minimizing environmental impact in an over-fertilization context. Field Crops Res. 116, 140–146.

Deng, X., Shan, L., Zhang, H., Turner, N., 2006. Improving agricultural water use efficiency in arid and semiarid areas of China. Agric. Water Manag. 80, 23–40.

Deubel, A., Hofmann, B., Orzessek, D., 2011. Long-term effects of tillage on stratification and plant availability of phosphate and potassium in a loess chernozem. Soil Tillage Res. 117, 85–92.

Dong, W.J., Chen, J., Zhang, B., Tian, Y.L., Zhang, W.J., 2011. Responses of biomass growth and grain yield of midseason rice to the anticipated warming with FATI facility in East China. Field Crops Res. 123, 259–265.

Erenstein, O., 2011. Cropping systems and crop residue management in the Trans-Gangetic Plains: Issues and challenges for conservation agriculture from village surveys. Agric. Syst. 104, 54–62.

Fang, S., Su, H., Liu, W., Tan, K., Ren, S., 2013. Infrared warming reduced winter wheat yields and some physiological parameters, which were mitigated by irrigation and worsened by delayed sowing. Plos one 8, e67518.

FAO, 2013. FAOSTAT database: agriculture production. Food and Agriculture Organization of the United Nations, Rome.

Fischer, R.A., Edmeades, G.O., 2010. Breeding and cereal yield progress. Crop Sci. 50, S-85–S-98.

GOV: The Central People's Government of the People's Republic of China, 2009. http://www.gov.cn/gzdt/2009-08/28/content_1403670.htm.

Griffith, D., 1984. Theory of spatial statistics. In: Gaile, G.L., Willmott, C.J. (Eds.), Spatial statistics and models. D. Reidel Publishing Company, Dordrecht, The Netherlands.

Guo, J.H., Liu, X.J., Zhang, Y., et al.,2010. Significant acidification in major Chinese croplands. Science 327, 1008–1010.

He, Z.H., Xia, X.C., Peng, S.B., Lumpkin, T.A., 2013. Meeting demands for increased cereal production in China. J. Cereal Sci., http://dx.doi.org/10.1016/j.jcs.2013.07.012.

Hou, R., Ouyang, Z., Li, Y., Wilson, G.V., Li, H., 2012. Is the change of winter wheat yield under warming caused by shortened reproductive period? Ecol. Evol. 2, 2999–3008.

Huang, S., Peng, X.X., Huang, Q.R., Zhang, W.J., 2010. Soil aggregation and organic carbon fractions affected by long-term fertilization in a red soil of subtropical China. Geoderma 154, 364–369.

Huang, S., Zeng, Y.J., Wu, J.F., Shi, Q.H., Pan, X.H., 2013. Effect of crop residue retention on rice yield in China: A meta-analysis. Field Crops Res. 154, 188–194.

IPCC, 2007. Climate change 2007: the physical science basis. Contribution of Working Group I to the Fourth Assessment Report of the Intergovernmental Panel on Climate Change. Cambridge University Press, Cambridge.

Li, D., Wang, L., Huang, G., Guo, P., 2009. Effect of conservation tillage on the water and soil loss in slope land of the Loess Plateau. J. Anhui Agric. Sci. 37, 6087–6088, (in Chinese with English abstract).

Li, Y., Ma, X.L., Wang, T.Y., et al.,2011. Increasing maize productivity in China by planting hybrids with germplasm that responds favorably to higher planting densities. Crop Sci. 51, 2391–2400.

Liu, Y., Wang, E., Yang, X., Wang, J., 2010. Contributions of climatic and crop varietal changes to crop production in the North China Plain, since 1980s. Glob. Change Biol. 16, 2287–2299.

Liu, L., Xu, X.L., Zhuang, D.F., Chen, X., Li, S., 2013. Changes in the potential multiple cropping system in response to climate change in China from 1960-2010. Plos one 8, e80990.

Lobell, D.B., Burke, M.B., Tebald, C., Mastrandrea, M.D., Falcon, W.P., Naylor, R.L., 2008. Prioritizing climate change adaptation needs for food security in 2030. Science 319, 607–610.

Lobell, D.B., Schlenker, W., Costa-Roberts, J., 2011. Climate trends and global crop production since 1980. Science 333, 616–620.

Lobell, D.B., Sibley, A., Ortiz-Monasterio, J.I., 2012. Extreme heat effects on wheat senescence in India. Nat. Clim. Change 2, 186–189.

National Bureau of Statistics of China., 2012. China Statistical Yearbook. China Statistics Press, Beijing (in Chinese).

Peng, S.B., Buresh, R.J., Huang, J., et al.,2010. Improving nitrogen fertilization in rice by site-specific N management. A review. Agron. Sustain. Dev. 30, 649–656.

Qian, C., Yu, Y., Zhao, Y., et al.,2012. Responses of spring corn growth and yield in a cold area of China to field warming at nighttime during pre-anthesis stage. Chinese J. Appl. Ecol. 23, 2483–2488, (in Chinese with English abstract).

Sommer, R., Glazirina, M., Yuldashev, T., et al.,2013. Impact of climate change on wheat productivity in Central Asia. Agric. Ecosyst. Environ. 178, 78–99.

Tang, K.L., 2004. Soil erosion in China. Science and Technology Press, p. 845 (in Chinese).

Tao, F.L., Zhang, Z., Shi, W.J., et al.,2013. Single rice growth period was prolonged by cultivars shifts but yield was damaged by climate change during 1981-2009 in China, and late rice was just opposite. Glob. Change Biol. 19, 3200–3209.

Tao, F.L., Yokozawa, M., Xu, Y.L., Hayashi, Y., Zhang, Z., 2006. Climate changes and trends in phenology and yields of field crops in China, 1981-2000. Agri. For. Meteorol. 138, 82–92.

Tian, Y.L., Chen, J., Chen, C.Q., et al.,2012. Warming impacts on winter wheat phenophase and grain yield under field conditions in Yangtze Delta Plain. China. Field Crops Res. 134, 193–199.

Wang, C.R., Tian, X.H., Li, S.X., 2004. Effects of plastic sheet-mulching on ridge for rainwater-harvesting cultivation on WUE and yield of winter wheat. Sci. Agric. Sin. 37, 208–214, (in Chinese with English abstract).

Wang, H.L., Zhang, X.C., Song, S.Y., Ma, Y.F., Yu, X.F., 2013. Regulation of whole field surface plastic mulching and double ridge-furrow planting on seasonal soil water loss and maize yield in rain-fed area of Northwest Loess Plateau. Sci. Agric. Sin. 46, 917–926, (in Chinese with English abstract).

Wang, H.L., Zhang, X.C., Song, S.Y., Ma, Y.F., Yu, X.F., Liu, Y.L., 2011. Effects of whole field-surface plastic mulching and planting in furrow on soil temperature, soil moisture, and corn yield in arid area of Gansu Province, Northwest China. Chin J. Appl. Ecol. 22, 2609–2614, (in Chinese with English abstract).

Wang, J., Wang, E., Yang, X., Zhang, F., Yin, H., 2012a. Increased yield potential of wheat-maize cropping system in the North China Plain by climate change adaptation. Clim. Change 113, 825–840.

Wang, C.J., Pan, G.X., Tian, Y.G., Li, L.Q., Xu, H.Z., Xiao, J.H., 2010. Changes in cropland topsoil organic carbon with different fertilizations under long-term agro-ecosystem experiments across mainland China. Sci China Life Sci. 53, 858–867.

Wang, X.B., Wu, H.J., Dai, K.A., Zhang, D.C., Feng, Z.H., Zhao, Q.S., Wu, X.P., Jin, K., Cai, D.X., Oenema, O., Hoogmoed, W.B., 2012b. Tillage and crop residue effects on rainfed wheat and maize production in northern China. Field Crops Res. 132, 106–116.

Xiao, G.J., Zhang, Q., Yao, Y.B., et al.,2008. Impact of recent climatic change on the yield of winter wheat at low and high altitudes in semi-arid northwestern China. Agric. Ecosyst. Environ. 127, 37–42.

Xiao, Y.G., Qian, Z.G., Wu, K., et al.,2012. Genetic gains in grain yield and physiological traits of winter wheat in Shandong Province, China, from 1969 to 2006. Crop Sci. 52, 44–56.

Xie, R., Li, S., Jin, Y., et al.,2008. The trends of crop yield responses to conservation tillage in China. Sci. Agric. Sin. 41, 397–404, (in Chinese with English abstract).

Xie, Z.J., Li, M.S., Xu, J.S., Zhang, S.H., 2009. Contributions of genetic improvement to yields of maize hybrids during different eras in North China. Sci. Agric. Sin. 42, 781–789, (in Chinese with English abstract).

Xue, Y., Duan, H., Liu, L., Wang, Z., Yang, J., Zhang, J., 2013. An improved crop management increases grain yield and nitrogen and water use efficiency in rice. Crop Sci. 53, 271–284.

Yang, X., Liu, Z., Chen, F., 2010. The possible effects of global warming on cropping systems in China I. The possible effects of climate warming on northern limits of cropping systems and crop yields in China. Sci. Agric. Sin. 43, 329–336, (in Chinese with English abstract).

Yong, T.W., Yang, W.Y., Xiang, D.B., Chen, X.R., Wan, Y., 2012. Production and N nutrient performance of wheat-maize-soybean relay strip intercropping system and evaluation of interspecies competition. Acta Pratac. Sin. 21, 50–58, (in Chinese with English abstract).

Zhang, S., Sadras, V.O., Chen, X., Zhang, F., 2013a. Water use efficiency of dryland wheat in the Loess Plateau in responseto soil and crop management. Field Crops Res. 151, 9–18.

Zhang, T., Huang, Y., Yang, X., 2013b. Climate warming over the past three decades has shortened rice growth duration in China and cultivar shifts have further accelerated the process for late rice. Glob. Change Biol. 19, 563–570.

Zheng, T.C., Zhang, X.K., Yin, G.H., et al.,2011. Genetic improvement of grain yield and associated traits in Henan Province, China, 1981 to 2008. Field Crops Res. 12, 225–233.

Zhou, Y., He, Z.H., Sui, X.X., Xia, X.C., Zhang, X.K., Zhang, G.S., 2007. Genetic improvement of grain yield and associated traits in the northern China winter wheat region from 1960 to 2000. Crop Sci. 47, 245–253.

CHAPTER

4

Improving farming systems in northern Europe

Pirjo Peltonen-Sainio, Ari Rajala, Hannu Känkänen, Kaija Hakala

MTT Agrifood Research Finland, Plant Production, Jokioinen, Finland

1 SPECIAL FEATURES OF NORTHERN EUROPEAN CONDITIONS FOR CROP PRODUCTION

Northern European growing conditions have particular features that differ markedly from those typically associated with the rest of Europe. These features are critical for crop development, growth and yield determination *per se*, as well as for the mechanisms and approaches needed and used in adapting to and managing such northern agricultural systems. Finland is the northernmost agricultural country (Fig. 4.1), having all of its arable land of some 2 million hectares above 60°N. Agriculture is possible at such northern latitudes due to the Gulf Stream. The Gulf Stream, together with its northern extension towards Europe, the North Atlantic Drift, is a powerful, warm, and swift Atlantic Ocean current that originates in the Gulf of Mexico and influences the climate of the west coast of Europe. Hence, due to the Gulf Stream, the northern European temperature regimens during the growing season differ from those elsewhere at comparable latitudes and from those in the more southern regions.

Finland represents the uppermost extreme for field crop production and is an interesting case with respect to how crop physiology needs to be taken into account when targeting improved, sustainable agricultural systems under exceptional conditions. In Finland, over 60% of the arable land is allocated to small-grain cereals and rapeseed. The second most important group is grass crops, occupying 30% of the total arable land. Other crops are grown on much smaller areas. This chapter addresses small-grain cereals and rapeseed as these are the crops that can be successfully grown under such northern conditions.

The most typical features of northern growing conditions are harsh winters, intensive, exceptionally rapid rate of development due to exposure of very early growth stages to long days and relatively rapidly increasing mean temperatures,

Crop Physiology. DOI: 10.1016/B978-0-12-417104-6.00004-2

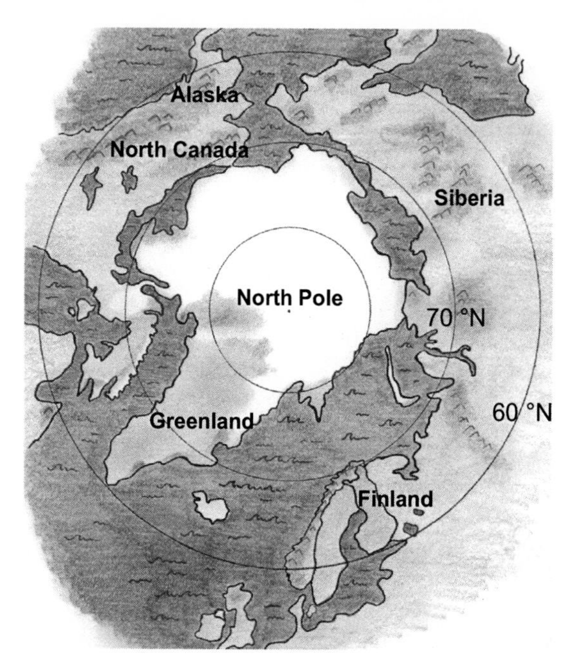

FIG. 4.1 Europe has the northernmost conditions for active field crop production with the largest areas of arable land. The Nordic countries have a total of 9.1 million hectares of arable land (FAO Statistics 2013 in http://faostat.fao.org). While all of the arable land and seed cropping in Finland is above 60°N, in Sweden 90% of seed cropping is at lower latitudes, in Norway 50% and in Denmark 100%. Evidently, in such northern extremes, differences in growing conditions and production risks change markedly according to latitude. (*Drawing: Jaana Nissi.*)

generally cool mean temperatures during the growing season, risk of night frosts, early summer drought and risk of abundant precipitation close to harvest and later at the time of sowing of winter cereals. Successful crop production under conditions with such a particular combination of features requires specific adaptation mechanisms. However, it is not only the degree of difficulty that such basic conditions represent but, particularly, the substantial fluctuations in conditions that occur within and among years and seasons (Peltonen-Sainio and Niemi, 2012) that determine the requirements of plant breeding and development of adaptive crop management practices. As our examples will indicate, such specific adaptation strategies are very vulnerable to changes in political and economic environments (Peltonen-Sainio et al., 2009a) as these are invariably dynamic and unpredictable, and their impacts are difficult to quantify (Richards, 2002).

1.1 Harsh winters

Winters in the north are harsh in many ways. When over-wintering conditions have been too harsh and resulted in problems of winter survival, or even total crop failure, it is typically due to a combination of unfavorable, critical conditions and stresses that are, when occurring together or when succeeding each other, lethal for the crop.

Under northern conditions, winter survival of crops is typically dependent on latitude, species and winter-hardiness of the cultivar. Low temperatures rarely cause crop death, as hardened over-wintering crops in the north usually tolerate freezing temperatures during winter (Hömmö, 1994; Antikainen, 1996). In addition, where the temperatures are the lowest, there is usually a protective snow cover. On the other hand, abundant snow cover often maintains the temperature at the plant stand close to zero, favoring infection by low-temperature parasitic fungi such as *Microdochium nivale, Sclerotinia* spp. and *Typhula* spp. that can cause serious economic damage to cereals, grasses and legumes (Hömmö and Pulli, 1993; Hömmö, 1994; Nissinen, 1996; Serenius et al., 2005). The typical period for permanent snow cover in southern Finland (Jokioinen) was from 16 December to 10 April and, in central Finland, (Jyväskylä) from 26 November to 22 April according to data covering the last three decades. Melting of snow can result in formation of hermetic ice cover and anoxia (Hofgaard et al., 2003). Cycles of melting during the day and freezing at night of the uppermost soil layers can cause frost heaving and damage such as root breakage, especially in spring. Furthermore, insufficient carbohydrate reserves for maintenance respiration are a problem when winters are long and plants are covered with snow or when winters are exceptionally warm and temperatures are above zero for long periods and enhance crop metabolism (Niemeläinen, 1990; Antikainen, 1996; Hakala and Pahkala, 2003). However, different combinations of these and other constraints represent a challenge for winter survival when they occur once or several times during winter (Hömmö, 1994; Peltonen-Sainio et al., 2011a).

1.2 Short and intensive growing seasons

The thermal growing season (i.e. the period when daily mean temperatures are permanently above a base temperature of +5°C) is short in the north. Typically, it lasted from 27 April till 15 October in southern Finland (Jokioinen) and from 2 May till 6 October in central Finland (Jyväskylä) according to data covering the last three decades. The proportion of the agronomically relevant and physiologically advantageous part of the season is likely to be less relative to that in central Europe (Trnka et al., 2011). This is because under northern growing conditions there is a period during the beginning of the thermal growing season that cannot be utilized for crop growth for various reasons, including too moist soils (melted snow that is directed from the fields with sub-soil drainage systems) that need to dry before sowing is possible with heavy machinery, and low temperatures that slow down germination, seedling establishment and early growth. Also risks of night frost are higher the earlier the sowing. However, the post-sowing period prior to canopy closure is likely to be relatively short under northern conditions with long days and rapidly increasing temperatures following sowing. The lost part of the early thermal growing season is dependent on local conditions but also on crop species. For example, rapeseed is especially sensitive to low temperatures at germination and night frosts during the early seedling stage and therefore, it is sown later and it has a longer lost thermal period early in the growing season than, for example, cereals (Peltonen-Sainio et al., 2006).

An extensive European study revealed that, in the northernmost regions, the number of effective growing days was strikingly low (Trnka et al., 2011). Under northern growing conditions, there is also a substantial period lost at the end of the thermal growing season that is characterized by higher rainfall challenging maturation and harvests, decreasing mean temperatures, increased risks of night frost and reduced light intensity (Rajala and Peltonen-Sainio, 2000; Peltonen-Sainio et al., 2009b; Saikkonen et al., 2012), as day length shortens rapidly from 12 h at the autumnal equinox in late September to 8 hours at the end of October until it reaches the minimum of 5.5 h around 21 December in southern Finland (e.g. www.gaisma.com, 9.7.2013).

Mean temperatures. Typically conditions during the northern European growing season are cool but, nonetheless, favor cereal and rapeseed growth and many other temperate crops that can cope with the short growing season. On the other hand, temperatures are even relatively high at crop establishment because it takes some time for the field to dry prior to sowing and, during late spring, mean daily temperatures increase quite quickly as days lengthen. The question of temperatures sustaining early growth is, however, dependent on crop species. For cereals adapted to northern growing conditions, the temperature regimen in early summer is always sufficient for good establishment in regions where they are commonly grown. However, because germination and early seedling establishment of rapeseed are favored by higher temperature after sowing (Peltonen-Sainio et al., 2006), rapeseed benefits from delayed sowing in comparison with cereals. Such differences between cereals and rapeseed in response to temperature in northern conditions differ from those described by Angus et al. (1981) for temperate cultivars grown in a Mediterranean climate.

The growing season in northern Europe does not end in increasingly high, stressful temperatures and terminal drought typical of Mediterranean (Acevedo et al., 1999) and continental climates of vast regions of North America and Asia (Johnson et al., 2004; Sadras and Angus, 2006). In Nordic countries, temperatures and potential evapotranspiration typically fall and precipitation increases towards the end of the growing season (Mukula and Rantanen, 1987). Despite generally cool conditions, elevated temperatures also hamper crop growth in the northernmost European regions, as it did across the rest of Europe according to an extensive study that observed how variation in yield and climate coincided (Peltonen-Sainio et al., 2010). For example, turnip rape (*Brassica rapa* L.) is very sensitive to elevated temperatures at seed set and fill (Peltonen-Sainio et al., 2007a): a 2–3°C increase in daily mean temperature during the seed filling phase reduced yield by up to 450 kg ha^{-1} depending on cultivar when grown in Finland. Similarly all spring cereals exhibited sensitivity to elevated temperatures at their critical yield determining phases contrary to autumn sown cereals (Table 4.1) (Peltonen-Sainio et al. 2011b).

The risk of night frost can hold farmers back from sowing very early in years when a warm spring begins exceptionally early. In the event of severe night frosts, below −5°C, occurring after seedling emergence, cereal stands are rarely completely destroyed as the developing apex is close to the soil surface and is well protected and insulated by the expanding leaves. However, rapeseed is sensitive to night frosts and, at −5 to −7°C, most seedlings are lost (Pahkala and Sovero, 1988; Pahkala et al., 1991). Hence, early sowing of rapeseed might necessitate re-sowing later in the season, and this can expose the plant stand to unfavorable conditions at maturation if done too late. Night frosts occur typically in spring and early summer. Mid-summer is the only period when crop-damaging frost does not occur. For example, in southern Finland such a frost-free period is between 20 June and 10 August (Solantie, 1987).

TABLE 4.1 Growing season climate of two locations in Finland

	Mean		Precipitation			
Month	**Temp (°C)**	**Radiation (MJ/m²)**	**Mean (mm)**	**Min (mm)**	**Max (mm)**	**SD**
JOKIOINEN, SOUTH FINLAND						
May	9.6	18.2	36.6	0.9	87.2	21.6
June	14.1	19.9	59.3	11.2	131.1	34.8
July	16.3	18.5	83.1	0.8	164.3	38.9
August	14.6	14.2	76.7	12.8	158.6	32.6
September	9.4	8.3	61.3	11.5	135.9	32.9
JYVÄSKYLÄ, CENTRAL FINLAND						
May	8.7	17.7	39.5	5.4	113.5	22.4
June	13.9	18.7	60.9	8.3	156.7	30.9
July	16.2	17.9	82.0	17.8	145.8	35.9
August	13.8	12.9	81.7	23.3	181.8	44.6
September	8.4	7.4	61.6	14.3	101.3	27.8

At the end of the growing season, not only do temperatures become lower, but there is also an increased risk of night frosts (Peltonen-Sainio et al., 2008). In general, the only likely economic risk related to late season night frost concerns rapeseed, for which the seed ripening processes are interrupted (Bonham-Smith et al., 2006) resulting in high chlorophyll content in the seed and reduced quality. Heavy rain rather than late season frost is the major driver for harvesting well before the end of the thermal growing season (Rajala and Peltonen-Sainio, 2000).

1.3 Early summer drought and uneven distribution of precipitation

Fluctuations in amount and timing of precipitation are substantial within and between the seasons in northern Europe, but there are some typical features. One of these is early summer drought resulting from low precipitation and rapid drying of soils in late spring (Peltonen-Sainio et al., 2011b). As this occurs during the reproductive pre-anthesis phase, it is the most critical stress reducing the yield potential of grain crops (Rajala et al., 2009, 2011). Limitations that early summer drought causes for floret set and viability is further emphasized as it associates with long-day induced accelerated developmental rate (Peltonen-Sainio and Rajala, 2007).

We compared the availability of water with that needed by a cereal crop at its most critical phase of yield determination (from 200 to 450°Cd from sowing under Finnish growing conditions) (Peltonen-Sainio et al., 2011b), and found that according to 30-year climate datasets, only 30–60% of the rainfall needed for undisturbed and non-limited formation of yield was achieved on a long-term average basis, depending on region. Interestingly, when assessing trends in the occurrence of early summer drought in the 20th century, it was found that the weather observations for this particular 30-year period suggest that the early summer drought has even slightly eased off compared to early parts of the century (Ylhäisi et al., 2010).

In northern Europe, rains are more common later in the growing season. Sufficient rains at grain filling, increasing grain weight, can somewhat compensate for the reduced number of grains resulting from early summer drought (Peltonen-Sainio et al., 2007b; Peltonen-Sainio and Jauhiainen, 2008). Compensation is, however, limited for many reasons, e.g. lower total sink capacity of reduced number of grains in the ear and/or lower potential grain size. Early, high temperature sensitive events, even before flowering, when carpels develop and their weight is determined, could impose an upper limit on grain size (Scott et al., 1983; Wardlaw, 1994; Calderini et al., 1999a, 1999b; Calderini and Reynolds, 2000), as could the events soon after pollination through endosperm cell division (Rajala and Peltonen-Sainio,2004, 2011; Egli, 2006). Due to only limited compensation ability, increasingly abundant rains at the end of the cereal and rapeseed growth cycles represent a risk for crop production, especially when heavy rains cause lodging, disease infestations, yield losses, quality deterioration and problems with the moist soils unable to carry heavy machinery (Rajala and Peltonen-Sainio, 2000).

2 ADAPTATION: A MATTER OF CROP RESPONSES WHEN COPING WITH NORTHERN CONDITIONS

Northern European growing conditions have many special features that are exceptional and challenging (Peltonen-Sainio and Niemi, 2012). This means that adaptation measures developed to cope with such special conditions are partly unique or at least tailored to meet the demands set by the northern climate. Success in northern field crop production requires interplay between breeding of cultivars adapted to grow and yield in such an exceptionally short and intensive growing season, and development of farming systems sustaining yield potential (Peltonen-Sainio, 2012).

2.1 Development and growth: the need to hurry

2.1.1 Phenophases

The life cycles of the cultivars adapted to northern European conditions must match the requirements of the short and intensive growing season. Both long days and relatively rapidly increasing mean temperatures are factors that enhance development rate (Peltonen-Sainio and Rajala, 2007), which is maintained at a steady, high pace throughout the period when spikelet, floret and floret organ primordia viability is determined, until pollination at 400–600°Cd depending on the cereal species. The day lengthens from 12 h at the vernal equinox in March to 16 h at the end of April, reaching a maximum of 19 h around midsummer in southern Finland (e.g. www.gaisma.com, 9.7.2013).

Evidently, rapid development during the pre-anthesis generative phase (Peltonen-Sainio and Rajala, 2007) causes yield penalties. Hence, enhanced developmental rate is a costly but vital adaptation mechanism to cope with short-season conditions at high latitudes. Early maturing cultivars specially bred for northern conditions are necessary for sufficient yield stability (Peltonen-Sainio et al., 2011c; Peltonen-Sainio and Niemi, 2012). Because the environment is so restrictive, growers have a reduced opportunity to compensate for poor management, and they cannot have alternative sowing windows during the season to match the most critical developmental phases with favorable conditions. Therefore, in addition to well-adapted cultivars, farmers in northern Europe need to have a fine-tuned management strategy to sustain their yield formation in an economically and environmentally feasible way.

Despite early developmental phases being hastened by long days in the north (Peltonen-Sainio and Rajala, 2007) and cereal stands heading earlier than at lower latitudes, the grain-filling phase is not markedly shorter when measured in days (Kivi, 1967). The balance between pre- and post-heading phases is critical for obtaining high yields with sufficient duration for both pre- and post-anthesis phases. This is emphasized by differences in how favorable the climatic conditions are: often yield is favored by precipitation at grain fill and therefore grain-filling rates remain high (Peltonen-Sainio, 1990, 1991), in contrast to typical early summer drought which reduces the number of developing florets. Early maturing cultivars tend to suffer most from this due to lower capacity for compensation (Peltonen-Sainio and Jauhiainen, 2013). On the other hand, the importance of flexibility between duration of main growth phases is highlighted by the recent finding that delay in sowing did not associate with comparable delay in ripening of cereals stands (Peltonen-Sainio and Jauhiainen, 2013).

2.1.2 Role of main yield-determining components and compensation between them

A rapidly progressing pre-anthesis generative phase results in a lower number of viable, grain-producing florets and therefore, associates with reduced grain number and yield penalties (Peltonen-Sainio and Rajala, 2007). Despite this, there seems to be excess production of florets at pre-anthesis, particularly under non-water- and non-nutrient-limiting conditions (Peltonen-Sainio and Peltonen, 1995) when compared with number of grains (Peltonen-Sainio et al., 2007b). As grain number predominates over grain weight in yield determination (Garcia del Moral et al., 2003; Borras et al., 2004; Sadras 2007), grain yields at high latitudes are modest as a consequence of low grain number.

At the critical window for yield determination, early summer drought is common, but there is also increasingly strict competition for resources between generative developing organs, elongating stems and tillers. Simultaneously, floret number per head no longer increases but starts to decline during the pre-anthesis phase, several apical developmental stages before pollination (Peltonen-Sainio and Peltonen, 1995). Such competition-induced reasonably low number of grains per unit land area cannot be sufficiently compensated for by, for example, tillering at high latitude conditions (Peltonen-Sainio et al., 2009c), which again emphasizes the critical role of early summer drought as a climatic constraint.

2.1.3 Dynamics of tillering

Although tillering is species and cultivar dependent, the environment also plays an important role in modifying tiller performance (Mela and Paatela, 1974; Simmons et al., 1982; Lafarge, 2000; Prystupa et al., 2003; Peltonen-Sainio et al., 2009c). During long days, tiller initiation, tiller development and growth, heading capacity of tillers and contribution of tiller heads to grain yield have special features. In contrast with many other regions in the world, tillers generally contribute only modestly to grain yield of spring cereals at high latitudes (Peltonen-Sainio and Järvinen, 1995; Peltonen-Sainio et al., 2009c). At long day conditions, tiller contribution to grain yield exceeds that of main shoot only in two-row barley (*Hordeum vulgare* L.) whereas, in other cereal species, main shoot dominance is so strong that the mean proportion of tiller-produced yield is typically below 20% (Peltonen-Sainio et al., 2009c). Vegetative tillers are produced despite the high seeding rates that are used to enhance the uniculm habit induced by long days and they sustain early canopy closure (Peltonen-Sainio, 1997). Long days contribute to low tillering directly, by enhancing apical dominance (Michael and Beringer, 1980), and indirectly through rapid main shoot stem elongation that leads to shading and changes in the red:far red ratio within the canopy (Evers et al., 2007). Such changes in light conditions signal increasing competition (Franklin and Whitelam, 2005) and thereby, imminent shortage of resources. In contrast to tillering in spring cereals, branching of spring sown turnip rape and oilseed rape (*Brassica napus* L.) is an important characteristic for the crop for optimizing the formation of yield components in relation to the resources available. Seeding rate and/or number of emerged seedlings or seedlings that survived early summer frost (Pahkala and Sovero, 1988) can vary considerably but, nevertheless, result in comparable yields. This indicates that branching successfully compensates for reduced plant numbers per unit land area in northern conditions. When number of harvested plants per square meter ranged from some 50 up to more than 500, the number of branches per plant and number of pods per branch differed greatly, but their compensation ability was largely dependent on availability of resources (Pahkala et al., 1994).

Early summer drought ceases tillering. Lauer and Simmons (1988) indicated that short-lived tillers are expensive for the crop energy balance. Early summer drought, typical for northern Europe, depresses tiller growth and hinders expression of yield potential of tillers. Suppressed tillering, however, increases the risk of tillering at late grain filling. Late tillering is disadvantageous as it maintains moisture in mature plant stands and hampers harvest. Risk for late tillering is increased by rains that become more frequent as the growing season advances. Furthermore, possible over-supply of nutrients that remained in the soil due to drought may pose an additional environmental risk.

In contrast to spring cereals, for autumn sown cereals, short days, low temperatures and high precipitation favor tillering. This is especially true for winter rye (*Secale cereale* L.), as 53–58% of total grain yield was produced by tillers, depending on cultivar and growing conditions (Hakala and Pahkala, 2003). For winter cereals, autumn weather favors not only tillering, but also the root system is well established to avoid harmful effects of early summer drought and elevated temperatures when the next growing season advances (Peltonen-Sainio et al., 2011b). Hence, their cultivation could be considered as an interesting adaptation mechanism to cope with constraints caused by early summer drought. Despite this, grain yields fluctuate far more in winter than in spring cereals, because of large variation in success of over-wintering (Peltonen-Sainio et al., 2009a, 2011a).

Although tillering is dependent on species, cultivars and environmental conditions, the uniculm growth habit sustained by strong apical dominance seems to be hard to manipulate, for example, through exogenous hormonal stimuli such as plant growth regulators (Rajala and Peltonen-Sainio, 2001; Peltonen-Sainio et al., 2003), or by using mechanical treatments to break them up (Peltonen and Peltonen-Sainio, 1997; Peltonen-Sainio and Peltonen, 1997). On the other hand, restricted-tillering cultivars, like those of wheat (*Triticum aestivum* L.) with a major *tin* gene (Richards, 1988; Motzo et al., 2004; Palta et al., 2007), may offer a way of avoiding harmful late tillering at grain filling, for example, after drought has suppressed tillering during early growth phases. Growth and yield of uniculm barley lines with the uc_2 gene were studied in Alaska (Dofing and Karlsson, 1993; Dofing and Knight, 1994; Dofing, 1996). Those experiments showed low grain yields in uniculm lines compared to conventionally tillering lines primarily due to lower grain number per head. Nevertheless, they considered that the uniculm phenotype results in several potential modifications for phenological development that may be of particular value in northern environments with the need to avoid plant stands ripening too late.

2.1.4 *Adaptation to over-wintering*

Winter conditions and their impacts cannot be reliably forecast (Hollins et al., 2004) to allow decisions to be taken well in advance as to whether to sow winter cereals in a particular year or not. Development of other adaptation strategies for harsh winters is also challenging. There are two major ways to adapt to northern winters: using spring types and completely avoiding the challenges of over-wintering or using a complex combination of adaptation mechanisms for the various risks that might result in crop failure. The first is realized by the limited number of winter cereal types grown above 60°N. Winter rye and wheat are the only winter cereals adapted to such northern conditions, while winter barley, oat (*Avena sativa* L.) and rapeseed are of no economic importance. There are also regional differences in cultivation of the two winter cereals (Peltonen-Sainio et al., 2011a). As winter rye is clearly more winter hardy, it is typically grown in more northern regions than winter wheat.

The cultivation area of winter cereals fluctuates from year to year but, as a trend, the cultivated area has been declining during the past decade (Peltonen-Sainio et al., 2011a). At present, the total area of winter cereals in Finland is only about 5% (www.mmmtike.fi, 10.7.2013), with even smaller areas being cultivated when autumn conditions are not favorable for sowing. For example, abundant autumn rainfall can prevent sowing, as fields are not able to carry the machinery. Nevertheless, an example of one adaptation strategy is that winter cereals are often sown on sloping fields to aid surface water runoff from the fields and thereby reduce risks of formation of hermetic ice cover. Late sowing to avoid dense canopies before winter and use of cultivars resistant to pathogens are important to avoid winter damages as well (Serenius et al., 2005). For susceptible winter cereals, especially in the northernmost areas with thick snow cover, plant protection measures against snow mold infections are used (Serenius et al., 2005).

3 GAPS BETWEEN POTENTIAL AND ACTUAL YIELDS

According to Evans and Fischer (1999), potential yield is 'the maximum yield which could be reached by a crop in given environments' as determined from physiological principles, and yield potential is the 'yield of a cultivar when grown in environments to which it is adapted, with nutrients and water non-limiting and with pests, diseases, weeds, lodging and other stresses effectively controlled'. Actual yield is again the realized part of the existing yield potential under prevailing conditions.

National, regional and global yields have increased due to genetic improvement and better crop management, but also due to their interaction (Evans, 2005). However, achievements *per se* and the roles of both of these components and the interplay affecting yield trends vary markedly from one region to another and with time, being dependent not only on prevailing growing conditions, but also on market and policy environments. Evidently challenges arising from the climatic environment are emphasized in the northern agricultural production areas.

3.1 Changes in yield trends in northern growing areas

Stagnated crop yields, reduced rates of improvement in crop yields or even decline in yield trends have recently been experienced features of agricultural production in many farming systems of the world (Bell et al., 1995; Calderini and Slafer, 1998; Peltonen-Sainio et al., 2007a, 2009a). Both timing of and reasons for such turning points differ according to region.

Evans (1993) reported examples of potential reasons for yield plateaus and declines including extension of increasing cultivation areas to less favorable regions, changes in socioeconomic policies, markets, costs, and input use, but also in ways to analyze and interpret the changes in annual yields. In addition to extensification of farming, in some production systems, one of the reasons for declining yield is in fact intensification of farming. For instance, when farmers try to grow more crops per year, they sacrifice yield of an individual crop (Egli, 2008), which again shows less yield progress per harvest. Multiple cropping is not practiced now or in the future in northern European conditions due to rigid climatic restrictions.

3.1.1 Steady increases in genetic gains

Keeping pace with genetic improvements requires continuous updating of crop physiology methodology in breeding for ever higher yielding capacity (Foulkes et al., 2007). Hence, it is possible that leveling-off in genetic yield potential in certain regions is a result of fully exploited traits such as shortening of stem length, thereby increasing resource allocation to grains, i.e. by increasing harvest index (HI) and grain number (Austin et al., 1980; Siddique et al., 1989; Foulkes et al., 2007). However, Peltonen-Sainio et al. (2008) indicated that, under northern conditions, HI of modern wheat cultivars, as opposed to barley and oat cultivars, was lower than that of other temperate spring cereals and also when compared with HI in many of the important wheat production regions. This example might indicate that such traits are not yet thoroughly exploited in breeding for yield enhancement in germplasm adapted to northern conditions.

In the northernmost European growing region, genetic gains were evident for cereal grain yields during the entire study period from 1961 till today (Peltonen-Sainio et al., 2007a, 2009a) (Fig. 4.2). This was the case for all spring and winter cereals as well as rapeseed. The mean genetic yield increases were 28 kg ha^{-1} $year^{-1}$ for barley and winter rye, 21 kg for oat, 36 kg for spring wheat, 29 kg for winter wheat and 17 kg for rapeseed. Quality traits have also been continuously improved. One example is rapeseed, for which oil content has increased with higher yields, and seed chlorophyll content, which reduces yield quality, has decreased (Peltonen-Sainio et al., 2011d). Furthermore, along with improved yields, genetic improvements in yield removed nitrogen (N) have also taken place with significant environmental impacts (Peltonen-Sainio and Jauhiainen, 2010). Such improvements highlight the central, multifold

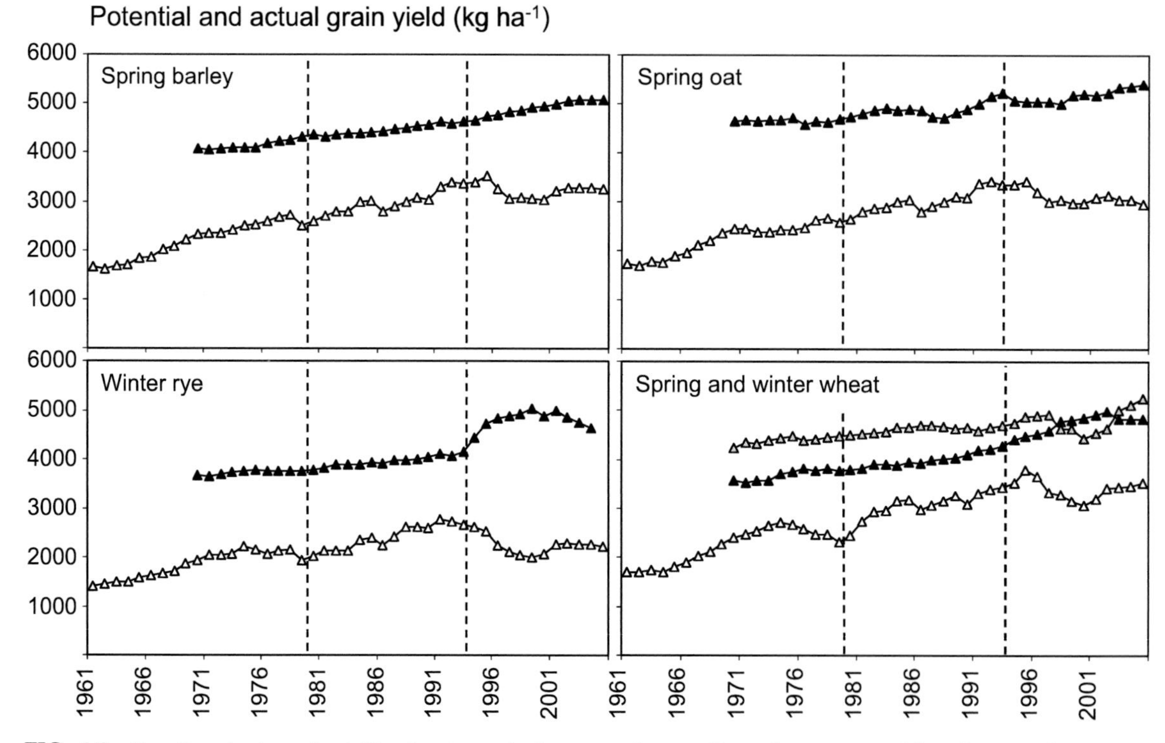

FIG. 4.2 Genetic gains in grain yields of spring and winter cereals according to long-term multilocation experiments carried out in Finland since 1970 and national yield trends in Finland since 1961. Genetic gains (increase in potential yields) are measured according to the year of introduction of a new line/cultivar to the MTT testing program. The solid triangles are potential yields and open triangles actual yields; both as 5-year moving averages. Potential yield of winter wheat is indicated with a gray triangle and that of spring wheat with a closed triangle. Vertical broken lines differentiate three agronomical subperiods: pre-modern and the first and the latter parts of modern periods. (*Source: Peltonen-Sainio et al., 2009a.*)

role of plant breeding in adapting cultivars to the northern climatic conditions.

In northern areas, breeding has been successful in increasing yield potential without any signs of slowing pace of increase in recent years for any of the temperate crops. In fact, achievements in genetic yield potential have been even higher during the last 10 years than earlier (Peltonen-Sainio et al., 2007a, 2009a). Interestingly, high yield plasticity of a cultivar associated with responsiveness to favorable conditions, without yield penalties under stressful conditions, were found in spring cereals, contrary to winter cereals, in which yields of plastic cultivars was particularly low in poor environments (Peltonen-Sainio et al., 2011c).

Breeding cultivars specifically to northern conditions is a major challenge. Before discussing the traits that contribute to genetic yield increases, it is important to underline that earliness has always been the principal trait in breeding programs for improving adaptation to northern conditions (Kivi, 1963; Aikasalo, 1988; Rekunen, 1988; Peltonen-Sainio, 1990). Comparison of historical and present day oat cultivars revealed that modern oat cultivars had a slightly shorter grain-filling period than their predecessors (Peltonen-Sainio and Rajala, 2007). This was also reflected in the growth period from sowing to maturity. This means that no trade-off between the length of pre- and post-anthesis main phases occurred. However, despite the shorter growing time, grain yields were improved. Earlier maturity in cereals – and even without any yield penalty – enhances yield stability and evidently indicates improved adaptation to the short northern growing season.

As reported for other regions, also in northern regions, increased yield potential of modern cultivars resulted from reduced plant height and increased HI and number of grains per head (Peltonen-Sainio, 1990). However, improved yields of oat were not merely a consequence of increased HI, but also of higher total phytomass (Peltonen-Sainio, 1991), which means increases in total biomass production per unit area.

3.1.2 Periodic fluctuation in realization of existing genetic yield potential

The successful use of management practices for controlling crop growth under prevailing conditions determines the extent to which genetic improvements in yield potential are realized. There have been major changes in the means for managing crop stands. During the 1961–1980 agronomical sub-period (later termed the pre-modern period), widespread agricultural mechanization occurred, farmers adopted improved agricultural practices and commercial inputs were introduced and applied. In the 1981–1994 period (the first part of the modern period), crop management was further intensified while, in 1994, Finland became a member of the European Union, and especially since then (the latter part of the modern period), northern field crop production has been directed towards greater sustainability, often supported by policy and economic incentives. Furthermore, changes in cereal pricing on global and regional markets have impacted on input use for cereal and rapeseed production.

In the first part of the modern period production was intensified, and the equation typical for the process was:

Actual yield = *f* (***Growing conditions*** × *Genetic gains*) × (*Growing conditions* × ***Crop management gains***) × (*Genetic gains* × *Crop management gains*)

In the latter part of the modern agricultural period, the equation gradually broadened to become:

Actual yield = *f* [(***Growing conditions*** × *Genetic gains*) × (*Genetic gains* × ***Crop management gains***) × (*Growing conditions* × ***Crop management gains*** × *Market effects* × *Environmental motives*)]

All these major elements, influencing rate and direction of changes in yield trends, comprise numerous sub-components. Fitting trend lines to data is relatively easy, but the complexity of components and changes in them over time makes interpretation of the causal effects on changes in yield trends challenging.

Peltonen-Sainio et al. (2007a) reported periodic fluctuation in rapeseed yield trends: during early stages of adaptation of rapeseed to northern conditions yields increased by up to 50 kg ha^{-1} $year^{-1}$, thereafter they leveled-off for about 20 years, increased again for a short period in the early 1990s by about 65 kg ha^{-1} $year^{-1}$ and thereafter markedly declined by 27 kg ha^{-1} $year^{-1}$ over the last 15–20 years, despite steady and marked improvements in genetic yield potential. The decline in national rapeseed yields was so dramatic in the 2000s that the yields approached already those typical for the very early stages of rapeseed production in Finland (Peltonen-Sainio et al., 2007a). This study indicated that the national yield trend for rapeseed did not coincide with genetic gains in yield potential, thereby emphasizing that crop management and the prevailing growing conditions were important contributors to the realization of actual yields. Identified challenges for rapeseed production, likely primarily contributing to yield decline, were inability of rapeseed taproot to penetrate through compacted subsoil (Peltonen-Sainio et al., 2011e), inadequate control of pests and diseases and elevated temperatures close to flowering (Peltonen-Sainio et al., 2007a). Kirkegaard et al. (2006) and Lisson et al. (2007) reported similar evidence for southern New South

Wales in Australia to associate with underperformance of oilseed rape.

Also some fluctuations were noted in national cereal yield trends in Finland, which share some of the features discussed for rapeseed. Realization of yield potential varied according to the time period. Furthermore, for all cereal species except wheat, despite the evident success of plant breeding, yields declined over the last 10 years (Peltonen-Sainio et al., 2009a). Recent decline in yield trends contrasted with periods prior to 1995, when national grain yields increased by 32 to 76 kg ha^{-1} $year^{-1}$ depending on species. In the pre-modern period, national yields of all cereals increased on average by 42 kg ha^{-1} $year^{-1}$. In the first part of the modern period, the mean rate of yield increase was even slightly higher, 53 kg ha^{-1} $year^{-1}$. At that time, higher rates of fertilizer, plant growth regulator and pesticide input were used to sustain improved yield potential of cultivars. Hence, both plant breeding and management practices were successful and contributed to increasing national yields more so than was solely due to plant breeding, which increased 1.7–5.4-fold (depending on species) in 1970–1980 and 1.7–2.8-fold in 1981–1994. In 1970–1980, national oat and winter rye yields were more enhanced by crop management than by breeding, in contrast to barley, while in 1981–1994 national yield increases were higher than genetic gains for all cereals other than winter rye. For all cereals other than wheat, national yields declined over the last 10 years. The decline was greatest in oat (−37 kg ha^{-1} $year^{-1}$) and winter rye (−36 kg ha^{-1} $year^{-1}$) and least in barley (−12 kg ha^{-1} $year^{-1}$) (Peltonen-Sainio et al., 2009a). Wheat yields might also have even continued to increase during the latter part of the modern period if cultivation had not extended into less favorable, northern regions.

The EU environmental program for agriculture seeks increased sustainability and a reduced environmental footprint for agriculture. Rates of fertilizer application for arable crops were restricted and economic incentives to produce efficiently decreased dramatically as producer prices for cereals fell by two-thirds while the prices of inputs remained the same or, recently, even increased. Consequently, there was a reduction in the use of inputs in the absence of efficient incentives. Since 1995, the average annual change in productivity (the relationship between production volume and the inputs used) has been negative for cereal farms (Niemi and Ahlstedt, 2007). At the same period of declined yield trends, growing conditions have fluctuated challenging the yield stability more. Furthermore, modern cultivars tend to have higher responsiveness to fluctuating conditions than the earlier cultivars (Öfversten et al., 2002; Peltonen-Sainio et al., 2011c).

Considering future prospects for cereal trends in northern areas, it is likely that marked changes will occur again, but for different reasons. Envisaged, substantial global changes will affect agriculture (Soussana et al., 2012) also in northernmost Europe. There will be increased demand for agricultural commodities to feed the expanding human population and to meet the needs driven by increased standard of living in highly populated regions. Competition for land will have substantial effects on global markets and pricing. In addition to this, climate change will influence global changes and alter the trends for, and balances among, many field crops in northern Europe (Olesen and Bindi, 2002; Ewert et al., 2005; Olesen et al., 2011).

4 CHALLENGES AND PRACTICES IN ATTEMPTING TO IMPROVE SUSTAINABILITY

Sustainability in northern Europe is particularly concerned with reduced leaching of N and phosphorus (P) and trade-offs with maintenance of soil fertility. Alternative tillage systems, cover crops, crop rotations, multiple cropping, and improved resource-use efficiency are increasingly important in this context. In recent years, increased concern over climate change has also led to assessing greenhouse gas emissions from agriculture, including land use changes and the effects of tillage. Improved sustainability also includes elements of floral and faunal diversity.

4.1 Nutrient leaching

In northern Europe in general, but especially in Finland with its more than 100000 lakes, 14000 km of Baltic coastline and substantial annual precipitation, the risks of nutrient leaching to natural water systems are high. Grasslands ensure continuous ground cover mainly in the central and northern parts of Finland with dairy production, whereas partial ground cover from annual crops in southern areas determines the potential for nutrient leaching, particularly in autumn. Leaching can also occur during mild winter periods when large volumes of water run from the fields and in spring when the snow melts. Nutrients, especially N and P, leach in runoff water and soil particles in uncovered soils are prone to erosion. Therefore, it is essential that most of the nutrients are available for crop growth and are also utilized for biomass production instead of remaining unused in the soil after harvest. This is ecologically and economically desirable.

One major challenge is to match the available nutrient supply to crop needs in variable high latitude weather conditions. Typically, in the short growing season characterized by intensive and early crop development and growth, nutrients are applied only at sowing and are expected to sustain growth for the entire period from sowing to maturity. However, the major constraint is early summer drought, which can result in problems with nutrient uptake early in the season.

The EU environmental program for agriculture, launched in Finland in 1995, included a special scheme aimed at increasing agricultural sustainability by reducing the associated environmental load. For example, fertilizer application rates for arable crops were restricted. At the same time, economic incentives to produce efficiently decreased dramatically as producer prices for cereals dropped by two-thirds while the prices of inputs remained the same (Niemi and Ahlstedt, 2007) and national cereal yields started to decline (Peltonen-Sainio et al., 2009a). However, when both fertilizer use and national yields declined, the effectiveness of this scheme depended on the changes in N and P load in the natural waterways. This indicated that reduction of nutrient load in the environment includes interesting crop physiology dimensions and challenges. Owing to the spatial variation in soils and yield (Rajala et al., 2007), precision farming methods can help to reduce nutrient leaching (Blackmore et al., 2003).

4.2 Maintenance of soil fertility and production

Leaching of nutrients is primarily reduced by cutting down on excessive fertilizer. The primary challenge for fertilizer dosage is highly fluctuating conditions. Over 90% of arable land in Finland is included in the EU environmental program, in which use of N and P is restricted according to the productivity (mean yield) of each field. If soil P is high, no additional fertilization is allowed. The most essential elements for determining nutrient needs are crop species and cultivars, how well they are adapted to conditions of the particular fields according to cropping history, success of crop establishment and use of plant protection according to the needs of the crop.

Changes in fertilizer use have resulted in yield declines in cereals but also changes in main yield components. Reduced N application rates were associated with lower test weight, grain weight and also grain protein concentration, while reduced rates of P application were associated with reduced grain weight and protein concentration in some cases (Salo et al., 2007). However, concomitant changes other than fertilizer use occurred as economic incentives to produce efficiently decreased dramatically (Niemi and Ahlstedt, 2007). Examples of such changes critical for yield formation include inadequate investments in drainage and liming, and increases in areas of reduced tillage and direct drilling.

The role of versatile crop rotation on soil fertility is increasingly recognized. Especially growing legumes is envisaged to increase in the near future (Peltonen-Sainio and Niemi, 2012; Peltonen-Sainio et al., 2013a), which would improve soil structure and lead to an increase in soil carbon content (Jensen et al., 2012). Känkänen et al. (2013) concluded that the potential of biological N fixation in Finnish conventional agriculture would reduce N fertilizer need by 60 % compared to the current situation and save fossil energy, taking account of energy use of machines, about 3700 terajoules per year.

4.3 Reductions in soil tillage

Traditionally, most fields sown to annual crops were autumn plowed, but this practice has declined recently. Reducing soil tillage is a means of decreasing erosion and thereby nutrient leaching. It is therefore favored in the EU environmental program, followed by a prominent increase in area of conservation tillage (Alakukku et al., 2010). Such systems, including shallow stubble tillage and no-till, have been attractive also because of the high cost of fossil fuels and labor. In northern Europe, the moldboard plow is usually replaced by chisel or disk implements. Both represent potential conservation tillage methods for spring cereals on clay and silty clay soils typical of these humid climates (Rydberg, 1987; Aura, 1999; Heinonen et al., 2002).

No-till, also called zero tillage or direct drilling, has been practiced on increasing areas in Finland until 2007, when 30% of the winter cereal, 10% of spring cereal and 17% of oilseed rape area was direct drilled. After that the no-till area has stagnated to 13% of total agricultural land. The success of reduced tillage and direct drilling depends on weather conditions and soil quality. According to Rasmussen (1999), this method proved to be most advantageous in Scandinavia on the heaviest clay soils, which are the most difficult soils to prepare using conventional soil tillage methods.

During the rapid increase of no-till area, one argument for replacing plowing was the expected increase in soil carbon, and thereby decreased emissions. However, effects on greenhouse gas emissions have been questioned (Reicosky, 2003) when considering the putative increase in NOx emissions (Regina et al., 2007). Sheehy et al. (2013) reported a risk of increased N_2O emissions in clayey soils in spring cereal systems in northern European boreal climate, based on studies on conventional and no-till fields in southwestern Finland. Furthermore, Regina and Alakukku (2010) found no consistent changes in CO_2 fluxes caused by no-till management during the first years of no-till practice, and CH_4 fluxes did not differ between the treatments. Baker et al. (2007) concluded that, in fact, the net effect of reduced tillage would be an increase in greenhouse gas emissions.

In spite of some set-backs in greenhouse gas emissions, the positive image of reduced tillage in the context of the environment remains mainly because of the reduced erosion and leaching. Some recent findings partly question also this as, for example, Alakukku et al. (2010) found that the mean area of ponded water that hampers crop growth as a consequence of high precipitation, tended to be larger in conservation tillage fields than after plowing. An additional drawback is also the often increased need for pesticides, which can be reduced to some extent through sound crop rotation and use of resistant cultivars. Recently, increased dependency on glyphosate in agricultural systems may pose many ecological and environmental risks that are particularly detrimental in northern ecosystems with long biologically inactive winters and short growing seasons (Helander et al., 2012). Field experiments in Scandinavia with plowless tillage of clay loams and clay soils compared to conventional autumn plowing usually show reductions in total P losses of 10–80% by both surface and subsurface runoff (Ulén et al., 2010). Contrary to what was expected with the EU environmental program, environmental gains have not been always reached. For example, Muukkonen et al. (2007) reported that no-till increased the loss of dissolved reactive P (DRP) in surface runoff. According to Ulén et al. (2010), effects of not plowing during the autumn on losses of DRP are frequently negative, since the DRP losses without plowing compared to conventional plowing have increased up to fourfold in experimental fields. They concluded that erosion control measures should be further evaluated for fields with high erosion risk since reduction in particulate-bound P losses may be low and DRP losses still high.

Differences in the amount and quality of yield from plowed and no-till soils vary between years, fields and crop species. Timing and implementation of drilling in no-till soil is especially challenging. When early summer drought occurs after drilling it may interfere with seedling establishment. On the other hand, if plant stand establishes well, prospects are good for successful pre-anthesis growth, even if there is scarcity of early summer precipitation, because of the relatively high water content in the no-till soil. Typically, yields are higher in rainy summers when soils are plowed, while under dry conditions, direct drilled crops may yield more (Känkänen, 2008). However, in rainy summers, the yield decrease can be quite marked when the crop is direct drilled, and especially so with spring barley and field pea (*Pisum sativum* L.), which are sensitive to wet soils. Direct-drilled crops are also more sensitive to dry and warm periods at grain filling when maturation processes are advanced.

Oat is a stable crop when direct drilled, while spring wheat often yields less except in dry summers. High levels of crop residues can cause problems for soil drying in spring and especially in the case of oat and wheat monocultures (Känkänen et al., 2011). Direct drilling also works for winter cereals. Establishment is better than on plowed soil in dry autumns contrary to the case in wet autumns that are more typical of northern conditions. Establishment of direct-drilled spring turnip rape is problematic, mainly because the fertilizer is placed too close to the seed, which in turnip rape markedly reduces germination. This can be avoided by increasing the amount of seed, but this is not economically feasible and may increase the risk of volunteer turnip rape in the subsequent crop. A higher seeding rate was also advantageous for spring cereals in no-till clay soil (Känkänen et al., 2011).

The protein content of spring cereal grains has been low in direct drilling (Känkänen et al., 2011), which is positive for malting barley but negative for other cereals, especially for bread wheat. This is obviously a consequence of low mineralization of N in no-till soil (Kristensen et al., 2000). Protein content could be improved by increasing N rates, but this is conflictive with general environmental aims. Alakukku (2006) found no decrease with zero-tillage in grain N concentration from crops grown on clay soil. However, no-till in clay loam soils reduced both N and P content more than grain yield. Also some effects of the tillage system, though not fully consistent, on grain size and variation in grain weight have been demonstrated recently (Lötjönen and Isolahti, 2010; Känkänen et al., 2011). These findings emphasize the need for evaluation of long-term effects of tillage systems to understand thoroughly likely interlinks with productivity and environmental impacts.

Crop rotation in no-till fields is critical. The amount and quality of crop residues affects drilling performance, and yield as well may be threatened by diseases originating from straw residues. Snails have caused some damage to winter cereals, especially when direct drilled. Monoculture is therefore a particular risk for no-till fields. Furthermore, direct drilling of small seeded oil crops is difficult after a crop with a high straw yield if the straw is not removed. The substantial decrease of yield and quality calls again for long-term trials based on understanding how to defeat the challenges typical for no-till systems.

The degree of sustainability of the methods that have been adopted in northern regions due to economic and environmental reasons depends on how successful reduced or no-till is, compared with the conventional methods with respect to plant stand establishment, determination of yield potential and yield realization during the post-anthesis phase. As described earlier, under northern European conditions, fields are often highly heterogeneous with respect to plant growth conditions and yield production capacity (Rajala et al., 2007). No-till in heterogeneous fields increases the risk of uneven growth and ripening. Furthermore, sowing often occurs later in untilled than in tilled soil. Although fossil fuel is used less without plowing, the advantage is partly lost as late harvest often results in higher grain moisture content and increased energy use for grain drying, which is the typical measure to secure high quality yields in northern areas.

4.4 Crops for nutrient uptake and soil coverage

Cover or catch crops are used for reducing nutrient leaching, transferring N to the next main crop, increasing biodiversity and maintaining or improving soil structure. The mode of action of the cover crop largely depends on climatic conditions (Schröder, 2001). In the short growing season of northern latitudes, the time after cereal harvest is too short for sowing the cover crop; hence undersowing in cereals is the only method for establishing a cover crop stand (Jensen, 1991; Alvenäs and Marstorp, 1993). Only early crops, like summer potato (*Solanum tuberosum* L.), enable sowing the cover crop after harvest of the main crop. On cereal farms, undersowing enables immediate uptake of residual N by the cover crop after harvest of the main crop (Breland, 1996). Cover crops can grow until late autumn or even the next spring before being either incorporated into the soil or eliminated with glyphosate.

Undersown grass crops catch available N from the soil, thus reducing N leaching. Italian ryegrass (*Lolium multiflorum* Lam.) uptakes N effectively in autumn and timothy (*Phleum pratense* L.) particularly in the following spring (Känkänen and Eriksson, 2007). Clovers fix atmospheric N and transfer it to the next crop. Undersown crops compete with the main crop for water, nutrients and space. However, these effects on the main crop are relatively low if the undersown crop species are chosen carefully, taking competition into account and using moderate seeding rates. Delayed inter-sowing does not benefit the yield formation of the main crop and greatly decreases the biomass production of the catch crop (Kvist, 1992; Ohlander et al., 1996). Thus, simultaneous undersowing with the main crop is used. The yield decrease and seed costs are, however, the main reasons for limited undersowing in Finland, despite its evident positive effects on soil, diversity and environment. In order to increase the profitability of cover crops, Känkänen (2010) outlined a model of an adaptive undersowing system, which emphasizes allocation of measures according to needs. By this means, the goal of undersowing is defined and, thereafter, many decisions are to be made by the farmer to reach the comprehensive benefit on environment and farming.

Buffer zones are uncultivated areas between fields and watercourses that have been established to slow runoff water from agriculture, enhance infiltration, and absorb P in soil and vegetation, thereby trapping sediment and nutrients. Their extensive use in Finnish farms is a successful outcome of the environmental program. Buffer zones and constructed wetlands are good at reducing soil and particulate P losses via surface runoff, but the effect on dissolved reactive P is more complicated (Uusi-Kämppä et al., 2000). The surface runoff losses of sediment decreased by more than 50%, total P by 30% and total N by 50% by using buffer zones on tilled soil (Uusi-Kämppä, 2010). In spring, the implementation of buffer zones even increased the losses of dissolved reactive P but mowing and removing the residue from the buffer zones effectively decreased the losses.

Recently, regional agro-environmental projects, where farmers and environmental authorities cooperate closely, have been established. Projects are a combination of research, advisory and practical farming, which helps knowledge transfer between different parties. Sustainable farming is promoted not only by introducing cover crops, buffer zones, and other profitable measures, but also by raising the environmental awareness of farmers and understanding of the important role of soil structure and fertility on both the environment and high yields.

4.5 Multiple and versatile cropping

Relay intercropping, which normally is carried out by sowing winter cereal under spring cereal in northern Europe, could improve nutrient usage and reduce negative effects of tillage in conventional autumn sowing. Relay cropping of spring barley and winter wheat has been studied in Finland and Sweden with some inconsistencies: in Finland the system was uncertain and resulted in yield decreases of both crops (Känkänen et al., 2004), while in Sweden it was more successful although further studies were needed (Roslon, 2003).

Crop rotation is a basic operation in sustainable agriculture; however, this has been neglected when short-term economic goals have been pursued in recent years. Appropriate crop rotations increase the diversity of farming systems, reduce risk for crop diseases and maintain and improve soil structure. Legumes likely have a most prominent role towards more diverse and sustainable agriculture. Biological N fixation can be increased greatly by more intensive use of legumes in grasslands, utilization of green manure and undersown crops, and enhanced growing of pulse crops (Känkänen et al., 2013).

The need for taking care of soil and other aspects of the environment will increase if climate change progresses as anticipated. Clay soils, common in Sweden and especially in southern Finland, are particularly vulnerable. A change from cold winters with soil frost and snow cover to wet and mild winters represents a risk of deteriorating soil structure. To improve and maintain soil structure and conditions, methods such as reduced tillage, versatile crop rotations and use of cover crops are further highlighted as the climate changes. The number of alternative crop species available for northern crop rotations will also increase due to longer growing seasons (Peltonen-Sainio et al., 2009b; Elsgaard et al., 2012). Many of the above examples not only describe how changes in farming systems and crop management practices can improve sustainability in northern agriculture, but they also highlight the marginality of environmental advantages when crop management is prone to highly variable climate and when economic incentives to produce efficiently decrease dramatically.

4.6 Improving resource-use efficiency

In addition to developing crop management systems as indicated, it is important to search for the potential solutions that plant breeding could offer through cultivar development. Improved resource-use efficiency is especially important today and will increase in importance in the future when fertilizer prices may rise due to increasing energy costs. Nitrogen-use efficiency (NUE) is defined as grain yield produced per unit of N available in the soil (Moll et al., 1982). Like grain yield itself, NUE is a complex trait determined by many physiological processes during the lifespan of the crop.

In intensive and short growing seasons typical of high latitudes, capacity to take up N according to crop growth requirements is essential to avoid both under-realization of yield potential and increased risks for nutrient leaching (Løes, 2003). According to Ortiz-Monasterio et al. (1997), N uptake is elemental for NUE especially when N is in short supply. Therefore, with current reduced fertilizer use rates, improved ability of the crop to take up N and P should be an aim of plant breeding.

Under northern conditions, cereal cultivars have not been specifically bred or screened for efficient nutrient uptake and ability to transport the N taken up into the harvestable yield. Despite this, significant NUE improvements for wheat and oat between 1909 and 2002 have been demonstrated (Muurinen et al., 2006) and most of the progress in improved NUE originated from improved N uptake. Plant breeding has also increased significantly the yield removed N in all seeds crops grown in northern Europe (Peltonen-Sainio and Jauhiainen, 2010). Further breeding for improved NUE by developing modern genomic tools and benefiting from variability in germplasm, represents a novel, underutilized way to improve cost efficiency of input use in agriculture, reduce nutrient losses to the environment and sustain formation and realization of yield potential.

5 THE FUTURE AND CLIMATE CHANGE

The global mean temperature has risen by 0.76°C (IPCC, 2007) during the last century (1906–2005). In Finland, the average increase has been about the same for this period (Tuomenvirta, 2004; Tietäväinen et al., 2010). However, the warming has increased since, with the latest finding for 1909–2008 being a 0.93°C increase in annual temperatures, and during the last 50 years (1959–2008), even as high as 0.3°C per decade, which is more than two times that observed for global annual temperatures for the same period (Tietäväinen et al., 2010). During the last century, the largest increase in temperatures has been in the spring, but lately the winter temperatures have started to increase more (Table 4.2) (Tuomenvirta, 2004; Tietäväinen et al., 2010).

A warmer climate will mitigate some of the climatic constraints that currently limit agriculture at high latitudes. One of the main constraints is the short growing season. Increases in average temperatures during the last century have already extended the growing season in Finland, especially in spring (Carter, 1998; Kaukoranta and Hakala, 2008). Between 1965 and 2007, the advancement of the start of growing season was 2.0–2.8 days per decade, and the corresponding advancement in the start of sowing was from 0.6 to 1.7 days per decade for cereals (depending on the region in Finland) to 2.5 days per decade for sugar beet (and 3.4 days per decade for potato [Kaukoranta and Hakala, 2008]). The same trend is projected to continue in the future even in the case of considerable reductions in greenhouse gas emissions, but especially if these reductions are not completely realized (Table 4.2) (IPCC, 2007).

In the future, the lengthening of the thermal growing season will occur more at the end than at the beginning of the growing season (Table 4.2), with a considerable fraction of the increased temperature sum accumulating late in the autumn. Because of the rapid reduction in sunlight hours and light intensity in the north in the autumn (Saikkonen et al., 2012), an increase in the length of the thermal growing season in the autumn would benefit crop production only marginally (Peltonen-Sainio et al., 2009b). Most of the increase in the thermal growing season in spring could be used for crop growth. However, sowing early in spring may not be possible because of too high soil moisture and the possibility of early season frosts. Therefore, cultivation of winter cereals and perennials will become increasingly crucial in the future, as they can start growing once the conditions become favorable, they can also make use of the wintertime moisture in the field, they may capture the nutrients left by the previous crop in rotation in autumn, they can escape the harmful effects of elevated temperatures and they have higher yield potentials (Peltonen-Sainio et al., 2009b, 2011b).

TABLE 4.2 Current growing season and climatic variables for Jokioinen, Finland (60°48′N, 23°30′E – average of 1971–2000, year 1986), and scenarios for the future [periods 2010–2039 (year 2025), 2040–2069 (year 2055) and 2070–2099 (year 2085)] according to the A2 scenario of IPCC (2000), and calculated as a mean of 19 climate models

Jokioinen 60°48′N 23°30′E	Growing season			Winter days (d)	Temp (sum dd)	Precipitation			
	Start	End	Length (d)			MAM (mm)	JJA (mm)	SON (mm)	DJF (mm)
1986	Apr 27	Oct 12	169	139	1193	106	209	195	139
2025	Apr 22	Oct 18	181	121	1371	108	215	201	145
2055	Apr 16	Oct 28	196	93	1577	112	221	210	155
2085	Apr 05	Nov 09	219	27	1859	119	228	224	170
Denmark 54–56°N, 8–12°E	Apr 1–15	Nov 1–15	>225	0-25	1400–1600	125–150	175–225	150–300	150–200

(Peltonen-Sainio et al., 2009b). MAM, spring (March, April, May); JJA, summer (June, July, August); SON, autumn (September, October, November); DJF, winter (December, January, February). Danish (54–58°N, 8–12°E) temperatures derived from Tveito et al. (2001) and precipitation from Frich et al. (1997) are averaged for 1961–1990. Temperature sum (Temp Sum) and growing season length are calculated at >5°C, and the number of winter days is calculated <0°C. Precipitation in mm per season.

Warmer and shorter winters in the future will probably result in diminished risk of freezing damage and better over-wintering conditions for autumn-sown crops and perennials. However, the benefits of milder autumns and winters may be compromised by several factors. Warmer autumns provide shorter periods for cold hardening of over-wintering plants, leading to higher vulnerability to winter injury by severe frosts (Bélanger et al., 2002), which will continue to occur regularly, despite the general warming of climate (Jylhä et al., 2008). Reduced snow cover may also cause increased risk of winter injury when periods of severe frosts coincide with periods of reduced or no snow cover, leaving the over-wintering plants unprotected. On the other hand, the projected increase in wintertime precipitation (IPCC, 2007) (Table 4.2) may also result in deeper snow cover, which under relatively mild winter conditions may melt and again freeze, resulting in soil heaving and injuries, especially to root tissue, and ice encasement leading to anoxia (Bélanger et al., 2002). According to the latest scenarios, the conditions for freezing and melting of snow during winter will become more frequent in the future as winter days with daily maximum temperatures passing 0°C are predicted to increase, although the number of days with average temperatures below 0°C will decrease (Jylhä et al., 2008). Especially fluctuating conditions during winters challenge over-wintering capacity even in relatively winter hardy crops (Peltonen-Sainio et al. 2011a), which may be a significant obstacle delaying introduction of autumn sown crops to a large extent to northern agricultural systems.

Trends in precipitation during the last century are hard to project (Tuomenvirta, 2004; IPCC, 2007). However, annual precipitation has increased in northern Europe during the last century and, according to IPCC (2007), the increase is 'very likely' to continue in the future. Such observed increases in June precipitation by 5–9 mm per decade may have facilitated 15–20% of attained cereal yield increases between 1961 and 2000 (Ylhäisi et al., 2010). Increase in annual precipitation is projected to take place more during winter than during the growing season. Whether the wintertime precipitation will increase as much as in the present scenarios is of course not certain, considering the difficulties in predicting rain in general and the great variation in precipitation on average (Peltonen-Sainio et al., 2009d). The scenarios show increases in precipitation by 13, 19, 29 and 31 mm for MAM, JJA, SON and DJF (Table 4.2), respectively, in southern Finland by the end of the next century, according to emission scenario A2 (greenhouse gas emissions are not efficiently restricted; IPCC, 2000). This would exceed the natural variation observed at present. In any case, even though the precipitation will increase in the North, the increase is likely to benefit only marginally agricultural production, as it will mainly take place outside the growing season and would not increase sufficiently during the growing season to sustain higher yield potentials that longer growing seasons *per se* may enable (Ylhäisi et al., 2010; Peltonen-Sainio et al., 2013b). Therefore, early season drought will thus remain a problem in the future, as the projected slight increases in growing season precipitation will probably not compensate for the increased evapotranspiration caused by higher growing season temperatures and higher canopy biomasses in the field. Moreover, a trend showing increase in heavy precipitation events has been detected for the past century, and this trend is expected to strengthen during the 21st century (IPCC, 2007, 2012). Heavy rain events tend to lead to surface flow of water, benefiting the crop water economy less than rain falling evenly over a longer time period.

Autumn-sown crops and perennials that can make use of winter moisture in spring would facilitate adaptation to early season drought. However, over-wintering problems will likely challenge the yields of autumn-sown crops during the next few decades and cause substantial variation within and between years (Peltonen-Sainio et al., 2009b, 2011a). In addition, not all crops can be sown in autumn, although the range of suitable cultivars will probably increase in the future. Another practical tool to cope with drought is irrigation. Irrigation is neglected in the North, but if the occurrence of early season drought increases and especially if the prices of agricultural commodities increase, irrigation could become an increasingly attractive method to reduce annual yield fluctuations, as there are large water reservoirs available for irrigation in northern Europe (Peltonen-Sainio et al., 2013b). Irrigation during the early growing season would favor grain set (Rajala et al., 2009, 2011), improve nutrient uptake and reduce the risk of nutrient leaching.

The future climatic conditions of southern Finland have often been compared with the current conditions in Denmark (Fig. 4.1). When climatic variables only are considered, the conditions in southern Finland for the period 2070–2099 (year 2085 in Table 4.2) according to scenario A2, indeed resemble those currently pertaining to Denmark. However, the most striking differences between the climates of Denmark and Finland are in photoperiod, the number of winter days and springtime precipitation, the latter two being likely more favorable in Denmark even in the case of prominent effects of climate change. Future projections indicate that agroclimatic conditions in northern Europe are going to change drastically due to climate change (Trnka et al., 2011), which again enables large-scale shifts in production areas of many important field crops towards northern European regions thereby also facilitating more diverse crop rotations in the future (Elsgaard et al., 2012; Peltonen-Sainio et al., 2013a).

In Finnish open top chamber experiments, meadow fescue (*Festuca pratensis* Hudson) benefited from increases in growing season temperatures, length of growing season and ambient CO_2 (Hakala and Mela, 1996), while the yield of spring wheat decreased in elevated growing season temperatures due to a higher developmental rate. Elevated CO_2 concentration did compensate for the yield loss, but only to the extent of bringing the yield to about the same level as at ambient temperatures (Hakala, 1998). In a warmer climate, achieving yield increases requires new, more productive crop cultivars adapted to the longer and warmer growing season (Peltonen-Sainio et al., 2009b, 2011b). The barley germplasm adapted to northern agroenvironments exhibited some capacity for adaptation to some climatic variables, like elevated temperatures that may be increasingly harmful in the future (Hakala et al., 2012). Breeding efforts and novel genetic resources are, however, needed to cope with extreme events such as prolonged droughts and excessive rains or increasing temperature sum accumulation prior to heading, which also at present cause yield penalties (Hakala et al., 2012). The forage grass cultivars seem to adapt better than cereals to the expected future growing conditions (Hakala and Mela, 1996). The increase in the concentration of CO_2 should also bring about yield benefits for both grasses and cereals, at least as suggested by the considerable increase in photosynthetic efficiency of both wheat and meadow fescue at higher CO_2 concentrations (Hakala et al., 1999).

While crop yields are likely to increase in the future in Finland, realizing the higher future yield potentials requires comprehensive adaptive measures. Elevated temperatures in association with long day conditions hasten the development of seed-producing crops and cause yield penalties (Peltonen-Sainio et al., 2011b). Thereby, long day conditions may hamper full exploitation of the potential advantages of a longer growing season in the future. Secondly, the actual future growing season will have to end earlier than the thermal one because of lowering light intensities in the season from September onwards in particular (Saikkonen et al., 2012; www.gaisma.com, 9.7.2013). To take full advantage of a longer growing season, new improved cultivars of autumn-sown crops should be taken into cultivation in order to promote crop growth early in the season. Investment in field drainage and irrigation systems should also be urged to avoid extreme soil moisture and subsequent difficulties during spring cultivation, to sustain higher realization of yield potentials and also to reduce nutrient leaching. New breeds and more precise use of inputs are likely needed. All such developments towards sustainably intensified cropping systems (Peltonen-Sainio, 2012) call for consistency in political preferences and decisions at national scale and the EU.

As demonstrated by recorded advancements in sowings due to earlier onset of growing seasons (Kaukoranta and Hakala, 2008) and evidenced by a switch between early and late maturing crop species and cultivars depending on farmers' experiences on climatic conditions (Himanen et al., 2013; Peltonen-Sainio et al., 2013c), farmers have valuable readiness to adapt to changes in conditions. They are also eager to test new cultivars of the commonly cultivated crops, and also completely new crops such as forage maize (*Zea mays* L.) and winter oilseed rape, whenever possible. Spring oilseed rape has already given reasonable yields and is forecasted to replace the less productive turnip rape in the northernmost European conditions within the next decades (Peltonen-Sainio et al., 2009e).

Until recently the northernmost European climate has not favored invasions of pests and diseases to the same extent as elsewhere in Europe. With longer growing seasons and milder winters, the pathogen and pest problems will probably increase (Carter et al., 1996; Kaukoranta, 1996; Hakala et al., 2011). Even though most problems with pests, pathogens and weeds will be manageable in the future, just as they are in warmer climates at present, the required control measures will bite off some of the potential profit that climate change would bring to Finnish agriculture. Higher fertilizer rates to match the increasing yield potential will also increase the input costs of farming. More importantly, higher inputs of pesticide and fertilizer will lead to new environmental challenges. For example, higher precipitation, especially heavy rains, and longer and milder winters may increase risk for soil erosion and leaching of both agrochemicals and fertilizer nutrients (Forsius et al., 2013).

6 CONCLUDING REMARKS

Both breeding cultivars adapted to northern growing conditions and developing crop management to sustain yield potential have represented exceptional challenges for northern Europe compared with the global challenges that crop production typically faces. Despite this, a crop physiology approach has not long been the typical foundation when tailoring and developing cultivars and crop management practices for these particular conditions. During earlier decades, development was largely based on testing management practices and application of methods for large-scale comparative trials, in which yield response *per se*, economic outcome and general applicability of the method were often sufficient for adopting a novel method or system. However, crop physiology will evidently play a far more important role in the future. Plant breeding and development of crop management practices and cropping systems assisted by crop physiological understanding are the avenues to cope with the complex interactions between agriculture, environment and changing climate. Namely, substantial challenges will be faced when northern agriculture has to balance between climate change-induced opportunities and risks and become more environmentally friendly in the highly variable and promptly changing conditions. And such processes will take place when global changes are comprehensive when related to fundamental questions such as increasing demand for food and reduced global production areas and capacities. Northern agroecosystems have interesting and challenging future prospects in the light of global changes that call for environmentally and socioeconomically sustainable intensification in order to cope with all the simultaneously mobile elements of future changes.

References

Acevedo, E.H., Silva, P.C., Silva, H.R., Solar, B.R., 1999. Wheat production in Mediterranean environments. In: Satorre, E.H., Slafer, G.A. (Eds.), Wheat ecology and physiology of yield determination. Food Products Press, New York, pp. 295–323.

Aikasalo, R., 1988. The results of six-row barley breeding and the genetic origin of varieties released. J. Agric. Sci. Finland 60, 293–305.

Alakukku, L., 2006. Structure of clay topsoil affected by tillage intensity. In: NJF Seminar 378; Tillage systems for the benefit of agriculture and the environment. Arranged by NJF section I: Soil, water and environment. Nordic Agricultural Academy, Odense, Denmark, 29–31 May 2006, p. 226 (Extended abstracts).

Alakukku, L., Ristolainen, A., Sarikka, I., Hurme, T., 2010. Surface water ponding on clayey soils managed by conventional and conservation tillage in boreal conditions. Agric. Food Sci. 19, 313–326.

Alvenäs, G., Marstorp, H., 1993. Effect of a ryegrass catch crop on soil inorganic-N content and simulated nitrate leaching. Swed. J. Agric. Res. 23, 3–14.

Angus, J.F., Cunningham, R.B., Moncur, M.W., MacKenzie, D.H., 1981. Phasic development in field crops I. Thermal response in the seedling phase. Field Crops Res. 3, 365–378.

Antikainen, M., 1996. Cold acclimation in winter rye (*Secale cereale* L.): Identification and characterization of proteins involved in freezing tolerance. Annales Universitatis Turkuensis SER. AII, TOM. 87. PhD Thesis of University of Turku.

Aura, E., 1999. Effect of shallow tillage on physical properties of clay soil and growth of spring cereals in dry and moist summers in southern Finland. Soil Tillage Res. 50, 169–176.

Austin, R.B., Bingham, J., Blackwell, R.D., et al.,1980. Genetic improvement in winter wheat yield since 1900 and associated physiological changes. J. Agric. Sci. (Camb.) 94, 675–689.

Baker, J.M., Ochsner, T.E., Venterea, R.T., Griffis, T.J., 2007. Tillage and soil carbon sequestration – What do we really know? Agric. Ecosyst. Environ. 118, 1–5.

Bélanger, G., Rochette, P., Castonguay, Y., Bootsma, A., Mongrain, D., Ryan, D.A.J., 2002. Climate change and winter survival of perennial forage crops in eastern Canada. Agron. J. 94, 1120–1130.

Bell, M.A., Fischer, R.A., Byerlee, D., Sayre, K., 1995. Genetic and agronomic contributions to yield gains: A case study for wheat. Field Crops Res. 44, 55–65.

Blackmore, S., Godwin, R.J., Fountas, S., 2003. The analysis of spatial and temporal trends in yield map data over six years. Biosyst. Eng. 84, 455–466.

Bonham-Smith, P.C., Gilmer, S., Shou, R., Galka, M., Abrams, S.R., 2006. Non-lethal freezing effects on seed degreening in *Brassica napus*. Planta 224, 145–154.

Borras, L., Slafer, G.A., Otegui, M.E., 2004. Seed dry weight response to source-sink manipulations in wheat, maize and soybean: a quantitative reappraisal. Field Crops Res. 86, 131–146.

Breland, T.A., 1996. Green manuring with clover and ryegrass catch crops undersown in small grains: Effects on soil mineral nitrogen in field and laboratory experiments. Acta Agric. Scand. Sect. B, Soil Plant Sci. 46, 178–185.

Calderini, D.F., Slafer, G.A., 1998. Changes in yield and yield stability in wheat during the 20th century. Field Crops Res. 57, 335–347.

Calderini, D.F., Reynolds, M.P., 2000. Changes in grain weight as a consequence of de-graining treatments at pre- and post-anthesis in synthetic hexaploid lines of wheat (*Triticum durum* x *T. tauschii*). Aust. J. Plant Physiol. 27, 183–191.

Calderini, D.F., Abeledo, L.G., Savin, R., Slafer, G.A., 1999a. Effect of temperature and carpel size during pre-anthesis on potential grain weight in wheat. J. Agric. Sci. (Camb.) 132, 453–459.

Calderini, D.F., Abeledo, L.G., Savin, R., Slafer, G.A., 1999b. Final grain weight in wheat as affected by short periods of high temperature during pre- and post-anthesis under field conditions. Aust. J. Plant Physiol. 26, 453–458.

Carter, T.R., 1998. Changes in the thermal growing season in Nordic countries during the past century and prospects for the future. Agric. Food Sci. Finland 7, 161–179.

Carter, T.R., Saarikko, R.A., Niemi, K.J., 1996. Assessing the risks and uncertainties of regional crop potential under a changing climate in Finland. Agric. Food Sci. Finland 5, 329–350.

Dofing, S.M., 1996. Near-isogenic analysis of uniculm and conventional-tillering barley lines. Crop Sci. 36, 1523–1526.

Dofing, S.M., Karlsson, M.G., 1993. Growth and development of uniculm and conventional-tillering barley lines. Agron. J. 85, 58–61.

Dofing, S.M., Knight, C.W., 1994. Yield component compensation in uniculm barley lines. Agron. J. 86, 273–276.

Egli, D.B., 2006. The role of seed in the determination of yield of grain crops. Aust. J. Agric. Res. 57, 1237–1247.

Egli, D.B., 2008. Soybean yield trends from 1972 to 2003 in mid-western USA. Field Crops Res. 106, 53–59.

Elsgaard, I., Børgesen, C.D., Olesen, J.E., et al., 2012. Shifts in comparative advantages for maize, oat and wheat cropping under climate change in Europe. Food Add. Contam. A 29, 1514–1526.

Evans, L.T., 1993. Crop evolution, adaptation and yield. Cambridge University Press, Cambridge.

Evans, L.T., 2005. The changing context for agricultural science. J. Agric. Sci. (Camb.) 143, 7–10.

Evans, L.T., Fischer, R.A., 1999. Yield potential: its definition, measurement, and significance. Crop Sci. 39, 1544–1551.

Evers, J.B., Vos, J., Chelle, M., Andrieu, B., Fournier, C., Struik, P.C., 2007. Simulation the effects of localized red:far red ratio on tillering in spring wheat (*Triticum aestivum*) using a three-dimensional virtual plant model. New Phytol. 176, 325–336.

Ewert, F., Rounsevell, M.D.A., Reginster, I., Metzger, M.J., Leemans, R., 2005. Future scenarios of European agricultural land use. I. Estimating changes in crop productivity. Agric. Ecosyst. Environ. 107, 101–116.

FAO Statistics, 2013. http://faostat.fao.org.

Forsius, M., Anttila, S., Arvola, L., et al., 2013. Impacts and adaptation options of climate change on ecosystem services in Finland: a model based assessment. Curr. Opin. Environ. Sustain. 5, http://dx.doi.org/10.1016/j.cpsist-2013.01.001.

Foulkes, M.J., Snape, J.W., Shearman, V.J., Reynolds, M.P., Gaju, O., Sylvester-Bradley, R., 2007. Genetic progress in yield potential in wheat: recent advances and future prospects. J. Agric. Sci. (Camb.) 145, 17–29.

Franklin, K.A., Whitelam, G.C., 2005. Phytochromes and shade-avoidance responses in plants. Ann. Bot. 96, 169–175.

Frich, P., Rosenorn, S., Madsen, H., Jensen, J.J., 1997. Observed Precipitation in Denmark, 1961-90. Danish Meteorological Institute Technical Report 97-8, Copenhagen.

Garcia del Moral, L.F., Garcia del Moral, M.B., Molina-Cano, J.L., Slafer, G.A., 2003. Yield stability and development in two- and six-rowed winter barleys under Mediterranean conditions. Field Crops Res. 81, 109–119.

Hakala, K., 1998. Growth and yield potential of spring wheat in a simulated changed climate with increased CO_2 and higher temperature. Eur. J. Agron. 9, 41–52.

Hakala, K., Mela, T., 1996. The effects of prolonged exposure to elevated temperatures and elevated CO2 levels on the growth, yield and dry matter partitioning of field-sown meadow fescue (*Festuca pratensis*, cv. *Kalevi*). Agric. Food Sci. Finland 5, 285–298.

Hakala, K., Pahkala, K., 2003. Comparison of central and northern European winter rye cultivars grown at high latitudes. J. Agric. Sci. (Camb.) 141, 169–178.

Hakala, K., Heliö, R., Tuhkanen, E., Kaukoranta, T., 1999. Photosynthesis and Rubisco kinetics in spring wheat and meadow fescue under conditions of simulated climate change with elevated CO_2 and increased temperatures. Agric. Food Sci. Finland 8, 441–457.

Hakala, K., Hannukkala, A., Huusela-Veistola, E., Jalli, M., Peltonen-Sainio, P., 2011. Pests and diseases in a changing climate: a major challenge for Finnish crop production. Agric. Food Sci. 20, 3–14.

Hakala, K., Jauhiainen, L., Himanen, S.J., Rötter, R., Salo, T., Kahiluoto, H., 2012. Sensitivity of barley varieties to weather in Finland. J. Agric. Sci. (Camb.) 150, 145–160.

Heinonen, M., Alakukku, L., Aura, E., 2002. Effects of reduced tillage and light tractor traffic on the growth and yield of oats (*Avena sativa*). Adv. Geoecol. 35, 367–378.

Helander, M., Saloniemi, I., Saikkonen, K., 2012. Glyphosate in northern ecosystems. Trends Plant Sci. 17, 569–574.

Himanen, S.J., Hakala, K., Kahiluoto, H., 2013. Crop responses to climate and socioeconomic change in northern regions. Region. Environ. Change 13, 17–32.

Hofgaard, I.S., Vollsnes, A.V., Marum, P., Larsen, A., Tronsmo, A.M., 2003. Variation in resistance to different winter stress factors within a full-sib family of perennial ryegrass. Euphytica 134, 61–75.

Hollins, P.D., Kettlewell, P.S., Peltonen-Sainio, P., Atkinson, M.D., 2004. Relationships between climate and Finnish winter cereal grain quality and their potential for forecasting. Agric. Food Sci. 13, 295–308.

Hömmö, L.M., 1994. Resistance of winter cereals to various winter stress factors – inter- and intraspecific variation and the role of cold acclimation. Agric. Sci. Finland 3, Suppl. 1 (PhD Thesis, Agricultural Research Centre of Finland).

Hömmö, L., Pulli, S., 1993. Winterhardiness of some winter wheat (*Triticum aestivum*), rye (*Secale cereale*), triticale (X *Triticosecale*) and winter barley (*Hordeum vulgare*) cultivars tested at six locations in Finland. Agric. Sci. Finland 2, 311–327.

IPCC, 2000. Emissions scenarios. In: Nakicenovic, N., Swart, R. (Eds.), Special Report of the Intergovernmental Panel on Climate Change. Cambridge University Press, Cambridge.

IPCC, 2007. Summary for policymakers. In: Solomon, S., Qin, D., Manning, M. et al., (Eds.), Climate change 2007: The physical science basis. Contribution of Working Group I to the Fourth Assessment Report of the Intergovernmental Panel on Climate Change. Cambridge University Press, Cambridge.

IPCC, 2012. Managing the risks of extreme events and disasters to advance climate change adaptation. In: Field, C.B., Barros, V., Stocker, T.F. et al., (Eds.), A Special Report of Working Groups I and II of the Intergovernmental Panel on Climate Change. Cambridge University Press, Cambridge.

Jensen, E.S., 1991. Nitrogen accumulation and residual effects of nitrogen catch crops. Acta Agric. Scand. 41, 333–344.

Jensen, E.S., Peoples, M.B., Boddey, R.M., et al., 2012. Legumes for mitigation of climate change and provision of feedstocks for biofuels and biorefineries. Agron. Sustain. Devel. 32, 329–364.

Johnson, E.N., Miller, P.R., Blackshaw, R.E., et al., 2004. Seeding date and polymer coating effects on plant establishment and yield of fall-seeded canola in the Northern Great Plains. Can. J. Plant Sci. 84, 955–963.

Jylhä, K., Fronzek, S., Tuomenvirta, H., Carter, T.R., Ruosteenoja, K., 2008. Changes in frost, snow and Baltic sea ice by the end of the twenty-first century based on climate model projections for Europe. Climat. Change 86, 441–462.

Känkänen, H., 2008. Effects of direct drilling on spring cereals, turnip rape and pea. In: NJF Seminar 418 New insights into sustainable cultivation methods in agriculture, Piikkiö, Finland, 17-19 September 2008. NJF Report 4 3, 10-11.

Känkänen, H., 2010. Undersowing in a northern climate: effects on spring cereal yield and risk of nitrate leaching. MTT Sci., *MTT Tiede* 8 93 Doctoral Dissertation.

Känkänen, H., Eriksson, C., 2007. Effects of undersown crops on soil mineral N and grain yield on spring barley. Eur. J. Agron. 27, 25–34.

Känkänen, H., Huusela-Veistola, E., Salo, Y., Kangas, A., Vuorinen, M., 2004. Päällekkäisviljely: Lupauksia ja pettymyksiä (Relay intercropping: Promises and disillusions). Maa- ja elintarviketalous 64: 35 (In Finnish, with English abstract).

Känkänen, H., Alakukku, L., Salo, Y., Pitkänen, T., 2011. Growth and yield of spring cereals during transition to zero tillage on clay soils. Eur. J. Agron. 34, 35–45.

Känkänen, H., Suokannas, A., Tiilikkala, K., Nykänen, A., 2013. Biologinen typensidonta fossiilisen energian säästäjänä: 2. korjattu painos (Reducing use of fossil energy by biological N fixation). MTT Raportti 76, 60, (In Finnish, with English abstract).

Kaukoranta, T., 1996. Impact of global warming on potato late blight: risk, yield loss and control. Agric. Food Sci. Finland 5, 311–327.

Kaukoranta, T., Hakala, K., 2008. Impact of spring warming on sowing times of cereal, potato and sugar beet in Finland. Agric. Food Sci. 17, 165–176.

Kirkegaard, J.A., Robertson, M.J., Hamblin, P., Sprague, S.J., 2006. Effect of blackleg and sclerotinia stem rot on canola yield in the high rainfall zone of southern New South Wales, Australia. Aust. J. Agric. Res. 57, 201–212.

Kivi, E.I., 1963. Domestic plant breeding for the improvement of spring cereal varieties in Finland. Acta Agric. Fenn. 100, 1–37.

Kivi, E.I., 1967. Ilmastotekijäin vaikutus mallasohrasadon määrään ja laatuun. Mallasjuomat 11, 1–10, (In Finnish).

Kristensen, H.L., McCarty, G.W., Meisinger, J.J., 2000. Effects of soil structure disturbance on mineralization of organic soil nitrogen. Soil Sci. Soc. Am. J. 64, 371–378.

Kvist, M., 1992. Catch crops undersown in spring barley. Competitive effects and cropping methods. Crop Product. Sci. 15, 210.

Lafarge, M., 2000. Phenotypes and the onset of competition in spring barley stands of one genotype: daylength and density effects on tillering. Eur. J. Agron. 12, 211–223.

Lauer, J.G., Simmons, S.R., 1988. Photoassimilate partitioning by tiller and individual tiller leaves in field-grown spring barley. Crop Sci. 28, 279–282.

Lisson, S.N., Kirkegaard, J.A., Robertson, M.J., Zwart, A., 2007. What is limiting canola yield in southern New South Wales? A diagnosis of causal factors. Aust. J. Exp. Agric. 47, 1435–1445.

Løes, A.K., 2003. Studies of the availability of soil phosphorus (P) and potassium (K) in organic farming systems, and of plant adaptations to low P- and K-availability. Ph.D. diss. Agricultural Univ. Norway, 2003, 29.

Lötjönen, T., Isolahti, M., 2010. Direct drilling of cereals after ley and slurry spreading. Acta Agric. Scand. B Soil Plant Sci. 60, 307–319.

Mela, T., Paatela, J., 1974. Grain yield of spring wheat and oats as affected by population density. Ann. Agric. Fenn. 13, 161–167.

Michael, G., Beringer, H., 1980. The role of hormones in yield formation. Physiological Aspects of Crop Productivity. Proceedings of the 15th Colloquium International Potash Institute. Pudoc, Wageningen, The Netherlands, pp. 85–116.

Moll, R.H., Kamprath, E.J., Jackson, W.A., 1982. Analysis and interpretation of factors which contribute to efficiency of nitrogen utilization. Agron. J. 74, 562–564.

Motzo, R., Guinta, F., Deidda, M., 2004. Expression of a tiller inhibitor gene in the progenies of interspecific crosses *Triticum aestivum* L. × T. *turgidum subsp. durum*. Field Crops Res. 85, 15–20.

Mukula, J., Rantanen, O., 1987. Climatic risks to the yield and quality of field crops in Finland. I. Basic facts about Finnish field crops production. Ann. Agric. Fenn. 26, 1–18.

Muukkonen, P., Hartikainen, H., Lahti, K., Särkelä, A., Puustinen, M., Alakukku, L., 2007. Influence of no-tillage on the distribution and lability of phosphorus in Finnish clay soils. Agric. Ecosyst. Environ. 120, 2–4, 299-306.

Muurinen, S., Slafer, G.A., Peltonen-Sainio, P., 2006. Breeding effects on nitrogen use efficiency of spring cereals under northern conditions. Crop Sci. 46, 561–568.

Niemeläinen, O., 1990. Factors affecting panicle production of cocksfoot (*Dactylis glomerata* L.) in Finland. III. Response to exhaustion of reserve carbohydrates and to freezing stress. Ann. Agric. Fenn. 29, 241–250.

Niemi, J., Ahlstedt, J. (eds), 2007. Finnish Agriculture and Rural Industries 2007. Agrifood Research Finland, Economic Research, Publications 103a.

Nissinen, O., 1996. Analysis of climatic factors affecting snow mold injury in first-year timothy (*Phleum pratence* L.) with special reference to *Sclerotinia borealis*. Acta Univ Ouluen A Sci Rerum Nat. 289. PhD Thesis of University of Oulu.

Öfversten, J., Jauhiainen, L., Nikander, H., Salo, Y., 2002. Assessing and predicting the local performance of spring wheat varieties. J. Agric. Sci. 139, 397–404.

Ohlander, L., Bergkvist, G., Stendahl, F., Kvist, M., 1996. Yield of catch crops and spring barley as affected by time of undersowing. Acta Agric. Scand. B Soil Plant Sci. 46, 161–168.

Olesen, J.E., Bindi, M., 2002. Consequences of climate change for European agricultural productivity, land use and policy. Eur. J. Agron. 16, 239–262.

Olesen, J.E., Trnka, M., Kersebaum, K.C., et al., 2011. Impacts and adaptation of European crop production systems to climate change. Eur. J. Agron. 34, 96–112.

Ortiz-Monasterio, J.I., Sayre, K.D., Rajaram, S., McMahon, M., 1997. Genetic progress in wheat yield and nitrogen use efficiency under four nitrogen rates. Crop Sci. 37, 898–904.

Pahkala, K., Sovero, M., 1988. The cultivation and breeding of oilseed crops in Finland. Ann. Agric. Fenn. 27, 199–207.

Pahkala, K., Laakso, I., Hovinen, S., 1991. The effect of frost treatment on turnip rape seedlings, ripening of seeds and their fatty acid composition. Proc. GCIRC Rapeseed Cong. 6, 1749–1753.

Pahkala, K., Sankari, H., Ketoja, E., 1994. The relation between stand density and the structure of spring rape (*Brassica napus* L.). J. Agron. Crop Sci. 172, 269–278.

Palta, J.A., Fillery, I.R.P., Rebetzke, G.J., 2007. Restricted-tillering wheat does not lead to greater investment in roots and early nitrogen uptake. Field Crops Res. 104, 52–59.

Peltonen, J., Peltonen-Sainio, P., 1997. Breaking uniculm growth habit of spring cereals at high latitudes by crop management. II Tillering, grain yield, and yield component differences. J. Agron. Crop Sci. 178, 87–95.

Peltonen-Sainio, P., 1990. Genetic improvements in the structure of oat stands in northern growing conditions during this century. Plant Breed. 104, 340–345.

Peltonen-Sainio, P., 1991. High phytomass producing oats for cultivation in the northern growing conditions. J. Agron. Crop Sci. 166, 90–95.

Peltonen-Sainio, P., 1997. Leaf area duration of oat at high latitudes. J. Agron. Crop Sci. 178, 149–155.

Peltonen-Sainio, P., 2012. Crop production in a northern climate. *In* : Meybeck, A., et al., (eds), Proceedings of a Joint FAO/OECD Workshop, Building Resilience to Climate Change in the Agriculture Sector, available online at http://www.fao.org/agriculture/crops/news-events-bulletins/detail/en/item/134976/, pp. 183-216.

Peltonen-Sainio, P., Järvinen, P., 1995. Seeding rate effects on tillering, grain yield, and yield components of oat at high latitude. Field Crops Res. 40, 49–56.

Peltonen-Sainio, P., Peltonen, J., 1995. Floret set and abortion in oat and wheat under high and low nitrogen regimes. Eur. J. Agron. 4, 253–262.

Peltonen-Sainio, P., Peltonen, J., 1997. Breaking uniculm growth habit of spring cereals at high latitudes by crop management. I Leaf area index and biomass accumulation. J. Agron. Crop Sci. 178, 79–86.

Peltonen-Sainio, P., Rajala, A., 2007. Duration of vegetative and generative development phases in oat cultivars released since 1921. Field Crops Res. 101, 72–79.

Peltonen-Sainio, P., Jauhiainen, L., 2008. Association of growth dynamics, yield components and seed quality in long-term trials covering rapeseed cultivation history at high latitudes. Field Crops Res. 108, 101–108.

Peltonen-Sainio, P., Jauhiainen, L., 2010. Cultivar improvement and environmental variability in yield removed nitrogen of spring cereals and rapeseed in northern growing conditions according to a long-term dataset. Agric. Food Sci. 19, 341–353.

Peltonen-Sainio, P., Niemi, J.K., 2012. Protein crop production at the northern margin of farming: To boost, or not to boost. Agric. Food Sci. 21, 370–383.

Peltonen-Sainio, P., Jauhiainen, L., 2013. Weather variability lessons from the past: sowing to ripening dynamics and yield penalties for northern agriculture in 1970–2012. Manuscript submitted.

Peltonen-Sainio, P., Rajala, A., Simmons, S., Caspers, R., Stuthman, D.D., 2003. Plant growth regulator and day-length effects on preanthesis main shoot and tiller growth in conventional and dwarf oat. Crop Sci. 43, 227–233.

Peltonen-Sainio, P., Känkänen, H., Pahkala, K., Salo, Y., Huusela-Veistola, E., Peltonen, J., 2006. Polymer coated turnip rape seed did not facilitate early broadcast sowing under Finnish growing conditions. Agric. Food Sci. 15, 152–165.

Peltonen-Sainio, P., Jauhiainen, L., Hannukkala, A., 2007a. Declining rapeseed yields in Finland: how, why and what next? J. Agric. Sci. (Camb.) 145, 587–598.

Peltonen-Sainio, P., Kangas, A., Salo, Y., Jauhiainen, L., 2007b. Grain number dominates grain weight in cereal yield determination: evidence basing on 30 years' multi-location trials. Field Crops Res. 100, 179–188.

Peltonen-Sainio, P., Muurinen, S., Rajala, A., Jauhiainen, L., 2008. Variation in harvest index of modern spring barley, oat and wheat under northern conditions. J. Agric. Sci. (Camb.) 146, 35–47.

Peltonen-Sainio, P., Jauhiainen, L., Laurila, I.P., 2009a. Cereal yield trends in northern European conditions: Changes in yield potential and its realisation. Field Crops Res. 110, 85–90.

Peltonen-Sainio, P., Jauhiainen, L., Hakala, K., Ojanen, H., 2009b. Climate change and prolongation of growing season: changes in regional potential for field crop production in Finland. Agric. Food Sci. 18, 171–190.

Peltonen-Sainio, P., Jauhiainen, L., Rajala, A., Muurinen, S., 2009c. Tiller traits of spring cereals in tiller-depressing long day conditions. Field Crops Res. 113, 82–89.

Peltonen-Sainio, P., Jauhiainen, L., Hakala, K., 2009d. Are there indications of climate change induced increases in variability of major field crops in the northernmost European conditions? Agric. Food Sci. 18, 206–226.

Peltonen-Sainio, P., Hakala, K., Jauhiainen, L., Ruosteenoja, K., 2009e. Comparing regional risks in producing turnip rape and oilseed rape – Impacts of climate change and breeding. Acta Agric. Scand. B Soil Plant Sci. 59, 129–138.

Peltonen-Sainio, P., Jauhiainen, L., Trnka, M., et al., 2010. Coincidence of variation in yield and climate in Europe. Agric. Ecosyst. Environ. 139, 483–489.

Peltonen-Sainio, P., Hakala, K., Jauhiainen, L., 2011a. Climate-induced overwintering challenges for wheat and rye in northern agriculture. Acta Agric. Scand. B Soil Plant Sci. 61, 75–83.

Peltonen-Sainio, P., Jauhiainen, L., Hakala, K., 2011b. Crop responses to temperature and precipitation according to long-term multi-location trials at high-latitude conditions. J. Agric. Sci. 149, 49–62.

Peltonen-Sainio, P., Jauhiainen, L., Sadras, V.O., 2011c. Phenotypic plasticity of yield and agronomic traits in spring cereals and rapeseed at high latitudes. Field Crops Res. 24, 261–269.

Peltonen-Sainio, P., Jauhiainen, L., Hyövelä, M., Nissilä, E., 2011d. Trade-off between oil and protein in rapeseed at high latitudes: means to consolidate protein crop status? Field Crops Res. 121, 248–255.

Peltonen-Sainio, P., Jauhiainen, L., Laitinen, P., Salopelto, J., Saastamoinen, M., Hannukkala, A., 2011e. Identifying difficulties in rapeseed root penetration in farmers' fields in northern European conditions. Soil Use Manag. 27, 229–237.

Peltonen-Sainio, P., Hannukkala, A., Huusela-Veistola, E., et al., 2013a. Potential and realities of enhancing rapeseed- and grain legume-based protein production in a northern climate. J. Agric. Sci. (Camb.) 151, 303–321.

Peltonen-Sainio, P., Jauhiainen, L., Hakala, K., Ruosteenoja, K., 2013. Rainfed production challenges future yields at high latitude conditions? Manuscript submitted.

Peltonen-Sainio, P., Jauhiainen, L., Niemi, J.K., Hakala, K., Sipiläinen, T., 2013c. Do farmers rapidly adapt to past growing conditions by sowing different proportions of early and late maturing cereals and cultivars? Manuscript submitted.

Prystupa, P., Slafer, G.A., Savin, R., 2003. Leaf appearance, tillering and their coordination in response to NxP fertilization in barley. Plant Soil 255, 587–594.

Rajala, A., Peltonen-Sainio, P., 2000. Manipulating yield potential in cereals by plant growth regulators. In: Basra, A.S. (Ed.), Growth regulators in crop production. Food Products Press, Binghamton, New York, pp. 27–70.

Rajala, A., Peltonen-Sainio, P., 2001. Plant growth regulator effects on spring cereal root and shoot growth. Agron. J. 93, 936–943.

Rajala, A., Peltonen-Sainio, P., 2004. Intra-panicle variation in progress of cell division in developing oat grains: A preliminary study. Agric. Food Sci. 13, 163–169.

Rajala, A., Peltonen-Sainio, P., 2011. Pollination dynamics, grain weight and grain cell number within the inflorescence and spikelet in oat and wheat. Agric. Sci. 2, 283–290.

Rajala, A., Peltonen-Sainio, P., Kauppila, R., Wilhelmson, A., Reinikainen, P., Kleemola, J., 2007. Within-field variation in grain yield, yield components and quality traits of two-row barley. J. Agric. Sci. (Camb.) 145, 445–454.

Rajala, A., Hakala, K., Mäkelä, P., Muurinen, S., Peltonen-Sainio, P., 2009. Spring wheat response to timing of water deficit through sink and grain filling capacity. Field Crops Res. 114, 263–271.

Rajala, A., Hakala, K., Mäkelä, P., Peltonen-Sainio, P., 2011. Drought effect on grain number and grain weight at spike and spikelet level in six-row spring barley. J. Agron. Crop Sci. 197, 103–112.

Rasmussen, K.J., 1999. Impact of ploughless soil tillage on yield and soil quality: A Scandinavian review. Soil Tillage Res. 53, 3–14.

Regina, K., Alakukku, L., 2010. Greenhouse gas fluxes in varying soils types under conventional and no-tillage practices. Soil Tillage Res. 109, 144–152.

Regina, K., Perälä, P., Alakukku, L., 2007. Greenhouse gas fluxes in boreal agricultural soils under conventional tillage and no-till practice. *In*: Jandl, R., Olsson, M. (eds), European cooperation in the field of scientific and technical research COST Action 639: Greenhouse-gas budget of soils under changing climate and land use (BurnOut), pp. 53-56.

Reicosky, D.C., 2003. Tillage-induced CO2 emissions and carbon sequestration: effect of secondary tillage and compaction. In: Garcia-Torres, L., Benites, J., Martinez-Vilela, A., Holgado-Gabrera, A. (Eds.), Conservation agriculture. Kluwer Acad. Pub, Dordrecht, The Netherlands, pp. 291–300.

Rekunen, M., 1988. Advances in the breeding of oats. Comparative trials with historical varieties in 1977–87. J. Agric. Sci. Finland 60, 307–321.

Richards, R., 1988. A tiller inhibitor gene in wheat and its effect on plant growth. Aust. J. Agric. Res. 38, 749–757.

Richards, R., 2002. Current and emerging environmental challenges in Australian agriculture – the role of plant breeding. Aust. J. Agric. Res. 53, 881–892.

Roslon, E., 2003. Relay cropping of spring barley and winter wheat. Doctoral thesis, Swedish University of Agricultural Sciences, Uppsala 2003. Acta Universitatis Agriculturae Sueciae, Agraria 427.

Rydberg, T., 1987. Studies in ploughless tillage in Sweden 1975–1986. Raport, Jordbearbetningsavdelning, Sveriges Lantbrukuniversitet, Uppsala, pp. 53–76.

Sadras, V.O., 2007. Evolutionary aspects of the trade-off between seed size and number in crops. Field Crops Res. 100, 125–138.

Sadras, V.O., Angus, J.F., 2006. Benchmarking water-use efficiency of rainfad wheat in dry environments. Aust. J. Agric. Res. 57, 847–856.

Saikkonen, K., Taulavuori, K., Hyvönen, T., et al., 2012. Climate change-driven species' range shifts filtered by photoperiodism. Nat. Clim. Change 2, 239–242.

Salo, T., Eskelinen, J., Jauhiainen, L., Kartio, M., 2007. Reduced fertilizer use and changes in cereal grain weight, test weight and protein content in Finland in 1990-2005. Agric. Food Sci. 16, 3–16.

Schröder, J.J., 2001. Reduction of nitrate leaching. The role of cover crops. Brochure of EU Concerted Action (AIR3) 2108.

Scott, R.W., Appleyard, M., Fellowes, G., Kirby, E.J.M., 1983. Effect of genotype and position in the ear on carpel and grain growth and mature grain weight of spring barley. J. Agric. Sci. (Camb.) 100, 383–391.

Serenius, M., Huusela-Veistola, E., Avikainen, H., Pahkala, K., Laine, A., 2005. Effects of sowing time on pink snow mold, leaf rust and winter damage in winter rye varieties in Finland. Agric. Food Sci. 14, 362–376.

Sheehy, J., Six, J., Alakukku, L., Regina, K., 2013. Fluxes of nitrous oxide in tilled and no-tilled boreal arable soils. Agric. Ecosyst. Environ. 164, 190–199.

Siddique, K.H.M., Kirby, E.J.M., Perry, M.W., 1989. Ear to stem ratio in old and modern wheats: relationship with improvement in number of grains per ear and yield. Field Crops Res. 21, 59–78.

Simmons, S.R., Rasmusson, D.C., Wiersma, J.V., 1982. Tillering in barley: genotype, row spacing and seeding rate effects. Crop Sci. 22, 201–215.

Solantie, R., 1987. Last spring frosts and first autumn frosts in Finland. Meteorol. Publ. 6, 60.

Soussana, J.F., Fereres, E., Long, S., et al., 2012. A European science plan to sustainably increase food security under climate change. Glob. Change Biol. 18, 3269–3271.

Tietäväinen, H., Tuomenvirta, H., Venäläinen, A., 2010. Annual and seasonal mean temperatures in Finland during the last 160 years based on gridded temperature data. Internatl. J. Climatol. 30, 2247–2256.

Trnka, M., Olesen, J.E., Kersebaum, K.C., et al., 2011. Agroclimatic conditions in Europe under climate change. Glob. Change Biol. 17, 2298–2318.

Tuomenvirta, H., 2004. Reliable estimation of climatic variations in Finland. Diss. University of Helsinki. Finnish Meteorological Institute Contributions No. 43, 79.

Tveito, O.E., Forland, E.J., Alexandersson, H., et al., 2001. Nordic climate maps. KLIMA report 6/2001.

Ulén, B., Aronsson, H., Bechmann, M., Krogstad, T., Øygarden, L., Stenberg, M., 2010. Soil tillage methods to control phosphorus loss and potential side-effects: a Scandinavian review. Soil Use Manag. 26, 94–107.

Uusi-Kämppä, J., 2010. Effect of outdoor production, slurry management and buffer zones on phosphorus and nitrogen runoff losses from Finnish cattle farms. MTT Sci. MTT Tiede 7, 45. Diss; Doctoral Dissertation.

Uusi-Kämppä, J., Braskerud, B., Jansson, H., Syversen, N., Uusitalo, R., 2000. Buffer zones and constructed wetlands as filters for agricultural phosphorus. J. Environ. Qual. 29, 151–158.

Wardlaw, I.F., 1994. The effect of high temperature on kernel development in wheat: variability related to pre-heading and post-anthesis conditions. Aust. J. Plant Physiol. 21, 731–739.

Ylhäisi, J., Tietäväinen, H., Peltonen-Sainio, P., et al., 2010. Growing season precipitation in Finland under recent and projected climate. Nat. Haz. Earth Syst. Sci. 10, 1563–1574.

CHAPTER

5

Raising productivity of maize-based cropping systems in eastern and southern Africa: Step-wise intensification options

John Dimes, Daniel Rodriguez, Andries Potgieter

Queensland Alliance for Agriculture and Food Innovation (QAAFI), The University of Queensland, Toowoomba, QLD, Australia

1 INTRODUCTION

Close to 400 million people live in eastern and southern Africa (ESA), with more than half living in extreme poverty and 75% residing in rural areas (Mulugetta et al., 2011). Food security is a major concern, as the region is barely self-sufficient in food grains with net imports of 10% if South Africa is excluded (FAOSTAT, 2009). Maize is the main and preferred food staple in ESA, with per capita consumption averaging 44 kg $year^{-1}$ in Ethiopia and exceeding 100 kg $year^{-1}$ in Malawi. In years of surplus, maize is also an important source of income to many farmers. Approximately 65% of the agricultural land in sub-Saharan Africa suffers from degradation (UNEP/ISRIC.1991; GEF, 2003) including low nutrient content, physical degradation and erosion which results in generally low yields of approximately 1 t ha^{-1} for the staple grains (Rockstrom and Falkenmark, 2000). In some cases, land degradation is irreversible. With growth of population and incomes, the demand for maize is projected to increase approximately 3–4% annually over the next 10 years, requiring at least 40% more maize grain to be accessed in a very short time horizon.

Rain-fed, smallholder agricultural systems prevail in ESA and sub-Saharan Africa, and farmers are highly vulnerable to biotic and abiotic stresses and price fluctuations. Intensification of these cropping systems is challenging, and much of the past growth in agricultural production has been through expansion of area cultivated, having severe environmental consequences. In these circumstances, increasing the productivity

Crop Physiology. DOI: 10.1016/B978-0-12-417104-6.00005-4

and resource-use efficiency of agriculture is an essential component to achieve food secured households. Together with the lack of functional markets, insidious nutrient depletion and significant season-to-season variation in weather, farmers find it difficult to justify the investment of scarce resources in new technologies.

The aims of this chapter are to describe key constraints to the sustainable intensification of maize-based cropping systems in ESA, and discuss the need for stepping-stone approaches to the increase in farmers' investment in technology.

2 MAIZE-BASED FARMING SYSTEMS IN EASTERN AND SOUTHERN AFRICA

2.1 Maize cropping systems

Maize production occupies between 40% (3.0 Mha, Tanzania) and 90% (1.2 Mha, Malawi) of cultivated lands across countries in ESA. Maize crops are intercropped, predominantly with legumes, but also with vegetable crops, generally in fields closer to farm homesteads. Intercropping is favored because of small farm size, manual systems of land preparation in association with animal draught power and problems of weed control. The majority of rural households pursue mixed farming enterprises and maize residues are an important feed source for cattle and goats during the dry seasons, either grazed *in situ* or stored close to night pens. In higher rainfall regions where dairy production is undertaken, green maize leaves are also an important feed source (Herrero et al., 2013).

The legume species vary across the region, although the most common species are beans (*Phaseolus vulgaris*), cowpeas (*Vigna unguiculata*), pigeon pea (*Cajanus cajan*), groundnuts (*Arachis hypogea*), chickpea (*Cicer arietinum*) and soybeans (*Glycine max*). Maize intercropped with beans, cowpeas and pigeon peas are more common in small land-holdings, whereas sole legume crops such as groundnuts and soybeans in rotation with maize are more common where there is less pressure on the land. The legumes are an important source of protein and are often valuable 'women crops', both for household consumption and for cash income.

2.2 Climate

We have examined maize production in five countries of ESA – Ethiopia, Kenya, Tanzania, Malawi and Mozambique, and the main analysis relates to 14 locations across the region (Table 5.1). These sites are the current focus of a maize–legume intensification project led by CIMMYT (Mulugetta et al., 2011). They have varying extents of climate similarity in the sub-region (Fig. 5.1a) and are highly representative of its most severe and persistent food insecurity (Fig. 5.1b). Semi-arid sites (Bulawayo and Katumani), where maize production is commonly practiced, are included in the analysis to compare maize response in drier cropping environments.

Average annual rainfall for the 14 locations is generally high, ranging from 750 to 1900 mm (Table 5.1). Malawi, Mozambique and Zimbabwe sites experience a well-defined unimodal rainfall pattern, with >94% of rainfall between October and May. In contrast, the Kenya sites experience a bimodal pattern in which both seasons (3–4 months rainfall duration) are fully exploited for maize production and together receive 81–87% of annual rainfall. In Tanzania and Ethiopia, there is substantial rainfall (20–44% of annual) outside the main cropping periods that is generally considered too unreliable for maize production.

Average in-crop rainfall for the 16 site-seasons range from 395 mm at Selian, Tanzania to 858 mm for the long rains season at Kakamega in western Kenya (Fig. 5.1). The in-crop rainfall could be considered mostly reliable (11 of 16 site-seasons have CV% ≤30%) and adequate for rain-fed maize. The variability of rainfall for bimodal seasons at Embu (CV = 35% and 42%)

TABLE 5.1 Climate characteristics of sites analyzed in this study. Sites in Kenya have a bimodal rainfall pattern, whereas the remaining sites have a unimodal pattern with >94% annual rain concentrated in a single season from October to May

Country	Region	Rainfall station	Climate record	Annual rainfall (mm)	For period of maize growth cycle	
					Daily average temperature (°C)	Daily radiation (MJ m^{-2})
BIMODAL						
Kenya	Eastern	Embu	1983–2013	1286	20	18
	Western	Kakamega	1980–2011	1927	25	21
	Semi-arid tropics	Katumani	1957–1998	683	20	18
UNIMODAL						
Ethiopia	Western	Baco	1986–2012	1292	20	24
	Rift Valley	Melkassa	1977–2011	812	21	19
		Hawassa	1982–2011	1013	20	23
Tanzania	Northern	Mbulu	1981–2010	823	22	22
		Selian	1992–2011	769	21	18
	Eastern	Ilonga	1981–2011	1013	25	17
Malawi	Mid-altitude	Chitedze	1949–2008	897	22	19
		Kasungu	1947–1998	797	23	21
	Low-altitude	Chitala	1947–1998	890	23	21
Mozambique	Manica Province	Sussundenga	1969–2005	1168	24	22
		Chimoio	1951–2011	1081	24	21
	Tete Province	Dedza (proxy for Angonia)	1958–1998	950	20	19
Zimbabwe	Semi-arid tropics	Bulawayo	1939–2007	577	22	24

and unimodal seasons at Mbulu (41%), Kasungu (37%) and Sussundenga (35%) suggest more frequent water deficits in maize crops. This is particularly the case at Kasungu and Sussundenga where deep (1.5 m), light textured sandy soils are more common. At the majority of sites, heavier textured clay loams to light clay soils predominate, with maize rooting depths of 0.9 to 1.5 m.

2.3 Maize management and crop performance

For the countries under consideration, maize is generally the most important crop for farmers' investment choices except in Ethiopia, where tef (*Eragrostis tef*), an indigenous species, is an equally important cash crop and food

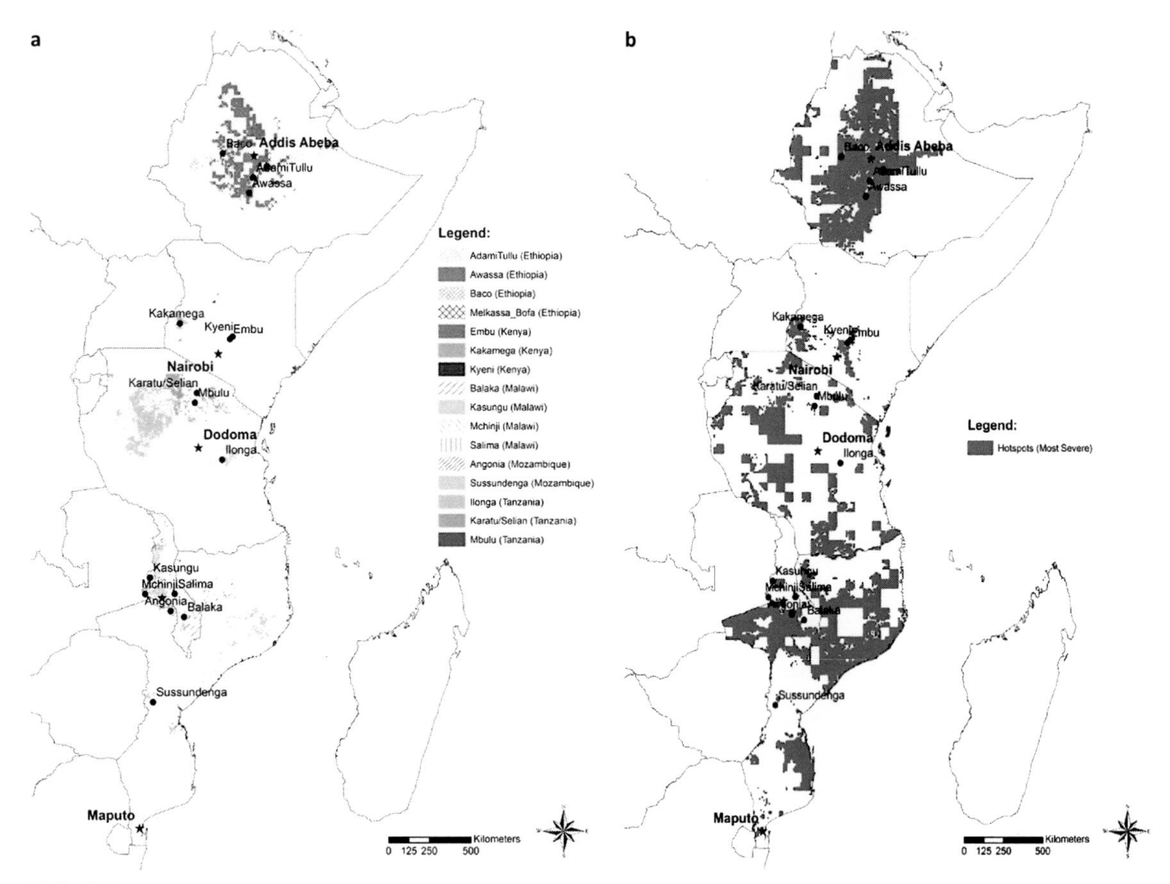

FIG. 5.1 (a) similarity of monthly temperature and rainfall patterns. (b) most severe clustering of food insecurity based on yield gap data (1999–2001) in 5 countries of eastern and southern Africa. Please see color plate at the back of the book.

staple. Even so, average maize yields across the sub-region are typically 1–2 t ha^{-1}, less than half the estimated attainable yield of 3–4 t ha^{-1} (Koo, 2012). Improved germplasm occupies less than 25% of the maize area and, although hybrids have been developed in the region since the 1960s, farmers' uptake has been low (Denning et al., 2009). Farmers' investment in soil fertility management has been even more limited (Wichelns, 2006), despite the widespread distribution of soils having inherently low soil organic matter content and associated low soil nutrient supply. Poor legume yield limits the contribution of N biological fixation (Giller and Cadisch, 1995; Giller et al., 2009) and farmyard manure is mostly of low nutrient content (Probert et al., 1995, 2005). Maize crops typically experience substantial weed competition due to labor shortages delaying and restricting weeding frequency – some crops will be weeded only once, most are weeded twice, up to flowering. There is generally substantial weed biomass in fields at maize maturity and dry season weed control is almost non-existent. In relation to the in-crop rainfall (Fig. 5.2), farmers' current agronomic practices result in chronically poor

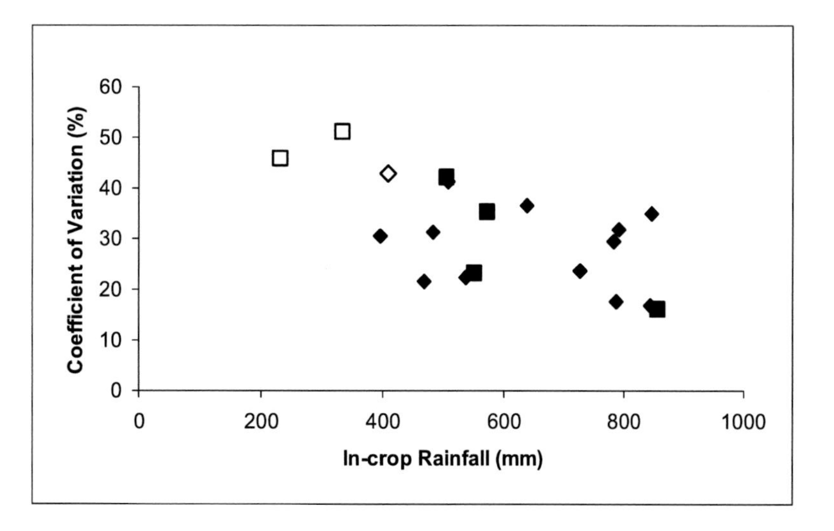

FIG. 5.2 Average in-crop rainfall and its coefficient of variation for the 16 site-seasons (solid symbols) examined in this study; semi-arid environments (open symbols) are shown for comparisons. Bimodal (squares) and unimodal (diamonds) cropping seasons are represented.

efficiency in the use of both rainfall and labor (Rockstrom et al., 2009).

3 SUSTAINABLE INTENSIFICATION OF SUB-SAHARAN AGRICULTURE

3.1 A stepping-stone approach for adoption of complex technological packages

Increasingly, conservation agriculture (CA) (Wall et al., 2013), defined as minimal or no soil disturbance, maintenance of a minimal soil cover (30% ground cover) and incorporating crop rotation, particularly with legumes, is being proposed as the basis of intensifying smallholder agriculture in Africa. Although its applicability in the African context has been contested by some (Giller et al., 2009), it does address the widespread issue of soil sustainability that was highlighted in the Introduction. The principles of CA have extremely wide applicability, being practiced on a wide range of environments and cropping systems (Wall, 2007). However, CA is knowledge intensive, equipment and expertise sensitive and requires high cash inputs including herbicides, seed of improved varieties and fertilizers (Wall, 2007; Thierfelder and Wall, 2011). For a comparison, Chapter 3 outlines the experience in the deployment of conservation agriculture in China which shares a dominance of small-holdings with Africa, albeit in a context of more developed infrastructure and better opportunities for technology adoption.

While significant energy and cost savings have been the basis of wide-scale adoption in mechanized systems, these are not so readily available in Africa's largely manual-based cropping systems. Further, farming systems in developed world agriculture are geared to exploiting the gains in soil moisture obtainable under CA (Harrington and Erenstein, 2005; Wall, 2007), whereas this is not the case in Africa's low input cropping systems where water productivity is perennially low (Rockstrom et al., 2009). Promotion of CA in Africa represents significant practice change for farmers in terms of their weed control and use of crop residues (Erenstein, 2002; Wall, 2007). It also assumes the financial capacity of farmers, along with well-developed market access and incentives, to invest in the requisite technologies of improved seed, herbicides and fertilizer. As Bolliger et al. (2006) noted, CA practices offer huge advantages for farmers with sufficient capital and opportunity, but this is a 'fortunate few' farmers in sub-Saharan Africa.

There are further technical issues affecting adoption rates of CA in Africa. The most reliable effects of CA (reduced soil loss and increased soil organic matter content) and benefits to farmers' yields accrue with long delays – mostly longer than 5 years (Derpsch, 2005; Erenstein et al., 2012). Giller et al. (2009) pointed out that farmers need immediate returns to investment when considering adoption of CA practices and therefore the long-term benefit is a major hurdle for adoption. More recently, Edmeades et al. (2012) observed that, in general, uptake of CA techniques has been quite limited in Africa, primarily because of competing uses for stover as forage, low stover production, external input requirements and lack of suitable machinery or machinery services. Earlier studies have suggested that CA increased labor inputs for weed control when herbicides are unavailable (Muliokela et al., 2001) and Andersson and Giller (2012) point out that, under these circumstances, CA shifts the peak labor demand from sowing to weeding.

Conservation agriculture does not work if the residues are not retained (Theirfelder and Wall, 2011). Hence the strong interdependence between reduced tillage and mulching becomes an important constraint to the adoption of residue retention practices in Africa given the overwhelmingly low N status of most maize-cropping systems. It has been acknowledged that better N management is required particularly in the first years of conversion from tillage-based agriculture to CA (Theirfelder and Wall, 2011). Wall (2007) estimated that an additional 20 kg N ha^{-1} needs to be applied to CA crops to offset the reduction in available nitrogen in untilled soil, although this would decline as soil organic matter and N mineralization increases. However, Africa is chronically low in fertilizer use, with an average annual below 10 kg N ha^{-1} (Wichelns, 2006), despite the generally infertile soils.

Given the large diversity in levels of farmers' endowment and livelihood strategies and the multiplicity of existing trade-offs (Giller et al., 2011), simplification and integration at the household or farming system is required. Here we propose that, depending on farmers' present performance and investment, three basic steps could be identified together with supporting options for farmers to engage and participate in markets to sustainable intensity productivity, increase food production and reduce poverty.

3.1.1 *Step 1: Improving agronomic practice*

The agronomic output of under-performing growers ('hanging-in farmers', Dorward et al., 2009) needs to be improved. This can be achieved by working with smallholder farmers so that they practice, learn and improve agronomic practices including the use of improved or good quality seed, improved sowing techniques, better match plant densities and arrangements to local conditions (rainfall, soil type, slope), the preparation and use of manures and composts using local ingredients, and in-crop weed control. These are just a few simple agronomic practices that would require no additional investment and are likely to lift yields and reduce yield variability and risk for approximately 30–40% of the population of farmers across eastern and southern Africa. Similar observations were made by Edmeades et al. (2012) who recommended that emphasis should be placed on promoting component technologies that provide increases in productivity and income in the short term, compared to technologies that are expected to provide benefits in the long term.

3.1.2 *Step 2: Increasing farmers' investment*

This step involves working with the better performing farmers ('stepping up farmers', Dorward et al., 2009) to evaluate the benefits and trade-offs (e.g. risks) from alternative investments given a limited availability of resources such as fertilizers, herbicides, stubble from previous crops. Gradually incorporating more CA component technologies might be an option for this group of farmers. This approach is likely to provide significant benefits to approximately 60–70% of the farming population across eastern and southern Africa.

3.1.3 Step 3: Identifying and supporting transformational changes and engagement with markets

Step 3 consists of supporting leading farmers, i.e. those that already apply significant investment in their production system. This will include the promotion of collective bargaining arrangements that improve access to input and output markets, explore options to access credit to invest in transformational changes, for example, machinery, new and more innovative practices and specialization in crops and livestock activities, identifying new products or value adding to existing products that are likely to increase returns with minimal additional risk. Even though this is likely only to provide significant benefits to 5–10% of the farming population, this fraction of the population can be expected to have the required capital and motivation to initiate small agribusinesses that will benefit larger numbers of farmers.

4 METHODS

4.1 Participatory approaches to identify relevant and actionable interventions in stepping stone approaches: Case studies from Tanzania and Mozambique

Following learnings from previous works (Carberry et al., 2004; Whitbread et al., 2010), surveys and participatory crop modeling workshops were conducted with farmers from three villages, one in Tanzania (Mandela), and two in Mozambique (Marera and Rotanda). Yields of crops from 13 recent cropping seasons (1999–2011 in Mandela, 2001–2013 in Mozambique) were collected – i.e. farmers' recall. At Mandela, 10 farmers provided a total of 90 maize yield estimates (five farmers all 13 seasons, three farmers, 5–7 seasons and two farmers, 4 and 3 seasons). At Marera, six farmers provided 67 maize yields (four farmers, 13 seasons) and at Rotanda, seven farmers provided 74 estimates (four farmers, 13 seasons). At each site, farmers cropped within an area no further away from each other than 10 km, therefore farmers' yields could be related to in-crop rainfall records from nearby weather stations for each of the seasons, i.e. Ilonga Research Station, 50 km from Mandela, Chimoio airport, 30 km from Marera and Rotanda Met station at a nearby village.

APSIM (Keating et al., 2003) was then used to simulate maize yields using representative management and soil descriptions obtained from farmers' resource allocation maps (Defoer and Budelman, 2000) to parameterize the model. At Mandela, we used farmer's management applied in 2011 to a maize field to calibrate the model to the farmer's yield achieved in that season. The 2011 management was then used to simulate previous seasons back to 1999; this means we tested the same crop management across seasons and assumed that no significant changes have occurred in crop management during that period. Parameters for an improved open pollinated variety were used in all cases. Yield distributions from farmer surveys and model outputs were compared to evaluate the performance of the model. Similar steps were adopted at Marera and Rotanda (data not shown).

4.2 Scaling-out stepping-stone approaches for the sustainable intensification of agriculture across contrasting environments in eastern and southern Africa

Given the large environmental variation (Waddington et al., 2010), APSIM was used to evaluate maize responses to alternative management interventions for 19 growing environments in Ethiopia, Kenya, Tanzania, Malawi, Mozambique and Zimbabwe. To assess farmers' investments in a stepping-stone approach, five management interventions were simulated:

1. Weed competition removed through effective use of knock-down and pre-emergent herbicides.
2. Application of a low rate of N fertilizer as starter (50 kg of 12:24:12 NPK ha^{-1}), and top-dressing of nitrogen at 9 leaves (25 kg urea ha^{-1}).
3. Application of a recommended rate of N fertilizer as starter (100 kg ha^{-1} of 12:24:12 NPK), and a top-dressing of nitrogen at 9 leaves (75 kg urea ha^{-1}).
4. Application of maize residues as a mulch (1500 kg ha^{-1}) with moderate N content (0.66%N, C:N ratio = 60:1) at sowing, and the low rate of N fertilizer. This amount of residue provides approximately 40% groundcover at sowing, and decomposes during the 120-day crop cycle to less than 100 kg ha^{-1} at maturity.
5. Application of maize residues (1500 kg ha^{-1}) with a moderate N content at sowing and the recommended rate of N fertilizer.

Fixed parameters included maximum rooting depth = 120 cm, maximum plant available water = 108 mm, soil organic carbon in the 0–10 cm layer = 1.1%, plant population = 4 plants m^{-2}, row spacing = 75 cm and crop duration = 120 d. For each season and site, maize was sown on the earliest rainfall opportunity within nominated sowing windows. Currently, farmers favor improved open pollinated varieties (OPV) and even hybrids, with less frequent use of landraces. To capture the interaction of cultivar with management and rainfall, three sets of genetic parameters related to yield potential were used. Maximum grain number per ear was set to 400 for landrace, 450 for OPV and 550 for hybrid. Maximum grain growth rate was set to 6, 8 and 9 mg $grain^{-1}$ d^{-1}, respectively. These have been married to four sets of crop growth and development coefficients required for constant crop duration across the temperature and stress conditions of the various sites (Dimes et al., 2013).

5 RESULTS

5.1 Model performance

Figure 5.3 compares the actual and simulated frequency distribution of maize yields for the period 1999–2011 at Mandela, Tanzania.

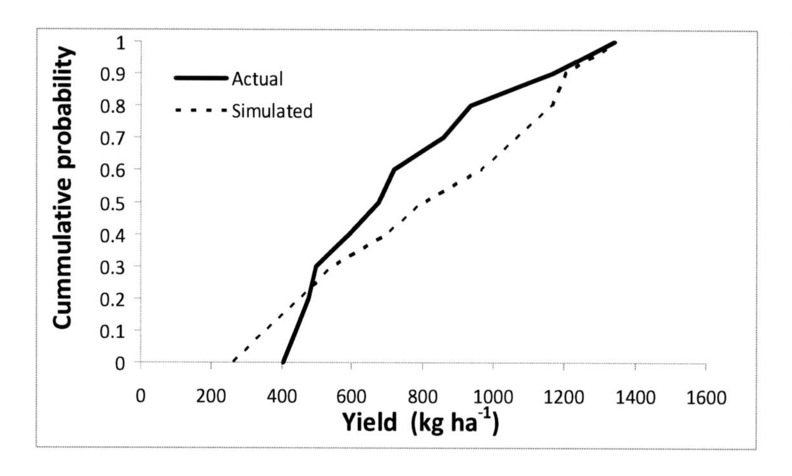

FIG. 5.3 Maize yield distributions derived from farmer surveyed data and simulated yields using APSIM for the period 1999 to 2011 at Mandela, Tanzania.

Simulated and observed yields were also discussed with the participating farmers (Fig. 5.4). At Mandela, farmers agreed that the simulated maize yield was very close to the farmer's yield in 2011, i.e. the year used for model calibration, and that the average for the field across years (solicited during the discussion, see asterisks in Fig. 5.4) was also close to the average of the simulated yields. Disagreements between simulated and actual yields were observed for some seasons. Possible causes of disagreement were discussed with the farmer, in particular for 1999, a year affected by a severe drought. Using the management described for 2011, the model simulated a higher than observed yield in 1999, and yield failure in 2000 (Fig. 5.4). The survey showed that two farmers from the same village recalled higher yields in 1999 than in 2000, while another two farmers reported the reverse, and one other farmer reported no difference between these seasons. We concluded that the discrepancy between the simulated and the farmer's recalled yield in 1999 was probably management related (e.g. unfavorable sowing date, poor weed control, etc.).

Figure 5.4 also shows modeled results for the alternative management practices where 50 kg urea ha^{-1} was applied in combination with two in-crop weeding events and a single weeding event scenario with no fertilizer applied. The combination of two in-crop weeding events and the application of urea increased maize yield across seasons by 140% compared to farmers' practice. In contrast, when only a single in-crop weeding event was simulated (indicated as a practice for some by both farmers and extension officers during the meeting) maize yield was zero in 4 seasons, and about 75% lower than farmers' practice across seasons.

Before presenting results, the farmer was asked to estimate the yield in 2011 if he had applied the urea fertilizer. His response of between 2000 and 2250 kg ha^{-1}, (8–9 bags $acre^{-1}$ in Fig. 5.2) was in agreement with the simulated 2250 kg ha^{-1}. This

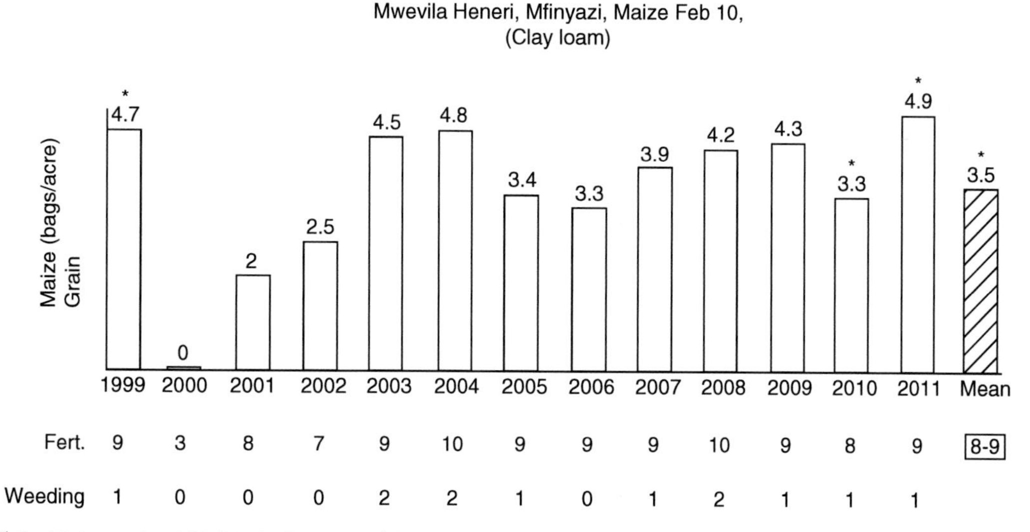

FIG. 5.4 Maize grain yield (100-kg bags $acre^{-1}$, bar chart) simulated for farmer's field and management at Mandela. Numbers at the bottom show simulated yield (bags $acre^{-1}$) for two alternative management practices: +Fert: 50 kg urea ha^{-1}, and one weeding in crop. Management related to an OPV sown on February 10, 2011 in a clay loam soil, with manual weed control on March 10 and 31 and no applied fertilizer or manure. The asterisks indicate yield data obtained from the farmer after simulations were performed (except in 2011 which was obtained initially from the farmer and used to calibrate the model soil inputs). The cross-hatched bar is the mean yield of 13 seasons simulated.

suggested that these farmers understood the yield benefits of fertilizer, but chose not to use it given cash constraints at the time of buying.

5.2 Farmers' performance and stepping-stone approaches for the sustainable intensification of agriculture

Figure 5.5 shows the relationship between farmers' yield estimates and seasonal rainfall for Mandela, Marera and Rotanda. Farmers at Mandela and Marera did not use fertilizer, whereas some growers used fertilizers in Rotanda. A farmer from Rotanda indicated seasons where he used and did not use fertilizer (highlighted in Fig. 5.5).

Mandela in Tanzania had lower cropping season average rainfall (607 mm) and large season-to-season variability (356 to 1033 mm). Marera and Rotanda in Mozambique had similar cropping season rainfall (i.e. 758 mm and 774 mm, respectively) and the lowest cropping season rainfall was in excess of 500 mm.

Fifty-six percent of all the yield estimates (n = 231) in Figure 5.5 were below 1000 kg ha^{-1}, and 73% showed rainfall water-use efficiencies (i.e. yield per unit seasonal rainfall) below 2 kg ha^{-1} mm^{-1}. Depending on cropping areas by individual landholders, such low water productivity is a major factor contributing to food insecurity. The highest rainfall-use efficiency was 9.5 kg ha^{-1} mm^{-1} and was achieved at Rotanda where fertilizer use is common. The benefits of fertilizer use on rainfall-use efficiency are apparent also from the farmer that did use fertilizer in some seasons (2.1–4.7 kg ha^{-1} mm^{-1}) but not in others (1.2–2.3 kg ha^{-1} mm^{-1}).

Yield–rainfall relationships in unfertilized crops on very similar soils in the same village were highly scattered (Fig. 5.6a). It is interesting to note that there are three distinct groups of farmers under the same agroecological conditions: (1) those who consistently achieve up to 4.1 kg grain ha^{-1} mm^{-1}; (2) those that consistently fail to produce a crop; and (3) those that consistently achieve average yields. This highlights the

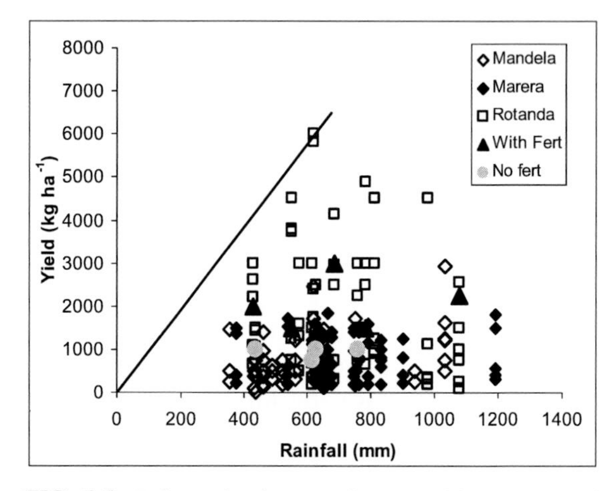

FIG. 5.5 Relationship between farmer yield estimates and rainfall for up to 13 consecutive cropping seasons at Mandela (eastern Tanzania), Marera and Rotanda (Mozambique). One farmer at Rotanda indicated seasons with and without fertilizer use. The line represents the maximum rainfall use efficiency (9.5 kg ha^{-1} mm^{-1}).

benefits to be obtained from improving the performance of the poor and average performing farmers and from working with the better farmers to increase investment in inputs and manage any associated risk.

Figure 5.6b shows the simulated maize yields using farmers' management at Mandela including open pollinated varieties, allowing for weed competition, and lack of fertilizers. In these simulations, we used recommended plant populations and row configurations. The envelope lines are for the highest and 'best of the rest' efficiencies drawn in Figure 5.6a.

As seasonal rainfall at Mandela extends to below 400 mm (Fig. 5.6), maize yields would benefit from moisture conservation practices such as mulching. However, application of maize residues with low N content depressed simulated yields significantly (Fig. 5.7), due to immobilization of soil N associated with decomposition of residues. In contrast, the same amount of legume residues (C:N ratio = 20:1) increased maize yields to above 2000 kg ha^{-1} in all seasons due to the input of organic N in these

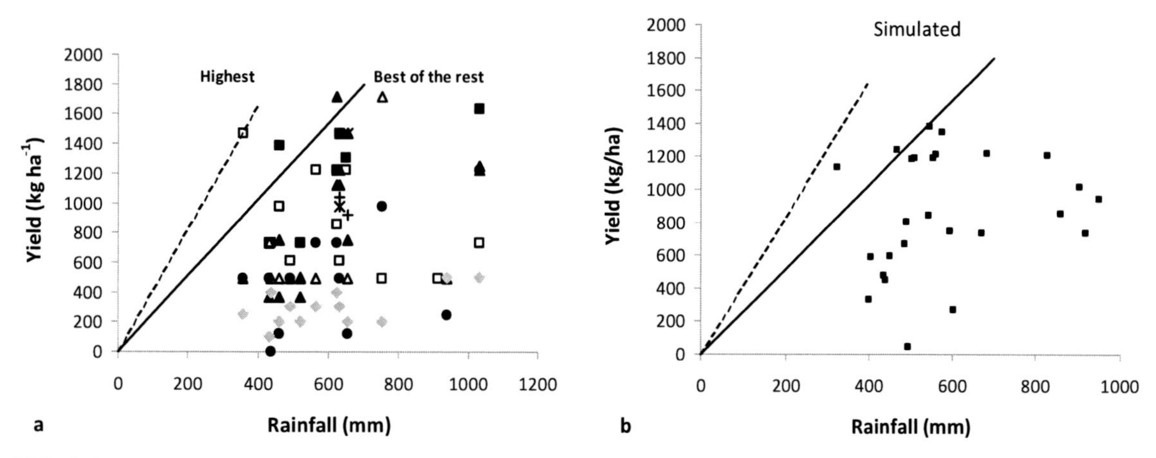

FIG. 5.6 Relationships between yield and seasonal rainfall at Mandela, eastern Tanzania. (a) Farmer yield estimates (1999–2011, 10 farmers), (b) simulated yield (1986–2011). The lines represent the boundary functions for the relationship between grain yield and annual rainfall for the best farmer, and for the best of the remaining farmers. The rainfall water-use efficiencies are 4.1 kg ha^{-1} $mm^{-1}$for highest and 2.5 kg ha^{-1} mm^{-1} for 'best of the rest'.

unfertilized maize crops. Application of 80 kg N ha^{-1} along with maize residues further increased yield and rainfall-use efficiency (Fig. 5.7), also showing benefits from moisture conservation during the driest seasons.

With a simulated yield ceiling of approximately 4 t ha^{-1}achieved with maize residue and N fertilizer, the average rainfall-use efficiency was 7.1 kg ha^{-1} mm^{-1} (Fig. 5.7). At Rotanda, maximum yield was 6 t ha^{-1} and rainfall-use efficiency of 9.5 kg ha^{-1} mm^{-1}. What are, therefore, the management steps to increase yield and rainfall-use efficiency to match those at Rotanda?

Simulations indicate that weed control with herbicides could increase yield from 860 kg ha^{-1} (baseline) to 1670 kg ha^{-1} and rainfall-use efficiency from 1.6 to 3.1 kg ha – $^{-1}$ mm^{-1} (Fig. 5.8). In the absence of weed competition, hybrid seed can increase the yield and efficiency by a further 32% (1950 kg ha^{-1}, 3.6 kg ha^{-1} mm^{-1}), but a small dose of fertilizer (20 kg N ha^{-1}) with open pollinated variety increased simulated yield and productivity by more than 50% (2620 kg ha^{-1} and 4.8 kg ha^{-1} mm^{-1}) compared to open pollinated variety without fertilizer.

If cash resources are not a constraint, a high dose of fertilizer (180 kg N ha^{-1}) with hybrid seed would more fully exploit the potential of the Mandela environment (yield mostly greater than 6000 kg ha^{-1} and rainfall-use efficiency = 10.7 kg ha^{-1} mm^{-1}, Fig. 5.8). However, unlike any of the

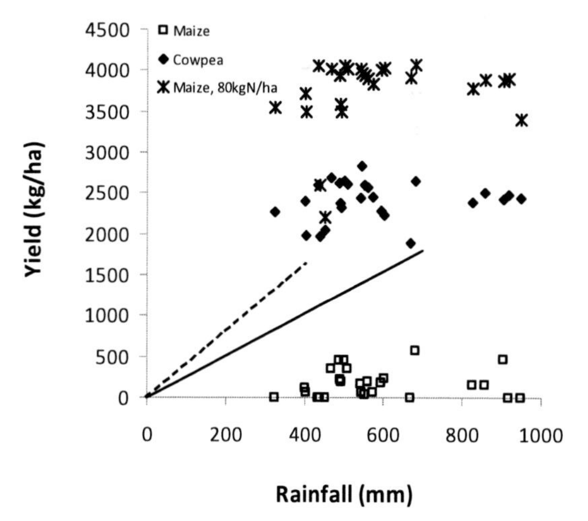

FIG. 5.7 Relationship between simulated maize yield and seasonal rainfall at Mandela for baseline farmer management. Comparison of yield responses to residues with low C:N ratio and no fertilizer (maize, C:N ratio = 80:1), high C:N ratio (cowpea, C:N ratio = 20:1) and high C:N ratio plus fertilizer (maize + 80 kg N ha^{-1}) Lines have slopes of 4.1 (dashed) and 2.5 (solid) kg ha^{-1} mm^{-1} as derived from Figure 5.6a.

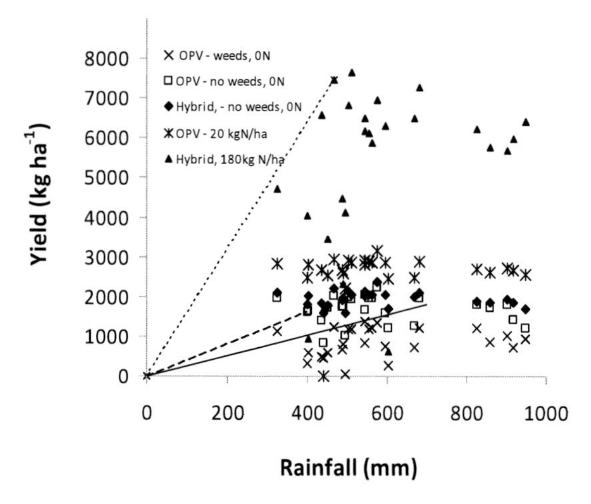

FIG. 5.8 Relationship between simulated maize yield and seasonal rainfall at Mandela for open pollinated varieties (OPV) and hybrids with and without weed competition and varying rates of N fertilizer (0, 20 and 180 kg N ha^{-1}). Lines have slopes of 4.1 (dashed), 2.5 (solid) and 15.9 (short dash) kg ha^{-1} mm^{-1}.

previous interventions, there are also seasonal yields much less than 6000 kg ha^{-1} and as low as 600 kg ha^{-1} in a 600 mm season, indicating higher yield variability and production risk at this level of investment.

In summary, the simulations showed that small additional investments (small additions of N fertilizer or legume residues, improved weed control) can improve the productivity of farmers while maintaining yield stability across the seasonal conditions at Mandela. Higher productivity is possible with higher investments (hybrid seed and high fertilizer N) but at the expense of increasing seasonal variability of yield and therefore returns on investment.

5.3 Scaling out the stepping-stone approach for the sustainable intensification of agriculture across contrasting agroecologies in eastern and southern Africa

5.3.1 Maize yield responses across rainfall environments

Maize response to yield improvement technologies interacts strongly with rainfall amount and distribution during the growing season as demonstrated by simulation of the high investment strategy at Mandela. Here, we examine how the step-wise uptake of crop management technologies applied to landrace, OPV and hybrid maize interacts solely with rainfall patterns (i.e. soil, plant population and spacing and crop duration held constant) across a range of semi-arid to wet humid tropical environments in eastern and southern Africa. The baseline management is maize OPV with weed competition, no fertility inputs and no surface residues.

The average yield across sites simulated for the baseline crop management was 630 kg ha^{-1} for the landrace[1], 940 kg ha^{-1} for the OPV and 1260 kg ha^{-1} for the hybrid (Fig. 5.9). These are in line with the yields reported by farmers at Mandela and Marera, where fertilizer use was almost non-existent. To be expected, magnitude of the maize yield response to improved weed control, fertilizer and mulch was much less in the semi-arid environments (Katumani and Bulawayo) compared to the more humid sites – approximately 40% less at the highest yield.

The average percentage increase above the baseline yields for the technology interventions was relatively constant across the three cultivars, except for the hybrid with N inputs, where proportional increases were lower because of the higher baseline yield and/or interactions with water supply limiting yield (Table 5.2). The interaction with water supply was more apparent in the drier environments where the proportional responses of the hybrid to N inputs are much lower than those of the lower yielding

[1]For this analysis, yields of the unimproved landrace were underestimated because a grain growth rate normally associated with longer crop duration (typically 140 days or longer for landraces) was married to a shorter crop cycle. This was necessitated by operational aspects of simulating the bimodal cropping systems as well as for control of the rainfall and crop growth period sampled across sites. Hence the results of the landrace here are to show how a low-yielding cultivar will respond to the yield improvement technologies and should not be used to compare the yield benefits of the improved seed.

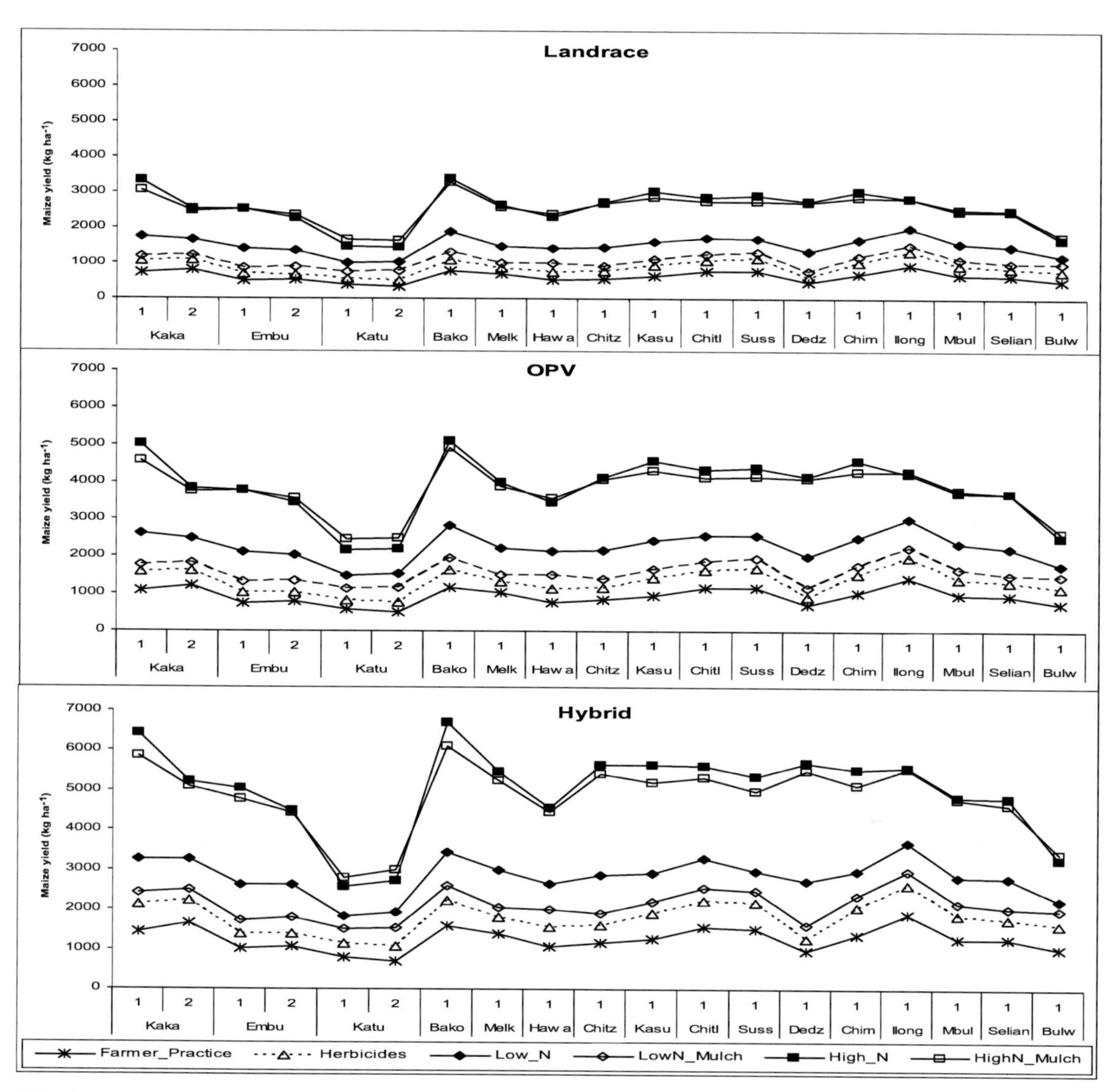

FIG. 5.9 Simulated average maize yields for 19 site-seasons using landrace, improved OPV and hybrid germplasm descriptions and their response to increasing inputs – herbicides to eliminate weed competition, N fertilizer and maize residues. 1 & 2 refers to seasons in bimodal rainfall environments. The baseline management is weed competition, no fertility inputs and no crop residues. Low N is 18 kg N ha^{-1}, high N is 81 kg N ha^{-1}, mulch = 1500 kg maize stubble ha^{-1} at sowing.

cultivars. On the other hand, eliminating weed competition in the drier climate was more beneficial (52% increase) compared to the wetter climates (40% increases). More notably, the proportional yield response to low N rate was higher in the drier (160%) than in the wetter environments (140%). This is explained by the interaction of water supply limiting soil N mineralization in the drier environments, thereby exacerbating the crop nitrogen deficit. The application of small N doses is therefore more beneficial in overcoming the N constraint in these environments.

TABLE 5.2 Average yield change (%) relative to baseline management for maize cultivars and technology interventions in humid tropic and semi-arid environments

	Environment					
	Humid tropic			Semi-arid		
Technology	**Landrace**	**OPV**	**Hybrid**	**Landrace**	**OPV**	**Hybrid**
Herbicides	40	40	41	52	52	52
Low N	139	139	124	162	163	140
High N**	314	315	304	277	277	243
Low N + Mulch	66	66	65	106	107	102
High N + Mulch	305	305	285	318	316	268

The baseline management is weed competition, no fertilizer and no surface residues. Low N is 18 kg N ha^{-1}, High N is 81 kg N ha^{-1}, mulch = 1500 kg maize stubble ha^{-1} at sowing.

*** Note the declining marginal response to fertilizer use – the high N rate is 4.5 times the input of the low rate, while the yield advantage is less than 3.2 times the yield of the baseline management (0N inputs).*

Overall, the tested technologies increased simulated maize yields across the environments and the biggest source of the yield gain was nitrogen fertilizer. However, with low nitrogen, even in the drier semi-arid environments, maize residues reduced crop yields due to N immobilization. With high nitrogen input, there were small improvements due to moisture conservation at the semi-arid sites but these were largely absent (on average) in the more humid locations.

6 DISCUSSION AND CONCLUSIONS

High water–nitrogen co-limitation is associated with smaller yield gaps and higher water-use efficiencies in cereal crops in Mediterranean environments (Sadras, 2004; Cossani et al., 2010; Chapter 7). The basis of this concept is the work of Bloom et al. (1985) in developing the analogy that efficient use of resources in plant systems is akin to efficient allocation of capital resources in the economic sphere. The principle is that plant growth is maximized when resource limitations are balanced. For rain-fed systems targeting high production and rainfall-use efficiency, this principle implies that nitrogen inputs need to be managed carefully to match water availability during crop growth cycle. Sinclair and Park (1993) also drew upon the work of Bloom et al. to explain why Liebig's Law of the Minimum does not apply at higher production levels.

The cropping systems analyzed by Sadras (2004), Cossani et al. (2010) and Chapter 7 typically have higher resource inputs and technical efficiencies (smaller yield gaps) than those of Africa's smallholder maize cropping systems (Carberry et al., 2013). The water–N co-limitation index ranges from 1 when both stresses are balanced to zero when a single stress dominates (Box 7.1 in Chapter 7). We have not explored co-limitation, but our results suggest that for the humid tropic sites in this study the water–N co-limitation index would be close to zero, i.e. crop growth is limited almost entirely by nitrogen availability. As Sinclair and Park (1993) concede, the limiting factor paradigm is applicable and useful where an extreme and obvious stress decreases yield. Hence, in the context of agricultural development in Africa's low input low output cropping systems, Liebig's Law of the Minimum does retain high relevance.

The surveys at Mandela and Marera demonstrate the extent of farmers' current low maize yields and rainfall-use efficiency in these wet environments. The benefits of fertilizer in raising efficiency are evident in the surveys from Rotanda where fertilizer use is more common, and in the simulations for Mandela. The broader simulation analysis across rainfall environments shows that, even in drier areas, maize will respond strongly to small N inputs in soils with organic carbon of 1.1% or lower. This is consistent with the empirical findings of Twomlow et al. (2010) in the semi-arid regions of southwest and western Zimbabwe.

The modeled results suggest that, in these environments, soil N supply mediated by biological activity is limited by the frequent drying of the soil surface layers. Irrespective of germplasm and rainfall, application of small doses of N fertilizer is likely to result in the largest payoff to farmers. This was the basis of the small dose research reported by Twomlow et al. (2010) for the semi-arid tropic and these results suggest that it is applicable to 'hanging in' farmers' fields in higher rainfall environments as well. However, improved weed management (and in many instances, plant stands) would need to be a first step, else the benefits of small fertilizer applications would be compromised. This is consistent with feedback from on-farm trial results, where farmers have responded to the labor saving and convenience of herbicides. Results reported here indicate that a 40% yield advantage could be possible if weed competition is largely eliminated in these N-starved cropping systems. Training of extension officers and farmers on effective use of herbicides and more widespread promotion of pre-emergent weedicide would be a priority in scaling-out this component of conservation agriculture.

Although crop residue retention improves soil organic matter in the long term and protects the soil against erosion, maize crop residues also have the capacity to immobilize soil nitrogen in the short term. The recent review by Grahmann et al. (2013) concluded that conservation agriculture has lower nitrogen-use efficiency than conventional systems, hence the need to adjust fertilization. The promotion of smaller doses of fertilizer without residues in Africa can be based on a number of logical arguments in pursuit of longer-term adoption of crop residue retention. First, resource-poor farmers are reluctant to invest in fertilizer; hence smaller doses will be more acceptable as a first step. Secondly, smaller doses of nitrogen and residue retention are incompatible in these N-poor cropping systems. Such negative effects and resultant lower returns on the fertilizer investment will discourage farmers from investing in this needed technology. Lastly, in low input mixed farming systems, stover production is low and there are competing uses for stover (Baudron et al., 2013). High yielding crops are therefore necessary to return adequate amounts of stover. Hence, it is logical that increasing the use of mineral fertilizers in Africa is the essential first step to wider adoption of residue retention. The alternative source of nitrogen would be biological fixation, but Africa's smallholder farmers seemingly have even less incentive to invest in legumes as cropped areas are much less and yield gaps even larger than those of maize.

Improved cultivars could raise farm production but, in the absence of fertilizer inputs, productivity gains will be limited (Fig. 5.8). Also, in rain-fed systems with yield above 3 t ha^{-1}, the returns to higher investments become more uncertain across the range of seasonal rainfall in the drier environments (e.g. Mandela, Fig. 5.8, Table 5.2). For resource-poor farmers, hybrid seed is expensive, increases production risk and produces crops with high harvest index. Accordingly, maize hybrids would occupy a more distant role in intensification of current production systems and as a stepping-stone in the scaling-out of conservation agriculture in Africa (Fig. 5.10).

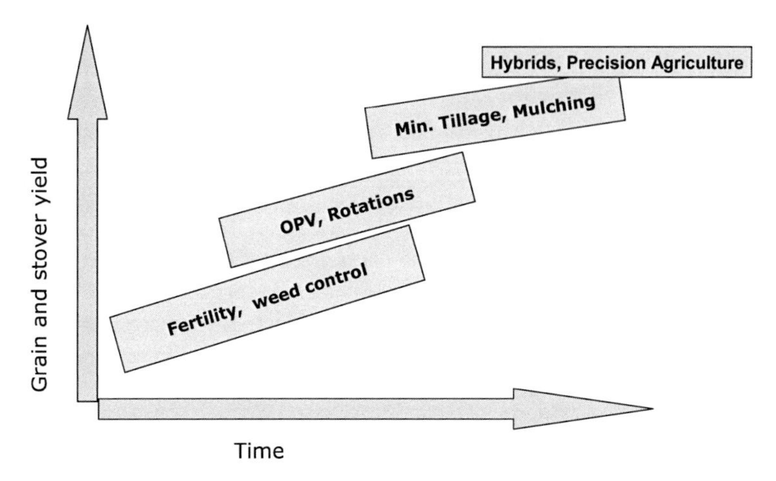

FIG. 5.10 A possible technology sequence for testing with farmers for step-wise intensification of maize production systems in Eastern and Southern Africa.

Our analysis points to nitrogen inputs as the crucial step to raising the productivity of maize production in eastern and southern Africa. While benefits of conservation agriculture for crop production take up to 5 years and longer to emerge under well-managed conditions (Derpsch 2005; Erenstein et al., 2012), the FAO statistics reported by Tittonell and Giller (2013) show that fertilizer use in Africa has been almost static since the 1960s. Therein lays the real intensification challenge for most of sub-Saharan Africa because, as Keating et al. (2010) pointed out, unless ways are found to relieve the soil fertility constraint in Africa, efficient use of other natural and human resources will remain low.

Acknowledgements

This work was part of the project "Sustainable Intensification of Maize-Legume Systems for Food Security in Eastern and Southern Africa (SIMLESA)", funded by Australian Centre for International Agricultural Research (ACIAR). Daniel Rodriguez and Andries Potgieter are supported by the Queensland Alliance for Agriculture and Food Innovation (QAAFI), The University of Queensland, Australia.

References

Andersson, J.A., Giller, K.E., 2012. On heretics and God's blanket salesmen: contested claims for Conservation Agriculture and the politics of its promotion in African smallholder farming. In: Sumberg, J., Thompson, J. (Eds.), Contested agronomy: agricultural research in a changing world. Earthscan, London.

Baudron, F., et al., 2013. Conservation agriculture in African mixed crop-livestock systems: Expanding the niche. Agric. Ecosyst. Environ. (in press). http://dx.doi.org/10.1016/j.agee.2013.08.020.

Bloom, A.J., Chapin, F.S.I., Mooney, H.A., 1985. Resource limitation in plants – an economic analogy. Annu. Rev. Ecology. Systemat. 16, 363–392.

Bolliger, A., Magid, J., Amado, T.J.C., et al., 2006. Taking stock of the Brazilian 'zero–till revolution': A review of landmark research and farmer's practice. Adv. Agron. 91, 49–110.

Carberry, P., Gladwin, C., Twomlow, S., 2004. Linking simulation modelling to participatory research in smallholder farming systems. In : Delve, R.J., Probert, M.E., (Eds), Modelling nutrient management in tropical cropping systems. ACIAR Proceedings No. 114, Canberra, Australia, pp. 32–46.

Carberry, P.S., Liang, W.L., Twomlow, S., et al., 2013. Scope for improved eco-efficiency varies among diverse cropping systems. Proc. Natl. Acad. Sci. USA. 110, 8381–8386.

Cossani, C.M., Savin, R., Slafer, G.A., 2010. Co-limitation of nitrogen and water on yield and resource-use efficiencies of wheat and barley. Crop Past. Sci. 61, 844–851.

Derpsch, R., 2005. The extent of conservation agriculture adoption worldwide: implications and impact. In: Proceedings on CD. III world congress on conservation agriculture: linking production, livelihoods and conservation. Nairobi, Kenya, 3-7 Oct, 2005.

Dimes, J., Hassen, S., Abebe, T., et al., 2013. Rainfall variability and maize production risk across agro-ecologies of Eastern and Southern Africa. SIMLESA Project Final Report.

Defoer, T., Budelman, A., 2000. Managing soil fertility in the tropics. A Resource Guide for participatory learning and action research. Royal Tropical Institute, Amsterdam, The Netherlands.

Denning, G., Kabambe, P., Sanchez, P., Malik, A., Flor, R., Harawa, R., Sachs, J., 2009. Input subsidies to improve smallholder maize productivity in Malawi: Toward an African Green Revolution. PLoS biology 7 (1), e1000023.

Dorward, A., Anderson, S., Nava Bernal, N., et al., 2009. Hanging in, stepping up and stepping out: livelihood aspirations and strategies of the poor. Develop. Pract. 19, 240–247.

Edmeades, G., Shumba, E., Wandschneider, T., Dixon, J., 2012. Sustainable Intensification of Maize-Legume Cropping Systems for Food Security in Eastern and Southern Africa (SIMLESA). Mid-term Review Report to ACIAR.

Erenstein, O., 2002. Crop residue mulching in tropical and semi-tropical countries: an evaluation of residue availability and other technological implications. Soil Till. Res. 67, 115–133.

Erenstein, O., Sayre, K., Wall, P., Hellin, J., Dixon, J., 2012. Conservation agriculture in maize- and wheat-based systems in the (sub)tropics: lessons from adaptation initiatives in South Asia, Mexico, and Southern Africa. J. Sustain. Agric. 36, 180–206.

FAOSTAT, 2009. http://faostat.fao.org/default.aspx.

GEF. 2003. What kind of world? The challenge of land degradation. Global Environment Facility (GEF), p.4.

Giller, K.E., Cadisch, G., 1995. Future benefits from biological nitrogen fixation: An ecological approach to agriculture. Plant Soil 174, 255–277.

Giller, K.E., Witter, E., Corbeels, M., Tittonell, P., 2009. Conservation agriculture and smallholder farming in Africa: The heretic's view. Field Crops Res. 114, 23–34.

Giller, K.E., Tittonell, P., Rufino, M.C., et al., 2011. Communicating complexity: Integrated assessment of trade-offs concerning soil fertility management within African farming systems to support innovation and development. Ag. Syst. 104, 191–203.

Grahmann, K., Verhulst, N., Buerkert, A., Ortiz-Monasterio, I., Govaerts, B., 2013. Nitrogen use efficiency and optimization of nitrogen fertilization in conservation agriculture. CAB Rev. 8, No. 053.

Harrington, L., Erenstein, O., 2005. Conservation agriculture and resource conserving technologies – A global perspective. Agromeridian. 1, 32–43.

Herrero, M., Havlik, P., Valin, H., et al., 2013. Biomass use, production, feed efficiencies, and greenhouse gas emissions from global livestock systems. Proc. Natl. Acad. Sci USA 110, 20888–20893.

Keating, B.A., Carberry, P.S., Bindraban, P.S., Asseng, S., Meinke, H., Dixon, J., 2010. Eco-efficient agriculture: concepts, challenges, and opportunities. Crop Sci. 50 (Suppl. 1), S-109.

Keating, B.A., Carberry, P.S., Hammer, G.L., et al., 2003. An overview of APSIM, a model designed for farming systems simulation. Eur. J. Agron. 18, 267–288.

Koo, J. 2012. http://harvestchoice.org/labs/2 Accessed 13/3/2014.

Muliokela, S.W., Hoogmoed, W.B., Stevens, P., Dibbits, H., 2001. Constraints and possibilities for conservation farming in Zambia. In: Garcia-Torres, L., Benites, J., Martínez-Vilela, A. (Eds.), Conservation agriculture, a worldwide challenge. First World Congress on Conservation Agriculture. Vol. II: Offered Contributions. XUL, Cordoba, Spain, pp. 61–65.

Mulugetta, M., Dimes, J., Dixon, J., et al., 2011. The sustainable intensification of maize-legume farming systems in eastern and southern Africa (SIMLESA) program. *In:* 5th World Congress on Conservation Agriculture and Farming Systems Design, Australian Centre for International Agricultural Research, pp. 1-5.

Probert, M.E., Delve, R.J., Kimani, S.K., Dimes, J.P., 2005. Modelling nitrogen mineralization from manures: representing quality aspects by varying C:N ratio of subpools. Soil Biol. Biochem. 37, 279–287.

Probert, M.E., Okalebo, J.R., Jones, R.K., 1995. The use of manure on small-holders farms in semi-arid Kenya. Exp. Agric. 31, 371–381.

Rockstrom, J., Falkenmark, M., 2000. Semiarid crop production from a hydrological perspective: gap between potential and actual yields. Crit. Rev. Plant Sci. 19, 319–346.

Rockstrom, J., Kaumbutho, P., Mwalley, J., et al., 2009. Conservation farming strategies in East and Southern Africa: Yields and rain water productivity from on-farm action research. Soil Till. Res. 103, 23–32.

Sadras, V.O., 2004. Yield and water-use efficiency of water- and nitrogen-stressed wheat crops increase with degree of co-limitation. Eur. J. Agron. 21, 455–464.

Sinclair, T.R., Park, W.I., 1993. Inadequacy of the Liebig limiting-factor paradigm for explaining varying crop yields. Agron. J. 85, 742–746.

Thierfelder, C., Wall, P.C., 2011. Reducing the risk of crop failure for smallholder farmers in Africa through the adoption of conservation agriculture. In: Bationo, A. et al., (Ed.), Innovations as key to the Green Revolution in Africa. Springer, Amsterdam, The Netherlands, pp. 1269–1277.

Tittonell, P., Giller, K.E., 2013. When yield gaps are poverty traps: The paradigm of ecological intensification in African smallholder agriculture. Field Crops Res. 143, 76–90.

Twomlow, S., Rohrbach, D., Dimes, J., et al., 2010. Microdosing as a pathway to Africa's Green Revolution: Evidence from broad-scale on-farm trials. Nutr. Cycl. Agroecosyst. 88, 3–15.

UNEP/ISRIC. 1991. World map of the status of human-induced soil degradation (GLASOD). An explanatory note, 2nd edn). UNEP: Nairobi, Kenya, and ISRIC: Wageningen, Netherlands.

Waddington, S.R., Li, X.Y., Dixon, J., Hyman, G., de Vicente, M.C., 2010. Getting the focus right: production constraints for six major food crops in Asian and African farming systems. Food Secur. 2, 27–48.

Wall, P.C., 2007. Tailoring conservation agriculture to the needs of small farmers in developing countries: An analysis of issues. J. Crop Improve. 19, 137–155.

Wall, P.C., Thierfelder, C., Ngwira, A., Govaerts, B., Nyagumbo, I., Baudron, F., 2013. Conservation agriculture in Eastern and Southern Africa. In: Jat, R.A., Graziano de Silva, J. (Eds.), Conservation agriculture: global prospects and challenges. CABI, Cambridge, USA.

Whitbread, A., Robertson, M., Carberry, P., Dimes, J., 2010. Applying farming systems simulation to the development of more sustainable smallholder farming systems in Southern Africa. Eur. J. Agron. 32, 51–58.

Wichelns, D., 2006. Improving water and fertilizer use in Africa: challenges, opportunities and policy recommendations. Background paper prepared for the African Fertilizer Summit, Abuja, Nigeria, 9–13 Jun, 2006.

CHAPTER

6

Cropping systems in environments with high yield potential of southern Chile

Mario Mera[1,2], *X. Carolina Lizana*[3], *Daniel F. Calderini*[3]

[1]Instituto de Investigaciones Agropecuarias (INIA)
[2]Universidad de La Frontera, Temuco, Chile
[3]Universidad Austral de Chile, Instituto de Produccion y Sanidad Vegetal, Valdivia, Chile

1 INTRODUCTION

1.1 Food production in Chile, a national aim

The Chilean food sector is the second most important export area after the mining industry. International trade agreements with more than 50 countries have opened markets to over two billion people. Under this scenario, Chile aims to become one of the ten main food exporters of the planet, a Food Power. It is already among the top twenty food exporters and, more importantly, one of the fastest growing food exporters in the world. The sector increases the value of its exports by one billion US dollars annually, and duplicated the export worth during the last decade. The Chilean food sector represents 10.3% of the GNP, a figure only surpassed by New Zealand and Belgium. Different crops are produced in the continental territory of the country, which ranges from almost 18° to 53°S, including the driest desert in the world (the Atacama Desert) and places where rainfall surpasses 4000 mm year^{-1}. Tropical, subtropical and temperate crops are grown across these latitudes where complex cropping systems are developed in a wide diversity of agro-climatic conditions in an intricate network involving the agri-food industry and exporters. Southern Chile would contribute to this national aim through the production of a wide variety of food, including cereals, potato, oilseeds and protein grains, fresh and dry fruit, beef, milk and salmon.

Crop Physiology. DOI: 10.1016/B978-0-12-417104-6.00006-6

This chapter analyzes the environmental conditions and cropping system of southern Chile. It focuses on key annual crops like temperate cereals, potato, rapeseed and lupin, highlighting the remarkable yield potential of this region and its opportunities to increase actual crop yields, contributing in turn to the national food production aim. Challenges such as the management of herbicide-resistant weeds and the likely impact of climate change are discussed.

1.2 National production and yield trends of key annual crops of southern Chile: temperate cereals and potato

Temperate cereals are sown along a wide range of latitudes in Chile, i.e. from 33° to 41°S, though most of the cereal production is between 36° and 40°S. Among temperate cereals, wheat is the most important in terms of both production (1.34 million t year^{-1}) and area (266 000 ha), followed by oat (36% of the wheat area), triticale (8%) and malting barley (5%). Bread wheat dominates, with pasta wheat sown in only 6% of the total wheat area (ODEPA, 2014).

Despite the importance of wheat, the harvested area steadily decreased over the last 40 years from 750 000 ha to the present 260 000 ha at a rate of 9700 ha year^{-1} (ODEPA, 2014). Similar to wheat, the area of malting barley decreased (1300 ha year^{-1}) over the last 40 years, but it has kept stable around 18 000 ha since 2000. Contrary to these crops, oat has shown a stable harvested area after the 1960s, averaging 82 000 ha throughout the last 50 years and increasing recently to 126 000 ha (ODEPA, 2014). Different causes have powered the reduction of wheat and barley areas but the main explanation has been the shift from commodity grain crops to forestry and more profitable apple, wine grape, and lately berries and hazelnut, of which Chile is a world recognized producer and exporter, while cereals are mainly grown for the domestic market. The stability of the oat cropped area, on the other hand, could be associated with the wheat–oat rotation until the mid 2000s and, mainly, the double purpose of oat grains for human food and animal feed, especially the latter in the light of the long area dedicated to cattle production (see below). The increase of the oat area in the last years has been fuelled by the production to overseas markets as 70% of the national production is exported, i.e. about US$ 250 million per year (García, 2011; SOFOFA, 2013).

Despite changes in acreage, wheat, oat and barley showed a sharp increase in grain yield between 1980 and 1995. Recently, Engler and del Pozo (2013) reviewed Chilean grain yield trends of these crops and their likely causes. The authors evaluated the period between 1929 and 2009 finding similar yield improvement for all these crops after the 1980s, i.e. 101 kg ha^{-1} year^{-1} as average. Different causes can explain these yield gains, but Engler and del Pozo (2013) highlighted, in addition to plant breeding (Matus et al., 2012), the trade liberalization policy in the 1980s which facilitated the availability of technology and improved the competitiveness of the agricultural sector, price bands reducing price fluctuations and an intensive outreach program led by agricultural government agencies, i.e. INIA and INDAP. For oat, additional causes contributed to the yield increase including higher oat prices in the last five years, better competitive conditions of Chilean farmers to supply the Latin American market, government policies to stimulate oat export, the release of shorter and high-yielding varieties (e.g. Supernova INIA released in 2010), the use of plant growth regulators to avoid lodging and contract farming between growers and processing industries/exporters.

Regarding potato, this crop historically covered between 6 and 8% of the annual cropping area of Chile. In 1980, the total area was 88 760 ha decreasing to 49 576 ha in 2013 (Fig. 6.1a). Tuber production increased from 903 100 Mg in 1980 to 1 159 022 Mg at the present (Fig. 6.1b) as a consequence of the balance between the reduction of area (756 ha year^{-1}) and the increased tuber

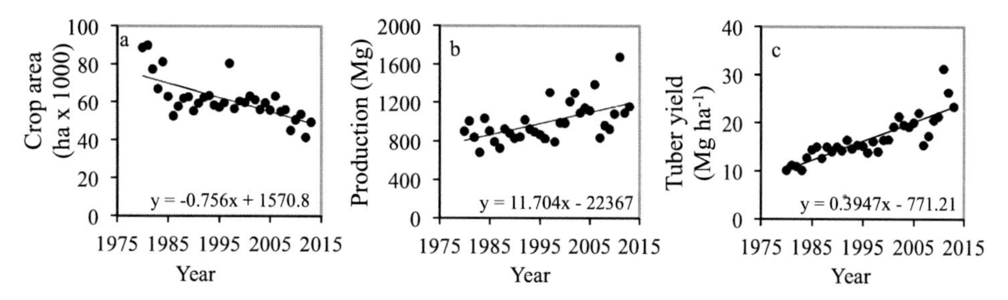

FIG. 6.1 Potato crop area, tuber production and tuber yield in Chile between 1980 and 2013.

yield (395 kg ha^{-1} year^{-1}; Fig. 6.1c). However, tuber production increased at a lower rate (1.3% year^{-1}) than other annual crops; for instance, wheat averaged 3.2% year^{-1} for the same period. This was due to lower potato yield gain (3.9% year^{-1}) between 1980 and 2013 in comparison with maize (7.9% year^{-1}), wheat (7.8% year^{-1}), other temperate cereals (5.8% year^{-1}), rapeseed (4.8% year^{-1}) and sugar beet (4.7% year^{-1}) (ODEPA, 2014). The slower gains in potato could be explained by the large technical gap among potato cropping systems coexisting in the country, i.e. ranging from large areas (>300 ha cropped by entrepreneurial farmers and mainly aimed at supplying the chip industry), to subsistence farming. In fact, there are 67000 small potato farmers in Chile (93% of total potato farmers) planting as average less than 1 ha year^{-1}.

2 ENVIRONMENTAL AND AGRICULTURAL FEATURES OF SOUTHERN CHILE

2.1 Climate and soils. High diversity in a narrow piece of land

From the agricultural viewpoint, southern Chile spans an area from about 37° to 41°S, which comprises three administrative regions spanning some 500 km (Fig. 6.2). The southern limit of La Araucanía (39°S) is where the Mediterranean climate (Acevedo et al., 1999; Chapter 7) that characterizes central Chile ends, southwards the climate is described as 'temperate warm with short dry season' and 'temperate rainy with Mediterranean influence' (Uribe et al., 2012).

Four natural agricultural segments are recognizable from east to west: the Andes foothills, the longitudinal central valley, the coastal range, and the coastal plains. The central valley and the coastal plains are more relevant to agriculture. Naturally, the coastal plains are under considerable oceanic influence, which moderates temperature fluctuations. The coastal range is conspicuous in the north of La Araucanía and produces a rain shadow on adjacent east drylands. This effect is less significant toward the south due to the lowering of the range. Annual rainfall increases north to south (Fig. 6.2b), varying from about 1000 mm in rain-shadowed areas with a dry period of roughly four months, to 2000 mm or even more in the southernmost central valley and sectors of the coastal plains (CONAMA, 2008; Uribe et al., 2012).

Temperature of the agricultural areas of southern Chile decreases to the south (Fig. 6.2c) and has relatively minor fluctuations throughout the year. The coldest month is July, where minimum monthly means, in the agricultural areas, varies between 2° and 4°C, and maximum monthly means between 10° and 12°C. January is normally the warmest month, with minimum monthly means between 9° and 10°C and maximum monthly means between 18° and 25°C, depending on the zone (Uribe et al., 2012). Frosts are common during winter and usually reach

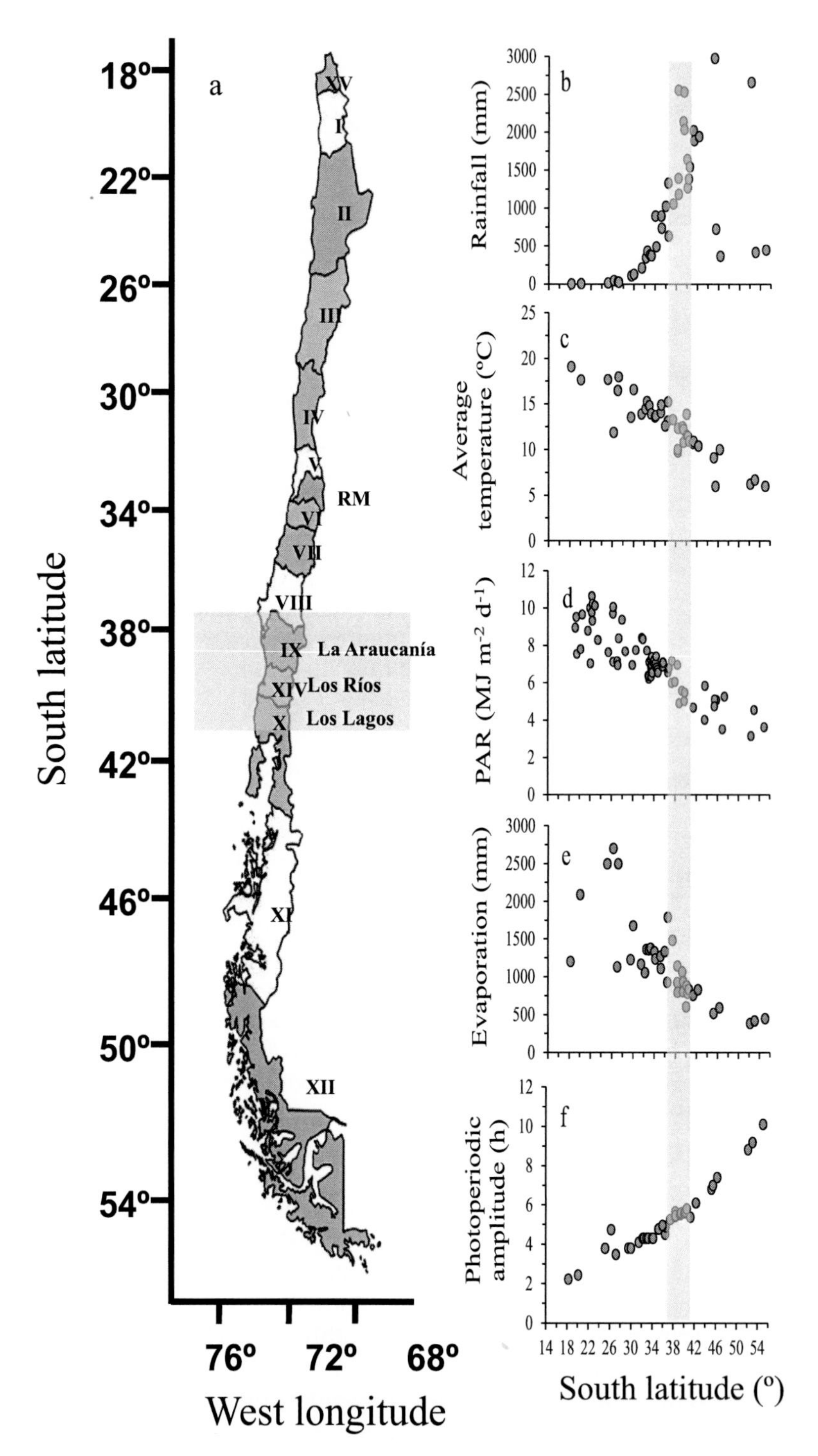

FIG. 6.2 (a) Map of Chile, (b) rainfall, (c) annual average temperature, (d) annual average photosynthetic active radiation (PAR), (e) cumulative pan evaporation and (f) photoperiod amplitude across the latitudes of Chile. Gray area highlights the range between 37° and 41°S comprising the regions of La Araucanía (37° 50′–39° 29′S), Los Ríos (39° 29′–40° 23′S) and Los Lagos (40° 23′–43° 44′S), the focus of this chapter.

down to −5°C in the agricultural areas, but may sporadically occur also late in spring and early summer. As with temperature, solar radiation decreases southwards (Fig. 6.2d), averaging 4.2 MJ m^{-2} d^{-1} in winter (July), and then increasing to peak at 18.8–20.9 MJ m^{-2} d^{-1} in January (Uribe et al., 2012). Additionally, pan evaporation shows a range between 650 and almost 1500 mm $year^{-1}$ in southern Chile, while the amplitude of photoperiod is by 5–6 hours contrasting the shortest and longest days during the year (Fig. 6.2e, f).

The soil commonly found in the central valley is a silty loam Andisol locally termed *trumao*, with medium depth or profound, derived from relatively recent volcanic ashes. The soils in the coastal hills are either red clay Ultisols derived from ancient volcanic ashes, or granitic Alfisols; and those in the coastal plains are deep sandy loams of metamorphic micaceous origin, with a *trumao*-like top layer (Rouanet et al., 1988). *Trumao* soils are high in both organic matter (>15%) and phosphorus-retention capacity; therefore, most crops are quite responsive to phosphate fertilization. On the other hand, the red clay soils are generally lower in organic matter (<10%), and several crops respond markedly to nitrogen fertilization. *Trumaos* store up to 200 mm of water; Huber and Trecaman (2004) and Dörner et al. (2013a) have shown that Andisols normally have maximum plant available water (PAW) between 22 and 35% vol^{-1} under a wide range of management systems. But the hydraulic conductivity of these soils is extremely high, both under saturated and unsaturated conditions, due to the bimodal pore system that could dry the soil even to the permanent wilting point (Dörner et al., 2009, 2012, 2013b). On the other hand, Ultisols and Alfisols normally have higher bulk density and lower water-holding capacity as compared to Andisols. For example, PAW of an Ultisol in southern Chile ranges between 20 and 25% vol^{-1}.

Soils generally have a pH below 6, and many below 5.5, a condition that hinders the growth and productivity of various crops. Abuse of ammoniac fertilizers has contributed to this condition increasing the risk of aluminum toxicity (Valle et al., 2009b); consequently, a number of entrepreneurial farmers are routinely liming prior to establishing wheat, barley or rapeseed crops.

2.2 The social context

From the socioeconomic viewpoint, two groups are readily identified, small farming and entrepreneurial farming. Small farming is frequently conducted by native ethnic groups, who usually have low education and scarce agricultural equipment and financial resources. They are considered essential in the aim of turning Chile into an 'agro-alimentary power' and policies to encourage associative farming are enhancing their contribution to Chilean agriculture. Entrepreneurial farming, on the other hand, is generally steered by a group with more access to equipment and capital, cultivating at least 200 ha of wheat each. However, the current gap between these groups is large and it is the crops of entrepreneurial farmers that place southern Chile on the map of the high-yielding environments.

2.3 Cereal, potato, lupin and oilseed crops

Rain-fed agriculture is predominant in southern Chile, with minor irrigated sectors devoted primarily to potato and fruit crops. The temperate climate is propitious for small grain cereals (wheat, *Triticum aestivum* L; triticale, *X Triticosecale* Wittmack; oat, *Avena sativa* L.; barley, *Hordeum vulgare* L.), cool-season grain legumes (pea, *Pisum sativum* L.; lentil, *Lens culinaris* Medik.; lupins (*Lupinus albus* L., *L. angustifolius* L., *L. luteus* L.), oilseeds such as rapeseed (*Brassica napus* L.) and linseed (*Linum usitatissimum* L.), potato (*Solanum tuberosum* L.) and sugar beet (*Beta vulgaris* subsp. *vulgaris* var. *altissima* Döll). However, due to its temperature regime, La Araucanía is

TABLE 6.1 Crop area and yield in three regions (La Araucanía, Los Ríos, Los Lagos) of southern Chile averaged for three cropping seasons (2009–10, 2010–11, 2011–12)

Crop	Area (ha)	Average yield (Mg ha^{-1})	Yield of entrepreneurial farmers[3] (Mg ha^{-1}) Estimated minimum	Mean yield of current variety trials (Mg ha^{-1})
Bread wheat	140915	6.34	7.7[4]	10.0[4]
Oat	68052	5.25	6.8[5]	9.0[5]
Potato	27016	32.80	35.0[6]	50.0[6]
Lupins[1]	24198	2.55	3.0[7]	4.5[7]
Triticale	19270	5.04	7.0[4]	12.0[4]
Rapeseed	15976	3.92	4.5[8]	5.5[8]
Barley[2]	10876	5.64	7.2[5]	9.5[5]

[1]*Comprises bitter Lupinus albus (≈50%), sweet L. albus (≈45%) and sweet L. angustifolius (≈5%);*
[2]*Mainly malting barley;*
[3]*Entrepreneurial farmers would be those sowing over 200 ha of wheat;*
[4]*C. Jobet, INIA, personal communication;*
[5]*J. Santander, SOFO, personal communication;*
[6]*rain-fed, J. Inostroza, INIA, personal communication;*
[7]*M. Mera, unpublished data;*
[8]*winter rapeseed, Mera & Espinoza (2012)*
Source of data: ODEPA (2014).

a frontier for commercial grain maize and common bean. Despite the diversity of cropping options, lack of profitable break crops limits crop rotations and favors cropping systems heavily relying on bread wheat (Table 6.1). Potato ranked the fourth crop in Chile over the last five years considering both area and production after cereals (wheat, maize) and sugar beet (ODEPA, 2014). Additionally, due to the attractive price of its oil, the acreage with rapeseed has increased in recent years, benefiting the cereals in the farming system. The area with lupins still remains limited.

The soils and climate (Fig. 6.2) of southern Chile favor outstanding crop productivity, as illustrated in Table 6.1. Wheat yields among entrepreneurial farmers are often over 8 Mg ha^{-1}. Wheat yields greater than 10 Mg ha^{-1} are not unusual, particularly among farmers that have adopted rapeseed in the rotation. High wheat yields are also achieved when potato is the preceding crop. In recent years, winter wheat trials are yielding about 10 Mg ha^{-1} (Table 6.1) and at some sites about 12 Mg ha^{-1} are achieved as an average of whole experiments comprising cultivars and breeding lines. Current oat varieties are reaching yields unseen a decade ago, when the average was about 4.5 Mg ha^{-1}. Currently, the average in southern Chile is greater than 5 Mg ha^{-1}, and yields of 8 Mg ha^{-1} have become quite normal for entrepreneurial farmers. A similar picture can be seen with triticale and barley (Table 6.1). Potatoes can be cultivated without irrigation in southern Chile but there is frequently a high yield penalty associated with water stress. Under irrigation, however, yields of 50 Mg ha^{-1} and more are common, and even 90 Mg ha^{-1} are not unusual among the best farmers. Rapeseed yields are generally greater than 3.5 Mg ha^{-1} and utilizing winter rapeseed hybrids good farmers are getting over 4.5 Mg ha^{-1} of grain yield, with 46–51% oil content. Cool season grain legumes, such as dry pea and large-seeded lentil, were once significant crops in Chile but as a result of international trade agreements have almost disappeared from southern agriculture. However, the yield potential for both crops is high.

Southern Chile has nearly 2.4 million hectares of prairies and around 40000ha of fodder crops, which sustain 80% of the country's beef and 85% of the dairy production (R. Demanet, Universidad de La Frontera, Temuco, personal communication). Some 220000ha are sown pastures, based predominantly on perennial ryegrass and white clover, some 660000ha are improved pastures, and the remaining, unsown natural pastures (ODEPA, 2014). Fodder crops are mainly oat, maize for silage (increasingly), and brassicas (turnip, rutabaga, rapeseed, kale). Some 820000ha with pastures are subjected to eventual rotation with annual crops (R. Demanet, Universidad de La Frontera, Temuco, personal communication).

The area with fruit crops in southern Chile is relatively small compared to the rest of the country but it is estimated to grow at a rate of 5% annually. Hazelnuts grew remarkably in recent years, reaching 2600ha in 2012 and sweet cherries reached 410ha (CIREN, 2012). Estimations for 2013 are 4800ha with hazelnuts and over 800ha with sweet cherries (A. Gonzalez, INIA, Temuco, personal communication). Berries are also relevant, particularly blueberries, which summed up to 4220ha in 2012. Red apples are a more traditional southern fruit crop with 2300ha concentrated in La Araucanía.

According to CONAF (2008a,b, 2009), 3.3 million hectares in these regions are forests and 2.4 million belong to native forest. Forestry in Chile is based primarily on pines (*Pinus radiata*) and eucalyptus (*Eucaliptus globulus, E. nitens*), with a planted area of around 805000ha. The remaining area is mixed forest.

2.4 Tillage practices

Traditional tillage, involving the inversion of the upper soil layer and burning of residues, was used for over a century in southern Chile. In the 1980s, no-tillage after burning of crop residues began to be utilized and, by the turn of the millennium, the vast majority of the entrepreneurial farms had adopted it. The incorporation of cereal residues is challenging, as a good crop may return 10Mg stubble ha^{-1}. The use of combine harvesters with straw chopper and spreader, set for low cutting to leave short standing stalks, has become essential and with equipment to manage large amount of crop residues, a growing number – now probably 20% – of the commercial farmers have stopped burning and are incorporating crop residues.

3 CEREAL-BASED CROPPING SYSTEMS AT HIGH YIELD POTENTIAL CONDITIONS

3.1 Temperate cereals in southern Chile

The reduction of the wheat cropped area commented on in section 1.2 was compounded by the higher (>60%) reduction that occurred between 35° and 36°S, concentrating this crop in southern Chile, which is, in turn, higher yielding as is shown in Figure 6.3 by the increase of average yield recorded in regions from Central Chile (under irrigation) to the south (mainly cropped at rain-fed conditions). At present, production of wheat, oat, barley and triticale in southern Chile

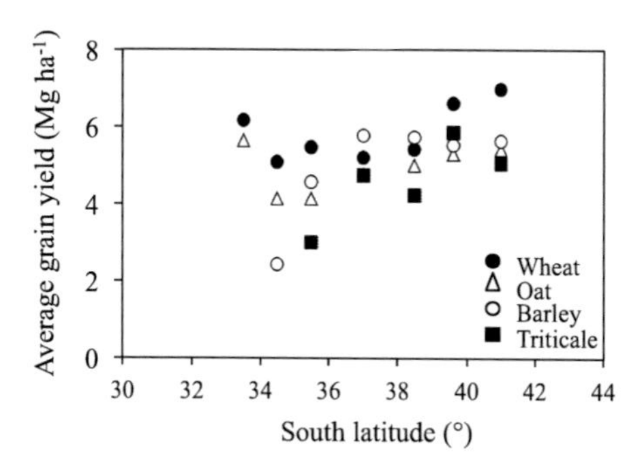

FIG. 6.3 Average grain yield of wheat, oat, barley and triticale from Central (33°S) to southern (41°S) Chile. *Source: ODEPA, 2014.*

accounts for more than 50% of the whole country (from 56% of wheat to 89% of triticale production), mainly in La Araucanía (ODEPA, 2014).

Three categories of temperate cereals are sown in southern Chile: (1) winter (with high vernalization sensitivity); (2) facultative (high photoperiod sensitivity); and (3) spring (low vernalization and photoperiod sensitivities). Sowing dates range from autumn (May) to late winter (August) depending on the category (the earlier for winter and the later for spring cultivars) and the latitude and altitude. Facultative cultivars, especially in wheat and oat, account for more than 60% of the area, since the high photoperiod sensitivity of this group allows sowing at a wide range of dates minimizing the frost risk from anthesis on (from almost as early as winter varieties until as late as spring varieties). Chapter 12 reviews the main environmental drivers of development in temperate cereals (temperature *per se*, photoperiod and vernalization) and their impact on crop phenophases.

3.2 Environmental and crop characteristics accounting for the high potential yield of temperate cereals

Southern Chile is one of the most favorable areas of the world for temperate cereals (e.g. Sandaña et al., 2009; Bustos et al., 2013). For instance, grain yield of spring wheat, barley and triticale can reach 12 Mg ha^{-1} or more (Fig. 6.4) associated, as expected, with high grain number per unit area (Sadras and Slafer, 2012). The elevated grain number potential of southern Chile is the consequence of the crop growth rate (CGR) (Fig. 6.5a) during the critical window for the determination of this yield component (Fischer, 1985; Savin and Slafer, 1991; Abatte et al., 1998; Chapter 12), which is in turn associated with the high photothermal quotient, i.e. the ratio between solar radiation and temperature bracketing anthesis (Fischer, 1985; Savin and Slafer, 1995). Interestingly, the relationship between grain number of wheat and both crop growth rate and

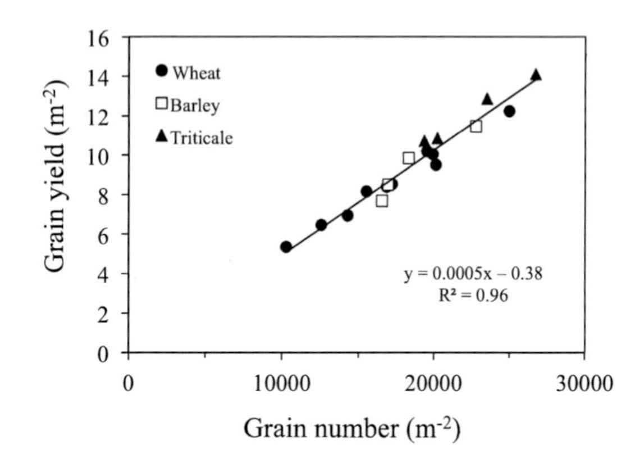

FIG. 6.4 Relationship between grain yield and grain number of wheat, barley and triticale under different sowing dates or phosphorus treatments in Valdivia (39° 47′S).

photothermal quotient recorded in southern Chile is similar to that in other environments, but reaching higher values (Fig. 6.5a, b). Similar to wheat, the favorable condition around anthesis is the main reason for the elevated yields of other temperate cereals (Ariznabarreta and Miralles, 2008; Estrada et al., 2008; Quiroz, 2010).

In addition to grain number, grain weight is an important trait to understand the high yields reached by temperate cereals in this area. Recently, García et al. (2013) analyzed experimental data from a set of 105 doubled haploid lines plus their parents (Bacanora and Weebil) evaluated in Ciudad Obregón (México), Buenos Aires (Argentina) and Valdivia (Chile), pointing out the higher thousand-grain weight recorded in Valdivia. High grain weight potential in southern Chile has been associated with favorable temperature during two periods: grain filling (e.g. García et al., 2013; Lizana and Calderini, 2013); and booting to anthesis, when flowers grow setting the upper limit of grain size (Calderini and Reynolds, 2000; Hasan et al., 2011). Indeed, average temperature during grain filling in southern Chile, i.e. between 15° and 18°C (Sandaña et al., 2009; Hasan et al., 2011), is close to the optimum (15°C) for grain weight of wheat and barley (Chowdhury and Wardlaw, 1978). In

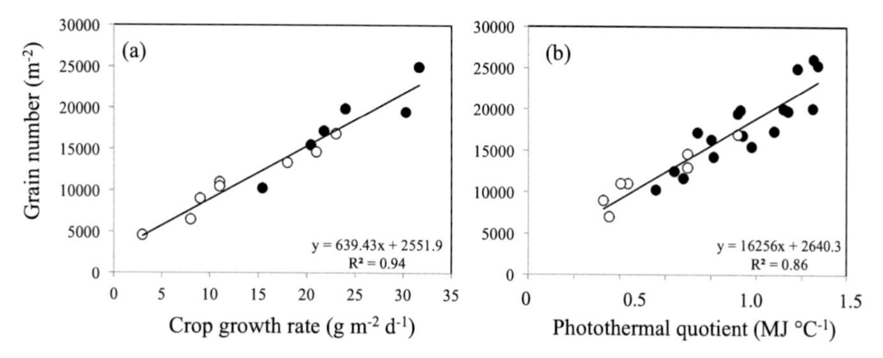

FIG. 6.5 Relationship between grain number of wheat and (a) crop growth rate and (b) the photothermal quotient during the critical window for grain number determination (see text) under different sowing dates or phosphorus treatments recorded in southern Chile (closed circles) and Argentina (open circles). *Sources: Sandaña and Pinochet (2011), Sandaña et al. (2012) and Quiroz (2010, only in b) for southern Chile, and Lázaro et al. (2009) for Argentina.*

recent evaluations of nine spring wheat genotypes, individual grain weight in Valdivia was 20% higher than in Ciudad Obregón despite larger grain number in the former site. Similar results have been found for malting barley, cv. Scarlet (Calderini et al., 2013). Moreover, improved genotypes could further raise potential yield in southern Chile, as was demonstrated by Bustos et al. (2013) and analyzed in section 6.1; the authors reported 16 Mg grain ha^{-1} in experiments assessing double haploid lines derived from the cross between spring wheat cultivars having high grain number vs high grain weight. Therefore, Chile's paradox of outstanding yield potential and the reduction of wheat and barley areas could be partially beaten through increasing yield potential by plant breeding and by bridging the gap between actual and potential yield (Table 6.1 and Fig. 6.4). However, investment in product diversification, alternative markets and government policies are as relevant as yield improvement to revitalize wheat and barley as it was proven for oat production (section 1.2).

3.3 Considerations for spring cereals in cropping systems

In a recent survey of 20 winter and 20 spring wheat cultivars in four sites from 36° 58′ to 40° 54′S, under 50 mm irrigation in the northern site and rain-fed conditions in the other three (from 38° 41′ to 40° 54′S), winter wheats over-yielded spring wheats by 18.4% (11.1 and 9.4 Mg ha^{-1}, respectively, averaged across the genotypes). In addition, winter wheats showed higher grain yield stability as indicated by the coefficient of variation, i.e. 15.9% and 30.9% in winter and spring wheats, respectively. These results are supported by crop simulation when changes in the rainfall regimen associated with phases of El Niño Southern Oscillation (ENSO) were assessed for rain-fed agricultural locations of Chile (Meza et al., 2003). The authors showed that spring wheat was more sensitive to weather variation than winter wheat and oat in Temuco (38° 46′S) and Valdivia (39° 47′S) and, also, the authors showed that an accurate and anticipated weather forecast would improve the stability of the crop yield and farmers' income by optimizing management practices such as sowing date and fertilization rate (Meza et al., 2003).

Both, grain yield and its stability are strongly affected by environmental and crop conditions during the critical period around anthesis. Spring cereals have a very short crop cycle in southern Chile as shown in Figure 6.6, reaching high crop grow rate (CGR), e.g. 300–320 kg dry matter ha^{-1} d^{-1} during the linear accumulation of above-ground biomass in wheat and barley (Calderini et al., 2013). Therefore, spring cultivars have lower chances than winter cultivars to compensate any biotic or abiotic constraint decreasing the ability of the crop to intercept fully the incident solar radiation around anthesis (Fig. 6.6), affecting in turn the CGR and grain number achieved by the crop (Fig. 6.5a). For example, 40 days delay in sowing of spring wheat and barley decreased the emergence–anthesis period by 22 days, reaching lower fraction of radiation interception at anthesis (by 75% across the crops). Accumulated solar radiation during the whole crop cycle of these crops decreased by 21% and grain yield by 13.2%. In this experiment, grain number reduction (18% across both crops) almost completely explained the impact of delayed sowing on grain yield (Quiroz, 2010).

3.4 Main features of cereal-based crop rotations and production

Cereal cropping systems of southern Chile are based on wheat as the main crop in agreement with its area (Table 6.1). Though generally wheat rotates with crops such as oat, rapeseed, potato, lupin, barley, triticale, and with pastures in Los Ríos and Los Lagos, the monoculture of wheat is occasionally practiced, particularly by small farmers. Monoculture of wheat has shown clear detrimental effects, reducing the attainable farm yields by 50% or more, mainly due to soil diseases (e.g. *Gaeumannomyces graminis*), weeds (e.g. *Lolium multiflorum*, *Avena fatua*) and nutrient availability to the crop (Rouanet et al., 2005). Sugar beet, grain legumes (lupin, pea and lentil), rapeseed and even oat were found to be the best previous crops for wheat in southern Chile (Rouanet et al., 2005). High availability of nutrients (N, P and K) in the soil after potato benefits the following wheat crop. Also, the increased rapeseed area accomplished recently (and described in section 5) is highly beneficial for the cereal-based cropping systems to precede wheat and barley, facilitating weed control and breaking soil diseases cycle (Kirkegaard et al., 2000). Therefore, these rotations contribute highly to the sustainability of the cereal-based cropping systems in the present and more in the near future if the rapeseed area continues increasing becoming, in turn, one of the keys for bridging the gap between actual and attainable yields (Fig. 6.3 and Table 6.1).

The availability of nutrients to the crop is also a constraint for cereal-based cropping systems,

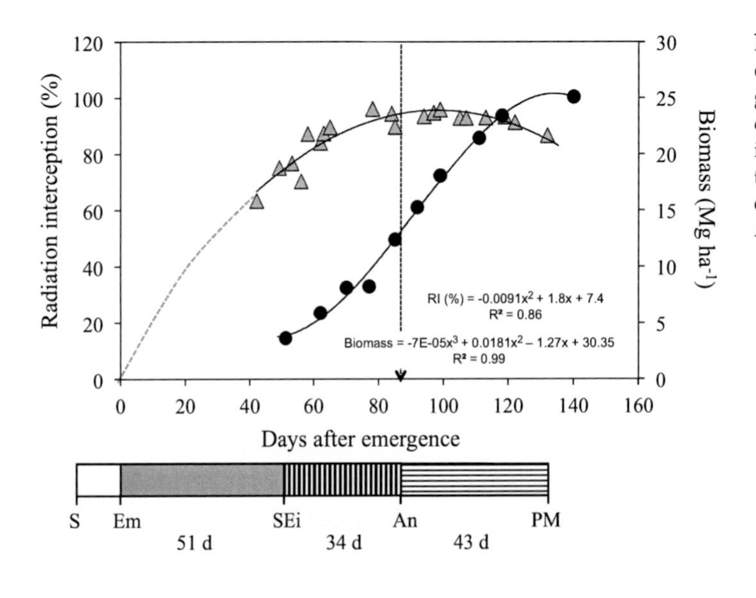

FIG. 6.6 Time-course of fraction of radiation interception (triangles) and above-ground biomass (circles) of a spring wheat cultivar growing under potential conditions in Valdivia (39° 47′S). The arrow shows the time of anthesis. S: sowing; Em: seedling emergence; Sei: stem elongation initiation; An: anthesis; PM: physiological maturity.

mainly nitrogen and phosphorus, the latter particularly in volcanic soils (section 2.1). The demand for nutrients to achieve a yield of 12 Mg ha^{-1} could reach 270 kg N ha^{-1} and 40 kg P ha^{-1}, showing a nutrient-uptake efficiency of 0.65 kg kg^{-1} and nutrient-use efficiency of 40 g g^{-1} for nitrogen (Chapter 8 presents an in-depth analysis of crop nitrogen uptake and efficiencies) and 0.15 kg kg^{-1} and 300 g g^{-1} for phosphorus (Rodriguez et al., 2001; Valle et al., 2011). Due to the characteristics of *trumao* soils, phosphorus shortage and soil aluminum toxicity are predominately important in southern Chile because of their negative effect on CGR and grain yield of cereals (e.g. Sandaña and Pinochet, 2011; Valle et al., 2009a). CGR is the result of accumulated radiation intercepted by the crop and radiation-use efficiency (Chapter 15). The latter is a more conservative trait in temperate cereals, while radiation interception is more sensitive (Fig. 6.7) and the main cause of CGR and grain yield decreases under low availability of nitrogen (Muurinen and Peltonen-Sainio, 2006), phosphorus (Sandaña and Pinochet, 2009) and other soil constraints such as aluminum toxicity (Valle et al., 2009b) and mechanical impedance (Sadras et al., 2005). This is highly important for winter, facultative and spring cereals but the timing for an appropriate fertilization is particularly critical for spring cultivars taking into account the short time to reach full radiation interception of this group (Fig. 6.6). The optimal economic nitrogen dose has been found to be 250 kg N ha^{-1} for high-yielding facultative wheats (Campillo et al., 2007). Entrepreneurs generally fertilize with phosphorus at sowing (ranging between 70 and 130 kg P_2O_5 ha^{-1}) and nitrogen (180–220 kg N ha^{-1}) split at sowing (by 25%), starting (45%) and ending (30%) tillering in winter wheat, while nitrogen is split twice (sowing and starting-mid tillering) in spring cultivars using the lower-range doses. As in other cropping areas, nitrogen fertilization is a key for barley to avoid grain protein concentration below or above the industry requirements (ranging between 9 and 12% for most of them).

Water availability could constrain the productivity of temperate cereals in southern Chile broadening the gap between attainable and potential yields (Table 6.1). Water deficit can result from an increase of evaporation and a decrease of rainfall, mainly after anthesis, i.e. between November and December (Fig. 6.8), but stored soil water can buffer this climatic deficit, particularly in *trumao* soils (Quiroz, 2010). Additionally, the tillage system could be a key for soil water content. Although the management of crop residues in high yield potential southern Chile is

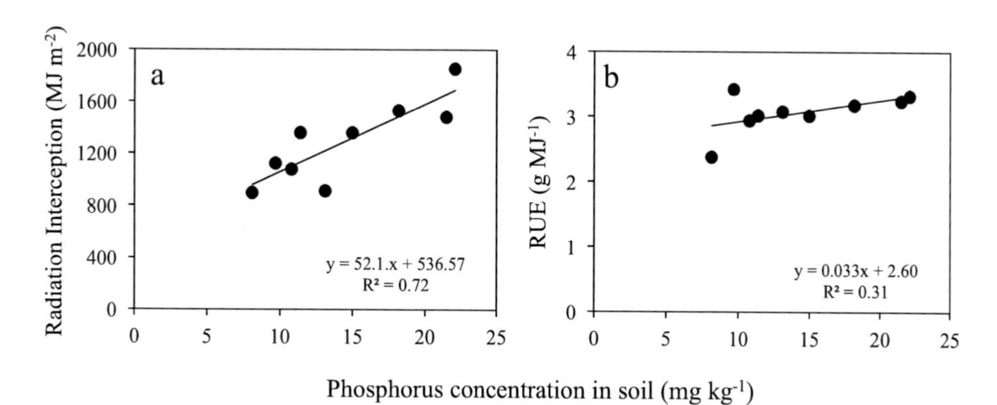

FIG. 6.7 Relationship between accumulated radiation interception (a) and radiation-use efficiency (PAR radiation) (b) of wheat and soil phosphorus concentration in southern Chile. *Adapted from Sandaña and Pinochet (2011) and Sandaña et al. (2012).*

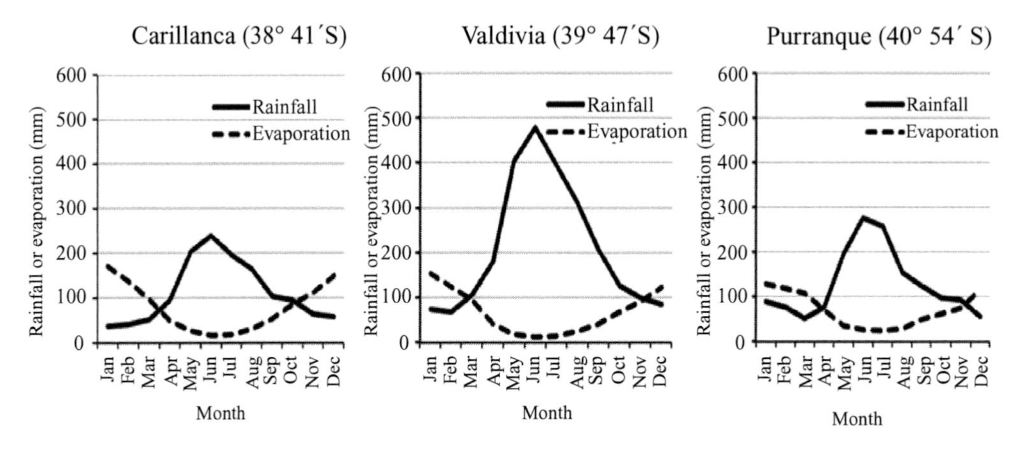

FIG. 6.8 Time-course of average rainfall (line) and pan evaporation (dashed line) in three sites of southern Chile.

challenging, the adoption of no-tillage without burning could preserve more soil water than conventional tillage as was shown for the oat–wheat–oat rotation in a compacted Alfisol of Central Chile (35° 97′S) (Martínez et al., 2011). However, these authors found that no-till reduced grain yield in the 3-year rotation as a consequence of soil compaction. Moreover, the negative impact of no-till on soil water infiltration was reported in a comparison between conventional tillage and no-till after 4 and 7 years in a Mollisol of central Chile (Martínez et al., 2008). High organic matter *trumaos* have elevated aggregate stability as well as resilience against mechanical stresses (Dörner et al., 2011); these soils are therefore less sensitive to compaction than Alfisols and Mollisols. Consequently, the use of low or no-till would improve the yield stability and the sustainability of the cereal-based cropping system of southern Chile.

4 THE POTATO-BASED CROPPING SYSTEMS; BETWEEN SUBSISTENCE AGRICULTURE AND HIGH INPUT PRODUCTION

4.1 Overview

Potato is grown in a wide range of agro-ecological zones of Chile between 18° and 53°S and from sea level to 3000 m altitude (Montaldo, 1974). However, over 95% of the potato production is concentrated between 27° and 41°S (ODEPA, 2014). In northern areas (27°–32°S), about 60% of potato is grown between March and October (late planting or autumn–winter crops), 15% is cultivated from July to December (early planting or spring crop) and 12% between December and May (late planting or summer crop). Autumn and summer plantings are considered off-season crops grown in river valleys with irrigation. The most common crop rotation in this zone comprises wheat, potato, grain legumes (beans or pea) under irrigated conditions. In central Chile (33°–34°S), potato is grown early (July–December, spring crops) or late (January–June, summer crops). A common crop rotation is similar to the northern one: wheat, potato and grain legumes (beans or pea under irrigation and lentil or chickpea in rain-fed conditions). In Central–South Chile (34°–37°S) rain-fed potato is grown from October to March; however, early (September) and late (November) plantings are also found in this zone. Crop rotations are diverse. For example, potato is grown after wheat, oat or barley and before sugar beet, maize or sunflower (very small area). Most of the seed potato production is in southern Chile, where entrepreneurs grow consumer and seed potatoes between October and March in a typical rotation wheat–oat–potato–rapeseed, mainly to avoid biotic constraints and fulfill the standard seed certification requirements, regarding a 4-year crop rotation. Entrepreneurs generally rent farmland to satisfy this requirement. On the other hand, subsistence farmers rotate wheat and potato or grow potato as a monoculture. Across the country, subsistence farms coexist with large, high-tech farms but, in general, at the north (specifically in La Serena, 29° 54′S) high-tech based agriculture of early potato production is performed to supply the high-price market of early potato, while in the south, potato-seed and industry production based in a favorable environment for plant health and high yield is dominant.

Average potato yield in Chile over the last 10 years has remained above 20 Mg ha^{-1} (Fig. 6.1c and Table 6.2), but tuber yield varies between 5 and 80 Mg ha^{-1} across cropping systems. This is not only due to differences in climate and soil conditions but also contrasting efficiency on the use of resources. Small farms plant low quality 'seed' in monocultures or under inadequate crop rotation. Furthermore, the lack of or deficient irrigation is usual among these farmers, who apply very low doses of fertilizers and pesticides. This hinders the control of late blight (*Phytophtora infestans*), which causes the greatest losses among small potato producers. Also, small farms are highly dependent on hand labor for planting, weed control, fertilization, and other practices, while entrepreneurs grow potato under intensive input systems, highly mechanized. Despite these problems southern Chile concentrates 77% of potato production and has the higher average tuber yield (Table 6.2).

4.2 Bridging the gap between actual and potential yields

The gap between actual and potential yield of potato is highly influenced by land ownership which, in turn, associates with the technology level. Figure 6.9a shows a wide dispersion of tuber yield in different counties, ranging between 2 and 34 Mg ha^{-1} around 38°S, where the average farm area ranges from less than 1 to 18 ha approximately (Fig. 6.9b). The dispersion of tuber yield and land ownership would be greater if data from individual farmers were available. Similarly, the irrigated area at each county ranges from nil to 100% (Fig. 6.9c), which significantly affects actual yields. The average tuber yield thus associates with both proportion of the

TABLE 6.2 Potato crop area, tuber production and tuber yield in different regions of Chile

Region	Latitude S (°)	Crop area (ha)	%	Production (Mg)	%	Tuber yield (Mg ha^{-1})
IV	29–32	2546	5.2	51 863	4.5	20.4
V	32–34	1103	2.3	16 392	1.4	14.9
RM	33–34	5104	10.4	112 644	9.8	22.1
VI	34–35	942	1.9	19 220	1.7	20.4
VII	35–36	3017	6.2	69 068	6.0	22.9
VIII	36–38	8372	17.1	152 632	13.2	18.2
La Araucanía	38–39	14 459	29.6	314 582	27.3	21.8
Los Ríos	39–40	3334	6.8	76 035	6.6	22.8
Los Lagos	40–43	10 012	20.5	340 220	29.5	34.0

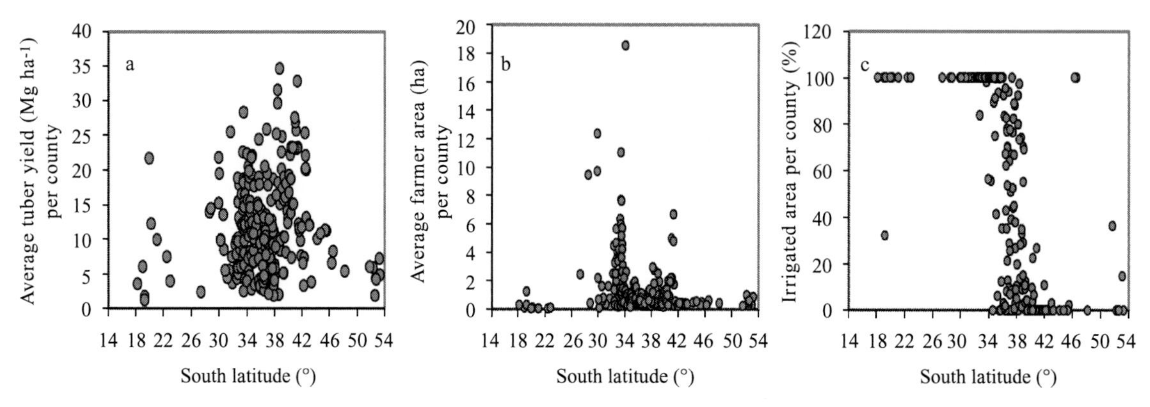

FIG. 6.9 Average (a) tuber yield, (b) farmer area and (c) irrigated area per county across Chile. *Data from National Agricultural Census 2007 (INE, 2007).*

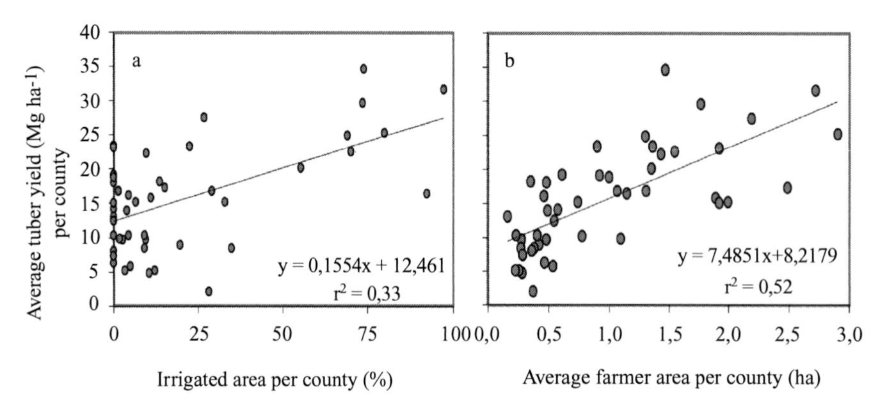

FIG. 6.10 Relationship between average tuber yield and (a) the proportion of potato irrigated area per county and (b) average farmer area per county in southern Chile (37–41°S). *Data from the National Agricultural Census 2007 (INE, 2007).*

potato-irrigated area per county (Fig. 6.10a) and average farmer area per county (Fig. 6.10b).

Assuming a simulated potential yield from 70 to 100 Mg ha^{-1}, and actual yield of farmers from 25 to 50 Mg ha^{-1}, the yield gap in southern Chile averages 61%. As shown in Figure 6.11, the gap is lower in the best agro-ecological zone for potato production (41° 28′S), but still, implies a loss of approximately 40 Mg ha^{-1} of yield. A large proportion of this gap can be overcome with an adjustment in management practices as genetic gain in yield seems to be small in recent years. For instance, new cultivars showed similar yield to those introduced many years ago, which are widely grow in southern Chile (e.g. cv. Desiree introduced in 1968). However, it should be noted that in rain-fed areas of La Araucanía, modern cultivars (e.g. Karu INIA) could reduce the yield gap by better performance under water-shortage conditions.

Irrigated experiments showed tuber yield over 80 Mg ha^{-1} in southern Chile (Kalazich et al., 2004). The main determinant of tuber yield was the number of tubers m^{-2}, while average tuber weight was not associated with yield (Fig. 6.12). Then, plant and environment conditions during the brief period of tuberization seem key to determine tuber yield. Although

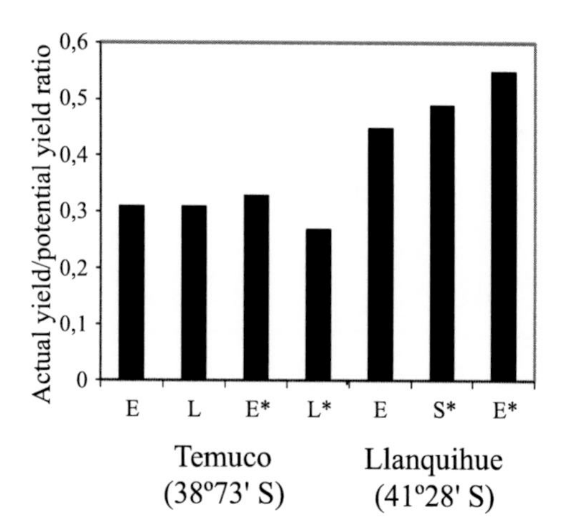

FIG. 6.11 Actual:potential yield ratio in different potato-cropping systems of southern Chile. Actual yield was recorded from farmers' data and potential yield was calculated with the LINTUL-potato simulation model. E: early planting; L: late planting; S: seed producers. * Potato producers with more than 10ha. *Adapted from Haverkort et al. (2014), using actual yields from survey carried out by the Agricultural Research Institute (INIA).*

no data are available to assess this association in southern Chile, simulations by Kooman et al. (1996a) identified the timing of this stage as critical to maximum tuber dry matter production.

High yield of potato has been associated with the modeled CGR and tuber growth rate (TGR) during tuber filling (Fig. 6.13; Table 6.3). This simulation (LINTUL potato simulation model) also showed that yield was unrelated to the duration between: (1) emergence and full radiation interception by the crop; (2) full radiation interception and harvest; and (3) the whole crop cycle (Fig. 6.13a–c). Haverkort et al. (2014) reported that average daily growth rate between emergence and harvest explained almost 70% of actual tuber yield measured in farmers' fields in Chile. Further, modeled crop growth rate declined under increasing temperature in southern Chile and increased with high radiation:temperature ratio (Fig. 6.14a, b). The rate of leaf appearance and leaf expansion of potato increases linearly in the range of 9–25°C (Benoit et al., 1983; Struik et al., 1989). Minimum and maximum temperatures (8–20°C) during potato emergence recorded in southern Chile are, therefore, within the range of linear increase. Low temperatures, especially at night (8–10°C) also stimulate tuberization leading to a high number of tubers per plant.

Solar radiation in southern Chile, around 20MJ m^{-2} d^{-1} during the crop cycle (October–February), is not limiting for potato yield and radiation-use efficiency (RUE) can reach 3.5g MJ^{-1}, which is close to the maximum reported for this crop species (Sinclair and Muchow, 1999). Field experiments in Valdivia showed that tuber yield was positively associated with the maximum fraction of radiation interception reached

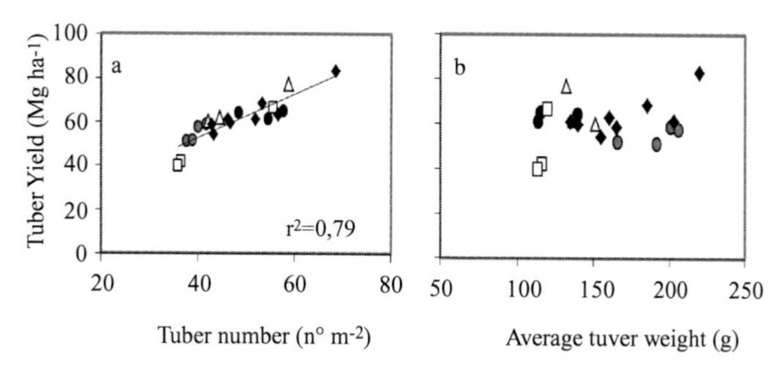

FIG. 6.12 Relationship between tuber yield and (a) tuber number and (b) average tuber weight. Cultivar Desiree (closed circles), Karu INIA (white triangles) and Yagana INIA (white squares) evaluated under different temperatures; cultivar Desiree cropped at different hill volume treatments (gray circles); cultivars Desiree and Yagana assessed under source–sink and UV-B treatments (closed diamonds) in Valdivia. Source: Lizana (unpublished).

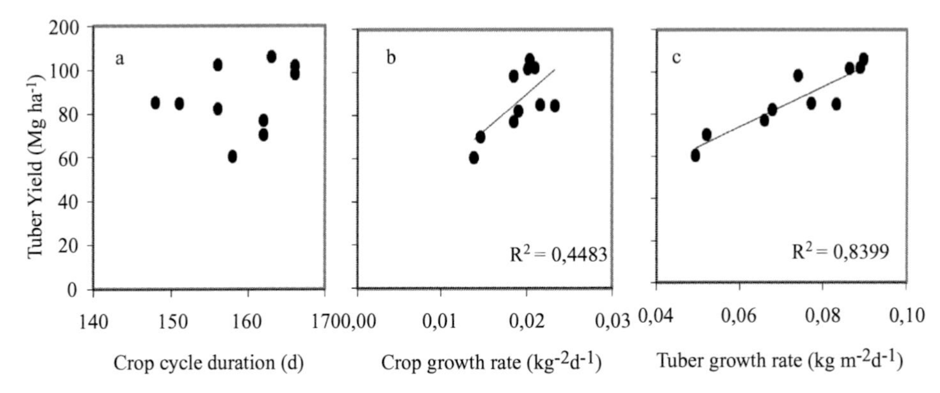

FIG. 6.13 Relationship between tuber yield and (a) the crop cycle duration, (b) crop growth rate and (c) tuber growth rate. Simulated data for 10 sites from 29° 54′ to 41° 54′S as described in Table 6.3.

TABLE 6.3 Modeled fresh tuber yield (FTY), crop cycle duration (CCD), crop (CGR) and tuber (TGR) growth rate in a transect of 10 cropping areas of Chile

Site	South latitude	FTY ($Mg\ ha^{-1}$)	CCD (days)	CGR ($kg\ m^{-2}\ d^{-1}$)	TGR ($kg\ m^{-2}\ d^{-1}$)
La Serena	29°54′	60.0	158	0.014	0.049
La Platina	33°36′	77.1	162	0.018	0.066
Rengo	34°24′	106.1	163	0.020	0.090
Talca	35°24′	70.2	162	0.015	0.052
Chillán	36°36′	98.3	166	0.018	0.074
Angol	37°48′	85.2	148	0.022	0.077
Carillanca	38°42′	84.8	151	0.023	0.083
Valdivia	39°48′	102.4	156	0.021	0.089
Purranque	40°54′	101.9	166	0.020	0.086
Puerto Montt	41°30′	82.2	156	0.019	0.068

Simulations performed with LINTUL potato model with historical climate data from INIA, 1989 and CNE/PNUD/UTFSM 2008.

by the crop (Fig. 6.15b), hence cultivars with a higher rate of leaf area expansion at the beginning of the crop cycle (e.g. Karu INIA) reached key phases (tuberization) with higher interception. Remarkably, a linear relationship was found in this experiment between tuber yield and the highest fraction of radiation interception reached by the crop (Fig. 6.15). Despite these promissory results, much more research should be accomplished in southern Chile to understand fully the physiological causes of potato yield gaps.

Water balance between precipitation and evaporation along the crop cycle is widely variable across southern Chile ranging between −108 mm in the north (37°S) and −28 mm in the south (41°S) but, in both sites, stored soil water at sowing is close to field capacity. The potato area under irrigation in southern Chile is growing,

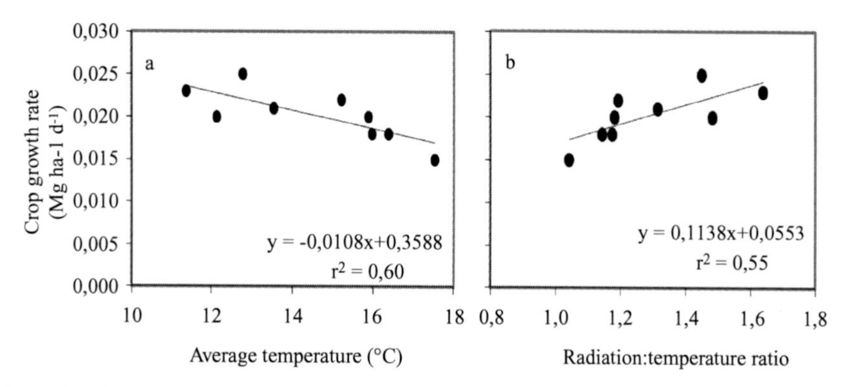

FIG. 6.14 Relationship between crop growth rate and (a) average temperature and (b) radiation:temperature ratio. Data are from 10 sites between 29° 54′ and 41° 54′ S as described in Table 6.3.

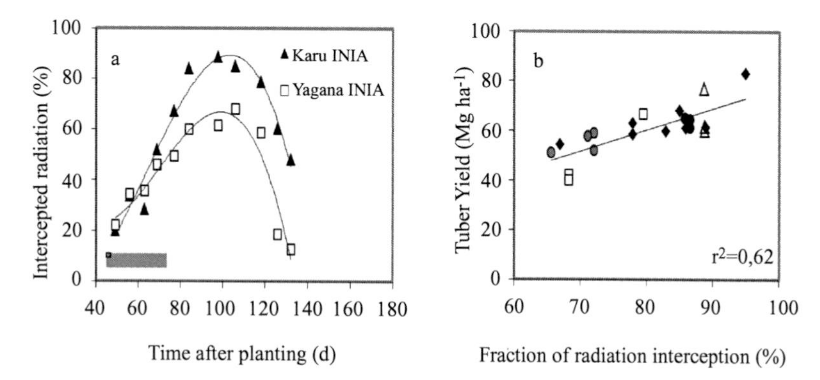

FIG. 6.15 (a) Time-course of the fraction of radiation interception during the crop cycle in two Chilean cultivars of potato: Karu INIA (closed triangles) and Yagana INIA (open squares). Gray bar indicates tuberization phase. (b) Relationship between tuber yield and the maximum fraction of radiation interception reached by the crop. Data are from different experiments carried out in Valdivia: cultivar Desiree (closed circles), Karu INIA (white triangles) and Yagana INIA (white squares) cropped at different temperatures; cultivar Desiree cropped at different hill volume treatments (gray circles); cultivars Desiree and Yagana at different shadow and UV-B treatments (closed diamonds). Source: Lizana (unpublished).

driven by large holdings using highly efficient irrigation systems. The average water-use efficiency calculated for southern Chile on the actual yield basis, is 9 g tuber l^{-1}. However, the borderline of water-use efficiency for the highest yield site (Llanquihue, 41° 28′S) has been estimated at 24 g l^{-1} (Haverkort et al., 2014). The significant difference in water-use efficiency could be reduced by improving the management of irrigation.

Nutrient availability is a key in potato-cropping systems of southern Chile. A significant ($P < 0.01$) relationship between actual yields and total amount of fertilizers applied during the crop cycle explained 45% of the yield variation in a survey across production systems (Haverkort et al., 2014). High rate of nutrient application is usual in southern Chile, especially high amount of phosphorus applied at planting (300–450 kg P_2O_5 ha^{-1}), because of the important phosphorus retention of *trumao* soils (section 2.1). On the other hand, no relationship was found between actual yield and nutrient-use efficiency of main macronutrients that ranged between 445 and 85 g potato g^{-1} N applied,

225 to 58 g potato g^{-1} P_2O_5 applied and 280 and 56 g potato g^{-1} K_2O applied (Haverkort et al., 2014). In this context, adjustment of fertilization rates is a way to reduce the costs to farmers more than to increase potato yield at the present.

Late blight caused by *Phytophthora infestans* is the main biotic constraint to potato production around the world, and also in southern Chile. The causal agent of this disease in Chile is the A1 mating type, which is considered to be more aggressive now than a decade ago. Medium and big farmers control this disease with chemicals but at a cost that small farmers cannot afford. When weather favors *Phytophtora*, crops in small farms can fail. The national breeding program of potato considers resistance to *Phytophtora* within its main breeding objectives.

5 RAPESEED AND LUPIN IN CURRENT FARMING SYSTEMS

5.1 Rapeseed

Over recent years, the area sown with rapeseed in southern Chile has rapidly increased, with some 40000 ha sown in 2012 and 2013 (Fig. 6.16). This is good news due to the well-known benefits of this crop in the rotation; indeed, many farmers think that higher wheat yields are in part due to the presence of rapeseed in the rotation. Most farmers chop and incorporate all rapeseed residues into the soil, without burning, and use no-till or reduced tillage to establish the subsequent crop, typically wheat.

Production of rapeseed is almost entirely based on hybrids of canola quality, over 95% originated in western Europe. Both winter and spring hybrids are available, but winter hybrids dominate. Most of the rapeseed is sown in early autumn (April). Early sowing is essential for a sufficiently large root system to resist uprooting by the time of the first frosts (May–June). Due to the vigor of the hybrid plants, around 35 plants m^{-2} are enough to ensure quick soil coverage and high yields. The rosette stage lasts through July and stem elongation starts in early August. Flowering takes about 50 days, from the beginning of October to the third week of November, grain-filling progresses until the end of December and ripening occurs during the first two weeks of January. Due to this phenology, rapeseed is one of the crops most well adapted to southern Chile conditions. Flowering proceeds during relatively cool days, since the maximum monthly mean temperatures of October and November are about 15 and 20°C respectively. This condition contributes to reduced flower drop and increases silique set. In addition, flowering usually elapses when soil water is not limiting. Between April and September, the crop receives generally around 1000–1500 mm of rainfall. During October and November another 150 mm are normally expected, although large variation may exist from one year to another. This is the period where evapotranspiration becomes gradually greater than precipitation (Fig. 6.6), but the soil still retains enough humidity for grain filling to progress with little or no water stress. December is typically a month where weather determines how good the yield will be. Another 50 mm rainfall is normal during December but, again, large annual variations occur.

The rapeseed crop is demanding of equipment, fertilizers and agrochemicals, hence the crop is grown almost exclusively by entrepreneurial farmers. The use of a pre-emergence herbicide (metazachlor, trifluralin), two sprayings of fungicide to prevent *Phoma* and *Sclerotinia*, plus one or two sprays of insecticide against leaf miners and aphids are generalized practices. Fertilizations of around 200 kg N ha^{-1}, 160 kg P_2O_5 ha^{-1}, 50 kg K_2O ha^{-1} and 50 kg S ha^{-1} are common. To reduce leaching during fall and winter, nitrogen fertilization is split between sowing, rosette with 7–9 leaves, and first flower buds. Some farmers over-use nitrogen and sulfur, so a more rational fertilization could make this crop more profitable as fertilizer represents 30–40% of costs.

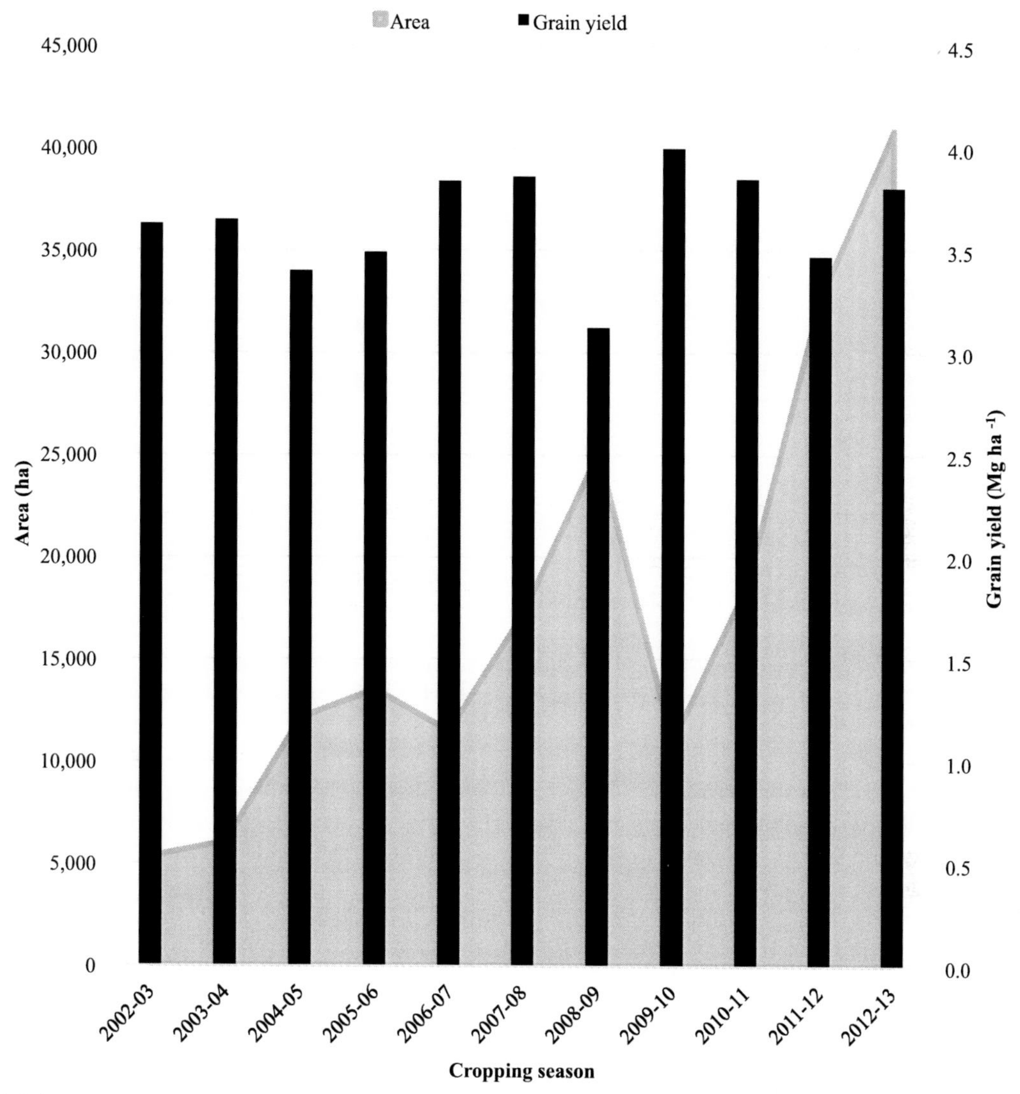

FIG. 6.16 Area sown and average grain yield of rapeseed in Chile during the period 2002–2012.

Figure 6.17a illustrates how critical phosphate fertilization could be for a rapeseed crop in a *trumao* soil with 8 g kg^{-1} of available P (Olsen) (Montenegro et al., 1992). Higher yields in Maquehue (Fig. 6.13b) may be ascribed to the particular weather of the cropping season and to greater yield potential of the varieties utilized, as the soil of Maquehue had 10 g kg^{-1} of available P (Olsen) (Ruiz, 2008).

Hybrids with greater yield potential have quickly replaced open pollinated varieties and, utilizing winter hybrids, good farmers are getting grain yields in the range of 4.5 and 5.5 Mg ha^{-1}, with 46–51% oil content. In yield trials, the best genotypes are exceeding 5.5 Mg ha^{-1}, with 51% oil (Mera and Espinoza, 2012). More recently, the best genotype exceeded 6 Mg ha^{-1}, with nearly 50% oil, i.e. a yield of 3 Mg ha^{-1} of oil, as

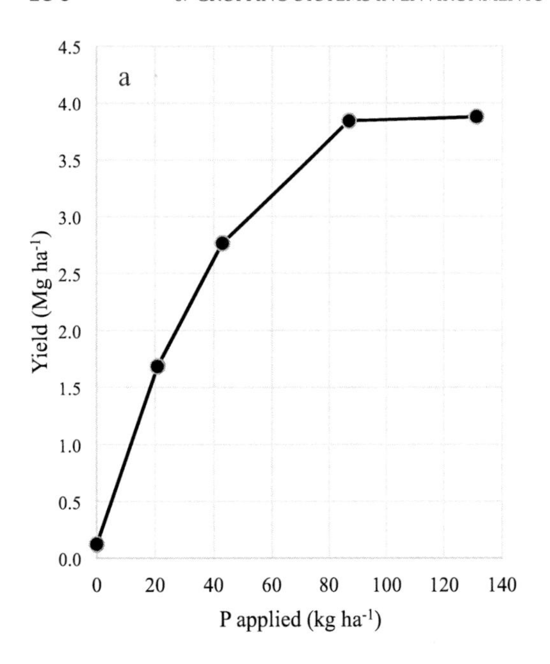

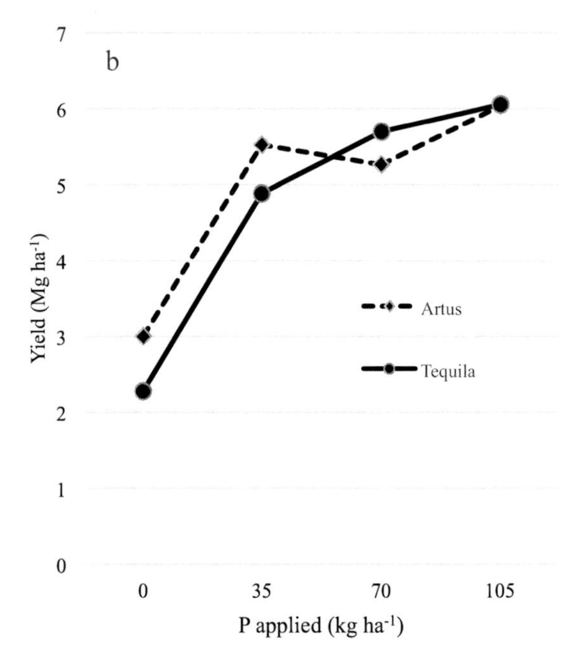

FIG. 6.17 Response of rapeseed rape to phosphate fertilization in two sites in La Araucanía. (a) Cv. Norin 16 in a silty loam Andisol of Lastarria (39° 11′S, 72° 29′W), adapted from Montenegro et al. (1992); (b) cvs. Artus and Tequila in a silty loam Andisol of Maquehue (38° 50′S, 72° 42′ W). *Adapted from Ruiz (2008).*

an average of two locations (Mera M, unpublished). Plant height may become excessive, particularly under rainy springs, making the crop prone to lodging during grain filling; some winter hybrids are about 1.75 m tall but many high-yielding hybrids may reach 1.9 and 2.0 m. These are huge plant sizes compared with the shorter canola varieties grown in Australia and Canada.

Spring and winter hybrids with the Clearfield® system have been introduced. The Clearfield hybrids carry two genes conferring resistance to imidazolinones, a group of broad-spectrum herbicides belonging to the acetolactate synthase (ALS) inhibitor family. In Chile, the resistant hybrids are used in combination with a mix of active ingredients, imazamox and imazapyr. The Clearfield hybrids have shown yields similar to the best non-Clearfield ones. This technology is significant for southern agriculture since the current Chilean legislation prohibits growing transgenic crops unless the entire harvest is exported; therefore, the use of glyphosate-resistant varieties is restricted (Chapter 18). The Clearfield rapeseed hybrids have proved useful to deal with high infestations of wild radish (*Raphanus sativus*) and wild turnip (*Brassica rapa*). However, the mix imazamox + imazapyr does not adequately control *Anthemis*, another defiant weed in rapeseed. The Clearfield wheat varieties give a good opportunity to control wild oat (*Avena fatua*), even biotypes resistant to ACCase and ALS inhibitors, because they are still susceptible or slightly resistant to the imazamox + imazapyr mix. This is particularly useful in farming systems based on tight wheat–oat rotations.

5.2 Lupins

Unlike wheat and other cereals, the average yield of lupins in southern Chile has not increased in the last decades. In 1984, the country mean yield for wheat, oat and lupin crops was close to 2 Mg ha^{-1}. Twenty-eight years later, the yields of wheat and oat are approaching 6 Mg ha^{-1} whereas that of lupins remains near 2 Mg ha^{-1}. Lupins are concentrated in La Araucanía and two species are grown commercially, the white lupin, *Lupinus albus* L. and the narrow-leafed lupin, *L. angustifolius* L. Both sweet and bitter types of white lupin are cultivated, whereas the cultivated narrow-leafed is exclusively sweet.

The bitter white is exported to Mediterranean Europe and Arabic countries for human consumption and is almost entirely in the hands of small farmers. Sweet white and narrow-leafed lupin grains are used domestically as animal feed. Due to its higher protein content, the fish food manufacturers prefer dehulled white lupin for salmon diets. The salmon industry is currently an important, and potentially a huge, demander of plant protein and oil and, consequently, the expansion of the lupin and rapeseed crops in Chile are closely linked to it. Lupin yield is about 2.5 Mg ha^{-1} (Table 6.1) but this average takes into account the lower yields of small farmers that grow bitter lupin applying little technology. The sweet types are a better indicator of the yield potential of the crop. As a rule of thumb, farmers in southern Chile consider 3 Mg ha^{-1} as a good yield and 4 Mg ha^{-1} as a very good yield for sweet lupin. Experimentally, 4 and 5 Mg ha^{-1} are frequently obtained with narrow-leafed lupin (Mera et al., 2011); up to 6 Mg ha^{-1} may be achieved with both sweet (Peñaloza and Tay, 2011) and bitter white lupin (Mera and Galdames, 2007).

White lupin varieties of both determinate and indeterminate growing types (Chapter 12) have been used in southern Chile. Determinacy has been defined as the condition by which eventually all the buds on the plant become floral (Huyghe, 1997). So far, the indeterminate and semi-determinate (pods on mainstem and first-order branches) growing habits have empirically proved more suited to get higher and more stable yields. Autumn-sown varieties of sweet white lupin are predominant. With a cycle of about 10 months they make full use of the growing period of southern Chile. One of the main areas of sweet white lupin production is the rain-shadowed drylands of La Araucanía. Historically, this area receives 993 mm rainfall in the 6 months of autumn and winter (April–September). White lupin may be severely damaged by waterlogging when rainfall is intense, particularly in soils with poor drainage. In October, November and December, the historic rainfall records are 68, 50 and 37 mm respectively, whereas evaporation records are 72, 90 and 117, and maximum mean temperatures 19.3°, 22.4° and 22.9°C, respectively (Rouanet, 1982). Lupin crops start flowering in early October and keep on until late November. Pod growing and grain filling are mostly during November and December. Leaf drop progresses during the first fortnight of January, physiological maturity occurs by mid-January, and the dry pod point is reached during the second fortnight. The reproductive stages of white lupin in these drylands often proceed in a period of quick loss of soil humidity and temperature rise. Timely autumn sowings allow drought escape. In addition, early autumn sowings allow the root systems of lupins to be more developed by the time of the frosts, which could cause tissue damage and uprooting.

Lupin crops in southern Chile face a far less extreme water stress than lupin crops in western Australia, where temperatures shorten the life cycle of autumn-sown narrow-leafed lupin to some 7 months, which is equivalent to the period that uses a spring-sown lupin in southern Chile. Consequently, entrepreneurial farmers in Chile may attain fairly regularly 2.5–3 Mg ha^{-1} in the rain-shadowed drylands, an environment that can be considered among the few in southern Chile that suffer water deficit during the final quarter of the cropping season, and 3–4 Mg ha^{-1} in the (non rain-shadowed) central valley.

In narrow-leafed lupin crops of southern Chile, a large proportion of total grain yield is on the branches, rather than the mainstem (Mera et al., 2004). The relative productivity of mainstem pods is more significant in Australia than in Chile. In cropping seasons with available soil moisture during late spring - early summer (December–January), lupin crops of southern Chile may have a significant proportion of the total yield on second-order branches.

Sweet lupins are barely fertilized, and bitter lupin is not fertilized at all. Weed control is the main production weakness. For early autumn-sown lupins, the slow growing phase may last 160 days. Then the task is to achieve the longest possible weed control with a pre-emergence herbicide and a prompt cover and soil shadowing by a good crop plant density. Triazines, particularly simazine, are widely used. The effectiveness of this herbicide group is enhanced under the humid conditions of southern Chile, and reduces drastically the presence of ubiquitous weeds, such as ryegrass (*Lolium* spp.). Compared to simazine, metribuzin better controls some broadleaf weeds and is used in lupin varieties that tolerate this herbicide. Several post-emergence grass-selective (ACCase-inhibiting) herbicides are also used when needed. Densities of about 40 seeds m^{-2} for narrow-leafed lupin and 30 seeds m^{-2} for white lupin, to end up with crop populations of 30 and 20 plants m^{-2}, respectively, allow achievement of a relatively fast soil coverage and high yields. Row spacings of 17 and 34 cm are used for narrow-leafed lupin, whereas 34 cm is preferred for white lupin. Pre-harvest applications of diquat or paraquat to prevent weed seed set are seldom used.

The narrow-leafed lupin is found mainly in the non rain-shadowed central valley of La Araucanía though its acreage is less than that of white lupin. Traditionally, Australian varieties have been used; the first local variety was released only recently (2010). It is a spring-sown crop (August mostly, or the second fortnight of July if the weather is favorable) and is harvested usually by mid-February. According to historical weather data (Rouanet, 1983), from August to January the crop receives between 552 and 579 mm rainfall. In addition, monthly mean maximum temperatures are 11.7°, 13.7°, 16.4°, 17.9° and 22.4°C for the period August–December (Rouanet, 1983). The Australian varieties of narrow-leafed lupin that have been used in Chile are all unresponsive to vernalization due to the presence of the dominant gene *Ku*, which brings forward flowering by 2–5 weeks in western Australia (Cowling et al., 1998). As far as we are aware, types responsive to vernalization have not been tested in Chile and their performance in our high-rainfall, longer season southern environment is unknown.

6 CHALLENGES AND OPPORTUNITIES FOR CROPPING SYSTEMS OF SOUTHERN CHILE

Several and diverse are the challenges to be faced by farming systems based on annual crops of southern Chile such as yield and economic sustainability, land competition from fruit, beef and dairy production, and even industrial and market demands in a highly globalized country. But the national aim of being a 'Food Power' gives, on the other hand, good opportunities for annual crops, mainly sustained by the remarkable yield potential of this area. This section focuses on a few issues, among many others, that could contribute to the cropping systems and those which should be beaten to exploit fully the outstanding agro-ecosystems of southern Chile.

6.1 Yield potential improvement of wheat for the advance of southern cropping systems

Despite the remarkable long trend increase of grain yield (section 1.2), the rate of yield gain of wheat did not maintain the same rate after the 1980s (1980s: 194.5 kg ha^{-1} $year^{-1}$, 1990s: 89.9 kg ha^{-1} $year^{-1}$ and 2000s: 135.3 kg ha^{-1} $year^{-1}$). This outcome is far from auspicious and makes us wonder whether the sharp increase in grain yield achieved between 1980 and 1995 (Engler and del Pozo, 2013) is really possible.

Yield potential remains the main target for breeders since it has been shown to increase both attainable and on-farm yields (Slafer and Calderini, 2005; Fischer and Edmeades, 2010; Chapter 16), even under conditions that are frequently stressful during grain filling (Acreche et al., 2008). The increase of yield potential is especially important in favorable environments such as that of southern Chile, where it is relatively easy to take advantage of high-yielding cultivars (Hall and Richards, 2013). Recently, Bustos et al. (2013) showed that local wheat cultivars sown by farmers are not fully exploiting the favorable environmental conditions of southern Chile. The authors reported that the local cultivars were out-yielded by two doubled haploid lines (DH) and two CIMMYT cultivars (Fig. 6.18a). The highest yield reached by the best DH was 16.6 Mg ha^{-1} of grain dry matter in experimental plots. Interestingly, 12% of the 105 double haploid lines assessed in two growing seasons yielded between 14 and 16 Mg ha^{-1}. DH lines reached higher grain yield due to both higher biomass and harvest index, while CIMMYT's varieties surpassed the local ones by improved harvest index with similar biomass (Fig. 6.18b). In the light of these results, a clear opportunity could be envisaged for both wheat breeding and cereal-based cropping system of southern Chile to increase potential and actual yield; this could be achieved even by harvest index, which would be easier to improve than biomass as was demonstrated in the past by the 'Green Revolution' or by improving spike dry matter through modifying phenology as proposed in Chapter 12. The strategy to improve grain yield by increasing biomass could imply higher demand of resources. Therefore, the challenge of improving nutrient- and water-use efficiency would become a central objective of wheat breeding programs aimed at taking advantage of the high-yielding conditions of southern Chile. However, more research is necessary to evaluate the demand and efficiencies of resources for these DH lines, the heritability of traits and interactions in cropping systems.

Additionally, lodging remains a challenge for wheat breeding aimed at increasing yield potential (Foulkes et al., 2011), especially in the favorable environments of Chile, due to the high weight of the spikes during grain filling and the likely occurrence of weather conditions for lodging (rain plus wind) at this phenophase. Lower plant rate and more equidistant distribution of plants could be considered in the future to prevent lodging according to the results of Bustos et al. (2013), who found crops at 44 plants m^{-2}

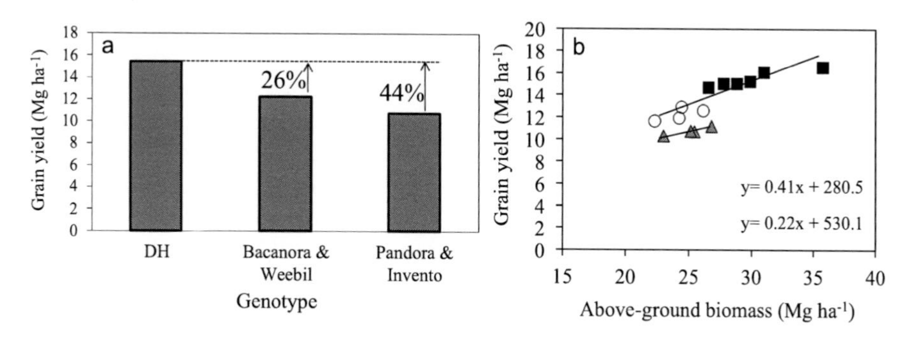

FIG. 6.18 (a) Averaged grain yield of two highest yielding doubled haploid lines, two CIMYT cultivars (Bacanora and Weebil) and two local cultivars (Pandora and Invento). (b). Relationship between grain yield and above-ground biomass of the two highest yielding doubled haploid lines (squares), two CIMMYT (circles) and two local (triangles) cultivars. Modified from Bustos et al. (2013).

were shorter and produced the same yield as crops at 350 plants m^{-2}.

Clearly, both breeding and crop management are essential to bridge the gap between actual, attainable and potential yields. Fortunately, the development of strategies to ameliorate the management of wheat and other crops has been boosted by initiatives as CropCheck Chile (http://www.fundacionchile.com/archivos/ManualTrigo_baja7853253.pdf) aimed at providing decision-making protocols available to farmers and overseeing the progress of the crop. This will facilitate seizing the opportunities brought by plant breeding.

6.2 The need to link breeding and physiology for potato improvement

Most potato breeding programs are oriented to improve one or more of the following attributes: tuber yield; table and processing quality; starch content; earliness and dormancy (CIP, 2006; Simakov et al., 2008; Bradshaw, 2009; Birch et al., 2012). These traits, in addition to regional specific objectives, such as disease-resistance, show variations among breeding programs and regions. In southern Chile, breeding for high yield, tolerance to water stress, peel and pulp color, dry matter content according to use (fresh market or industry) and resistance to late blight, potato leafroll virus (PLRV), potato virus Y (PVY) and potato cyst nematode (*Globodera rostochiensis*) have been the main targets (J. Kalazich, INIA, personal communication). However, potato breeding objectives do not consider either an ideotype approach or physiological associated traits as in other crops, e.g. wheat, maize, sunflower (Chapters 14, 15, 16 and 17). For instance, no phenological adjustment studies have been carried out to improve yield as has been researched in wheat (e.g. Chapter 12). Kooman (1995) postulated that the period from emergence to tuberization should be adequately long to generate a critical leaf area and stolon number to induce high numbers of tubers. The length of this phase, however, is highly correlated with the duration of subsequent periods (Kooman et al., 1996b) and opposes the goal of improving earliness. For this reason, the duration of specific stages of the crop should be a research objective of physiologists and breeders to explore their impact on tuber yield. Khan et al. (2013) proposed the use of simulation models to identify breeding targets and to evaluate the effective tuber bulking duration (ETBD), which is currently assumed to end at harvest. Assessments like this would provide important information to potato breeding programs (Chapter 14) in addition to the present research in markers and other molecular tools (Chapter 18). An example of the gap between breeding and physiology to improve potato performance is that current temperature, radiation and photoperiod are favorable for cultivars of subspecies *tuberosum*, which are better adapted to long days, but the phenological drivers and responses to photoperiod are still little known in our environment. Therefore, the development of potato breeding programs supported by a physiological approach is at once a challenge and an opportunity for potato-based cropping systems of southern Chile.

6.3 Lupin improvement is a key to consolidate this crop in southern cropping systems

Increased lupin breeding and crop management research is needed for this crop to consolidate as the choice for plant protein in Chile. Lupin yields should be high (over 3.5 Mg ha^{-1}) for this grain to be consistently competitive with imported soybean meal. The minimum protein yield for any protein crop to become an option is thought to be 1 Mg ha^{-1}, but a yield over 1.5 Mg ha^{-1} of protein would definitively make lupins a solid protein crop. A more aggressive breeding program is required to take full advantage of the privileged environment of southern Chile. Current Chilean varieties and lines of the albus species have significantly out-yielded recent dwarf determinate French varieties in experiments, in spite of the fact that in the country there is a relatively small-budget breeding program carried out by the private sector and very little breeding and research activities in the public sector. More intense work on lupin breeding and agronomy might have a profound effect on this crop and the agricultural systems of southern Chile.

6.4 The challenge of herbicide-resistant weeds

Intensive use of the soil with tight rotations based on cereals allowed grass weed populations to proliferate, and their control has become particularly challenging after the arising of herbicide-resistant biotypes. Chile is now one of the many countries affected by the presence of herbicide-resistant weeds. The first case (*Lolium rigidum* resistant to ACCase inhibitors) was reported in 1997 and, ever since, resistant biotypes have been reported in another 10 weed species (Heap, 2013). Seven of them are relevant for farming systems of southern Chile: *Lolium rigidum*, *Lolium multiflorum*, *Avena fatua*, *Cynosurus echinatus*, *Raphanus sativus*, *Anthemis cotula*, and *Anthemis arvensis*. During the last decade, resistant populations of grass weeds (*L. rigidum*, *L. multiflorum*, *A. fatua* and *C. echinatus*) grew quickly in wheat crops, raising the cost associated with weed control from 7% to more than 20% of the total cost (Espinoza et al., 2011). The case of *Lolium* is particularly serious as multiple resistances have evolved and some biotypes are resistant to both ACCase inhibitors and ALS inhibitors, and even a biotype of *L. multiflorum* has acquired resistance to these two herbicides plus resistance to glycine herbicides, meaning that some *L. multiflorum* populations are also resistant to glyphosate (Díaz and Galdames, 2012). Populations of this resistant biotype have already spread across La Araucanía and the region immediately northward (Espinoza et al., 2013), particularly in fields with no-tillage (Espinoza et al., 2012). Like *L. rigidum*, *C. echinatus* has evolved resistance to both ACCase and ALS inhibitors. Luckily, *A. fatua* biotypes only resistant to ACCase inhibitors have been found so far.

To add complexity to the situation, broadleaf weed biotypes (*R. sativus*, *A. cotula* and *A. arvensis*) resistant to ALS inhibitors, including the mix imazamox + imazapyr of the Clearfield system, were reported in 2010 (Espinoza and Rodriguez, 2012). Fortunately, the biotype of *R. sativus* proved susceptible to MCPA, and the biotypes of *Anthemis* were susceptible to clopyralid.

Indeed, the scenario is challenging and case-specific strategies must be enforced to prevent worsening of the situation. For resistant *Lolium* biotypes, a pre-sowing strategy already adopted by farmers is the burning of crop residues and the application of either paraquat or diquat after weed emergence. A post-sowing strategy is based on soil-applied herbicides, for example, the mix flufenacet, flurtamone and diflufenican in winter cereal crops, metazachlor in rapeseed crops, and metribuzin or simazine in lupin crops. Alternatively, a growing number of farmers are going back to conventional tillage in fields where no-tillage was practiced for years as a way to promote a faster and greater germination of *Lolium*, then weeds are mechanically eliminated. Indeed, crop rotation expands the choice of chemicals and the use of alternative herbicides is an essential mitigation measure in any case of weed resistance.

6.5 Scenarios of global change

Current knowledge of climate change is described in Chapter 20; therefore, this section will highlight briefly the likely effects of global warming and the increase of UV-B radiation on cropping systems of southern Chile. Projections for the southern hemisphere indicate that changes in average temperature will be lower than in the northern hemisphere (Easterling, 1997; Chapter 20). Current scenarios of moderate climate change (B2) for southern Chile forecast for

2100 an average temperature increase of 1–2°C for winter and between 2 and 3°C in summer (DGF, 2006). Precipitation is estimated to drop around 20% in winter and by 40% in summer. Under these projections, spring–summer crops (i.e. spring wheat or potato) are more vulnerable than winter cereals or rapeseed, especially in areas where water availability is relatively low (i.e. La Araucanía). Less change in rainfall is expected for Los Ríos and Los Lagos. A key aspect will be the timing of high temperature events, taking into account that the degree of impact of heat stress on grain yield of wheat depends on the crop phenophase as was found in field experiments (Lizana and Calderini 2013). The most sensitive stage has been the booting–anthesis period as grain yield decreased 18% under current scenarios, without impact on grain protein concentration (Lizana and Calderini 2013). Preliminary results did not show impact on potato tuber yield when temperature was raised 2.6°C over the environment during tuberization and at tuber filling, but a significant loss of tuber commercial quality (mainly deformation) was found (Lizana et al., unpublished data).

UV-B increases have been registered in southern Chile, especially in spring–summer periods (Lovengreen et al., 2000), raising stress pressure on summer crops. UV-B/PAR ratio and crop phenophase were identified as the key determinants of UV-B impact on wheat yield (Lizana et al., 2009).

Simulation assessment based on long-term climate records showed that spring wheat and potato have a strong dependence on annual weather conditions in Temuco and Valdivia, but management practices such as planting date and rate, and fertilization strategies seem promising to mitigate global change in southern Chile (Meza et al., 2003).

Little information is available at present and more knowledge, integrating climatic, soil, crop and rotation characteristics, is necessary to have a complete picture of the possible impact of climate change in this area.

7 CONCLUDING REMARKS

On-farm crop yields in southern Chile can be exceptional as a result of (1) favorable environments due to the solar radiation:temperature ratio, rainfall and soil characteristics, (2) adequate genetics and (3) high-input crop management. Despite this, annual crops like wheat, barley and potato have shown a significant decrease in their acreages mainly due to the expansion of fruit crops and pastures. On the other hand, oat has had a remarkable increment of area and yield, the former especially over the last 5 years. The development of this crop is an example of successful interaction between farmers, breeding, industry and export policies. In the case of crops, such as wheat and potato, Chilean breeding programs could continue improving yield potential to take advantage of the environment. In the case of rapeseed, western European hybrids have made this crop reborn in southern cropping systems and it is becoming a stable oilseed crop. Lupins, on the other hand, are crops with insufficient genetic materials; the local breeding is not providing the array of varieties needed to fit different environments, and research on agronomy is scarce. As a result, while cereal, potato and oilseed yields have steadily increased, lupin yields have remained stagnant for 20 years.

Tolerance to drought is increasingly a concern for crop improvement as drier seasons are becoming more frequent and the climate change scenarios forecast hotter and dryer conditions for southern Chile. The expanding area of no-till soil management could be helpful to mitigate global change, but little research has been carried out to size the impacts of global change and more knowledge is required taking into account the unique characteristics of *trumao* soils that prevent simple extrapolations of studies performed in other environments. Input-use efficiency has also been a rather neglected issue, weakly addressed by research and technology transfer. In the case of wheat, potato and rapeseed, abuse of fertilizer doses, particularly nitrogen, is widespread and herbicide-resistant biotypes of weeds have become particularly challenging for southern cropping systems.

References

Abbate, P.E., Andrade, F.H., Lázaro, L., et al.,1998. Grain yield increase in recent Argentine wheat cultivars. Crop Sci. 38, 1203–1209.

Acevedo, E., Silva, P., Silva, H., Solar, B., 1999. Wheat production in Mediterranean environments. In: E.H., Satorre, E.J., Slafer, G.A., (Eds.), Wheat ecology and physiology of yield determination, The Haworth Press Inc., New York, pp. 295-331.

Acreche, M.M., Briceño-Félix, G., Sánchez, J.A.M., Slafer, G.A., 2008. Physiological bases of genetic gains in Mediterranean bread wheat yield in Spain. Eur. J. Agron. 28, 162–170.

Arisnabarreta, S., Miralles, D.M., 2008. Critical period for grain number establishment of near isogenic lines of two- and six-rowed barley. Field Crops Res. 107, 196–202.

Benoit, G.R., Stanley, C.D., Grant, W.J., Torrey, D.B., 1983. Potato top growth as influenced by temperatures. Am. Potato J. 60, 489–502.

Birch, P.R.J., Bryan, G., Fenton, B., et al.,2012. Crops that feed the world 8: Potato: are the trends of increased global production sustainable? Food Secur. 4, 477–508.

Bradshaw, J., 2009. Potato breeding at the Scottish plant breeding station and the Scottish Crop Research Institute: 1920-2008. Potato Res. 52, 141–172.

Bustos, D.V., Hasan, A.K., Reynolds, M.P., Calderini, D.F., 2013. Combining high grain number and weight through a DH-population to improve grain yield potential of wheat in high-yielding environments. Field Crops Res. 145, 106–115.

Calderini, D.F., Lizana, C., Sandaña, P., Riegel, R., 2013. Productividad de biomasa, captura de recursos y sustentabilidad de trigo y cebada en ambientes de alto potencial de rendimiento del sur de Chile. In: Valle, S., Lizana, C., Calderini, D. (Eds.), Sistemas de producción de trigo y cebada: decisiones de manejo en base a conceptos ecofisiológicos para optimizar el rendimiento, la calidad y el uso de los recursos. Universidad Austral de Chile, Chile, pp. 63–79.

Calderini, D.F., Reynolds, M.P., 2000. Changes in grain weight as a consequence of de-graining treatments at pre- and post-anthesis in synthetic hexaploid lines of wheat (*Triticum durum* × *T. tauschii*). Aust. J. Plant Physiol. 27, 183–191.

Campillo, R., Jobet, C., Undurraga, P., 2007. Optimization of nitrogen fertilization for high-yielding potential wheat on Andisols at the Araucanía Region, Chile (in Spanish). Agric. Téc. 67, 281–291.

Chowdhury, S.I., Wardlaw, I., 1978. The effect of temperature on kernel development in cereals. Aust. J. Agric. Res. 29, 205–223.

CIP, 2006. Procedures for standard evaluation trials of advanced potato clones. Report, International Potato Center: La Molina, Peru.

CIREN, Centro de Información de Recursos Naturales, 2012. Catastro frutícola. Principales resultados. Región de La Araucanía, Julio 2012. ODEPA, CIREN: Santiago, Chile. [http://icet.odepa.cl/tmp/obj_699434/6117_catastro-IXRegion-2012.pdf]

CNE/PNUD/UTFSM, 2008. Irradiancia solar en los territorios de la república de Chile. Ed. Proyecto CHI/00/G32 Chile: Remoción de Barreras para la Electrificación Rural con Energías Renovables. Online:Available at http://www.termic.cl/descargas/RegistroSolarimetrico%2013%20MB%20prof.%20Sarmientos.pdf

CONAF, Corporación Nacional Forestal, 2008a. Catastro de uso del suelo y vegetación, período 1993-2007: monitoreo y actualización. Región de Los Ríos. MINAGRI-CONAF-CONAMA-UACH, Santiago, Chile.

CONAF, Corporación Nacional Forestal, 2008b. Catastro de uso del suelo y vegetación, período 1993-2007: monitoreo y actualización. Región de Los Lagos. MINAGRI-CONAF-CONAMA-UACH, Santiago, Chile.

CONAF, Corporación Nacional Forestal, 2009. Catastro de uso del suelo y vegetación, período 1993-2007: monitoreo y actualización. Región de La Araucanía. MINAGRI-CONAF-CONAMA-UACH, Santiago, Chile.

CONAMA, Corporación Nacional del Medio Ambiente, 2008. Análisis de vulnerabilidad del sector silvoagropecuario, recursos hídricos y edáficos de Chile frente a escenarios de cambio climático. Capítulo Impactos productivos en el sector silvoagropecuario de Chile frente a escenarios de Cambio Climático. Centro AGRIMED, Universidad de Chile.

Cowling, W.A., Huyghe, C., Swiecicki, W., 1998. Lupin breeding. In: Gladstones, J.S., Atkins, C.A., Hamblin, J. (Eds.), Lupin as crop plants. CAB International, Wallingford, pp. 93–120.

DGF-University of Chile, 2006. Estudio de la Variabilidad Climática en Chile para el Siglo XXI. Informe Final. Available from: http://www.sinia.cl/1292/articles-50188_recurso_8.pdf (verified April, 22 2014).

Díaz, J., Galdames, R., 2012. A new biotype of Lolium multiflorum resistant to glyphosate in cereal production. *In*: De Prado, R., (ed.), International workshop on glyphosate weed resistance: European status and solutions, 79. Cordoba, Spain, 3-4 May, 2012.

Dörner, J., Dec, D., Feest, E., Díaz, M., Vásquez, N., 2012. Dynamics of soil structure and pore functions of a volcanic ash soil under tillage. Soil Till. Res. 125, 52–60.

Dörner, J., Dec, D., Peng, X., Horn, R., 2009. Efecto del cambio de uso en la estabilidad de la estructura y la función de los poros de un Andisol (Typic Hapludand) del sur de Chile. J. Soil Sci. Plant Nutr. 9, 190–209.

Dörner, J., Dec, D., Zúñiga, F., Sandoval, P., Horn, R., 2011. Effect of the land use change on Andosol's pore functions and their functional resilience after mechanical and hydraulic stresses. Soil Till. Res. 115-116, 71–79.

Dörner, J., Zúñiga, F., López, I., 2013a. Short-term effects of different pasture improvement treatments on the physical quality of an Andisol. J. Soil Sci. Plant Nutr. 13, 381–399.

Dörner, J., Dec, D., Zúñiga, F., et al.,2013b. Soil changes in the physical quality of an Andosol under different management intensities in southern Chile. In: Krümmelbein, J., Horn, R., Pagliai, M. (Eds.), Soil degradation, advances in Geoecology 42. Catena Verlag GMBH, Reiskirchen, Germany.

Easterling, D.R., Horton, B., Jones, P.D., et al.,1997. Maximum and minimum temperature trends for the globe. Science 277, 364–367.

Engler, A., del Pozo, A., 2013. Assessing long- and short-term trends in cereal yields: the case of Chile between 1929 and 2009. Cie. Invest. Agrar. 40, 55–67.

Espinoza, N., Palma, J.C., Rodríguez, C., 2011. Malezas resistentes: Impacto en los costos de producción de trigo. Crops Land 4, 3–8.

Espinoza, N., Rodríguez, C., 2012. Aparecen las primeras malezas de hoja ancha resistentes a herbicidas en el país. Crops Land 5, 3–6.

Espinoza, N., Rodríguez, C., Contreras, G., 2012. Strategies to manage resistance to glyphosate in Lolium multiflorum in southern Chile. *In*: De Prado, R., (ed.), International workshop on glyphosate weed resistance: European status and solutions, 85. Cordoba, Spain, 3-4 May, 2012.

Espinoza, N., Llancaqueo, F., Mera, M., Contreras, G., De Prado, R., 2013. Mapa de la resistencia a glifosato, inhibidores de ACCasa e inhibidores de ALS en biotipos de ballica (Lolium multiflorum) de Chile. XXI Congreso de la Asociación Latinoamericana de Malezas (ALAM) y XXXIV Congreso de la Asociación Mexicana de la Ciencia de la Maleza (ASOMECIMA). Cancún, México, 11-15 Noviembre, 2013.

Estrada-Campuzano, G., Miralles, D., Slafer, G., 2008. Yield determination in triticale as affected by radiation in different development phases. Eur. J. Agron. 28, 597–605.

Fischer, R.A., 1985. Number of kernels in wheat crops and the influence of solar radiation and temperature. J. Agric. Sci. (Camb.) 105, 447–461.

Fischer, R.A., Edmeades, G.O., 2010. Breeding and cereal yield progress. Crop Sci. 50, S85–S98.

Foulkes, M.J., Slafer, G.A., Davies, W.J., et al.,2011. Raising yield potential in wheat: III. Optimizing partitioning to grain while maintaining lodging resistance. J. Exp. Bot. 62, 469–486.

García, J.C., 2011. Producción de avena (*Avena sativa* L.). Aspectos generales. Primer seminario de avena: Avena de la Araucanía: Aporte en la dieta alimenticia cultivos. Universidad de la Frontera. http://www.slideshare.net/vincficaUFRO/produccin-de-avena-juan-carlos-garca.

García, G.A., Hasan, A.K., Puhl, L.E., Reynolds, M.P., Calderini, D.F., Miralles, D.J., 2013. Grain yield potential strategies in an elite wheat double-haploid population grown in contrasting environments. Crop Sci. 53, 2577–2587.

Hall, A.J., Richards, R.A., 2013. Prognosis for genetic improvement of yield potential and water-limited yield of major grain crops. Field Crops Res. 143, 18–33.

Hasan, A.K., Herrera, J., Lizana, X.C., Calderini, D.F., 2011. Carpel weight, grain length and stabilized grain water content are physiological drivers of grain weight determination of wheat. Field Crops Res. 123, 241–247.

Haverkort, A.J., Sandaña, P., Kalazich, J., 2014. Yield gaps and ecological footprints of potato production systems in Chile. Potato Res., Published online March 2014.

Heap I., The International Survey of Herbicide Resistant Weeds. Online. Internet. Tuesday, August 27, 2013. Available www.weedscience.org.

Huber, A., Trecaman, R., 2004. Eficiencia del uso del agua en plantaciones de Pinus radiata en Chile. Bosque (Valdivia) 25, 33–43.

Huyghe, C., 1997. White lupin (*Lupinus albus* L.). Field Crops Res. 53, 147–160.

INE, Instituto Nacional de Estadísticas. 2007. Censo agropecuario y forestal 2007. Estadísticas por comuna. Online. Available http://www.ine.cl/canales/chile_estadistico/censos_agropecuarios/censo_agropecuario_07.php.

INIA, 1989. Mapa agroclimático de Chile. Instituto Nacional de Investigaciones Agropecuarias. Ministerio de Agricultura, Chile.

IPCC, 2007. Climate change 2007. Cambridge Press, Cambridge.

Kalazich, J., López, H., Rojas, J., et al.,2004. Karu-INIA, new potato cultivar for Chile. Agric. Téc. 64, 409–413.

Khan, M., van Eck, H., Struik, P., 2013. Model-based evaluation of maturity type of potato using a diverse set of standard cultivars and a segregating diploid population. Potato Res. 56, 127–146.

Kierkegaard, J.A., Sarwar, M., Wong, P.T.W., Mead, A., Howe, G., Newell, M., 2000. Field studies on the biofumigation of take-all by brassica break crops. Aust. J. Agric. Res. 51, 445–456.

Kooman, P.L., 1995. Yielding ability of potato crops as influenced by temperature and daylength. Thesis Landbouw Universiteit Wageningen.

Kooman, P.L., Fahem, M., Tegera, P., Haverkort, A.J., 1996a. Effects of climate on different potato genotypes 1. Radiation interception, total and tuber dry matter production. Eur. J. Agron. 5, 193–205.

Kooman, P.L., Fahem, M., Tegera, P., Haverkort, A.J., 1996b. Effects of climate on different potato genotypes 2. Dry matter allocation and duration of the growth cycle. Eur. J. Agron. 5, 207–217.

Lázaro, L., Abbate, P., Cogliatti, D., Andrade, F., 2009. Relationship between yield, growth and spike weight in wheat under phosphorus deficiency and shading. J. Agric. Sci. (Camb.) 148, 83–93.

Lizana, X.C., Hess, S., Calderini, D., 2009. Crop phenology modifies wheat responses to increased UV-B radiation. Agric. For. Meteorol. 149, 1964–1974.

Lizana, X.C., Calderini, D.F., 2013. Yield and grain quality of wheat in response to increased temperatures at key periods for grain number and grain weight determination: considerations for the climatic change scenarios of Chile. J. Agric. Sci. (Camb.) 151, 209–221.

Lovengreen, C.H., Fuenzalida, H., Villanueva, L., 2000. Ultraviolet solar radiation at Valdivia, Chile (39,88S). Atmos. Environ. 34, 4051–4061.

Martínez, G.I., Ovalle, C., Del Pozo, A., et al.,2011. Influence of conservation tillage and soil water content on crop yield in dryland compacted Alfisol of Central Chile. Chil. J. Agric. Res. 71, 615–622.

Martínez, E., Fuentes, J.P., Silva, P., Valle, S., Acevedo, E., 2008. Soil physical properties and wheat root growth as affected by no-tillage and conventional tillage systems in a Mediterranean environment of Chile. Soil Till. Res. 99, 232–244.

Matus, I., Mellado, M., Pinares, M., Madariaga, R., del Pozo, A., 2012. Genetic progress in winter wheat cultivars released in Chile from 1920 to 2000. Chil. J. Agric. Res. 72, 303–308.

Mera, M., Harcha, C., Miranda, H., Rouanet, J.L., 2004. Genotypic and environmental effects on pod wall proportion and pod wall specific weight in *Lupinus angustifolius*. Aust. J. Agric. Res. 54, 397–406.

Mera, M., Galdames, R., 2007. Boroa-INIA, first bitter lupin (*Lupinus albus*) cultivar released in Chile for export. Chil. J. Agric. Res. 67, 320–324.

Mera, M., Alcalde, J.M., Buirchell, B., 2011. Lupino: las variedades australianas recientes. Tierra Adentro 92, 41–43.

Mera, M., Espinoza, N., 2012. Los primeros híbridos invernales de raps Clearfield. Crops Land 6, 10–14.

Meza, F.J., Wilks, D.S., Riha, S.J., Stedinger, J.R., 2003. Value of perfect forecasts of sea surface temperature anomalies for selected rain-fed agricultural locations of Chile. Agric. For. Meteorol. 116, 117–135.

Montaldo, P., 1974. Regiones ecológicas del cultivo de la papa (*Solanum tuberosum*) en Chile. Agro. Sur. 2, 25–27.

Montenegro, A., Besoain, M., Toro, C., 1992. Evaluación del efecto de las rocas fosfóricas en el cultivo del raps. *In*: Campillo, R., (ed.), 1er Seminario nacional sobre uso de rocas fosfóricas en agricultura. Serie Carillanca N°29. Temuco, Chile, pp. 347-371.

Muurinen, S., Peltonen-Sainio, P., 2006. Radiation-use efficiency of modern and old spring cereal cultivars and its response to nitrogen in northern growing conditions. Field Crops Res. 96, 363–373.

ODEPA. Oficina de estudios y políticas agrarias., 2014. Estadísticas por Macro rubros Agrícolas. Online. Available http://www.odepa.cl/estadisticas/productivas.

Peñaloza, E., Tay, J., 2011. Fertilización del cultivo de lupino. *In* : Hirzel, J., (ed.), Fertilización de cultivos en Chile. Colección libros INIA 28. Instituto de Investigaciones Agropecuarias: Chillán, Chile, pp. 351-367.

Quiroz, J., 2010. Rendimiento y producción de biomasa de trigo, cebada y triticale bajo riego y secano durante el llenado de grano en sur de Chile. MSc. Thesis. Universidad Austral de Chile.

Rodríguez, J., Pinochet, D., Matus, F., 2001. Fertilización de los cultivos. Registro propiedad intelectual: 118.251. Santiago, Chile.

Rouanet, J.L., 1982. Clasificación agroclimática IX región, 2ª aproximación; Macroárea I. IPA Carillanca 2 (1), 22–26.

Rouanet, J.L., 1983. Clasificación agroclimática IX región, 2ª aproximación; Macroárea II. IPA Carillanca 2 (2), 23–25.

Rouanet, J.L., Romero, O., Demanet, R., 1988. Áreas agroecológicas en la IX región: Descripción. IPA Carillanca 7 (1), 18–23.

Rouanet, J.L., Mera, M., Acevedo, E., Silva, P., 2005. Rotaciones y sus efectos sobre la productividad de los cultivos y sobre la calidad del suelo. In: Rouanet, J.L. (Ed.), Rotaciones de cultivos y sus beneficios para la agricultura del sur., Fundación Chile. Instituto de Investigaciones Agropecuarias, Santiago Chile, pp. 29–46.

Ruiz, M.J., 2008. Efecto del fósforo en la producción de raps canola (*Brassica napus*) en un suelo andisol de la región de La Araucanía. Thesis Ingeniero Agrónomo, Universidad de La Frontera., Temuco, Chile.

Sadras, V.O., Slafer, G.A., 2012. Environmental modulation of yield components in cereals: Heritabilities reveal a hierarchy of phenotypic plasticities. Field Crops Res. 127, 215–224.

Sadras, V.O., O'Leary, G.J., Roget, D.K., 2005. Crop responses to compacted soil, capture and efficiency in the use of water and radiation. Field Crops Res. 91, 131–148.

Sandaña, P., Ramírez, M., Pinochet, D., 2012. Radiation interception and radiation use efficiency of wheat and pea under different P availabilities. Field Crops Res. 127, 44–50.

Sandaña, P., Pinochet, D., 2011. Ecophysiological determinants of biomass and grain yield of wheat under P deficiency. Field Crops Res. 120, 311–319.

Sandaña, P.A., Harcha, C.I., Calderini, D.F., 2009. Sensitivity of yield and grain nitrogen concentration of wheat, lupin and pea to source reduction during grain filling. A comparative survey under high yielding conditions. Field Crops Res. 114, 233–243.

Savin, R., Slafer, G.A., 1991. Shading effects on the yield of an Argentinian wheat cultivar. J. Agric. Sci. (Camb.) 116, 1–7.

Slafer, G.A., Calderini, D.F., 2005. Importance of breeding for further improving durum wheat yield. In: Royo, C., Nachit, M.N., Di Fonzo, N., Araus, J.L., Pfeiffer, W.H., Slafer, G.A. (Eds.), Durum wheat breeding. Current approaches and future strategies. Food Products Press, The Haworth Press, New York, pp. 87–95.

Simakov, E., Anisimov, B., Yashina, I., Uskov, A., Yurlova, S., Oves, E., 2008. Potato breeding and seed production system development in Russia. Potato Res. 51, 313–326.

Sinclair, T.R., Muchow, R.C., 1999. Radiation use efficiency. Adv. Agron. 65, 215–265.

Struik, P., Geertsema, C., Custers, C.H.M.G., 1989. Effects of shoot, root and stolon temperature on the development of the potato (*Solanum tuberosum* L.) I. Development of the haulm. Potato Res. 32, 133–141.

SOFOFA, 2013. Sociedad de Fomento Fabril. http://web.sofofa.cl

Uribe, J.M., Cabrera, R., de la Fuente, A., Paneque, M., 2012. Atlas Bioclimático de Chile. Universidad de Chile, Santiago de Chile.

Valle, S.R., Pinochet, D., Calderini, D.F., 2009a. Al toxicity effects on radiation interception and radiation use efficiency of Al-tolerant and Al-sensitive wheat cultivars under field conditions. Field Crops Res. 114, 343–350.

Valle, S.R., Carrasco, J., Pinochet, D., Calderini, D.F., 2009b. Grain yield, above-ground and root biomass of Al-tolerant and Al-sensitive wheat cultivars under different soil aluminum concentrations at field conditions. Plant Soil 318, 299–310.

Valle, S.R., Pinochet, D., Calderini, D.F., 2011. Uptake, partitioning and use efficiency of N, P, K, Ca and Al by wheat Al-sensitive and Al-tolerant cultivars under a wide range of soil Al concentrations. Field Crops Res. 121, 392–400.

Van Dam, J., Kooman, P.L., Struik, P.C., 1996. Effects of temperature and photoperiod on early growth and final number of tubers in potato (*Solanum tuberosum* L.). Potato Res. 39, 51–62.

CHAPTER

7

Cereal yield in Mediterranean-type environments: challenging the paradigms on terminal drought, the adaptability of barley vs wheat and the role of nitrogen fertilization

Roxana Savin[1], *Gustavo A. Slafer*[1,2], *C. Mariano Cossani*[3], *L. Gabriela Abeledo*[4], *Victor O. Sadras*[5]

[1]Department of Crop and Forest Sciences & AGROTECNIO (*Center for Research in Agrotechnology*), University of Lleida, Lleida, Spain

[2]ICREA (Catalonian Institution for Research and Advanced Studies)

[3]CIMMYT (International Maize and Wheat Improvement Center), Mexico, DF, Mexico

[4]Department of Crop Production, University of Buenos Aires, Buenos Aires, Argentina

[5]South Australian Research and Development Institute, Waite Campus, Adelaide, Australia

1 INTRODUCTION

The Mediterranean agricultural region comprises the lands surrounding the Mediterranean Sea: the southern strip of Europe, northern lands of Africa and west Asia (Fig. 7.1). This region has relatively cold and wet winters, dry and hot spring-summers (Aschmann, 1973) and soils with typically low fertility (Ryan et al., 2009). Aschmann (1973) highlighted the concentration of rainfall in the winter half year as the most distinctive element of the Mediterranean climate, and proposed 65% of annual rainfall in this period as a boundary in his definition. Annual rainfall is frequently scarce, broadly ranging from 200 to 600 mm (Fig. 7.1b), with high inter-annual

Crop Physiology. DOI: 10.1016/B978-0-12-417104-6.00007-8

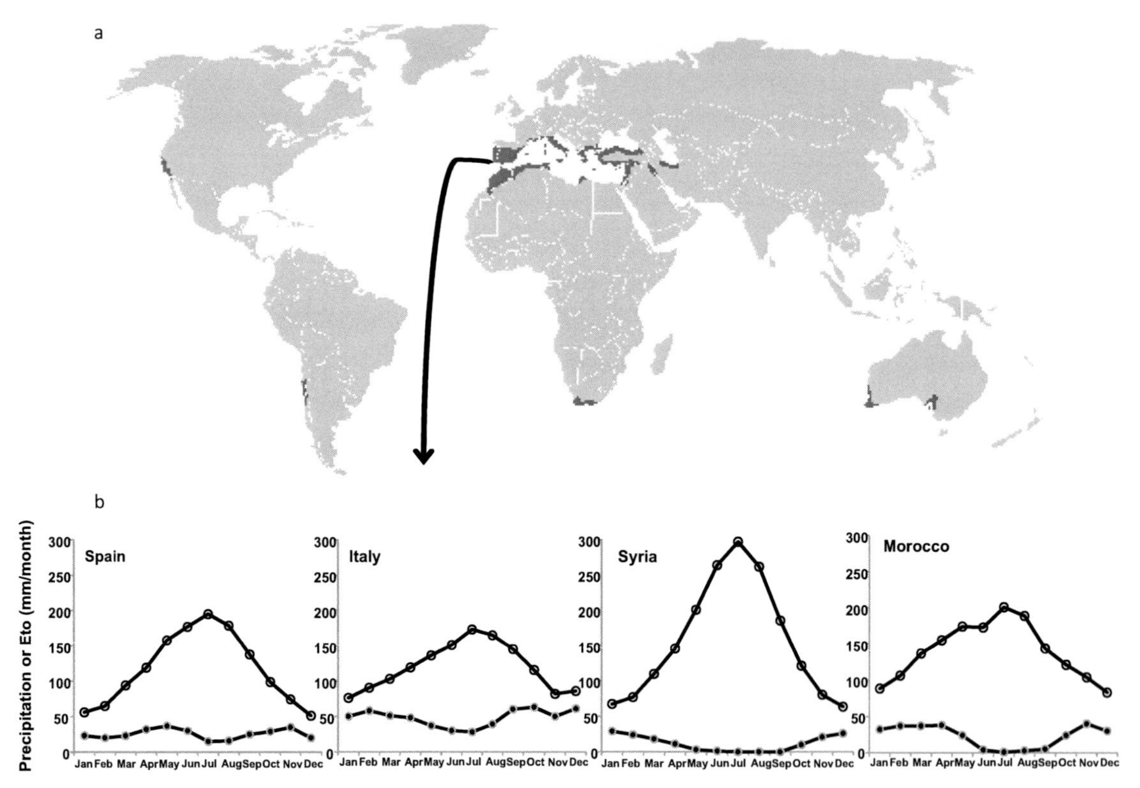

FIG. 7.1 (a) Areas with Mediterranean-type environments; and (b) long-term precipitation (closed symbols) and reference evapotranspiration (open symbols) at selected locations. *Data from www.allmetsat.comand http://badc.nerc.ac.uk/mybadc.*

variability. Other agricultural lands in the world share these characteristics and, by extension, are considered to have Mediterranean-type climates (Fig. 7.1a).

Rainfall seasonality restricts agriculture to winter-spring crops and the amount and variation of rainfall drives relatively low and highly variable yield. The contemporary cereal-based systems of Mediterranean regions are similar to those described by Pliny in ancient Rome and Greece: wheat and barley are the backbone, with grain legumes, pastures, and/or *Brassica* species used as break-crops. These cropping systems have therefore stood the test of time but, alongside this long history, a core paradigm has grown that has rarely been challenged. In this chapter, we revise three elements of this paradigm. First, the notion that annual crops in these environments are exposed to 'terminal drought'. Using physiological and environmental analysis, we show that this label over-emphasizes grain filling, and neglects the most critical pre-flowering period; this has implications for both crop management and breeding. Second, the notion that barley performs better than wheat at the lowest end of the rainfall range. We emphasize the lack of solid scientific evidence supporting this view, and briefly consider the implications for patterns of land allocation. Third, the notion that nitrogen fertilization may reduce yield in low rainfall sites and seasons. We argue that, again in this case, the scientific evidence does not fully support this viewpoint, and discuss the opportunities to improve yield and water-use efficiency by better management of nitrogen fertilization.

2 TERMINAL DROUGHT?

Drought and heat stress, particularly during flowering and grain filling, are trademarks of the Mediterranean region (Loss and Siddique, 1994; Acevedo et al., 1999). Stresses are more severe in the southern (N Africa – W Asia) than in the northern (S. Europe) part of the Mediterranean basin (Fig. 7.1b). The combination of less severe stress and higher technological inputs leads to higher and less variable yield in the northern than in the southern Mediterranean region (Fig. 7.2).

The term 'terminal stress' has been used to characterize these environments, hence the emphasis on agronomic practices and crop traits to balance water use, before and after anthesis to reduce the likelihood of extreme water deficit during grain filling (Fischer, 1979; Mitchell et al., 1996; van Herwaarden et al., 1998b; Leport et al., 1999; Sadras, 2002; Palta et al., 2004). However, physiological and environmental analyses have challenged the appropriateness of this characterization. From a physiological viewpoint, the strong correlation between cereal yield and grain number and the modest contribution of grain size to yield variation in Mediterranean environments of Europe and Australia (Fig. 7.3a,b), highlights the importance of growing conditions between stem elongation and few days after flowering, when grain number is determined (Fig. 7.3c). Management and environmental conditions that favor growth during this developmental window have a direct impact on grain number and yield (Fig. 7.4). Selective pressure for yield and agronomic adaptation in Australian wheat over the last five decades has resulted in enhanced pre-flowering growth (Sadras et al., 2012b). Environmental analysis in Mediterranean parts of Europe and Australia reinforce the notion that the more damaging stress pattern starts to develop well before flowering, thus compromising grain number rather than grain size (Chapter 13).

We thus suggest that the tag 'terminal drought' has overemphasized management and breeding

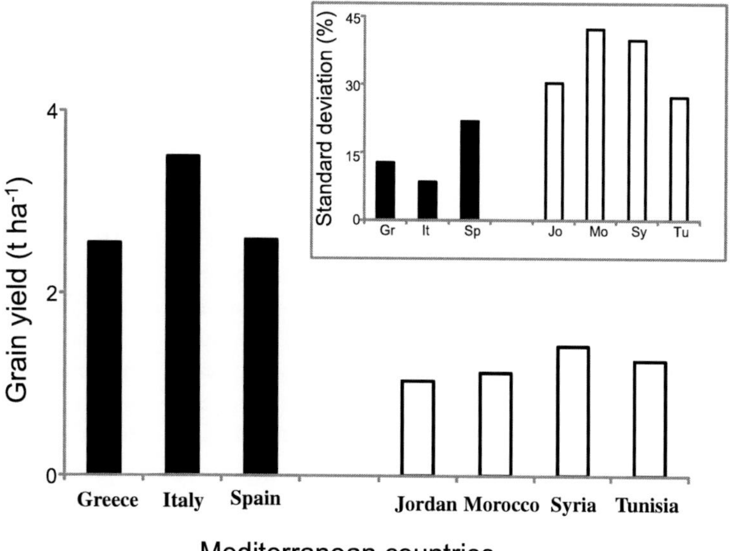

FIG. 7.2 Wheat yield in selected countries in the North (closed bars) and South (open bars) regions of Mediterranean region. Inset shows the standard deviation of yields. *Data from FAOSTAT for the period 1988 to 2012.*

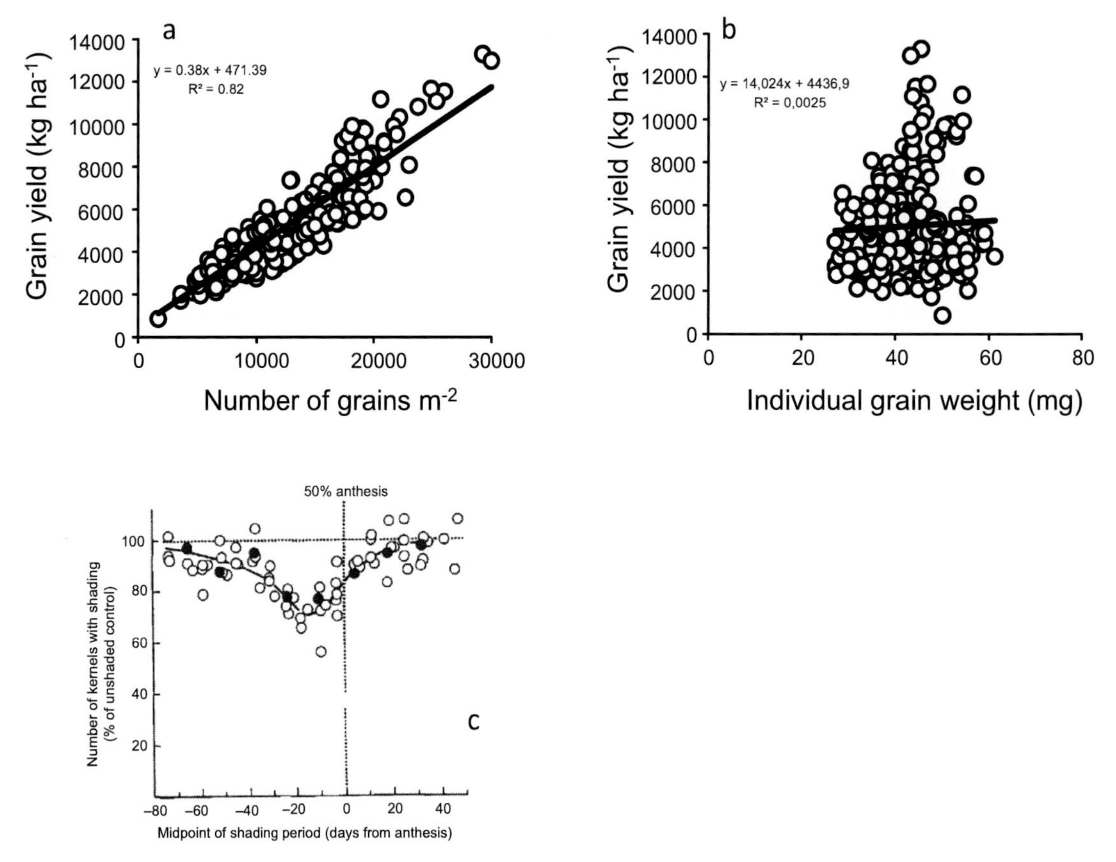

FIG. 7.3 Wheat yield in Mediterranean-type environments is mostly related with (a) grain number; (b) grain size is a minor component of the variation in yield. (c) Grain number is determined in a developmental window between stem elongation and a few days after flowering. Collectively, this analysis supports the proposition that the pre-anthesis period, rather than post-anthesis, is the most critical for yield determination of wheat in Mediterranean environments. *Sources: (a) Slafer et al. (2014), only Mediterranean data* and *(b) Fischer (1985).*

perspectives that focus on grain filling at the expense of the more critical pre-flowering stage. Of course, poor grain filling is undesirable for quality reasons, but grain size only contributes to small adjustments of yield (say 10%), whereas large variations in yield (say twofold) are necessarily associated with grain number and its components, heads m^{-2} and grains per head (Slafer et al., 2014). Agronomic practices and breeding strategies targeting crops in Mediterranean environments may therefore need to reassess the relative weight placed on pre- and post-anthesis growth.

3 DOES BARLEY OUT-YIELD WHEAT UNDER SEVERE WATER DEFICIT?

Land allocation to wheat and barley is based on a presumed differential sensitivity to stresses between these species. In southern Europe, the driest areas (less than 250 mm $year^{-1}$) are dominated by barley monoculture while agricultural systems become more complex and barley tends to be replaced by durum or bread wheat (depending on the country) with

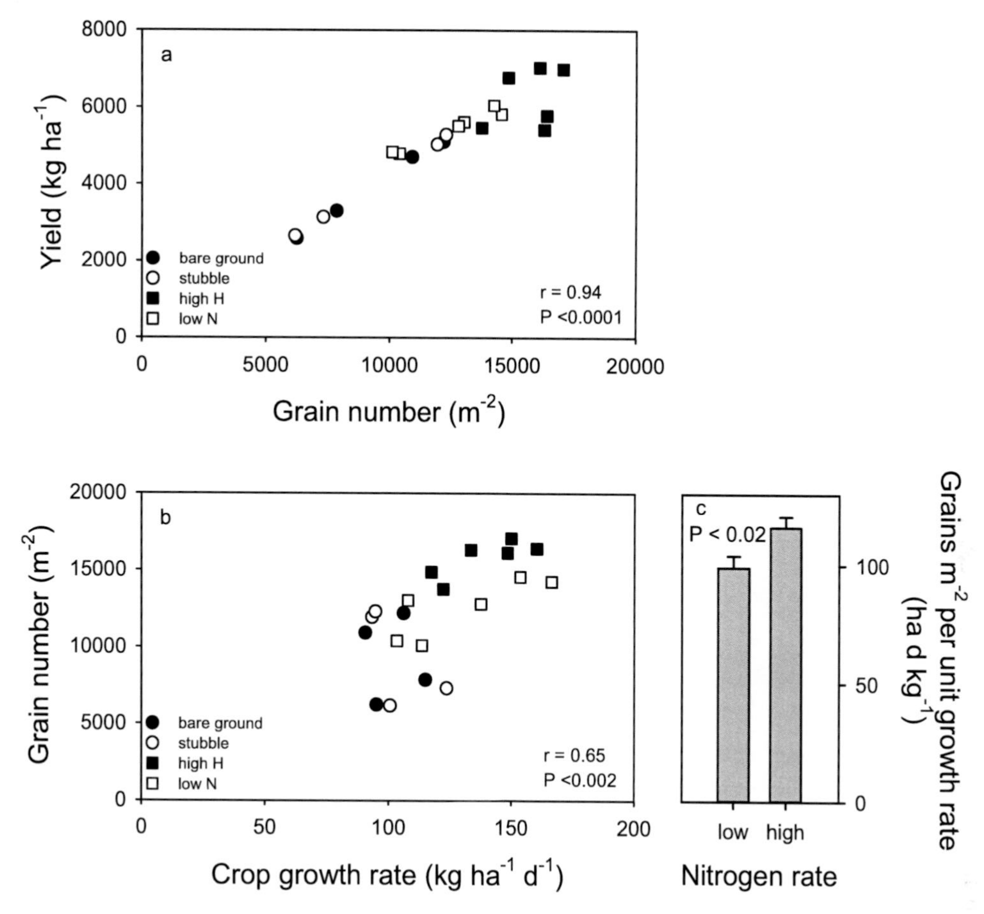

FIG. 7.4 Physiological analysis of wheat yield as affected by management (N fertilizer, stubble) and environment (location, season, rainfall) in a Mediterranean-type environment of South Australia. (a) Yield is a primary function of grain number and (b) grain number is proportional to the growth rate between stem elongation and a few days after flowering. (c) For the same growth rate, N-deficient crops set fewer grains. *Source: Sadras et al. (2012a).*

increasing precipitation (Lopez-Bellido, 1992; Anderson and Impiglia, 2002; Cossani, 2010). A similar pattern is evident in the West Africa and North Asia Mediterranean region where barley/livestock systems dominate below 250 mm year^{-1} (Fig. 7.5a).

A consequence of this pattern of land allocation is that, in general, wheat yields are higher than those of barley in the region (Fig. 7.5b). Barley generally has a larger embryo than wheat, and this contributes to traits with putative role in drought adaptation, i.e. barley has more seminal roots, longer roots and larger leaf area during crop establishment (Richards, 2006). There are, however, not many comparative studies of wheat and barley under field conditions in which barley has consistently outperformed wheat in low-yielding conditions, and *vice versa*. Recently, a set of comparative field experiments showed that yield of both crops tended to be similar under both low- and high-productivity conditions (Fig. 7.6). In these experiments, averaged grain weight ranged from 23.8 to 47.7 mg grain^{-1}, being higher for durum wheat than

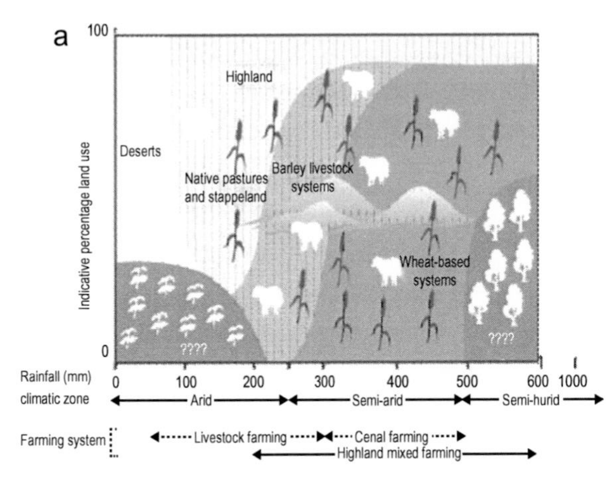

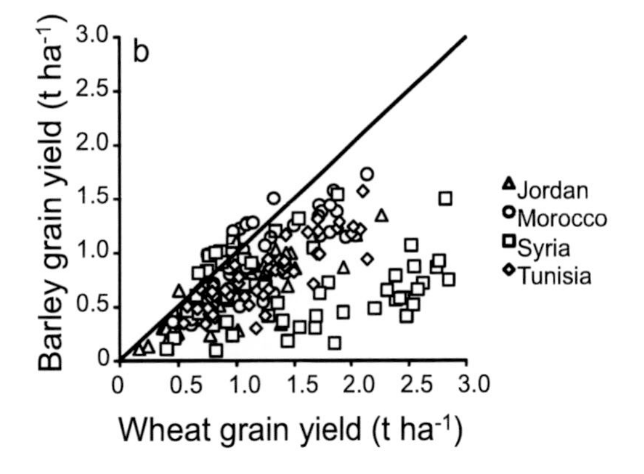

FIG. 7.5 (a) Patterns of land use in the West Africa and North Asia Mediterranean region as a function of annual rainfall. (b) Grain yield of barley and wheat in Jordan, Morocco, Syria and Tunisia between 1988 and 2012. The line is y = x. *Sources: (a) Ryan et al. (2008), (b) FAOSTAT (www.faostat.org).*

barley and bread wheat. The three species responded similarly in terms of grain nitrogen content to changes in the environmental conditions explored and, in terms of grain weight, barley was as stable as bread wheat (Cossani et al., 2011a). Thus, the monoculture of barley in harsh environments does not unequivocally improve the productivity of these fields (Cossani et al., 2007), tending to decrease the diversification of production and thereby increasing the uncertainty in the system. More comparisons of wheat and barleys of similar phenology under realistic field conditions are necessary to test the proposition that barley out-yields wheat under severe stress.

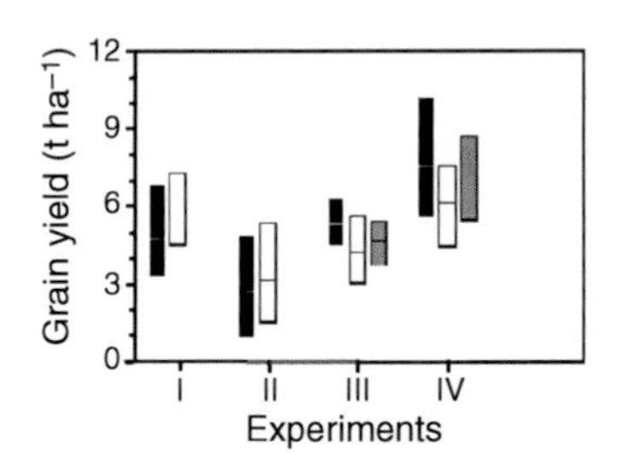

FIG. 7.6 Barley, bread wheat and durum wheat grain yield (black, white and gray bars, respectively) for experiments carried out in NE Spain between 2003 and 2007. The vertical bars represent the range between minimum and maximum yield, while the horizontal lines the average yield for each experiment. *Source: Cossani et al. (2009).*

4 DOES NITROGEN FERTILIZATION REDUCE YIELD IN LOW-RAINFALL CONDITIONS?

Rain-fed crops rely on two sources of water: the water stored in the soil at sowing, and the rainfall during the season. Owing to the strong rainfall seasonality in Mediterranean-type environments, soil water content at sowing is often low (Fig. 7.7). Seasonal rainfall is therefore the main source of water as well as a key driver of yield (Keatinge et al., 1986; Garabet et al., 1998). Water stress has thus been considered the dominant yield-limiting factor in these environments (Loss and Siddique 1994; Bennet et al., 1998; Acevedo et al., 1999). Water availability, however, only accounts for part of the variation in yield in these environments, typically about one-third (French and Schultz, 1984a,b; Sadras and Angus, 2006). Other factors, such as nutrient deficit, weeds, insects, diseases, untimely sowing, rain distribution during the crop cycle, and extreme weather, account for the remaining variation in yield; this is reflected in a common and often large gap between the actual and

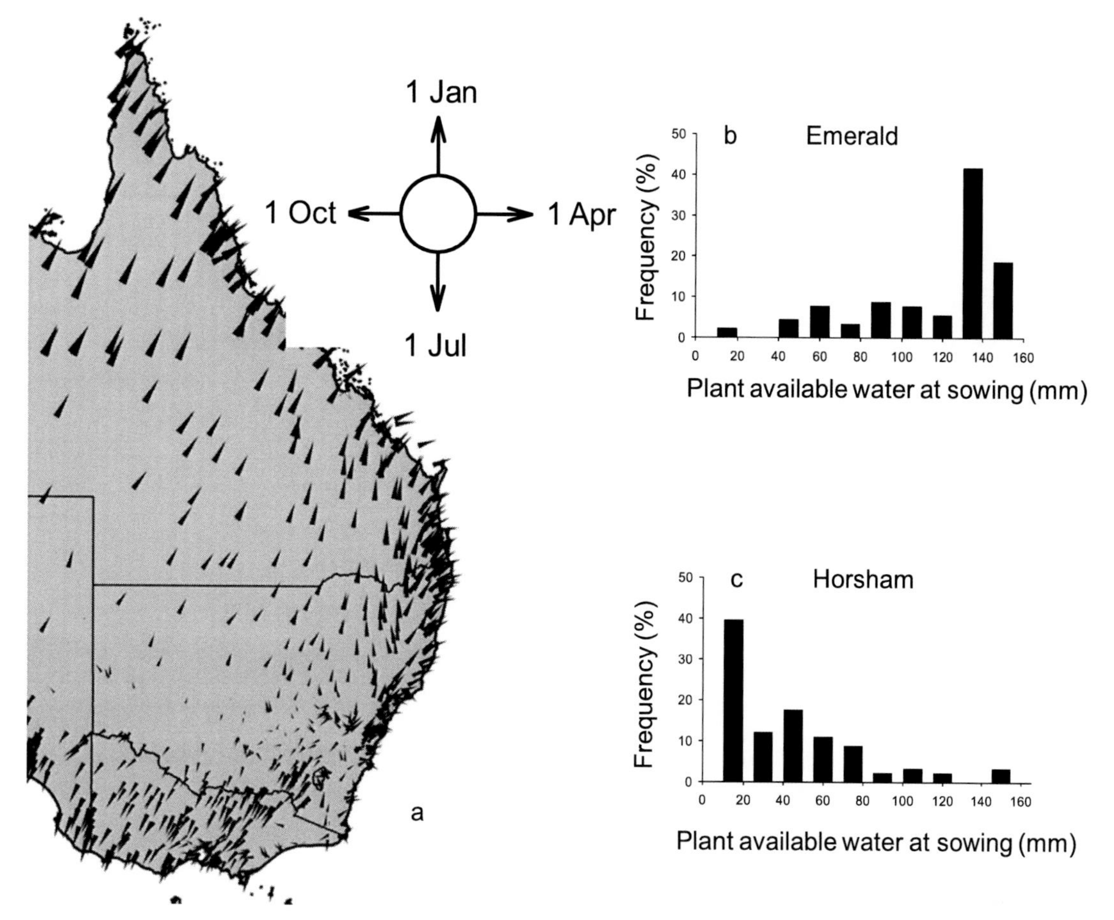

FIG. 7.7 (a) Rainfall seasonality in eastern Australia shifts from winter-dominant in the Mediterranean-type environments of the south, to summer-dominant in the north. The magnitude of the vectors represent the intensity of seasonality, and their direction represents the time of year with the largest rainfall concentration, e.g. a vertical upward vector would indicate maximum concentration of rain on 1 January. The probability of starting the wheat growing season with a full profile is high in (b) Emerald, a northern location with summer-dominant rainfall and low in (c) Horsham, a southern location with winter-dominant rainfall. *Sources: (a) Williamson (2007) (b,c) Sadras and Rodriguez (2007).*

maximum yield that can be achieved with a given evapotranspiration (Fig. 7.8).

Soils in Mediterranean environments have low fertility (Ryan et al., 2009); when this combines with low rates of fertilizer, cereal crops are likely to be nitrogen deficient. In the case of south-eastern Australia, where growers pay the full price of fertilizer, risk aversion accounts for the low usage of fertilizer (Sadras and Roget, 2004). This risk stems from low initial water content in soils (Fig. 7.7c) and limited practical capacity to predict seasonal rainfall (Anwar et al., 2008). In the Mediterranean region of West Asia and North Africa (WANA), crop management is far more conservative, with rain-fed cereals commonly grown, year after year, under nitrogen deficiency (e.g. Mossedaq and Smith, 1994; Oweis et al., 1998; Ryan, 2000; Ryan et al., 2008), even though the relevance of increasing nitrogen fertilization to increase yields in these dry lands

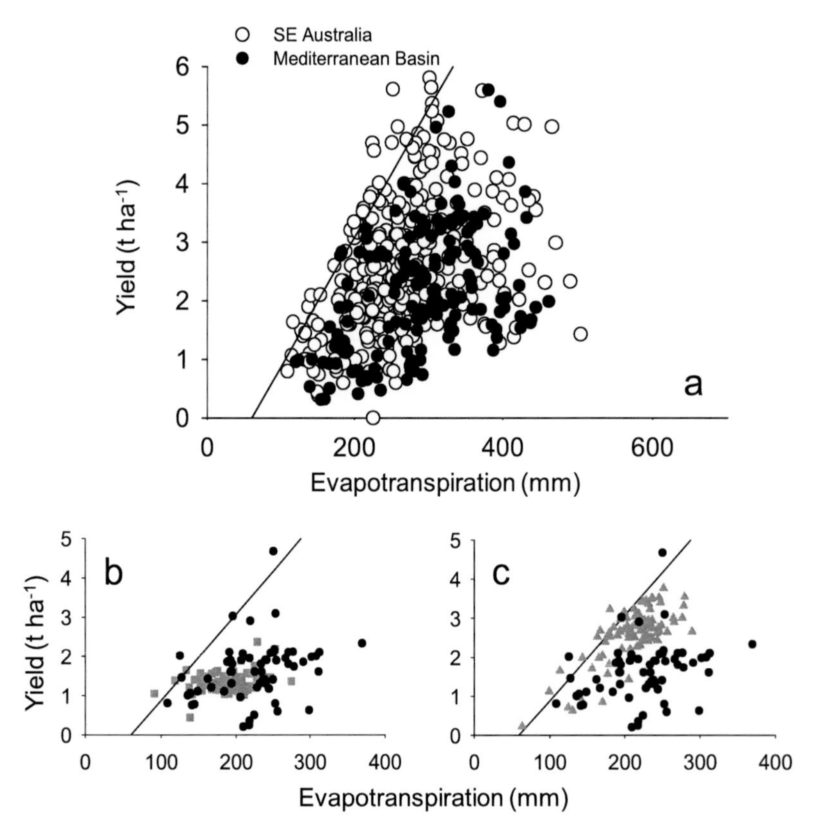

FIG. 7.8 (a) Relationship between actual grain yield and evapotranspiration for wheat crops in the Mediterranean Basin and SE Australia. (b, c) Relationship between actual grain yield and evapotranspiration for wheat crops in SE Australia where black points are actual measurements in grower fields and gray points are modeled assuming (b) low N supply as in growers' fields and (c) high rates of N supply. In all three figures, lines are boundary functions representing the maximum yield that can be achieved for a given evapotranspiration. *Sources: (a) Sadras and Angus (2006), (b,c) Sadras and Roget (2004).*

systems has been reported (Oweis et al., 1998; Ryan et al., 2008, 2009; Cossani et al., 2011b). Thus, the conservative attitude of WANA farmers being reluctant to fertilize their cereals may mainly reflect socioeconomic rather than technological limitations (Cossani et al., 2011b); Chapter 5 expands on these concepts for other regions of Africa. In addition, some studies have shown that excess nitrogen in dry conditions can be detrimental for yield (van Herwaarden et al., 1998a), further adding to the perception that nitrogen fertilization is unnecessary or undesirable in dry conditions. 'Haying-off', whereby high availability of nitrogen favors early growth at the expense of reproductive growth, leads to relatively high biomass but low harvest index and low yield (van Herwaarden et al., 1998a). This phenomenon has been mostly observed in central New South Wales, where soils are relatively fertile, and rainfall is uniformly distributed during the year (small vectors in Figure 7.7a), hence not conforming to the main features of Mediterranean environments. Indeed, haying-off is unlikely in the more typical Mediterranean environments of western and south-eastern Australia (Palta and Fillery, 1995; Asseng et al., 1998; Asseng and van Herwaarden, 2003; Sadras et al., 2012a) and the Mediterranean Basin, where Keatinge et al. (1985) concluded that nitrogen fertilization may reduce grain weight, but not yield. Our conclusion is that haying-off is scientifically plausible, but the concept has been unduly extrapolated and might have influenced farmers' practices in true Mediterranean environments where it is unlikely.

The interaction between water and nitrogen has been investigated from several perspectives.

Hooper and Johnson (1999) used a literature survey of fertilization experiments in arid, semi-arid, and sub-humid ecosystems to test the hypotheses that water limits above-ground net primary productivity at lower levels of annual precipitation, whereas nitrogen availability is the primary limitation with higher precipitation. Their analysis did not support this but the alternative hypothesis that plant growth was co-limited by water and nitrogen throughout the range of precipitation from 200 to 1100 mm year^{-1}. Water and nitrogen co-limitation has been demonstrated for wheat in Mediterranean environments (Box 7.1), and further supports the need to revisit common fertilizer practices in the context of increasing food demand and water scarcity. We thus critically assess the role of nitrogen fertilization to improve the capture

BOX 7.1

WATER AND NITROGEN CO-LIMITATION OF WHEAT YIELD IN SE AUSTRALIA AND NE SPAIN

Limitations imposed by single soil resources to crop productivity have been widely investigated but interactions between resources have received less attention. Bloom et al. (1985) proposed the use of analogies from economy in understanding the economy of resource use in plants. Thus, a plant must acquire resources (water, nitrogen, etc.) to be stocked, invested or used in several ways to construct several main products (leaves, stems, roots). In this context, Bloom et al. (1985) hypothesized that plants modify resource allocation so that their limitation of growth is nearly equal for all resources. Consequently, plants or crops would maximize their growth when all resources are equally limiting, rather than when growth is severely limited by a single resource. Therefore, yield would be positively related to the degree of co-limitation of key resources, chiefly water and nitrogen in dry, unfertile environments.

The theory of co-limitation was tested in Mediterranean areas for water and N (Sadras, 2004, 2005; Cossani et al., 2010). Simulation exercises supported the hypothesis that the gap between attainable and actual wheat yield of water- and N-stressed crops was negatively related to the degree of water and N co-limitation (Sadras, 2005). Cossani et al. (2010) used experimental data to test the same hypothesis in wheat and barley crops in NE Spain. The empirical data supported conclusions from the modeling outputs: yield, as well as water-use efficiency (WUE), is positively related to the degree of co-limitation by water and nitrogen. This is in line with findings by de Wit (1992) who reported several cases of interactions between nutrients (including those between N and water) determining that resources are used more efficiently with increasing availability of other resources. We suggest that resources are used more efficiently when there is a balance in their availability, reflected by a higher degree of co-limitation. Furthermore, this statement is in line with the reported inadequacy of Liebig's law of the minimum (Sinclair and Park, 1993), suggesting compensations among resources through plant acclimations, in agreement with the economic analogy (Bloom et al., 1985).

Therefore, a better adjustment of estimates of N requirements (which is a cornerstone of the decision on fertilization) would emerge from an accurate assessment of availability of other major resources (particularly those which cannot be modified by management) in order to maximize the degree of co-limitation.

Source: Sadras et al. (2012a)

and efficiency of use of available water to cereals in Mediterranean environments.

4.1 Cereal yield and nitrogen in the Mediterranean Basin

Here we revise a series of detailed experimental and modeling studies carried out in northern Spain, and studies testing the role of nitrogen in the broader regions of the Basin to assess the role of nitrogen, particularly in low yielding, water-stressed crops.

4.1.1 Cereal responses to nitrogen in Northern Spain

Table 7.1 summarizes experiments in farmers' fields where wheat and barley were grown with local practices to investigate responses to nitrogen supply. Year-to-year variability in the experiments was similar to the average variation for farmers of the region (Fig. 7.9a). These experiments showed neutral to positive, but not negative, effects of nitrogen on yield of cereals in realistic farming conditions including yields down to 2 t ha^{-1}(Fig. 7.9b). No evidence of 'haying-off' was found under these conditions (see above).

To expand this analysis over a time series of 17 years, the CERES-Wheat model was calibrated and validated for the genotypes and soil–weather conditions of the region, returning a range of yield from virtually zero to more than 7 t ha^{-1} in well fertilized crops (Fig. 7.10). Within the range of yields measured in different field experiments (from 0.7 to 7.6 t ha^{-1}), the model predicted performance with an acceptable error (Abeledo et al., 2008), even though it tended to underestimate yield in the most stressful conditions due to an over-sensitivity to water stress (Chipanshi et al., 1997, 1999). In agreement with field experiments, increasing nitrogen supply was neutral for the severely water-stressed crops (yield <1.5 t ha^{-1}) and increased yield above this threshold (Fig. 7.10). The model did not return cases of haying-off under severe stress. In an average year, the difference between the potential and attainable yield was minimized when the nitrogen available for the crop was at least ≈100 kg N ha^{-1} (Fig. 7.11); in those years with intermediate or low stress (and, therefore, less difference between actual and potential yield), the amount of nitrogen that minimized the difference was higher (150–200 kg N ha^{-1}) and in years with severe stress, the yield difference was maximum, and 50 kg N ha^{-1} was enough for the poor crop growth. In addition, more nitrogen availability reduced grain yield variability within each particular environmental situation (i.e. the greater the availability of nitrogen, the lower absolute value of the standard error of each point), irrespective of the environmental situation (i.e. average, good or poor years).

4.1.2 Cereal responses to nitrogen in the Mediterranean basin

Sixteen field experiments compared yield of fertilized and unfertilized rain-fed wheat and barley in Morocco, Jordan, Italy and Spain with the aim of testing whether farmers' reluctance to fertilize regularly rain-fed cereals is agronomically justified. Weeds and pests were controlled and local cultivars and practices were used. Treatments consisted of a range of nitrogen rates and timings of application (up to the onset of stem elongation). Consistent responses to fertilization were obtained even in environments with very low yields (Fig. 7.12). On average, grain yield increased more than 25% in response to fertilization and the response was similar in both cereals (Fig. 7.12). The magnitude of the response to fertilization increased with the yield of non-fertilized crops (i.e. lower water restriction). However, even many cases of crops exposed to extreme stress (yields lower than 1.5 t ha^{-1}) also responded positively to nitrogen fertilization (Fig. 7.12). In none of the conditions did fertilization significantly reduced yield compared with the unfertilized crops, confirming that haying-off is unlikely in typical Mediterranean conditions.

TABLE 7.1 Experiments performed at NE Spain comparing wheat and barley production under different nitrogen and water availabilities (rain-fed, r and irrigated, i)

Exp.	Species	Sowing date	First Detectable Node (DC 3.1)	Anthesis (DC 6.5)	Harvest (DC 9.0)	Environment	Nitrogen treatment (KgN ha^{-1})	Initial water content (mm)	Initial N-NO3- content (KgN ha^{-1})	Irrigation treatments (mm)	Rainfall (sowing -maturity) (mm)
I	Barley	21-Nov-03	13-apr-04	11-may-04	21-Jun-04	rainfed	0-40-80-120-160-200	288	94	0	284
	Bread wheat			31-may-04	12-Jul-04						
II	Barley	11-Nov-04	12-apr-05	08-may-05	20-Jun-05	rainfed and irrigated	0 and 200	83	34	222	159
	Bread Wheat			10-may-05	20d-28i Jun 05					222	160 r; 168 i
III	Barley	28-Nov-05	11-apr-06	03-may-06	06-Jun-06	rainfed and irrigated	0-50-100 and 150	240	115	76	92
	Durum wheat		17-apr-06	09-may-06	13d -20i Jun 06					95	92 r; 95 i
	Bread wheat		17-apr-06	16-may-06	13d -20i Jun 06					95	92 r; 95 i
IV	Barley	06-Nov-06	26-mar-07	30-apr-07	18-Jun-07	rainfed and irrigated	0-75 and 150	201	150	318	331
	Durum wheat			07-may-07	25-Jun-07					336	332
	Bread wheat			15-may-07	25-Jun-07					336	332

Data presented in Cossani et al. (2009).

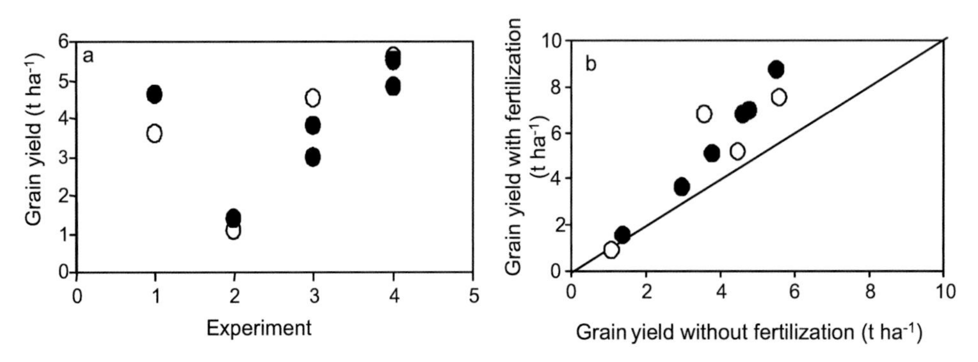

FIG. 7.9 (a) Grain yield of rain-fed, unfertilized barley (open symbols) and wheat (closed symbols) in four experiments conducted in farmers' fields at Urgell, NE Spain. (b) Grain yield from fertilized and unfertilized crops under rain-fed conditions (as there were different fertilized treatments, the greatest response to the addition of fertilizer was considered). *Adapted from Cossani et al. (2009).*

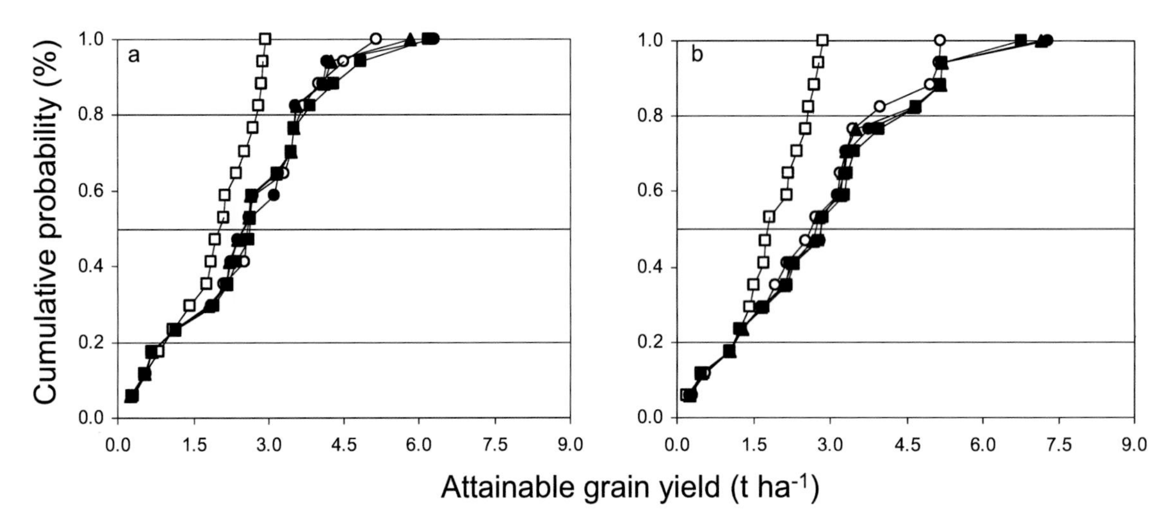

FIG. 7.10 Cumulative frequency distribution of simulated attainable yield for the wheat cultivars (a) Anza and (b) Soissons in response to N availability at sowing using a climate series of 17 years (1989–2006) for Agramunt, NE Spain. N availability was 50 kg N ha^{-1} (open squares), 100 kg N ha^{-1} (open circles), 150 kg N ha^{-1} (closed circles), 200 kg N ha^{-1} (closed triangles) and 250 kg N ha^{-1} (open triangles). *Source: Abeledo et al. (2008).*

These findings were further tested in fields of 20 farmers of Béja (sub-humid) and Siliana (semi-arid) regions of Tunisia (Cossani et al., 2011b). Local practices and tools were used except for nitrogen fertilization; three fertilization strategies were compared. The main comparison was between yields of unfertilized and fertilized crops following a recommendation derived from an *ad-hoc* model (Cossani et al., 2011b). As in Tunisia, the straw also has market value, both grain and 'total yield', which represents the overall marketable yield in terms of grain-equivalent yield, were considered. Grain yield with no fertilization ranged from 1.0 to 3.5 t ha^{-1}, and increased up to 7 t ha^{-1} with additional nitrogen (Fig. 7.13). Therefore, the response to fertilization was positive in most cases for both grain and total yield (Fig. 7.13a,b).

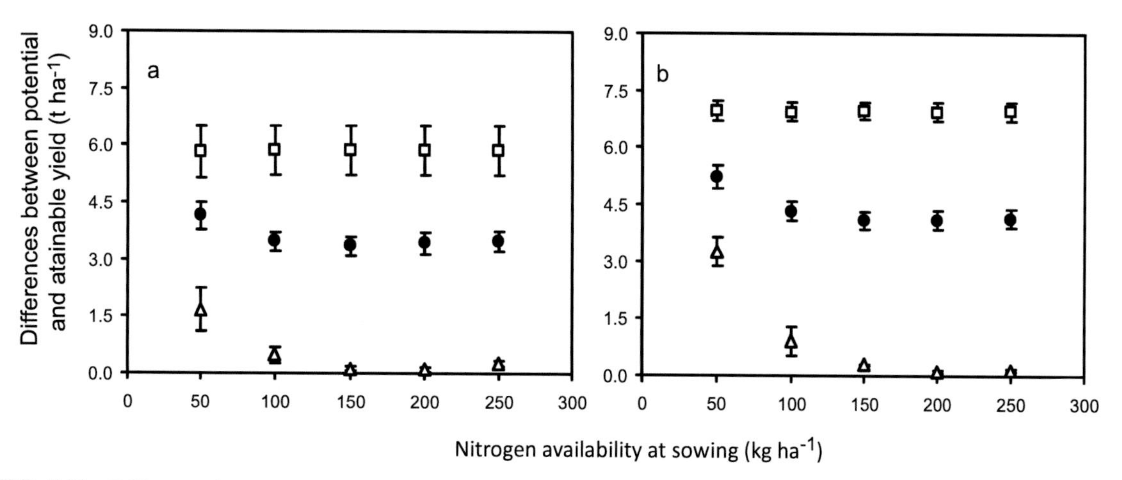

FIG. 7.11 Differences between potential and attainable yield for wheat cultivars (a) Anza and (b) Soissons as a function of nitrogen availability for the 3 years with the higher gap (open squares), a mean gap (closed circles), and the 3 years with the lower gap (open triangles) using a long-term simulation (17 years, 1989–2006) for Agramunt, NE Spain. *Source: Abeledo et al. (2008).*

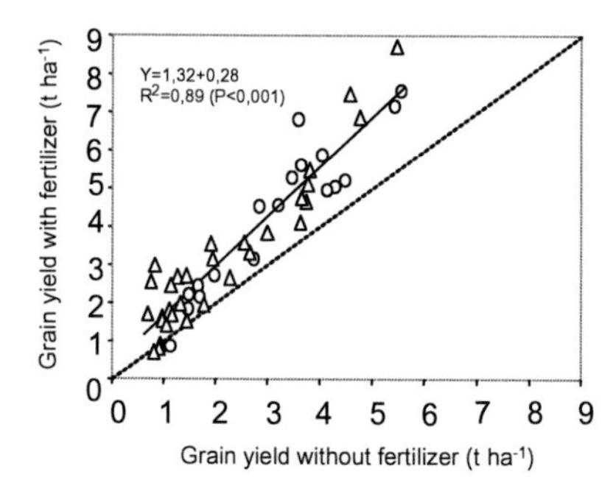

FIG. 7.12 Relationship between grain yield with and without fertilizer under rain-fed conditions for wheat (triangles) and barley (circles) in Morocco, Jordan, Italy and Spain. Dotted line is the relationship 1:1. *Source: Savin et al. (under revision).*

In summary, there is strong empirical evidence for a positive or neutral effect of nitrogen on the yield of water-stressed cereals in typical Mediterranean environments. This is consistent with physiological principles. In dry, nitrogen-poor environments, crop yield is favored by nitrogen and water co-limitation (Box 7.1), and high nitrogen supply favors both water uptake and water-use efficiency (Box 7.2). However, the increased yield and water-use efficiency of well-fertilized crops is associated with relatively low nitrogen-use efficiency.

4.2 Wheat yield and nitrogen in Australia

Rain-fed crops in the south-eastern and western Mediterranean regions of Australia grow under recurrent water stress, which combines with generalized low soil fertility and local soil problems such as salinity (Fischer, 2009). Until the early 1990s, farmers were reluctant to fertilize their cereals for many reasons, including severe root disease problems. Wider adaptation of break crops improved the health of wheat crops, thus leading to more consistent responses to nitrogen, even though the background water limitations did not change or worsened in western Australia. In environments with recurrent water stress, therefore, cereals yield can be consistently improved with appropriate nitrogen fertilization (Fig. 7.14).

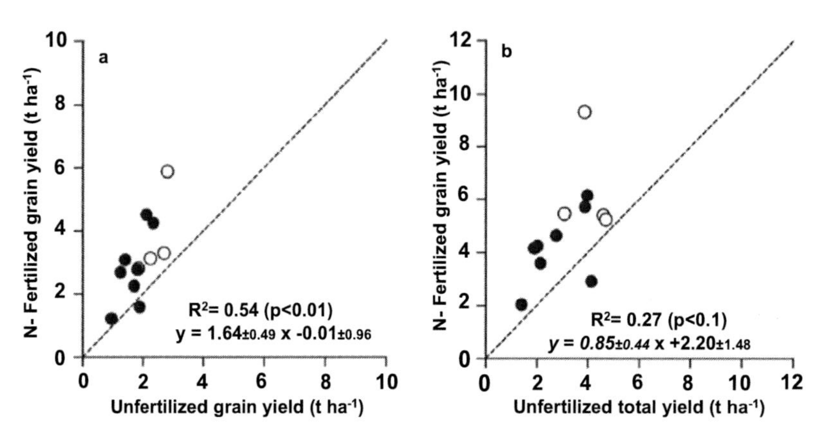

FIG. 7.13 Relationship between (a) grain yield and (b) total biomass at maturity in fertilized (closed symbols) and unfertilized (open symbols) rain-fed wheat crops in farmers' fields in Tunisia. Dotted line is the relationship 1:1. *Source: Cossani et al. (2011b).*

BOX 7.2

THE NITROGEN-DRIVEN TRADE-OFF BETWEEN THE EFFICIENCY IN THE USE OF WATER AND NITROGEN

Agronomically, water-use efficiency (WUE) can be defined as the ratio of grain yield and seasonal evapotranspiration, and is conveniently disaggregated in the following components (Cooper et al., 1987):

$$WUE = \frac{TE}{1+\frac{E}{T}} \cdot HI$$

where *TE* is shoot biomass per unit seasonal crop transpiration, *T* is crop transpiration, *E* is evaporation from the soil surface, and *HI* is harvest index. A nitrogen-deficient crop has low biomass/transpiration ratio (Brueck, 2008) and high evaporation/transpiration ratio associated with small canopies (Cooper et al., 1987; Caviglia and Sadras, 2001; Norton and Wachsmann, 2006). Nitrogen-deficiency may also impair water uptake from deeper soil layers (Figure Box 7.1), hence further contributing to higher *E/T*. The response of harvest index to nitrogen supply could be positive, negative or neutral (van Herwaarden et al., 1998a) but the consistent decrease of *TE* and increase of *E/T* drive the reduction in WUE characteristic of nitrogen-deficient crops (Sadras and Rodriguez, 2010). Owing to the law of diminishing returns, yield per unit nitrogen supply declines with increasing input of nitrogen

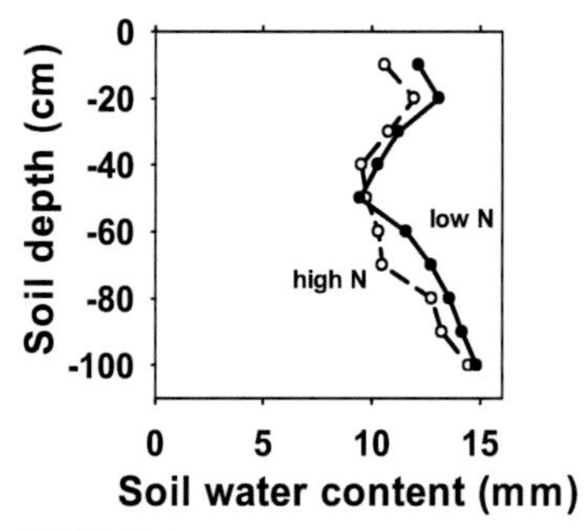

FIGURE BOX 7.1 Nitrogen deficient wheat crops are unable to extract water from deep soil layers. *Source: Sadras et al. (2012a)*

BOX 7.2 (cont.)

(Chapter 8). In consequence, crops require high nitrogen input to achieve high water-use efficiency, but this leads to low yield per unit nitrogen supply; hence the trade-off between these resources (Figure Box 7.2). This trade-off is high-wired in the physiology of crops and therefore applies irrespective of species (except legumes?), soil, climate and management (Belder et al., 2005; Kim et al., 2008; Sadras and Rodriguez, 2010). In making fertilization decisions, growers might be addressing this trade-off when they chose to under-fertilize their crops, hence achieving an economically sensible compromise between the efficiency of both resources.

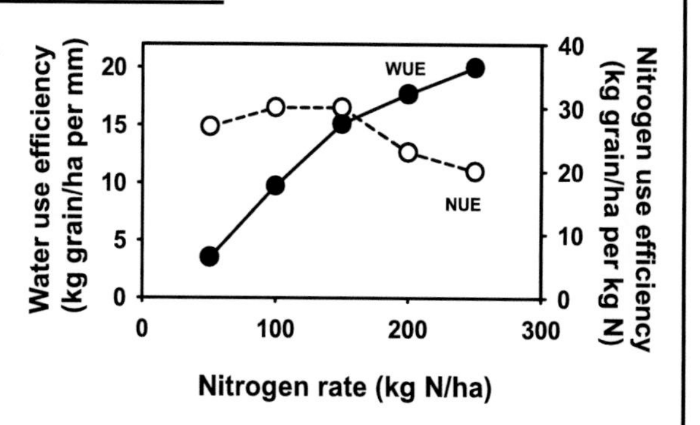

FIGURE BOX 7.2 The nitrogen-driven trade-off between yield per unit water use and yield per unit nitrogen supply. *Adapted from Sadras and Rodriguez (2010).*

Source: Sadras and Rodriguez, 2010

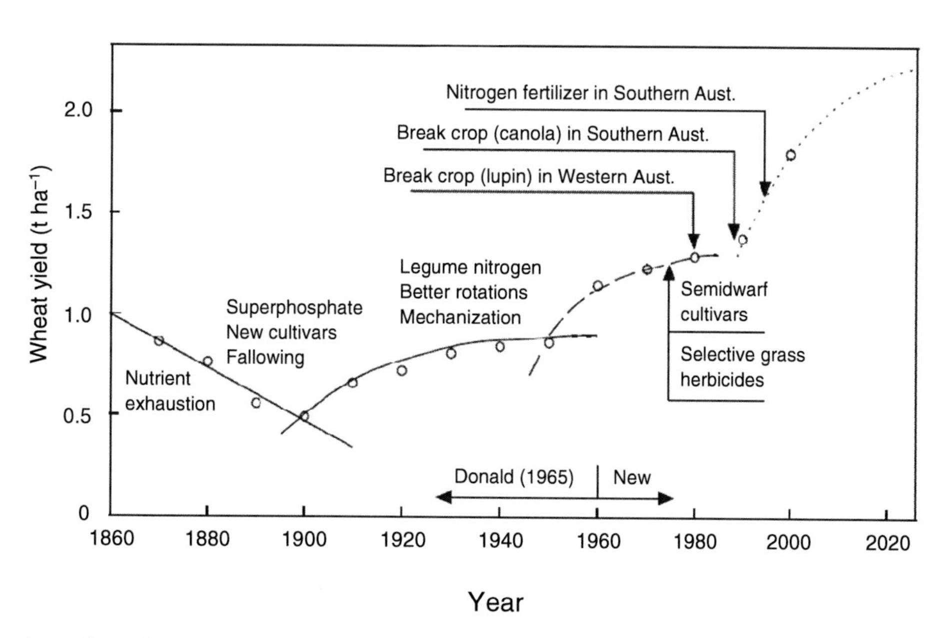

FIG. 7.14 Australian wheat yields with likely explanations for the trends. *Source: Angus (2001) and Passioura (2002).*

Experiments and modeling provide reinforcing evidence for the need of higher rates of fertilization in south-eastern Australia. Experiments manipulating water supply, stubble management and nitrogen input, showed that nitrogen supply has a twofold effect on grain number and hence yield: increasing nitrogen supply increases the rate of crop growth during the critical stage from stem elongation to flowering, and increases grain set per unit growth rate (Fig. 7.4). In the dry Mallee region, yields in growers' fields tended to level at about 2 t ha^{-1} with increasing rainfall and evapotranspiration (see black points, Fig. 7.8b). Modeling this production system reproduced this pattern (see gray points, Fig. 7.8b). In comparison, the gap between actual and attainable yield for a given evapotranspiration was dramatically reduced when yield was modeled under the assumption of abundant supply of nitrogen (see gray points, Fig. 7.8c). A good part of the yield gap in these environments is therefore associated with nitrogen deficiency. Growers are aware of the benefits of nitrogen fertilization, but they also have to deal with risks and trade-offs between yield per unit water use and yield per unit nitrogen uptake (Box 7.1).

Selection for yield and agronomic adaptation has dramatically changed the phenotype of Australian wheat (Sadras and Lawson, 2011, 2013; Sadras et al., 2012b). Yield per unit transpiration increased at a rate of 0.012 kg ha^{-1} mm^{-1} year^{-1} during most of the last century. This resulted from improved harvest index and, during the last five decades, improved biomass. The improvement in biomass, associated with enhanced radiation-use efficiency before flowering, resulted from a significant increase in nitrogen uptake, and changes in the profiles of nitrogen distribution in the canopy. Breeding with a dual focus on water and nitrogen may thus help to improve the adaptation of cereals to dry, nitrogen-poor environments (Sadras and Richards, 2014).

5 CONCLUDING REMARKS

We revised three main statements usually accepted for cereals crops under Mediterranean environments. These were: crops are exposed to terminal drought; barley out-yields wheat in dry conditions; and water deficit overrides the benefit of nitrogen fertilization. We argued that the scientific evidence does not fully support these statements, and that agronomic practices and breeding strategies targeting crops in Mediterranean environments may therefore need to reassess the relative weight placed on pre- and post-anthesis growth, the allocation of land to different species with putatively different drought adaptation and the opportunity for wider diversity in the drier environments, and the opportunities to improve yield and water-use efficiency by better management of nitrogen fertilization accounting for both agronomic responses and economic risk.

References

Abeledo, L.G., Savin, R., Slafer, G.A., 2008. Wheat productivity in the Mediterranean Ebro Valley: analyzing the gap between attainable and potential yield with a simulation model. Eur. J. Agron. 28, 541–550.

Acevedo, H.E., Silva, P.C., Silva, H.R., Solar, B.R., 1999. Wheat production under Mediterranean environments. In: Satorre, E.H., Slafer, G.A. (Eds.), Wheat: ecology and physiology of yield determination. Haworth Press, Inc, New York, p. 503.

Anderson, W.K., Impiglia, A., 2002. Management of dryland wheat. In: Curtis, B.C., Rajaram, S., Gómez-Macpherson, H. (Eds.), Wheat: improvement and production. Plant production and protection series. FAO, Rome, p. 567.

Angus, J.F., 2001. Nitrogen supply and demand in Australian agriculture. Aust. J. Exp. Agric. 41, 277–288.

Anwar, M.R., Rodriguez, D., Liu, D.L., Power, S., O'Leary, G.J., 2008. Quality and potential utility of ENSO-based forecasts of spring rainfall and wheat yield in south-eastern Australia. Aust. J. Agric. Res. 59, 112–126.

Araus, J.L., Slafer, G.A., Royo, C., Serret, M.D., 2008. Breeding for yield potential and stress adaptation in cereals. Crit. Rev. Plant Sci. 27, 377–412.

Aschmann, H., 1973. Distribution and peculiarity of Mediterranean ecosystems. In: di Castri, F., Mooney, H.A.

(Eds.), Mediterranean type ecosystems. Springer-Verlag, Heildeberg, pp. 11–19.

Asseng, S., Keating, B.A., Fillery, I.R.P., et al., 1998. Performance of the APSIM-wheat model in Western Australia. Field Crops Res. 57, 163–179.

Asseng, S., van Herwaarden, A.F., 2003. Analysis of the benefits to wheat yield from assimilates stored prior to grain filling in a range of environments. Plant Soil 256, 217–229.

Belder, P., Bouman, B.A.M., Spiertz, J.H.J., Peng, S., Castañeda, A.R., Visperas, R.M., 2005. Crop performance, nitrogen and water use in flooded and aerobic rice. Plant Soil 273, 167–182.

Bennet, S.J., Saidi, N., Enneking, D., 1998. Modelling climatic similarities in Mediterranean areas: a potential tool for plant genetic resources and breeding programmes. Agric. Ecosyst. Environ. 70, 129–143.

Bloom, A.J., Chapin, F.S.I., Mooney, H.A., 1985. Resource limitation in plants – an economic analogy. Annu. Rev. Ecol. Systemat. 16, 363–392.

Brueck, H., 2008. Effects of nitrogen supply on water-use efficiency of higher plants. J. Plant Nutr. Soil Sci. 171, 210–219.

Caviglia, O.P., Sadras, V.O., 2001. Effect of nitrogen supply on crop conductance, water- and radiation-use efficiency of wheat. Field Crops Res. 69, 259–266.

Chipanshi, A.C., Ripley, E.A., Lawford, R.G., 1997. Early prediction of spring wheat yields in Saskatchewan from current and historical weather data using the Ceres-Wheat model. Agric. For. Meteorol. 84, 223–232.

Chipanshi, A.C., Ripley, E.A., Lawford, R.G., 1999. Large-scale simulation of wheat yields in a semi-arid environment using a crop-growth model. Agric. Syst. 59, 57–66.

Cossani, C.M., 2010. Grain yield and resource use efficiency of bread wheat, barley and durum wheat under Mediterranean environments. Thesis doctoral. Universidad de Lleida.

Cossani, C.M., Savin, R., Slafer, G.A., 2007. Contrasting performance of barley and wheat in a wide range of conditions in Mediterranean Catalonia (Spain). Ann. Appl. Biol. 151, 167–173.

Cossani, C.M., Slafer, G.A., Savin, R., 2009. Yield and biomass in wheat and barley under a range of conditions in a Mediterranean site. Field Crops Res. 112, 205–213.

Cossani, C.M., Slafer, G.A., Savin, R., 2010. Co-limitation of nitrogen and water on yield and resource-use efficiencies of wheat and barley. Crop Past. Sci. 61, 844–851.

Cossani, C.M., Slafer, G.A., Savin, R., 2011a. Do barley and wheat (bread and durum) differ in grain weight stability through seasons and water-nitrogen treatments in a Mediterranean location? Field Crops Res. 121, 240–247.

Cossani, C.M., Thabet, C., Mellouli, H.J., Slafer, G.A., 2011b. Improving wheat yields through N fertilization in Mediterranean Tunisia. Exp. Agric. 47, 459–475.

Cooper, P.J.M., Gregory, P.J., Tully, D., Harris, H.C., 1987. Improving water use efficiency of annual crops in rainfed systems of west Asia and North Africa. Exp. Agric. 23, 113–158.

de Wit, C.T., 1992. Resource use efficiency in agriculture. Agric. Sys. 40, 125–151.

FAOSTAT, 2012. FAOSTAT Data. http://faostat.fao.org/default.aspx.

Fischer, R.A., 1979. Growth and water limitation of dryland wheat yield in Australia: a physiological framework. J. Aust. Inst. Agric. Sci. 45, 83–94.

Fischer, R.A., 1985. Number of kernels in wheat crops and the influence of solar radiation and temperature. J. Agric. Sci. 105, 447–461.

Fischer, R.A., 2009. Farming systems of Australia: exploiting the synergy between genetic improvement and agronomy. In: Sadras, V.O., Calderini, D.F. (Eds.), Crop physiology: applications for genetic improvement and agronomy. Academic Press, San Diego, pp. 23–54.

French, R.J., Schultz, J.E., 1984a. Water use efficiency of wheat in a Mediterranean-type environment. I. The relation between yield, water use and climate. Aust. J. Agric. Res. 35, 743–764.

French, R.J., Schultz, J.E., 1984b. Water use efficiency of wheat in a Mediterranean-type environment. II. Some limitations to efficiency. Aust. J. Agric. Res. 35, 765–775.

Garabet, S., Wood, M., Ryan, J., 1998. Nitrogen and water effects in a Mediterranean-type climate. I. Dry matter yield and nitrogen accumulation. Field Crops Res. 57, 309–318.

Hooper, D.U., Johnson, L., 1999. Nitrogen limitation in dryland ecosystems: responses to geographical and temporal variation in precipitation. Biogeochemistry 46, 247–293.

Keatinge, J.D.H., Dennett, M.D., Rodgers, J., 1986. The influence of precipitation regime on the crop management of dry areas in northern Syria. Field Crops Res. 13, 239–249.

Keatinge, J.D.H., Neate, P.J.H., Shepherd, K.D., 1985. The role of fertilizer management in the development and expression of crop drought stress in cereals under Mediterranean environmental conditions. Exp. Agric. 21, 209–222.

Kim, K.I., Clay, D.E., Carlson, C.G., Clay, S.A., Trooien, T., 2008. Do synergistic relationships between nitrogen and water influence the ability of corn to use nitrogen derived from fertilizer and soil? Agron. J. 100, 551–556.

Leport, L., Turner, N.C., French, R.J., et al., 1999. Physiological responses of chickpea genotypes to terminal drought in a Mediterranean-type environment. Eur. J. Agron. 11, 279–291.

López-Bellido, L., 1992. Mediterranean cropping systems. In: Pearson, C.J. (Ed.), Ecosystems of the world. Elsevier, The Netherlands, pp. 311–356.

Loss, S.P., Siddique, K.H.M., 1994. Morphological and physiological traits associated with wheat yield increases in Mediterranean environments. Adv. Agron. 52, 229–275.

Mitchell, J.H., Fukai, S., Cooper, M., 1996. Influence of phenology on grain yield variation among barley cultivars grown under terminal drought. Aust. J. Agric. Res. 47, 757–774.

Mossedaq, F., Smith, D.H., 1994. Timing nitrogen application to enhance spring wheat yields in a Mediterranean climate. Agron. J. 86, 221–226.

Norton, R.M., Wachsmann, N.G., 2006. Nitrogen use and crop type affect the water use of annual crops in south-eastern Australia. Aust. J. Agric. Res. 57, 257–267.

Oweis, T., Pala, M., Ryan, J., 1998. Stabilizing rainfed wheat yields with supplemental irrigation and nitrogen in a Mediterranean climate. Agron. J. 90, 672–681.

Palta, J.A., Fillery, I.R.P., 1995. N application increases pre-anthesis contribution of dry matter to grain yield in wheat grown on a duplex soil. Aust. J. Agric. Res. 46, 507–518.

Palta, J.A., Turner, N.C., French, R.J., 2004. The yield performance of lupin genotypes under terminal drought in a Mediterranean-type environment. Aust. J. Agric. Res. 55, 449–459.

Passioura, J.B., 2002. Environmental biology and crop improvement. Funct. Plant Biol. 29, 537–546.

Richards, R.A., 2006. Physiological traits used in the breeding of new cultivars for water-scarce environments. Agric. Water Manag. 80, 197–211.

Ryan, J., 2000. Soil and plant analysis in the Mediterranean region: limitations and potential. Commun. Soil Sci. Plant Anal. 31, 2147–2154.

Ryan, J., Ibrikci, H., Sommer, R., McNeill, A., 2009. Nitrogen in rainfed and irrigated cropping systems in the Mediterranean region. Adv. Agron. 104, 53–136.

Ryan, J., Singh, M., Pala, M., 2008. Long-term cereal-based rotation trials in the Mediterranean region: implications for cropping sustainability. Adv. Agron. 97, 273–319.

Sadras, V.O., 2002. Interaction between rainfall and nitrogen fertilisation of wheat in environments prone to terminal drought: economic and environmental risk analysis. Field Crops Res. 77, 201–215.

Sadras, V.O., 2004. Yield and water-use efficiency of water- and nitrogen-stressed wheat crops increase with degree of co-limitation. Eur. J. Agron. 21, 455–464.

Sadras, V.O., 2005. A quantitative top-down view of interactions between stresses: theory and analysis of nitrogen –water co-limitations in Mediterranean agro-ecosystems. Aust. J. Agric. Res. 56, 1151–1157.

Sadras, V.O., Angus, J.F., 2006. Benchmarking water-use efficiency of rainfed wheat in dry environments. Aust. J. Agric. Res. 57, 847–856.

Sadras, V.O., Lawson, C., 2011. Genetic gain in yield and associated changes in phenotype and competitive ability of SA wheat varieties released between 1958 and 2007. Wheat Breeding Assembly, Perth, 24-26 August, 2011.

Sadras, V.O., Lawson, C., 2013. Nitrogen and water-use efficiency of Australian wheat varieties released between 1958 and 2007. Eur. J. Agron. 46, 34–41.

Sadras, V.O., Lawson, C., Hooper, P., McDonald, G.K., 2012a. Contribution of summer rainfall and nitrogen to the yield and water use efficiency of wheat in Mediterranean-type environments of South Australia. Eur. J. Agron. 36, 41–54.

Sadras, V.O., Lawson, C., Montoro, A., 2012b. Photosynthetic traits of Australian wheat varieties released between 1958 and 2007. Field Crops Res. 134, 19–29.

Sadras, V.O., Richards, R.A., 2014. Improvement of crop yield in dry environments: benchmarks, levels of organisation and the role of nitrogen. J. Exp. Bot. 65, 1981-95.

Sadras, V.O., Rodriguez, D., 2007. The limit to wheat water use efficiency in eastern Australia. II. Influence of rainfall patterns. Aust. J. Agric. Res. 58, 657–669.

Sadras, V.O., Rodriguez, D., 2010. Modelling the nitrogen-driven trade-off between nitrogen utilisation efficiency and water use efficiency of wheat in eastern Australia. Field Crops Res. 118, 297–305.

Sadras, V.O., Roget, D.K., 2004. Production and environmental aspects of cropping intensification in a semiarid environment of southeastern Australia. Agron. J. 96, 236–246.

Sinclair, T.R., Park, W.I., 1993. Inadequacy of the Liebig limiting-factor paradigm for explaining varying crop yields. Agron. J. 85, 742–746.

Slafer, G., Savin, R., Sadras, V.O., 2014. Coarse and fine regulation of wheat yield components in response to genotype and environment. Field Crops Res. 157, 71–83.

van Herwaarden, A.F., Farquhar, G.D., Angus, J.F., Richards, R.A., Howe, G.N., 1998a. 'Haying-off', the negative grain yield response of dryland wheat to nitrogen fertilizer. I. Biomass, grain yield, and water use. Aust. J. Agric. Res. 49, 1067–1082.

van Herwaarden, A.F., Richards, R.A., Farquhar, G.D., Angus, J.F., 1998b. 'Haying-off', the negative grain yield response of dryland wheat to nitrogen fertiliser. III The influence of water deficit and heat shock. Aust. J. Agric. Res. 49, 1095–1110.

Williamson, G., 2007. Climate and root distribution in Australian perennial grasses; implications for salinity mitigation. School of Earth and Environmental Sciences, The University of Adelaide, Adelaide.

PART II

CARBON, WATER AND NUTRIENT ECONOMIES OF CROPS

CHAPTER

8

Quantifying crop responses to nitrogen and avenues to improve nitrogen-use efficiency

François Gastal, Gilles Lemaire, Jean-Louis Durand, Gaëtan Louarn

Institut National de la Recherche Agronomique (INRA), Lusignan, France

1 INTRODUCTION

Soil N availability is generally a major limiting factor for productivity of agricultural systems. Supply of N to crops and cropping systems, through N fertilizers and utilization of legumes, is therefore one of the key elements for producing sufficient food to meet the demand of increasing human population (Angus, 2001; Eickout et al., 2006). Over the last 40 years, the worldwide use of mineral N fertilizers increased sevenfold in parallel with the doubling of agricultural food production. Nevertheless, production of N fertilizers through the Haber–Bosch process is extremely consuming in fossil energy and produces large emissions of greenhouse gases. Moreover, misuse of N in intensive agricultural systems has led, through nitrogen cascades (Galloway and Cowling, 2002), to important environmental impacts such as eutrophication of freshwater (London, 2005) and marine ecosystems (Beman et al., 2005), and gaseous emissions of N oxides and ammonia into the atmosphere (Ramos, 1996; Stulen et al., 1998). Nevertheless, emissions per unit production could be lower in high input systems, as indicated by Grassini and Cassman (2012), and Chapter 2. For many years, a relatively high grain-to-fertilizer price ratio, particularly in subsidized agricultural systems, incited farmers to apply excess N to allow for high yield and profit. This insurance strategy led to a progressive accumulation of N in soils and risk of N leaching (Addiscott et al., 1991). This effect, combined with the long residence time of N within the soil organic matter, suggests that the pollution of ground water that we observe today could well be the delayed consequence of the intensification of cropping systems one or two decades ago (Mariotti, 1997). Problems associated with climate change, biofuel production

Crop Physiology. DOI: 10.1016/B978-0-12-417104-6.00008-X

and global food security are also questioning the efficiency of use of N fertilizer in agricultural systems (Cassman, 2007).

As a result of these environmental and economical aspects, there is an important need to improve nitrogen efficiency of cropping systems. Nitrogen-use efficiency depends on agronomic practices including management of mineral and organic nitrogen fertilization and the use of legumes in cropping systems, and genetic progress in nitrogen-use efficiency. Adoption of a more restricted strategy for supply and timing of N fertilizers on crops is difficult because we have a limited capacity to predict weather, which determines both soil N mineralization and crop growth potential. In many agricultural areas, rainfall uncertainty is an important factor on decisions related to fertilizer application, either for lack of rainfall reducing crop growth and soil N mineralization, or for excess rainfall increasing the risk of nitrate leaching (Chapter 7; Sadras, 2002; de Koeijer et al., 2003; Sadras and Roget, 2004; Cabrera et al., 2007). A systematic reduction of N application to avoid excess of N in soils would increase the likelihood of temporary crop N deficiency, i.e. when soil N availability does not meet plant N demand. Instead, tactical strategies as well as breeding approaches are required to improve crop N-use efficiency under both conditions that favor over-fertilization (subsidized economies, intrinsically high fertility, high grain-to-nitrogen price), and systems where shortage of nitrogen or other pedoclimatic constraints are chronic (Chapter 2). Understanding the processes and traits that govern N uptake, N distribution and growth responses to N is important to maximize crop production with minimum N input to reduce environmental hazards (Cassman et al., 2002).

The objective of this chapter is to develop a framework of the principles governing regulation of N economy (i.e. N uptake and allocation) and growth of plants and crops, and to apply these principles in tools for improving both fertilization management and breeding strategies. First, we develop a theoretical analysis of the dynamics of plant and crop N demand in relation to growth potential during the crop cycle, and discuss agronomical tools to evaluate nitrogen status of crops. Second, we analyze the physiological and morphological responses of plants and crops to N deficiency. In a final section, we examine the concept of nitrogen-use efficiency in the light of the principles developed previously, and discuss avenues for improving nitrogen-use efficiency through plant breeding and agronomy.

For most crop species, plant life cycle can be simplified into two main phases: (1) the pre-flowering phase, when plants develop foliage and roots, and when young leaves and roots behave as sinks; and (2) the post-flowering phase, when senescing leaves are sources of carbohydrates and reduced N for developing storage organs such as seeds, fruits or tubers. From a complementary perspective, Chapter 10 discusses senescence as the point of transition from the carbon capture phase of crop development to the period of remobilization of nitrogen and other nutrients. The present approach of N uptake and N allocation considers distinctly these two phases. In addition, we consider explicitly two levels of organization, the individual plant and the plant population, i.e. the crop. The aim is to identify the buffering effect on some physiological traits when scaling up from organ to whole plant and to crop (Chapter 1).

The chapter develops the case of the ecophysiological understanding of the drivers of plant and crop N nutrition. This focus means that important aspects of the nitrogen economy of crops are beyond the scope of this chapter. Biological fixation is not treated *per se*, but it is shown that once within the plant, biologically fixed N behaves similarly to N originating from mineral N absorption (Oti-Boateng et al., 1994; Peoples and Baldock, 2001; Kahn et al., 2002). Chapter 9 deals with N_2-fixation from the viewpoint of evolutionary conflict, i.e. traits that maximize microbe fitness are not necessarily those that maximize host–plant fitness involving microbial mutualisms. The dynamics of N in soil is also not directly considered here (Powlson et al., 1992; Mary and Recous, 1994; Garnier et al., 2003; Murphy et al., 2003). Another important point not discussed within this chapter is the interactions between plant N nutrition and biotic stresses. On this topic readers can refer to White (1993) and Waring and Cobb (1992) for plant–insect interactions and to Delin et al. (2008) for fungal diseases.

2 CROP N DEMAND: ITS REGULATION AT PLANT AND CROP LEVELS

In this section, we develop a framework based on both empirical relationships and theory that allows for the regulation of plant N uptake in relation to both plant growth potential and soil N supply. From this general framework, a diagnostic tool of plant and crop N status is derived to quantify the intensity and the timing of N deficiency during the crop developmental cycle.

Crop N demand at any time of the crop cycle can be defined as the amount of N necessary to sustain growth potential, i.e. maximal plant growth and plant mass. Crop N demand results from growth potential and critical plant N concentration. Critical plant N concentration is defined as the minimum plant N concentration necessary to achieve maximum crop mass (Greenwood et al., 1990). Thus, crop N demand corresponds to critical N uptake that is the minimum crop N uptake necessary to achieve maximum crop mass. If N uptake is higher than critical N uptake (correspondingly if crop N concentration is higher than critical N concentration), then the crop experiences 'luxury' N consumption. Conversely, if N uptake is lower than critical N uptake (correspondingly if crop N concentration is lower than critical N concentration), then the crop experiences N deficiency.

This concept of critical N can be applied in dynamic terms, so the daily crop N demand (or critical N uptake rate) is the quantity of N required each day for the crop to maintain its potential growth rate over a given period of time. This dynamic approach of crop N demand has been used extensively on perennial forage crops such as alfalfa and grasses (Lemaire and Salette, 1984a,b; Lemaire et al., 1985) and further extended to annual crops such as wheat (Justes et al., 1994), maize (Plénet and Lemaire, 1999) and canola (Colnenne et al., 1998). All these studies bring convergent results that have been assembled within consistent theory (Greenwood et al., 1990; Lemaire and Gastal, 1997; Gastal and Lemaire, 2002).

2.1 N dilution and N-uptake dynamics

The N dilution process within a growing crop is empirically approached by plotting plant (or crop) N concentration with plant (or crop) mass. Further, a theoretical framework is presented to provide ecophysiological evidence supporting such empirical relationships.

2.1.1 Empirical approach

The actual plant N concentration in a crop stand declines even under favorable N supply as the crop mass increases (Greenwood et al., 1986). This decline can be described empirically by a negative power function (Lemaire and Salette, 1984) relating plant N concentration (%N) to crop mass (W, t ha^{-1}) during vegetative growth:

$$\%N = aW^{-b} \quad (8.1)$$

The coefficient a represents plant N concentration for $W = 1\,t\,ha^{-1}$; it depends on the rate of steady state N supply and may also be affected by the species. Coefficient b is dimensionless and represents the ratio between the relative decline in plant %N and the relative crop growth rate. From Eq. 8.1 it is possible to derive the relationship between crop N uptake (N) and crop mass (W) during vegetative growth:

$$N = a'W^{1-b} \quad (8.2)$$

The coefficient a' corresponds to the crop N uptake for $W = 1\,\mathrm{t\,ha^{-1}}$. Its value is $10a$ when N is expressed in $\mathrm{kg\,ha^{-1}}$. So, as for a, a' depends on the rate of the steady state N supply of the crop. The allometric coefficient $1-b$ is the ratio between the relative N uptake rate and the relative growth rate of the crop.

When N supply is the minimum necessary for achieving the maximum crop growth rate during the entire growth period, then it is possible to define the critical N dilution curve:

$$\%Nc = a_c W^b \quad (8.3)$$

where a_c is the critical plant %N for a crop mass of $1\,\mathrm{t\,ha^{-1}}$. An example of the determination of the critical dilution curve for a maize crop is presented in Figure 8.1. In the same way, it is possible to define the dynamics of critical crop N uptake:

$$N = a'_c W^{1-b} \quad (8.4)$$

where a'_c is the critical crop N uptake for a crop mass of $1\,\mathrm{t\,ha^{-1}}$.

The values of coefficients a_c or a'_c and b have been established for the main cultivated species according to the method developed by Justes et al. (1994) (Table 8.1). For a given species, coefficients a_c (or a'_c) and b remain constant in a large range of climatic conditions. Lemaire et al. (2007) showed that for wheat, maize, canola, sorghum and sunflower the coefficients of Eq. 8.4 established in temperate conditions hold in sub-tropical situations. So, coefficients a_c and b can be considered as crop species characteristics. Coefficient b varies between 0.25 for canola and 0.50 for rice with a majority of values around 0.30–0.40 as for perennial grasses, alfalfa, pea, maize and sorghum (Table 8.1). Due to uncertainty in the determination of b, a common average value of 0.35 is proposed. No clear differences in the value of b appear between C_3 and C_4 species neither between monocots and dicots. The coefficient a_c is lower in C_4 than in C_3 species, in correspondence with the difference in leaf N content according to their respective photosynthetic pathway (Brown, 1978). Within each metabolic group it is difficult to observe clear differences between species, as pointed out by Greenwood et al. (1990). Due to the tendency to have a

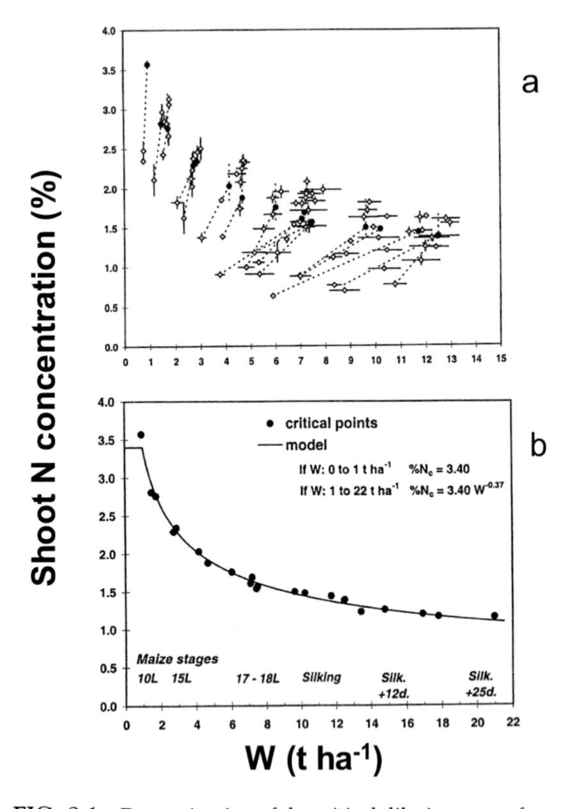

FIG. 8.1 Determination of the critical dilution curve from data obtained with different N fertilization rate experiments in maize. (a) Open symbols are the different experiments. Dark symbols are the critical N concentration calculated as the intersection of the oblique line representing response of crop mass to increased plant N concentration, and the vertical line representing the increases in plant N concentration without any increase in crop mass (accumulation of excess of N). (b) The critical dilution curve of maize. *From Plénet and Lemaire (1999).*

TABLE 8.1 Coefficients a_c and b of the critical dilution curve $\%N = a_c W^b$ for different crop species

Crop species	a_c (kgN ha^{-1})	b (dimensionless)	References
Temperate grasses (C_3)	4.8	0.32	Lemaire and Salette (1984a,b)
Lucerne (C_3)	4.8	0.33	Lemaire et al. (1985)
Pea (C_3)	5.1	0.32	Ney et al. (1997)
Wheat (C_3)	5.3	0.44	Justes et al. (1994)
Canola (C_3)	4.5	0.25	Colnenne et al. (1998)
Rice (C_3)	5.2	0.52	Sheehy et al. (1998)
Tomato (C_3)	4.5	0.33	Tei et al. (2002)
Sunflower (C_3)	4.5	0.42	Debaeke et al. (2012)
Cabbage (C_3)	5.1	0.33	Ekbladh and Witter (2010)
Potato (C_3)	4.6	0.42	Bélanger et al. (2001)
Maize (C_4)	3.4	0.37	Plénet and Lemaire (1999)
Sorghum (C_4)	3.9	0.39	Plénet and Cruz (1997)
Tropical grasses (C_4)	3.6	0.34	Duru et al. (1997)

correlation between a_c and b, parameters of Eqs 8.3 and 8.4 are relatively close together for species of the same metabolic group (Gastal and Lemaire, 2002).

Crop N demand can be analyzed in dynamic terms on a daily basis. The derivative of Eq. 8.4 relates the daily crop N demand to the daily crop growth rate:

$$\frac{dN}{dt} = \frac{dN}{dW}\frac{dW}{dt} = a_c(1-b)W^{-b}\left(\frac{dW}{dt}\right) \quad (8.5)$$

Hence the daily crop N demand follows the daily crop growth rate, but for a similar daily crop growth rate, the daily crop N demand declines as the crop mass W increases.

In conclusion, for a given species at any moment of its vegetative period:

1. the dynamics of the crop dry matter accumulation determines both the critical plant %N and critical N uptake
2. the relationship between critical crop N uptake (i.e. crop N demand) and crop growth rate is fairly independent of external variables such as soil (other than N supply) and climate
3. this relationship is slightly variable among species of the same metabolic group, and reflects the metabolic differences between C_3 and C_4.

2.1.2 Physiological principles

Greenwood et al. (1990) and Lemaire and Gastal (1997) developed a theory to account for the empirical observations in section 2.1.1. Following Caloin and Yu (1984) assumption, plant mass W is composed of two compartments: (1) Wm, the metabolic tissues directly involved in growth processes (photosynthesis and meristematic activity) with a high N concentration (%Nm); and (2) Ws, the structural tissues necessary for plant architecture with a low N concentration (%Ns). Then:

$$W = Wm + Ws \quad (8.6)$$

and the plant N concentration (%N) is:

$$\%N = \frac{1}{W}(\%NmWm + \%NsWs) \quad (8.7)$$

If we assume that Wm increases allometrically with W, then:

$$Wm = kW^{\alpha} \quad (8.8)$$

and:

$$\%N = k(\%Nm - \%Ns)W^{\alpha-1} + \%Ns \quad (8.9)$$

Eq. 8.9 represents a negative power function between plant N concentration and crop mass provided that $\alpha < 1$, and is equivalent to the empirical relationship of Eq. 8.2. The main difference between the empirical and theoretical models is the small positive asymptote (%Ns ≈ 0.8%) in Eq. 8.9, but the difference between the two equations is small (Lemaire and Gastal, 1997). Hence $\alpha - 1$ in Eq. 8.9 is mathematically close to $-b$ in Eq. 8.1.

Following Hardwick (1987) and considering that Wm is dominated by photosynthetic tissues, it is postulated that Wm scales with plant area:

$$Wm = pLAI \quad (8.10)$$

where p is a metabolic leaf density (t ha^{-1}). It is then possible to establish an allometric relationship between leaf area index (LAI) and W, according to Eq. 8.8:

$$LAI = \frac{k}{p}W^{\alpha} \quad (8.11)$$

The coefficient k/p is the LAI for W = 1 t ha^{-1} that corresponds also to the leaf area ratio (LAR) at this crop stage. It has been called 'leafiness' coefficient (Lemaire et al. 2007), representing the intrinsic plant shape and architecture through its LAR. When the crop grows, Eq. 8.11 predicts a decrease of the plant LAR with W; this decline is more or less rapid depending on plant density. Lemaire et al. (2007) compared α in Eq. 8.11 and $1 - b$ in Eq. 8.4 for different crops such as maize, sorghum, wheat, canola, sunflower, growing in subtropical and temperate environments near the critical N status. They experimentally confirmed that α is approximately equal to $1-b$. With this assumption, a relationship between crop N uptake dynamics and LAI expansion was derived from Eqs 8.2 and 8.11:

$$N = \frac{a'p}{k}LAI \quad (8.12)$$

Such proportionality between N uptake and LAI has been empirically observed for wheat (Sylvester-Bradley et al., 1990; Grindlay et al., 1993) and alfalfa (Lemaire et al., 2005). Jamieson and Semenov (2000) used such a relation for modeling N uptake of wheat crops. Lemaire et al. (2007, 2008) showed that the coefficient of proportionality between crop N uptake and LAI depends on crop N status through coefficient a', but also on plant shape and architecture through the leafiness coefficient k/p that is very variable among crop species. Eq. 8.12 therefore is less general than Eq. 8.2 for simulation of crop N demand. Lemaire et al. (2007) wondered if the dynamics of N uptake (or of N dilution) is dominated by crop mass accumulation (Eq. 8.2) or LAI expansion (Eq. 8.12). They concluded that both are incomplete representations of the same global process. For the same crop mass increment, the corresponding increment in N uptake should not be exactly the same if this biomass is mainly composed of leaf (high Wm) or stem tissue (high Ws). This accounts for the finding that phenologic stages that govern leaf/stem ratio in wheat have an important influence on N dilution curves when comparing cultivars with different timing of stem elongation (Angus, 2007).

Lemaire et al. (2005, 2007) showed that the dilution of N with plant growth in isolated plants was less marked than for plants in dense stands. These authors showed that $1-b$ and α from seedling emergence until LAI ≈ 1, i.e. the period where the plants can be considered as near isolated, are in a range 0.90–0.95, indicating a low decrease in plant %N and LAR with increasing plant mass. For LAI >1, as competition for light increases, both $1-b$ and α decline rapidly to a common value of 0.60–0.70 indicating a parallel

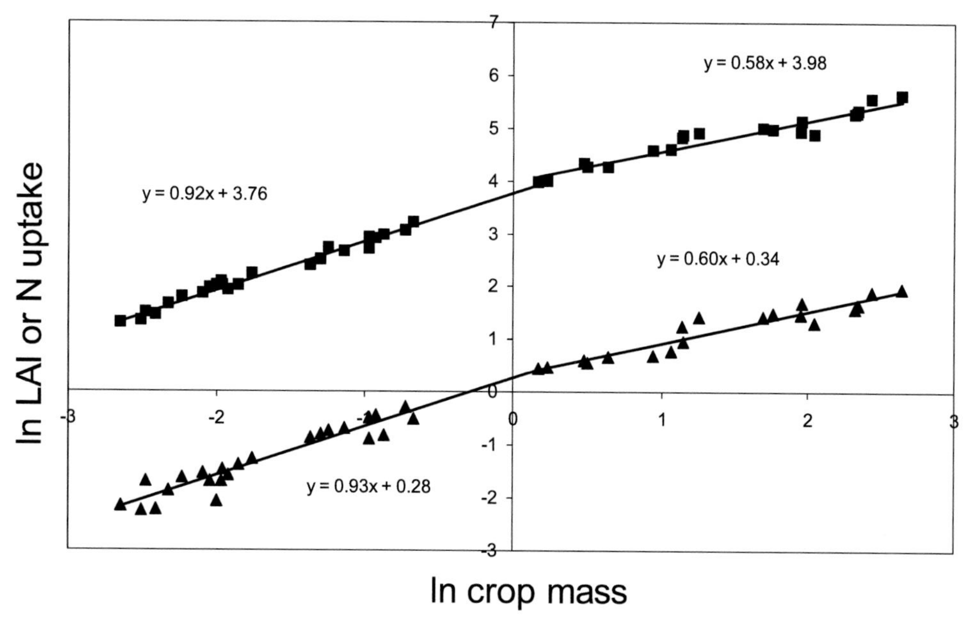

FIG. 8.2 Relationships between LAI (▲) or N uptake (■) and crop mass in a ln-ln scale (R^2 >0.91). Data for vegetative wheat crops grown under non-limiting N supply. *From Lemaire et al. (2007).*

acceleration in the decrease of both plant %N and LAR (Fig. 8.2). Theoretically, coefficient α in Eq. 8.11 should be 2/3 if three conditions are met: (1) plant mass W scales with plant volume; (2) plant mass per unit volume is constant; and (3) plant growth is isometric, i.e. the plant grows with the same relative rate in the three dimensions (Niklas, 1994). Lemaire et al. (2007) provided evidence favoring the notion of $\alpha \approx 2/3$ for the majority of crop species.

For isolated plants α is slightly lower than 1; $\alpha = 1$ would have implied that the plants grew only in two dimensions, i.e. in area but not in height or in thickness. The value of 0.90–0.95 indicates that plants tend to optimize their leaf area expansion when they are near isolated, while in a dense stand, plants invest more in the third dimension allowing for the placement of their leaves within the illuminated layers at the top of the canopy. As the growth in the third dimension, i.e. in height, implies more structural tissues with low N concentration, the decline in plant %N is accelerated as the light environment changes in growing canopies (Pons et al., 1989; Tremmel and Bazzaz, 1995). So, one of the causes of the plant %N decrease with increasing crop mass is the architectural and morphological changes of plants in response to the presence of neighbors.

Early plant-to-plant interactions are related to the perception of neighboring plants via changes in light spectrum including sensing red:far red ratio (R:FR) of light by phytochrome and blue light by cryptochrome (Ballaré et al., 1995, 1997). This allows plants to anticipate competition for radiation through modifications of their architecture (tillering, internodes and petiole elongation, size of sheath and lamina) and strategies for shade avoidance (Aphalo and Ballaré, 1995; Gautier et al., 1998; Hérault-Bron et al., 1999; Bahmani et al., 2000). This response leads to an increase in the proportion of structural tissues (Ws), and then to an acceleration of the N dilution process as crop mass increases.

Another cause of N dilution is the decrease of N concentration of shaded leaves as the canopy develops. The vertical pattern of foliar N is associated with the light extinction in the canopy and has been interpreted in terms of optimization of canopy photosynthesis, as discussed in section 3.3.3.

In conclusion, this theoretical framework shows that N dilution operates at different levels of organization: the organ (i.e. leaf, stem), the whole plant and the canopy. N dilution has to be considered as an ontogenetic process and its expression in terms of crop mass accumulation allows an overall simple representation of complex processes. In this way, critical N dilution curves generically capture most of the environmental sources of variation as well as variations among the metabolic groups C_3 and C_4.

2.2 Co-regulation of N uptake by both N soil availability and plant growth rate potential

The approach described above seems to indicate that plant growth itself regulates plant N uptake. This appears contradictory with physiological processes indicating that plant growth is the consequence of plant N uptake and not the reverse. To reconcile these two points of view we have to consider that (1) plant N uptake is directly related to N availability to roots, and thus to N supply in soil, and (2) plant N uptake is also feedback controlled by the crop growth rate. As reviewed in Lemaire and Millard (1999), plant N uptake is feedback regulated by shoot N and C signaling irrespective of the source of soil N, i.e. nitrate, ammonium or N_2 fixation (Ryle et al., 1986; Gastal and Saugier, 1989; Oti-Boateng et al., 1994; Tourraine et al., 1994). A positive regulation comes from a C signal corresponding to photosynthetic assimilate transported by phloem from leaves to roots, and a negative signal comes from organic N recirculated from shoots to roots (Lejay et al., 1999; Forde, 2002). This later regulation acts as an N satiety signal, repressing the N uptake of the plant when its capacity to store organic N compounds within new growing organs becomes saturated.

So an increase in plant growth rate increases both (1) leaf area and then plant photosynthesis, leading to a positive C signal to the root transport system for nitrate absorption, and (2) sequestration of N organic compounds in new growing organs (leaves, stems and roots) leading to a reduction of the negative N signaling to root nitrate transport systems. Conversely, a decrease in plant growth rate leads to a decrease of the positive C-signaling and to an increase in the organic N-compounds recirculating in the phloem and repressing N absorption by the roots. So the actual N absorption rate of a plant is co-regulated by both soil N supply and its own growth rate potential. Two groups of transport systems, respectively with low and high affinity for nitrate operate in plants (Glass et al., 2002). Devienne-Baret et al. (2000) proposed a model accounting for the co-regulation of plant N uptake by soil nitrate concentration and plant growth rate:

$$\frac{dN}{dt} = a_c(1-b)W^{-b}\left(\frac{dW}{dt}\right)_{\max}\left[V_H\frac{C}{K_H+C}+V_L\frac{C}{K_L+C}\right] \quad (8.13)$$

where V and K are the coefficients of the Michaelis–Menten formula and subscripts describe the high (H) and the low (L) affinity transport systems for nitrate; C is the actual NO_3^- concentration in soil solution, W is the crop mass, and $1-b$ is the allometric coefficient of Eq. 8.2.

Lemaire et al. (2007) discussed whether leaf expansion or crop mass accumulation is the variable that best describes the feedback control of crop N uptake. They concluded that biomass accumulation and LAI expansion are two faces of the same regulation process. Nevertheless, the slope of the N uptake vs LAI relationship (Eq. 8.12) is more variable across species than the slope of the N uptake vs crop mass relationship (Eq. 8.4) when crops are maintained near the critical N supply. The variation of plant morphology and leafiness across crop species induces variation in N uptake plant capacity per unit of LAI indicating that leaf area expansion is not the only way by which the plant can sequestrate organic nitrogen, but that stem growth or leaf thickness can also be a means for the plant to store reduced N, delaying the feedback control of N absorption N by roots.

Eq. 8.13 represents the crop N uptake vs crop mass trajectories for different steady-state N supply conditions, i.e. C = constant. The four trajectories in Figure 8.3 correspond to rates of N supply for the agronomic range between highly-limiting N to saturated plant capacity to sequestrate organic N. Each trajectory represents a 'virtual' constant plant N status maintained by a steady state N supply, whereas a 'real' crop in a changing environment could move from one trajectory to another according to changes in soil supply with time. The dotted lines As-Af-Ac-Am and Bs-Bf-Bc-Bm represent two response curves of the crop to increased N supply for either (1) different time during crop growth, (2) different genotypes at the same time having different growth potential, or (3) different environments (i.e. temperature) modulating crop growth rate. So, whatever the cause of the variation in crop mass provided that N supply remains at steady state, any increase in crop mass (ΔW) is accompanied by a corresponding increase in crop

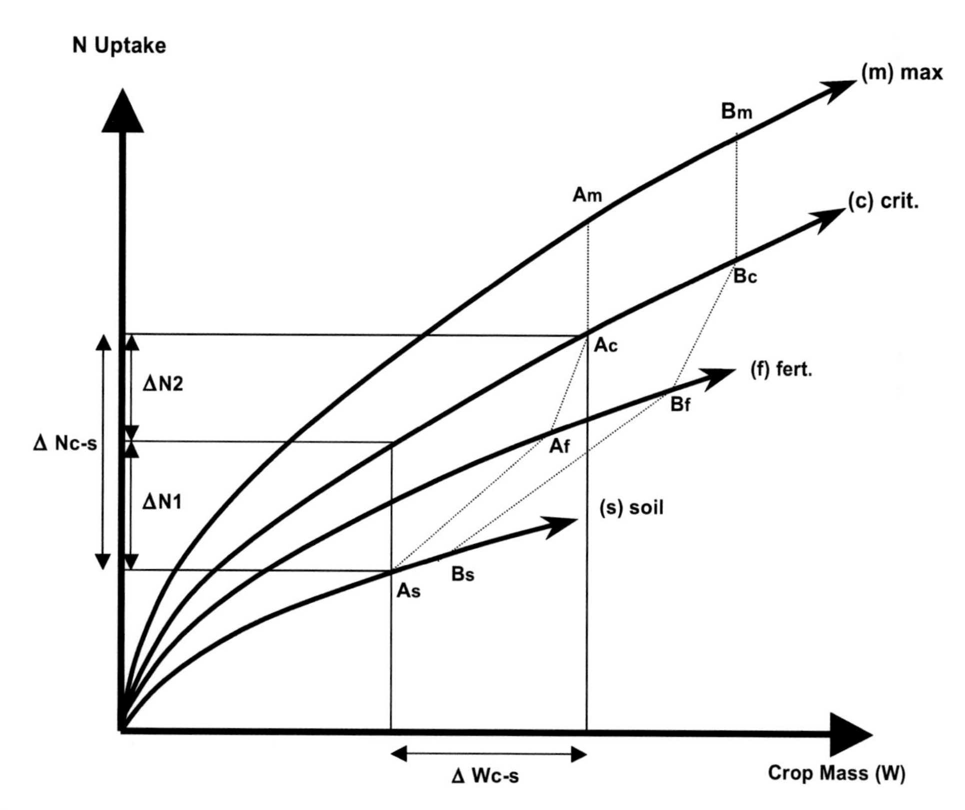

FIG. 8.3 N uptake vs crop mass trajectories for different steady-state N supply corresponding to: (s) non-fertilized crops representing a highly limiting N condition, (f) a suboptimal fertilization rate, (c) the critical N uptake, and (m) the maximum N uptake where the capacity of the plant to sequestrate organic N according to its growth rate is saturated. (A) and (B) represent either (1) two growth stages of the same crop, or (2) two crops having different growth rate, or (3) an environmental effect. The dotted lines represent the response curves to increased N supply for the situations A and B.

N uptake (ΔN) according to Eq. 8.13 (C constant). So the corresponding $\Delta N/\Delta W$ represents the effect of the increment in crop mass (W) on the corresponding increment in crop N uptake (N) along a given N uptake vs crop mass trajectory; this reflects the feedback effect of plant growth on plant N uptake. Thus, an increase in plant N uptake can be considered as a consequence of the increment in plant growth as stated by Eq. 8.2. When N supply increases from (s) to (c) then, at a given time, we can observe an increment of N uptake, $\Delta Nc\text{–}s$, that can be related to the corresponding increment in crop mass $\Delta Wc-s$ considered as the direct consequence of the increment in crop N uptake. But, in fact, as shown in Figure 8.3, $\Delta Nc-s$ can be seen as the sum of two components: (1) ΔN_1, the increment in N uptake necessary for the plant to achieve its critical N status and hence its potential growth rate; and (2) ΔN_2, the supplement of N uptake corresponding to the increase in crop growth rate. This is an illustration of the co-regulation of plant N uptake by both soil N supply and plant growth capacity. As soil N supply increases, plant N uptake increases according to the increase of C in Eq. 8.13 and, as a result of the increase of plant N status, plant growth rate is accelerated and then, as a feedback consequence, a supplement quantity of N is taken up according to Eq. 8.2. So, at any moment, crop growth rate is the consequence of crop N uptake and vice versa.

In conclusion, the close relationship between N uptake and crop mass accumulation as expressed by Eq. 8.2 reflects the feedback regulation of root mineral N absorption by shoot growth dynamics. So the co-regulation of N uptake by both soil N supply and plant or crop growth rate as stated by Eq. 8.13 is an emerging property of the whole plant–soil system.

2.3 Diagnostic of plant N status in crops

The main consequence of the theory developed above is that neither the plant N concentration nor the crop N uptake *per se* can indicate unequivocally the crop N nutrition status. Eqs 8.1 and 8.2 indicate that both plant N concentration and crop N uptake have to be interpreted in relation with crop mass. Following the concept of N dilution curves, a nitrogen nutrition index has been proposed, derived from measurements of N concentration and crop biomass. Indirect methods of evaluation of nitrogen nutrition index are also an alternative.

2.3.1 *Nitrogen nutrition index*

The critical N dilution curve as determined by Eq. 8.3 for each crop species (Table 8.1) allows the separation of the actual N status of crops in two situations (Fig. 8.4): above the critical N curve the crops are in N luxury consumption, and below the critical curve the crops are in deficient N supply. It is necessary to measure simultaneously plant N concentration and crop mass at any moment of the vegetative growth period of the crop. Lemaire and Gastal (1997) proposed a Nitrogen Nutrition Index (NNI) to quantify the intensity of both N deficiency and luxury consumption of a given crop. NNI is calculated as the ratio between the actual plant N concentration of the crop (%Na) and the critical plant N concentration (%Nc) corresponding to the actual crop mass (Wa):

$$NNI = \frac{\%Na}{\%Nc} \tag{8.14}$$

When NNI is close to 1, the plant N status is considered as near optimum. Departures from 1 indicate deficiency (NNI <1; the intensity of deficiency is then equal to $1-\text{NNI}$) or excess nitrogen (NNI >1, the intensity of excess is then equal to $\text{NNI}-1$).

Nevertheless, this approach does not take into account that the minimum plant N concentration is not 0 but %Ns, the N concentration of the structural compartment of the plant mass (Eq. 8.9). As stated, %Ns should correspond to the minimum N content for the plant to stay alive. Angus and Moncur (1985) postulated the

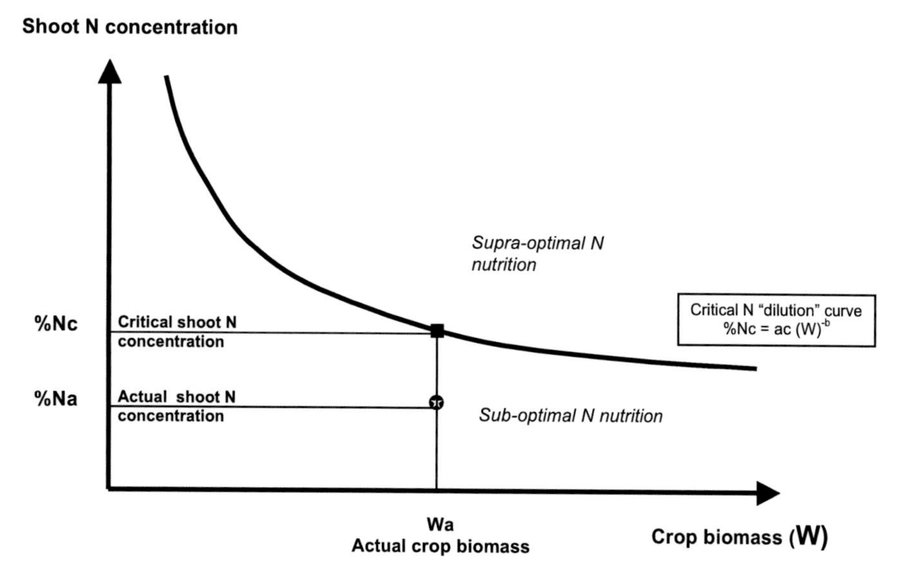

FIG. 8.4 The critical N-dilution curve and the calculation of the Nitrogen Nutrition Index (NNI) (see text for further details on NNI).

existence of such a minimum plant N concentration with a parallel evolution of %Ns with %Nc. Determination of %Ns in field conditions is difficult because even under very severe N deficiency the minimum plant N concentration is not achieved. Since N deficiency is generally lower at early growth stages because growth rate and plant N demand are low, such estimation tends to indicate a decrease in %Ns with plant mass (Plénet and Lemaire, 1999). Lemaire and Gastal (1997) used Eq. 8.9 to derive %Ns of 0.77% for wheat and 0.82% for maize. Therefore, a common value of about 0.8% can be used. The theory developed above postulates that %Ns should be stable along the crop growth cycle. As a default, we can use a constant value of 0.8% for %Ns. Then a Nitrogen Nutrition Index that reflects this physiological attribute can be calculated:

$$NNI' = \frac{\%Na - \%Ns}{\%Nc - \%Ns} \tag{8.15}$$

NNI′ is therefore physiologically more robust than NNI, but involves a greater degree of uncertainty related to the actual value of %Ns, which has not been documented for many crop species.

2.3.2 Assessment of crop N status in field experiments: a prerequisite for interpretation of agronomic data

NNI allows for the interpretation of the response of crops to N nutrition in agronomic experiments. Lemaire and Meynard (1997) showed that the same N rate could lead to different plant N status, depending on plant growth potential, soil N mineralization and overall soil N availability (dry soil, for example). As a consequence, the classical response curves approach of crop yield vs N fertilizer rate leads to variable results with great difficulties of generalization. Each year-site experiment provides a particular fitted equation $Y = f(N)$ where Y is the crop yield and N is the nitrogen application rate. In consequence, the determination of Nopt, i.e. the N application rate necessary for maximizing yield (or economic return) can be obtained only statistically with a very large uncertainty.

The determination of the NNI of a crop at target stages allows a more generic approach:

- the reference treatment becomes the non-limiting N treatment (when NNI ≥ 1 along the growth period) instead of the nil-N rate which corresponds to an unpredictable crop N status
- instead of crop response to N rates, it is then possible to study a more generic response of crops to N where the reduction of crop yield can be directly related to the intensity of N deficiency experienced by crops as expressed by 1−NNI: $Y_a/Y_{max} = f(1-NNI)$, where Y_a is the actual yield and Y_{max} is the maximum yield achievable under the conditions of experiment.

Y_{max} can be estimated directly from experiment or calculated from crop models using local soil and climatic parameters. The relative yield (Y_a/Y_{max}) accounts for the variations due to local and temporal conditions. Thus the response curve $Y_a/Y_{max} = f(1-NNI)$ is more generic than the classical response curve relating Y_a to N application rates. Moreover, such an approach allows the calculation of the relationship between 'crop N status' (NNI) and N application rates. By this way, it is possible to analyze how various N treatments (rate, timing and chemical forms of N application) allow fitting crop N demand. In particular, the crop NNI corresponding to the nil-N treatment (NNI_0) can be considered as an indicator of the capacity of soil to provide N to crops and to fit its N demand. Hence this parameter is useful for interpretation of multiannual and multilocal experimental networks.

Determination of crop NNI is necessary not only for interpreting data from N fertilizer experiments, but also for interpreting all other agronomic data where other field conditions have been manipulated. Soil N availability depends on many other factors than N, such as soil water content, availability of other nutrients, soil structure, pest and diseases, etc. (section 3.5). Thus any manipulation of these factors through crop management could indirectly affect crop N nutrition status. The crop NNI allows quantification of the indirect effect of these other agronomical factors (irrigation, soil tillage, PK fertilization, etc.) on crop N nutrition. For example, this approach allowed Barro et al. (2012) to show that under shading (50% PAR) tropical grasses grew better and legumes grew less than in full light. For grasses, shading contributed to maintaining soil water content, hence improving soil mineral N availability and leading to a higher NNI. For legumes, shading reduced N_2 fixation and NNI. The NNI thus helped to unveil these unexpected links and contrasting patterns between grasses and legumes.

NNI estimates the instantaneous crop N status when plant %Na and actual crop mass (Wa) have been determined. But under changing N supply in the field it is necessary to determine NNI several times during the growth period of the crop. Jeuffroy and Bouchard (1999) used NNI to characterize the N deficiency of wheat crops in terms of intensity, duration and its timing (Fig. 8.5). They showed that an integrated NNI explained a higher percentage of the variation in grain number of wheat within a large experimental data set than a single NNI observed at flowering (96% and 92% respectively; Justes et al., 1997). On oilseed rape, a strong relationship between seed number and crop NNI during seed set was also observed (Jeuffroy et al., 2003). On maize, Plénet and Cruz (1997) showed a high correlation between grain number and NNI averaged from seedling emergence to 20 days after silking. Grain weight was also correlated to NNI. The slope of the regression between the relative grain yield (*Ya/Ymax*) and the intensity of nitrogen deficiency was −1.15, implying that grain yield decreased 15% when crop N deficiency intensifies 10%. Using NNI, Doré and Meynard (1995) showed that pea N deficiency could be caused either by insect (*Sitona lineatus L.*) damage on nodules or by deteriorated soil structure. In all cases, a strong relationship was

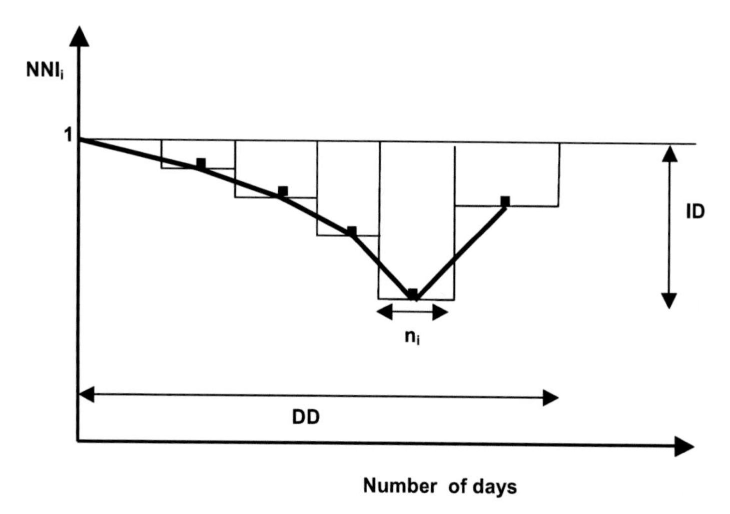

FIG. 8.5 Evolution of instantaneous NNI (NNIi) during growth period and estimation of integrated NNI. DD represents the duration of the period of N deficit, and ID represents the intensity of N deficit.

observed between seed number and crop NNI at flowering.

Many experiments compared the yield of cultivars for evaluation of breeding progress or for determination of adapted variety within a regional context. Such experiments are carried out over several years and on several sites for analyzing genotype–environment interactions (G × E). Part of these G × E interactions can be generated by variations in crop N status according to differences among genotypes in their own N demand associated to their growth capacity and to differences in N availability due to soil–climate–management conditions. Here the use of crop NNI at a given growth stage as co-variable should help in the agronomical interpretation of differences observed among cultivars and G × E interactions.

All these examples illustrate the potential of NNI for interpreting agronomic data from field experiments and for detecting possible indirect effects of N induced by manipulation of other factors. One important application of this approach is the use of the NNI for fertilizer management (Lemaire et al., 2008). The possibility to predict crop NNI evolution through the use of dynamic crop models combined with *in situ* crop measurements is promising (Naud et al., 2008).

2.3.3 Simplified methods for evaluating nitrogen nutrition index

Despite its robustness, NNI remains more a research than a management tool. NNI determination is time consuming because it requires both actual determination of both crop mass and plant N concentration. Therefore it is necessary to develop non-invasive, cost-effective methods for a rapid determination of plant N status and use NNI as a reference for calibration.

The theory developed above for explaining N dilution with plant growth in dense canopy shows that the decrease in plant N concentration is the result of two processes: (1) the decline in plant leaf area ratio as crop mass increases; and (2) the preferential allocation of N to well illuminated leaves in the upper layer of the canopy as it develops (see section 3.3.3 below). Therefore, Lemaire et al. (1997) suggested that while plant N concentration declines with crop mass accumulation, the N concentration of the upper leaves of the canopy is more stable and well correlated with the NNI. This assumption was evaluated on several C_3 and C_4 grasses, showing that N concentration of the upper leaves of the canopy is a good proxy of the NNI of a crop (Fig. 8.6). A higher correlation and a better stability between site-years were observed when N

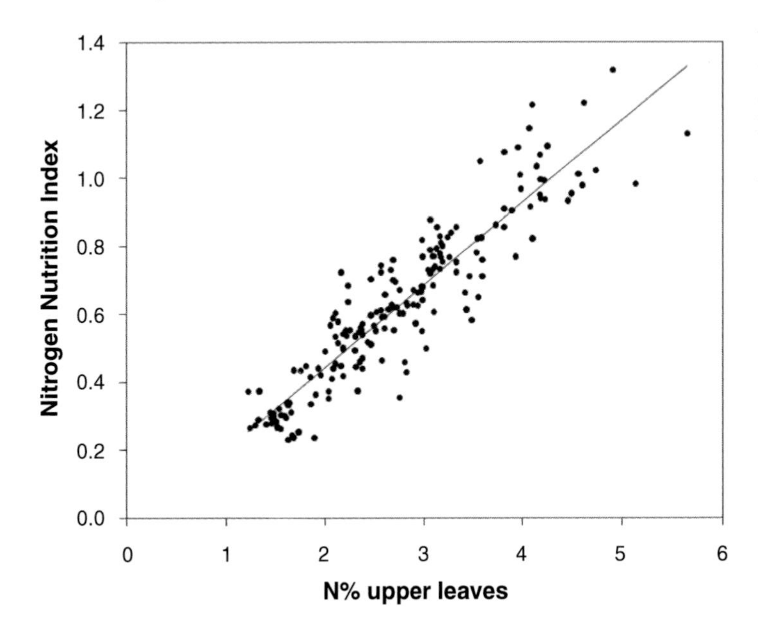

FIG. 8.6 Relationship between N concentration of the upper leaf layer of the canopy and Nitrogen Nutrition Index, evaluated on perennial forage C_3 grasses (*Festuca aundinacea, Lolium perenne, Dactylis glomerata*). *From Farrugia et al. (2004) and Gastal et al. (2002).*

concentration of the upper leaves was expressed on a dry weight (DW) rather than on a leaf area basis (Farrugia et al., 2004; Ziadi et al., 2009). Data from Farrugia et al. (2002, 2004) allowed the following relationship between N concentration of upper leaves on a DW basis (%Nup) and NNI for perennial C_3 grasses to be derived (Fig. 8.6):

$$NNI = 0.242\%Nup - 0.041 \qquad (8.16)$$

For these perennial C_3 grasses, a constant N concentration of the upper leaves of the canopy (%Nup = 4.3) allowed achievement of an NNI of 1 for the crop, independently of its mass during growth cycle. For maize, a lower critical value in %Nup = 3.7 was found (Ziadi et al., 2010).

Chlorophyll measurement with SPAD is also a practical, commonly used method for the estimation of the upper leaf N concentration (Feibo et al., 1988; Piekelek and Fox, 1992; Reeves et al., 1993; Matsunaka et al., 1997, among many others). Peng et al. (1996) showed that SPAD measurements are well correlated with visual leaf color estimations. Caviglia and Sadras (2001) showed that NNI and SPAD measurements provide consistent measures of crop nitrogen deficiency in wheat. However, these methods frequently show significant variation of calibration according to genotypes and environments. Normalization of SPAD readings has been proposed to overcome these difficulties (Ziadi et al., 2010). Promising methods to assess crop nitrogen status are being developed with the aid of remote-sensing tools (e.g. Babar et al., 2006; Houlès et al., 2007).

2.4 Intra- and inter-specific interactions within plant stands

The theory developed above stands for pure crops and for plants considered as the average plant of a population. Important questions arise from mixed crops about how plant–plant interactions impact plant N acquisition and allocation. Variability between individual plants and species components of mixed crops also matters and has then to be considered.

Light competition affects plant %N dilution (Lemaire and Gastal, 1997; Seginer, 2004).

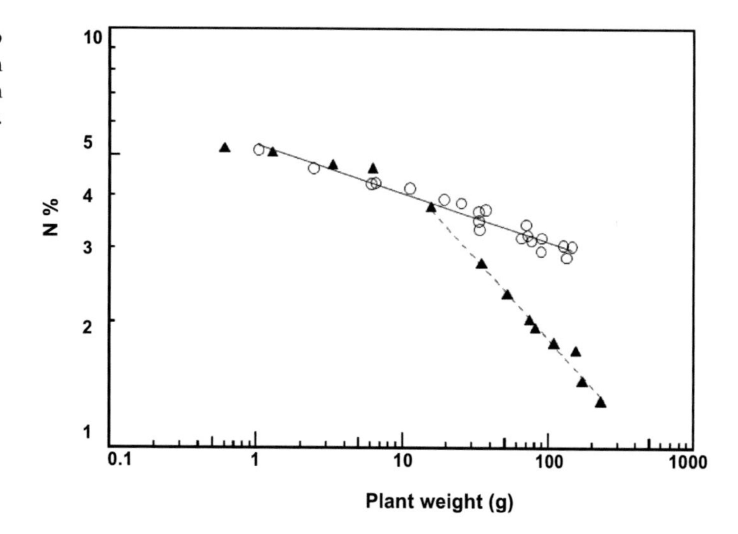

FIG. 8.7 Comparison of the relationship between plant %N and plant mass between isolated plants (○) and the average plant in a dense canopy (▲) for sweet sorghum cv. Keller. *From Lemaire and Gastal, 1997.*

Critical dilution curves differ between isolated plants and the average plant in a dense canopy (Fig. 8.7). Within a dense alfalfa stand, Lemaire et al. (2005) showed that plants dominated for light had a lower N concentration than dominant plants at similar plant mass. Two different processes contribute to the reduction in plant %N arising from competition for light in dense canopies (Lemaire et al., 2005): (1) the proportion of metabolic vs structural N pool is altered by light-induced plant plasticity, associated with a decreased in leaf:stem ratio (i.e. higher biomass allocation to stems); and (2) the average %N of each pool is affected by plant light microclimate (Hirose et al., 1988; Lemaire et al., 1991), and subordinate plants display systematically lower N concentration, particularly in stems. Irrespective of light-induced photomorphogenetic adaptations, the most conserved relationship was between leaf mass and plant N content. Thus, allocation of N among individual plants reflects the contribution of individual plants to the crop leaf mass, and probably to light interception, rather than their contribution to the whole crop mass.

Mixed crops are often grown expecting better resource-use efficiency. Therefore, it is of interest to analyze the N nutrition of each species separately. Nevertheless, specific constraints occur when it comes to analyze N dilution and diagnose N nutrition under these conditions. For a given shoot mass and leaf area, light interception differs between pure and mixed crops (Sinoquet et al., 1990). As a consequence, the critical N dilution curve is likely to change between pure and mixed stands (Cruz and Soussana, 1997). This critical N dilution curve is expected to stand in between the critical curves of isolated plants and average plants in pure dense canopies for the dominant species, and could even be lower for the subordinate species. Experimental designs have been proposed to provide critical N dilution curves per plant or species and assist in the interpretation of the intra- and inter-specific effects of light-mediated interferences on N nutrition (Cruz and Lemaire, 1986). Mixtures in alternate rows were analyzed with respect to both single- and double-row pure stands of each species in an alfalfa–cocksfoot mixture (Fig. 8.8). These row-based analyses showed that the presence of alfalfa did not benefit cocksfoot N nutrition when compared to double-row pure stands. Similarly, shoot N dilution remained the same for double-row and mixed alfalfa.

A difficulty in expressing N dilution with respect to dry mass on a ground area basis (or an

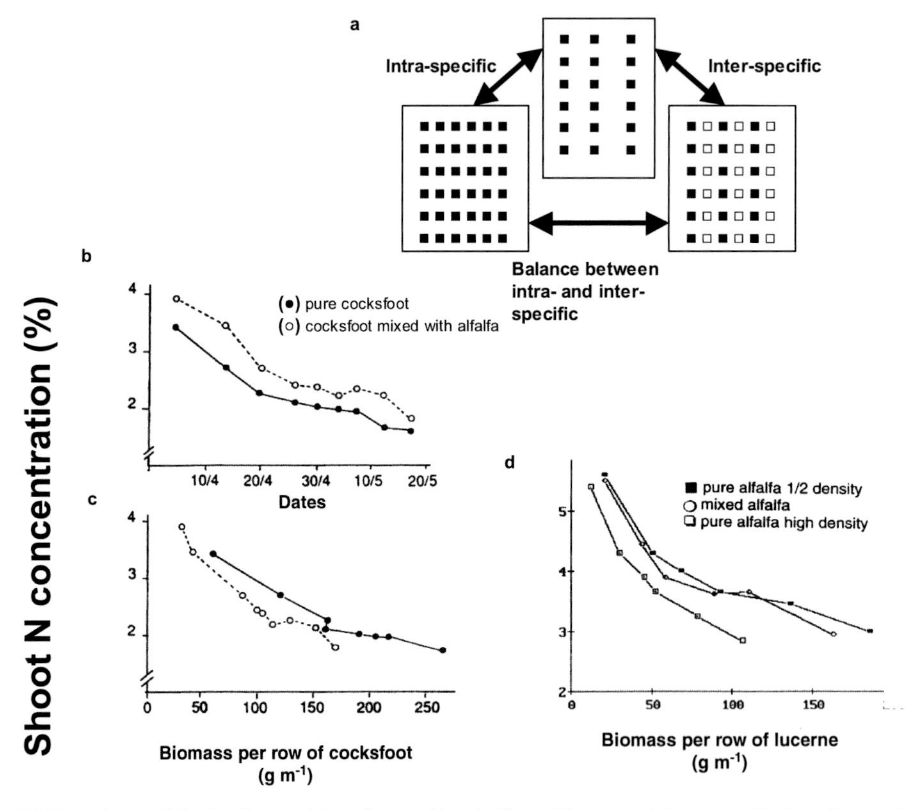

FIG. 8.8 (a) Experimental design for studying the combined effect of intra- and inter-specific interference in mixed crops. The row of one species in double-spaced row plots can be compared with row within pure crop in single-spaced row (intra-specific effect) or with row within mixture (inter-specific effect). Evolution during spring growth of plant N concentration of a pure double-spaced row of cocksfoot compared with row of cocksfoot in mixture with alfalfa either on (b) time basis or (c) dry matter basis. (d) Evolution of plant N concentration of alfalfa rows in pure stand with single-spaced rows, in pure stand in double-spaced rows, and in mixture with cocksfoot. *From Cruz and Lemaire (1986).*

average plant/row basis) is that the space available for each species has to be assessed. In clover-based pastures for instance, neither rows nor individual plants can clearly be defined. To overcome this limitation, Soussana and Arregui (1995) proposed to relate the shoot %N of each species to the whole canopy biomass. The rationale behind this is that whole biomass could be a good proxy for total canopy leaf area and could account for neighbor shading, assuming that species have similar heights/influences upon each other. They also compared the diagnosis of N nutrition from this method with a row-based method as mentioned above, and found slightly higher average nitrogen nutrition index in the row-based method. Contrary to alfalfa, however, white clover improved the N nutrition of the associated grasses irrespective of the method of calculation.

Both methods present limitations. To avoid the uncertainty related to the partitioning of light between the species in the mixture and its effects on the expression of the critical N curves, Cruz and Soussana (1997) suggested plotting the %N against the total radiation intercepted by each component rather than whole canopy biomass. According to the biophysical theory of Monteith (1972), this variable represents the potential dry matter production of a crop. But estimation of light partitioning in mixed stands can be tedious, even with simplified approaches (Barillot et al., 2011; Louarn et al., 2012). Alternatives include the use of shoot N relationships stable across light environment (e.g. plant N content vs leaf mass as proposed in Lemaire et al. (2005), analogous to Eq. 8.12), or simplified diagnostic tools independent of the light and soil surface (e.g. N concentration of the upper leaves, see section 2.3.3). All these options remain to be assessed.

Overall, intra- or inter-specific interactions between plants lead to hierarchical N resource sharing among plants according to their hierarchical position for light interception. These plant–plant interactions have to be considered for understanding the competition for soil N resources. In complex plant communities, N resource sharing might be dominated by competition for light in conditions of high N supply, and by intrinsic N uptake capacity of plants under low N supply (Dybzinski and Tilman, 2007).

3 RESPONSE OF PLANTS AND CROPS TO N DEFICIENCY

Once the timing and the intensity of N deficiency of a crop has been quantified, it is possible to analyze its responses to a wide range of N deficiency through the different functions leading to crop mass accumulation: (1) the interception of light (leaf expansion processes and leaf senescence); and (2) the radiation-use efficiency (photosynthesis and respiration), and allocation patterns within plants (i.e. shoot vs root, harvest index).

3.1 Crop life cycle and plant N economy

Here we present selected comparisons between grain crops (wheat and rice vs maize) and between grain and forage crops to illustrate differences and similarities between species. In most crop species, the plant life cycle with regard to N nutrition can be divided into two main phases: the vegetative phase, from seedling emergence until flowering; and the reproductive phase, from fecundation to maturity. In unstressed crops, the end of the first phase often coincides with peak leaf area, stem height and branching (tillering), but peak leaf area is anticipated in relation to flowering under stressful conditions. In the reproductive stage, leaves, stems and roots behave predominantly as net sources of N organic compounds for seed development and grain filling. This N recycling is associated with leaf senescence that leads to a progressive decline in crop photosynthesis and C supply (Chapter 10). During this post-flowering phase, plant N uptake can provide an additional source of N for grains.

In wheat (Palta and Fillery, 1995; Habash et al., 2006) and rice (Mae, 1997; Tabuchi et al., 2007), 60–95% of the grain N at harvest comes from remobilization of N stored in roots and shoots before flowering. In maize, this proportion is only 45–65% (Ta and Weiland, 1992; Rajcan and Tollenaar, 1999a,b; Gallais and Coque, 2005). With good water supply, wheat and rice can develop a larger LAI (7–8) than maize (4–6); this, according to Eq. 8.12, allows a greater capacity of storage of organic N compounds in shoots. That corresponds to the double function of leaves, as organs of C assimilation, and as temporary storage of N in Rubisco (Sinclair and Sheehy, 1999; Thomas and Sadras, 2001; Chapter 10). Lemaire et al. (2007) showed that the capacity of shoot N storage per unit LAI at optimum N supply is 37 kg N ha^{-1} for wheat and 30 kg N ha^{-1} for maize, reflecting differences in both metabolic pathway and intrinsic LAR between the two species. Then the total capacity of shoot N storage at flowering can reach 250 kg N ha^{-1} for wheat and only 140 kg N ha^{-1} for maize. As a result, maize grain yield is more dependent on plant

N uptake capacity after flowering than wheat or rice.

Canola can accumulate a large quantity of N in vegetative parts at the beginning of flowering (Lainé et al., 1993). However, grain yield of canola is low in comparison with the total crop mass, and then an important part of the N stored in the vegetative organs is not used for grain production (Rossato et al., 2001).

For perennial vegetative plants like forage grasses or legumes, the plant life cycle is interrupted by cutting before flowering and then the plant N economy is entirely dominated by the accumulation of N within leaves and stems. Remobilization of N from shaded leaves at the bottom of the canopy to well illuminated leaves at the top occurs as the crop LAI develops (Lemaire et al., 1991; section 2.1.2). For regrowth after defoliation, plants remobilize organic N stored in roots and stubbles (Avice et al., 1996, 1997; Ourry et al., 1996). In these species, perenniality is related to the capacity to restore N root reserve during regrowth (Lemaire et al., 1992). As a consequence, the dynamics of shoot N accumulation during regrowth does not reflect exactly the dynamics of N absorption and assimilation.

The selected comparisons of wheat and rice vs maize and between annual grain crops and perennial forage crops demonstrate that plant N economy differs substantially according to crop species and management. But, despite these differences, a common feature can be pointed out across crop species: the necessity to store sufficient quantities of N in vegetative organs either (1) to produce maximum herbage mass for forage crops and ensure perenniality, or (2) to meet as much as possible the N demand of grains. The more the grains depend on post-flowering plant N uptake the more the crop is susceptible to terminal stress such as the exhaustion of N in soil and/or reduced soil water content and then N availability. Stay-green types are particularly able to maintain leaf chlorophyll and photosynthetic capacity during grain filling (section 3.4).

3.2 Effects of N deficiency on crop mass accumulation

Maximizing harvested production of a crop corresponds in general to maximizing crop mass accumulation. This is obvious for forage crops where shoot mass is harvested. It is also the case for most of the grain crops such as cereals where grain number is correlated to crop mass at flowering (Uhart and Andrade, 1995; Desmotes-Mainard et al., 1999; Desmotes-Mainard and Jeuffroy, 2004).

3.2.1 Radiation-use efficiency and PAR interception

Crop biomass accumulation is related to the quantity of photosynthetic active radiation (PAR) intercepted by the crop during its life cycle (Monteith, 1972); in a first approximation this relationship is linear (Gallagher and Biscoe, 1978; Gosse et al., 1984, 1986; Sinclair et al., 1992). The slope of this relationship represents the average conversion of energy intercepted by the crop in biomass, and is commonly called radiation-use efficiency (RUE). Bélanger et al. (1992) showed that N deficiency reduces both components of crop mass accumulation in a tall fescue sward: the quantity of intercepted PAR and the RUE (Fig. 8.9a). Further, these authors analyzed the responses of RUE and intercepted PAR to N deficiency using NNI. As shown in Figure 8.9b, in relative terms, RUE was more affected by moderate N deficiency (NNI = 0.6–0.8) than the quantity of PAR intercepted by the crop, but for very severe N deficiency (NNI around 0.3), the response of the two variables converged. Similar results have been obtained by Trápani and Hall (1996) on sunflower and Muchow and Davies (1988) on sorghum and maize. The response of LAI expansion to N deficiency follows a similar pattern as RUE, but with a slightly higher sensitivity: for NNI = 0.6, RUE was reduced by 30% while LAI was reduced by 40% as compared to NNI = 1. The difference in response between LAI and PAR was due to the asymptotic

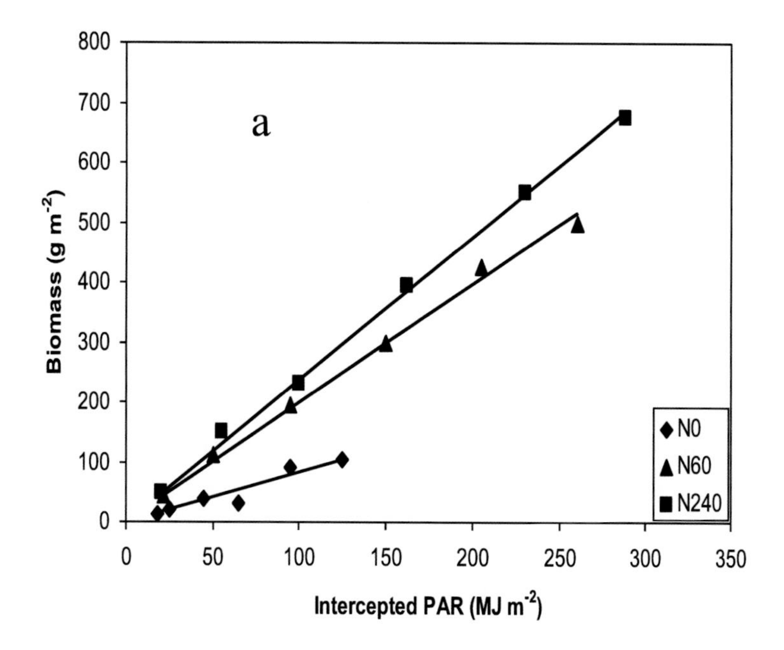

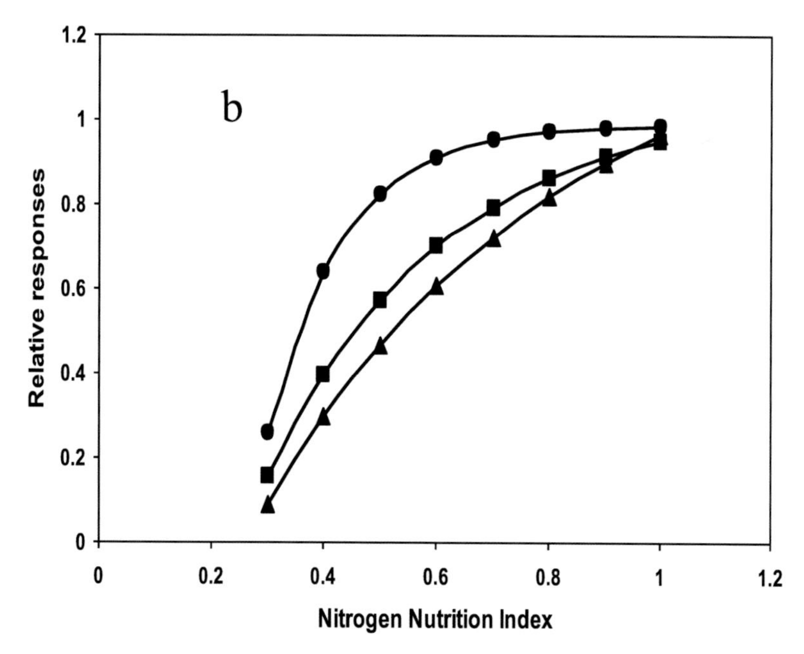

FIG. 8.9 Relationships between shoot biomass accumulation and cumulative intercepted PAR during regrowth of tall fescue swards receiving nil, 60 or 240 kg N ha^{-1} (a) and effects of crop N status determined by the nitrogen nutrition index (NNI) of tall fescue swards receiving different N rates (b) on (1) the relative quantity of intercepted PAR, PAR_{act}/PAR_{max} (●), (2) the relative-use efficiency for shoots, RUE_{act}/RUE_{max} (■), and the relative LAI, LAI_{act}/LAI_{max} (▲). *From Bélanger et al. (1992).*

relationship between the radiation-interception efficiency (i.e. the ratio between the visible radiation intercepted by a crop and the incident visible radiation) of the crop and the LAI. The reduction of the quantity of intercepted PAR is important when LAI declines below 3, under severe N deficiency.

The relative effect of N deficiency on both RUE and the quantity of intercepted PAR depends on the timing of the N deficiency. Deficiency before the crop reaches LAI = 3 is likely to reduce the quantity of PAR intercepted by the crop, while N deficiency after this stage would only have a small effect. This partially accounts for the divergent reports on the relative sensitivity of RUE and light interception in response to N deficiency.

RUE is an integrated variable accounting for photosynthesis and respiration. However, because RUE is generally estimated on a shoot biomass basis, it also depends on assimilate allocation to roots. Bélanger et al. (1994) compared the effect of N nutrition on the components of RUE of tall fescue with RUE calculated on the basis of (1) canopy gross photosynthesis, (2) total (shoot + roots) biomass, and (3) shoot biomass (Table 8.2). Canopy gross photosynthesis was less affected by N deficiency than accumulation of total biomass: in high N condition, about 50% of assimilated C was used for plant growth while in deficient N conditions this proportion is only 37%, reflecting higher relative respiration in N-deficient plants (Table 8.2). The accumulation of shoot biomass was more affected by N supply than the accumulation of total biomass, reflecting a change in dry matter allocation to roots, which increased from 17% in crops with 180 kg N ha^{-1} to 37% with no fertilizer (Table 8.2). The lower shoot:root ratio in N-deficient plants and crops is widely documented (Barta, 1975; Robson and Parsons, 1978; Jarvis and MacDuff, 1989). The higher relative root growth could be one of the reasons why the respiration losses are higher in relative terms in N-deficient swards than under high N supply.

3.2.2 *Effect of N deficiency on canopy size and radiation interception*

The response of leaf area to N deficiency is mediated by a decline in both the expansion of individual leaves and branching or tillering (Wilman and Pearse, 1984; Gastal and Lemaire, 1988; Vos and Biemond, 1992; Trápani and Hall, 1996). These authors showed that N had little effect on the rate of leaf appearance and the duration of individual leaf expansion in many species. The accumulation of non-structural carbohydrates in N-deficient leaves suggests that carbohydrate supply is not the cause of reduced leaf expansion (Gastal and Lemaire, 2002). N deficiency alters the rates of cell division and cell expansion, while the final cell length is little affected (MacAdam et al., 1989; Gastal and Nelson, 1994; Fricke et al., 1997; Trápani et al., 1999). In all instances, the impact of N on leaf expansion rate of monocots was related more to the effect of

TABLE 8.2 Effects of nitrogen supply on radiation-use efficiency (RUE) of tall fescue swards

N rate (kg ha^{-1})	(1) RUEgp (g DM MJ^{-1})	(2) RUEt (g DM MJ^{-1})	(2)/(1)	(3) RUEs (g DM MJ^{-1})	[(2 − (3)]/(2)
0	4.82	1.78	0.37	1.12	0.37
60	5.03	2.28	0.45	1.73	0.24
120	5.05	2.56	0.51	2.07	0.19
180	5.62	2.65	0.47	2.21	0.17

RUE is calculated on the basis of (1) canopy gross photosynthesis (RUEgp), (2) total biomass (RUEt), and (3) shoot biomass (RUEs). The ratio [(2)−(3)]/(2) indicates fractional allocation of dry matter to roots. After Bélanger et al. (1994).

N on cell production than cell expansion rate. In dicots, early studies concluded that the shortage of N primarily reduced cell growth rate. Nevertheless, whether the response of leaf growth to N deficiency differs between monocots and dicots, as suggested by Radin (1983), remains an open question.

Gastal et al. (1992) proposed a quantitative relationship between the leaf elongation rate (LER) of tall fescue and the *NNI* of the sward:

$$\frac{LER_{actual}}{LER_{critical}} = 1.39 - 1.9e^{-1.49NNI} \quad (8.17)$$

where subscript 'actual' refers to suboptimal N condition while 'critical' refers to non-limiting N. This response curve crosses the X-axis at NNI = 0.21, corresponding to the plant N status at which leaf growth ceases. For a severe N deficiency, NNI = 0.4, the actual LER is reduced to about 30% of its maximum.

Lemaire et al. (2008) analyzed the response of LAI of different crop species to N deficiency using the allometric relationships between LAI and crop mass W (Eq. 8.11). This approach allows the separation of the reduction of LAI directly related to the reduction of crop mass from a specific reduction of LAI at similar crop mass: i.e. a reduction in plant leafiness coefficient corresponding to (1) a decrease in leaf/stem ratio, and/or (2) an increase in specific leaf weight (SLW). Figure 8.10 shows the response of LAI of maize and wheat to N supply. For about the same level of N deficiency (Fig. 8.10a,d), the two crop species behave differently: the reduction in LAI of maize is entirely and allometrically linked to the reduction in crop mass (Fig. 8.10b), while wheat also drastically reduces its LAI at a given crop mass in response to N deficit (Fig. 8.10e). Lemaire et al. (2008) showed that tall fescue behaves as maize while canola behaves as wheat, with sorghum and sunflower having intermediate responses. As a consequence of the response of leaf area per unit of crop mass to N deficiency, the accumulation of N per unit of LAI can be either reduced in N-deficient condition as for maize (Fig. 8.10c) or maintained as for wheat (Fig. 8.10f).

Hence, classification of crop species in either metabolic group (C_3 vs C_4) or botanical groups (monocots vs dicots) does not allow the determination of crop response type: under N deficiency some species such as maize or tall fescue and, to a lesser extent, sorghum and sunflower tend to maximize light interception by minimizing the reduction in crop LAI, while other species such as wheat and canola do not. Do these two opposite strategies mean that a trade-off exists between photosynthesis rate per unit leaf area and leaf area expansion? Does maize drop photosynthesis per unit leaf area more than wheat? Answers to these questions require more analytical studies on response of leaf and canopy photosynthesis to N deficiency.

3.2.3 *Effect of N deficiency on leaf photosynthesis*

The response of leaf photosynthesis to irradiance is largely dependent on the leaf N concentration. Photosynthetic proteins, including large amounts of Rubisco and, to a lesser extent, light harvesting complex proteins, represent about 60% of the leaf N content (Evans, 1983; Field and Mooney, 1986; Lawlor, 2002). Leaf photosynthesis at saturating light (A_{max}) increases asymptotically with leaf N content (see Grindlay, 1997, for review). This relationship shows a positive intercept on the leaf N concentration axis, indicating that when leaf photosynthesis rate becomes zero, leaves would still contain a significant amount of N, corresponding to structural N (Ns) in Eq. 8.9 (Field and Mooney, 1986). The variation in the A_{max}/SLN (leaf N content per unit of leaf area) relationship seems relatively limited among cultivated species of the same metabolic group (Evans, 1989; van Keulen et al., 1989). However, Sinclair and Horie (1989) showed a lower A_{max} at similar leaf N per unit area for soybean compared to non-legume C_3 species. This lower A_{max} was associated with higher

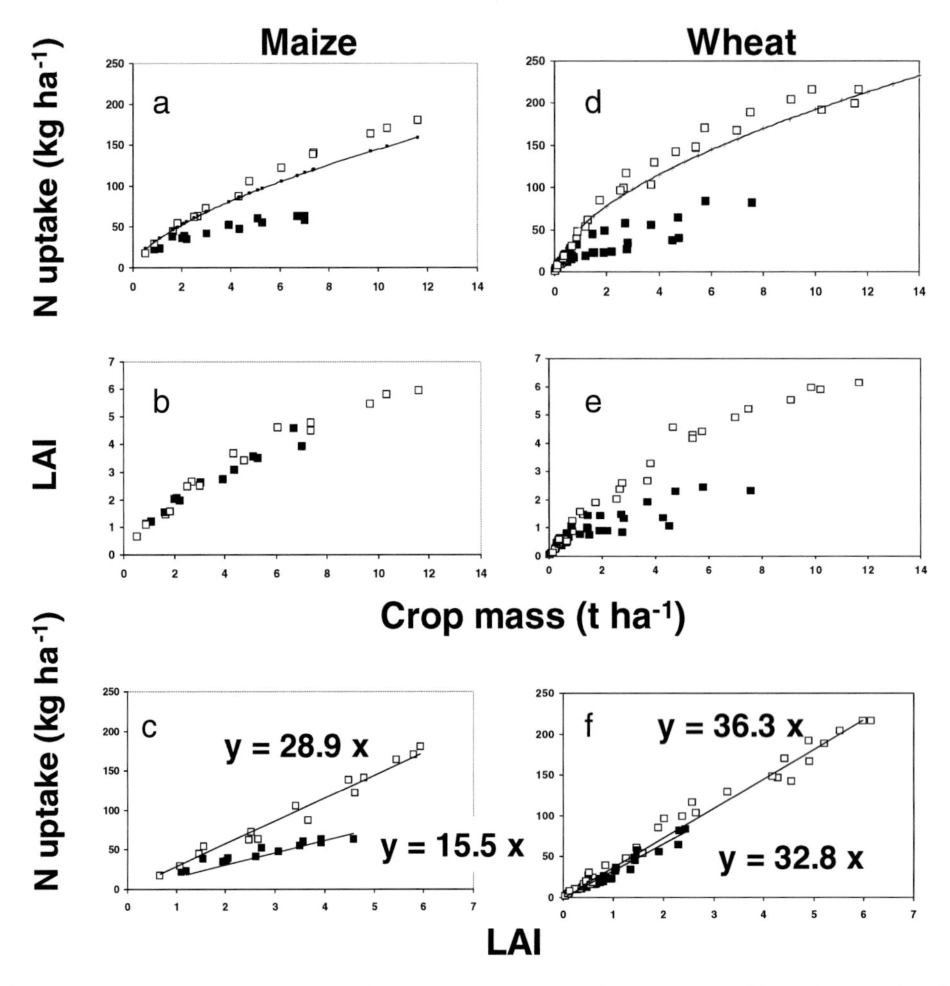

FIG. 8.10 Comparison of (a–c) maize and (d–f) wheat responses to non-limiting (□) and limited N supply (■) measured in field crops in France. Relationships between (a, d) N uptake and crop mass, (b, e) LAI and crop mass, and (c, f) N uptake and LAI. *From Lemaire et al. (2008).*

structural N concentration of leaves. The variations in A_{max}/SLN relationship between species could be due to (1) differences in nitrogen costs of PEP-carboxylase and Rubisco and the relative amount of these two enzymes per unit leaf area, and/or (2) the possible accumulation of vegetative storage proteins in legume species.

There is a discussion among crop scientists whether A_{max} has to be related to leaf N content per unit leaf mass or per unit leaf area basis (Grindlay, 1997). Our opinion is that none of these two approaches is completely right. The two processes of leaf photosynthesis, light harvesting by chlorophyll and CO_2 reduction by Rubisco, are affected by leaf N content. Both processes have to be expressed on a volume basis and then leaf thickness should be considered. But the correlation between leaf thickness and SLA is weak as SLA varies with non-structural carbohydrate concentrations. A measure of leaf N concentration per unit of leaf volume could be based on leaf water content at saturation (Thornton et al., 1999).

This problem reinforces the necessity for scaling metabolic plant processes and for relating them to plant form and size (Enquist et al., 2007).

Data concerning the response of leaf photosynthesis to N at low irradiance, i.e. photosynthetic light efficiency, are relatively scarce. Gastal and Bélanger (1993) indicated that for tall fescue the effect of leaf N content on photosynthetic light efficiency was low or not significant and always much lower than the effect on A_{max}. Connor et al. (1993) showed no correlation between photosynthetic light efficiency and N in sunflower. However, Dreccer (1999) found contradictory results. Another component of the net C exchange at low light is dark respiration, which seems to increase with increasing N, in association with maintenance (Connor et al., 1993). Hence, as the leaves are progressively shaded, the effect of N deficiency on leaf N net photosynthesis becomes lower.

Leaf photosynthesis and responses to light and CO_2 concentration have been formalized in the biochemical model originally proposed by Farquhar et al. (1980) where leaf photosynthesis is determined by the carboxylation/oxygenation capacity of Rubisco (Vcmax) or by electron transport and regeneration of ribulose 1,5 biphosphate (Jmax). Linear correlation between leaf N concentration per unit leaf area and both Vcmax and Jmax have been observed in many studies (Niinemets and Tenhunen, 1997; Prieto et al., 2012) and allows the responses of leaf photosynthesis to irradiance and to leaf N content described above to be represented in a more mechanistic way. Alternatively, a coordination hypothesis between Vcmax and Jmax has been proposed (Chen et al., 1993) and has been the basis for a more general model linking light gradient within canopies, leaf nitrogen content and leaf photosynthesis (Maire et al., 2012).

3.2.4 *Integrating from leaf to canopy photosynthesis*

Gastal and Bélanger (1993) proposed a quantitative relationship between canopy gross photosynthesis at high irradiance (CGP_{max}) and NNI for tall fescue swards: a reduction in NNI from 1 to 0.4 reduced the relative CGP_{max} from 1 to 0.6. This relatively low responsiveness of CGP_{max} to N deficiency was due to the greater number of shaded leaves with increasing canopy size and the decoupling of photosynthesis and leaf N content at low irradiance. The same authors showed that the maximum light yield of the canopy, i.e. the photosynthetic light efficiency of the canopy at low irradiance, was not related to the sward NNI. So, when canopy photosynthesis is integrated over a day or over a longer period, there is only a limited response to N during the periods of low irradiance. For this reason, the effect of N deficiency on integrated canopy gross photosynthesis appears relatively limited (Table 8.2). Genetic variation in A_{max}, when it exists, is in great part buffered when integrated at the level of canopy (Sinclair et al., 2004).

3.2.5 *Trade-off between light capture and radiation-use efficiency*

Crops can respond to N deficiency through a reduction in light interception, a reduction in canopy photosynthesis or both. Sinclair and Horie (1989), Grindlay (1997) and Vos et al. (2005) proposed that these two types of responses represent a trade-off between maintaining resource capture or resource-use efficiency. In conditions of limited N supply, maintenance of leaf area must lead to a decreased leaf N concentration per unit leaf area (SLN) as in maize (Fig. 8.10b,c), whereas maintenance of SLN is expected to be associated with reduced leaf area as in wheat (Fig. 8.10e,f). Differences in critical SLN for LAI expansion and RUE can underlie the differences among species in response to N deficit (Lemaire et al., 2008). For sorghum and maize, the critical SLN for leaf expansion is about 1.0 g m^{-2} (Muchow, 1988; van Oosterom et al., 2001), while the critical SLN for RUE is 1.4–1.5 g m^{-2} (Muchow and Davis, 1988; Muchow and Sinclair, 1994). As a consequence, N deficiency affects RUE before LAI expansion and maize and sorghum maintain resource capture rather than resource-use efficiency. For sunflower, SLN thresholds are 2.0 g m^{-2} for leaf size and 1.5 g m^{-2} for RUE (Bange et al., 1997). For canola, the quantity of PAR intercepted by the crop is severely decreased by N deficiency while RUE is less affected (Wright et al., 1988). For wheat, there is also ample evidence that N deficit affects LAI before RUE (Gallagher and Biscoe, 1978; Meinke et al., 1997). Variation between species in strategy of radiation capture vs leaf N and radiation-use efficiency was also confirmed by Fletcher et al. (2013).

Lemaire et al. (2008) showed consistent differences among species in their response to N deficiency in terms of resource capture, mediated by the ability to maintain LAI expansion despite N deficiency. Because the specific leaf area (SLA) is not substantially affected by N deficiency when comparisons are made at same LAI, the capacity for maintaining LAI expansion is mainly determined by resource allocation between leaf and stem. These differences among species are not related to either metabolic (C_3 vs C_4) or botanical type (monocot vs dicot). C_4 species, however, tend to reduce RUE more than C_3 species. As a consequence, there is not a clear trade-off between resource capture and resource-use efficiency among species as postulated.

3.3 N deficiency effects on C and N allocation within plants and canopies

3.3.1 C and N allocation to roots

Nitrogen deficit increases the root:shoot ratio (Table 8.2) according to the concept of functional equilibrium (Brouwer, 1962; Thornley, 1977). However, variation exists between species in the intensity of the root:shoot response to nitrogen deficiency (Robinson, 1994). This response implies a feedback whereby the increase in root:shoot ratio in response to N deficiency would lead to an increase in plant N uptake that could contribute to limiting the crop N deficiency. Local patches of mineral N in soil generally induce root proliferation while root elongation is enhanced in free mineral N soil zones, leading to a foraging development of roots (Samuelson et al., 1992; Robinson, 1994; Zhang and Forde, 2000).

Root density, architecture, morphology and root growth responses to localized mineral N supply differ between species and a large range of root morphological plasticity exists (Fitter, 1991; Grime et al., 1991; Wasson et al., 2012; Trachsel et al., 2013). Lemaire et al. (1996) stated that differences in root architecture accounted for the higher N uptake of sorghum under low soil N in comparison to maize. Intra-specific differences in root architecture have been poorly investigated (Manschadi et al., 2006, 2008). Oyanagi et al. (1993) and Oyanagi (1994) investigated genetic differences in vertical root distribution and root angle in wheat. A series of studies reported a weak but significant genetic correlation between several root traits, biomass production and yield of maize and wheat under N limiting N supply (Guingo et al., 1998; Coque and Gallais, 2006; Laperche et al., 2006; Herrera et al., 2013; Lynch, 2013).

Plant below-ground N may represent up to 30–60% of the total N for legumes at pod filling and around 35% for cereals at grain filling (McNeil et al., 1997; Peoples and Baldock, 2001; Kahn et al., 2002). Hence, variation in shoot:root allocation of N organic compounds could explain differences in shoot N recovery between species and genotypes. Total root length and root area per unit of soil volume are important traits related to N absorption. The difference in root length density of cereals and dicots is about twofold, and this is partially associated with a large compensating difference in hydraulic conductivity (Hamblin and Tennant, 1987). Improving our understanding of the relationship between plant growth, plant productivity, and root architecture and dynamics under soil conditions is of major importance (Whu et al., 2005).

3.3.2 *C and N allocation to stems*

As discussed above, crop species respond differently to N deficiency in terms of allocation of dry matter to photosynthetic and non-photosynthetic tissues. The effect of N deficiency on the allocation pattern of N is the result of the responses of (1) leaf:stem ratio, and (2) stem N concentration (%Nstem). Species such as wheat and canola reduce proportionally more leaf growth than total shoot growth in response to N deficiency, while maize and tall fescue reduce leaf growth in the same proportion to shoot growth, and sorghum and sunflower have an intermediate response (Lemaire et al., 2008). Owing to the stability of specific leaf area with N deficiency when compared at similar LAI, the leaf:stem ratio of wheat and canola is sharply reduced by N deficiency while it is not affected in maize. The leaf:stem ratio generally decreases as crop mass increases (Lemaire et al., 1992). Therefore, a progressively greater proportion of C is allocated to the stem as the crop ages. Stems are mainly composed of structural tissues (Ws in Eq. 8.6) having a low N concentration (%Ns).

As indicated by Eq. 8.12, N accumulates within the canopy proportionally to LAI. The capacity of a crop to accumulate N in canopy per unit LAI (N_{LAI}) can then be decomposed in two components, N accumulation in leaves (Nleaves) and N accumulation in stem (Nstems):

$$N_{LAI} = \frac{Nleaves}{LAI} + \frac{Nstems}{LAI} \tag{8.18}$$

or

$$N_{LAI} = SLN + \frac{Nstems}{LAI} \tag{8.19}$$

Figure 8.11 shows that sorghum and maize responded to N deficiency through a decrease in N allocation to stems per unit LAI, while wheat and canola maintained their N allocation to stems. Reduction of allocation of N to stem should allow the plant to maintain a high LAI, a high SLN or both, thus favoring light capture and/or radiation-use efficiency under N deficiency. However, as stated by Lemaire et al. (2008), the reduction in N allocation to stem could reduce later N transfer to reproductive meristems, leading to a reduced grain number and grain yield (Demotes-Mainard et al, 1999; Vega et al., 2001). If such a hypothesis were confirmed, it would lead to a trade-off between light capture and light-use efficiency on one hand, and reproductive growth on the other hand.

3.3.3 *N allocation within canopies*

Numerous studies indicate that N distribution within canopies is not uniform (see review from Grindlay, 1997). This non-uniform distribution can be due to a non-uniform distribution of irradiance within canopies (Hirose and Werger, 1987; Lemaire et al., 1991; Schieving et al., 1992; Shiraiwa and Sinclair, 1993; Anten et al., 1995). Parallel to the extinction of light, there is a gradient in leaf age in the canopy. Hikosaka et al. (1994) and Lemaire et al. (1991) showed that the age effect is less important than the irradiance effect. The relationships between light and foliar N profiles in canopies have been analyzed in several species: lucerne (Lemaire et al., 1991); sunflower (Sadras et al., 1993; Connor et al., 1995), wheat and oilseed rape (Dreccer, 1999), cotton (Milroy et al., 2001), soybean (Sinclair and Shiraiwa, 1993), *Lysimachia* spp. (Hirose et al., 1988), maize (Drouet and Bonhomme, 1999), *Carex* spp. (Aerts and de Caluwe, 1994).

Rousseaux et al. (1996, 1999, 2000) tried to distinguish the effect of light quality (R:FR ratio) from the effect of light intensity on remobilization of N from shaded leaves, and found that both signals played a role in determining the onset of remobilization (see also Chapter 10). The adjustment of SLN with irradiance within the canopy has been interpreted in terms of optimal canopy photosynthesis (Field 1983; Anten et al., 1995). From a physiological point of view, adjustment of SLN with irradiance operates through acclimation of the photosynthesis

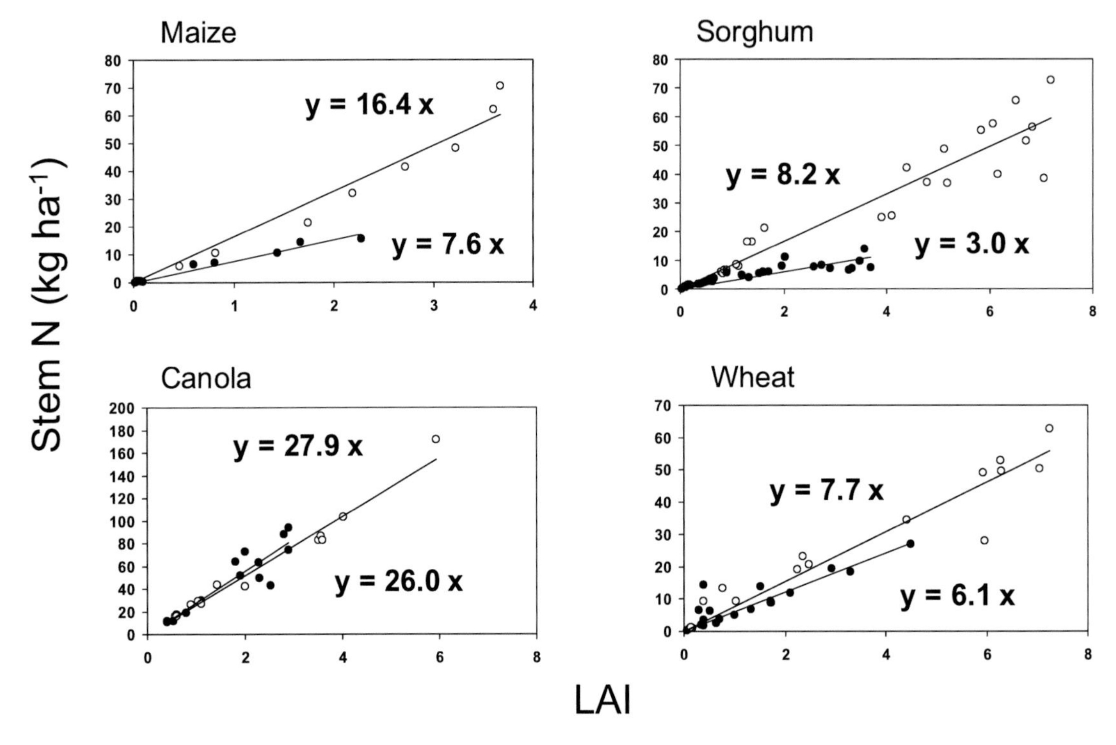

FIG. 8.11 Relationship between nitrogen accumulation in stems and LAI for maize, sorghum, canola and wheat grown under non-limiting (○) or limiting (●) N supply. *From Lemaire et al. (2008).*

apparatus to the light intensity experienced by individual leaves during their growth (Prioul et al., 1980a). As proposed by Hirose and Bazzaz (1998), the preferential allocation of N to the top of the canopy would correspond to a trade-off between radiation-use efficiency and nitrogen-use efficiency, i.e. the increase in dry matter production per unit of supplemental absorbed nitrogen.

This light acclimation of leaves is largely reversible within a few days (Prioul et al., 1980b). So young leaves at the top of the canopy with typically high SLN and photosynthesis rate recycle N to growing leaves and drop their SLN as they are progressively shaded by newer leaves. In lucerne (Lemaire et al., 1991) and sunflower (Sadras et al., 1993), leaf N concentration per unit mass decreases to a smaller extent than SLN accounting for variations in specific leaf area. The gradient in SLN is more or less steep within the canopy depending on the light extinction profile that varies between crop species. Some species show an important decline in crop SLN (the average leaf SLN of the canopy) with increasing LAI such as wheat or canola, while species such as maize have a relatively constant crop SLN due to their more erect leaf inclination and, subsequently, a more uniform distribution of light and leaf N throughout the canopy (Lemaire et al., 2008). This inter-specific variation in leaf N profile suggests that species may differ in their acclimation responses to light (Gastal and Lemaire, 2002) leading to inter-specific variation in the response of canopy photosynthesis to N deficiency. Moreover, it is often observed that SLN declines less rapidly in canopies than light (Hirose and Werger 1987; Wright et al., 2006; Dewar et al. 2012).

3.4 Harvest index and components of grain yield

For grain crops, yield is closely correlated to grain number per unit area of soil, which depends on the growth rate in a critical window around flowering. This period coincides with the maximum rate of crop N uptake (Fischer, 1993). So, limitation of crop growth rate by N deficiency at this period decreases grain number and yield in maize (Uhart and Andrade, 1995), wheat (Desmotes-Mainard et al., 1999; Martre et al., 2003; Sadras et al., 2012) and rice (Mae, 1997). Jeuffroy and Bouchard (1999) established for wheat a relationship between grain number and the severity and duration of the N deficiency before flowering as calculated by NNI. In fact, N deficiency during the vegetative period has two consequences for the determination of grain number: (1) a lower crop growth rate at flowering, restricting C supply to spike primordia, decreasing spike survival (Arisnabarreta and Miralles, 2004; Desmotes-Mainard and Jeuffroy, 2004); and (2) a decrease in the N content in the spike stems (Desmotes-Mainard et al., 1999; Desmotes-Mainard and Jeuffroy, 2004) corresponding to a direct effect of N deficiency on floret fertility (Abbate et al., 1995).

The other grain yield component, grain weight, is generally less affected by N deficiency at flowering than the grain number (Plénet and Cruz, 1997). But it is necessary to take into account the negative correlation between grain weight and grain number: a reduced grain number resulting from pre-flowering N deficiency can lead to a more favorable source:sink ratio during the grain-filling period. Grain filling in both carbohydrates and proteins depends on (1) recycling C and N compounds from vegetative plant organs, and (2) post-flowering photosynthesis and root N absorption. The relative importance of these two components depends on plant species (Borras et al., 2004) and probably on genotypes of the same species and their capacity to store of C and N compounds in their vegetative organs before flowering as discussed above.

Delaying leaf senescence should allow not only provision for more carbohydrate for grain filling by prolonged photosynthesis, but also provision for C to roots for maintaining N absorption (Dreccer, 2005). So stay-green mutants of maize show a larger source:sink ratio during grain filling with an increased proportion of N derived from soil (Rajcan and Tollenaar, 1999a,b). Using stay-green genotypes in low N supply conditions seem to be beneficial for both grain sorghum (Borell and Hammer, 2000; Jordan et al. 2012) and maize (Mi et al., 2003). Chapter 10 further develops aspects of the stay-green syndrome from the perspective of plant senescence.

Recovery of N in grain (nitrogen harvest index, NHI) can be considered as a component of N-use efficiency of grain crops (Sinclair, 1998). However, breeding progress in NHI has been limited because of the apparent inverse genetic relationship between yield and grain protein content (Feil et al., 1990; Canevara et al., 1994; Simmonds, 1995). However, studies in maize show that this inverse relationship can be genetically modulated (Uribelarrea et al., 2007).

3.5 Interactions of nitrogen deficit with other resources and stresses

3.5.1 *Co-limitation of N and other resources*

Crop growth depends on the capture of resources including CO_2, radiation, water, N and other nutrients. So the response of crops to the availability of one particular resource such as N depends on the availability of other resources. In agricultural systems, plant growth is often simultaneously limited by several factors (Sinclair and Park, 1993; Sadras, 2005). Kho (2000) developed the concept of limitation index L_i of one resource i, N for example, as being the ratio between the elementary response of crop mass δW to elementary increase of the resource δN, and the use efficiency of this resource W/N:

$$L_i = \frac{\delta W/\delta N}{W/N} \tag{8.20}$$

This limitation index varies from a maximum of 1 when the N resource is the only limiting factor, to 0 when N becomes non-limiting. This author shows that when several factors are limiting crop growth, then the sum of the limitation index of each factor is equal to 1. Sadras (2005) used the concept of co-limitation following Bloom et al. (1985) who hypothesized that the plant maximizes its growth when growth is equally limited by all resources. For two resources such as nitrogen and water a co-limitation index CI can then be calculated (Sadras, 2005):

$$CI = 1 - L_{nitrogen} - L_{water} \tag{8.21}$$

and then maximizing crop growth corresponds to maximum CI, i.e. water and nitrogen are equally limiting. Sadras et al. (2004) showed that, under a given level of stress, crops in SE Australia perform better under conditions of high co-limitation between water and N.

Hooper and Johnson (1999) and Sadras (2005) stressed that co-limitation may arise from variable mechanisms operating at different time scales and at different levels of organization from molecular to ecosystem. For instance, at the ecosystem level, N × water and N × P interactions reflect the close links between resources mediated by biogeochemical feedbacks (Hooper and Johnson, 1999). The early proposition that, at the same level of stress, water–nitrogen co-limitation favors wheat yield (Sadras 2005) has received further support in independent studies in European Mediterranean environments (Cossani et al. 2010) and is further discussed in Chapter 7.

3.5.2 *N deficiency–water deficit interactions*

In addition to the direct impact of water deficit on primary productivity, Sadras (2004) showed that yield and water-use efficiency of water and nitrogen stressed wheat crops increase with the degree of water and nitrogen co-limitation. Several studies showed the effect of drought on plant or crop N status (Lemaire and Denoix, 1987; Bassiri Rad and Caldwell, 1992; Karrow and Maranville, 1994; Onillon et al., 1995; Pratersak and Fukaï, 1997); Gonzalez-Dugo et al. (2010) have recently reviewed this topic. Under water deficit, a reduced crop N uptake has been observed for many species: perennial ryegrass, Italian ryegrass and cocksfoot (Colman and Lazemby, 1975; Gonzalez-Dugo et al., 2005); rice (Pirmoradian et al. 2004); wheat (Larsson, 1992); maize (Pandley et al. 2000); and sunflower (Gonzalez Dugo et al., 2010). The reduction in N uptake under water deficit is due to (1) a reduction in N demand by diminution in plant growth at optimal N concentration, especially for shoots vs roots (Spollen et al., 1993), (2) a reduction in optimal plant N demand per unit biomass, and (3) a reduction in soil N availability. Any decrease in growth rate reduces N demand defined by the dilution curve, and so does water deficit (Lemaire and Meynard, 1997). Changes in N demand per unit mass were studied by Gonzalez-Dugo et al. (2012) by manipulating mineral nitrogen concentration and water potential of the nutrient solution in a split-root design in three grass species. That study showed that plant N status also depended on the capacity of stressed plants to absorb and allocate nitrogen to leaves. Lemaire and Denoix (1987) showed that during a prolonged drought when plants were forced to extract water from deeper soil horizons where there was a lack of mineral N, the water-use efficiency dropped, indicating that N deficiency increased. This phenomenon has been also observed in maize and sorghum (Lemaire et al., 1996). Water deficit may hence decrease the nitrogen nutrition index (Fig. 8.12). The effect appears reversible and recovery from drought also brings about a rapid recovery of the crop nitrogen status. Taking into account soil root density, soil water content and mineral N concentration profiles in tall fescue

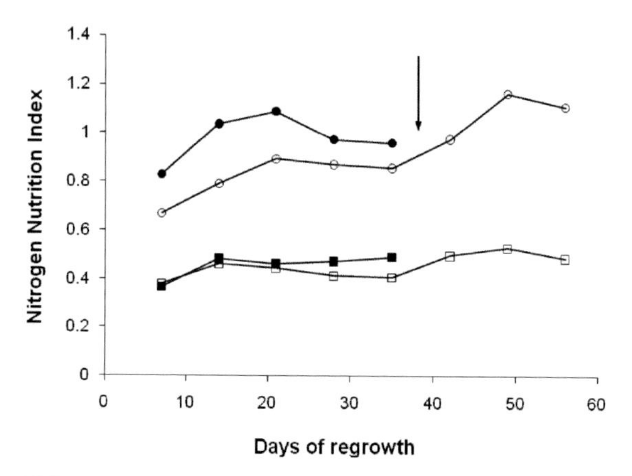

FIG. 8.12 Evolution of the nitrogen nutrition index of tall fescue swards during growth at two nitrogen fertilization rates (high N: circles, low N: squares) and under two contrasted water supply regimens (irrigated: closed symbols; rain-fed: opened symbols). The arrow indicates rewatering of rain-fed plots. *From Gonzalez-Dugo et al., 2010.*

and Italian ryegrass subjected or not to water deficit, a single linear relationship was found between N uptake and mineral N flux to the rhizosphere (Durand et al., 2010). However, a recent field study on tall fescue showed that the critical nitrogen concentration is reduced under water deficit (Errecart et al., 2013), which is consistent with the controlled condition study by Gonzalez-Dugo et al. (2012). In the long term, extreme water deficits could even strongly alter the ability of the root system to supply the shoot N demand following water deficit recovery (Poirier et al., 2012).

Another factor contributing to co-limitation between water and nitrogen is the effect of soil moisture on soil N net mineralization (Pastor and Post, 1985; Smolander, 2005). Birch (1958), Bloem et al. (1992), Franzluebbers et al. (2000) and Austin et al. (2004) showed that soil N mineralization is rapidly activated during soil re-watering after drought, leading to increased plant N uptake.

3.5.3 *N × P × S interactions*

Soil phosphorus (P) deficiency and hence the importance of N × P interaction is widespread (Kho, 2000). The effect of P deficiency on the N nutrition of crop can result from two processes: (1) a direct effect of P deficiency on plant growth, leading to a decrease in crop mass, thus to a limited N dilution and hence to an increase in plant N concentration; and (2) an indirect effect of plant P nutrition on soil N availability through mineralization of soil organic matter and/or stimulation of root growth. As proposed by Sinclair and Park (1993), the concept of pseudo-substitution of inputs contributes to explaining nutrient co-limitation at the plant and population levels. In soil with low N availability, plants may respond to P fertilizer by increasing root growth and enhancing N uptake. Therefore, part of the crop response to P application can be attributed to an indirect N nutrition effect.

Duru et al. (1997) used NNI for analyzing the interaction between N and P deficiency. The application of P-fertilizer on a P-deficient permanent grassland led to two different effects (Fig. 8.13): (1) an increase in the NNI of the sward from about 0.6 to 0.75 for the unfertilized N treatment and from 0.9 to 1.0 for the N fertilized treatment; this increase in NNI was associated with a corresponding increase in sward biomass, that is an indirect effect of P application; and (2) an increase in the sward biomass accumulation at similar NNI, that is a direct effect of P application. The NNI method therefore seems suitable for detecting and analyzing interactions between nitrogen and phosphorus.

Sterner and Elser (2002), Ågren (2004) and Matzek and Vitousek (2009) revised the concept of ecological stoichiometry, with emphasis on C:N:P ratios. Several authors proposed the use of plant N:P ratios as tools for diagnostic and fertilizer management in wetlands and other types of natural vegetation (Verhoeven et al., 1996; Koerselman and Meuleman, 1996; Güsewell and Koerselman, 2002; Güsewell et al., 2003; Drenovsky and Richards, 2004). However, in crop species, the large capacity of plants to store excess N or excess P in organic compounds (i.e. proteins, phytate) depending on the relative availability of soil nutrients, leads to a large

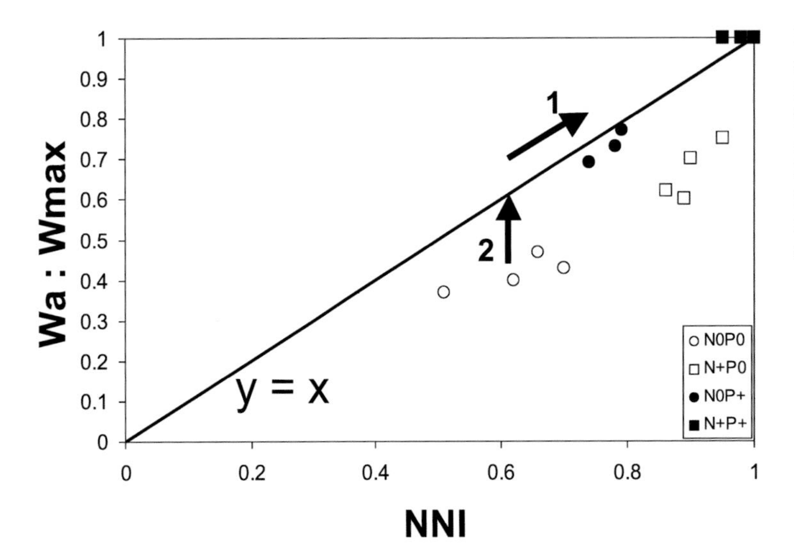

FIG. 8.13 The ratio of actual crop mass (Wa) and maximum crop mass (Wmax) as a function of nitrogen nutrition index (NNI) captures the interactions between N and P in natural grasslands. P deficiency can be decomposed in two effects: (1) an indirect effect through N availability, and (2) a direct effect. *From Duru et al. (1997).*

variability in N:P ratio, limiting the value of this indicator for practical purposes (Sadras, 2006; Greenwood et al., 2008).

Plant nitrogen nutrition interacts greatly with S nutrition because several amino acids and proteins involved in plant metabolism have S bonds. Sulfur mineralization in soil is also intimately coupled with N and P mineralization through stoichiometric relationships. Recently, Salvagiotti and Miralles (2008) reported the effect of the interaction between N and S fertilization on biomass and grain yield of wheat, showing that, under some conditions, S deficiency may limit the yield response to N supply.

4 NITROGEN-USE EFFICIENCY

Improving nitrogen-use efficiency is a major goal in plant breeding for sustainable agriculture (Hirel et al., 2007). From an agronomic point of view, nitrogen-use efficiency (NUE) is the increment of yield (ΔY) for each added unit of N fertilizer (ΔNf), i.e. the slope of the crop yield response curve to the rate of N supply. As this response curve is asymptotic, NUE declines as the rate of N application increases (Fig. 8.14).

From an ecophysiological and plant-breeding perspective, such a global definition of NUE does not allow the identification of plant traits controlling NUE because plants respond to total N supply (N_{tot}) including fertilizer (N_{fert}) and soil N supply (N_{soil}) (Fig. 8.14). Thus, according to variation in N_{soil} due to soil, climate and management, different N_{tot} can be obtained with

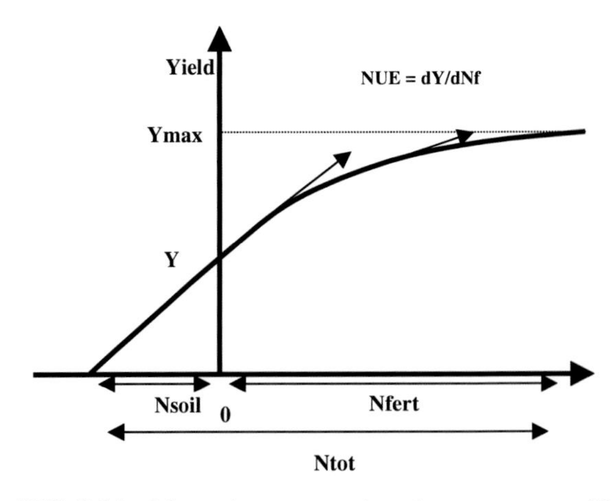

FIG. 8.14 Schematic representation of crop response to N supply, and determination of nitrogen-use efficiency. Nsoil is the soil N supply through mineralization of organic matter, Nfert is the fertilizer N supply, Ntot is the total N supply for the crop.

the same N_{fert}, then leading to large differences in NUE not directly attributable to species or genotypes.

Moll et al. (1982) proposed to define NUE as the yield produced per unit of available N in the soil. Hence, NUE can be divided into two components:

1. the N-uptake efficiency (NuptE) as the increment in N uptake by the crop per unit of increment in N supply in the soil
2. the N-utilization efficiency (NutE) as the capacity of the plant to produce a supplement of biomass per unit of N uptake.

According to crop species, yield represents either the above-ground biomass, as for forage crops, or the grain biomass as for cereals, grain legumes or oilseed crops. In grain crops, it is then necessary to take into account the harvest index (HI). Thus, NUE can be expressed as:

$$NUE = \frac{\Delta Nupt}{\Delta Nt} * \frac{\Delta W}{\Delta Nupt} * \frac{\Delta Y}{\Delta W} \tag{8.22}$$

$$NUE = NuptE * NutE * HI \tag{8.23}$$

Therefore, NUE can be considered as resulting from three types of processes: (1) the capacity of the crop to capture soil N; (2) the capacity of the crop to use N for biomass elaboration; and (3) the capacity of the crop to allocate C and N to grains.

4.1 N-uptake efficiency

Genotypic variation in N uptake at different N supply has been investigated in rice (Borell et al., 1998), wheat (Le Gouis et al., 2000) and maize (Bertin and Gallais, 2001). Some studies show that modern cultivars have a higher N-uptake capacity than older ones (Ortiz-Monasterio et al., 1997; Brancourt-Humel et al., 2003) because they have a higher biomass. This effect corresponds to a shift from points A to points B in Figure 8.3 and is quantitatively accounted for by Eq. 8.13. So the first way to improve N-uptake capacity of a crop is to increase its growth rate. More interesting would be to increase the N-uptake capacity at a similar biomass. Lemaire and Meynard (1997) showed that cocksfoot was more able to capture N than tall fescue when N supply was limited (60 kg N ha^{-1}), while the two species had similar N-uptake capacities when N supply increased to 120 kg N ha^{-1} (Fig. 8.15). Lemaire et al. (1996) showed that grain sorghum has a higher N-uptake capacity than maize when both crops

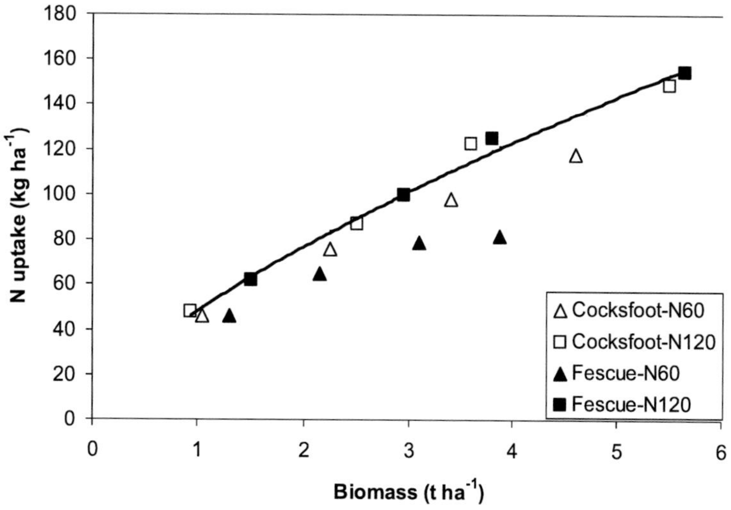

FIG. 8.15 Comparison of N uptake of tall fescue and cocksfoot at two rates of N supply. The curve is the critical N uptake for temperate grass species with the parameters in Table 8.1. *From Lemaire and Meynard (1997).*

were grown under low N supply, but the two species removed the same quantity of N from soil under conditions of high N supply. As for cocksfoot vs tall fescue, sorghum appears to have a greater root-length density than maize. This would confer to sorghum a better capacity of interception of mineral N before its immobilization by soil microbes within N mineralization sites. Soil N recovery by crops is the result of the nitrogen balance between crop uptake rate, immobilization by soil microbial communities and losses by leaching, denitrification and volatilization (Lemaire et al., 2004). Root architecture and perhaps root exudates could play a role in this balance and hence on plant N-uptake efficiency. So, because of a permanent turnover of N in soil, plants having a dense root system are able to compete more efficiently against microbes for capturing mineral N when the principal source of N supply is mineralization of organic matter. When there is ample mineral N in soil, the immobilization capacity of the microbial community is saturated and then the difference between species for N mineral capture greatly disappears. Nevertheless, Dybzinski and Tilman (2007) showed that differences between species in N uptake capacity can persist even under high N.

Laperche et al. (2007) detected coincidences between quantitative trait loci (QTL) for root architecture and QTL for traits associated with N-uptake efficiency in wheat. Moreover, Camus-Kulandaivelu et al. (2006) identified a large genotypic variation across maize lines for the density and length of lateral roots. Hence, assuming no trade-offs, it should be possible to breed maize for improved N uptake in low N supply conditions that would lead to a better use of soil N and then a reduction of the N fertilizer required for a target yield. Genetic studies showed that N-uptake efficiency was the most important component of NUE in rice (Singh et al., 1998) and wheat (Le Gouis et al., 2000).

Improved ability of a plant to capture N under low N supply could thus be related to several mechanisms:

1. denser root systems which would allow a more complete and rapid exhaustion of soil mineral N and a greater competition of plant against microbes or against neighbors (weeds or companion species in mixtures)
2. a more efficient nitrate and ammonium transport system in roots which would also favor the plant in the competition with microbes for N released by gross mineralization
3. a variation in the quantity and quality of C exudates which could alter the gross N fluxes in the rhizosphere (Paterson, 2003).

All these mechanisms can operate simultaneously and all could involve significant trade-offs. They provide keys for interpreting genotype $\times$ environment interactions and relevant objectives for functional genomic and genetic approaches with the aim of improving N recovery by crops.

The main limitations in interpreting experimental data on comparison of NuptE across genotypes is that at least two levels of N application rates (Nfert) are required for estimating ΔNt (Eq. 8.21), and the assumption that soil N (Nsoil) remains similar under the two N treatments, supposing the additivity of Nsoil and Nfert which is probably not the case because of the 'priming effect'. Moreover, the response curve Nupt = f(Nt) is asymptotic, so its derivative dNupt/dNt declines as Nt increases, and the estimation of NuptE depends on the range of ΔNt experienced, leading to difficulties for generalizations. To overcome this difficulty, we propose to determine systematically the NNI for each genotype and to compare NuptE across genotypes at similar NNI. Moreover, we can suppose that ranking genotypes on their NNI under low N supply could be a way to rank their capacity to fit their own N demand. Following this approach, an analysis of breeding improvement on Australian wheat cultivars allowed Sadras and Lawson (2013) to show that NNI, accounting for the effect of biomass on nitrogen uptake, increased linearly with year of cultivar release, hence supporting the conclusion that breeding for yield improved the nutrition status of wheat in association with an increased capacity to take up nitrogen. In a similar way, Ciampitti and Vyn (2012) showed that modern maize cultivars have a higher N-uptake capacity than old ones because they have a higher biomass accumulation potential. However, in the absence of any information on the ranking of NNI across 'old' and 'modern' cultivars, it is impossible to know whether this difference is only due to the increase in biomass or whether there is also a genetic shift in N uptake efficiency *per se* (i.e. at similar crop mass). This study illustrates the necessity to use NNI diagnostic for a better understanding of nitrogen-use efficiency in crops.

4.2 Nitrogen-utilization efficiency

As crop mass increases during growth (Fig. 8.3, from points A to B), the response of biomass to supplemental N fertilization and N uptake increases (Fig. 8.3, dotted line from As–Af to Bs–Bf and from As–Ac to Bs–Bc). Hence, N-utilization efficiency (NutE = $\Delta W/\Delta Nupt$) increases with crop growth. This approach allows a dynamic analysis of NutE where time has to be implicitly taken into account through the crop growth rate: the higher the crop growth rate, the higher the crop mass at a given time and then the higher the NutE. So when comparisons of NutE between crop species or genotypes are made, it is important to consider that smaller crops should have a lower NutE than the more productive crop. In consequence, breeding for a high NutE should lead to the selection of genotypes having a higher crop mass potential. More interesting is detecting differences in NutE among genotypes when comparisons are made at the same crop mass by plotting the ratio of actual crop mass (Wa) to maximum crop mass (Wmax) vs NNI (Fig. 8.13). To our knowledge, such a comparison between crop species and genotypes has not been done and we have no clear indications whether variation in NutE persists when genotypes are compared at same growth potential except when we compare C_3 with C_4 crops. Moreover, as shown in Figure 8.3, $\Delta W/\Delta Nupt$ declines as the crop N status increases owing to the asymptotic response of W to N uptake. As a consequence, the interpretation of differences in NutE across genotypes must take into account the possibility of differences in plant N status. Thus, a genotype having a high N-uptake efficiency leading to a near optimum NNI would be classified with a lower NutE than a genotype with low NuptE. Therefore, there is a fundamental trade-off between NutE and NuptE that requires comparisons of NutE among genotypes at similar NNI. Caju et al. (2011) reported an analysis of nitrogen-use efficiency across European wheat cultivars showing large variations in both NuptE and NutE but, in the absence of NNI, it is impossible to know to which extent these differences are linked to trivial effects of crop mass, or crop N status or are intrinsic (i.e. difference remaining for comparisons at similar crop mass). Semenov et al. (2007) used a modeling approach for deconvoluting nitrogen-use efficiency components on wheat. However, as the identification of crop N status was not considered in their analysis, the trade-off between NutE and NuptE was not clearly explicated.

4.3 Harvest index and protein concentration in grain

The effect of N deficiency on harvest index (HI) of grain crops has been analyzed in section 3.4. Eqs 8.22–8.23 indicate that HI is an important source of variation in grain yield per unit of N uptake. In addition to agronomic NUE, the nitrogen harvest index (NHI) is an important consideration for grain crops (section 3.4). NHI reflects the grain protein content (Sinclair, 1998) and thus an important aspect of the grain nutritional quality. Desai and Bhatia (1978) showed that NHI among durum wheat cultivars was positively correlated with HI, thus indicating that the distribution of N between straw and grain, to a large extent, depends upon the partitioning of dry matter between these two compartments. These authors showed also that grain protein concentration was not correlated to NHI. Grain produced per unit of N uptake and grain N concentration are inversely related (Fig. 8.16). A variable proportion of the variation in yield per unit N uptake is accounted for by grain protein concentration or by NHI according to the crop type (Fig. 8.16). Charmet et al. (2005) analyzed genetic control of protein accumulation in wheat grain. Correlation analyses revealed that dry matter and N accumulation rates in grain were repeatedly and closely correlated over several years, which was not the case for duration of dry matter and N accumulation, and that protein composition was primarily influenced by the N accumulation rate. Five QTL were significantly associated with the kinetics of dry matter and N accumulation, and two of them also influenced protein composition. Chapter 17 examines the associations between yield and quality traits, including trade-offs between yield and grain oil concentration in sunflower and yield and grain protein in cereals.

4.4 Breeding for Nitrogen-Use Efficiency?

Engineering nitrogen-efficient crops is a worldwide objective for breeding programs on most crop species. Nevertheless, as shown above, NUE is a complex trait including many elemental processes with several trade-offs leading to high genotype × environment interactions and relatively low heritability. Literature is abundant reporting G × E studies on nitrogen-use efficiency for different crop species. Nevertheless, few of them allow decomposition of NUE into its three components (Eq. 8.22). Three important conclusions have to be drawn from the above approach:

1. As plant mass increases, both N-uptake and N-utilization efficiency increase as the result of the N dilution process and the feedback control of N uptake by plant growth rate.

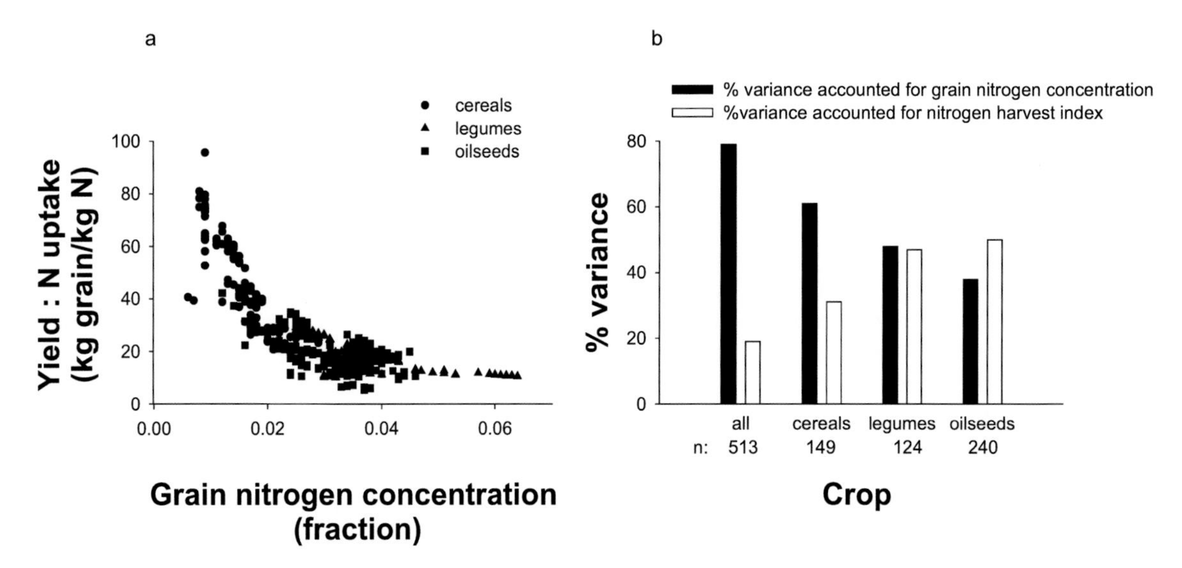

FIG. 8.16 (a) Relationship between grain yield per unit nitrogen uptake and grain N concentration across different crop species. (b) Percentage of variance in grain yield per unit N uptake accounted for grain nitrogen concentration and nitrogen harvest. *From Sadras (2006).*

As a consequence, a large plant will have a higher NuptE and a higher NutE than a small one, leading to an increase in NUE with crop mass. An increase in NUE is the indirect result of breeding for high crop mass.
2. More interesting would be to select for higher nitrogen-use efficiency at similar crop mass, corresponding to an intrinsically more efficient cultivar. The low variability of critical N dilution curves across species (except C_3 vs C_4) indicates that intra-specific variability in NutE is probably low and difficult to target in breeding programs. However, the variability in NuptE allows breeding for the capacity of crops to fit their own N demand under lower N supply conditions. Ranking germplasm in their capacity to maintain high NNI in low N conditions should be relevant for breeding.
3. For grain crops, NUE depends not only on the capacity of crops to accumulate N at flowering but also to have an efficient transfer of C and N during grain filling. More detailed analysis is necessary for the determination of genetic control of these processes, as outlined in Chapter 10. However, these studies need to include unambiguous assessment of crop N status at flowering.

5 CONCLUSIONS

The first conclusion of this chapter is the necessity to integrate all elementary processes and to scale them up to the level of whole plant and crop where they are agronomically relevant. Nitrogen metabolism of plants is controlled by physiological processes such as nitrate or ammonium transport through cell membranes in roots, nitrate reduction in roots and leaves, N_2 fixation within nodules for legumes, and ammonium assimilation. Each of these metabolic processes is regulated at molecular level with some degree of genetic variability. However, when all these processes are integrated by scaling up to whole plant and plant population or community level, the overall integrated regulation of N uptake and N-use efficiency can be summarized by general rules, for which inter-specific variability is more limited. For example, whatever the source of N, i.e. nitrate, ammonium or biological N_2 fixation, the feedback control of N acquisition by shoot growth appears to operate similarly and dominate the overall N-uptake capacity of the plant despite different intrinsic capacity of N uptake per unit of root area or root mass. So the genetic variability in N-uptake capacity potentially observed at organ level or lower seems to be 'buffered' at whole plant level where feedback mechanisms are operating. Moreover, when scaling-up from individual plant to plant population, competition for light and photomorphogenic responses leading to isometric growth impose a strong regulation of N-uptake dynamics at crop level. Isometric growth of plants and N dilution curves are emerging properties at the plant population level and therefore are poorly related to molecular and genetic controls. Hence, N acquisition and allocation in a vegetative crop is dominated by the plant–plant interactions for light acquisition: sharing of N among individual plants parallels sharing of light.

The second conclusion deals with the possibilities of application of this integrated knowledge for improving N-use efficiency through breeding and agronomy. It appears that improving the ability of plants to absorb and accumulate N efficiently from soil (N-uptake efficiency) is the first objective for high N-use efficiency in cropping systems. This ability of plants to take up mineral soil N has to be investigated both under low and high N conditions. As demonstrated in this chapter, in both conditions, N-uptake efficiency of a plant is directly dependent upon its growth capacity depending on (1) its own genetic potential, (2) external conditions such as soil and climate, (3) management techniques, and (4) interactions between these three components. So, any improvement in crop growth capacity by both breeding and crop management (e.g. irrigation, P, K and S fertilization, and planting density) will increase N-uptake efficiency. More important is to improve the N-uptake capacity at similar crop mass. This would allow achievement of a given yield with less N fertilizer. Such an objective requires a more efficient root system (root length density) to increase the competitive ability of the plant for using soil mineral N against soil microbes and weeds. Breeding for root architecture is not easy, but several studies demonstrated that it is possible. Crop management techniques to enhance root development have to be more deeply tested through soil tillage and soil structure preservation, and their quantitative effects on crop N uptake must be investigated.

Timing of N fertilizer application is an important way for improving N fertilizer recovery by crops and hence N-use efficiency. As crop N-uptake capacity is largely determined by crop growth rate, application of N is ideally timed just before the acceleration of crop growth. Progress has to be made with non-invasive and operational tools for rapid diagnostics on crop N status, calibrated with the NNI method. These tools allow monitoring of the evolution of crop N status along its growth period. For most crops, studies on the effect of timing and intensity of short periods of N starvation along the growth cycle would allow the determination of periods of high and low responsiveness in terms of grain yield and grain quality. Such information could then be included within decision-making tools to reduce fertilizer supply at the minimum to achieve target yield (Rodriguez et al., 2009).

Breeding to increase N-utilization efficiency for dry matter production (forage or energy crops) is not easy because, apart from the difference between C_4 and C_3 species, no clear interspecific differences are generally observed as attested by the similarity of the critical N dilution curves. However, the situation is likely to be different if we consider grain production and grain protein content owing to the great differences observed in harvest index and nitrogen harvest index among genotypes. The identification of the key elements responding to N deficiency in terms of grain development will provide us with more information on the extent to which the whole plant N status has localized effects on grain formation, grain development and grain-filling processes. Since the N status of a plant on its own does not allow for determining the N status of reproductive meristems, the development of more refined models describing both partitioning and translocation within plant is necessary to take into account the intimate relation between N and C fluxes at the organ level. This will require, as reviewed by Hammer et al. (2004), a cooperative effort between molecular biologists, crop physiologists, geneticists, agronomists and specialists in bioinformatics.

References

Abbate, P., Andrade, F.H., Culot, J.P., 1995. The effects of radiation and nitrogen on number of grains in wheat. J. Agric. Sci. (Camb.) 124, 351–360.

Addiscott, T.M., Withmore, A.P., Powlson, D.S., 1991. Farming, fertilisers and the nitrate problem. CAB International, Wallingford, UK.

Aerts, R., de Caluwe, H., 1994. Effects of nitrogen supply on canopy structure and leaf nitrogen distribution in Carex species. Ecology 75, 1482–1490.

Ågren, G.I., 2004. The C:N:P stoichiometry of autotrophs – Theory and observations. Ecol. Lett. 7, 185–191.

Angus, J.F., Moncur, M.W., 1985. Models of growth and development of wheat in relation to plant nitrogen. Aust. J. Agric. Res. 36, 537–544.

Angus, J.F., 2001. Nitrogen supply and demand in Australian agriculture. Aust. J. Exp. Agric. 41, 277–288.

Angus, J.F., 2007. Should nitrogen dilution curves be expressed in relation to biomass or development? In: Bosch Serra, A.D., Teira Esmatges, M.R., Vilar Mir, J.M. (Eds.), Towards a better efficiency in N use. Proceedings of the 15th Nitrogen Workshop. Editorial Milenio, Lleida, Spain, pp. 305–308.

Anten, N.P.R., Schieving, F., Werger, M.J.A., 1995. Patterns of light and nitrogen distribution in relation to whole canopy carbon gain in C_3 and C_4 mono- and dicotyledonous species. Oecologia 101, 504–513.

Aphalo, P.J., Ballaré, C.L., 1995. On the importance of information-acquiring systems in plant-plant interactions. Funct. Ecol. 9, 5–14.

Arisnaberrata, S., Miralles, D.J., 2004. The influence of fertiliser nitrogen application on development and number of reproductive primordia in field grown two- and six-rowed barleys. Aust. J. Agric. Res. 55, 357–366.

Austin, A.T., Yahdjian, L., Stark, J.M., et al., 2004. Water pulses and biogeochemical cycles in arid and semi-arid ecosystems. Oecologia 141, 221–235.

Avice, J.C., Ourry, A., Lemaire, G., Boucaud, J., 1996. Nitrogen and carbon flows estimated by ^{15}N and ^{13}C pulse chase labelling during regrowth of Medicago sativa L. Plant Physiol. 112, 281–290.

Avice, J.C., Ourry, A., Lemaire, G., Volenec, J.J., Boucaud, J., 1997. Root protein and vegetative storage protein are key organic nutrients for alfalfa shoot regrowth. Crop Sci. 37, 1187–1193.

Babar, M.A., Reynolds, M.P., van Ginkel, M., Klatt, A.R., Raun, W.R., Stone, M.L., 2006. Spectral reflectance to estimate genetic variation for in-season biomass, leaf chlorophyll, and canopy temperature in wheat. Crop Sci. 46, 1046–1057.

Bahmani, I., Hazard, L., Varlet-Grancher, C., et al., 2000. Differences in tillering of long- and short-leaved perennial ryegrass genetic lines under full light and shade treatments. Crop Sci. 40, 1095–1102.

Ballaré, C.L., Scopel, A.L., Sanchez, R.A., 1995. Plant photomorphogenesis in canopies, crop growth and yield. Hort. Sci. 30, 1172–1181.

Ballaré, C.L., Scopel, A.L., Sanchez, R.A., 1997. Foraging for light: photosensory ecology and agricultural implications. Plant Cell Environ. 20, 820–825.

Bange, M.P., Hammer, G.L., Rickert, K.G., 1997. Effect of specific leaf nitrogen on radiation use efficiency and growth of sunflower. Crop Sci. 37, 1201–1207.

Barillot, R., Louarn, G., Escobar-Gutiérrez, A.J., Huynh, P., Combes, D., 2011. How good is the turbid medium-based approach for accounting for light partitioning in contrasted grass–legume intercropping systems? Ann. Bot. 108, 1013–1024.

Barro, R.S., Varella, A.C., Lemaire, G., et al., 2012. Forage yield and nitrogen nutrition dynamics of warm-season native forage genotypes under two shading levels and in full light. Rev. Braz. Zootec. 41, 1589–1597.

Barta, A.L., 1975. Effect of nitrogen nutrition on distribution of photosynthetically incorporated $^{14}CO_2$ in *Lolium perenne* L. Can. J. Bot. 53, 237–242.

Bassiri Rad, H., Caldwell, M.M., 1992. Temporal changes in root growth and ^{15}N uptake and water relations in two tussock grasses species recovering from water stress. Physiol. Plant. 86, 525–531.

Bélanger, G., Gastal, F., Lemaire, G., 1992. Growth analysis of a tall fescue sward fertilised with different rates of nitrogen. Crop Sci. 6, 1371–1376.

Bélanger, G., Gastal, F., Warembourg, F., 1994. Carbon balance on tall fescue: effects of nitrogen and growing season. Ann. Bot. 74, 653–659.

Bélanger, G., Walsh, J.R., Richards, J.E., Milburn, P.H., Ziadi, N., 2001. Critical nitrogen curve and nitrogen nutrition index for potato in eastern Canada. Am. J. Potato Res. 78 (5), 355–364.

Beman, J.M., Arrigo, K., Matson, P.M., 2005. Agricultural runoff fuels large phytoplankton blooms in vulnerable areas of the ocean. Nature 434, 211–214.

Bertin, P., Gallais, A., 2001. Physiological and genetic basis of nitrogen use efficiency in maize. 2. QTL detection and coincidences. Maydica 46, 53–68.

Birch, H.F., 1958. The effect of soil drying on humus decomposition and nitrogen availability. Plant Soil 10, 9–31.

Bloem, J., Ruiter, P.D., Koopman, G., Lebbink, G., Brussard, L., 1992. Microbial numbers and activity in dried and rewetted arable soil under integrated and conventional management. Soil Biol. Biochem. 24, 655–665.

Bloom, A.J., Chapin, F.S.I., Mooney, H.A., 1985. Resource limitation in plants, an economic analogy. Ann. Rev. Ecol. System. 16, 363–392.

Borell, A., Garside, A.L., Fukaï, S., Reid, D.J., 1998. Season nitrogen rate and plant type affect nitrogen uptake and nitrogen use efficiency in rice. Aust. J. Agric. Res. 49, 829–843.

Borell, A., Hammer, G.L., 2000. Nitrogen dynamics and the physiological basis of stay-green in sorghum. Crop Sci. 40, 1295–1307.

Borras, L., Slafer, G.A., Otegui, M.E., 2004. Seed dry weight response to source-sink manipulations in wheat, maize and soybean: a quantitative reappraisal. Field Crops Res. 86, 131–146.

Brancourt-Humel, M., Doussinault, G., Lecomte, C., Brérard, P., Le Buanec, B., Trottet, M., 2003. Genetic improvement in agronomic traits of winter wheat cultivars released in France from 1946 to 1992. Crop Sci. 43, 37–45.

Brouwer, R., 1962. Distribution of dry matter in the plant. Neth. J. Agric. Sci. 19, 361–370.

Brown, R.H., 1978. A difference in N use efficiency in C_3 and C_4 plants and its implication in adaptation and evolution. Crop Sci. 18, 93–98.

Cabrera, V.E., Jagtap, S.S., Hildebrand, 2007. Strategies to limit (minimize) nitrogen leaching on dairy farms driven by seasonal climate forecasts. Agric. Ecosyst. Environ. 122, 479–489.

Caju, O., Allard, V., Martre, P., et al., 2011. Identification of traits to improve the nitrogen-use efficiency of wheat genotypes. Field Crops Res. 123, 139–152.

Caloin, M., Yu, O., 1984. Analysis of the time course change in nitrogen content of *Dactylis glomerata* L. using a model of plant growth. Ann. Bot. 54, 69–76.

Camus-Kulandaivelu, L., Veyreiras, J.B., Madur, D., et al., 2006. Maize adaptation to temperate climate: relationship between population structure and polymorphism in the Dwarf 8 gene. Genetics 172, 2449–2463.

Canevara, M.G., Romani, M., Corbellini, M., Perenzin, M., Borghi, B., 1994. Evolutionary trends in morphological, physiological, agronomical and qualitative traits of *Triticum aestivum* L. cultivars bred in Italy since 1990. Eur. J. Agron. 3, 175–185.

Cassman, K.G., Dobermann, A.D., Walters, D., 2002. Agroecosystems, nitrogen-use efficiency, and nitrogen management. Ambio 31, 132–140.

Cassman, K.G., 2007. Climate change, biofuels, and global food security. Environ. Res. Lett. 2, 11–12.

Caviglia, O.P., Sadras, V.O., 2001. Effect of nitrogen supply on crop conductance, water- and radiation-use efficiency on wheat. Field Crops Res. 69, 259–266.

Charmet, G., Robert, N., Branlard, G., Linossier, L., Martre, P., Triboi, E., 2005. Genetic analysis of dry matter and nitrogen accumulation and protein composition in wheat kernels. Theor. Appl. Genet. 111, 540–550.

Chen, J.L., Reynolds, J.F., Harley, P.C., Tenhunen, J.D., 1993. Coordination theory of leaf nitrogen distribution in a canopy. Oecologia 93 (1), 63–69.

Ciampitti, I.A., Vyn, T.J., 2012. Physiological perspectives of changes over time in maize yield dependency on nitrogen uptake and associated nitrogen use efficiency: a review. Field Crops Res. 43, 48–67.

Colman, R.L., Lazemby, A., 1975. Effect of moisture on growth and nitrogen response by *Lolium perenne* L. Plant Soil 42, 1–13.

Colnenne, C., Meynard, J.M., Reau, R., Justes, E., Merrien, A., 1998. Determination of a critical nitrogen dilution curve for winter oilseed rape. Ann. Bot. 81, 311–317.

Connor, D.J., Hall, A.J., Sadras, V.O., 1993. Effects of nitrogen content on the photosynthetic characteristics of sunflower leaves. Aust. J. Plant Physiol. 20, 251–263.

Connor, D.J., Sadras, V.O., Hall, A.J., 1995. Canopy nitrogen distribution and the photosynthetic performance of sunflower crops during grain filling. A quantitative analysis. Oecologia 101, 274–281.

Coque, M., Gallais, A., 2006. Genomic regions involved in response to grain yield selection at high and low nitrogen fertilisation in maize. Theor. Appl. Genet. 112, 1205–1220.

Cossani, C.M., Slafer, G.A., Savin, R., 2010. Co-limitation of nitrogen and water, and yield and resource-use efficiencies of wheat and barley. Crop Past. Sci. 61, 844–851.

Cruz, P., Lemaire, G., 1986. Analyse des relations de compétition dans une association de luzerne (Medicago sativa L.) et de dactyle (Dactylis glomerata L.). II – Effets sur la nutrition azotée des deux espèces. Agronomie 6, 735–742.

Cruz, P., Soussana, J.F., 1997. Mixed crops. In: Lemaire, G. (Ed.), Diagnosis of the nitrogen status in crops. Springer-Verlag, Heidelberg. pp. 131–144.

Debaeke, P., van Oosterom, E.J., Justes, E., et al., 2012. A species-specific critical nitrogen dilution curve for sunflower (*Helianthus annuus* L.). Field Crops Res. 136, 76–84.

de Koeijer, T.J., de Buck, A.J., Wossink, G.A.A., Oenna, J., Renkema, J.A., Struik, P.C., 2003. Annual variation in weather: its implications for sustainability in the case of optimising nitrogen input in sugar beet. Eur. J. Agron. 19, 251–264.

Delin, S., Nyberg, A., Linden, B., et al. (2008). Impact of crop protection on nitrogen utilisation and losses in winter wheat production. Eur. J. Agron. 28 (3), 361–370.

Desai, R.M., Bhatia, C.R., 1978. Nitrogen uptake and nitrogen harvest index in durum wheat cultivars varying in their grain protein concentration. Euphytica 27, 561–566.

Desmotes-Mainard, S., Jeuffroy, M.H., 2004. Effects of nitrogen and radiation on dry matter and nitrogen accumulation in the spike of winter wheat. Field Crops Res. 87, 221–233.

Desmotes-Mainard, S., Jeuffroy, M.H., Robin, S., 1999. Spike dry matter and nitrogen accumulation before anthesis in wheat as affected by nitrogen fertiliser: relationship to kernels per spike. Field Crops Res. 64, 249–259.

Devienne-Baret, F., Justes, E., Machet, J.M., Mary, B., 2000. Integrated control of nitrate uptake by crop growth rate and soil nitrate availability under field conditions. Ann. Bot. 86, 995–1005.

Dewar, R.C., Tarvainen, L., Parker, K., Wallin, G., McMurtrie, R.E., 2012. Why does leaf nitrogen decline within tree canopies less rapidly than light? An explanation from optimization subject to a lower bound on leaf mass per area. Tree Physiol. 32 (5), 520–534.

Doré, T., Meynard, J.M., 1995. On-farm approach of attacks by the pea weevil (*Sitona lineatus* L.) and the resulting damage to pea (*Pisum sativum* L.) crops. J. Appl. Entomol. 119, 49–54.

Dreccer, M.F., 1999. Radiation and nitrogen use in wheat and oilseed rape crops. Wageningen Universiteit, Wageningen, PhD Thesis.

Dreccer, M.F., 2005. Nitrogen use at the leaf and canopy level: a framework to improve N use efficiency. J. Crop Impr. 15, 97–125.

Drenovsky, R.E., Richards, J.H., 2004. Critical N:P values: predicting nutrient deficiencies in desert shrublands. Plant Soil 259, 59–69.

Drouet, J.-L., Bonhomme, R., 1999. Do variations in local leaf irradiance explain changes to leaf nitrogen within row maize canopies? Ann. Bot. 84, 61–69.

Durand, J.L., Gonzalez-Dugo, V., Gastal, F., 2010. How much do water deficits alter the nitrogen nutrition status of forage crops? Nutr. Cycl. Agroecosyst. 88, 231–243.

Duru, M., Lemaire, G., Cruz, P., 1997. The nitrogen requirement for major agricultural crops: grasslands. In: Lemaire, G. (Ed.), Diagnosis on the nitrogen status in crops. Springer-Verlag, Heidelberg. pp. 56–72.

Dybzinski, R., Tilman, D., 2007. Resource use patterns predict long-term outcomes of plant competition for nutrients and light. Am. Nat. 170, 305–318.

Eickhout, B., Bouwman, A.F., Zeijts, V.H., 2006. The role of nitrogen in world food production and food sustainability. Agric. Ecosyst. Environ. 116, 4–14.

Ekbladh, G., Witter, E., 2010. Determination of the critical nitrogen concentration of white cabbage. Eur. J. Agron. 33, 276–284.

Enquist, B.J., Tiffney, B.H., Niklas, K.J., 2007. Metabolic scaling and the evolutionary dynamics of plant size, form, and diversity: toward a synthesis of ecology, evolution, and paleontology. Internatl. J. Plant Sci. 168, 729–749.

Errecart, PM, Agnusdei, MG, Lattanzi, FA, Marino, MA, Berone, GD. 2013. Critical N concentration declines with soil water availability in tall fescue. Crop Sci. in press.

Evans, J.R., 1983. Nitrogen and photosynthesis in the flag leaf of wheat (*Triticum aestivum* L.). Plant Physiol. 72, 297–302.

Evans, J.R., 1989. Photosynthesis and nitrogen relationships in leaves of C_3 plants. Oecologia 78, 9–19.

Farquhar, G.D., Caemmerer, S., Berry, J.A., 1980. A biochemical model of photosynthetic CO2 assimilation in leaves of C_3 species. Planta 149, 78–90.

Farrugia, A., Gastal, F., Duru, M., Scholefield, D., 2002. Simplified sward nitrogen status assessment from N concentration of upper leaves. Multifunction grasslands. In: Durand, J.L., Emile, J.C., Huyghe, C., Lemaire, G. (Eds), Proceeding of the 19th General Meeting of the European Grassland Federation, La Rochelle, France, 294-295.

Farrugia, A., Gastal, F., Scholefield, D., 2004. Assesment of nitrogen status of grassland. Grass Forage Sci. 59, 113–120.

Feibo, W., Lianghuam, W., Fuha, X., 1998. Chlorophyll meter to predict nitrogen side-dress requirement for short-season cotton. Field Crops Res. 56, 309–314.

Feil, B., Thiraporn, R., Geisler, G., 1990. Genotypic variation in grain nutrient concentration in tropical maize grown during a rainy and a dry season. Agronomie 10, 717–725.

Field, C., 1983. Allocating leaf nitrogen for the maximization of carbon gain – Leaf age as a control on the allocation program. Oecologia 56 (2–3), 341–347.

Field, C., Mooney, H.A., 1986. The photosynthesis-nitrogen relationship in wild plants. In: Givnish, T.J. (Ed.), On the economy of plant form. Cambridge University Press, Cambridge, pp. 25–53.

Fischer, R.A., 1993. Irrigated spring wheat and timing and amount of nitrogen fertiliser. II – Physiology of grain yield response. Field Crops Res. 33, 57–80.

Fitter, A.H., 1991. The ecological significance of root system architecture: an economic approach. In: Atkinson, D. (Ed.), Plant root growth. An ecological perspective. Blackwell Scientific Publications, Oxford, pp. 229–242.

Fletcher, A.L., Johnstone, P.R., Chakwizira, E., Brown, H.E., 2013. Radiation capture and radiation use efficiency in response to N supply for crop species with contrasting canopies. Field Crops Res. 150, 126–134.

Forde, B.G., 2002. The role of long distance signalling in plant responses to nitrate and other nutrients. J. Exp. Bot. 53, 39–43.

Franzluebbers, A., Haney, R., Honeycutt, C., Schomberg, H., Hons, F., 2000. Flush of carbon dioxide following rewetting of dried soil relates to active organic pools. Soil Sci. Soc. Am. J. 64, 613–623.

Fricke, W., McDonald, A.J.S., Matson-Djos, L., 1997. Why do leaves and leaf cells of N-limited barley elongate at reduced rates? Planta 202, 522–530.

Gallagher, J.N., Biscoe, P.V., 1978. Radiation absorption, growth and yield of cereals. J. Agric. Sci. (Camb.) 91, 47–60.

Gallais, A., Coque, M., 2005. Genetic variation and selection for nitrogen use efficiency in maize: a synthesis. Maydica 50, 531–537.

Galloway, J.N., Cowling, E.B., 2002. Reactive nitrogen and the world: 200 years of change. Ambio 31, 64–71.

Garnier, P., Néel, C., Atia, C., Recous, S., Mary, B., Lafolie, F., 2003. Modelling of carbon and nitrogen dynamics with and without straw incorporation. Eur. J. Soil Sci. 54, 555–568.

Gastal, F., Lemaire, G., 1988. Study of a tall fescue sward grown under nitrogen deficiency conditions. In: Proceedings of the XIIth General meeting of the European Grassland Federation. Dublin, Ireland. pp. 323–327.

Gastal, F., Saugier, B., 1989. Relationships between nitrogen uptake and carbon assimilation in whole plant of tall fescue. Plant Cell Environ. 12, 407–418.

Gastal, G., Bélanger, G., Lemaire, G., 1992. A model of the leaf extension rate of tall fescue in response to nitrogen and temperature. Ann. Bot. 70, 437–442.

Gastal, F., Bélanger, G., 1993. The effect of nitrogen fertilization and the growing season on photosynthesis of field-grown tall fescue canopies. Ann. Bot. 72, 401–408.

Gastal, F., Nelson, C.J., 1994. Nitrogen use within the growing leaf blade of tall fescue. Plant Physiol. 105, 191–197.

Gastal, F., Lemaire, G., 2002. N uptake and distribution in crops: an agronomical and ecophysiological perspective. J. Exp. Bot. 53, 789–799.

Gautier, H., Varlet-Grancher, C., Baudry, N., 1998. Comparison of horizontal spread of white clover (*Trifolium repens* L.) grown under two artificial light sources differing in their content in blue light. Ann. Bot. 82, 41–48.

Glass, A.D.M., Britto, D.T., Kaiser, B.N., et al., 2002. The regulation of nitrate and ammonium transport systems in plants. J. Exp. Bot. 53, 855–864.

Gonzalez-Dugo, V., Durand, J.L., Gastal, F., Picon-Cochard, C., 2005. Short-term response of the nitrogen nutrition status of tall fescue and Italian ryegrass swards under water deficit. Aust. J. Agric. Res. 56, 1269–1276.

Gonzalez-Dugo, V., Durand, J.L., Gastal, F., 2010. Water deficit and nitrogen nutrition of crops: a review. Agron. Sustain. Devel. 30, 529–544.

Gonzalez-Dugo, V., Durand, J.L., Gastal, F., Bariac, T., Poincheval, J., 2012. Restricted root-to-shoot translocation and decreased sink size are responsible for limited nitrogen uptake in three grass species under water deficit. Exp. Environ. Bot. 75, 258–267.

Gosse, G., Chartier, M., Lemaire, G., 1984. Mise au point d'un modèle de prévision de production pour une culture de luzerne. Comp. Rend. Acad. Sci. Paris III 18, 541–544.

Gosse, G., Varlet-Grancher, C., Bonhomme, R., Chartier, M., Allirand, J.M., Lemaire, G., 1986. Production maximale de matière sèche et rayonnement solaire intercepté par un couvert végétal. Agronomie 6, 47–56.

Grassini, P., Cassman, K.G., 2012. High-yield maize with large net energy yield and small global warming intensity. Proc. Natl. Acad. Sci. USA 109 (4), 1074–1079.

Greenwood, D.J., Neeteson, J.J., Draycott, A., 1986. Quantitative relationships for the dependence of growth rate of arable crops on their nitrogen content, dry weight and aerial environment. Plant Soil 91, 281–301.

Greenwood, D.J., Lemaire, G., Gosse, G., Cruz, P., Draycott, A., Neeteson, J.J., 1990. Decline in percentage N of C_3 and C_4 crops with increasing plant mass. Ann. Bot. 66, 425–436.

Greenwood, D.J., Karpinets, T.V., Zhang, K., Bosh-Serra, A., Boldrini, A., Karawulova, L., 2008. A unifying concept for the dependence of whole-crop N:P ratio on biomass: theory and experiment. Ann. Bot. 102, 967–977.

Grime, J.P., Campbell, B.D., Mackey, J.M.L., Crick, J.C., 1991. Root plasticity, nitrogen capture and competitive ability. In: Atkinson, D. (Ed.), Plant root growth. An ecological perspective. Blackwell Scientific Publications, Oxford, pp. 381–397.

Grindlay, D.J.C., 1997. Towards an explanation of crop nitrogen demand based on leaf nitrogen per unit leaf area. J. Agric. Sci. (Camb.) 128, 377–396.

Grindlay, D.J.C., Sylvester-Bradley, R., Scott, R.K., 1993. Nitrogen uptake of young vegetative plants in relation to green area. J. Sci. Food Agric. 63, 116–123.

Guingo, E., Hébert, Y., Charcosset, A., 1998. Genetic analysis of root traits in maize. Agronomie 18, 225–235.

Güsewell, S., Koerselman, W., 2002. Variation in nitrogen and phosphorus concentrations of wetland plants. Persp. Plant Ecol. Evol. Syst. 5, 37–61.

Güsewell, S., Koerselman, W., Verhoeven, J.T.A., 2003. Biomass N:P ratios as indicators of nutrient limitation for plant populations in wetlands. Ecol. Appl. 13, 372–384.

Habash, D.Z., Bernard, S., Shondelmaier, J., Weyen, Y., Quarrie, S.A., 2006. The genetics of nitrogen use on hexaploid wheat: N utilization, development and yield. Theor. Appl. Genet. 114, 403–419.

Hamblin, A.P., Tennant, D., 1987. Root length density and water uptake in cereals and grain legumes: how well are they correlated. Aust. J. Agric. Res. 38, 513–527.

Hammer, G.L., Sinclair, T.R., Chapman, S.C., van Oosterom, E., 2004. On system thinking, systems biology, and the in silico plant. Plant Physiol. 134, 909–911.

Hardwick, R.C., 1987. The nitrogen content of plants and the self-thinning rule in plant ecology: a test of the core-skin hypothesis. Ann. Bot. 60, 439–446.

Héraut-Bron, V., Robin, C., Varlet-Grancher, C., Afif, C., Gückert, A., 1999. Light quality (red:far red ratio): does it affect photosynthetic activity, net CO2 assimilation and morphology of young white clover leaves? Can. J. Bot. 77, 1425–1431.

Herrera, J.M., Noulas, C., Feil, C., Stamp, C., Liedgens, M., 2013. Nitrogen and genotype effects on root growth and root survivorship of spring wheat. J. Plant Nutr. Soil Sci. 176, 561–571.

Hikosaka, K., Terashima, I., Katoh, S., 1994. Effects of leaf age, nitrogen nutrition and photon flux density on the distribution of nitrogen among leaves of a vine (Ipomoea tricolor Cav) grown horizontally to avoid mutual shading of leaves. Oecologia 97, 451–457.

Hirel, B., Le Gouis, J., Ney, B., Gallais, A., 2007. The challenge of improving nitrogen use efficiency in crop plants: towards a more central role of genetic variability and quantitative genetics within integrated approaches. J. Exp. Bot. 58, 2369–2387.

Hirose, T., Werger, M.J.A., 1987. Maximising daily canopy photosynthesis with respect to the leaf-nitrogen allocation pattern in the canopy. Oecologia 72, 520–526.

Hirose, T., Werger, M.J.A., Pons, T.L., van Rheenen, J.W.A., 1988. Canopy structure and leaf nitrogen distribution in a stand of *Lysimachia vulgaris* L. as influenced by stand density. Oecologia 77, 145–150.

Hirose, T., Bazzaz, F.A., 1998. Trade-off between light- and nitrogen-use efficiency in canopy photosynthesis. Ann. Bot. 82, 195–202.

Hooper, D.U., Johnson, L., 1999. Nitrogen limitation in dryland ecosystems: responses to geographical and temporal variation in precipitation. Biogeochemistry 46, 247–293.

Houlès, V., Guérif, M., Mary, B., 2007. Elaboration of a nutrition indicator for winter wheat blades on leaf area index and chlorophyll meter content for making nitrogen recommendations. Eur. J. Agron. 27, 1–11.

Jamieson, P.D., Semenov, M.A., 2000. Modelling nitrogen uptake and redistribution in wheat. Field Crops Res. 68, 21–29.

Jarvis, S.C., Macduff, J.H., 1989. Nitrate nutrition of grasses from steady-state supplies in flowing solution culture following nitrate deprivation and/or defoliation. J. Exp. Bot. 40, 695–975.

Jeuffroy, M.H., Bouchard, C., 1999. Intensity and duration of nitrogen deficiency on wheat grain number. Crop Sci. 39, 1385–1393.

Jeuffroy, M.H., Colnenne, C., Reau, R., 2003. Nitrogen nutrition and rapeseed seed number. Towards enhanced value of cruciferous oilseed crops by optimal production and use of the high quality seed components. In: Proceedings 11th International Rapeseed Congress. Copenhagen, Denmark. pp. 6–10.

Jordan, D.R., Hunt, C.H., Cruickshank, A.W., Borrell, A.K., Henzell, R.G., 2012. The relationship between the stay-green trait and grain yield in elite sorghum hybrids grown in a range of environments. Crop Sci. 52, 1153–1161.

Justes, E., Mary, B., Meynard, J.M., Machet, J.M., Thellier-Huché, L., 1994. Determination of a critical nitrogen dilution curve for winter wheat crops. Ann. Bot. 74, 397–407.

Justes, E., Jeuffroy, M.H., Mary, B., 1997. The nitrogen requirement for major agricultural crops: wheat, barley and durum wheat. In: Lemaire, G. (Ed.), Diagnosis on the nitrogen status in crops. Springer-Verlag, Heidelberg, pp. 73–91.

Kahn, D.F., Peoples, M.B., Chalk, P.M., Herridge, D.F., 2002. Quantifying below-ground nitrogen of legumes. 2. Comparison of 15N and non isotopic methods. Plant Soil 239, 277–289.

Karrow, M., Maranville, J.W., 1994. Response of wheat cultivars to different soil-nitrogen and moisture regimes. 2 – Nitrogen uptake, partitioning and influx. J. Plant Nutr. 17, 745–761.

Kho, R.M., 2000. On crop production and the balance of available resources. Agric. Ecosyst. Environ. 80, 71–85.

Koerselman, W., Meuleman, A.F.M., 1996. The vegetation N:P ratio: a new tool to detect the nature of nutrient limitation. J. Appl. Ecol. 33, 1441–1450.

Lainé, P., Ourry, A., Macdujj, J.H., Boucaud, J., Salette, J., 1993. Kinetic parameters of nitrate uptake by different catch crop species: effect of low temperatures or previous nitrate starvation. Physiol. Plant. 88, 85–92.

Laperche, A., Brancourt-Humel, M., Heumez, E., Gardet, O., Le Gouis, J., 2006. Estimation of genetic parameters of a DH wheat population grown at different N stress levels characterized by probe genotypes. Theor. Appl. Genet. 112, 787–807.

Laperche, A., Devienne-Baret, F., Maury, O., Le Gouis, J., Ney, B., 2007. A simplified conceptual model of carbon and nitrogen functioning for QTL analysis of winter wheat adaptation to nitrogen deficiency. Theor. Appl. Genet. 113, 1131–1146.

Larsson, M., 1992. Translocation of nitrogen in osmotically stressed wheat seedlings. Plant Cell Environ. 15, 447–453.

Lawlor, D.W., 2002. Carbon and nitrogen assimilation in relation to yield: mechanisms are the key to understanding production systems. J. Exp. Bot. 53, 421–429.

Le Gouis, J., Béghin, D., Heumez, E., Pluchard, P., 2000. Genetic differences for nitrogen uptake and nitrogen utilization efficiencies in winter wheat. Eur. J. Agron. 12, 163–173.

Lejay, L., Tillard, P., Lepetit, M., et al., 1999. Molecular and functional regulation of two nitrate uptake systems by N- and C-status of Arabidopsis plants. Plant J. 18, 509–519.

Lemaire, G., Salette, J., 1984a. Relation entre dynamique de croissance et dynamique de prélèvement d'azote par un peuplement de graminées fourragères. 1 – Etude de l'effet du milieu. Agronomie 4, 423–430.

Lemaire, G., Salette, J., 1984b. Relation entre dynamique de croissance et dynamique de prélèvement d'azote par un peuplement de graminées fourragères. 2 – Etude de la variabilité entre génotypes. Agronomie 4, 431–436.

Lemaire, G., Cruz, P., Gosse, G., Chartier, M., 1985. Etude des relations entre la dynamique de prélèvement d'azote et la dynamique de croissance en matière sèche d'un peuplement de luzerne (*Medicago sativa* L.). Agronomie 5, 685–692.

Lemaire, G., Denoix, A., 1987. Croissance estivale en matière sèche de peuplements de fétuque élevée et de dactyle dans l'Ouest de la France. II – Interactions entre les niveaux d'alimentation hydrique et de nutrition azotée. Agronomie 7, 381–389.

Lemaire, G., Onillon, B., Gosse, G., Chartier, M., Allirand, J.M., 1991. Nitrogen distribution within a lucerne canopy during regrowth: relation with light distribution. Ann. Bot. 68, 483–488.

Lemaire, G., Khaithy, M., Onillon, B., Allirand, J.M., Chartier, M., Gosse, G., 1992. Dynamics of accumulation and partitioning of N in leaves, stems and roots of lucerne (*Medicago sativa* L.) in a dense canopy. Ann. Bot. 70, 429–435.

Lemaire, G., Charrier, X., Hébert, Y., 1996. Nitrogen uptake capacities of maize and sorghum crops in different nitrogen and water supply conditions. Agronomie 16, 231–246.

Lemaire, G., Gastal, F., 1997. N uptake and distribution in plant canopies. In: Lemaire, G. (Ed.), Diagnosis of the nitrogen status in crops. Springer-Verlag, Heidelberg, pp. 3–43.

Lemaire, G., Meynard, J.M., 1997. Use of the nitrogen nutrition index for the analysis of agronomical data. In: Lemaire, G. (Ed.), Diagnosis of the nitrogen status in crops. Springer-Verlag, Heidelberg, pp. 45–56.

Lemaire, G., Plénet, D., Grindlay, D., 1997. Leaf N content as an indicator of crop N nutrition status. In: Lemaire, G. (Ed.), Diagnosis of the nitrogen status in crops. Springer-Verlag, Heidelberg, pp. 189-199.

Lemaire, G., Millard, P., 1999. An ecophysiological approach to modelling resources fluxes in competing plants. J. Exp. Bot. 50, 15–28.

Lemaire, G., Recous, S., Mary, B., 2004. Managing residues and nitrogen in intensive cropping systems. New understanding for efficient recovery by crops. In: Proceedings of the 4th International Crop Science Congress. Brisbane, Australia September, 2004.

Lemaire, G., Avice, J.C., Kim, T.H., Ourry, A., 2005. Developmental changes in shoot N dynamics of lucerne in relation to leaf growth dynamics as a function of plant density and hierarchical position within the canopy. J. Exp. Bot. 56, 935–943.

Lemaire, G., van Oosterom, E., Sheehy, J., Jeuffroy, M.H., Massignam, A., Rossato, L., 2007. Is crop demand more closely related to dry matter accumulation of leaf area expansion during vegetative growth? Field Crops Res. 100, 91–106.

Lemaire, G., van Oosterom, E., Jeuffroy, M.H., Gastal, F., Massignam, A., 2008. Crop species present different qualitative types of response to N deficiency during their vegetative growth. Field Crops Res. 105, 253–265.

London, J.G., 2005. Nitrogen study fertilises fears of pollution. Nature 433, 791.

Louarn, G., Da Silva, D., Godin, C., Combes, D., 2012. Simple envelope-based reconstruction methods can infer light partitioning among individual plants in sparse and dense herbaceous canopies. Agric. For. Meteorol. 166-167, 98–112.

Lynch, J.P., 2013. Steep, cheap and deep: an ideotype to optimize water and N acquisition by maize root systems. Ann. Bot. 112, 347–357.

MacAdam, J.W., Volenec, J.J., Nelson, C.J., 1989. Effects of nitrogen on mesophyll cell division and epidermal cell elongation in tall fescue leaf blades. Plant Physiol. 89, 549–556.

McNeil, A.M., Zhu, C., Fillery, I.R.P., 1997. Use in situ 15N-labelling to estimate the total below ground nitrogen on pasture legumes in intact soil-plant systems. Aust. J. Agric. Res. 48, 295–304.

Mae, T., 1997. Physiological nitrogen efficiency in rice: nitrogen utilization, photosynthesis, and yield potential. In: Ando, T. (Ed.), Plant nutrition for sustainable food production and environment. Kluver Academic Publishers, Dordrecht, The Netherlands, pp. 51–60.

Maire, V., Martre, P., Kattge, J., et al., 2012. The coordination of leaf photosynthesis links C and N fluxes in C-3 plant species. Plos One 7, e38345.

Manschadi, A.M., Christopher, J., deVoil, P., Hammer, G.L., 2006. The role of root architectural traits in adaptation of wheat to water-limited environments. Funct. Plant Biol. 33, 823–837.

Manschadi, A.M., Hammer, G.L., Christopher, J.T., deVoil, P., 2008. Genotypic variation in seedling root architectural traits and implications for drought adaptation in wheat (*Triticum aestivum* L.). Plant Soil 303, 115–129.

Mariotti, A., 1997. Quelques reflexions sur le cycle biogéochimique de l'azote dans les agrosystèmes. Lemaire, G., Nicolardot, B. (Eds.), Maîtrise de l'azote dans les agrosystemes, 83, INRA Editions, Paris, pp. 9–24.

Martre, P., Porter, J.R., Jamieson, P.D., Triboï, E., 2003. Modelling grain nitrogen accumulation and protein composition to understand the sink/source regulations of nitrogen utilization in wheat. Plant Physiol. 133, 1959–1967.

Mary, B., Recous, S., 1994. Measurement of nitrogen mineralization and immobilization fluxes in soil as a means of predicting net mineralization. Eur. J. Agron. 3, 291–300.

Matsunaka, T., Watanabe, Y., Miyawaki, T., Ichiwaka, N., 1997. Prediction of grain protein content in winter wheat through leaf color measurements using a chlorophyll meter. Soil Sci. Plant Nutr. 43, 127–134.

Matzek, V., Vitousek, P.M., 2009. N:P stoichiometry and protein:RNA ratios in vascular plants: an evaluation of the growth-rate hypothesis. Ecol. Lett. 12, 765–771.

Meinke, H., Hammer, G.L., van Keulen, H., Rabbinje, R., Keating, B.A., 1997. Improving wheat simulation capabilities in Australia from a cropping systems perspective: water and nitrogen effects on spring wheat in a semi-arid environment. Eur. J. Agron. 7, 75–88.

Mi, C., Liu, J., Chen, F., Zhang, F., Cui, Z., Liu, X., 2003. Nitrogen uptake and remobilization in maize hybrids differing in leaf senescence. J. Plant Nutr. 26, 447–459.

Milroy, S.P., Bange, M.P., Sadras, V.O., 2001. Profiles of leaf nitrogen and light in reproductive canopies of cotton (*Gossypium hirsutum*). Ann. Bot. 87, 325–333.

Moll, R.H., Kamprath, E.J., Jackson, W.A., 1982. Analysis and interpretation of factors which contribute to efficiency of nitrogen utilization. Agron. J. 74, 562–564.

Monteith, J.L., 1972. Solar radiation and productivity in tropical ecosystems. J. Appl. Ecol. 9, 747–766.

Muchow, R.C., 1988. Effect of nitrogen supply on the comparative productivity of maize and sorghum in a semi-arid tropical environment. I – Leaf growth and leaf nitrogen. Field Crops Res. 18, 1–16.

Muchow, R.C., Davis, R., 1988. Effect of nitrogen supply on the comparative productivity of maize and sorghum in a semi-arid tropical environment. II – Radiation interception and biomass accumulation. Field Crops Res. 18, 17–30.

Muchow, R.C., Sinclair, T.R., 1994. Nitrogen response of leaf photosynthesis and canopy radiation use efficiency in field grown maize and sorghum. Crop Sci. 34, 721–727.

Murphy, D.V., Recous, S., Stockdale, E.A., et al., 2003. Gross nitrogen fluxes in soil: theory, measurement and application of 15N pool dilution techniques. Adv. Agron. 79, 69–119.

Naud, C., Makowski, D., Jeuffroy, M.H., 2008. Is it useful to combine measurements taken during growing season with dynamic model to predict the nitrogen status of winter wheat? Eur. J. Agron. 28 (3), 291–300.

Ney, B., Doré, T., Sagan, M., 1997. The nitrogen requirement of major agricultural crops: grain legumes. In: Lemaire, G. (Ed.), Diagnosis of the nitrogen status in crops. Springer-Verlag, Heidelberg, pp. 107–117.

Niinemets, U., Tenhunen, J.D., 1997. A model separating leaf structural and physiological effects on carbon gain along light gradients for the shade-tolerant species *Acer saccharum*. Plant Cell Environ. 20, 845–866.

Niklas, K.J., 1994. Plant allometry: the scaling of form and process. University of Chicago Press, Chicago.

Onillon, B., Durand, J.L., Gastal, F., Tournebize, R., 1995. Drought effects on growth and carbon partitioning in a tall fescue sward grown at different rates of nitrogen fertilization. Eur. J. Agron. 4, 91–99.

Ortiz-Monasterio, J.I., Satre, K.D., Rajaram, S., McMahon, M., 1997. Genetic progress in wheat yield and nitrogen use efficiency under four nitrogen rates. Crop Sci. 37, 898–904.

Oti-Boateng, C., Wallace, W., Sisbury, J.H., 1994. The effect of the accumulation of carbohydrate and organic nitrogen on N_2 fixation of faba bean cv. Fiord. Ann. Bot. 73, 143–149.

Ourry, A., Macduff, J.H., Ougham, H., 1996. The relationship between mobilization of N reserves and changes in translatable messages following defoliation in *Lolium temulentum* L. and *Lolium perenne* L. J. Exp. Bot. 47, 739–747.

Oyanagi, A., Nakamoto, T., Wada, M., 1993. Relationship between root growth angle of seedlings and vertical distribution of roots in the field in wheat cultivars. Jap. J. Crop Sci. 62, 565–570.

Oyanagi, A., 1994. Gravitropic response, growth angle and vertical distribution of roots of wheat (*Triticum aestivum* L). Plant Soil 165, 323–326.

Palta, J.A., Fillery, I.R.P., 1995. N application increases pre-anthesis contribution of dry matter to grain yield in wheat grown on a duplex soil. Aust. J. Agric. Res. 46, 507–518.

Pandley, R.K., Maranville, J.W., Chetima, M.M., 2000. Deficit irrigation and nitrogen effects on maize in a Sahelian environment. II – Shoot growth, nitrogen uptake and water extraction. Agric. Water Manag. 46, 15–27.

Pastor, J., Post, W.M., 1985. Development of a linked-forest productivity-soil process model. ORNL/TM-5919. Oak Ridge National Laboratory, Oak Ridge, Tenessee, USA.

Paterson, E., 2003. Importance of rhizodeposition in the coupling of plant and microbial productivity. Eur. J. Soil Sci. 54, 741–750.

Peng, S., Garcia, F.V., Laza, R.C., Sanico, A.L., Visperas, R.M., Cassman, K.G., 1996. Increased N-use efficiency using a chlorophyll meter on high-yielding irrigated rice. Field Crops Res. 47, 243–252.

Peoples, M.B., Baldock, J.A., 2001. Nitrogen dynamics of pastures: nitrogen fixation inputs, the impact of legumes on soil nitrogen fertility, and the contributions of fixed nitrogen to Australian farming systems. Aust. J. Exp. Agric. 41, 327–346.

Piekelek, W.P., Fox, R.H., 1992. Use of a chlorophyll meter to predict side-dress nitrogen for maize. Agron. J. 84, 59–65.

Pirmoradian, N., Sepashkhah, A.R., Maftoun, M., 2004. Deficit irrigation and nitrogen effects on nitrogen-use efficiency and grain protein on rice. Agronomie 24, 143–153.

Plénet, D., Cruz, P., 1997. The nitrogen requirement for major agricultural crops: maize and sorghum. In: Lemaire, G. (Ed.), Diagnosis of the nitrogen status in crops. Springer-Verlag, Heidelberg, pp. 93–106.

Plénet, D., Lemaire, G., 1999. Relationships between dynamics of nitrogen uptake and dry matter accumulation in maize crops. Determination of critical N concentration. Plant Soil 216, 65–82.

Poirier, M., Durand, J.L., Volaire, F., 2012. Persistence and production of perennial grasses under water deficits and extreme temperatures: importance of intraspecific *vs.* interspecific variability. Glob. Change Biol. 18, 3632–3646.

Pons, T.L., Schieving, F., Hirose, T., Werger, M.J.A., 1989. Optimization of leaf nitrogen allocation for canopy photosynthesis in *Lysimachia vulgaris* L. In: Lambers, H. (Ed.), Causes and consequences of variation in growth rate and productivity in higher plants. SPB Academic, The Hague, pp. 175–186.

Powlson, D.S., Hart, P.B.S., Johnston, A.E., Poulton, P.R., Jenkinson, D.S., 1992. Influence of soil type, crop management and weather on the recovery of 15N-labelled fertiliser applied to winter wheat in spring. J. Agric. Sci. (Camb.) 118, 83–110.

Pratersak, A., Fukaï, S., 1997. Nitrogen availability and water stress interaction on rice grown in field. Field Crops Res. 52, 249–260.

Prieto, J.A., Louarn, G., Perez Peña, J., Ojeda, H., Simonneau, T., Lebon, E., 2012. A leaf gas exchange model that accounts for intra-canopy variability by considering leaf nitrogen content and local acclimation to radiation in grapevine (*Vitis vinifera* L.). Plant Cell Environ. 35, 1313–1328.

Prioul, J.L., Brangeon, J., Reyss, A., 1980a. Interaction between external and internal conditions in the development of photosynthetic features in grass leaf. I. Regional responses along a leaf during and after low-light or high-light. Plant Physiol. 66, 762–769.

Prioul, J.L., Brangeon, J., Reyss, A., 1980b. Interaction between external and internal conditions in the development of photosynthetic features in grass leaf. II. Reversibility of light-induced responses as a function of development. Plant Physiol. 66, 770–774.

Radin, J.W., 1983. Control of plant growth by nitrogen: differences between cereals and broadleaf species. Plant Cell Environ. 6, 65–68.

Rajcan, I., Tollenaar, M., 1999a. Source:sink ratio and leaf senescence in maize. I. Dry matter accumulation and partitioning during grain filling. Field Crops Res. 60, 245–253.

Rajcan, I., Tollenaar, M., 1999b. Source:sink ratio and leaf senescence in maize. II. Nitrogen metabolism during grain filling. Field Crops Res. 60, 255–265.

Ramos, C., 1996. Effect of agricultural practices on the nitrogen losses in environment. In: Rodriguez-Barrueco, C. (Ed.), Fertiliser and environment. Kluver Academic Publishers, Dordrecht, The Netherlands, pp. 355–361.

Reeves, D.W., Mask, P.L., Wood, C.W., Delaney, D.P., 1993. Determination of wheat nitrogen status with a hand-held chlorophyll meter: influence of management practices. J. Plant Nutr. 16, 781–796.

Robinson, D., 1994. The responses of plants to non-uniform supplies of nutrients. New Phytol. 127, 635–674.

Robson, M.J., Parsons, A.J., 1978. Nitrogen deficiency in small closed communities of S24 ryegrass. I – Photosynthesis, respiration, dry matter production and partition. Ann. Bot. 42, 1185–1197.

Rodriguez, D., Robson, A., Belford, R., 2009. Dynamic and functional monitoring technologies for applications in crop management. In: Sadras, V.O., Calderini, D.F. (Eds.), Crop physiology: applications for genetic improvement and agronomy. Academic Press, San Diego, pp. 489–510.

Rossato, L., Lainé, P., Ourry, A., 2001. Nitrogen storage and remobilization in *Brassica napus* L. during the growth cycle: nitrogen fluxes within the plant and changes in soluble protein pattern. J. Exp. Bot. 52, 1655–1663.

Rousseaux, M.C., Hall, A.J., Sanchez, R.A., 1996. Far-red enrichment and photosynthetically active radiation level influence leaf senescence in field-grown sunflower. Physiol. Plant. 96, 217–224.

Rousseaux, M.C., Hall, A.J., Sanchez, R.A., 1999. Light environment, nitrogen content, and carbon balance of basal leaves of sunflower canopies. Crop Sci. 39, 1093–1100.

Rousseaux, M.C., Hall, A.J., Sanchez, R.A., 2000. Basal leaf senescence in a sunflower (*Helianthus annuus*) canopy: responses to increased R/FR ratio. Physiol. Plant. 110, 477–482.

Ryle, G.J.A., Powell, C.E., Gordon, A.J., 1986. Defoliation in white clover: nodule metabolism, nodule growth and maintenance, and nitrogenase functioning during growth and regrowth. Ann. Bot. 57, 263–271.

Sadras, V.O., Hall, A.J., Connor, D.J., 1993. Light-associated nitrogen distribution profile in flowering canopies of sunflower (*Helianthus annuus* L.) altered during grain filling. Oecologia 95, 488–494.

Sadras, V.O., 2002. Interaction between rainfall and nitrogen fertilisation of wheat in environments prone to terminal drought: economic and environmental risk analysis. Field Crops Res. 77, 201–215.

Sadras, V.O., Roget, D.K., 2004. Production and environmental aspects of cropping intensification in a semiarid environment of South Eastern. Aust. Agron. J. 96, 236–246.

Sadras, V.O., 2004. Yield and water use efficiency of water and nitrogen stressed wheat crops increase with the degree of water and nitrogen co-limitation. Eur. J. Agron. 21, 455–464.

Sadras, V.O., Baldock, J., Cox, J., Bellotti, B., 2004. Crop rotation effect on wheat grain yield as mediated by changes in the degree of water and nitrogen co-limitation. Aust. J. Agric. Res. 55, 599–607.

Sadras, V.O., 2005. A quantitative top-down view of interactions between stresses: theory and analysis of nitrogen-water co-limitation in Mediterranean agro-ecosystems. Aust. J. Agric. Sci. 56, 1151–1157.

Sadras, V.O., 2006. The N:P stoichiometry of cereal, grain legume and oilseed crops. Field Crops Res. 95, 13–29.

Sadras, V.O., Lawson, C., 2013. Nitrogen and water use efficiency of Australian wheat variety released between 1958 and 2007. Eur. J. Agron. 46, 34–41.

Sadras, V.O., Lawson, C., Hooper, P., McDonald, G.K., 2012. Contribution of summer rainfall and nitrogen to the yield and water use efficiency of wheat in Mediterranean-type environments of South Australia. Eur. J. Agron. 36, 41–54.

Salvagiotti, F., Miralles, D.J., 2008. Radiation interception, biomass production and grain yield as affected by the interaction of nitrogen and sulphur fertilisation in wheat. Eur. J. Agron. 28 (3), 282–290.

Samuelson, M.E., Eliasson, L., Larsson, C.M., 1992. Nitrate-regulated growth and cytokinin responses in seminal roots of barley. Plant Physiol. 98, 309–315.

Schieving, F., Pons, T.L., Werger, M.J.A., Hirose, T., 1992. The vertical distribution of nitrogen and photosynthetic activity at different plant densities in Carex acutiformis. Plant Soil 14, 9–17.

Seginer, I., 2004. Plant spacing effect on the nitrogen concentration of a crop. Eur. J. Agron. 21 (3), 369–377.

Semenov, M.A., Jamieson, P.D., Martre, P., 2007. Deconvoluting nitrogen use efficiency in wheat: a simulation study. Eur. J. Agron. 26, 283–294.

Sheehy, J.E., Dionara, M.J.A., Mitchell, P.L., et al., 1998. Critical concentrations: implications for high-yielding rice (*Oryza sativa* L.) cultivars in tropics. Field Crops Res. 59, 31–41.

Shiraiwa, T., Sinclair, T.R., 1993. Distribution of nitrogen among leaves in soybean canopies. Crop Sci. 33, 804–808.

Simmonds, N.W., 1995. The relation between yield and protein in cereal grains. J. Sci. Food Agric. 67, 309–315.

Sinclair, T.R., Horie, T., 1989. Leaf nitrogen, photosynthesis and crop radiation use efficiency: a review. Crop Sci. 29, 90–98.

Sinclair, T.R., Shiraiwa, T., Hammer, G.L., 1992. Variation in radiation-use efficiency with increased diffuse radiation. Crop Sci. 32, 1281–1284.

Sinclair, T., Shiraiwa, T., 1993. Soybean radiation-use efficiency as influenced with by non uniform specific leaf nitrogen and diffuse radiation. Crop Sci. 32, 1281–1284.

Sinclair, T.R., 1998. Historical changes in harvest index crop N accumulation. Crop Sci. 38, 638–643.

Sinclair, T.R., Park, W.I., 1993. Inadequacy of the Liebig limiting-factor paradigm for explaining varying crop yields. Agron. J. 85, 742–746.

Sinclair, T.R., Sheehy, J.E., 1999. Erect leaves and photosynthesis in rice. Science 283, 1456–1457.

Sinclair, T.R., Purcell, L.C., Sneller, C.H., 2004. Crop transformation and the challenge to increase yield potential. Trends Plant Sci. 9, 70–75.

Singh, U., Ladha, J.K., Castillo, I.E., Punzalan, G., Tirol-Padre, A., Duqueza, M., 1998. Genotypic variation in nitrogen use efficiency. I. Medium- and long-duration rice. Field Crops Res. 58, 35–53.

Sinoquet, H., Moulia, B., Gastal, F., Bonhomme, R., Varlet-Grancher, C., 1990. Modelling the radiative balance of the components of a well-mixed canopy: application to a white clover - tall fescue mixture. Acta Oecol. 11, 469–486.

Smolander, A., Barnette, L., Kitunen, V., Lumme, I., 2005. N and C transformations in long term N-fertilised forest soils in response to seasonal drought. Appl. Soil Ecol. 29, 225–235.

Soussana, J., Arregui, M., 1995. Impact de l'association sur le niveau de nutrition azotée et la croissance du ray-grass anglais et du trèfle blanc. Agronomie 15 (2), 81–96.

Spollen, W.G., Sharp, R.E., Saab, I.N., Wu, Y., 1993. Regulation of cell expansion in roots and shoots at low water potentials. In: Smith, J.A.C., Griffiths, H. (Eds.), Water deficits. Plant responses from cell to community. BIOS Scientific Publishers, Oxford, pp. 37–52.

Sterner, R.W., Elser, J.J., 2002. Ecological stoichiometry: the biology of elements from molecules to the biosphere. Princeton University Press, Princeton, NJ.

Stulen, I., Perez-Soba, M., De Kok, L.J., Van Der Eerden, L., 1998. Impact of gazeous nitrogen deposition on plant functioning. New Phytol. 139, 61–70.

Sylvester-Bradley, R., Stokes, D.T., Scott, R.K., Willington, V.B.A., 1990. A physiological analysis of the diminishing responses of winter wheat to applied nitrogen. 2. Evidence. Asp. Appl. Biol. II Cereal Qual. 25, 289–300.

Ta, C.T., Weiland, R.T., 1992. Nitrogen partitioning in maize during ear development. Crop Sci. 32, 443–451.

Tabuchi, M., Abiko, T., Yamaya, T., 2007. Assimilation of ammonium-ions and re-utilization of nitrogen in rice (*Oryza sativa* L.). J. Exp. Bot. 58, 2319–2328.

Tei, F., Benincasa, P., Guidici, M., 2002. Critical nitrogen concentration in processing tomato. Eur. J. Agron. 18, 45–56.

Thomas, H., Sadras, V.O., 2001. The capture and gratuitous disposal of resources by plants. Funct. Ecol. 15, 3–12.

Thornley, J.H.M., 1977. Root:shoot interactions. In: Jennings, D.H. (Ed.), Integration of activity in higher plants. Cambridge University Press, Cambridge, pp. 367–389.

Thornton, B., Lemaire, G., Millard, P., Duff, E.I., 1999. Relationships between nitrogen and water concentration in shoot tissue of *Molinia caerulea* L. during shoot development. Ann. Bot. 83, 631–636.

Tourraine, B., Clarkson, D.T., Muller, B., 1994. Regulation of nitrate uptake at the whole plant level. In: Roy, J., Garnier, E. (Eds.), A whole plant perspective on carbon-nitrogen interactions. SPB Academic Publishing, The Hague, pp. 11–30.

Trachsel, S., Kaeppler, S.M., Brown, K.M., Lynch, J.P., 2013. Maize root growth angles become steeper under low N conditions. Field Crops Res. 140, 18–31.

Trápani, N., Hall, A.J., 1996. Effects of leaf position and nitrogen supply on the expansion of leaves of field-grown sunflower. Plant Soil 184, 331–340.

Trápani, N., Hall, A.J., Weber, M., 1999. Effects of constant and variable nitrogen supply on sunflower (*Helianthus annuus* L.) leaf cell number and size. Ann. Bot. 84, 599–606.

Tremmel, D., Bazzaz, F.A., 1995. Plant architecture and allocation in different neighborhoods: implications for competitive success. Ecology 76, 262–271.

Uhart, S.A., Andrade, F.H., 1995. Nitrogen deficiency in maize. I. Effects on crop growth, development, dry matter partitioning and kernel set. Crop Sci. 35, 1376–1383.

Uribelarrea, M., Below, F.E., Moose, S.P., 2007. Divergent selection for grain protein affects nitrogen use in maize. Field Crops Res. 100, 82–90.

van Keulen, H., Goudrian, J., Seligman, N.G., 1989. Modelling the effects of nitrogen on canopy development and crop growth. In: Russell, G., Marshall, L.B., Jarvis, P.G. (Eds.), Plant canopies: their growth, form and function. Cambridge University Press, Cambridge, pp. 83–104.

van Oosterom, E.J., Carberry, P.S., Muchow, R.C., 2001. Critical and minimum N contents for development and growth of grain sorghum. Field Crops Res. 70, 55–73.

Vega, C.R.C., Andrade, F.H., Sadras, V.O., 2001. Reproductive partitioning and seed efficiency in soybean, sunflower and maize. Field Crops Res. 72, 163–175.

Verhoeven, J.T.A., Koerselman, W., Meuleman, A.F.M., 1996. Nitrogen- or phosphorus-limited growth in herbaceous, wet vegetation: relations with atmospheric inputs and management regimes. Trends Ecol. Evol. 11, 494–497.

Vos, J., Biemond, H., 1992. Effects of nitrogen on the development and growth of the potato plant. I. Leaf appearance, expansion growth, life span of leaves and stem branching. Ann. Bot. 70, 27–35.

Vos, J., van der Putten, P.E.L., Birch, C.J., 2005. Effect of nitrogen supply on leaf appearance, leaf growth, leaf nitrogen economy and photosynthetic capacity in maize (*Zea mays* L.). Field Crops Res. 93, 64–73.

Waring, G.L., Cobb, N.S., 1992. The impact of plant stress on herbivore population dynamics. In: Bernays, E.A. (Ed.), Insect-plant interactions. CRC Press, Boca Raton, pp. 167–226.

Wasson, A.P., Richards, R.A., Chatrath, R., et al., 2012. Traits and selection strategies to improve root systems and water uptake in water-limited wheat crops. J. Exp. Bot. 63, 3485–3498.

White, T., 1993. The inadequate environment-nitrogen and the abundance of animals. Springer-Verlag , Berlin, Germany.

Whu, L., McGechan, M.B., Watson, C.A., Baddeley, J.A., 2005. Developing existing plant root system architecture models to meet future agricultural challenges. Adv. Agron. 85, 181–219.

Wilman, D., Pearse, P.J., 1984. Effect of applied nitrogen on grass yield, nitrogen content, tillers and leaves in field swards. J. Agric. Sci. (Camb.) 103, 201–211.

Wright, G.C., Smith, C.J., Woodroofe, M.R., 1988. The effect of irrigation and nitrogen fertiliser on rapeseed (*Brassica napus* L.) production in south-eastern Australia. I – Growth and seed yield. Irrig. Sci. 9, 1–13.

Wright, I.J., Leishman, M.R., Read, C., Westoby, M., 2006. Gradients of light availability and leaf traits with leaf age and canopy position in 28 Australian shrubs and trees. Funct. Plant Biol. 33, 407–419.

Zhang, H., Forde, B.G., 2000. Regulation of *Arabidopsis* root development by nitrate availability. J. Exp. Bot. 51, 51–59.

Ziadi, N., Bélanger, G., Gastal, F., Claessens, A., Lemaire, G., Tremblay, N., 2009. Leaf nitrogen concentration as an indicator of corn nitrogen status. Agron. J. 101, 947–957.

Ziadi, N., Bélanger, G., Claessens, A., et al., 2010. Plant-based diagnostic tools for evaluating wheat nitrogen status. Crop Sci. 50 (6), 2580–2590.

CHAPTER

9

A Darwinian perspective on improving nitrogen-fixation efficiency of legume crops and forages

R. Ford Denison

Department of Ecology, Evolution and Behavior, University of Minnesota, USA

1 NITROGEN FIXATION'S ROLE IN AGRICULTURE

Plants and animals need nitrogen, mostly for proteins, but they are unable to use N_2 gas, which comprises 80% of the atmosphere. Only bacteria and archaea can convert atmospheric nitrogen into biologically useful forms, a process known as biological nitrogen fixation. In agriculture, symbiotic nitrogen fixation by rhizobia bacteria in legume root nodules is particularly important.

Even before recent increases in fertilizer prices, it was calculated that a 10% increase in nitrogen fixation by one crop (soybean) would provide hundreds of millions of dollars in economic benefits (Herridge and Rose, 2000). An implicit assumption was that this increase in nitrogen fixation could be achieved without costly side effects. Given the energy cost to legumes of supporting nitrogen fixation, this assumption seems risky unless, (1) legumes have enough surplus photosynthate that more allocation to nitrogen fixation does not detract from other beneficial uses, or (2) that increased fixation was achieved through an increase in efficiency (gN/gC), without increased photosynthate use. For reasons explained below, I will focus on the latter option.

The overall role of nitrogen fixation in agriculture is best understood in the context of total nitrogen budgets for agricultural ecosystems (Denison and Kiers, 2005). A key difference between natural and agricultural ecosystems is that only the latter are required to export vast quantities of food to distant cities. Nitrogen cycling occurs within both types of ecosystems. In agriculture, however, internal cycling can supply plant nitrogen needs only temporarily, because so much nitrogen is exported from the

Crop Physiology. DOI: 10.1016/B978-0-12-417104-6.00009-1

ecosystem each year as protein in grain, meat, or milk. For long-term sustainability, conservation of matter requires that nitrogen exports be balanced by external inputs, such as manure, synthetic fertilizer, or biological nitrogen fixation (Denison, 2012). Currently, little of that nitrogen returns from cities to farms, because the economic cost of transporting manure or human waste long distances exceeds its benefits.

Increasing energy prices would not necessarily favor more recycling of nitrogen or other nutrients. Synthetic nitrogen fertilizer is typically made from atmospheric nitrogen and natural gas (methane), so higher energy costs increase the cost of nitrogen fertilizer. Hydrogen produced by electrolysis of water can substitute for methane, so synthetic fertilizer production can be powered by wind (Nelson, 2007) or other renewable energy sources, although these may be more expensive than natural gas. One problem with recycling is that manure and other organic materials have much less nitrogen per kilogram than nitrogen fertilizer does. Calculations based on published parameters (Pimentel et al., 1984) show that, beyond a few kilometers, the higher transport-energy costs for manure more than compensate for the energy cost of making the fertilizer.

Unless we develop much more efficient ways of transporting organic nitrogen sources or abandon cities, high energy prices alone would not be enough to make closing the nitrogen cycle a viable option. Even farms that can afford to pay transport costs for manure, because they receive premium prices for organic products, compete for a limited supply. In California, the statewide manure supply is roughly 2000 kg irrigated ha^{-1} (Chaney et al., 1992), a fraction of the amount needed for most organic crops. Also, that manure often comes from animals fed grain grown with synthetic fertilizer, which was the ultimate source of the nitrogen in the manure.

Synthetic nitrogen fertilizer is not the only alternative to recycling nitrogen in organic forms, however. Currently, about two-thirds of global nitrogen inputs to agriculture come from nitrogen fertilizer and one-third from biological nitrogen fixation (Connor et al., 2011). Soybeans and alfalfa, two of our most widely grown crops, obtain much of their nitrogen from rhizobia in their root nodules. Other nitrogen-fixing legume crops and forages are important in some regions or could become so if increasing energy prices make synthetic nitrogen fertilizers less affordable. Increased reliance on biological nitrogen fixation could come from increasing use of legumes or, to some extent, from increasing the amount of nitrogen fixed by current legume crops and forages. This chapter explores the prospects for increasing nitrogen-fixation rates, but also the prospects for increasing nitrogen-fixation efficiency, reducing the metabolic cost of nitrogen fixation to legumes.

Greater reliance on biological nitrogen fixation could reduce water pollution from nitrogen fertilizer. Nitrogen-fixing legumes match nitrogen supply to demand. When a legume plant has enough available nitrogen (in soil or its own tissues), it prevents formation of new nitrogen-fixing root nodules (Streeter, 1988) and reduces the nitrogen-fixation rate of its existing nodules (Denison and Harter, 1995). This sophisticated regulation of supply by demand contrasts with over-application of fertilizer or manure by some farmers where determining optimal fertilizer rates is difficult or where subsidies favor excessive usage of nitrogen (Li et al., 2013; Chapter 3). This also contrasts with the under-application of fertilizer where water supply is uncertain and nitrogen management involves significant economic risk (Chapter 7) or where infrastructure, markets and opportunity costs limit the use of soluble fertilizer (Chapter 5).

Much of the nitrogen fixed by legumes is harvested in grain or hay. The nitrogen that remains is potentially available to subsequent crops grown in rotation, but is also subject to potential loss by leaching or denitrification, contributing to water or air pollution.

Substituting biological nitrogen fixation for some of our current fertilizer use would have only a small effect on total world greenhouse gas production. Fertilizer production accounts for a significant fraction (40–50%) of agricultural energy use, but agricultural production accounts for less than 5% of total energy use in industrial countries (Connor et al., 2011). For example, more energy is consumed in transporting wheat bread home and toasting it than was used to grow the wheat, including the energy cost of the nitrogen fertilizer (Avlani and Chancellor, 1977). Greater reliance on nitrogen fixation on farms would not solve the problem of nitrogen pollution from feedlots and cities.

Where nitrogen fertilizer is available, but costly, the main potential benefits from nitrogen fixation are reduced costs to farmers and reduced water pollution from farms. Where nitrogen fertilizer is not available or unaffordable, biological nitrogen fixation may be the only option for increasing nitrogen supply to crops. In these typically low-input cropping systems, however, poor legume yield limits the contribution of nitrogen biological fixation (Chapter 5).

2 A DARWINIAN PERSPECTIVE ON IMPROVING N_2 FIXATION

The wild ancestors of our crops and their rhizobial symbionts co-evolved for millions of years prior to domestication. Over that entire period, any trait that enhanced a plant's or microbe's fitness tended to increase in frequency, under natural selection. What opportunities remain for further improvements by humans (Denison et al., 2003)?

One way to improve nitrogen fixation might be to select legume crops or forages for earlier nodulation, more abundant nodulation, faster nodule growth, longer nodule duration, or increased nitrogen fixation per gram of nodule. It is worth asking, however, why past natural selection has not increased these beyond their current levels. Similarly, we can identify genes key to nitrogen fixation and could increase their expression. This approach would implicitly assume that past natural selection, followed by domestication and subsequent plant breeding, somehow resulted in suboptimal expression of key genes.

Again, it is worth considering specific reasons why expression of genes key to nitrogen fixation might be suboptimal. It seems safe to assume that, for any given gene, there are at least five possible single-base mutations that would increase its expression. With a mutation rate of 10^{-8} per base (Koch et al., 2000) and 2×10^7 soybean plants km^{-2}, every 10 square kilometers would include a plant with increased expression of that gene. Multiply by millions of years and it is clear that natural selection had ample opportunity to increase expression of any given gene, if greater expression caused a major increase in fitness (Denison, 2012).

Genes can have pleiotropic effects, of course. Maybe natural selection repeatedly rejected increased expression of the gene because, although effects on nitrogen fixation were beneficial, there were negative effects on root growth, flowering, or some other important process. But if those processes are still important in crops today, that may still be a good reason not to increase that gene's expression.

More broadly, I have argued that past natural selection is unlikely to have missed 'simple, trade-off-free improvements' (Denison, 2012), where 'simple' is defined as arising frequently via mutation, and 'trade-off-free' means increasing individual fitness across a wide range of past environments. For simplicity, I will call this the 'evolutionary-trade-off hypothesis'.

For example, I have argued that accepting some trade-offs will be key to improving crop productivity under drought, because the wild ancestors of our crops were exposed to drought reasonably often. The failure of recent high-profile effort to improve the performance of maize under drought (Roth et al., 2013) is consistent with this hypothesis. I predict that similar independent testing of other maize lines bred or engineered for drought tolerance would also show that improved performance under drought, if any, comes at some cost to other useful traits.

The trade-offs that constrained improvement by past natural selection need not always constrain us, however. For example, we would readily sacrifice adaptation to past CO_2 concentrations if it enhanced adaptation to current or future CO_2. This might be achieved by modifying the photosynthetic enzyme, Rubisco (Zhu et al., 2004), although the modifications needed might not be obvious.

What other trade-offs rejected by past natural selection might be acceptable to us today? Many opportunities involve trade-offs between individual fitness and the collective performance of crop communities (Denison et al., 2003). Yield increases from Green Revolution wheat and rice depended on reallocating resources from stem to grain (Chapter 16). The resulting high-yield plants are shorter and quickly out-competed by their lower-yielding predecessors (Jennings and de Jesus, 1968).

Maize tassel size is a less-known example. Tassels consume resources directly, but they also shade leaves, reducing photosynthesis (Duncan et al., 1967). Natural selection favored large tassels because individual plants could increase their fitness by pollinating more of their neighbors. It helped that the shadows cast by tassels usually fell on the leaves of competing neighbors, rather than a plant's own leaves. Plant breeders have gradually decreased tassel size as a side effect of breeding for yield, rather than as a deliberate objective (Duvick and Cassman, 1999). There are many other examples of trade-offs between individual-plant fitness and the overall productivity and resource-use efficiency of crop communities (Donald, 1968; Denison, 2012).

What about N_2 fixation? Nitrogen fixation is a metabolically costly process, so natural selection will not have favored legumes fixing more nitrogen than a plant itself needs, even if that would benefit other plants in the community (e.g. in intercrops) or future generations (e.g. in crop rotation). In nature, future generations would include a plant's own offspring, but also the competitors of those offspring.

If we breed legume crops to fix more nitrogen than an individual plant needs, perhaps by reducing their tendency to downregulate N_2 fixation when soil nitrate is abundant, the higher cost of nitrogen fixation relative to nitrate uptake will reduce legume yield, unless photosynthate is in excess. Increased nitrogen fixation could, however, leave more nitrogen in the soil for subsequent, non-legume crops (Herridge and Rose, 2000). It is not clear whether the economic benefits of savings on fertilizer would outweigh the lower legume yield. Some of the residual nitrogen from fixing more nitrogen than the legume uses would usually be lost before it could be taken up by the next crop, reducing economic benefits and increasing pollution.

In contrast to legumes, rhizobia fix much more nitrogen in symbiosis than they need themselves, sharing most of it with their legume hosts. Why have rhizobial mutants that divert resources from N_2 fixation to their own reproduction not displaced those that fix more nitrogen?

The rhizobia infecting an individual plant may benefit collectively, receiving more photosynthate from their host, if they provide the plant with more nitrogen. This is because an increase in nitrogen fixation increases host-plant photosynthesis (Bethlenfalvay et al., 1978).

If there were only one strain of rhizobia per individual plant, that would be the end of the story. But each plant is typically infected by four to twelve different strains (West et al., 2002a). These are each other's most likely competitors for the next host. What would happen if the plant allocated photosynthate to nodules equally, regardless of each nodule's nitrogen contribution? Mathematical analysis showed that, under those conditions, 'free rider' strains of rhizobia that divert all available resources from nitrogen fixation to their own reproduction would have displaced nitrogen-fixing strains (West et al., 2002a). Why has this not happened?

The evolutionary persistence of rhizobial nitrogen fixation, we hypothesized, must be due to 'host sanctions', whereby legume responses to less beneficial nodules reduce the reproduction of the rhizobia in those nodules (Denison, 2000; West et al., 2002a). Natural selection among legumes would favor such responses, which would reduce waste of resources on nodules that fix less nitrogen (West et al., 2002b), but not necessarily to the level that would be optimal in agriculture (Denison, 2000).

Subsequent experiments confirmed that such sanctions actually exist. Rhizobia prevented from fixing nitrogen, by exposure to an N_2-free ($Ar:O_2$) atmosphere, reproduced less than rhizobia on the same soybean plant that were allowed to fix N_2 (Kiers et al. 2003). Similar results have since been seen with wild lupines (Simms et al. 2006) and with pea and alfalfa (Oono et al., 2011). The physiological mechanism of legume-imposed sanctions is unknown, although we did find that nodule gas permeability and nodule-interior O_2 were lower in the non-fixing soybean nodules (Kiers et al., 2003), which would have limited rhizobial respiration and perhaps rhizobial reproduction.

A decrease in nodule gas permeability is a common response to stress, from drought to defoliation (Weisz et al., 1985; Hartwig et al., 1987; Denison et al., 1992a; Serraj and Sinclair, 1996; Denison, 1998). The change in nodule gas permeability is not oxygen-specific as permeability to hydrogen (Witty and Minchin, 1998) and ethylene (Weisz and Sinclair, 1988) occur at the same time. Because gases diffuse much more slowly in the aqueous phase, the overall gas permeability of nodules could be consistent with either a thin but continuous aqueous shell or a thicker aqueous shell penetrated by a very few gas-filled channels. The hydrogen gas concentration in the nodule interior and the response to a change in gas composition (which would affect diffusion in the gas phase but not the aqueous phase) are both consistent with significant diffusion through air-filled channels (Denison et al., 1992b; Witty and Minchin, 1994). Various physical mechanisms for reversibly blocking these channels have been proposed, each supported by some empirical evidence (De Lorenzo et al., 1993; Denison and Kinraide, 1995; Serraj et al., 1995).

Some plant or rhizobial genes linked to nitrogen fixation may apparently be expressed at suboptimal levels, but low expression of a given enzyme is more likely due to trade-offs than to prolonged failure of natural selection to favor simple genetic changes that consistently enhance fitness. Significant improvements will either require major genetic changes – not the simple ones already tested by natural selection – or accepting trade-offs that were rejected by past natural selection.

A focus on trade-offs leads to two questions. First, how do conditions on farms today differ from the conditions where the legumes and rhizobia spent most of their recent evolutionary history? Second, how do the interests of an individual rhizobial strain and its legume host differ, with respect to this trait or this gene? Differences between present and past conditions and conflicts of interest between legumes and rhizobia represent opportunities for humans to improve nitrogen fixation beyond what past natural selection has already done.

3 RATIONALE FOR FOCUS ON EFFICIENCY OF N_2 FIXATION RATHER THAN RATE

Total agricultural nitrogen fixation could be increased by increasing the use of legumes or by increasing the amount of nitrogen fixed by current legume crops and pastures. It is interesting to contemplate a world that relies mostly on legumes for protein and calories, rather than mostly on grasses like maize, rice, and wheat. Major changes in diets would be required and total food production would probably be less. For example, maize yields are often roughly triple soybean yields (Johnson et al., 1992). Contributing factors for this difference include C_4 photosynthesis in maize and the greater synthesis cost for soybean's high protein and high lipid seeds (Penning de Vries et al., 1974; Sinclair and de Wit, 1975). Also, the metabolic cost to plants of supporting biological nitrogen fixation by rhizobia is greater than the cost of taking up and using soil nitrogen (Penning de Vries et al. 1974), assuming the latter is available.

I will argue in the next section that evolutionary trade-offs between the fitness of rhizobia versus their legume hosts have resulted in certain inefficiencies and that reversing those inefficiencies can increase nitrogen fixation without increasing the photosynthate cost to legumes. Alternatively, a legume plant could fix more nitrogen by allocating more photosynthate to N_2 fixation. This could involve making more nodules or supplying more resources to each nodule. Either approach would raise the question of whether the additional nitrogen fixed would contribute enough to yield to make up for the photosynthate consumed. The evolutionary-trade-off hypothesis predicts that increasing photosynthate allocation to nodules is more likely to reduce yields, unless N_2 fixation is more valuable on farms today than it was in the lower-fertility soils where legumes evolved. That seems unlikely.

Nodules already consume a large fraction of a legume's photosynthate, although estimates vary with species and conditions. Pea plants were estimated to send 32% of photosynthate to nodules, although almost half of that returned to the shoot in the products of fixation (Minchin and Pate, 1973). One review estimated, based on published data, that nodules used 16–22% of photosynthate during vegetative growth, but only 3–10% during pod fill (Atkins, 1984). During pod fill, however, the same review estimated that N_2 fixation only cost 1.4 gC gN^{-1} (Atkins 1984), less than half the minimal theoretical cost from the same article. So the apparent high efficiency of N_2 fixation during pod fill may be an error, perhaps reflecting use of stored resources that were not measured. Carbon costs of N_2 fixation in other studies include 4.1 gC gN^{-1} for pea (Minchin and Pate, 1973), 5.1–8.1 gC gN^{-1} for alfalfa (Twary and Heichel, 1991) and 4.5–5.8 gC gN^{-1} for the N_2-fixing non-legume, *Alnus incana* (Lundquist, 2005).

This discussion implicitly assumes that plants are carbon (photosynthate) limited often enough that greater carbon allocation to N_2 fixation could significantly limit other plant functions that require photosynthate. If increased demand for photosynthates stimulates photosynthesis enough to meet that demand, without imposing some other cost, then symbiotic nitrogen fixation (and any other carbon-consuming process) would be essentially free. Effective rhizobia can certainly increase photosynthesis more than enough to offset their carbon costs. The question is whether this is due to improved nitrogen supply, an increase in demand for carbon, or some combination of the two.

A recent paper attempted to resolve this question by comparing the effects of rhizobial inoculation on photosynthesis with its effects on leaf nitrogen, using published data (Kaschuk et al., 2009). In the published experiments selected for analysis, the increase in photosynthesis with rhizobial inoculation was greater than the increase in leaf nitrogen. Because photosynthesis shows diminishing returns with nitrogen, these results suggest that some of the stimulation of photosynthesis could be due to increased demand for photosynthate, rather than increased nitrogen supply (Kaschuk et al., 2009). This conclusion was based on only three studies, however, and in two of them (those that reported data for multiple dates), the range of values for stimulation of photosynthesis always overlapped with the range of values for increased leaf nitrogen.

Photosynthate surpluses in plants must sometimes occur and their frequency may deserve more research (Thomas and Sadras, 2001), but experiments that increase nitrogen supply and sink demand for carbon simultaneously may not be the best approach. How much does photosynthesis increase in response to treatments (like heating to increase maintenance respiration or inoculation with ineffective rhizobia) that increase demand without the complication of simultaneous benefits? Why do plants have so many photosynthesis increasing adaptations, such as moving nitrogen from lower to upper leaves, if photosynthate is in surplus?

As for stimulation of photosynthesis by inoculation with effective rhizobia, is the increase perhaps due to greater stomatal opening? A plant might keep stomata partially closed, thus conserving scarce soil water (Chapter 11), when its potential growth was limited by nitrogen supply anyway. With nitrogen supply assured by rhizobia, the risk/benefit ratio of stomatal opening changes, in ways that may not be proportional to current leaf nitrogen.

Meanwhile, additional evidence that photosynthate is often limiting comes from research with super-nodulating legume genotypes. Legumes actively regulate nodule number, based on their need for nitrogen and on the availability of nitrate or ammonium in the soil. Some rhizobial strains subvert this regulation for their own benefit. Strains that produce rhizobitoxine, which interferes with plant ethylene signaling, can increase the number of nodules per plant (Yuhashi et al., 2000). Rhizobitoxine-producing rhizobia support less plant growth than strains that do not make rhizobitoxine, but acquire more resources from the plant (Ratcliff and Denison, 2009).

Researchers have developed 'super-nodulating' legume genotypes, which have weaker regulation of nodule number and therefore produce more nodules. As predicted by the photosynthate-limitation and evolutionary-trade-off hypotheses, these genotypes almost always have lower, not higher, yield (Novák, 2010). Yields of these genotypes can be restored if over-consumption of photosynthate is alleviated by reducing nodule number, for example, by using low inoculation rates (Gresshoff et al., 1988).

So increasing nodulation beyond the number of nodules set by legumes' evolution-tuned regulatory mechanisms seems to be counterproductive. What about increasing photosynthate allocation to nodules, rather than increasing nodule number?

Legumes already have sophisticated mechanisms for allocating photosynthate to nodules, based on how much nitrogen they need and on how much nitrogen each nodule is supplying. Decreased oxygen supply to the nodule interior limits photosynthate consumption. Note that a decrease in nodule gas permeability can be triggered either by an increase in the availability of nitrate in the soil (Denison and Harter, 1995) or by a decrease in nitrogen fixation by the rhizobia inside (Kiers et al., 2003). These similar responses to conditions that have opposite effects on overall nitrogen availability may seem strange, but both responses plausibly increase a plant's evolutionary fitness.

An indiscriminate increase in photosynthate supply to nodules would undermine sanctions and waste resources on underperforming nodules. Under many conditions, this would decrease yield rather than increase it. We should focus instead on increasing efficiency, i.e. gN fixed gC^{-1} consumed.

Nitrogen-fixation efficiency could be achieved either by plants that enhance the nitrogen-fixation efficiency of whatever rhizobia are in their nodules, or by increasing nodule occupancy of more efficient rhizobial strains. These approaches will be discussed sequentially in the next two sections.

3.1 Increasing N_2-fixation efficiency of current rhizobial strains

The evolutionary-trade-off hypothesis argues that prolonged natural selection is unlikely to have missed simple, trade-off-free improvements (Denison, 2012). That still leaves at least three reasons why the efficiency of N_2 fixation might be less than optimal.

First, there are probably some potential improvements that are so complex that they have not arisen often enough, via mutation, for adequate testing by natural selection. Imagine some hypothetical protein, unrelated to existing nitrogenases, that fixes N_2 more efficiently than they do. Natural selection mostly works by tinkering with existing mechanisms (Jacob, 1977), so there is no guarantee that such an enzyme would evolve, whatever its benefits. But do humans understand enzyme function well enough to design a better nitrogenase from scratch? Maybe someday.

A second possibility is that, although sophisticated legume mechanisms to increase nitrogen-fixation efficiency evolved under natural selection, they were lost during domestication bottlenecks or perhaps during plant breeding in high-nitrogen soils (Kiers et al., 2007). This seems possible and might justify additional research on the efficiency of nitrogen fixation in wild legume species or in landraces of crop or forage legumes.

A third reason that natural selection might have failed to improve some aspects of efficiency is that conflicts of interest between legumes and rhizobia can reduce overall efficiency. In such cases, the fitness of one partner or the other may be increased by behavior that undermines overall efficiency.

For example, many rhizobia accumulate energy-rich polyhydroxybutyrate (PHB) in nodules. The most obvious benefit to rhizobia from PHB is that it can support rhizobial survival for months, or even reproduction, without an external carbon source (Ratcliff et al. 2008). PHB accumulation appears to represent a net cost to an individual plant, however. Conservation of matter and energy imply a trade-off between PHB accumulation and nitrogen fixation. Empirical evidence for this trade-off includes greater nitrogen fixation by a PHB(−) mutant (Cevallos et al., 1996) and greater PHB accumulation by a non-fixing mutant (Hahn and Studer, 1986).

What about collective benefits? It was once argued that PHB is used by rhizobia for 'continuation of N_2 fixation at high rates until the last stages of seed development' (Bergersen et al., 1991). It is true that rhizobia would benefit collectively from ensuring the survival of their legume host species. Collective benefits have little effect on evolution, however (Denison, 2012). The descendants of rhizobia infecting an isolated plant might benefit from ensuring that their host plant produced seed. Despite such collective benefits, multiple strains per plant would still create a tragedy-of-the-commons situation (Hardin, 1968). This is because, in the absence of host sanctions, strains that contribute to their host's seed production would not have any benefit relative to those that did not (West et al., 2002a).

Similarly, PHB accumulation by rhizobia might seem to offer a collective benefit to legumes – and to farmers – by enhancing rhizobial survival and reproduction. But agricultural soils typically have many more rhizobia than are needed for optimal nodulation. High rhizobial population densities may even suppress nodulation somewhat, perhaps via quorum sensing (Jitacksorn and Sadowsky, 2008; Denison and Kiers, 2011).

It appears, therefore, that natural selection among rhizobia would favor greater PHB accumulation, whereas natural selection among legumes would favor suppression of PHB accumulation by rhizobia, if that were possible. For example, legumes could benefit from producing a chemical that would diffuse into rhizobia and block PHB synthesis (Oono et al., 2009), but rhizobia might be expected to evolve resistance to this manipulation by their host.

Some legume hosts do manipulate their rhizobial symbionts in dramatic ways. They produce peptides that cause rhizobia to swell, losing the ability for further cell division, as they differentiate into the nitrogen-fixing, bacteroid form (Van de Velde et al., 2010). Not all legumes do this, but ancestral state reconstruction, based on characterization of bacteroids in nodules of 40 legume species, showed that legumes have evolved this trait at least five times (Oono et al., 2010). This repeated evolution suggests that there is some benefit to the legume host from bacteroid swelling or from effects on rhizobial physiology that are correlated with swelling.

Imposing bacteroid swelling could benefit legumes in two different ways. There could be some immediate effect, such as an increase in nitrogen-fixation efficiency (gN gC^{-1}). There could also be evolutionary effects, perhaps benefiting future generations of legumes, without necessarily providing the individual-fitness benefit needed for the trait to evolve by natural selection (Oono et al., 2009).

Only a few rhizobial strains can infect both types of host: those that impose bacteroid swelling and those that do not. *Rhizobium leguminosarum* A34 makes swollen bacteroids in pea nodules and non-swollen bacteroids in bean nodules. *Bradyrhizobium* sp. 32H1 makes swollen bacteroids in peanut nodules and non-swollen ones in cowpea nodules. Their nitrogen-fixation efficiency (N fixed per C respired) was compared between hosts, based on the marginal increase in nitrogen fixation with an increase in fixation-linked respiration (Witty et al., 1983). In each case, efficiency was greater in the host that imposed bacteroid swelling (Oono and Denison, 2010). Swollen bacteroids also supported more shoot growth g^{-1} nodule.

Why would swollen bacteroids be more efficient? Their surface:volume ratio is less, which could perhaps reduce maintenance–respiration costs linked to maintaining gradients across membranes (Penning de Vries, 1975). In addition, *R. leguminosarum* A34 bacteroids made two to four times as much PHB per cell in bean nodules, where the bacteroids were not swollen, than in pea nodules (Oono and Denison, 2010). Decreased resource allocation to PHB could help explain the greater efficiency of swollen bacteroids in pea nodules, although this difference was not seen in the peanut-vs-cowpea comparison. It is not clear why rhizobial growth in the pea-nodule environment would have an immediate effect on PHB synthesis.

Over longer periods, decreased PHB synthesis could be an evolutionary effect of bacteroid swelling. Because swollen bacteroids are unable to reproduce, at least in species that have been tested (Sutton and Paterson, 1980, Zhou et al., 1985), they would no longer enhance their fitness by diverting resources from nitrogen fixation to PHB (Oono et al., 2009).

There may be ways for non-reproductive bacteroids to increase their inclusive fitness (Hamilton, 1963), however, by enhancing reproduction of undifferentiated clonemates, which share all of their genes. For example, non-reproductive bacteroids of some rhizobial strains excrete chemicals collectively termed rhizopines (such as l-3-O-methyl-scyllo-inosamine) (Murphy et al., 1995; Wexler et al., 1995), using resources that could otherwise have been respired in support of nitrogen fixation. Rhizopines can support the growth and reproduction of undifferentiated rhizobia of the same strain and are presumably produced for their benefit.

Rhizopine beneficiaries could be related rhizobia in the soil near rhizopine-producing nodules (Olivieri and Frank, 1994; Simms and Bever, 1998), but there they would have to compete with genotypes that did not contribute to rhizopine production, including non-rhizobia (Gardener and de Bruijn, 1998). It seems more likely that the main beneficiaries are not-yet-differentiated (hence, still-reproductive) clonemates of the rhizopine-producing bacteroids in the same nodule (Denison, 2000). Either way, nitrogen-fixation efficiency might be enhanced if legumes could suppress rhizopine production by bacteroids in their nodules. Research to test these hypotheses has been proposed but not yet funded.

Hydrogen production by bacteroids is another source of inefficiency. Even under ideal conditions, at least 25% of electron transfer to nitrogenase supports H_2 production rather than N_2 fixation (Evans et al., 1987). Some rhizobia recapture some of this hydrogen, which can enhance efficiency (Albrecht et al., 1979), although strains with an uptake hydrogenase may not be consistently more beneficial (Fuhrmann, 1990). Even if they were, ensuring that nodules are occupied mostly by desirable (e.g. H_2-recycling) strains would be difficult under field conditions, as discussed below.

Fortunately, the host legume has some control over how much energy is lost to hydrogen production in the first place. Among the factors that can increase allocation to hydrogen production is hydrogen itself. Because hydrogen is produced by nitrogenase, nodule-interior hydrogen increases with nitrogenase activity, while it decreases with hydrogen diffusion out of the nodule. Nitrogenase activity depends on nodule-interior respiration, which depends on inward diffusion of oxygen. The ratio of hydrogen production to its outward diffusion (and hence the hydrogen concentration) decreases if a larger fraction of diffusion occurs through gas-filled pores, which boost the ratio of hydrogen diffusion to oxygen diffusion (Denison et al., 1992b; Moloney et al., 1994). Therefore, nodules that maintain an air-filled path from the nodule interior to the atmosphere should have greater nitrogen-fixation efficiency. The diffusion resistance of that path must remain low enough to protect nitrogenase from inactivation by oxygen, but that only occurs at concentrations well above those needed for optimal respiration in support of nitrogen fixation (Denison et al., 1992b).

Nitrogen fixation is particularly sensitive to drought (Sinclair et al., 1987), so reducing this drought sensitivity could be as useful as an overall increase in nitrogen-fixation rate or efficiency. Crossing a drought-tolerant soybean line with a high-yielding one led to one line that out-yielded check varieties under moderate drought and another that out-yielded checks under more severe drought (Sinclair et al., 2007), in preliminary field trials.

Because drought affects nitrogen fixation before it affects photosynthesis (Durand et al., 1987), drought-stressed plants may be more limited by nitrogen than by carbon (Muller et al., 2011). Therefore, increasing photosynthate allocation to nitrogen fixation under drought could be beneficial, even without an improvement in efficiency.

An increase in efficiency (relative to drought-sensitive genotypes) is not inconceivable, however. Decreased nitrogen fixation under drought is partly due to decreased nodule gas-phase permeability (Serraj and Sinclair, 1996), which could result in increased nodule-interior hydrogen concentrations and lower efficiency, as discussed previously. So, it is possible that lines that maintain nitrogen fixation under drought do so partly by slowing the decrease in nodule permeability, thereby maintaining both the rate and efficiency of nitrogen fixation.

It is not clear how useful this and other physiological information could be in crop-improvement programs. The mechanism by which some legumes impose bacteroid swelling may be well enough understood (Van de Velde et al., 2010) that this trait could someday be transferred to other legumes, perhaps increasing their nitrogen-fixation efficiency.

In general, it may be easier to select for yield on low-nitrogen soils (Herridge and Rose, 2000) than to select for physiological traits that may be difficult to assess in breeding programs. Selection or testing in both low- and higher-nitrogen conditions may be needed, however. Alfalfa selected only in the absence of soil nitrogen did not have higher yield either with or without inorganic nitrogen, whereas alternating selection with and without nitrogen increased yield under both conditions (Teuber and Phillips, 1988).

Another problem with simply breeding for yield on low-nitrogen soils is that the relative performance of different legume-crop genotypes may depend on which rhizobial strains are present in the soils. In one field test, three modern cultivars suffered much more yield loss from the presence of non-fixing rhizobia in the soil, than did three older cultivars (Kiers et al., 2007). If this difference was due to the older cultivars' better allocation of resources to the best-performing nodules, then similar results might be expected on other soils. But some legumes also prevent initial infection by a subset of less beneficial strains (Devine and Kuykendall, 1996). Both this identity-based discrimination ('partner choice') and performance-based discrimination ('sanctions') may have implications for future crops grown on the same soil, as well as for the current crop. This will be discussed in the final section.

3.2 Increasing nodule occupancy by the most efficient strains

Rhizobial strains vary greatly in net benefits to their legume hosts. If nodule occupancy by the most beneficial strains could be increased, that would probably be more effective than simply increasing overall resource allocation to nitrogen fixation. The latter could actually decrease yields, as seen with supernodulating legume genotypes (Novák, 2010).

Inoculum strains are often more efficient than indigenous ones, apparently, although assays based on growth of singly inoculated plants can confound efficiency and some aspects of competitiveness (Kiers et al., 2013). In any case, inoculum strains often occupy only a minority of nodules, under field conditions (Novák, 2010).

Increasing nodule occupancy by inoculum strains might be achieved in various ways, such as enhanced growth on root exudates (Archana, 2010) or production of bacteriocins, which kill other rhizobia (Schwinghamer and Brockwell, 1978; Robleto et al., 1998). Similarly, legume crops and forages can be selected to nodulate preferentially with a given inoculum strain (Rosas et al., 1998).

An engineered rhizobial strain that overexpressed ACC (1-aminocyclopropane-1-carboxylate) deaminase, thereby interfering with ethylene-mediated regulation of nodule number by its legume host, formed significantly more nodules (Conforte et al., 2010). As noted above, forming more nodules than the plant's normal regulatory mechanisms allow is not necessarily beneficial. But the engineered strain also formed 67% of nodules even when inoculated (as 22% of the total) with a second strain, so this approach might be used to increase nodule occupancy by more beneficial strains. Greater competitiveness is more difficult to understand than increased nodule number. If two rhizobial cells are equally close to a nodulation-receptive root, it seems unlikely that the cell interfering with host ethylene signaling would have any particular advantage over the other. Maybe competitive benefits come at a later stage, such as preventing abortion of nodules early in development. From the viewpoint of senescence, Chapter 10 explores the lifespan of nodules with emphasis on species such as lucerne with indeterminate nodules, which continuously produce new cells from a persistent meristem that replace older senescent cells, and species with determinate nodules such as soybean and bean which lack persistent meristems and have more or less simultaneous senescence of all nodule cells.

One problem with engineering inoculum strains to be more competitive (even if there is no trade-off with their nitrogen-fixation efficiency) is that inoculum strains exchange genes with indigenous strains (Sullivan et al., 1995), which are often less beneficial. Even without horizontal gene transfer, the potential fitness benefits to indigenous rhizobia from reproduction in nodules would impose strong selection for mimicking the recognition signals of inoculum strains, resisting any toxins they produce, and using any root-provided resources that were initially available only to the inoculum strain. Is there an evolution-proof way to enhance the nodulation competitiveness of an inoculum strain?

The evolutionary-trade-off hypothesis suggests a possible solution. If an inoculum strain had a trait that enhanced nodulation competitiveness, but reduced fitness in the soil, it would tend to die out during years when its legume host is not grown. It would therefore need to be applied in high numbers each time the legume is sown. This would represent a small cost to farmers, but additional profit for inoculum companies. The advantage, however, would be that horizontal transfer of this trait to less beneficial indigenous strains would reduce, rather than increase, the latter's frequency in the soil.

What traits might enhance competitiveness for nodulation, while reducing survival in the soil? Consider a rhizobial cell that detects signals from a nodulation-receptive root. Would the cell always enhance its fitness by swimming towards the root? Certainly, if it succeeds in forming a million-rhizobia nodule (Denison and Kiers, 2011). But how likely is that?

The chance of nodulation depends on how far away the root is and on whether there are other rhizobia closer to the root. Rhizobia might assess distance to the root from the strength of the root signal, while the abundance of other rhizobia can be assessed from quorum-sensing signals. In addition to competing for nodulation, other rhizobia may attract rhizobia-eating protozoa (Danso et al., 1975). If the root is distant, surrounded by other rhizobia, and patrolled by protozoa, then a rhizobial cell might enhance its fitness by waiting for a better opportunity (Denison and Kiers, 2011).

Decreased nodulation at very high rhizobial densities (Lohrke et al., 2000) and increased nodulation by strains that ignore quorum-sensing signals (Rosemeyer et al., 1998) are both consistent with the hypothesis that rhizobia may avoid roots crowded with competitors. It is not clear, however, whether there are changes to quorum sensing that would enhance competitiveness for nodulation by an inoculum strain, while reducing fitness between hosts enough to prevent the trait from increasing in indigenous rhizobia, after a few of them acquire it via horizontal gene transfer.

A better approach might be to develop legume crops that mimic the cues released by protozoa that feed on rhizobia. Those crops could then be paired with inoculum strains engineered or selected to ignore those false predator cues. Indigenous rhizobia would be deterred from approaching the crop's roots, giving the nodulation advantage to the inoculum strain. Any indigenous rhizobia that ignored the predator-mimic cues would fall prey to predators during the years between legume crops, along with the inoculum strain.

The evolutionary-trade-off hypothesis suggests that host sanctions might also be improved (Denison, 2000). Plants greatly reduce rhizobial reproduction in nodules that fail to fix any nitrogen, but rhizobia fixing N_2 at 50% of their potential do not always trigger these host sanctions (Kiers et al., 2006). Would stricter sanctions be better?

Legumes benefit from sanctions in two different ways. Plants enhance their individual fitness (assuming photosynthate limits fitness) by not wasting photosynthate on nodules that provide little nitrogen. There are also potential benefits to future generations of the same legume species growing in the same soil. Assuming that soil populations of rhizobia are dominated by rhizobia released from nodules, sanctions this year will increase nodule occupancy by more beneficial rhizobia in future years.

Natural selection, however, is almost blind to benefits to future generations, even if some of the beneficiaries would be a plant's own descendants. The immediate benefits to an individual plant from the nitrogen supplied by even a mediocre strain of rhizobia will often outweigh the cost to future generations of allowing that strain to persist in the soil (Denison, 2000). Individual legume plants are analogous to people who consider the individual benefits and costs of getting vaccinated, but not the 'herd immunity' benefits (Althouse et al., 2010).

4 CONCLUSION

Plant breeders have often selected for traits that benefit this year's crop community, while reducing individual fitness in competition. Reducing the height of wheat and rice and the size of tassels in maize are two examples, among many (Denison et al., 2003). Selection for benefits to future crops has never been a plant breeding objective, nor was it favored by past natural selection, so it may have much more untapped potential than traits that have already been subject to natural or human selection. Future crop benefits could come from plant suppression of soil-borne pathogens (beyond the point where the suppressive plant benefits itself), crop residues with improved decomposition (García-Palacios et al., 2013), beneficial effects on soil physical properties (Cresswell and Kirkegaard, 1995), inhibition of harmful nitrogen transformations in the rhizosphere (Pariasca Tanaka et al., 2010), preferential enhancement of mycorrhizal species beneficial to the next crop in a rotation (Johnson et al., 1992), or an increase in relative abundance of the most beneficial indigenous rhizobia (Denison, 2012). Breeding for benefits to future crops seems more promising than approaches that implicitly ignore past improvement by natural selection.

References

Albrecht, S.L., Maier, R.J., Hanus, F.J., Russell, S.A., Emerich, D.W., Evans, H.J., 1979. Hydrogenase in *Rhizobium japonicum* increases nitrogen fixation by nodulated soybeans. Science 203, 1255–1257.

Althouse, B.M., Bergstrom, T.C., Bergstrom, C.T., 2010. A public choice framework for controlling transmissible and evolving diseases. Proc. Natl. Acad. Sci. USA 107, 1696–1701.

Archana, G., 2010. Engineering nodulation competitiveness of rhizobialbioinoculants in soils. In: Khan, M.S. (Ed.), Microbes for legume improvement. Vienna, Springer-Verlag, p. 157.

Atkins, C.A., 1984. Efficiencies and inefficiencies in the legume/Rhizobium symbiosis – a review. Plant Soil 82, 273–284.

Avlani, P.K., Chancellor, W.J., 1977. Energy requirements for wheat production and use in California. Transact. ASAE 20, 429–437.

Bergersen, F.J., Peoples, M.B., Turner, G.L., 1991. A role for poly-β-hydroxybutyrate in bacteroids of soybean nodules. Proc. R. Soc. B 245, 59–64.

Bethlenfalvay, G.J., Abu-Shakra, S.S., Phillips, D.A., 1978. Interdependence of nitrogen nutrition and photosynthesis in *Pisum sativum* L. II. Host plant response to nitrogen fixation by Rhizobium strains. Plant Physiol. 62, 131–133.

Cevallos, M.A., Encarnación, S., Leija, A., Mora, Y., Mora, J., 1996. Genetic and physiological characterization of a *Rhizobium etli* mutant strain unable to synthesize poly-β-hydroxybutyrate. J. Bacteriol. 178, 1646–1654.

Chaney, D.E., Drinkwater, L.E., Pettygrove, G.S., 1992. Organic soil amendments and fertilizers. Oakland, University of California.

Conforte, V.P., Echeverria, M., Sanchez, C., Ugalde, R.A., Menendez, A.B., Lepel, V.C., 2010. Engineered ACC deaminase-expressing free-living cells of *Mesorhizobium loti* show increased nodulation efficiency and competitiveness on *Lotus* sp. J. Gen. Appl. Microbiol. 56, 331–338.

Connor, D.J., Loomis, R.S., Cassman, K.G., 2011. Crop ecology: productivity and management in agricultural systems. Cambridge, Cambridge University Press.

Cresswell, H.P., Kirkegaard, J.A., 1995. Subsoil amelioration by plant roots – the process and the evidence. Aust. J. Soil Res. 33, 221–239.

Danso, S.K.A., Keya, S.O., Alexander, M., 1975. Protozoa and the decline of *Rhizobium* populations added to soil. Can. J. Microbiol. 21, 884–895.

De Lorenzo, C., Iannetta, P.P.M., Fernandez-Pascual, M., et al., 1993. Oxygen diffusion in lupin nodules. II. Mechanisms of diffusion barrier operation. J. Exp. Bot. 44, 1469–1474.

Denison, R.F., 2012. Darwinian agriculture: how understanding evolution can improve agriculture. Princeton, Princeton University Press.

Denison, R.F., 2000. Legume sanctions and the evolution of symbiotic cooperation by rhizobia. Am. Nat. 156, 567–576.

Denison, R.F., 1998. Decreased oxygen permeability: a universal stress response in legume root nodules. Bot. Acta. 111, 191–192.

Denison, R.F., Kiers, E.T., 2011. Life-histories of rhizobia and mycorrhizal fungi. Curr. Biol. 21, R775–R785.

Denison, R.F., Kiers, E.T., 2005. Sustainable crop nutrition: constraints and opportunities. In: Broadley, M. (Ed.), Plant nutritional genomics. Oxford, Blackwell Publishing, pp. 242–264.

Denison, R.F., Kinraide, T.B., 1995. Oxygen-induced depolarizations in legume root nodules. Possible evidence for an osmoelectrical mechanism controlling nodule gas permeability. Plant Physiol. 108, 235–240.

Denison, R.F., Harter, B.L., 1995. Nitrate effects on nodule oxygen permeability and leghemoglobin. Nodule oximetry and computer modeling. Plant Physiol. 107, 1355–1364.

Denison, R.F., Kiers, E.T., West, S.A., 2003. Darwinian agriculture: when can humans find solutions beyond the reach of natural selection? Q. Rev. Biol. 78, 145–168.

Denison, R.F., Hunt, S., Layzell, D.B., 1992a. Nitrogenase activity, nodule respiration and O_2 permeability following detopping of alfalfa and birdsfoot trefoil. Plant Physiol. 98, 894–900.

Denison, R.F., Witty, J.F., Minchin, F.R., 1992b. Reversible O_2 inhibition of nitrogenase activity in attached soybean nodules. Plant Physiol. 100, 1863–1868.

Devine, T.E., Kuykendall, L.D., 1996. Host genetic control of symbiosis in soybean (*Glycine max* L). Plant Soil 186, 173–187.

Donald, C.M., 1968. The breeding of crop ideotypes. Euphytica 17, 385–403.

Duncan, W.G., Williams, W.A., Loomis, R.S., 1967. Tassels and productivity of maize. Crop Sci. 7, 37–39.

Durand, J., Sheehy, J.E., Minchin, F.R., 1987. Nitrogenase activity, photosynthesis and nodule water potential in soyabean plants experiencing water deprivation. J. Exp. Bot. 38, 311–321.

Duvick, D.N., Cassman, K.G., 1999. Post-green-revolution trends in yield potential of temperate maize in the north-central United States. Crop Sci. 39, 1622–1630.

Evans, H.J., Harker, A.R., Papen, H., Russell, S.A., Hanus, F.J., Zuber, M., 1987. Physiology, biochemistry, and genetics of the uptake hydrogenase in rhizobia. Annu. Rev. Microbiol. 41, 335–361.

Fuhrmann, J., 1990. Symbiotic effectiveness of indigenous soybean bradyrhizobia as related to serological, morphological, rhizobitoxine, and hydrogenase phenotypes. Appl. Environ. Microbiol. 56, 224–229.

García-Palacios, P., Milla, R., Delgado-Baquerizo, M., Martín-Robles, N., Álvaro-Sánchez, M., Wall, D.H., 2013. Side-effects of plant domestication: ecosystem impacts of changes in litter quality. New Phytol. 198, 504–513.

Gardener, B.B., de Bruijn, F.J., 1998. Detection and isolation of novel rhizopine-catabolizing bacteria from the environment. Appl. Environ. Microbiol. 64, 4944–4949.

Gresshoff, P.M., Krotzky, A., Mathews, A., et al., 1988. Suppression of the symbiotic supernodulation symptoms of soybean. J. Plant Physiol. 132, 417–423.

Hahn, M., Studer, D., 1986. Competitiveness of a *nif-Bradyrhizobium japonicum* mutant against the wild-type strain. FEMS Microbiol. Lett. 33, 143–148.

Hamilton, W.D., 1963. The evolution of altruistic behavior. Am. Nat. 97, 354–356.

Hardin, G., 1968. The tragedy of the commons. Science 162, 1243–1248.

Hartwig, U., Boller, B., Nösberger, J., 1987. Oxygen supply limits nitrogenase activity of clover nodules after defoliation. Ann. Bot. 59, 285–291.

Herridge, D., Rose, I., 2000. Breeding for enhanced nitrogen fixation in crop legumes. Field Crops Res. 65, 229–248.

Jacob, F., 1977. Evolution and tinkering. Science 196, 1161–1166.

Jennings, P.R., de Jesus, J., 1968. Studies on competition in rice. I. Competition in mixtures of varieties. Evolution 22, 119–124.

Jitacksorn, S., Sadowsky, M.J., 2008. Nodulation gene regulation and quorum sensing control density-dependent suppression and restriction of nodulation in the *Bradyrhizobium japonicum*-soybean symbiosis. Appl. Environ. Microbiol. 74, 3749–3756.

Johnson, N.C., Copeland, P.J., Crookston, R.K., Pfleger, F.L., 1992. Mycorrhizae: possible explanation for yield decline with continuous corn and soybean. Agron. J. 84, 387–390.

Kaschuk, G., Kuyper, T.W., Leffelaar, P.A., Hungria, M., Giller, K.E., 2009. Are the rates of photosynthesis stimulated by the carbon sink strength of rhizobial and arbuscular mycorrhizal symbioses? Soil Biol. Biochem. 41, 1233–1244.

Kiers, E.T., Ratcliff, W.C., Denison, R.F., 2013. Single-strain inoculation may create spurious correlations between legume fitness and rhizobial fitness. New Phytol. 198, 4–6.

Kiers, E.T., Hutton, M.G., Denison, R.F., 2007. Human selection and the relaxation of legume defenses against ineffective rhizobia. Proc. R. Soc. B. 274, 3119–3126.

Kiers, E.T., Rousseau, R.A., Denison, R.F., 2006. Measured sanctions: legume hosts detect quantitative variation in rhizobium cooperation and punish accordingly. Evol. Ecol. Res. 8, 1077–1086.

Kiers, E.T., Rousseau, R.A., West, S.A., Denison, R.F., 2003. Host sanctions and the legume-rhizobium mutualism. Nature 425, 78–81.

Koch, M.A., Haubold, B., Mitchell-Olds, T., 2000. Comparative evolutionary analysis of chalcone synthase and alcohol dehydrogenase loci in Arabidopsis, Arabis, and related genera *(Brassicaceae)*. Mol. Biol. Evol. 17, 1483–1498.

Li, Y.X., et al., 2013. An analysis of China's fertilizer policies: impacts on the industry, food security, and the environment. J. Environ. Qual. 42, 972–981.

Lohrke, S.M., Madrzak, C.J., Hur, H.G., Judd, A.K., Orf, J.H., Sadowsky, M.J., 2000. Inoculum density-dependent restriction of nodulation in the soybean–*Bradyrhizobium* symbiosis. Symbiosis 29, 59–70.

Lundquist, P.O., 2005. Carbon cost of nitrogenase activity in *Frankia–Alnusincana* root nodules. Plant Soil. 273, 235–244.

Minchin, F.R., Pate, J.S., 1973. The carbon balance of a legume and the functional economy of its root nodules. J. Exp. Bot. 24, 259–271.

Moloney, A.H., Guy, R.D., Layzell, D.B., 1994. A model of the regulation of nitrogenase electron allocation in legume nodules: ii. comparison of empirical and theoretical studies in soybean. Plant Physiol. 104, 541–550.

Muller, B., et al., 2011. Water deficits uncouple growth from photosynthesis, increase C content, and modify the relationships between C and growth in sink organs. J. Exp. Bot. 62, 1715–1729.

Murphy, P.J., Wexler, W., Grzemski, W., Rao, J.P., Gordon, D., 1995. Rhizopines – their role in symbiosis and competition. Soil Biol. Biochem. 27, 525–529.

Nelson, T., 2007. Prairie project may reshape the renewable fuel landscape. Solutions 1, 14–16.

Novák, K., 2010. On the efficiency of legume supernodulating mutants. Ann. Appl. Biol. 157, 321–342.

Olivieri, I., Frank, S.A., 1994. The evolution of altruism in rhizobium: altruism in the rhizosphere. J. Hered. 85, 46–47.

Oono, R., Denison, R.F., 2010. Comparing symbiotic efficiency between swollen versus nonswollen rhizobial bacteroids. Plant Physiol. 154, 1541–1548.

Oono, R., Anderson, C.G., Denison, R.F., 2011. Failure to fix nitrogen by non-reproductive symbiotic rhizobia triggers host sanctions that reduce fitness of their reproductive clonemates. Proc. R. Soc. B 278, 2698–2703.

Oono, R., Denison, R.F., Kiers, E.T., 2009. Tansley review: Controlling the reproductive fate of rhizobia: how universal are legume sanctions? New Phytol. 183, 967–979.

Oono, R., Schmitt, I., Sprent, J.I., Denison, R.F., 2010. Multiple evolutionary origins of legume traits leading to extreme rhizobial differentiation. New Phytol. 187, 508–520.

Pariasca Tanaka, J., Nardi, P., Wissuwa, M., 2010. Nitrification inhibition activity, a novel trait in root exudates of rice. AoB Plants, 2010:plq014, doi:10.1093/aobpla/plq014.

Penning de Vries, F.W.T., 1975. The cost of maintenance processes in plant cells. Ann. Bot. 39, 77–92.

Penning de Vries, F.W.T., Brunsting, A.H.M., Van Laar, H.H., 1974. Products, requirements and efficiency of biosynthesis: A quantitative approach. J. Theor. Biol. 45, 339–377.

Pimentel, D., Berardi, G., Fast, S., 1984. Energy efficiencies of farming wheat, corn, and potatoes organically. In: Bezdicek, D.F., Power, J.F., Keeney, D.R., Wright, M.J. (Eds.), Organic farming: current technology and its role in a sustainable agriculture. Madison, American Society of Agronomy.

Ratcliff, W.C., Denison, R.F., 2009. Rhizobitoxine producers gain more poly-3-hydroxybutyrate in symbiosis than do competing rhizobia, but reduce plant growth. ISME J. 3, 870–872.

Ratcliff, W.C., Kadam, S.V., Denison, R.F., 2008. Polyhydroxybutyrate supports survival and reproduction in starving rhizobia. FEMS Microbiol. Ecol. 65, 391–399.

Robleto, E.A., Kmiecik, K., Oplinger, E.S., Nienhuis, J., Triplett, E.W., 1998. Trifolitoxin production increases nodulation competitiveness of *Rhizobium etli* CE3 under agricultural conditions. Appl. Environ. Microbiol. 64, 2630–2633.

Rosas, J.C., Castro, J.A., Robleto, E.A., Handelsman, J., 1998. A method for screening *Phaseolus vulgaris* L. germplasm for preferential nodulation with a selected *Rhizobium etli* strain. Plant Soil 203, 71–78.

Rosemeyer, V., Michiels, J., Verreth, C., Vanderleyden, J., 1998. *luxI*- and *luxR*- homologous genes of *Rhizobium etli* CNPAF512 contribute to synthesis of autoinducer molecules and nodulation of *Phaseolus vulgaris*. J. Bacteriol. 180, 815–821.

Roth, J.A., Caimpitti, I.A., Vyn, T.J., 2013. Physiological evaluations of recent drought-tolerant maize hybrids at varying stress levels. Agron. J. 105, 1129–1141.

Schwinghamer, E.A., Brockwell, J., 1978. Competitive advantage of bacteriocin and phage-producing strains of *Rhizobium trifolii* in mixed culture. Soil Biol. Biochem. 10, 383–387.

Serraj, R., Sinclair, T.R., 1996. Inhibition of nitrogenase activity and nodule oxygen permeability by water deficit. J. Exp. Bot. 47, 1067–1073.

Serraj, R., Fleurat-Lessard, P., Jaillard, B., Drevon, J.J., 1995. Structural changes in the inner-cortex cells of soybean root nodules are induced by short-term exposure to high salt or oxygen concentrations. Plant Cell Environ. 18, 455–462.

Simms, E.L., Bever, J.D., 1998. Evolutionary dynamics of rhizopine within spatially structured rhizobium populations. Proc. R. Soc. B 265, 1713–1719.

Simms, E.L., Taylor, D.L., Povich, J., et al., 2006. An empirical test of partner choice mechanisms in a wild legume-rhizobium interaction. Proc. R. Soc. B 273, 77–81.

Sinclair, T.R., de Wit, C.T., 1975. Photosynthate and nitrogen requirements for seed production by various crops. Science 189, 565–567.

Sinclair, T.R., Muchow, R.C., Bennett, J.M., Hammond, L.C., 1987. Relative sensitivity of nitrogen and biomass accumulation to drought in field-grown soybean. Agron. J. 79, 986–991.

Sinclair, T.R., Purcell, L.C., King, C.A., Sneller, C.H., Chen, P., Vadez, V., 2007. Drought tolerance and yield increase of soybean resulting from improved symbiotic N2 fixation. Field Crops Res. 101, 68–71.

Streeter, J., 1988. Inhibition of legume nodule formation and N_2 fixation by nitrate. CRC Crit. Rev. Plant Sci. 7, 1–23.

Sullivan, J.T., Patrick, H.N., Lowther, W.L., Scott, D.B., Ronson, C.W., 1995. Nodulating strains of *Rhizobium loti* arise through chromosomal symbiotic gene transfer in the environment. Proc. Natl. Acad. Sci. USA 92, 8985–8989.

Sutton, W.D., Paterson, A.D., 1980. Effects of the host plant on the detergent sensitivity and viability of *Rhizobium* bacteroids. Planta 148, 287–292.

Teuber, L.R., Phillips, D.A., 1988. Influences of selection method and nitrogen environment on breeding alfalfa for increased forage yield and quality. Crop Sci. 28, 599–604.

Thomas, H., Sadras, V.O., 2001. The capture and gratuitous disposal of resources by plants. Funct. Ecol. 15, 3–12.

Twary, S.N., Heichel, G.H., 1991. Carbon costs of dinitrogen fixation associated with dry matter accumulation in alfalfa. Crop Sci. 31, 985–992.

Van de Velde, W., Zehirov, G., Szatmari, A., et al., 2010. Plant peptides govern terminal differentiation of bacteria in symbiosis. Science 327, 1122–1126.

Weisz, P.R., Sinclair, T.R., 1988. A rapid non-destructive assay to quantify soybean nodule gas permeability. Plant Soil 105, 69–78.

Weisz, P.R., Denison, R.F., Sinclair, T.R., 1985. Response to drought stress of nitrogen fixation (acetylene reduction) rates by field-grown soybeans. Plant Physiol. 78, 525–530.

West, S.A., Kiers, E.T., Simms, E.L., Denison, R.F., 2002a. Sanctions and mutualism stability: why do rhizobia fix nitrogen? *Proc.* R. Soc. B 269, 685–694.

West, S.A., Kiers, E.T., Pen, I., Denison, R.F., 2002b. Sanctions and mutualism stability: when should less beneficial mutualists be tolerated? *J.* Evol. Biol. 15, 830–837.

Wexler, M., Gordon, D., Murphy, P.J., 1995. The distribution of inositol rhizopine genes in *Rhizobium* populations. Soil Biol. Biochem. 27, 531–537.

Witty, J.F., Minchin, F.R., 1998. Hydrogen measurements provide direct evidence for a variable physical barrier to gas diffusion in legume nodules. J. Exp. Bot. 49, 1015–1020.

Witty, J.F., Minchin, F.R., 1994. A new method to detect the presence of continuous gas-filled pathways for oxygen diffusion in legume nodules. J. Exp. Bot. 45, 967–978.

Witty, J.F., Minchin, F.R., Sheehy, J.E., 1983. Carbon costs of nitrogenase activity in legume root nodules determined using acetylene and oxygen. J. Exp. Bot. 34, 951–963.

Yuhashi, K.I., Ichikawa, N., Ezura, H., et al., 2000. Rhizobitoxine production by *Bradyrhizobium elkanii* enhances nodulation and competitiveness on *Macroptilium atropurpureum*. Appl. Environ. Microbiol. 66, 2658–2663.

Zhou, J.C., Tchan, Y.T., Vincent, J.M., 1985. Reproductive capacity of bacteroids in nodules of *Trifolium repens* (L.) and Glycine max (L.) Merr. Planta 163, 473–482.

Zhu, X.G., Portis, A.R., Long, S.P., 2004. Would transformation of C3 crop plants with foreign Rubisco increase productivity? A computational analysis extrapolating from kinetic properties to canopy photosynthesis. Plant Cell Environ. 27, 155–165.

CHAPTER

10

Senescence and crop performance

Howard Thomas, Helen Ougham

IBERS, Edward Llwyd Building, Aberystwyth University, Ceredigion, UK

1 INTRODUCTION

There is a graphic on the home page of the website for IRRI, the International Rice Research Institute in the Philippines (www.irri.org), entitled 'Inconvenient Divergence'. It consists of two constantly updated counters, one showing world population, the other the total global area of productive land. The numbers are going in different directions, relentlessly, minute by minute. For each hectare of growing area lost, the world adds about 20 new mouths to feed. Malthus is out of fashion with policy-makers (Dorling, 2013). Nevertheless, every day around the world, men and women are in fields, laboring to solve this unfavorable population–crop equation, and to do this they need every tool that agricultural science can provide.

As well as farmers, agronomists and researchers, our crops have to work harder. We need agricultural species to capture and convert resources ever more efficiently, offsetting the dwindling area of cultivatable land by extending the growing season and exploiting stressful environments. To do this, the crop has to establish early and quickly, through rapid growth and development of the whole plant and its parts. The crop must be adapted and adaptable to non-optimal environments. And it must be prolific, yielding usable products of high pre- and post-harvest quality. Senescence is an essential element in each of these aspects of crop performance.

The following discussion begins with the role of senescence in the development of crop and other plants, and considers how a senescence phase is programmed into the fate of plant structures at all levels of organization. There follows an examination of senescence as a strategic and tactical measure in the face of biotic and abiotic environmental challenge. Finally, the contribution of senescence to crop yield and quality is discussed, with particular emphasis on carbon capture, nitrogen recycling and the mechanisms of resource allocation.

2 SENESCENCE AND DEVELOPMENT

Senescence and development interact at different levels. Senescence is part of the program that specifies cell fate. It is triggered differentially in tissues and organs, resulting in complex anatomies and morphologies that change

Crop Physiology. DOI: 10.1016/B978-0-12-417104-6.00010-8

and adapt over time (Gunawardena, 2008). Senescence is the means by which resources are recycled from obsolete body parts to new developing structures (Feller et al., 2007; Guiboileau et al., 2010). And variations on the senescence program theme have been shaped by evolution to give rise to a diversity of structures within the angiosperm life cycle (Thomas et al., 2009).

2.1 Anatomical development through localized senescence

The senescence and death of the right cells at the right time is necessary for normal plant development and survival. The purposeful nature of developmental cell elimination implies tight regulation. Programmed cell death is often used as the general term for any process by which plant protoplasm, and sometimes the associated cell wall, is selectively removed as part of a genetically determined event (Lord and Gunawardena, 2012). Death is not the inevitable outcome of senescence in plant cells (Thomas et al., 2003). The cell senescence program is executed through propagating metabolic cascades that require biological energy and are under the control of specific genes. There are probably almost as many variants of programmed senescence and death as there are cell types and terminal events in the life cycles of plants and their parts (Reape et al., 2008; van Doorn et al., 2011). A plant cell might be induced to senesce and die by a neighbor, and a severe external stress can cause fatal trauma but, in most cases, cell elimination is a form of suicide, an autolytic process in which cytoplasm is both the source and the location of the degradative activities that ultimately bring about its own destruction. Vacuoles have an essential function in most kinds of autolytic cell senescence (Matile, 1997; Hatsugai et al., 2006; Müntz, 2007). Macromolecules are either engulfed by, or transported across, the tonoplast and degraded by peptidases, nucleases, peroxidases and other hydrolytic and oxidative enzymes in the vacuolar space, or else the tonoplast ruptures, flooding the cytosol with lytic enzymes and rapidly killing the cell. Table 10.1 lists the varieties of autolytic plant cell senescence, and includes two major forms of animal cell death for comparison.

To exchange resources with the environment and transport them to where they are needed for growth and development, plants differentiate systems of holes and tubes by means of localized, developmentally-regulated senescence and death of groups of cells (Kozela and Regan, 2003). Similarly, plant shapes and adaptations for different purposes can be achieved by controlled cell death and hollowing out or shedding of parts. Lysigeny, the disintegration of cells to form glands, channels and secretory ducts, is often accompanied by schizogeny, cell separation (Turner, 1999; Pickard, 2008; Liu et al., 2012). The tracheary elements and fibers of xylem are formed by distinctive autolytic cell death processes accompanied by lignification (Courtois et al., 2009; Ohashi-Ito and Fukuda, 2010); Chapter 11 further examines the anatomy of the xylem, its implications for water delivery in roots and leaves, and the links between water supply for transpiration and carbon assimilation. Abscission leading to the shedding of organs, and dehiscence of dry fruits to disperse seeds and anthers to release pollen, are cell-separation processes resembling schizogeny (Roberts et al., 2002).

2.2 Leaf senescence

Perception and mechanistic understanding of foliar senescence are largely conditioned by the behavior of mesophyll, the dominant tissue of leaves. Of the modifications that mesophyll cells undergo during senescence, those affecting the chloroplasts are the most dramatic. Leaf chloroplasts lose thylakoid membranes and form large lipid-rich plastoglobules as they redifferentiate into gerontoplasts (Parthier, 1988; Besagni and Kessler, 2013). Other mesophyll cell organelles, notably peroxisomes and vacuoles, also experience structural and functional alterations.

TABLE 10.1 Modes of autolytic cell senescence in plants. Apoptosis and necrosis, the two principal forms of cell death in animals, are listed for comparison

Type of cell senescence	Characteristics
Autophagy	A form of cell senescence characterized by the regulated assembly of specific lytic structures that break down cytoplasm under the control of signal cascades and differential gene expression. Autophagy also occurs in animals and is distinct from apoptosis
Transdifferentiation	Remodeling of structure and function of cytoplasmic organelles in post-mitotic cells. Only in the final stages is there loss of integrity and viability as lytic processes take control. Examples include senescence of green cells of leaves or pigmented tissues of fruits
Hypersensitivity	A kind of cauterization or containment reaction to attempted infection by a pathogen. Related to autophagy and to some kinds of spontaneous lesion formation in mutants
Lysigeny	The formation of glands, channels and secretory ducts by the disintegration of cytoplasm. Air spaces formed in roots in response to low oxygen stress are lysigenous in origin
Schizogeny	Senescence of schizogenous cells is a process of cell separation during which the middle lamella of the cell wall breaks down
Pseudosenescence	Light-dependent bleaching response of green cells to stress, superficially resembling senescence but expressing changes that are symptoms of physiological distress and declining viability
Apoptosis	Type of programmed cell death in animals characterized by blebbing, cell shrinkage, nuclear fragmentation, chromatin condensation, and DNA fragmentation
Necrosis	Traumatic cell death in animals resulting from acute cellular injury. Non-physiological mortality of plant cells in response to trauma resembles necrosis in animals

These modifications to mesophyll ultrastructure are accompanied by distinctive and highly visible changes in pigment complement, as the pathway of chlorophyll catabolism is activated (Fig. 10.1; Hörtensteiner and Kräutler, 2011). Salvaging metabolites and structural components, particularly those that are reserves of nitrogen and phosphorus, becomes a priority. As a consequence of the shift from chloroplasts to gerontoplasts, current photosynthesis becomes a declining source of the energy required for remobilization and other metabolic processes in senescing mesophyll cells (Hidema et al., 1991), and energy requirements are increasingly met by catabolism, associated with significant changes to respiratory and oxidative pathways (Keskitalo et al., 2005). In many species, there is also synthesis or modification of antibiotic or antiherbivore compounds that protect potentially vulnerable senescing leaf tissues against pathogens or pests (Watanabe et al., 2013).

Initiation and execution of the leaf senescence program are directed by *SAGs* (senescence-associated genes). Mutations in *SAGs* interfere with normal senescence, frequently revealing themselves by abnormal retention of chlorophyll, a trait referred to as stay-green (Thomas and Howarth, 2000; Barry, 2009; Thomas and Ougham, 2014); in a crop context, the stay-green syndrome has a different meaning, as outlined in section 4. The activation and progression of senescence and the expression of *SAGs* are under hormonal control. Cytokinins are senescence antagonists. Leaf lifespan in the *Arabidopsis* mutant *ore12-1* is extended as a result of interference with the cytokinin signal transduction pathway for the antisenescence hormone cytokinin (Kim et al., 2006). Plants transformed with a gene encoding isopentenyl transferase, a limiting step in cytokinin biosynthesis, fused with the promoter region of a *SAG*, make endogenous cytokinin in an autoregulated fashion. The result is

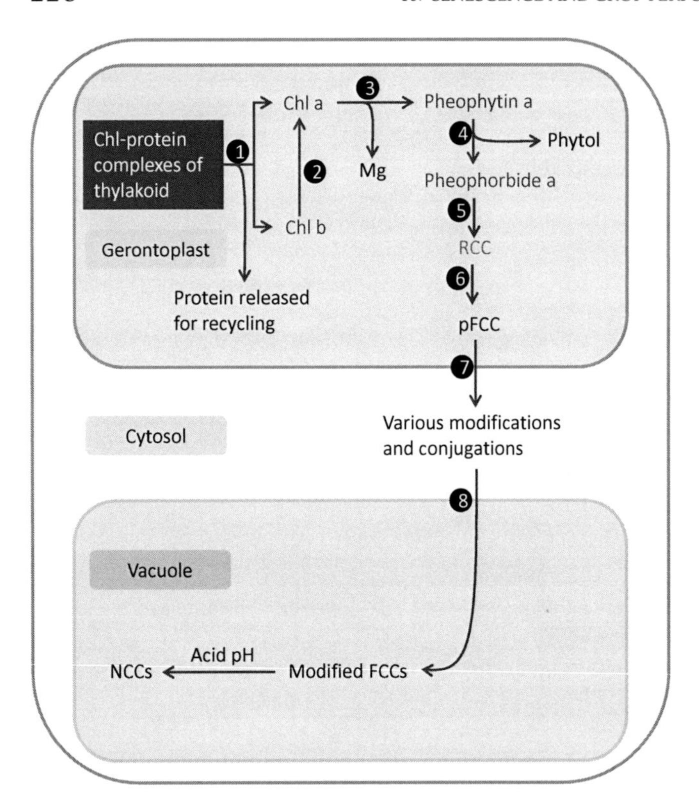

FIG. 10.1 The pathway and subcellular organization of chlorophyll breakdown during senescence. Enzymes and activities are identified as follows: (1) STAY-GREEN (encoded by the gene *SGR*); (2) chlorophyll b reductase; (3) dechelation reaction; (4) pheophytinase; (5) pheophorbide a oxygenase (*ACD1*); (6) RCC reductase (*ACD2*); (7) ATP-dependent catabolite transporter; (8) ABC transporter. Chl, chlorophyll; RCC, red Chl catabolite; (p)FCC, (primary) fluorescent Chl catabolite; NCC, non-fluorescent Chl catabolite.

delayed leaf senescence, an extension of photosynthetic activity, increased seed production, and improvements in stress tolerance in controlled environments (Gan and Amasino, 1995; Robson et al., 2004; Ha et al., 2012). In the absence of evidence with crop species under realistic field conditions, the importance of these traits for agriculture is as yet uncertain.

Other hormones have been implicated in foliar senescence in some species and tissues. Ethylene is a promoter of leaf senescence in *Arabidopsis* and a number of other plants. Senescence in these species is altered by chemical and genetic interference with ethylene physiology (Pierik et al., 2006). Of many other compounds with hormone-like influences on leaf senescence, jasmonic acid (JA) and salicylic acid (SA) are of particular interest, since they connect senescence with responses to herbivores and microbial pathogens (Morris et al., 2000; He et al., 2002; Thomas et al., 2013).

Cellular redox conditions are common factors in the signaling pathways by which leaf senescence responds to hormones and environmental factors. Reactive oxygen species (ROS) are of particular significance for redox control. The amount of ROS generated during plant metabolism often increases with tissue age. H_2O_2 is an ROS that originates as a product of normal enzymic reactions in peroxisomes, chloroplasts and other organelles (Zimmermann and Zentgraf, 2005). Among the redox-sensitive genes is *WRKY53*, which encodes a transcription factor. WRKY53 is induced by H_2O_2 and autoregulates its own synthesis by feedback inhibition. It interacts with a large number of genes of various kinds, including those encoding various SAGs and components of the SA and JA signaling networks (Miao and Zentgraf, 2007). If ROS are allowed to propagate without restraint, they cease to act as finely modulated signaling components and instead have the potential to damage and even kill cells. As the tight regulation of oxidation and ROS propagation relaxes during the final stages of senescence, viability is lost and finally cell death occurs (Van Breusegem and Dat, 2006).

2.3 Root senescence

Because the rhizosphere is largely hidden from sight, it is not readily apparent that senescence of root systems is at least as agriculturally and ecologically significant as senescence of above-ground parts. Root lifespan and the dynamics of subterranean tissue turnover have become a focus for research studies of carbon sequestration and global climate change (McCormack et al., 2013). Senescence of individual tissues contributes to building root systems. For example, cell differentiation in root vascular tissue follows the same route towards lignification and elimination of cytoplasm as that seen during xylogenesis in the shoot. It has sometimes been claimed that cells of the root cap undergo programmed death as they are shed during movement of the growing root through the soil. There are reports, however, that such cells remain viable and can be cultured to form callus tissue *in vitro* (Hawes and Pueppke, 1986).

Roots can slough off entire epidermal and cortical regions as they get older, but transport functions are retained in the stele and there may even be initiation of new lateral roots from the pericycle (Spaeth and Cortes, 1995). In a study of the patterns of senescence in maize (*Zea mays*) root tissues, the cortex of the main root was found to be alive for the whole lifespan of the plant, whereas the lifespan of the root hairs was only 2–3 days. Senescence of laterals in the older part of the first order root commences when the plant is at the six-leaf stage (Fusseder, 1987). Association with mycorrhizas tends to enhance root lifespan, whereas there is a negative correlation between root longevity and N content. Environmental stresses promote root cell senescence under some circumstances (Jovanovic et al., 2008). Drought often invokes an autophagy-like response in cells of the root tip, and flooding and hypoxia induce lysigenous air space formation in many species (Yamauchi et al., 2013).

The N-fixing root nodules of legumes can be determinate or indeterminate, depending on the host plant species. In most monocarpic legumes, the end of the nodule's lifespan coincides with pod filling. Indeterminate nodules, which continuously produce new cells from a persistent meristem that replaces older senescent cells, are found in species such as alfalfa (*Medicago sativa*) and pea (*Pisum sativum*). Developmental cell senescence in indeterminate nodules is a two-stage process, during which various genes related to defense and stress are activated, including cysteine proteases and enzymes of ROS metabolism. It begins with degradation and resorption of bacteroids and is followed by autolysis of host cells and salvage of materials (Guerra et al., 2010). The nodules of soybean (*Glycine max*) and bean (*Phaseolus vulgaris*) are determinate: they lack persistent meristems and senescence of all nodule cells occurs more or less simultaneously (Alesandrini et al., 2003). Depressed N fixation and accelerated senescence in determinate and indeterminate nodules are enhanced by stresses such as exposure to prolonged darkness, drought or high soil nitrate (Puppo et al., 2005).

2.4 Senescence of reproductive structures

Selective senescence of cells and tissues contributes to all stages of sexual reproduction (Fig. 10.2), from differentiation of male and female floral parts and gametes, through development of structures attractive to pollinators, to embryogenesis, seed maturation and fruit ripening. Like leaves, floral parts undergo senescence as a terminal phase of development, during which macromolecules break down and organelle structures and functions are modified (Rogers, 2013). Ethylene promotes flower senescence in many species, but not all. In some cases,

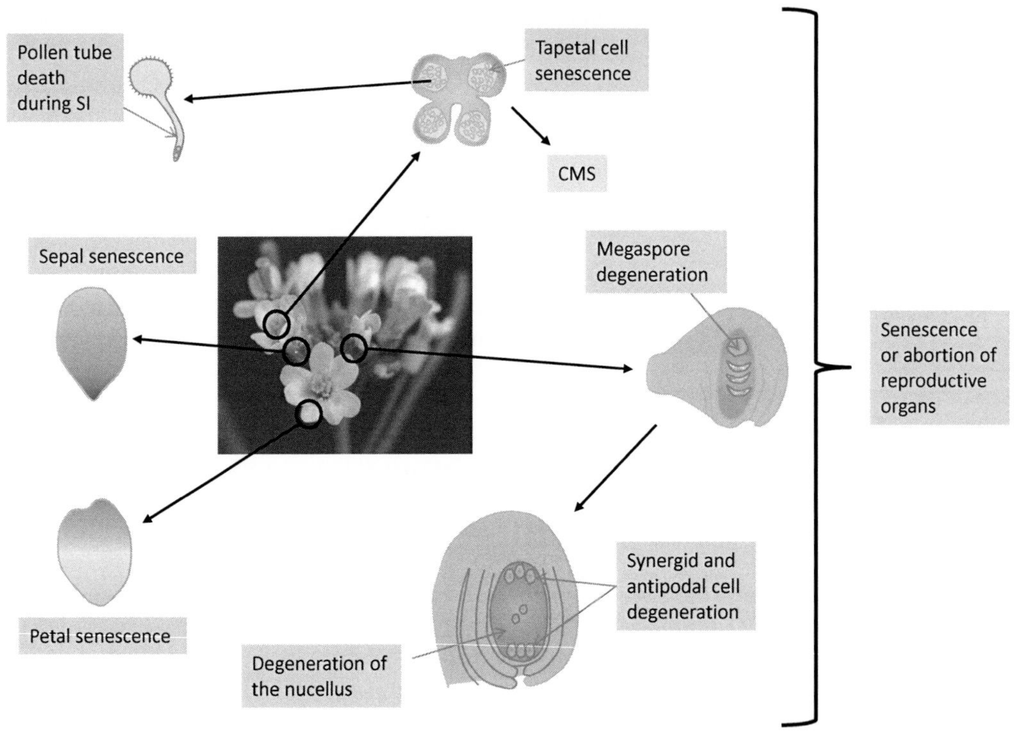

FIG. 10.2 Sites of autolytic cell death during development and senescence of floral organs. SI, self-incompatibility; CMS, cytoplasmic male sterility.

a surge of ethylene production follows pollination. Elimination of specific cells is essential for the formation of separate male and female flowers in sexually polymorphic plants. In maize and other monoecious and dioecious species, flowers at an early stage of development are indistinguishable, each bearing primordia for both male and female organs. Subsequently cell senescence results in the death of male or female tissues, giving rise to, respectively, a female or a male inflorescence (Diggle et al., 2011).

The development of tissues in the haploid phase of the angiosperm lifecycle is influenced by programmed cell senescence and death. Three of the four megaspores formed after meiosis of the megaspore mother cell are removed, leaving one megaspore that gives rise to the egg and other components of the embryo sac (Bell, 1996). Senescence leading to cell elimination is also an essential factor in pollen grain development. The tapetum undergoes a distinctive type of cell senescence resembling animal-type apoptotic cell death (Parish and Li, 2010). Fertilization in most angiosperms is followed by mitotic division of the zygote to produce one cell destined to develop into the embryo and another that undergoes a few rounds of mitosis to form the suspensor. Eventually, senescence is initiated in suspensor cells and they are deleted (Bozhkov et al., 2005). The endosperm of cereal grains consists of two cell types, starchy endosperm and aleurone, development of which culminates in distinctive developmentally-regulated cell senescence and death (Fath et al., 2000; Young and Gallie, 2000). Localized cell death is also believed to play a part in the development of the embryo (Li et al., 2004; Bozhkov et al., 2005; Sabelli and Larkins, 2009).

Failure of flowers to develop mature seeds may be the consequence of predation, adverse weather or defective pollination (Stephenson, 1981); but of particular significance for crop species is floral overproduction and the death of flowers within inflorescences, resulting in failure to realize full productive capacity. Abortion of pollinated flowers and fruits is selective and is thought to have arisen in evolution as a fitness attribute that optimizes resource allocation to reproduction, such that there is broadly a reciprocal relationship between number of seeds per plant and the extent of variability in seed size (Stephenson, 1981; Sadras, 2007; Sakai, 2007). Competition for resources between genetically different plant parts is under strong maternal control, which favors uniformity of progeny size and partially counteracts genomic conflict (Sadras and Denison, 2009). In wheat (*Triticum*), rice (*Oryza sativa*), soybean and other annual crops, grain number (a highly plastic trait with relatively low heritability), and grain size (which varies over a narrow range and has a higher heritability) are among the traits that contribute to yield (Sadras, 2007; Sadras and Slafer, 2012). The morphologies of species such as maize and sunflower (*Helianthus annuus*) where, unlike small-grain cereals, there has been agronomic selection for one or few inflorescences, are associated with increased variability of seed size and reduced plasticity in seed number (Sadras, 2007). A study of the developing wheat spikelet (Ghiglione et al., 2008) revealed that cessation of cell division, disruption of normal development and initiation of autophagic degeneration underlie the failure of floret primordia to reach fertility at anthesis. The trigger for floret cell death was proposed to be a combination of extended day length and competition with accelerated plant development for limited carbohydrate supply. Cell death is also implicated in shriveling of grape (*Vitis*) berry mesocarp, a variety- and environment-related characteristic with implications for wine quality (Fuentes et al., 2010; Bonada et al., 2013).

2.5 Fruit ripening

Ripening is the terminal stage in fruit development and shares a number of physiological and molecular features with senescence processes in other organs. At the same time, fruits display enhanced development of novel mechanisms for signaling to animal dispersers (including humans). For obvious practical reasons, there has been much research on shelf-life and post-harvest deterioration, but here we focus on ripening as a developmental event (Giovannoni, 2004). As in leaf senescence, transitions in plastid structure and function underlie the characteristic changes in color that occur during fruit ripening. Chlorophyll is broken down, chloroplasts become chromoplasts, and carotenoids form fibrils, crystals or globules in association with fibrillin proteins, the genes for which are strongly expressed during fruit ripening, leaf senescence and the development of floral parts, as well as in response to various environmental stresses (Egea et al., 2010; Piller at al., 2012).

A distinctive feature of ripening fruits is that they do not export significant amounts of products salvaged from the catabolism of macromolecules. Autolytic activities during ripening are primarily associated with shifts in fruit texture, flavor and aroma (Seymour et al., 2013). Hydrolytic enzymes become activated during ripening and attack cell wall carbohydrates, de-esterifying and depolymerizing polysaccharides, changing the physical properties of the wall matrix and, in soft fruits like tomato (*Solanum lycopersicum*) and peach (*Prunus persica*), loosening cell–cell adhesion. Fruits generally become sweeter as they ripen, and invertases play a vital part in the import and metabolism of translocated sucrose. The complex mix of tartness, astringency and fragrance that characterizes the mature fruit is also a consequence of the activation of lytic and oxidative metabolism during ripening (Kader, 2008; Schwab et al., 2008).

Fruits are classified according to whether or not they are climacteric, that is, exhibit a ripening-associated respiratory burst (Plaxton and Podestá, 2006). The function of the respiratory climacteric is unclear, but it is known to be associated with a surge in ethylene production. Ethylene regulates metabolism, including respiration, during ripening and also controls its own biosynthesis. Our understanding of ripening mechanisms has been greatly extended by analysis of a number of mutants (principally in tomato) with defects in ethylene synthesis or perception (Klee and Giovannoni, 2011). Genes for climacteric ethylene synthesis are under the control of the transcription factor RIN and other global regulators. Ripening in the tomato mutants *Never-ripe* and *Green-ripe* is incomplete as a consequence of a selective reduction in ethylene responsiveness. As in leaf senescence, ripening mutants with lesions in chlorophyll breakdown are stay-green (Barry, 2009).

3 SENESCENCE AND CROP ADAPTABILITY

Senescence is a tactic deployed when a random, unforeseeable stress (abiotic or biotic) is experienced. Seasonal or otherwise predictable environmental cues can trigger senescence as part of an adaptive strategy. When the speed and severity of a developing environmental stress outrun physiological capacity to resist or avoid pathological harm, tissues forgo the senescence phase of development and divert more or less directly to the more rapid pathways leading to cell death.

3.1 Seasonal influences on plant senescence

At latitudes away from the equator, plants use the seasonal change in day-length as a reliable environmental cue to prepare for the likely stresses (temperature, drought and so on) to come. In making such preparations, senescence is a major developmental event. Proper timing of leaf senescence is vital if a plant is to balance its carbon and nitrogen demands as photosynthetic carbon acquisition declines and nutrients are salvaged. Senescence is influenced by photoperiod in species with a day-length requirement for floral initiation (Chapter 12), but senescence is also under the control of a pathway independent of flowering (Wingler et al., 2009). A quantitative trait locus for delayed flag leaf senescence maps onto wheat chromosome 2D, close to an allele of the *Ppd-D1* locus for photoperiod insensitivity and the stature gene *Rht8* (Pestsova and Röder, 2002; Verma et al., 2004). This is consistent with extended greenness, dwarf habit and day-length insensitivity moving together as linked phenotypes during the selection of modern highly productive, short-stemmed, non-lodging bread wheat varieties. Studies of senescence in cereals are beginning to reveal in molecular detail the cross-talk between the regulatory pathways for photoperiodism, flowering, nutrient remobilization and grain fill (Parrot et al., 2012). Seasonal senescence in the perennial bioenergy grass *Miscanthus* is associated with floral induction and seems to be regulated by a combination of photoperiod and thermal time (Robson et al., 2012).

Detailed analyses of environmental factors controlling autumnal foliar senescence have been carried out on European aspen (*Populus tremula*). In this species, senescence is initiated strictly according to photoperiod, and can therefore be predicted by date, but once the process is underway, color changes are influenced by temperature (Fig. 10.3; Keskitalo et al., 2005). There is evidence that competence to commence leaf senescence in response to day-length is developed only after a separate photoregulatory pathway has triggered bud set and growth arrest (Fracheboud et al., 2009). Phytochrome plays an important role in day-length perception and autumnal responses. For example, overexpression of the phytochrome A gene in hybrid aspen (*Populus tremula* × *tremuloides*) resulted in day-length-insensitive plants that, unlike wild-type aspen, did not cease growth, acclimate to cold, develop dormancy or undergo leaf senescence and abscission in response to short days (Olsen

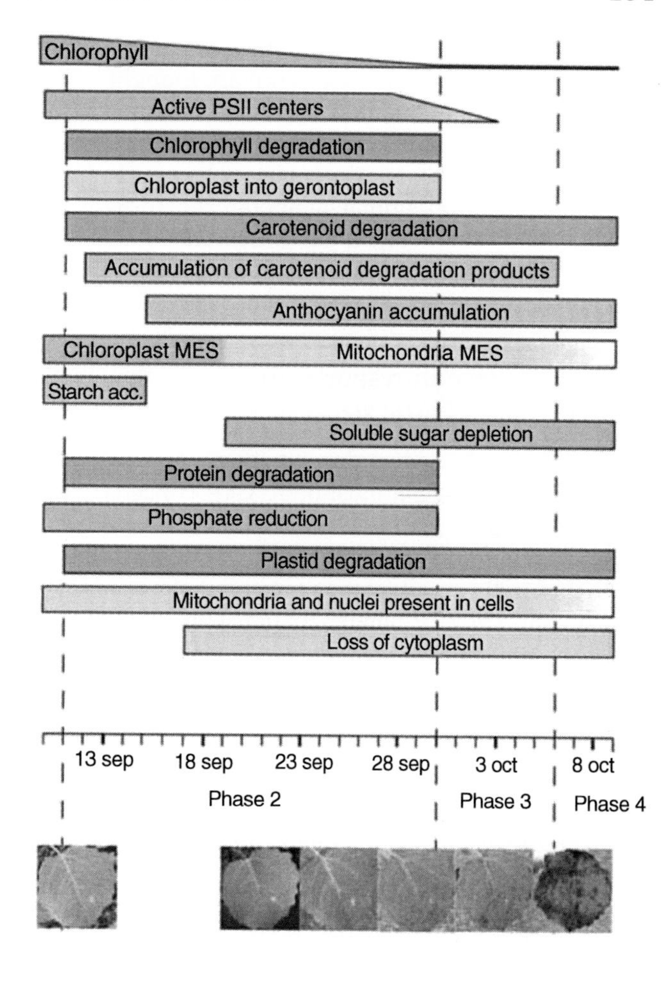

FIG. 10.3 The timetable of events during autumnal senescence in aspen leaves. Senescence is divided into four phases. Phase 1 (the mature, presenescent stage; not shown here) is followed by Phase 2, during which chloroplasts are converted to gerontoplasts, major pigmentation changes occur, N and P are mobilized and sugars metabolized. During Phase 2, the major energy source (MES) switches from chloroplasts to mitochondria. By Phase 3 less than 5% of original chlorophyll remains, cell contents are severely depleted but metabolism continues and viability is sustained in some cells. Phase 4 is the stage at which cell death is complete and few structures are recognizable within residual cell walls. *(From Keskitalo et al., 2005, Figure 10. http://bit.ly/15mqVcH).* Please see color plate at the back of the book.

et al., 1997). Fracheboud et al. (2009) point out that the strict photoperiodic control of senescence initiation seen in *Populus* may be exceptional, because remote-sensing measurements of mixed forests and whole biomes indicate that much of the phenological variation in canopy senescence can be accounted for by temperature variation (an observation of significance in the context of climate change). Apple (*Malus*), pear (*Pyrus*) and some other fruit species undergo foliar senescence, shed their leaves, and spend winter in a dormant state in response to chilling but not photoperiod (Heide and Prestrud, 2005). Curiously, however, induction of the autumn syndrome in apple is reported to have been rendered sensitive to short days by constitutive overexpression of a transcriptional regulator of cold tolerance from peach (Wisniewski et al., 2011). The underlying regulatory mechanism remains to be determined.

Chlorophyll degradation in *Populus* is enhanced at low fall temperatures, probably because of increased photo-oxidative stress. This means that the time from the onset of senescence until the leaves start to be visibly yellow could differ by up to 2 weeks, depending on how low the temperature falls during this period (Keskitalo et al., 2005). Anthocyanin production is also stimulated by the cold, and in years when the weeks immediately following senescence initiation are colder, decreased chlorophyll content and increased anthocyanin levels make the display of fall colors especially striking. Anthocyanins accumulated during foliar senescence are thought to have multiple functions. They may act as sunblockers or antioxidants, preventing light energy from damaging cells in which metabolic rate is constrained by suboptimal temperature. Additionally, or alternatively, there is evidence that fall colors act as visual warning signals to potentially predatory insects (Archetti et al., 2009).

3.2 Senescence in response to unpredictable abiotic stresses

Unseasonal flooding, drought, high or low temperatures, rising atmospheric CO_2 concentration, or elevated ultraviolet radiation (stresses recognized as likely to intensify as the global climate changes) can all lead to increased expression of *SAG*s, and to altered patterns of tissue senescence and death (Weaver et al., 1998; Guo and Gan, 2012). In addition to its role in photoperiodic regulation of seasonal senescence, light can directly invoke senescence and death in illuminated tissues because it is stressful when in excess. The light absorbed by the photosynthetic pigments powers C fixation, and normally any excess light energy is dissipated harmlessly through various protective biochemical mechanisms. But the balance between the supply of and demand for light energy may be disturbed by, for example, inhibition of C fixation by low or high temperature, or by water limitation or by interference from pathogens (Murata et al., 2007). Mild, short-term disruption of the energy balance results in photoinhibition but, under severe and extended light stress, photosynthetic tissues become bleached and cells die. Photobleaching, in contrast to senescence, is a rapid process involving retrograde signaling between the plastid, nucleus and cytosol, the propagation of ROS and free radicals, oxidation and decolorization of pigments and, in severe cases, destruction of the metabolic integrity and compartmentation of cells (Galvez-Valdivieso and Mullineaux, 2010). During such senescence-like pathological conditions (pseudosenescence; Table 10.1), neither chlorophyll catabolites nor the salvage products of proteins and nucleic acids are detectable, and senescence biomarkers often appear to be invoked independently of the senescence syndrome as a whole (Juvany et al., 2013).

Responses to dim light or darkness are complex. Plants may be subject to shading, or they get exposed to light gradients within canopies, or they experience sunflecking in forest understorey layers (Pearcy, 1990). As incident light undergoes gradual extinction within the plant canopy, due to absorption by chlorophyll, both the ratio of red to far-red wavelengths (perceived by phytochrome) and the flux of photosynthetically active radiation progressively decrease. These light signals cooperate to accelerate leaf senescence (Rousseaux et al., 1996). Such photoreceptor systems allow plants to detect potentially competitive neighbors by sensing changes in the wavelengths of reflected and transmitted light – the so-called shade avoidance syndrome (SAS; Ruberti et al., 2012). There is evidence that selection has attenuated the SAS during cereal domestication (Kebrom and Brutnel, 2007). Ectopic expression of an *Arabidopsis* phytochrome B gene in potato (*Solanum tuberosum*) allows higher optimum planting densities in the field, leading to increased tuber yield (Boccalandro et al., 2003). As well as photoreceptor and photosynthetic regulation, preferential translocation of cytokinin away from shaded foliage toward leaves higher in the canopy also contributes to gradients of senescence along shoot axes (Pons et al., 2001).

Darkness promotes senescence in individual attached or detached leaves. Darkening and/or detaching leaves alters assimilate supply–demand relations and influences other aspects of the senescence syndrome. For example, dark-incubated detached leaf tissue and senescing leaves of intact plants developing normally in the light share a broadly similar *SAG* expression profile, but there are many differences too. Comparative studies in *Arabidopsis* have shown that ethylene, JA and SA signaling networks regulate gene expression in natural developmental senescence, but the SA pathway does not operate in dark-induced senescence (Buchanan-Wollaston et al., 2005). Leaf senescence under prolonged darkness is delayed in rice carrying the submergence tolerance gene *SUB1A*, indicating that hormonal pathways regulating senescence responses to darkness and waterlogging are connected (Fukao et al., 2012).

The inescapable succession of day and night means that senescence must have an inherent tolerance of diel variation in temperature as it cycles above and below the optimum. But extremes of temperature at the wrong time in the crop growth cycle will trigger untimely and detrimental senescence. For example, high temperature during and after flowering in cereals often induces premature foliar senescence, resulting in poor grain quality and loss of yield (Vijayalakshmi et al., 2010; Lobell et al., 2012). Plants subject to high temperature stress frequently experience drought, which also promotes leaf senescence (Cairns et al., 2012). Drought-induction of senescence is an intrinsic acclimatory measure that aids water conservation by reducing the area of transpirational surface. Suboptimal temperatures can also be harmful, particularly in combination with high light intensity, promoting senescence, or ROS propagation and pseudosenescence (Masclaux-Daubresse et al., 2007). Many studies have shown that QTL for temperature and drought response coincide with loci for leaf senescence, and that selection targeting such loci can simultaneously improve stress tolerance (Ougham et al., 2007; Vijayalakshmi et al., 2010; Jordan et al., 2012; Emebiri, 2013; Thomas and Ougham, 2014).

If the supply of an essential mineral element drops below the level required for optimal growth, the shoot may develop deficiency symptoms, including senescence and cell death, and will allocate the resources thus released to enhancing root biomass and function (Hermans et al., 2006). Storing high concentrations of nitrate in cell vacuoles allows the immediate impact of a deficiency of soil N to be offset by first draining these reservoirs (van der Leij et al., 1998). Deficiency of a relatively immobile element such as calcium is apparent in the decreased growth and development of young organs, sometimes leading to cell death (White and Broadley, 2003). On the other hand, nitrogen is highly mobile and inadequate supplies trigger senescence in older organs first, as mobile N is salvaged from metabolites and macromolecules through catabolism and transported to younger tissues (Hill, 1980). Phosphorus is also mobile, but there are contradictory observations concerning the extent to which senescence is controlled by P status (Crafts-Brandner, 1992; Usuda, 1995; Colomb et al., 2000). *SAG*s are prominent among the genes upregulated in response to nutrient deficiency (Guo and Gan, 2012).

3.3 Diseases and plant senescence

Plants deploy senescence and death of cells, tissues or organs when threatened by, or exploiting, other organisms. Color changes associated with fall senescence, fruit ripening and floral display send invitations or warnings through the visual systems of potential pollinators, dispersers or predators (Archetti et al., 2009; Lomáscolo and Schaefer, 2010; Sheehan et al., 2013). These interactions are the finely balanced outcome of coevolutionary adaptations dating back to the origins of flowering plants. Plants must wage ceaseless war against disease, and resort to senescence as one of their most powerful weapons in the evolutionary arms race with pathogenic organisms.

Pathogens are classified into necrotrophs, biotrophs and hemibiotrophs. Necrotrophs kill host cells, usually by producing lethal toxins, and live on the resulting dead tissue. An example is *Botrytis cinerea*, the highly virulent causal agent of gray mold disease of soft fruits, which secretes the toxic terpenoid botrydial (Deighton et al., 2001). On the other hand, biotrophs, such as bacteria of the genus *Pseudomonas*, are pathological organisms that feed off tissue that must remain alive. Some pathogens, for example

Phytophthora infestans, the causal agent of potato late blight disease, are hemibiotrophs, that is, an initial biotrophic infection phase is followed by a necrotrophic phase. Host plants suffering attack by biotrophic pathogens often initiate hypersensitivity, a sacrificial autolytic cell death response to ensure survival of the whole organ or plant (Mur et al., 2008; Table 10.1).

Mutations in genes contributing to programmed senescence and autolytic cell elimination lead to plants in which the death of cells is abolished, reduced, premature or spontaneous. Spontaneous cell death is visible as tissue lesions which, in some cases, are associated with increased pathogen resistance (Lenk and Thordal-Christensen, 2009). Mutations that block the pathway of chlorophyll degradation (Fig. 10.1) represent points of contact between spontaneous lesion formation and leaf senescence. In the *Arabidopsis accelerated cell death* (*acd*) mutants *acd1* and *acd2*, the pheophorbide a oxygenase and RCC reductase steps, respectively, in the pathway of chlorophyll catabolism are knocked out. The mutant *lls1* is the maize equivalent of *acd1* (Mur et al., 2008). Blockages in the chlorophyll breakdown pathway in these mutants result in accumulation of pheophorbide or RCC, free photodynamic intermediates that stimulate the formation of harmful ROS and propagating lesions when illuminated (Tanaka et al., 2003; Pružinská et al., 2007). The enzymology and subcellular organization of the chlorophyll pathway normally ensure that the pigment can be disposed of safely without the risk of photosensitivity. This is necessary to sustain the viability of senescing cells during the phase of nutrient salvage (Thomas, 2010).

Chlorosis, or yellowing of leaf tissue, is a common feature of plant diseases and has close biochemical and genetic similarity to senescence. Some biotrophs produce chemicals that mimic senescence-promoting hormones, or toxins that disturb metabolism and trigger yellowing. For example, some forms of *Pseudomonas syringae*, a pathogen of several dicot species, secrete coronatine, a compound that induces chlorosis by acting as a JA analog (Uppalapati et al., 2005). By initiating senescence, such pathogens ensure that infected plant cells remain viable and become a source of sugars, and other nutrients, that support pathogen growth. Senescence, a stage when resources are being remobilized for use elsewhere in the plant, could offer a feast to potential pathogens, which may explain why plants upregulate defense pathways as they initiate normal senescence and why certain senescence hormones, particularly SA and JA, are also mediators of resistance against pathogens (Vlot et al., 2009). Cytokinins are another class of hormone implicated in host–pathogen interactions. Some pathogens that infect leaves stimulate cytokinin production and delayed senescence in a zone surrounding the infection site. The result is a 'green island', an area of tissue that retains chlorophyll against a background of senescing, yellowing leaf tissue (Walters and McRoberts, 2006). The extended source activity of the zombified green zone benefits the pathogen, which acts as a sink for photosynthate.

4 SENESCENCE AND CROP PRODUCTION

This chapter has discussed the principles and processes underlying senescence at all levels of biological organization and across a wide range of domesticated and wild plants. The present section focuses specifically on the crop and the central part played by senescence in the capture and allocation of resources. Assuming the establishment and geometry of the canopy is optimal, it is the onset and rate of senescence that will set a limit on how much assimilate the crop will accrue and, provided it has somewhere to put it, the crop will direct the extra resource that delayed senescence allows it to capture into increased yield (Fig. 10.4; Gregersen et al., 2013). But with this broad generalization come many qualifications and contradictions. As Fischer

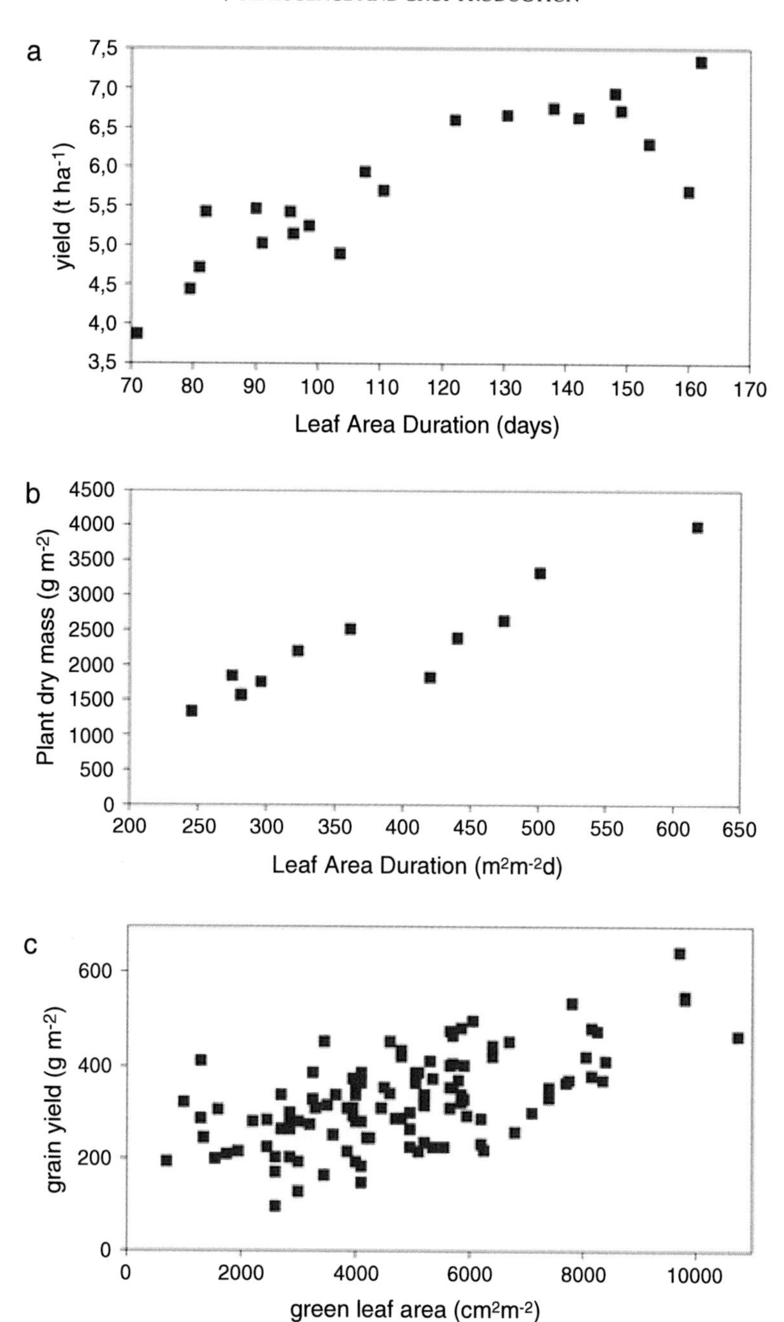

FIG. 10.4 Correlation of leaf senescence and productivity in three crop species. (a) Relationship between leaf area duration and grain yield in 10 maize hybrids. (b) Correlation of leaf area duration with plant dry mass in 11 genotypes of oilseed rape. (c) Correlation between grain yield and green leaf area at 25 days after anthesis in 11 inbred sorghum lines. *(From Gregersen et al., 2013, Figure 2. http://bit.ly/1593ZcC)*

and Edmeades (2010) show, stay-green is only one of a large number of traits that have combined to deliver the increases in productivity achieved for major grain crops in the modern era. Delayed senescence has brought yield benefits in only some, not all, crops and then only under certain agronomic conditions. The following discussion examines these matters in terms of the factors determining supply and demand in the developing crop, and the functional roles of senescence and autolytic mechanisms.

4.1 The carbon capture phase of crop development

Figure 10.5 shows the pattern of carbon assimilation and utilization over the life of a *Phaseolus vulgaris* leaf. Of the carbon fixed by photosynthesis, some is respired, some converted into leaf material (dry matter), and the remainder is available for export to support growth elsewhere in the plant (Thomas et al., 2013). Scaling from the maximal carbon capture capacity of a single leaf to the whole plant is problematical. Gay and Thomas (1995) combined growth analysis, gas exchange measurements and carbon content data to calculate the carbon balance of the fourth leaf of *Lolium temulentum*. Net gas exchange became positive about 7 days before full leaf expansion and leaf weight continued to increase so that the carbon balance with the rest of the plant did not become positive until just after final leaf size had been attained. Based on the model from this study, Thomas and Howarth (2000) showed that the net carbon contribution of each leaf of *L. temulentum* is sufficient to make about 3.7 more leaves. Extrapolating to the point of absurdity, if each of these leaves in turn were able without limit to make a further 3.7 leaves and so on and on, it would require only about 33 leaves to achieve the estimated net primary productivity of the entire biosphere! The serious point of this exercise is to emphasize the degree to which carbon capture is constrained by scale, ontogeny and environment, so that, in reality, it operates at no more than a minute fraction of capacity; but breeders and agronomists can take comfort from the fact that there is plenty of unexploited potential left to aim at.

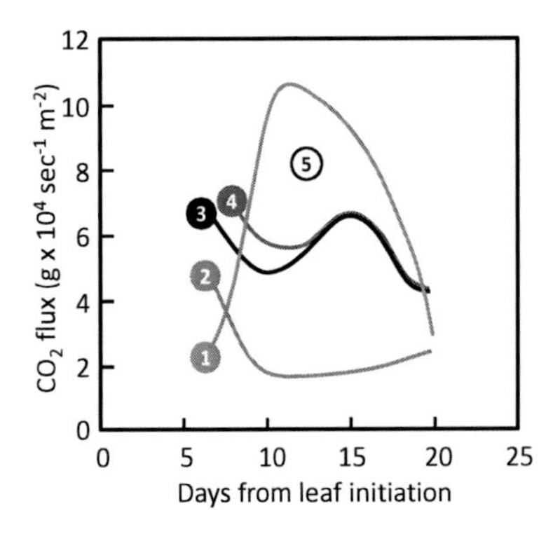

FIG. 10.5 Over the life of a *Phaseolus vulgaris* leaf, the carbon it assimilates through photosynthesis (1) can be respired (2 = dark respiration, 3 = dark + photorespiration), converted into leaf structure (4 = respiration + dry matter), or exported to support growth elsewhere in the plant (5 = area enclosed by 1 and 4). Leaf senescence, which begins about 10 days following leaf initiation, sets a limit on the leaf's carbon export capacity.

The point at which photosynthetic capacity begins to decline from its ontogenetic maximum marks the functional onset of senescence. Figure 10.5 shows that bean leaf senescence commences about 10 days following leaf initiation. The precise nature of the rate-determining metabolic step that accounts for the decrease in photosynthesis depends upon the species, and the developmental and environmental conditions. Often, declining efficiency of the light capture or electron transport systems of chloroplast thylakoids is the limitation. In other circumstances, the CO_2-fixation activity of Rubisco may set the pace for falling rates of photosynthesis as leaves senesce (see, for example, Mae et al., 1993).

4.2 Supply and demand during growth and senescence

A sink is defined as a net importer of nutrients (N, P, K, S and other minerals) and assimilates

(C derived directly or indirectly from photosynthesis). Developing seeds, bulbs, tubers and other structures that accumulate storage compounds are strong sinks, as are expanding leaves and branches during vigorous vegetative growth. Sources are organs that supply the precursors for sink metabolism through the vascular system. Sources and sinks are functionally coupled. The principal food crops are monocarpic, annual plants reproducing once at the end of the life cycle; but, in general, their wild relatives are perennial polycarps, reproducing and regrowing repeatedly over several years. The annual habit is thus inferred to be a loss-of-function trait, part of a suite of agronomically and nutritionally desirable characters selected for during domestication (Cox et al., 2006; Allaby et al., 2008).

Senescence is often considered to be the price paid for sex, reflecting the trade-off between survival and reproductive investment (e.g. Stearns, 1989). But the physiological and genetic determinants of plant habit and lifespan are more complex than this, comprising factors specifying maturity, structural integration, patterns of apical and primordium determinacy, the reproductive cycle and whole organism aging (Thomas et al., 2000; Davies and Gan, 2012; Thomas, 2013). Whole plant death in a monocarp is an extreme example of source–sink interaction. During development of the endosperm of cereal grains and the tuber parenchyma of potato, large amounts of starch are accumulated. These organs are supplied via the phloem with assimilated carbon mostly fixed by current photosynthesis. Some of the amino acid precursors for the synthesis of proteins with structural, reserve or enzymic functions in sinks may be the products of newly-assimilated inorganic nitrogen; but, in general, most amino acids imported by the sink are the recycled products of protein degradation occurring during senescence of source tissues (Xu et al., 2012). Chapter 8 deals with detail on the role of storage and current nitrogen assimilation in grain production. For example, nitrogen storage at flowering can reach 140 kg N ha^{-1} for maize compared to 250 kg N ha^{-1} for wheat; maize is thus more dependent on plant nitrogen uptake after flowering than wheat or rice. In canola, the combination of a large capacity to store nitrogen at flowering and low grain yield leads to a significant fraction of stored nitrogen that is not used for grain production (Chapter 8).

How is sink demand communicated to the source, or source capacity signaled to a potential sink? To put it crudely: does the source blow, or the sink suck? Answers to these questions, which are critical for understanding how senescence and resource allocation are regulated in the whole crop plant, are various and depend on species and circumstances. But recent studies of cell death in relation to nutrition have revealed a biologically conserved mechanism by which sugars and other factors might exert a regulatory influence over source senescence (Rolland et al., 2006; Wingler and Roitsch, 2008).

4.3 Sugar sensing and source–sink communication

It has long been accepted that sucrose traveling between sources and sinks has a central role in long-distance signaling in addition to its function in distributing carbon from current photosynthesis and reserves (Farrar et al., 2000). Plant cells have separate sensors for sucrose and the hexose products of sucrose hydrolysis. Changes in the sucrose–hexose ratio lead to different transduction pathways and inductive or repressive effects on gene transcription (Fig. 10.6; Smeekens et al., 2010). An extracellular invertase, which is associated with the plant cell wall and which hydrolyzes sucrose to glucose and fructose, facilitates the unloading of sugar from the vascular system (Lara et al., 2004; Rolland et al., 2006). The hexose products of enzymic inversion of sucrose are negative regulators of SnRK1, a global post-translational inhibitor and inducer of transcription with wide-ranging influence on development and environmental responses (Baena-Gonzalez et al., 2007).

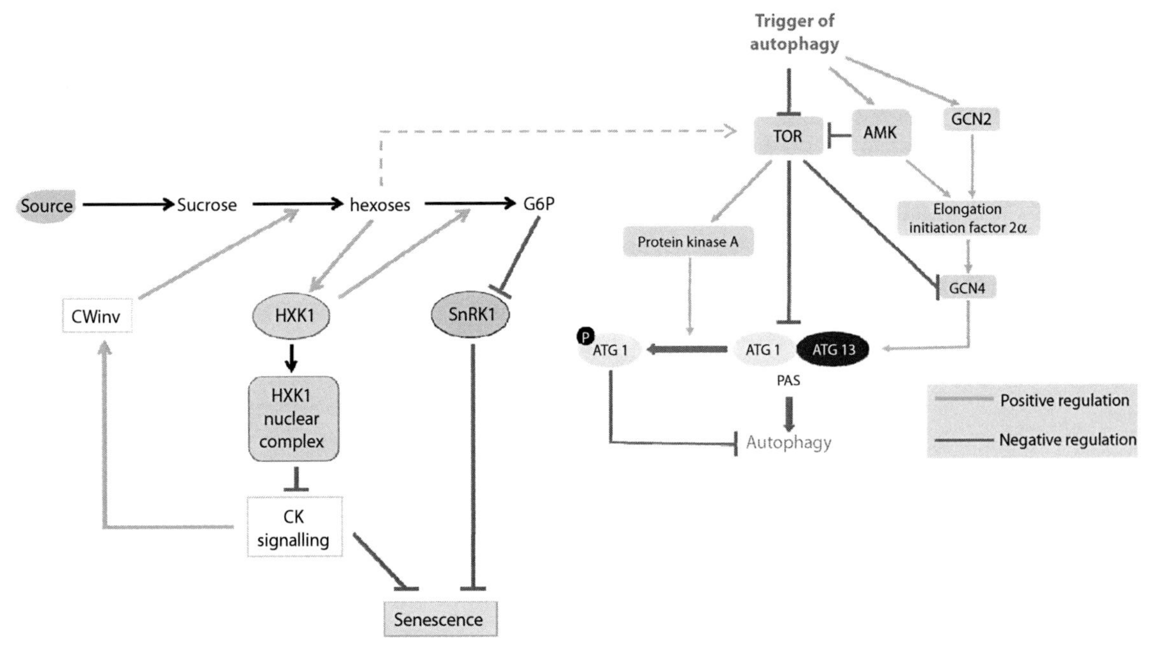

FIG. 10.6 Source–sink regulation of senescence involving global regulators Snf1-related kinase 1 (SnRK1), hexokinase 1 (HXK1) and target of rapamycin (TOR). SnRK1 delays senescence. HXK1 is part of a nuclear complex that promotes senescence by repressing cytokinin (CK) signaling. The products of hydrolysis of sucrose by cell wall invertase (CWinv) are positive regulators of HXK1 and TOR. Glucose-6-phosphate (G6P), the product of HXK1-catalyzed phosphorylation of glucose, is a negative regulator of SnRK1. TOR, protein kinase A, AMK (AMP-activated kinase) and GCN2 (general control nonderepressible 2) are kinases operating in autophagy signaling pathways. Elongation initiation factor 2α and the transcription factor GCN4 regulate expression of ATG1 and ATG13. PAS (pre-autophagosomal structure), the complex containing ATG1 and ATG13, is the precursor of the autophagosome. Please see color plate at the back of the book.

SnRK1 is a protein kinase activated by darkness, nutrient starvation, and high cellular concentrations of sucrose or low glucose or both: conditions associated with the induction of senescence (Jongebloed et al., 2004; Parrott et al., 2007). SnRK1 inhibits several key reactions of carbon and nitrogen metabolism by phosphorylating the corresponding enzymes. It also stimulates transcription of genes that encode enzymes of carbon mobilization. Plants in which *SnRK1* expression has been experimentally downregulated display premature senescence and other developmental irregularities (Thelander et al., 2004). The SnRK1 network in turn interacts with a second senescence-regulating pathway modulated by sugars (Fig. 10.6). The hexokinases encoded by the *Arabidopsis HXK1* and rice *HXK5* and *HXK6* genes function as glucose sensors. A fraction of HXK exists in the nucleus in high-molecular-weight complexes which repress transcription of photosynthetic genes and promote proteolytic degradation of transcription factors that function in plant hormone signaling pathways (Smeekens et al., 2010).

4.4 Autophagy, a universal integrative cell senescence mechanism

Autophagy (Table 10.1) participates in repair and maintenance activities in eukaryotic cells by degrading and recycling damaged macromolecules and organelles. It has been implicated in the incidence of diverse age-related physiological and pathological changes (Vellai et al., 2009), including nutrient-mediated source–sink regulation and senescence (Guiboileau et al., 2010; Thomas, 2013). A network of regulatory kinases controls autophagy gene (*ATG*) expression and the activities of ATG proteins they encode. Several different ATG proteins combine to make pre-autophagosomal structures that coalesce into an autophagosome. The autophagosome vesicle captures a portion of the cytoplasm and undergoes expansion with the aid of additional ATG proteins. The autophagosome then interacts with cytoskeletal microtubules and is transferred to the vacuole where lysis of the contents takes place (Liu and Bassham, 2012). Autophagosomes have been observed in an ever-increasing range of plant processes and structures, including senescing flower cells (Shibuya et al., 2013), during the HR (Mur et al., 2008) and in cell death caused by sugar starvation leading to decreased fertile floret numbers in wheat (Ghiglione et al., 2008).

Formation of the autophagosome is inhibited by protein kinase A, a negative regulator of autophagy. Protein kinase A is itself regulated by another kinase, TOR (target of rapamycin); TOR in turn is inactivated by the starvation-induced kinase AMK. AMK is also part of the kinase pathway that controls transcription of *ATG* genes. Figure 10.6 summarizes the network of interactions regulating autophagy. TOR is now thought to be a critical point of convergence for the regulatory network that coordinates energy status, sugar content, nitrogen availability, cell fate, and longevity (Baena-González and Sheen, 2008). It is absolutely essential for mRNA translation during cell growth and proliferation (Oldham and Hafen, 2003); but when differentiation is complete, TOR then becomes responsible for age-related deterioration, and lifespan can be increased by interfering with TOR signaling (Mair and Dillin, 2008). Senescence and cell death are accelerated in plant autophagy mutants (Thompson et al., 2005), and silencing TOR kinase induces early leaf yellowing and sugar accumulation (Deprost et al., 2007), probably through the interaction of TOR with the HXK and SnRK1 signaling pathways. The case for TOR as a nutrient-sensing regulator of plant growth, source–sink integration, and senescence is getting stronger (Guiboileau et al., 2010; Liu and Bassham, 2012).

4.5 Hormonal regulation of sources and sinks

There is good evidence for the positive participation of the senescence antagonist cytokinin in source–sink regulation (Davies and Gan, 2012). Cytokinins are normally produced in roots and transported to leaves. Tissues with the highest cytokinin levels are the strongest metabolic sinks and attract the majority of nutrients by out-competing less active structures. Post-flowering leaf senescence may be explained by redirection of cytokinins from the root away from foliage and into the developing seeds, which therefore become stronger sinks for nutrients diverted from leaves that have been triggered to senesce. Cytokinin regulation integrates neatly into the sugar-sensing/autophagy mechanism (Fig. 10.6). Downregulating the extracellular invertase associated with transfer of translocated carbon from the vascular system to the sink results in the inhibition of cytokinin-mediated delay of leaf senescence (Lara et al., 2004).

4.6 Leaves as storage organs

The investment of nitrogen, phosphorus, sulfur and other elements in building vegetative tissues is liquidated during reproductive development and canopy senescence. In this sense, growing leaves are storage sinks before they mature into sources of fixed carbon and, ultimately, recycled N, S and P (see also Chapter 8).

The initiation of senescence may be considered to be the point of transition from carbon to

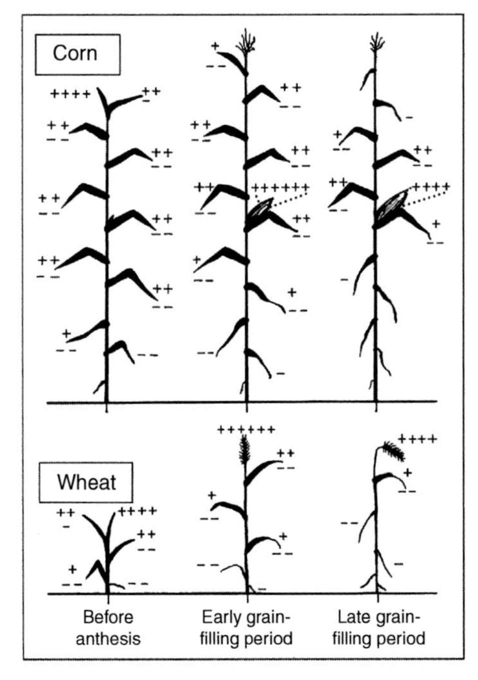

FIG. 10.7 Schematic overview of nitrogen transport at three stages of wheat and maize development. Minor (+) and major (++++++) fluxes into sinks are indicated as well as minor (−) and major (−−) nitrogen fluxes out of leaves. *(From Feller et al., 2007, Figure 3. http://bit.ly/1593D5S)*

nitrogen source. Young, actively growing vegetative and reproductive sinks accumulating proteins have a voracious appetite for nitrogen. When the demand cannot be met by import from the rhizosphere alone, nitrogen is withdrawn from older tissues. This is reflected in the fluxes into and out of organs at different stages in crop development (Fig. 10.7; Feller et al., 2007).

Internal redistribution of nitrogen between sources and sinks accounts, at least in part, for variations in the pattern of senescence. Progressive or sequential senescence is an example. It occurs during vegetative growth in herbaceous species when continued leaf production at the stem apex is frequently at the expense of the senescence of and nitrogen salvage from preceding leaves on the shoot. For instance, each vegetative grass tiller generally carries a fixed number of mature leaves at any given time and, by means of a self-regulatory feedback mechanism (Verdenal et al., 2008), every new leaf that appears is balanced by senescence of the lowermost leaf (Yang et al., 1998).

The foliage of soybean and other monocarpic species can be induced to initiate accumulation of vegetative storage proteins (VSPs) by removing developing pods. Depodding releases uncommitted nitrogen already assimilated by the plant. Nitrogen fixation by root nodules normally declines during fruiting in soybean, but is maintained in depodded plants. Elevated levels of nitrogen alter carbohydrate metabolism and sugar signaling in source leaves, resulting in a buildup of novel glycoprotein VSPs as a consequence of specific induction of new genes (Feller et al., 2008). Leaves of depodded plants also accumulate starch (Nakano et al., 2000), another metabolic sink for sugars. Taken together, these findings show that the leaves of depodded soybean plants have developed into secondary sinks.

4.7 Rubiscolytics

The changing status of the CO_2-fixing enzyme, Rubisco, the largest single protein repository of recoverable nitrogen in vegetative tissues, reflects the functional transitions undergone by the developing and senescing canopy. Rubisco is synthesized at a high rate in young growing leaves, and only when synthesis has stopped, at around full expansion, does breakdown of the protein start (Makino et al., 1984). Incidentally, the second most abundant protein fraction in leaves, the light-harvesting complexes (LHCs) of chloroplast thylakoids, has a similar turnover pattern (Hidema et al., 1992). By undergoing little or no simultaneous synthesis and breakdown at any time in the life of green tissue, Rubisco has the characteristics of a storage protein as well as an enzyme. Feller et al. (2007) coined the term 'Rubiscolytics' to describe the fate of Rubisco after its enzymic functions in carbon capture are terminated.

Rubisco is the most abundant protein in the terrestrial biosphere, and probably in marine biota as well (Raven, 2013). The seasonality of photosynthetic biomass revealed by remote imaging (see, for example, Grace et al., 2007) implies a corresponding global cycle of Rubisco synthesis and breakdown. We know in considerable detail how Rubisco proteins are synthesized and assembled into the functional holoenzyme (Whitney et al., 2011). But, despite its biosphere-wide scale, the enzymology and cell biology of Rubisco breakdown are a continuing and frustrating mystery (Feller et al., 2007). It seems likely that the early stages of proteolysis in senescence take place within the gerontoplast and subsequently the focus of degradative activity moves to the vacuole as remobilization of plastid nitrogen is succeeded by the autolytic events of terminal phase. Many plastid-localized proteases are known, but it remains unclear whether Rubisco is an *in vivo* substrate for any of them. Degradation of LHC is closely associated with the release and further catabolism of chlorophyll (Hörtensteiner and Kräutler, 2011); it is not known whether LHC apoproteins and Rubisco share the same proteolytic pathway. When the organization of the senescing mesophyll cell is examined using high-resolution microscopical and immunofluorescent methods, a dynamic picture emerges in which there is exchange of vesicles, containing proteases or stroma proteins or both, between the plastid and the vacuole (Chiba et al., 2003; Epimashko et al., 2004; Otegui et al., 2005; Martinez et al., 2008). It is to be hoped that increased application of such approaches will enlighten our ignorance.

Although there is wide variation between, and even within, species, proteolysis of Rubisco and other plastid proteins during senescence is, on the whole, rather insensitive to sink demand for nitrogen, to judge from surgical and steam-girdling experiments (Feller et al., 2007). Source tissues unable to export the products of proteolysis to the usual sinks either re-route them into new sinks, such as tillers in the case of cereals from which the developing ears have been removed, or simply accumulate them as free amino acids or VSPs. There is, however, growing evidence that nitrogen remobilization is attuned to leaf carbon status in a way that integrates it with the sugar/autophagy mechanism of source–sink regulation (Izumi et al., 2010). Ishida et al. (2008) implicated ATG-dependent autophagic transfer to the vacuole in the mobilization of stroma proteins, including Rubisco, and Parrott et al. (2010) linked senescence-related expression of a cysteine protease in barley (*Hordeum vulgare*) to high carbohydrate and relatively low nitrogen levels in the leaf.

The amino acid products of proteolysis frequently undergo further metabolism before being exported from the senescing leaf. Many species synthesize amides, glutamine and asparagine in particular. Enzymes of amino acid and amide metabolism are active during senescence as a consequence of enhanced gene transcription (Masclaux-Daubresse et al., 2008). Remobilization of nitrogen from Rubisco and other leaf proteins is accompanied by, and may be coregulated with, salvage of sulfur from the amino acids cysteine and methionine (Dubousset et al., 2009).

4.8 Genetic manipulation of the transition from carbon capture to nitrogen mobilization

The 'master switch' that, once identified and analyzed, will enable cessation of carbon acquisition and initiation of nitrogen salvage to be manipulated at will, is the philosopher's stone of crop senescence research. But if the promise of observations made on model systems in the laboratory is not to go the way of the dreams of alchemy, it must survive transfer to the crop in the field. Very few regulatory mechanisms have yet been shown to cross over in this way.

The turnover patterns of Rubisco and LHCs indicate that to make the transition from C source to N source, it is not sufficient simply to turn off biosynthesis of plastid components – catabolic pathways must be actively upregulated. A global regulatory regimen for senescence, such as that shown in Figure 10.6, is expected to have both down- and up-regulation capabilities. Genomics and systems biology are powerful tools that are being used to search for candidates (Breeze et al., 2011; Guo, 2013; Hickman et al., 2013). Transcriptomic and proteomic analyses of *SAG* expression networks, almost exclusively carried out on senescing *Arabidopsis* leaves, have revealed nodes of interaction centering on particular transcription factors that occupy strategic positions in pathways of developmental and environmental regulation. We have already discussed WRKY53, a member of a family of transcription factors that function in response to ROS and other stresses. Many of the regulators of senescence transcription revealed by studies in *Arabidopsis* can be identified in other species, including wheat (Guo, 2013).

Transcriptional regulators of the NAM/NAP/NAC family are of particular interest for crop production because they give an insight into genetic changes associated with domestication of wheat and direct evidence in realistic field conditions supports their role in altering senescence, yield and quality in this species (Uauy et al., 2006; Brevis et al., 2010). Wild emmer (*Triticum turgidum ssp dicoccoides*) is an ancestor of durum (pasta) wheat, *T. turgidum ssp durum*. Inbred lines of durum wheat in which chromosome 6 was substituted with emmer chromosome 6 have higher grain protein content (GPC) than non-recombinant durum wheat (Uauy et al., 2006). This is associated with more rapid leaf senescence and mobilization of foliar nitrogen in emmer. GPC maps to a single genetic locus on chromosome 6, encoding NAM-B1, a transcription factor of the NAC family. The protein encoded by a related *Arabidopsis* gene, *AtNAP*, is a positive regulator of senescence initiation (Guo and Gan, 2006). The *NAM-B1* alleles of durum wheat and hexaploid bread wheat (*T. aestivum ssp aestivum*) are non-functional as a consequence of sequence changes that became fixed during cereal domestication. When transcript levels of all *NAM*s in hexaploid wheat were knocked down, leaf senescence was delayed by 24 days, GPC was reduced by 5.8% and grains were 30% deficient in zinc and 24% in iron. There was no significant effect on thousand-kernel weight (Uauy et al., 2006). The GPC phenotype in barley has also been shown to be due to allelic variation in a *NAM*, and interaction between the *NAM*-regulated senescence pathway and the network controlling vernalization, photoinduction and flowering time has been established for this species (Parrott et al., 2012). *NAM* genes clearly have a critical role in regulating cereal leaf senescence and determining the partitioning of nitrogen and minerals between the grain and crop residue.

4.9 Post-harvest Phase

The topic of losses between harvest and the point of consumption is an enormous one and here we confine ourselves to one or two comments on the significance of senescence. About one-third of the edible fractions of food produced for human consumption is wasted, amounting to 1.3 billion tons per annum (FAO, 2011). Forty percent of food losses in developing countries occur at the post-harvest and processing stage, whereas in industrialized countries, more than 40% of losses occur at retail and consumer levels (Parfitt et al., 2010). Senescence and associated deteriorative processes contribute to both kinds of wastage. Senescence is often initiated and sustained by harvest and storage conditions. Post-harvest technologies such as refrigeration, chemical treatments, storage in inert atmospheres and rapid transportation to the consumer can slow deterioration. Senescence and ripening from the point of retail onward are important factors in consumer perception and the implementation of commercial and legal quality standards (Wills et al., 2007).

5 CONCLUSION: SENESCENCE AND ITS IMPLICATIONS FOR CROP IMPROVEMENT

Senescence as discussed in this chapter has a practical bearing on crops of all kinds: not just the major grain species that feed the world (wheat, rice, maize, sorghum), but green vegetables, root crops, fruits, timber, fiber and bioenergy crops, ornamentals and forages. Crop breeding and management techniques have enhanced plant resistance to early- and late-season stresses such as chilling and short day lengths. Improvement strategies for cereals have included altering the timing and rate of senescence, and this is considered to have made a significant contribution to performance and productivity over the last century or so (Gregersen et al., 2013).

In maize, rice and probably sorghum, breeding for delayed senescence has gone about as far as it can for now, and current idiotypes focus more on sink capacity, plant architecture and resistance to pests, diseases and stress (Lee and Tollenaar, 2007; Wu, 2009; Fischer and Edmeades, 2010). In modern high-performing varieties, senescence traits alone are only weakly linked with yield. This may in part be a scaling effect: the productivity benefits of selecting for single phenotypic characters at the individual plant level frequently do not translate into crop performance in the field (Pedró et al., 2012). Moreover, stay-green can have the effect of immobilizing N and P in the foliage and, consequently, it has been reported to be of limited or even negative value for some species, notably grain legumes such as soybean (Kumudini, 2002). Despite these limitations, QTL analysis shows that, in general, delayed senescence retains its effectiveness as an improvement trait if it is associated with selection for stress tolerance (Ougham et al., 2007). For example, sorghum lines exhibiting the stay-green phenotype often combine good grain yield with other desirable characters such as disease and insect resistance, wide environmental adaptiveness, appropriate time to maturity, good tillering, a juicy, palatable and digestible residue suitable for fodder, as well as potential for bioenergy use (Thomas and Howarth, 2000; Calviño and Messing, 2012; Jordan et al., 2012; Thomas and Ougham, 2014; Fig. 10.4c). Stay-green also turns out to have some surprising, and potentially useful, side effects. For instance, an insertion mutation in the *Medicago truncatula* *SGR* gene (Fig. 10.1) results not only in retention of leaf greenness, but also in delayed nodule senescence (Zhou et al., 2011).

In the three months since this chapter was started, the IRRI population counter has clocked up more than 19 million more mouths to feed, while the area of productive land has fallen by over a million hectares. Modern high-yielding agriculture is reliant on intensive and expensive fertilizers, pesticides, irrigation and advanced cultivars: they are required to keep leaves alive longer (Thomas, 1992). Understanding the genetic and environmental control of senescence as described in this chapter is essential if we are to sustain and increase the efficiency of crop production in a hungry world with shrinking land resources.

References

Alesandrini, F., Mathis, R., Van de Sype, G., Hérouart, D., Puppo, A., 2003. Possible roles for a cysteine protease and hydrogen peroxide in soybean nodule development and senescence. New Phytol. 158, 131–138.

Allaby, R.G., Fuller, D.Q., Brown, T.A., 2008. The genetic expectations of a protracted model for the origins of domesticated crops. Proc. Natl. Acad. Sci. USA 105, 13982–13986.

Archetti, M., Döring, T.F., Hagen, S.B., et al., 2009. Unravelling the evolution of autumn colors – an interdisciplinary approach. Trends Ecol. Evol. 24, 166–173.

Baena-Gonzalez, E., Rolland, F., Thevelein, J.M., Sheen, J., 2007. A central integrator of transcription networks in plant stress and energy signalling. Nature 448, 938–942.

Baena-González, E., Sheen, J., 2008. Convergent energy and stress signaling. Trends Plant Sci. 13, 474–482.

Barry, C.S., 2009. The stay-green revolution: Recent progress in deciphering the mechanisms of chlorophyll degradation in higher plants. Plant Sci. 176, 325–333.

Bell, P.R., 1996. Megaspore abortion: A consequence of selective apoptosis? Internatl. J. Plant Sci. 157, 1–7.

Besagni, C., Kessler, F., 2013. A mechanism implicating plastoglobules in thylakoid disassembly during senescence and nitrogen starvation. Planta 237, 463–470.

Bonada, M., Sadras, V.O., Fuentes, S., 2013. Effect of elevated temperature on the onset and rate of mesocarp cell death in berries of Shiraz and Chardonnay and its relationship with berry shrivel. Aust. J. Grape Wine Res. 19, 87–94.

Boccalandro, H.E., Ploschuk, E.L., Yanovsky, M.J., Sánchez, R.A., Gatz, C., Casal, J.J., 2003. Increased phytochrome B alleviates density effects on tuber yield of field potato crops. Plant Physiol. 133, 1539–1546.

Bozhkov, P.V., Filonova, L.H., Suarez, M.F., 2005. Programmed cell death in plant embryogenesis. Curr. Top. Devel. Biol. 67, 135–179.

Breeze, E., Harrison, E., McHattie, S., et al., 2011. High-resolution temporal profiling of transcripts during Arabidopsis leaf senescence reveals a distinct chronology of processes and regulation. Plant Cell 23, 873–894.

Brevis, J.C., Morris, C.F., Manthey, F., Dubcovsky, J., 2010. Effect of the grain protein content locus Gpc-B1 on bread and pasta quality. J. Cereal Sci. 51, 357–365.

Buchanan-Wollaston, V., Page, T., Harrison, E., et al., 2005. Comparative transcriptome analysis reveals significant differences in gene expression and signalling pathways between developmental and dark/starvation-induced senescence in Arabidopsis. Plant J. 42, 567–585.

Cairns, J.E., Sanchez, C., Vargas, M., Ordoñez, R., Araus, J.L., 2012. Dissecting maize productivity: ideotypes associated with grain yield under drought stress and well-watered conditions. J. Integ. Plant Biol. 54, 1007–1020.

Calviño, M., Messing, J., 2012. Sweet sorghum as a model system for bioenergy crops. Curr. Opin. Biotechnol. 23, 323–329.

Chiba, A., Ishida, H., Nishizawa, N.K., Makino, A., Mae, T., 2003. Exclusion of ribulose-1,5-bisphosphate carboxylase/oxygenase from chloroplasts by specific bodies in naturally senescing leaves of wheat. Plant Cell Physiol. 44, 914–921.

Colomb, B., Kiniry, J.R., Debaeke, P., 2000. Effect of soil phosphorus on leaf development and senescence dynamics of field-grown maize. Agron. J. 92, 428–435.

Courtois-Moreau, C.L., Pesquet, E., Sjodin, A., et al., 2009. A unique program for cell death in xylem fibers of Populus stem. Plant J. 58, 260–274.

Cox, T.S., Glover, J.D., van Tassel, D.L., Cox, C.M., DeHaan, L.R., 2006. Prospects for developing perennial grain crops. BioScience 56, 649–659.

Crafts-Brandner, S.J., 1992. Phosphorus nutrition influence on leaf senescence in soybean. Plant Physiol. 98, 1128–1132.

Davies, P.J., Gan, S., 2012. Towards an integrated view of monocarpic plant senescence. Russ. J. Plant Physiol. 59, 467–478.

Deighton, N., Muckenschnabel, I., Colmenares, A.J., Collado, I.G., Williamson, B., 2001. Botrydial is produced in plant tissues infected by Botrytis cinerea. Phytochemistry 57, 689–692.

Deprost, D., Yao, L., Sormani, R., et al., 2007. The Arabidopsis TOR kinase links plant growth, yield, stress resistance and mRNA translation. EMBO Rep. 8, 864–870.

Diggle, P.K., Di Stilio, V.S., Gschwend, A.R., et al., 2011. Multiple developmental processes underlie sex differentiation in angiosperms. Trends Genet. 27, 368–376.

Dorling, D., 2013. Population 10 billion. Constable and Robinson, London.

Dubousset, L., Abdallah, M., Desfeux, A-S., et al., 2009. Remobilization of leaf S compounds and senescence in response to restricted sulphate supply during the vegetative stage of oilseed rape are affected by mineral N availability. J. Exp. Bot. 60, 3239–3253.

Egea, I., Barsan, C., Bian, W., et al., 2010. Chromoplast differentiation: current status and perspectives. Plant Cell Physiol. 51, 1601–1611.

Emebiri, L.C., 2013. QTL dissection of the loss of green colour during post-anthesis grain maturation in two-rowed barley. Theor. Appl. Genet. 126, 1873–1884.

Epimashko, S., Meckel, T., Fischer-Schliebs, E., Lüttge, U., Thiel, G., 2004. Two functional different vacuoles for static and dynamic purposes in one plant mesophyll leaf cell. Plant J. 37, 294–300.

FAO., 2011. Global food losses and food waste. Extent, causes and prevention. Food and Agriculture Organization of the United Nations, Rome, http://www.fao.org/docrep/014/mb060e/mb060e00.pdf (accessed 18 August 2013).

Farrar, J., Pollock, C., Gallagher, J., 2000. Sucrose and the integration of metabolism in vascular plants. Plant Sci. 154, 1–11.

Fath, A., Bethke, P., Lonsdale, J., Meza-Romero, R., Jones, R., 2000. Programmed cell death in cereal aleurone. Plant Mol. Biol. 44, 255–266.

Feller, U., Anders, I., Mae, T., 2007. Rubiscolytics: fate of Rubisco after its enzymatic function in a cell is terminated. J. Exp. Bot. 59, 1615–1624.

Fischer, R.A., Edmeades, G.O., 2010. Breeding and cereal yield progress. Crop Sci. 50, S85–S98.

Fracheboud, Y., Luquez, V., Björkén, L., Sjödin, A., Tuominen, H., Jansson, S., 2009. The control of autumn senescence in European aspen. Plant Physiol. 149, 1982–1991.

Fuentes, S., Sullivan, W., Tilbrook, J., Tyerman, S., 2010. A novel analysis of grapevine berry tissue demonstrates a variety-dependent correlation between tissue vitality and berry shrivel. Aust. J. Grape Wine Res. 16, 327–336.

Fukao, T., Yeung, E., Bailey-Serres, J., 2012. The submergence tolerance gene SUB1A delays leaf senescence under prolonged darkness through hormonal regulation in rice. Plant Physiol. 160, 1795–1807.

Fusseder, A., 1987. The longevity and activity of the primary root of maize. Plant Soil 101, 257–265.

Galvez-Valdivieso, G., Mullineaux, P.M., 2010. The role of reactive oxygen species in signalling from chloroplasts to the nucleus. Physiol. Plant. 138, 430–439.

Gan, S., Amasino, R.M., 1995. Inhibition of leaf senescence by autoregulated production of cytokinin. Science 270, 1986–1988.

Gay, A.P., Thomas, H., 1995. Leaf development in *Lolium temulentum*: photosynthesis in relation to growth and senescence. New Phytol. 130, 159–168.

Ghiglione, H.O., Gonzalez, F.G., Serrago, R., et al., 2008. Autophagy regulated by day length determines the number of fertile florets in wheat. Plant J. 55, 1010–1024.

Giovannoni, J.J., 2004. Genetic regulation of fruit development and ripening. Plant Cell 16, S170–S180.

Grace, J., Nichol, C., Disney, M., Lewis, P., Quaife, T., Bowyer, P., 2007. Can we measure terrestrial photosynthesis from space directly, using spectral reflectance and fluorescence? Glob Change Biol. 13, 1484–1497.

Gregersen, P.L., Culetic, A., Boschian, L., Krupinska, K., 2013. Plant senescence and crop productivity. Plant Mol. Biol. 82, 603–622.

Guerra, J.C.P., Coussens, G., De Keyser, A., et al., 2010. Comparison of developmental and stress-induced nodule senescence in Medicago truncatula. Plant Physiol. 152, 1574–1584.

Guiboileau, A., Sormani, R., Meyer, C., Masclaux-Daubresse, C., 2010. Senescence and death of plant organs: Nutrient recycling and developmental regulation. Comp. Rend. Biol. 333, 382–391.

Gunawardena, A.H.L.A.N., 2008. Programmed cell death and tissue remodeling in plants. J. Exp. Bot. 59, 445–451.

Guo, Y., 2013. Towards systems biological understanding of leaf senescence. Plant Mol. Biol. 82, 519–528.

Guo, Y., Gan, S., 2006. AtNAP, a NAC family transcription factor, has an important role in leaf senescence. Plant J. 46, 601–612.

Guo, Y., Gan, S., 2012. Convergence and divergence in gene expression profiles induced by leaf senescence and 27 senescence-promoting hormonal, pathological and environmental stress treatments. Plant Cell Environ. 35, 644–655.

Ha, S., Vankova, R., Yamaguchi-Shinozaki, K., Shinozaki, K., Phan Tran, L-S.P., 2012. Cytokinins: metabolism and function in plant adaptation to environmental stresses. Trends Plant Sci. 17, 172–179.

Hatsugai, N., Kuroyanagi, M., Nishimura, M., Hara-Nishimura, I., 2006. A cellular suicide strategy of plants: vacuole-mediated cell death. Apoptosis 11, 905–911.

Hawes, M.C., Pueppke, S.G., 1986. Sloughed peripheral root cap cells: yield from different species and callus formation from single cells. Am. J. Bot. 73, 1466–1473.

He, Y., Fukushige, H., Hildebrand, D.F., Gan, S., 2002. Evidence supporting a role of jasmonic acid in Arabidopsis leaf senescence. Plant Physiol. 128, 876–884.

Heide, O.M., Prestrud, A.K., 2005. Low temperature, but not photoperiod, controls growth cessation and dormancy induction and release in apple and pear. Tree Physiol. 25, 109–114.

Hermans, C., Hammond, J.P., White, P.J., Verbruggen, N., 2006. How do plants respond to nutrient shortage by biomass allocation? Trends Plant Sci. 11, 610–617.

Hickman, R., Penfold, C.A., Breeze, E., et al., 2013. A local regulatory network around three NAC transcription factors in stress responses and senescence in Arabidopsis leaves. Plant J. 75, 26–39.

Hidema, J., Makino, A., Kurita, Y., Mae, T., Ojima, K., 1992. Changes in the levels of chlorophyll and light-harvesting chlorophyll a/b protein of PS II in rice leaves aged under different irradiances from full expansion through senescence. Plant Cell Physiol. 33, 1209–1214.

Hidema, J., Makino, A., Mae, T., Ojima, K., 1991. Photosynthetic characteristics of rice leaves aged under different irradiances from full expansion through senescence. Plant Physiol. 97, 1287–1293.

Hill, J., 1980. The remobilization of nutrients from leaves. J. Plant Nutr. 2, 407–444.

Hörtensteiner, S., Kräutler, B., 2011. Chlorophyll breakdown in higher plants. Biochim. Biophys. Acta Bioenerg. 1807, 977–988.

Ishida, H., Yoshimoto, K., Izumi, M., et al., 2008. Mobilization of Rubisco and stroma-localized fluorescent proteins of chloroplast to the vacuole by an ATG gene-dependent autophagic process. Plant Physiol. 148, 142–155.

Izumi, M., Wada, S., Makino, A., Ishida, H., 2010. The autophagic degradation of chloroplasts via Rubisco-containing bodies is specifically linked to leaf carbon status but not nitrogen status in Arabidopsis. Plant Physiol. 154, 1196–1209.

Jongebloed, U., Szederkenyi, J., Hartig, K., Schobert, C., Komor, E., 2004. Sequence of morphological and physiological events during natural ageing and senescence of a castor bean leaf: sieve tube occlusion and carbohydrate back-up precede chlorophyll degradation. Physiol. Plant. 120, 338–346.

Jordan, D.R., Hunt, C.H., Cruickshank, A.W., Borrell, A.K., Henzell, R.G., 2012. The relationship between the stay-green trait and grain yield in elite sorghum hybrids grown in a range of environments. Crop Sci. 52, 1153–1161.

Jovanovic, M., Lefebvre, V., Laporte, P., et al., 2008. How the environment regulates root architecture in dicots. Adv. Bot. Res. 46, 35–74.

Juvany, M., Müller, M., Munné-Bosch, S., 2013. Photo-oxidative stress in emerging and senescing leaves: a mirror image? J. Exp. Bot. 64, 3087–3098.

Kader, A.A., 2008. Flavor quality of fruits and vegetables. J. Sci. Food Agric. 88, 1863–1868.

Kebrom, T.H., Brutnell, T.P., 2007. The molecular analysis of the shade avoidance syndrome in the grasses has begun. J. Exp. Bot. 58, 3079–3089.

Keskitalo, J., Bergquist, G., Gardeström, P., Jansson, S., 2005. A cellular timetable of autumnal senescence. Plant Physiol. 139, 1635–1648.

Kim, H.J., Ryu, H., Hong, S.H., et al., 2006. Cytokinin-mediated control of leaf longevity by AHK3 through phosphorylation of ARR2 in Arabidopsis. Proc. Natl. Acad. Sci. USA 103, 814–819.

Klee, H.J., Giovannoni, J.J., 2011. Genetics and control of tomato fruit ripening and quality attributes. Annu. Rev. Genet. 45, 41–59.

Kozela, C., Regan, S., 2003. How plants make tubes. Trends Plant Sci. 8, 159–164.

Kumudini, S., 2002. Trials and tribulations: a review of the role of assimilate supply in soybean genetic yield improvement. Field Crops Res. 75, 211–222.

Lara, M.E.B., Garcia, M-C.G., Fatima, T., et al., 2004. Extracellular invertase is an essential component of cytokinin-mediated delay of senescence. Plant Cell 16, 1276–1287.

Lee, E.A., Tollenaar, M., 2007. Physiological basis of successful breeding strategies for maize grain yield. Crop Sci. 47, S202–S215.

Lenk, A., Thordal-Christensen, H., 2009. From nonhost resistance to lesion-mimic mutants: useful for studies of defense signaling. Adv. Bot. Res. 51, 91–121.

Li, R., Lan, S.Y., Xu, Z.X., 2004. Programmed cell death in wheat during starchy endosperm development. J. Plant Physiol. Mol. Biol. 30, 183–188.

Liu, P., Liang, S., Yao, N., Wu, H., 2012. Programmed cell death of secretory cavity cells in fruits of Citrus grandis cv. Tomentosa is associated with activation of caspase 3-like protease. Trees 26, 1821–1835.

Liu, Y., Bassham, D.C., 2012. Autophagy: pathways for self-eating in plant cells. Annu. Rev. Plant Biol. 63, 215–237.

Lobell, D.B., Sibley, A., Ortiz-Monasterio, J.I., 2012. Extreme heat effects on wheat senescence in India. Nat. Clim. Change 2, 186–189.

Lomáscolo, S.B., Schaefer, H.M., 2010. Signal convergence in fruits: a result of selection by frugivores? J. Evol. Biol. 23, 614–624.

Lord, C.E.N., Gunawardena, A.H.L.A.N., 2012. Programmed cell death in C. elegans, mammals and plants. Eur. J. Cell Biol. 91, 603–613.

Mae, T., Thomas, H., Gay, A.P., Makino, A., Hidema, J., 1993. Leaf development in *Lolium temulentum*: photosynthesis and photosynthetic proteins in leaves senescing under different irradiances. Plant Cell Physiol. 34, 391–399.

Mair, W., Dillin, A., 2008. Aging and survival: the genetics of life span extension by dietary restriction. Annu. Rev. Biochem. 77, 727–754.

Makino, A., Mae, T., Ohiro, K., 1984. Relation between nitrogen and ribulose-1,5-bisphosphate carboxylase in rice leaves from emergence through senescence. Plant Cell Physiol. 25, 429–437.

Martínez, D.E., Costa, M.L., Gomez, F.M., Otegui, M.S., Guiamet, J.J., 2008. "Senescence-associated vacuoles" are involved in the degradation of chloroplast proteins in tobacco leaves. Plant J. 56, 196–206.

Masclaux-Daubresse, C., Reisdorf-Cren, M., Orsel, M., 2008. Leaf nitrogen remobilisation for plant development and grain filling. Plant Biol. 10, 23–36.

Masclaux-Daubresse, C., Purdy, S., Lemaitre, T., et al., 2007. Genetic variation suggests interaction between cold acclimation and metabolic regulation of leaf senescence. Plant Physiol. 143, 434–446.

Matile, P., 1997. The vacuole and cell senescence. Adv. Bot. Res. 25, 87–112.

McCormack, M.L., Eissenstat, D.M., Prasad, A.M., Smithwick, E.A.H., 2013. Regional scale patterns of fine root lifespan and turnover under current and future climate. Glob. Change Biol. 19, 1697–1708.

Miao, Y., Zentgraf, U., 2007. The antagonist function of Arabidopsis WRKY53 and ESR/ESP in leaf senescence is modulated by the jasmonic and salicylic acid equilibrium. Plant Cell 19, 819–830.

Morris, K., Mackerness, S.A.H., Page, T., et al., 2000. Salicylic acid has a role in regulating gene expression during leaf senescence. Plant J. 23, 677–685.

Müntz, K., 2007. Protein dynamics and proteolysis in plant vacuoles. J. Exp. Bot. 58, 2391–2407.

Mur, L.A.J., Kenton, P., Lloyd, A.J., Ougham, H., Prats, E., 2008. The hypersensitive response; the centenary is upon us but how much do we know? J. Exp. Bot. 59, 501–520.

Murata, N., Takahashi, S., Nishiyama, Y.N., Allakhverdiev, S.I., 2007. Photoinhibition of photosystem II under environmental stress. Biochim. Biophys. Acta Bioenerg. 1767, 414–421.

Nakano, H., Muramatsu, S., Makino, A., Mae, T., 2000. Relationship between the suppression of photosynthesis and starch accumulation in the pod-removed bean. Aust. J. Plant Physiol. 27, 167–173.

Ohashi-Ito, K., Fukuda, H., 2010. Transcriptional regulation of vascular cell fates. Curr. Opin. Plant Biol. 13, 670–676.

Oldham, S., Hafen, E., 2003. Insulin/IGF and target of rapamycin signaling: a TOR de force in growth control. Trends Cell Biol. 13, 79–85.

Olsen, J.E., Junttila, O., Nilsen, J., et al., 1997. Ectopic expression of oat phytochrome A in hybrid aspen changes critical daylength for growth and prevents cold acclimatization. Plant J. 12, 1339–1350.

Otegui, M., Noh, Y.-S., Martínez, D., et al., 2005. Senescence-associated vacuoles with intense proteolytic activity develop in senescing leaves of Arabidopsis and soybean. Plant J. 41, 831–844.

Ougham, H., Armstead, I., Howarth, C., Galyuon, I., Donnison, I., Thomas, H., 2007. The genetic control of senescence revealed by mapping quantitative trait loci. Annu. Plant Rev. 26, 171–201.

Parfitt, J., Barthel, M., Macnaughton, S., 2010. Food waste within food supply chains: quantification and potential for change to 2050. Phil. Transact. R. Soc. Lond. B 365, 3065–3081.

Parish, R.W., Li, S.F., 2010. Death of a tapetum: a programme of developmental altruism. Plant Sci. 178, 73–89.

Parrott, D.L., Downs, E.P., Fischer, A.M., 2012. Control of barley (*Hordeum vulgare* L.) development and senescence by the interaction between a chromosome six grain protein content locus, day length, and vernalization. J. Exp. Bot. 63, 1329–1339.

Parrott, D.L., Martin, J.M., Fischer, A.M., 2010. Analysis of barley (*Hordeum vulgare* L.) leaf senescence and protease gene expression: a family C1A cysteine protease is specifically induced under conditions characterized by high carbohydrate, but low to moderate nitrogen levels. New Phytol. 187, 313–331.

Parrott, D.L., McInnerney, K., Feller, U., Fischer, A.M., 2007. Steam-girdling of barley (*Hordeum vulgare*) leaves leads to carbohydrate accumulation and accelerated leaf senescence, facilitating transcriptomic analysis of senescence-associated genes. New Phytol. 176, 56–69.

Parthier, B., 1988. Gerontoplasts: the yellow end in the ontogenesis of chloroplasts. Endocytobiol. Cell Res. 5, 163–190.

Pearcy, R.W., 1990. Sunflecks and photosynthesis in plant canopies. Annu. Rev. Plant Physiol. Plant Mol. Biol. 41, 421–453.

Pedró, A., Savina, R., Slafer, G.A., 2012. Crop productivity as related to single-plant traits at key phenological stages in durum wheat. Field Crops Res. 138, 42–51.

Pestsova, E., Röder, M., 2002. Microsatellite analysis of wheat chromosome 2D allows the reconstruction of chromosomal inheritance in pedigrees of breeding programmes. Theor. Appl. Genet. 106, 84–91.

Pickard, W.F., 2008. Laticifers and secretory ducts: two other tube systems in plants. New Phytol. 177, 877–888.

Pierik, R., Tholen, D., Poorter, H., Visser, E.J.W., Voesenek, L.A.C.J., 2006. The Janus face of ethylene: growth inhibition and stimulation. Trends Plant Sci. 11, 176–183.

Piller, L.E., Abraham, M., Dörmann, P., Kessler, F., Besagni, C., 2012. Plastid lipid droplets at the crossroads of prenylquinone metabolism. J. Exp. Bot. 63, 1609–1618.

Plaxton, W.C., Podestá, F.E., 2006. The functional organization and control of plant respiration. Crit. Rev. Plant Sci. 25, 159–198.

Pons, T.L., Jordi, W., Kuiper, D., 2001. Acclimation of plants to light gradients in leaf canopies: evidence for a possible role for cytokinins transported in the transpiration stream. J. Exp. Bot. 52, 1563–1574.

Pružinská, A., Anders, I., Aubry, S., et al., 2007. *In vivo* participation of red chlorophyll catabolite reductase in chlorophyll breakdown. Plant Cell 19, 369–387.

Puppo, A., Groten, K., Bastian, F., et al., 2005. Legume nodule senescence: roles for redox and hormone signalling in the orchestration of the natural aging process. New Phytol. 165, 683–701.

Raven, J.A., 2013. Rubisco: still the most abundant protein of Earth? New Phytol. 198, 1–3.

Reape, T.J., Molony, E.M., McCabe, P.F., 2008. Programmed cell death in plants: distinguishing between different modes. J. Exp. Bot. 59, 435–444.

Roberts, J.A., Elliott, K.A., Gonzalez-Carranza, Z.H., 2002. Abscission, dehiscence, and other cell separation processes. Annu. Rev. Plant Biol. 53, 131–158.

Robson, P.R.H., Donnison, I.S., Wang, K., et al., 2004. Leaf senescence is delayed in maize expressing the Agrobacterium IPT gene under the control of a novel maize senescence-enhanced promoter. Plant Biotechnol. J. 2, 101–112.

Robson, P.R.H., Mos, M., Clifton-Brown, J., Donnison, I., 2012. Phenotypic variation in senescence in Miscanthus: towards optimising biomass quality and quantity. BioEnerg. Res. 5, 95–105.

Rogers, H.J., 2013. From models to ornamentals: how is flower senescence regulated? Plant Mol. Biol. 82, 563–574.

Rolland, F., Baena-Gonzalez, E., Sheen, J., 2006. Sugar sensing and signaling in plants: conserved and novel mechanisms. Annu. Rev. Plant Biol. 57, 675–709.

Rousseaux, M.C., Hall, A.J., Sánchez, R.A., 1996. Far-red enrichment and photosynthetically active radiation level influence leaf senescence in field-grown sunflower. Physiol. Plant. 96, 217–224.

Ruberti, I., Sessa, G., Ciolfi, A., Possenti, M., Carabelli, M., Morelli, G., 2012. Plant adaptation to dynamically changing environment: The shade avoidance response. Biotechnol. Adv. 30, 1047–1058.

Sabelli, P.A., Larkins, B.A., 2009. The development of endosperm in grasses. Plant Physiol. 149, 14–26.

Sadras, V.O., 2007. Evolutionary aspects of the trade-off between seed size and number in crops. Field Crops Res. 100, 125–138.

Sadras, V.O., Denison, R.F., 2009. Do plant parts compete for resources? An evolutionary viewpoint. New Phytol. 183, 565–574.

Sadras, V.O., Slafer, G.A., 2012. Environmental modulation of yield components in cereals: Heritabilities reveal a hierarchy of phenotypic plasticities. Field Crops Res. 127, 215–224.

Sakai, S., 2007. A new hypothesis for the evolution of overproduction of ovules: an advantage of selective abortion for females not associated with variation in genetic quality of the resulting seeds. Evolution 61, 984–993.

Schwab, W., Davidovich-Rikanati, R., Lewinsohn, E., 2008. Biosynthesis of plant-derived flavor compounds. Plant J. 54, 712–732.

Seymour, G., Tucker, G.A., Poole, M., Giovannoni, J. (Eds.), 2013. The molecular biology and biochemistry of fruit ripening. Wiley, Chichester.

Sheehan, H., Hermann, K., Kuhlemeier, C., 2013. Color and scent: how single genes influence pollinator attraction. Cold Spring Harbor Symp. Quant. Biol. 77, 117–133.

Shibuya, K., Niki, T., Ichimura, K., 2013. Pollination induces autophagy in petunia petals via ethylene. J. Exp. Bot. 64, 1111–1120.

Smeekens, S., Ma, J., Hanson, J., Rolland, F., 2010. Sugar signals and molecular networks controlling plant growth. Curr. Opin. Plant Biol. 13, 274–279.

Spaeth, S.C., Cortes, P.H., 1995. Root cortex death and subsequent initiation and growth of lateral roots from bare steles of chickpeas. Can. J. Bot. 73, 253–261.

Stearns, S.C., 1989. Trade-offs in life-history evolution. Funct. Ecol. 3, 259–268.

Stephenson, A.G., 1981. Flower and fruit abortion: Proximate causes and ultimate functions. Annu. Rev. Ecol. System 12, 253–279.

Tanaka, R., Hirashima, M., Satoh, S., Tanaka, A., 2003. The Arabidopsis-accelerated cell death gene ACD1 is involved in oxygenation of pheophorbide a: inhibition of the pheophorbide a oxygenase activity does not lead to the Stay-Green phenotype in Arabidopsis. Plant Cell Physiol. 44, 1266–1274.

Thelander, M., Olsson, T., Ronne, H., 2004. Snf1-related protein kinase 1 is needed for growth in a normal day-night light cycle. EMBO J 23, 1900–1910.

Thomas, H., 1992. Canopy survival. In: Baker, N.R., Thomas, H. (Eds.), Crop photosynthesis: spatial and temporal determinants. Elsevier, Amsterdam, pp. 1–41.

Thomas, H., 2010. Leaf senescence and autumn leaf coloration. McGraw Hill 2010 Yearbook of Science and Technology, 211–214.

Thomas, H., 2013. Senescence, ageing and death of the whole plant. New Phytol. 197, 696–711.

Thomas, H., Howarth, C.J., 2000. Five ways to stay green. J. Exp. Bot. 51, 329–337.

Thomas, H., Huang, L., Young, M., Ougham, H., 2009. Evolution of plant senescence. BMC Evol. Biol. 9, 163.

Thomas, H., Ougham, H.J., 2014. The stay-green trait. J. Exp. Bot.doi: 10.1093/jxb/eru037.

Thomas, H., Ougham, H., Mur, L.A.J., Jansson, S., 2013. Senescence and programmed cell death. In: Buchanan, B., Gruissem, W., Jones, R. (Eds.), Biochemistry and molecular biology of plants. 2nd edn. Wiley, New York.

Thomas, H., Ougham, H.J., Wagstaff, C., Stead, A.J., 2003. Defining senescence and death. J. Exp. Bot. 54, 1127–1132.

Thomas, H., Thomas, H.M., Ougham, H., 2000. Annuality, perenniality and cell death. J. Exp. Bot. 51, 1781–1788.

Thompson, A.R., Doelling, J.H., Suttangkakul, A., Vierstra, R.D., 2005. Autophagic nutrient recycling in Arabidopsis directed by the ATG8 and ATG12 conjugation pathways. Plant Physiol. 138, 2097–2110.

Turner, G.W., 1999. A brief history of the lysigenous gland hypothesis. Bot. Rev. 65, 76–88.

Uauy, C., Distelfeld, A., Fahima, T., Blechl, A., Dubcovsky, J., 2006. A NAC gene regulating senescence improves grain protein, zinc, and iron content in wheat. Science 314, 1298–1301.

Uppalapati, S.R., Ayoubi, P., Weng, H., et al., 2005. The phytotoxincoronatine and methyl jasmonate impact multiple phytohormone pathways in tomato. Plant J. 42, 201–217.

Usuda, H., 1995. Phosphate deficiency in maize. V. Mobilization of nitrogen and phosphorus within shoots of young plants and its relationship to senescence. Plant Cell Physiol. 36, 1041–1049.

Van Breusegem, F., Dat, J.F., 2006. Reactive oxygen species in plant cell death. Plant Physiol. 141, 384–390.

van der Leij, M., Smith, S.J., Miller, A.J., 1998. Remobilisation of vacuolar stored nitrate in barley root cells. Planta 205, 64–72.

van Doorn, W.G., Beers, E.P., Dangl, J.L., et al., 2011. Morphological classification of plant cell deaths. Cell Death Different. 18, 1–6.

Vellai, T., Takács-Vellai, K., Sass, M., Klionsky, D.J., 2009. The regulation of aging: does autophagy underlie longevity? Trends Cell Biol. 19, 487–494.

Verdenal, A., Combes, D., Escobar-Gutiérrez, A.J., 2008. A study of ryegrass architecture as a self-regulated system, using functional–structural plant modelling. Funct. Plant Biol. 35, 911–924.

Verma, V., Foulkes, M.J., Worland, A.J., Sylvester-Bradley, R., Caligari, P.D.S., Snape, J.W., 2004. Mapping quantitative trait loci for flag leaf senescence as a yield determinant in winter wheat under optimal and drought-stressed environments. Euphytica 135, 255–263.

Vijayalakshmi, K., Fritz, A.K., Paulsen, G.M., Bai, G., Pandravada, S., Gill, B.S., 2010. Modeling and mapping QTL for senescence-related traits in winter wheat under high temperature. Mol. Breed. 26, 163–175.

Vlot, A.C., Dempsey, D.A., Klessig, D.F., 2009. Salicylic acid, a multifaceted hormone to combat disease. Annu. Rev. Phytopathol. 47, 177–206.

Walters, D.R., McRoberts, N., 2006. Plants and biotrophs: a pivotal role for cytokinins? Trends Plant Sci. 11, 581–586.

Watanabe, M., Balazadeh, S., Tohge, T., et al.,2013. Comprehensive dissection of spatio-temporal metabolic shifts in primary, secondary and lipid metabolism during developmental senescence in Arabidopsis thaliana. Plant Physiol. 162, 1290–1310.

Weaver, L.M., Gan, S., Quirino, B., Amasino, R.M., 1998. A comparison of the expression patterns of several senescence-associated genes in response to stress and hormone treatment. Plant Mol. Biol. 37, 455–469.

White, P.J., Broadley, M.R., 2003. Calcium in plants. Ann. Bot. 92, 487–511.

Whitney, S.M., Houtz, R.L., Alonso, H., 2011. Advancing our understanding and capacity to engineer Nature's CO_2-sequestering enzyme. Rubisco. Plant Physiol. 155, 27–35.

Wills, R.B.H., McGlasson, W.B., Graham, D., Joyce, D., 2007. Postharvest: an introduction to the physiology and handling of fruit, vegetables and ornamentals, 5th edn. Cabi Publishing, Wallingford.

Wingler, A., Purdy, S.J., Edwards, S.A., Chardon, F., Masclaux-Daubresse, C., 2009. QTL analysis for sugar-regulated leaf senescence supports flowering-dependent and -independent senescence pathways. New Phytol. 185, 420–433.

Wingler, A., Roitsch, T., 2008. Metabolic regulation of leaf senescence: interactions of sugar signaling with biotic and abiotic stress responses. Plant Biol. 10 (Suppl. 1), 50–62.

Wisniewski, M., Norelli, J., Bassett, C., Artlip, T., Macarisin, D., 2011. Ectopic expression of a novel peach (*Prunus persica*) CBF transcription factor in apple (Malus × domestica) results in short-day induced dormancy and increased cold hardiness. Planta 233, 971–983.

Wu, X., 2009. Prospects of developing hybrid rice with super high yield. Agron. J. 101, 688–695.

Xu, G., Fan, X., Miller, A.J., 2012. Plant nitrogen assimilation and use efficiency. Annu. Rev. Plant Biol. 63, 153–182.

Yamauchi, T., Shimamura, S., Nakazono, M., Mochizuki, T., 2013. Aerenchyma formation in crop species: A review. Field Crops Res. 152, 8–16.

Yang, J.Z., Matthew, C., Rowland, R.E., 1998. Tiller axis observations for perennial ryegrass (*Lolium perenne*) and tall fescue (*Festuca arundinacea*): number of active phytomers, probability of tiller appearance, and frequency of root appearance per phytomer for three cutting heights. NZ J. Agric. Res. 41, 11–17.

Young, T.E., Gallie, D.R., 2000. Programmed cell death during endosperm development. Plant Mol. Biol. 44, 283–301.

Zhou, C., Han, L., Pislariu, C., et al., 2011. From model to crop: functional analysis of a STAY-GREEN gene in the model legume Medicago truncatula and effective use of the gene for alfalfa improvement. Plant Physiol. 157, 1483–1496.

Zimmermann, P., Zentgraf, U., 2005. The correlation between oxidative stress and leaf senescence during plant development. Cell. Mol. Biol. Lett. 10, 515–534.

CHAPTER

11

Improving water transport for carbon gain in crops

Timothy J. Brodribb[1], *Meisha-Marika Holloway-Phillips*[2], *Helen Bramley*[3]

[1]University of Tasmania, Australia
[2]Australian National University, Australia
[3]University of Western Australia, Crawley, Australia

1 INTRODUCTION

The most intuitive approach to achieving the goal of improving the rate at which crops produce food must be to increase the rate at which plants fix CO_2 into carbohydrates. In pursuit of this goal, much research focuses on the photosystems and carboxylation enzymes in crop plants as possible regions for improving plant performance, e.g. the development of C_4 rice (von Caemmerer et al., 2012); wheat yield consortium (Reynolds et al., 2011). Fortunately for plants, and unfortunately for plant breeders, this same goal has been the target of hundreds of millions of years of evolution that have converged upon an efficient set of solutions for capturing solar energy and transferring it to the fixation of atmospheric CO_2. As a result there are certain optima in core processes such as the efficiency of light harvesting (Björkman and Demmig, 1987), and the efficiency of carboxylation enzymes (Tcherkez et al., 2006) that may prove difficult to improve significantly by human tampering.

Evolution does provide other options for increasing plant output; in particular, the multitude of trade-offs that constrain variation in plant function in the natural environment also present opportunities for increasing the maximum performance of plants beyond the evolutionary limits expressed in existing plant species (Denison, 2009; Chapter 9). Evolutionary trade-offs effectively place a governor on the rate and amplitude of natural variation because mutations that improve one aspect of plant function typically come with attendant costs. One classic example is that modifications increasing the supply of CO_2 into leaves for photosynthesis inevitably increase the parallel flow of water out of the leaf, increasing the risk of desiccation and damage of the leaf. Hence natural selection tends to constrain the porosity of the epidermis (determined

Crop Physiology. DOI: 10.1016/B978-0-12-417104-6.00011-X

by stomatal characters) that maintains a rather conservative balance between the rates of photosynthesis and transpiration (Wong et al., 1979). Domestication of plants allows us to relax some of the selective pressures existent in nature, therefore providing scope to explore a wider range of functional possibilities where increased costs or risks to the plant can be artificially managed.

The water transport system in plants provides a great, and largely unexplored potential for improving plant performance beyond the limitations imposed by evolutionary selection. This is because plant water status, which critically affects two major functions of plant leaves linked to production – stomata-mediated photosynthetic gas exchange and expansion growth – is governed by the balance between the supply of water from the root into the leaf and its loss by transpiration. Understanding the limitations to water transport and what regulates the dynamic permeability of plant tissues provides an opportunity to maximize carbon accumulation and growth by ensuring sufficient water supply. In this chapter, we therefore examine the principles of plant water transport with particular focus on how the anatomy of the xylem water-delivery system in the roots and leaves affects the efficiency with which water is supplied for transpiration under ample and water-deficit soil conditions, and present a vascular-centric view of the potential for improving crop yield.

2 WATER TRANSPORT AND CARBON GAIN

2.1 Water transport as a limitation

The greatest single factor limiting the productivity of rain-fed crops, growing on fertile soil, is water. This is most clearly demonstrated by the fact that crop productivity can generally be predicted by water use (Fig. 11.1). There is a multitude of traits that affect transpiration and therefore, crop production. However, surprisingly there have been few traits identified that directly influence the physiological drivers of transpiration rate (see Box 11.1 for a historical

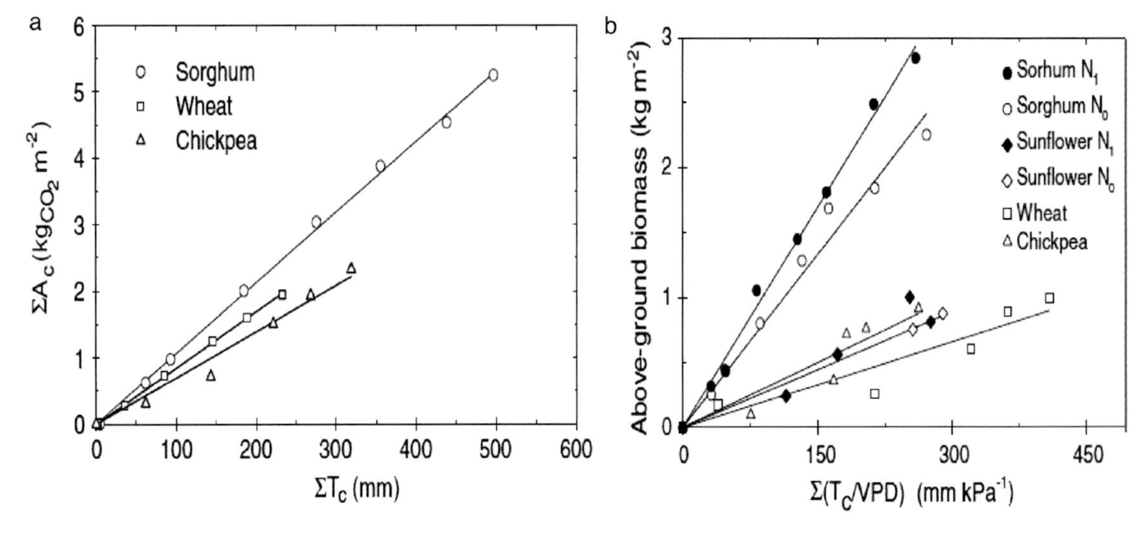

FIG. 11.1 The fundamental link between the amount of water used by a crop for transpiration and its growth and yield. (a) Relationship between cumulative daytime canopy net assimilation (A_c) and cumulative daytime canopy transpiration (T_c) for sorghum, wheat and chickpea; (b) relationships between above-ground biomass and cumulative crop transpiration normalized for day-time saturation vapor pressure deficit of the atmosphere (VPD). *Source: Steduto et al. (2007).*

BOX 11.1

VASCULAR-CENTRIC VIEW TO EXPLAINING WATER USE: NEW TARGETS FOR TRANSPIRATION REGULATION

The early analyses of crop water use (deWit, 1958; Fischer and Turner, 1978; Tanner and Sinclair, 1983) established the existence of a robust link between the amount of water used by a crop for transpiration and its growth and yield. Such a relationship is the consequence of the inextricable link between diffusion of CO_2 and water through the stomata during gas phase conductance. At the crop scale, the link between water use and biomass is commonly represented as the amount of water transpired multiplied by the efficiency by which water is converted to carbon – water-use efficiency – a crop specific factor. Passioura (1977, 1996) included an additional multiplier – harvest index – to account for the fact that not all carbon assimilated is directed to the harvestable component of the plant:

$$Y = W \times WUE \times HI$$

where, Y = yield, W = water transpired, WUE = water-use efficiency and HI = harvest index.

With respect to water, the equation suggests that production can be increased by either increasing plant water use and/or by increasing biomass production for a given unit of water consumed. However, because the two main components of interest in production systems tend to be growing season rainfall and yield, delineation between plant water use (the proportion of the environmental water supply that is transpired) and WUE (the translation of the water transpired to biomass) are often blurred into a single objective, popularly coined 'more crop per drop' (Giordano et al., 2006).

Targets for improving the efficiency of water use in the past have focused on crop or leaf level WUE including carbon isotope discrimination (Condon et al., 2004), and tolerance to cellular dehydration (Araus et al., 2003; Sinclair, 2011), as measured by survival on exposure to severe water stress. However, conditions where plant survival is required are unlikely to be economically viable and conflicting results have been found between carbon isotope discrimination and yield in water-limited environments, most likely because high WUE is often associated with plant traits that limit crop water use, such as early flowering and smaller leaf area (Blum, 2005 and references therein). This led Blum (2005) and others (Passioura and Angus, 2010; Sinclair, 2012) to suggest a reappraisal of the components of biomass production, emphasizing the need to maximize soil water use in order to maintain transpiration throughout the growing season.

The response has been a greater focus on water extraction capacities of the root to improve soil water mining, synchronizing growth duration with respect to seasonal patterns of rainfall, and regulation of transpiration to spread environmental water supply across the growing season (refer to reviews by Richards et al. (2010), Salekdeh et al. (2009), Sinclair (2012)). With regards to the latter, restricted transpiration is being achieved through selection for traits (e.g. low root and leaf hydraulic conductivity) that essentially reduce 'water wastage' by concentrating photosynthesis to periods of the day when WUE is likely to be at its greatest, e.g. by reducing night-time and midday transpiration (Sinclair et al., 2008; Sadok and Sinclair, 2011). With this shift in thinking, understanding the physiological drivers of transpiration has gained traction.

Transpiration at the plant level can be described according to two components: (1) the

BOX 11.1 *(cont.)*

driving force; and (2) the resistance of the pathway to water flow (which is represented as the reciprocal, conductance):

$$E = (\psi_{soil} - \psi_{leaf}).K_{plant}$$

where E is the leaf evapotranspiration rate, ψ is water potential, and K_{plant} is the hydraulic conductance of the plant.

K_{plant} in turn is a product of a number of anatomical features that define the flow capacity of the xylem and the integrity of the pipes under tension; (suggested reviews for further reading: Brodribb et al. (2010); Sack and Holbrook (2006); see section 3 for more details). These properties can be manipulated either to maintain water transport under decreasing ψ_{leaf} so that stomata can operate to much lower (more negative) leaf water potentials without detriment to recovery, or with increased sensitivity to vapor pressure deficit and leaf water potential to limit water use. The importance of stomatal closure to reduce the rate of ψ_{leaf} decline becomes apparent when viewed in the context of hydraulics, as damage to the vascular system by drought-induced cavitation (the formation of air bubbles within the xylem) impairs water transport and hence recovery of plants when water does arrive (Tyree and Zimmermann, 2002; Blackman et al., 2009); see more in section 4.3).

The limitations the water transport system places on gas exchange, however, are not restricted to the case of water-limited environments and, in this respect, nor the usefulness of considering yield potential as a function of available water. Yield potential is defined as the yield of a cultivar when grown in environments to which it is adapted, with nutrients and water non-limiting and with pests, diseases, weeds, lodging, and other stresses effectively controlled (Evans and Fischer 1999). However, water use has been paid minimal attention for the application of yield potential, except for canopy temperature depression (Fischer et al., 1998; Fischer, 2007), rather with emphasis placed on increasing transpiration efficiency (assimilation/transpiration) or the biochemical processes of assimilation (Foulkes et al., 2007; Murchie et al., 2009; Parry et al., 2011). This has also been the suggested approach to breeding for a warming environment (Ainsworth and Ort, 2010). However, hotter temperatures are usually associated with increased evaporative demand and therefore there will be an even greater need to improve the transport capacity of the plant to keep leaves hydrated (Roberts et al., 2012; Lobell et al., 2013). Maximum water transport efficiency perhaps is the greatest under-utilized property of the water transport system and therefore an exciting avenue for crop productivity.

It is by this virtue that we view a vascular-centric approach to plant water use as central to understanding opportunities and ramifications for biomass improvement under water-limited and water-ample conditions. Within this context, we define the water transport system as including the anatomical attributes and network assembly of the xylem which determine the physical limits to water flow, and the size and architecture of the root system which dictates the accessibility of soil derived water for transporting.

perspective on traits for transpiration). But soil water availability is only one potential source of dehydration, and hence reduced productivity, of crop plants. The capacity of the water transport network (xylem tissue) to deliver water from the soil to the sites of evaporation is finite, leading to an increase in water deficit in leaves as transpiration increases.

Perhaps the most easily observable demonstration of hydraulically-limited productivity occurs when photosynthesis is depressed during midday (Rouschal, 1938). In rice, photosynthetic depressions at peak irradiances are estimated to reduce daily carbon gain by up to 70% (Black et al., 1995; Murchie et al., 1999). These 'midday depressions' reflect the difference between the maximum observed value (morning peak) and the midday minimum, which gives rise to the well-accepted trough pattern in diurnal photosynthetic rates (Fig. 11.2). However, this calculation is likely to underestimate the reduction in photosynthetic potential at midday because photosynthetic limitation occurs well before the observation of a 'trough' in midday assimilation rates. In fact, photosynthetic depression begins at the point where assimilation rates do not continue to increase with increasing light intensity during the day. In maize, Hirasawa and Hsiao (1999) suggested that the midday depression was only minimal as assimilation plateaued mid-morning and then did not decline during midday. However, the maximum assimilation rate observed was between 5 and 10 μmol CO_2 m^{-2} s^{-1}, lower than the potential rate indicated by the response of assimilation to light intensity performed under low vapor pressure deficit (VPD),

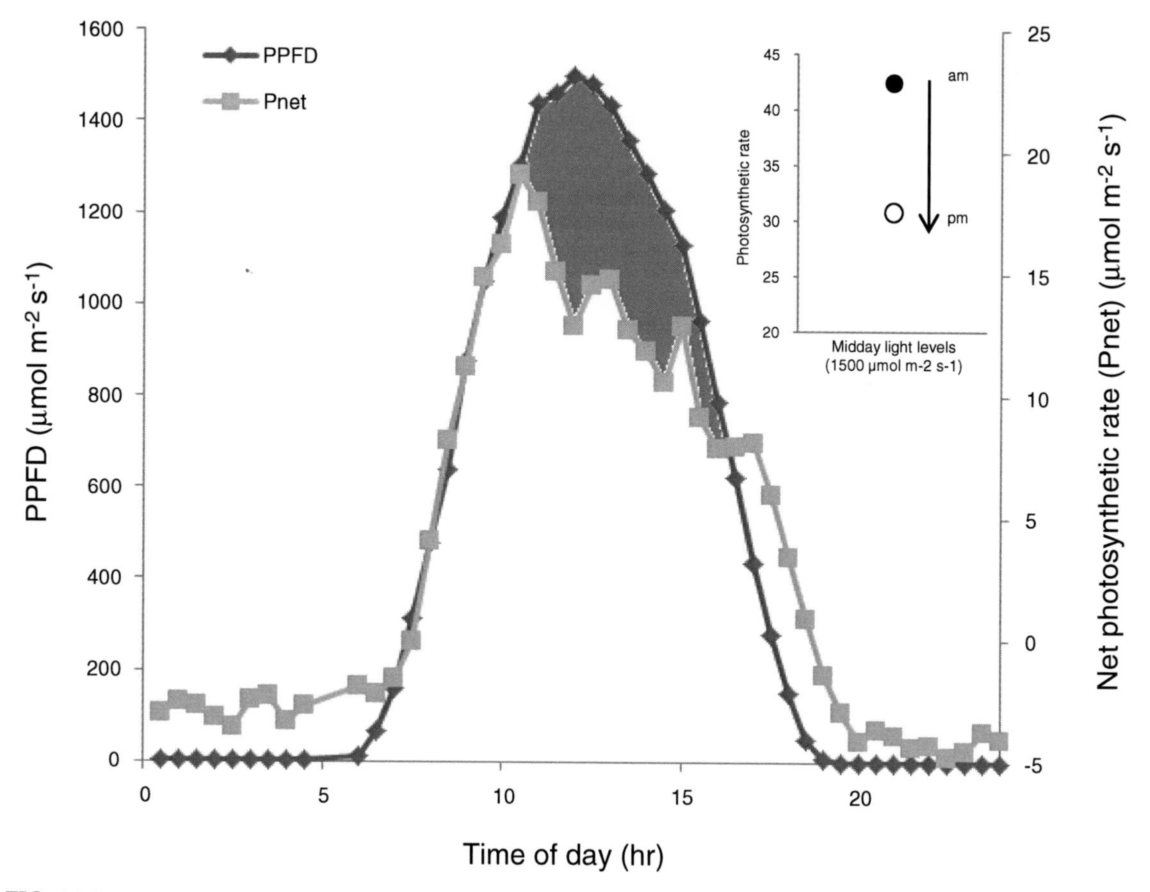

FIG. 11.2 Modeled diurnal progression of net photosynthesis and photosynthetic photon flux density (PPFD). The shaded area indicates the loss of potential carbon gain as vapor pressure deficit (VPD) increases during the day. Inset shows the reduction in photosynthetic rate under midday light intensity in the afternoon compared with the estimated rates under lower VPD in the morning. *Source: Hirasawa and Hsiao (1999).*

meaning high humidity, suggesting that radiation-use efficiency (RUE; dry-mass produced per unit intercepted solar radiation – g MJ^{-1}) beyond mid-morning was reduced as VPD increased (inset Fig. 11.2). While much work has been dedicated to understanding the reason of midday depressions, there are few estimates on the magnitude of the effect in terms of overall reductions in carbon gain and therefore the likely impact increasing RUE beyond midday is likely to have on productivity gains.

A principal cause of midday depression of photosynthesis in moist soil is the partial desiccation of leaves when the plant water transport system cannot fully replace water lost during rapid midday transpiration (Schulze and Hall, 1982; Brodribb and Holbrook, 2004a). The resultant water deficit causes stomata to close, thus limiting not only transpiration, but also photosynthesis (Brodribb and Holbrook, 2007). Owing to the reduction in transpirational cooling that occurs upon stomatal closure, canopy temperature has been used since the 1960s as a reliable indicator of water stress (Gates, 1964; Wiegand and Namken, 1966). This technique was formalized in the application as a crop water stress index by comparing relative differences in air and canopy temperatures of crops at different soil water contents (Idso et al., 1981; Jackson et al., 1981). Canopy temperature depression (CTD; Ta − Tc) has since been used to schedule irrigation and to evaluate cultivars for water use and tolerance to heat and drought, with greater CTD (cooler canopies) shown to be associated with greater yields in some situations, e.g. where water is limiting but present at depth (Lopes and Reynolds, 2010). This mirrors similar results found for the application of carbon isotope discrimination as a selection tool for transpiration efficiency (carbon assimilation/transpiration) (Condon et al., 2002), which is unsurprising given that both traits reflect stomatal limitations and a positive relationship between the two traits has been found (e.g. Fischer et al., 1998; Royo et al., 2002; Balota et al., 2008).

Midday depression provides an excellent example of transpiration outstripping the capacity of the xylem to maintain leaf hydration, and highlights the daily limits placed on plant gas exchange by the capacity of the xylem to transport water. Only in the last 15 years has the link between water transport and productivity been realized, and hence there is great potential for hydraulic improvement in crop plants.

2.2 Link between water transport and photosynthesis

Water transport and photosynthetic productivity are highly distinct processes that become tightly linked in terrestrial plants. Photosynthesis in air has the unavoidable consequence that, in exchange for CO_2, leaves lose vast quantities of water as transpiration. Failure to replace this transpired water vapor with liquid water transported from the soil would lead to desiccation and destruction of the photosynthetic apparatus. Thus all vascular plants have a dedicated water transport system that extracts water from soil and moves it to the sites of photosynthesis and transpiration in the leaves. As with any flowing fluid system, water traveling through the vascular system of a plant encounters a resistance to flow, meaning that even when soils are hydrated, leaves must exert a tension to pull water from the soil through the xylem into the leaf (Zimmermann, 1983).

The resultant tension is measured as a decrease in leaf water potential (Ψ_{leaf}) below zero. During transpiration, the water potential of leaves falls below that of the soil, and water flows passively along this water potential gradient in the non-living tubes of the xylem. Water potentials are analogous to electrical potentials (Dixon and Joly, 1895) in the sense that Ψ gradients determine water flow through the resistive hydraulic system in the same way that electrical resistors affect current flow produced by electrical potential gradients (Box 11.2, Eq. 3). In plants, more evaporation from leaves drives leaf

BOX 11.2

STOMATAL RESPONSIVENESS TO Ψ_{LEAF} ESTABLISHES THE CONNECTION BETWEEN HYDRAULIC AND PHOTOSYNTHETIC SYSTEMS

This connection can be represented by this series of equations:

1. $A = G_{sw}/1.6 \times \Delta CO_2$
2. $E = G_{sw} \times \Delta H_2O$
3. $\Delta\Psi = E/\ K_{plant}$
4. $\Psi_{leaf} = \Psi_{soil} - \Delta\Psi$
5. $G_{sw} = f(\Psi_{leaf})$

Starting with the rate of photosynthetic carbon assimilation (A), Eq. (1) relates CO_2 assimilation to the stomatal conductance to water vapor (G_{sw}); the factor 1.6 indicates that water vapor diffuses 1.6 times faster than CO_2 in air. Eq. 2 indicates the corresponding impact of G_{sw} upon transpiration. Transpiration rate (E) then determines the water potential gradient ($\Delta\Psi$) and leaf water potential (Ψ_{leaf}) inside the plant as a function of plant hydraulic conductance (K_{leaf}) (3 and 4). Because stomatal conductance is ultimately a function of Ψ_{leaf} (5) the final G_{sw} is constrained and regulates the assimilation rate (1).

water potential lower (more negative), which maintains an increased flux of water from the soil through the xylem to the leaf. However, the ability of leaves to sustain hydraulic tension is finite, and stomata close when leaf water potential falls below a threshold with a narrow range between −1MPa and −2MPa (Box 11.2, Eq. 5). Thus the conductivity of the xylem constrains maximum stomatal conductance and transpiration. The final link with photosynthesis occurs because water and CO_2 share a diffusional pathway through the stomata, so if xylem conductivity limits transpiration, then it also limits photosynthesis.

Confirmation of the mechanistic link between plant hydraulic conductivity and assimilation comes from the observation that variation among species in xylem conductivity is strongly associated with variation in maximum photosynthetic rate and growth of woody species. Originally, these correlations were based upon hydraulic conductivity measurements made in stems (Brodribb and Feild, 2000; Hubbard et al., 2001) but, more recently, it has been shown that the water supply characteristics of leaves are also strongly linked to photosynthetic rates (Brodribb et al., 2007; Flexas et al., 2013). Given that most food crop plants are non-woody, these recent advances in understanding hydraulics beyond the stem provide new perspectives with considerable relevance for modifying or breeding for hydraulic conductivity to enhance crop productivity.

Despite the potential for characteristics of the water transport system to be used in breeding, there has been little work establishing the magnitude of genetic variation in hydraulic conductivity in crop plants, and how this variation affects performance. A likely reason for this is the slow evolution of methods for accurately measuring xylem hydraulic conductivity in crop plants. Hydraulic conductivity in woody plants can be simply measured by measuring the resistance to the flow of water as it is pushed through excised branches (Sperry et al., 1988), but this technique is less suitable for herbaceous crop plants or the model plant *Arabidopsis thaliana* (Tixier et al., 2013), with little or no stem to work with. Unlike woody stems, pushing water through leaves and roots has the disadvantage that the flow pathway may not entirely follow the transpiration stream so different techniques are required.

One approach in monocots with linear leaves is simply to treat the leaf as a stem and to measure the resistance to pushing water through the large parallel veins as a means of measuring the axial part of the water transport network in leaves. A pot study employing this technique not only showed genetic variation in hydraulic conductivity among recombinant lines of *Avena barbarta,* but also that lines with higher hydraulic conductivity also had higher photosynthetic rates (Maherali et al., 2008). More recently, techniques suitable for measuring the hydraulic properties of crop plants have focused on whole leaves (Sack and Holbrook, 2006), where the rates of water uptake are used to quantify hydraulic conductivity. These studies suggest that for some of the principal crop species, variation in leaf water transport properties are primary determinants of photosynthetic rate (Holloway-Phillips and Brodribb, 2011b), biomass accumulation (Holloway-Phillips et al., in review) and the sensitivity of transpiration to changes in humidity (Ocheltree et al., 2014). Enhancing leaf hydraulic conductivity could hold the key to producing higher yielding crops while, conversely, it may be possible to reduce hydraulic conductivity as a means of increasing stomatal sensitivity to temperature and humidity, thus enhancing the efficiency of water use (Sinclair et al., 2008; Devi et al., 2012; Sinclair, 2012). Although the techniques employed for measuring leaf hydraulics are still under development, the data emerging from this approach indicate the leaf water transport system to be a primary source of functional variation in crop plants.

In short, the plant water transport system offers an opportunity to regulate plant hydration by manipulating the internal supply of water to plants rather than the external application of irrigation. Table 11.1 summarizes hydraulic-related traits that may provide new opportunities for improving crop performance; the majority of the suggested traits are based, however,

TABLE 11.1 Summary of hydraulic-related traits with potential relevance to the improvement of crop production under well-watered and water-limited conditions

Breeding potential	Trait	Function	Outcome	References
Biomass production under well-watered conditions	High hydraulic conductivity	Hydraulic conductivity sets the water supply capacity of the plant	(1) K_{plant} is linked to leaf level photosynthetic rate and stomatal conductance, (2) whole-plant growth rate and (3) biomass production	(1) e.g. in woody species (Brodribb and Feild, 2000; Hubbard et al., 2001), (2) e.g. in woody species (Domec and Gartner, 2003; Ducrey et al., 2008; Zhang and Cao, 2009; Fan et al., 2012), (3) in wheat (Holloway-Phillips et al., 2013 unpublished)
	Large vessel diameter	Water flux through the xylem is proportional to the diameter of the vessel raised to the power of 4. Therefore small changes in vessel width have a substantial effect on water flux	Hydraulic conductivity increases with vessel diameter	Based on the Hagen–Poiseuille law of fluid dynamics (Blizzard and Boyer, 1980; Woodhouse and Nobel, 1982; Zimmermann, 1983; Sperry and Hacke, 2004)

TABLE 11.1 Summary of hydraulic-related traits with potential relevance to the improvement of crop production under well-watered and water-limited conditions *(cont.)*

Breeding potential	Trait	Function	Outcome	References
	High vein density	The greater the vein density, the smaller the distance water must travel from vein endings to the sites of evaporation within the leaf	Reducing the resistance to water transport in the xylem and indirectly in the extra-xylary pathway increases hydraulic conductivity in the leaf	In woody species (Sack and Frole, 2006; Brodribb et al., 2007; Brodribb and Feild, 2010)
	Aquaporins	Aquaporins gate liquid water transport in the symplastic pathway	Reduces the resistance to water movement in the extra-xylary pathway to increase hydraulic conductivity	General review (Maurel et al., 2008); roots and aquaporin review (Bramley et al., 2013); grapevine (Perrone et al., 2012a), peanut (Devi et al., 2012)
	Increased root branching and root length density	Water uptake by roots is dependent on total surface area	Increases root hydraulic conductivity	
	High ratio of vein density to stomatal density	Water supply by the vascular system (vein density) must replace that which is lost during gas exchange via stomata (stomatal density). If supply exceeds the demand set by stomata, then leaf water potential and hence gas exchange may be buffered to a greater extent under high evaporative conditions	Lower sensitivity of gas exchange to perturbations in humidity in woody species. The trade-off, however, is that more water is lost for a given carbon gain	Brodribb and Jordan (2008), Carins Murphy et al. (2012)
	High root pressure	To refill emboli in the xylem when the plant is not transpiring	Efficient xylem repair should reduce the impacts of soil or atmospheric water stress on photosynthesis	Commonly observed in monocots (Milburn, 1973; Cochard et al., 1994; Stiller et al., 2003; Cao et al., 2012)
	Increased frequency of stomata on upper leaf surface	Higher maximum gas exchange and productivity	Greater stomatal conductance without a linked increase in vein density potentially improves the efficiency of resource use	Brodribb et al. (2013), Milla et al. (2013)

(continued)

TABLE 11.1 Summary of hydraulic-related traits with potential relevance to the improvement of crop production under well-watered and water-limited conditions *(cont.)*

Breeding potential	Trait	Function	Outcome	References
	Reduced genome size or cell size	Linked with higher densities of leaf veins and stomata	Higher density of conducting and photosynthetic tissue producing higher productivity	Beaulieu et al. (2008), Franks and Beerling (2009), Brodribb et al. (2013), Feild and Brodribb (2013)
Biomass production under water-limited conditions	Decreasing the vulnerability of xylem to cavitation by selecting for increased vessel wall thickness and/or decreased lumen breath	Emboli in the xylem restrict water transport causing photosynthetic depression and ultimately death	Greater resistance to cavitation is linked with increased drought tolerance. The collapse pressure of vessels has also been attributed to the ratio between vessel wall thickness and lumen breath	Woody species (Blackman et al., 2009, 2012; Brodribb and Cochard, 2009).
	Large hydraulic safety margin	Stomata can slow the rate of decline in leaf water potential and hydraulic conductivity under water stress by closing early during water stress exposure	A positive safety margin, where stomata close well before any substantial decline in hydraulic conductivity has been shown to be an important mechanism for certain woody species to survive soil water deficits	Woody species (Brodribb and Holbrook, 2004b; Choat et al., 2012). Forage grasses (Holloway-Phillips and Brodribb, 2011b)
	High major vein density in leaves	Provide for alternative routes for water transport if certain veins cavitate under drought	Increases P50, i.e. the leaf water potential at which 50% loss of hydraulic conductivity has occurred, and therefore drought tolerance	Scoffoni et al. (2012)
	Small vessel diameter	Increases the resistance of the xylem water pathway	Helps to reduce plant water use (conserve soil water) and therefore longer term water stress	Passioura (1972), Richards and Passioura (1989b)
	Aquaporins	Short-term regulation of water transport through the symplastic pathways in leaves and root	Reduces hydraulic conductivity in response to abiotic stress	Azaizeh et al. (1992), Gloser et al. (2007), Bramley et al. (2010)

TABLE 11.1 Summary of hydraulic-related traits with potential relevance to the improvement of crop production under well-watered and water-limited conditions *(cont.)*

Breeding potential	Trait	Function	Outcome	References
	Low ratio of vein to stomatal density	If supply is insufficient to meet demand set by stomata, under high evaporative conditions, LWP declines triggering stomatal closure and hence reducing transpiration rate	Restricted transpiration rate has been characterized in (1) soybean, (2) sorghum, (3) maize and (4) forage grasses and (5) show to be related to reduced leaf hydraulic conductivity)	(1) Sadok and Sinclair (2010), (2) Gholipoor et al. (2013), Choudhary and Sinclair (2013), (3) Gholipoor et al. (2013), (4) Sinclair et al. (2008), (5) Wherley and Sinclair (2009)

on research at the leaf-level in non-crop species. Therefore, there is still much research required for their validation in field crops. The functional characteristics of xylem systems vary enormously among species. At the scale of the individual plant, this variation results in two major constraints to carbon gain – via a limit on the maximum rates water can be transported from soil to leaves (section 3) and via the responsiveness of the water transport system to changing water supply/demand as a result of a finite capacity of the system to transport water, which results in impairment of transport via cavitation if the supply capacity is exceeded (section 4).

3 DETERMINANTS OF WATER TRANSPORT

3.1 Maximum efficiency in leaves

Much of the formative measurement and experimentation in plant hydraulics was carried out on the stems of woody species, focusing on one-dimensional flow along the plant axis (Tyree and Zimmermann, 2002). More recently, leaves have been the subject of new research providing insights that have more relevance for the typically non-woody cast of crop plants. In stems, hydraulic conductance is proportional to the size of xylem conduits, due to the physical law that relates fluid conductivity of a tube to the fourth power of its radius (Sperry and Hacke, 2004). In leaves, the situation is rather different because the transpiration stream not only flows through the non-living xylem 'plumbing', but is also required to pass through the living mesophyll, thus replacing water evaporating from mesophyll tissue near the stomata (Brodribb et al., 2010). Because the pathway for water flow through living cells is extremely slow and tortuous compared with the flow within xylem tissue, the xylem vein network delivers water close to the sites of evaporation in the leaf, thereby minimizing the distance water must travel through living cells outside the xylem (Fig. 11.3a; Raven, 1977). The 'extra-vascular' pathway between the veins and the sites of evaporation constitutes a major hydraulic resistance in the leaf, and the length of this pathway is largely determined by the density of minor veins in the leaf lamina (Sack and Frole, 2006; Brodribb et al., 2007). Thus it appears that a key adaptive mechanism for plants to modify the hydraulic conductivity of the leaf is by variation in the density of leaf venation (Boyce et al., 2009; Brodribb and Feild, 2010). In

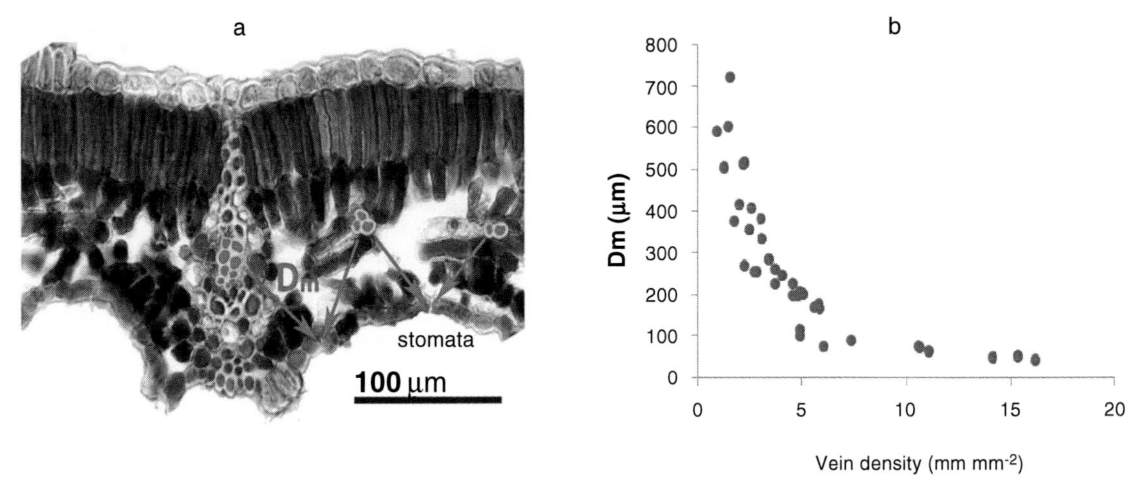

FIG. 11.3 (a) Leaf cross-section from *Curatela americana* showing the pathway for water movement (D_m) from the vein ending (vascular bundle) through the living tissue of the mesophyll to the stomata, where water finally exits as vapor. (b) Mesophyll path length (D_m) decreases as the pathway resistance for water decreases as a function of vein density (D_v) across a range of taxa. *Source: (b) Brodribb et al. (unpublished).* Please see color plate at the back of the book.

addition to this, there is growing evidence that plants are able to modify the conductivity of the hydraulic pathway outside the xylem network by the regulated expression of protein channels in the plasma membrane that facilitate water passage through the membrane (called aquaporins) in the leaf (section 4). Both vein density and aquaporin expression constitute leaf characters that could be manipulated by breeding or genetic manipulation either by direct measurement (methods technically difficult and time consuming), linked traits or using molecular markers, thus providing potential targets for crop improvement.

3.1.1 *Vein density – a key determinant of leaf hydraulic efficiency*

The density of vein branching in the leaf lamina is fundamentally important in determining the hydraulic conductivity of leaves, and hence their photosynthetic and water-use characteristics. Broadly, among land plants, there is a strong relationship between the length of the hydraulic pathway from veins to stomata, and the hydraulic conductivity (K_{leaf}) of the whole leaf (Brodribb et al., 2007). Longer flow pathways in species such as ferns and primitive angiosperms with low vein density are associated with low K_{leaf} and rates of assimilation, while species with high vein density clearly enjoy higher K_{leaf} and assimilation rates (Fig. 11.4; Brodribb and Feild, 2010; Flexas et al., 2013). In this context, it is impressive to note the extent of variation in vein density that exists among plant species. Studies of living plant species show that vein density ranges from less than 1 mm of vein length mm^{-2} of leaf area in many ferns and conifers, to more than 25 mm mm^{-2} in tropical angiosperm trees (Sack and Scoffoni, 2013). Notably, it was demonstrated that a substantial rise in vein density seems to have occurred during the evolution of angiosperms in the Cretaceous epoch (Brodribb and Feild, 2010; Feild et al., 2011) and this has been proposed as a significant contributing factor to the very high photosynthetic and productivity maxima among flowering plants.

While high vein density is important for high rates of C_3 photosynthesis, it also appears to be a prerequisite for the evolution of C_4 photosynthesis (Sage et al., 2012). In C_3 plants, high vein

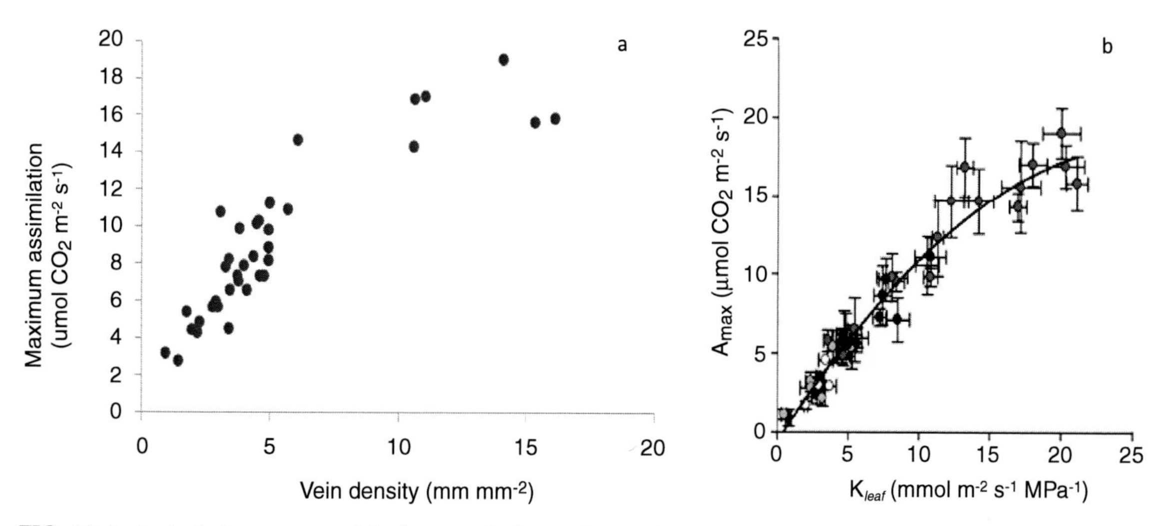

FIG. 11.4 Assimilation rate per unit leaf area can be boosted through increasing the water supply capacity of leaves as represented by (a) vein density and (b) the maximum hydraulic conductivity of the leaves (K_{leaf}) in bryophytes (black), lycopods (white), ferns (green), conifers (red), angiosperms (blue), and gymnosperms (brown) evaluated in the field under conditions of high soil water availability. *Source: Brodribb et al. (2007).* Please see color plate at the back of the book.

density enhances stomatal conductance due to more efficient water delivery, but in C_4 plants a higher density of veins also provides more surface area of photosynthetic bundle sheath cells, and more contact with mesophyll cells (Sage et al., 2013). Vein density thus appears to be a rather universal trait for improving photosynthetic yield regardless of the photosynthetic pathway or taxonomic affinity.

The water-delivery system of veins in leaves varies strongly across major plant lineages but, for angiosperms, consists of a hierarchy of vein orders which, among other developmental aspects, decrease in vessel diameter with increasing vein order, forming a mostly closed system with in-built redundancy (alternative pathways for water movement) (Wylie, 1939). In dicots, the patterning is termed reticulate, with a hierarchical organization of closed vein loops whereas, in monocots, the pattern is striate with longitudinal veins of several orders running parallel to the leaf, interconnected with small transverse veins (Fig. 11.5). The lower order or major veins having the largest diameter vessels are largely responsible for the rapid transport of water along the length of the leaf, as well as for mechanical support (Price et al., 2013). However, the vast majority of the total vein length is comprised of the higher order, minor veins which are responsible for leaking water out into mesophyll. Thus, as the bulk of transpired water is drawn through these veins, high minor vein densities contribute to high water supply capacity through increasing the surface area for exchange of xylem water with surrounding mesophyll tissue and reducing the distance through which water travels outside the xylem. Minor veins are also the sites of assimilate loading for distribution within the phloem to the rest of the plant and, as such, have been considered to be similarly important in setting maximum assimilation rates through reducing the build-up of sugars which can inhibit photosynthesis via feedback regulation (Adams et al., 2007).

Vein density is typically counted in leaves that have been cleared with a bleaching agent and treated with a lignin stain to highlight the xylem conduits. Counting vein density remains

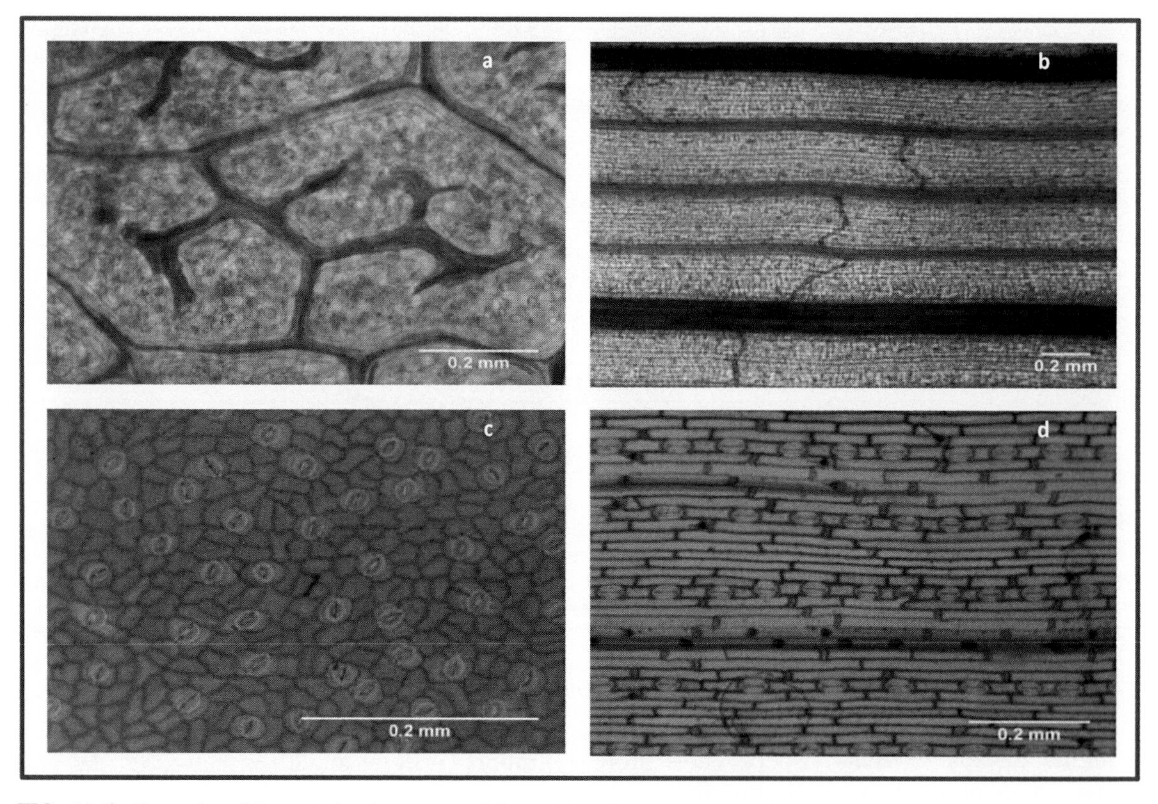

FIG. 11.5 Examples of the reticulated venation of dicots (a) and striate pattern of monocots with longitudinal veins cross-linked by smaller transverse veins (b), and corresponding stomatal patterning (c) dicot and (d) monocot. Source: (a and c) Carins-Murphy (unpublished); (b and d). *Source:Holloway-Phillips et al., unpublished*. Please see color plate at the back of the book.

an arduous process of tracing vein pathways from microscope images of prepared leaves, a method that is unsuitable to high-throughput phenotyping. However, developments using image tracing software and pattern recognition are providing promising new avenues for the automation of this process (Price et al., 2011; Larese et al., 2014), although issues with sample preparation and magnification currently prevent accurate estimation of the critical minor vein density at a whole leaf scale (Sack et al., 2012). These technical difficulties should be resolved by improved software and it should only be a matter of time before venation parameters can be extracted relatively quickly from trial plants.

The majority of crop species where leaf vasculature has been described include monocot crop and forage species which occupy the lower half of the reported angiosperm range for vein density (Fig. 11.6). The potential for vein density to be targeted as a means of modifying productivity or water-use efficiency by conventional breeding depends largely on the degree of genetic variation expressed in vein density, and the relative size of genetic, environmental and interaction controls of the trait. Studies in tree species show considerable variation both within and between trees (Sack and Holbrook, 2006). Within-tree variation in vein density is largely associated with light climate, with examples of >40% lower vein density in shade

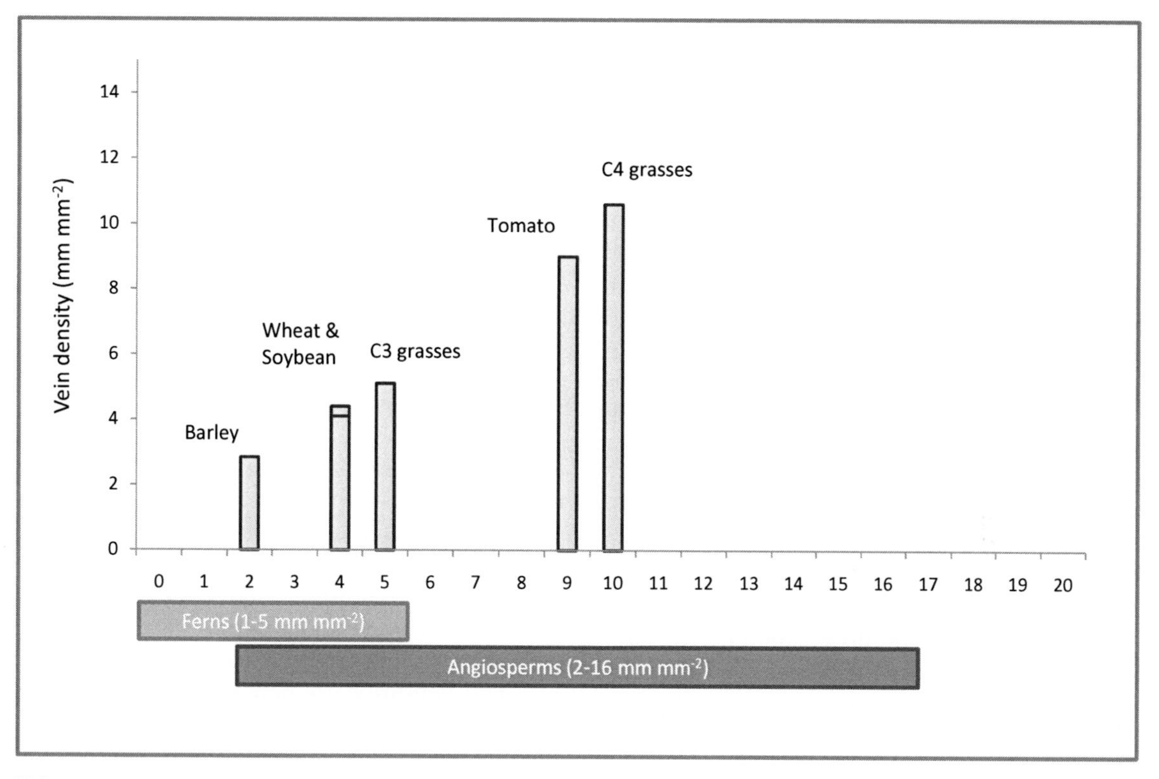

FIG. 11.6 Average vein densities (Dv) of common crop plants sit within the lower half of the reported angiosperm range. *Sources: barley (Dannenhoffer et al., 1990), wheat (3.78–4.50 mm mm^{-2}) Holloway-Phillips et al., 2013, unpublished; soybean (3.54–5.25 mm mm^{-2}) (Tanaka et al., 2010); C_3 and C_4 grasses (4.5–14.3 mm mm^{-2}) (Ueno, 2006) and tomato (Giraldo et al., 2013). Ranges of Dv for crop plants are averages across varietal comparisons. Fern and angiosperm ranges are those reported in Brodribb et al. (2007).*

versus sun leaves from species inhabiting forest communities (Zwieniecki et al., 2004; Carins Murphy et al., 2012). This within tree variation has been explained in terms of plants optimal investment in veins by building high vein density in sunlit leaves with high potential for photosynthesis, while minimizing vein investment in light-limited shade leaves (Brodribb and Jordan, 2011). The cost of veins occurs for two reasons – veins are constructed mainly of lignin, which is carbon dense (Lambers and Poorter, 1992), and vascular tissue displaces photosynthetic tissue, which therefore has the potential to reduce carbon return (Feild and Brodribb, 2013). Herbaceous species appear to exhibit much lower plasticity in vein density as might be expected by their limited capacity to experience both sun and shade conditions in the same plant (Uhl and Mosbrugger, 1999). However, in glasshouse-grown wheat, variation in vein density was correlated with shoot and ear biomass at anthesis in 13 Australian commercial wheat varieties, but there was a weak genetic component to this variation (Holloway-Phillips et al., in review). Field studies in soybean show considerable variation in vein density among inbred lines, and that this variation is associated with growth type and stomatal density (Tanaka and Shiraiwa, 2009). Much more work is needed to understand the genetic control and environmental modulation of vein density variation before the potential for modifying crop function can be assessed.

Testament to the centrality of leaf vein density to the photosynthetic performance of leaves is the very tight correlation between the density of stomata and leaf minor veins. Recent studies indicate that a high level of coordination between these traits exists both within and between species (Carins Murphy et al., 2012; Brodribb et al., 2013; Li et al., 2014). This coordination makes sense in terms of optimization of investment by plants in the manufacture of leaf veins and stomata because vein density determines the conductance of liquid water through the leaf, while the size and density of stomata limit the diffusive conductance of water vapor escaping the leaf as transpiration. The magnitude of these conductances determines maximum photosynthesis and transpiration, while the ratio of these conductances dictates the sensitivity of leaves to changes in humidity (Brodribb and Jordan, 2008), temperature and crop boundary layer. Observations of plants growing in natural systems indicate that the optimum ratio between veins and stomata occurs when hydraulic conductivity of the leaf is high enough to maintain stomata open under optimal photosynthetic conditions, but low enough to ensure sensitivity of stomatal conductance to changes in evaporative demand (Brodribb et al., 2013). Thus, plants grown under one environment, say shade, will perform suboptimally in terms of stomata remaining substantially closed when moved to full sun. In this context, it is clear that there is potential to manipulate the sensitivity of crop plants to humidity and temperature by modifying the ratio of vein density to stomatal density. This adds an important dimension to work on stomatal density and size in crop plants (Milla et al., 2013), with scope to produce plants where photosynthesis is insensitive to humidity due to a high ratio of vein to stomatal density. Conversely, it should be possible to find crop varieties with relatively low vein density relative to stomatal density resulting in lower productivity but high water-use efficiency.

3.2 Maximum efficiency in roots

There are few studies incorporating root hydraulic properties of crop plants in their trait-based selection and breeding programs, despite recognition for at least the past century of the importance of root systems in water and nutrient uptake (Weaver et al., 1922). Direct selection of root-based traits is considered 'impractical' (Reynolds and Tuberosa, 2008), because roots in the field are largely inaccessible and there are insufficient techniques to quantify rapidly suitable features for selection. Canopy temperature depression is often used as a surrogate measurement implying water uptake by plant roots (Reynolds and Tuberosa, 2008), but CTD is only useful in some environments and for some species (Bramley et al., 2013). Moreover, root-based traits are not the only factor contributing to CTD (discussed above) (Bramley et al., 2013).

Modern phenotyping programs generally use artificial conditions (Chen et al., 2011; Gowda et al., 2011; Wasson et al., 2012) to identify root architectural traits, such as angle of root axes (Manschadi et al., 2006) and root length density (Manschadi et al., 2006; Palta et al., 2011; Comas et al., 2013), which are believed to be beneficial in water-limited environments. Smaller root angle leads to deeper roots in some species (Manschadi et al., 2006; Gowda et al., 2011) and greater root length density will improve water capture, but vigorous root systems could deplete soil moisture too rapidly in crops growing on predominantly stored soil moisture (Palta et al., 2011). Root growth diverts assimilates away from productive organs, so ratios of leaf area to root surface area or root length are more functionally descriptive (Comas et al., 2013) and may be more useful allometry measurements. Root traits characterized under artificial conditions need verification in the field (Wasson et al., 2012).

Root systems account for up to half of the total plant resistance to water flow (Steudle, 2000; Javot and Maurel, 2002). Therefore, a fundamental understanding of root hydraulic properties and how these relate to plant performance is needed to improve crop production.

3.2.1 *Root hydraulic conductivity*

The efficiency of a root system in capturing water from the soil and transporting it to the shoot is determined by the hydraulic properties of its components, which are determined by root morphology, anatomy and aquaporin activity. Roots are comprised of a series of concentric tissue layers of varying size and structure surrounding a central cylinder containing the vascular tissue. Water taken up by a root travels radially through the apoplast via intercellular spaces and cell walls or through the symplast via plasmodesmata and crossing cell membranes. On reaching the vasculature, water travels axially to the shoot through metaxylem vessels. Axial resistance to water flow is low in comparison to the radial path (Frensch and Steudle, 1989; Amodeo et al., 1999; Bramley et al., 2009), but the focus in crop roots has been to reduce xylem diameter to increase root hydraulic resistance (Richards and Passioura, 1989a; Koizumi et al., 2007; Schoppach et al., 2013). This is considered to be a water-conserving strategy enabling crops growing on stored soil moisture to have sufficient water remaining in the soil profile for the grain-filling stage (Richards and Passioura, 1989a). Incorporating a higher root hydraulic resistance to conserve water is also now being considered as a way to increase yields in other crop species growing under water-limited environments (Choudhary and Sinclair, 2013). However, higher root resistance to water flow reduces the rate of water supply to leaves, increases the likelihood of water stress and cavitation and also probably results in slower growth, even when the soil water supply is sufficient. In addition, slower growth or smaller canopies could result in less shading of soil surface (Fischer, 2007; Mullan and Reynolds, 2010) so the water conserved through reduced transpiration may be offset by higher evaporation (Chapter 7). Higher root resistance to water flow is also associated with stomatal closure and midday depression in photosynthesis (Adachi et al., 2010). In rice, the most productive cultivars are able to maintain midday photosynthesis through better maintenance of leaf hydration due to higher root hydraulic conductance (lower resistance) (Adachi et al., 2010; Taylaran et al., 2011). Higher root hydraulic conductance is also correlated with shoot dry weight in rice genotypes under soil water deficit (Matsuo et al., 2009), implying that low root resistance may also be a beneficial trait in this crop species under drought.

The root radius has a high resistance to water flow because the anatomical structures are significant obstacles to cross. Suberization and lignification of cell walls in the endodermis and exodermis (if present) block water flow through the apoplast, forcing it to cross cell membranes to enter the symplast. Membrane transport of water is generally facilitated by aquaporins whose abundance and gating provide fine regulation of the rate of water flow through tissue and organs (Javot and Maurel, 2002; Tyerman et al., 2002). Manipulation of aquaporin expression may be a way genetically to improve root hydraulic efficiency and could be selected more easily than architectural traits in breeding programs through identification of molecular markers. Bramley et al. (2013) have recently discussed genetic variation in root anatomy, their hydraulic properties and aquaporin expression in crop species.

There is a widely held belief that smaller root diameters convey a higher hydraulic conductivity (Comas et al., 2013). Based on physical geometry and what we know about the axial and radial hydraulic resistances, this is supported theoretically because the path length from the exterior to the interior of the root is shorter and the absorbing area per unit root volume is larger in smaller diameter roots. However, inter- and intra-specific comparisons have demonstrated that root diameter and hydraulic conductivity are not always directly related (Bramley et al., 2009; Schoppach et al., 2013). Of greater importance is the route that water takes crossing the root radius and the obstacles it encounters (Bramley et al., 2009) and the diameter of the xylem as narrower roots tend to have narrower xylem vessels (Gowda et al., 2011; Schoppach et al., 2013). The most efficient structure should therefore be a root with wider or more abundant xylem vessels and a thin cortex.

Size of the root system is an important trait for effective water use, but it is not necessarily size (total length) that determines its hydraulic efficiency and both of these components should be considered when targeting crop adaptation to specific environments. Larger root systems should have greater hydraulic conductance, as has been observed for kiwifruit vines (Black et al., 2011) and lupins (Bramley et al., 2009), but the region of root involved in water uptake will influence total root hydraulic conductance and depends on the species (Bramley et al., 2009). For example, the hydraulic conductivity of a wheat root system may be more dependent on the amount of branching and lateral root production than total root length because the rate of water uptake is greater near the root apices. Smaller root systems have been inadvertently selected in breeding programs for greater yield in wheat (Waines and Ehdaie, 2007), probably because less carbon investment in root biomass has resulted in greater carbon allocation into grains, as indicated by higher harvest indices in modern cultivars (Siddique et al., 1990). The decrease in time to anthesis in modern cultivars (Siddique et al., 1990) could affect partitioning, but shorter growing season through later sowing tended to increase rather than decrease root biomass in near isogenic lines for the dwarfing gene *Rht* (Miralles et al., 1997). The differences in root architectural traits or hydraulic properties of modern and older wheat cultivars are not known, but recent interest in vigorous wheat genotypes has shown that they produce more branches and have high root length density, i.e. root length per unit soil volume (Liao et al., 2006). Modern cultivars were considered more opportunistic in their water use compared with the older cultivars as they had higher stomatal conductance when water was available, but rapidly closed stomata when water was limited (Siddique et al., 1990). Genetic variation in stomatal regulation (section 4) and water-use strategies exists in modern wheat cultivars and has been attributed to potential differences in root hydraulic resistance (Saradedevi et al. under review).

4 MAINTENANCE AND REGULATION OF WATER TRANSPORT

Although the maximum hydraulic conductivity of stems is largely static, being determined by the anatomy of the xylem, the great majority of the resistance to water flow in crop plants occurs in leaves and roots, and these tissues can display considerable dynamism in their hydraulic conductivity in response to abiotic stress. This dynamism is a critical determinant of the performance of crop varieties under limiting water supply or highly evaporative demand. Crop varieties that produce higher yields under water-limited conditions have been approached through breeding for higher transpiration efficiency (Condon et al., 2004) and other adaptations including stem carbohydrate remobilization, delayed leaf senescence and rooting depth (Richards et al., 2010). The physiology of the xylem lies at the core of plant water management, and should thus provide avenues for profoundly modifying the way plants respond to water stress. The following section examines how this dynamism in crop plant water transport can be attributed to rapid changes in membrane permeability or rapid cavitation and refilling of xylem embolisms.

4.1 Hydraulic regulation in roots

Hydraulic conductivity in many crop species changes diurnally (Henzler et al., 1999; Hachez et al., 2012; Henry et al., 2012) in tune with the demands of transpiration set by the environment (Vandeleur et al., 2013). Without this regulation, leaf water status would be more negative during the day (Tsuda and Tyree, 2000), potentially reaching critical xylem tensions that induce cavitation even under sufficient soil water availability. Roots appear to be the most flexible in terms of regulating hydraulic conductivity in response to their environment. The region of root or part of the root system involved in water uptake may also vary temporally (Lawlor, 1973; Bramley et al., 2010). Abiotic perturbation such as hypoxia, salinity, nutrient status, temperature and light can induce changes in root hydraulic conductivity within minutes or hours (Azaizeh et al., 1992; Lee et al., 2005; Gloser et al., 2007; Bramley et al., 2010). All these short-term responses are generally reversible and correlate with aquaporin activity (Tournaire-Roux et al., 2003; Boursiac et al., 2005; Hachez et al., 2012). Most crop species have at least 30 individual aquaporin isoforms and many PIPs (plasma-membrane intrinsic proteins) are only expressed in roots (Bramley et al., 2007). Some aquaporins are predominantly expressed in the endodermis (Perrone et al., 2012) and it is this tissue that is most likely responsible for regulating root hydraulic conductivity (Bramley et al., 2007).

As expected, high PIP gene expression, especially PIP2s, is often correlated with high root hydraulic conductivity (Perrone et al., 2012a; Gambetta et al., 2013). Within *Arabidopsis* accessions, root hydraulic conductance varied two-fold, which was related to expression of certain PIPs but not the level of suberization (Sutka et al., 2011). Although the functional roles of most aquaporins are still unclear, including the identification of specific aquaporins responsible for these hydraulic changes, advances have been made in characterizing expression patterns (Boursiac et al., 2005; Di Pietro et al., 2013; Li et al., 2013) in relation to root anatomy (Gambetta et al., 2013), post-translation regulation that controls pore gating (Törnroth-Horsefield et al., 2006; Di Pietro et al., 2013) and the role of signaling molecules induced by stress (Benabdellah et al., 2009; Di Pietro et al., 2013).

Root hydraulic conductivity also changes seasonally, with plant development and long-term responses to abiotic stress through changes in anatomy and morphology (Barrowclough et al., 2000; Henry et al., 2012; Vandeleur et al., 2013). In a study where a drought-sensitive Merlot variety was grafted onto two different rootstocks, the more drought tolerant root system had greater root hydraulic conductance during summer, and Merlot clones with the drought tolerant root system maintained higher stomatal conductance (Alsina et al., 2011). Greater root hydraulic conductance was associated with new roots and larger root system rather than changes in anatomy or conductivity of individual roots (Alsina et al., 2011). Root systems typically increase in size with plant development, but there also appears to be a link between root surface area and aquaporin expression as root size increased in antisense *Arabidopsis* and tobacco with inhibited expression of specific aquaporin isoforms (Kaldenhoff et al., 1998; Martre et al., 2002; Siefritz et al., 2002). Changes in anatomy in response to abiotic stress tend to increase the radial resistance to water flow such as increased deposition of suberin in response to soil water deficit (Vandeleur et al., 2013) and aerenchyma development in response to hypoxia (Ranathunge et al., 2003). Increased suberization is likely to aid water conservation, preventing loss of water from the root into dry soils. Under severe drought, root hydraulic resistance universally increases, whether due to changes in root anatomy, reduced contact with the soil due to root shrinkage and root death or reduced aquaporin abundance. Most of these changes are irreversible and reliant on new root growth when water supply returns.

Examination of root architectural traits alone will not identify the most suited root system for particular environments. Root hydraulic efficiency and root hydraulic properties need to be considered along with phenotypic plasticity (i.e. the change in phenotype in response to the environment) when selecting suitable crop germplasm for breeding programs.

4.2 Hydraulic regulation in leaves

The function of the venation network is to minimize the water flow pathway outside the xylem, through living cells. This 'extra-xylary pathway' constitutes a water transport bottleneck, explaining why leaves with a high vein density and hence a short extra-xylary pathway have high K_{leaf}. However, the conductivity of the extra-xylary pathway can be dynamically modified by the expression of aquaporins in the xylem bundle sheath (Nardini et al., 2005; Shatil-Cohen et al., 2011; Pantin et al., 2012), thus increasing K_{leaf} without changing the anatomy of the leaf, but rather by dynamic gene expression. The importance of aquaporin expression in regulating root hydraulic conductance has been known for some time (section 4.1), but their function in leaves is less clear.

There remains uncertainty about the magnitude of the aquaporin effect in leaves, but aquaporin inhibitors (mercury and silver) can reduce K_{leaf} by 30–40% (Aasamaa et al., 2005; Nardini and Salleo, 2005). The appeal of aquaporins for manipulating K_{leaf} in commercial plants has been recognized, and apparent varietal variation in aquaporin activity (Devi et al., 2012; Perrone et al., 2012b; Sinclair, 2012) suggests manipulation of this trait could lead to varieties with desirable hydraulic characteristics. In addition, the potential dynamism of aquaporin expression provides an opportunity to regulate water transport in response to short-term growing conditions such as temperature, CO_2, VPD and light (Cochard et al., 2007; Flexas et al., 2013). Downregulation of aquaporin during the early stages of water stress has been linked with incipient declines in K_{leaf} but, under sustained water stress, the effects of cavitation appear to outweigh those of aquaporin (Pou et al., 2013). Furthermore, it is possible that aquaporins may contribute to the coordination of water transport and photosynthetic activity in leaves given that aquaporins are not only responsible for facilitating water movement into living cells, but are also critical for the transfer of CO_2 (Flexas et al., 2013).

The pathway whereby CO_2 travels from the intercellular airspaces beneath stomata, through the mesophyll, to the sites of carboxylation in the chloroplast, has been identified as a key limitation to the uptake of CO_2 during photosynthesis (e.g. Flexas et al., 2008). For this reason, there has been a concentrated effort to investigate the potential for enhancing crop productivity by targeting improvements in mesophyll conductance. Interestingly, there appears to be a high level of coordination between mesophyll conductance and K_{leaf} in a diversity of plants, further reinforcing the fundamental principle of coordination between the separate systems of water transport and photosynthesis in leaves (Flexas et al., 2013). The reasons for this coordination are not well understood, and may involve common use of aquaporins by both pathways, or perhaps a common dependence on leaf anatomical characters (Syvertsen et al., 1995; Flexas et al., 2013). Regardless of the reasons behind the linkage, it means water transport and mesophyll conductance probably need to be considered as co-dependent traits in the leaf.

4.3 Water stress and cavitation

The preceding discussion examines how maximum water transport in leaves and roots is constrained by the anatomy of the xylem network and the metabolic activity of membrane water channels. Modification of these characteristics holds promise for producing crop plants with enhanced maximum productivity under optimum conditions of soil moisture and humidity, but another critical consideration when assessing the performance of crop varieties is their performance under dry soil and atmosphere. Under water stress, the failure of the water transport system by cavitation becomes a critical determinant of plant success.

All plant water transport systems are sensitive to dysfunction and damage under certain conditions of drying soil or atmosphere. Declining hydraulic conductivity under water stress is one of the trade-offs involved with transporting water under high tension, and occurs as a result of cavitation. Xylem cavitation is initiated when hydraulic tension in the xylem exceeds the ability of the xylem membranes to prevent air from being sucked into the water column. Once air bubbles are sucked into xylem conduits they block the passage of water, rendering these cavitated conduits non-functional. Most work on xylem cavitation has been done on woody plants, where cavitation occurs under stressful conditions, but cavitation may be a common, even diurnal, process in the leaves and roots of crop plants (Stiller et al., 2003). The most important anatomical determinants of vulnerability to cavitation in woody plants are at the scale of the pit membranes between xylem vessels as this is where cavitation is thought to be 'seeded' (Sperry and Ikeda, 1997).

The water potential at which air breaches the xylem pit membrane and enters the conduit is called the 'air-seeding threshold', and the cavitation vulnerability of the xylem of a species can be quantified by measuring hydraulic conductivity as water stress is imposed either by drying or centrifuging branches (Cochard et al., 2013). The most common quantification of xylem vulnerability to cavitation is the P50, which represents the water potential at which 50% of the xylem capacity to transport water has been lost due to cavitation. In woody plants, P50 exhibits low variation within species, while among species, the range extends from near −1 MPa in soybean (Sperry, 2000) and forage grasses (Holloway-Phillips and Brodribb, 2011b) to close to −10 MPa in drought tolerant conifers (Willson et al., 2008; Pittermann et al., 2012). The P50 in the stems of woody plants is ecologically significant as an indicator of the limit of water stress a species can endure, but among crop plants indications are that P50 is often reached, after which embolisms are rapidly repaired (Buchard et al., 1999; Kaufmann et al., 2009).

4.3.1 *Xylem vulnerability in crops*

Crop plants fall at the very vulnerable end of the xylem cavitation threshold. Although few measurements are available, relative to woody plants, the P50s of both roots and leaves in crop plants, such as rice, sugarcane, soybean and maize, are all less negative than −2 MPa (Fig. 11.7 and Table 11.2). Such highly vulnerable xylem is likely to be exposed to cavitating water potentials on a regular basis, and several studies have demonstrated the formation of xylem embolisms in crop species such as sugarcane (Neufeld et al., 1992) and rice (Stiller et al., 2003) despite a high availability of soil water under field conditions. Losses in hydraulic conductivity of crop and forage species due to cavitation are associated with reductions in photosynthesis (Stiller et al., 2003; Holloway-Phillips and Brodribb, 2011b), but this appears to be repairable nocturnally, provided that roots are sufficiently hydrated to develop root pressure (Tyree et al., 1986; Stiller et al., 2003). The analysis of cavitation events in crop plants has received little attention despite the encouragement of early research and, as such, it is difficult to conclude whether there is significant genetic variation for P50 or of its association with drought resistance. In grapevine, P50 ranged from −0.9 to −2.63 Mpa (Alsina et al., 2007). However, in rice (Stiller et al., 2003) and forage grasses (Holloway-Phillips and Brodribb, 2011a), there was little variation in P50 detected even though varieties were chosen for their adaptation to different environments. Instead, across the four varieties of forage grass investigated, the relationship between the leaf water potential at stomatal closure and P50 was much more insightful in terms of the amount of leaf death incurred under soil water deficits.

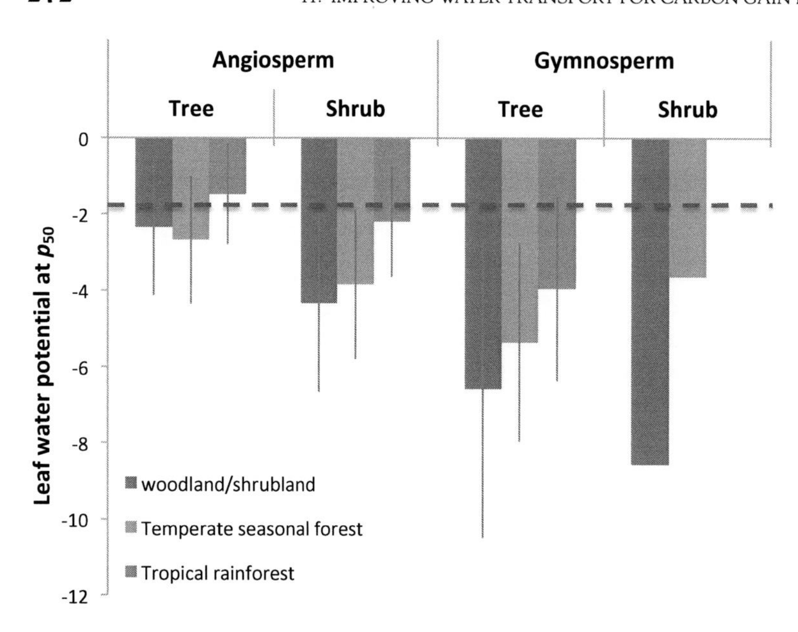

FIG. 11.7 Leaf water potential at 50% loss of hydraulic conductivity (P50) for a range of angiosperms and gymnosperms. Data redrawn from the supplementary information in Choat et al. (2012). The dashed line represents the average P50 for herbaceous crops.

TABLE 11.2 P50 (water potential at which 50% of the xylem capacity to transport water has been lost due to cavitation), common midday water potentials (ψ) in the field, and estimated cavitation under well-watered conditions for a range of crop species

Species	P50 (−MPa)	MD ψ (−MPa)	Field cavitation
Sunflower (s)	1.88	NA	NA
Rice (l)	1.6	>1.5	>60%
Sugarcane (l)	1.17	0.95	<50%
Forage grass (l)	1.10	1.10	<50%
Maize (s)	1.67	0.5	4.5%
Grapevine (s)	2.97	<1.8	<30%
Grapevine (p)	0.95	0.8	<30%
Olive (l)	7	2	<10%

Letters next to species indicate leaf (l), stem (s) and petiole (p). Data sources: sunflower (Stiller and Sperry, 2002); rice (Stiller et al., 2003); sugarcane (Neufeld et al., 1992); forage grasses (P50: Holloway-Phillips & Brodribb 2011b; MD ψ: Holloway-Phillips unpublished); maize (P50: Li et al. 2009; MD ψ: Nissanka et al., 1997); grapevine stem (P50: Choat et al., 2010; MD ψ: Barbe et al., 2005); grapevine petiole (Zufferey et al., 2011); olive (P50: Ennajeh et al., 2008; MD ψ: Angelopoulos et al., 1996).

Due to the high vulnerability of crop plants to xylem cavitation, stomatal behavior plays a key role in determining the degree to which species are exposed to losses of conductivity during the day. Stomatal closure at midday can occur in some crop species exposed to high evaporative demand at midday, and the timing of stomatal closure determines the minimum water potential, and hence the degree of cavitation that plants are likely to sustain. The so-called 'hydraulic safety margin' describes the difference in water potential between the point of stomatal closure and P50 (Fig. 11.8). In conifers, stomatal control is highly conservative with respect to xylem cavitation, leading to large safety margins (Choat et al., 2012) in comparison with woody angiosperms which tend to have much smaller, but still positive safety margins. The small amount of information from commercial herbaceous plants grown in controlled environment conditions suggests that they often have negative safety margins, meaning that stomatal closure occurs after the onset of cavitation (Holloway-Phillips and Brodribb, 2011b), exposing plants to significant diurnal xylem cavitation. The close overlap between stomatal and xylem dynamics in crops mean that the impact of changes in P50 in different crop varieties must be considered in combination with characters of stomatal behavior and plant turgor relations.

Inter- and intra-specific variation in stomatal response to soil moisture and VPD has been reported (Cox and Jolliff, 1987; Soar et al., 2006; Hopper et al., 2014); loosely termed 'isohydric' (maintaining approximate homeostasis in leaf hydration) and 'anisohydric' (allowing variation in leaf hydration as soil water diminishes). Co-existence of contrasting stomatal physiologies relates to trade-offs, e.g. between carbon assimilation and regulation of canopy temperature (Soar et al., 2009), and between utilizing the carbon investment in constructing resistant xylem and the risk of permanent damage from regulating too close to the functional limits. For example woody trees that display anisohydric stomatal responses tend also to have xylem less vulnerable to cavitation (McDowell et al., 2008); physiological constraints, e.g. the ability to reverse diurnal cavitation (section 4.3.2), and environmental variation (Chapter13).

As summarized by Tardieu (2012), 'any trait or trait-related allele can confer drought tolerance: just design the right drought scenario'. In this regard, Sperry et al. (2002) suggest that plants coordinate their hydraulic capacity to match their mode of water-use regulation and, furthermore, that this is achieved with minimum investment in roots and xylem cavitation

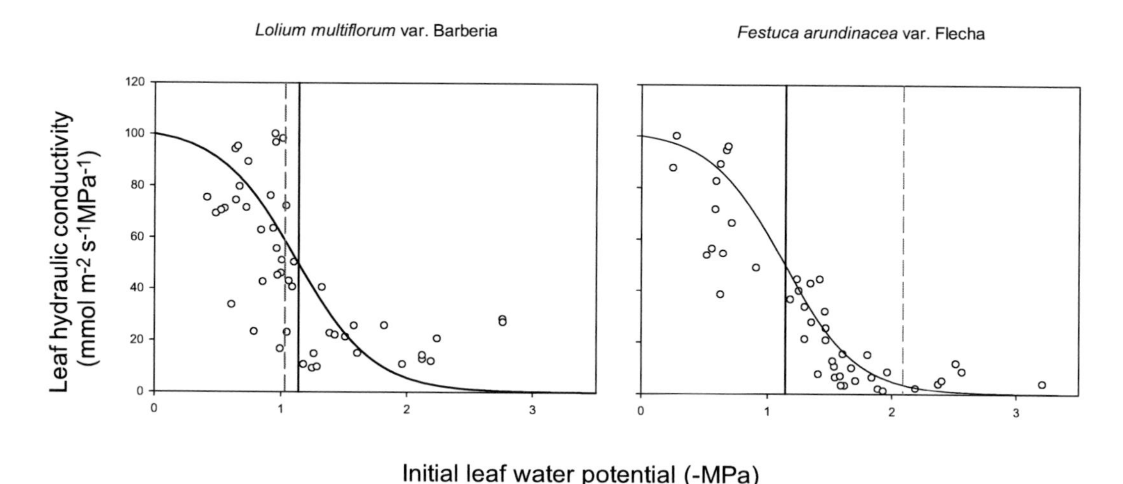

FIG. 11.8 Leaf hydraulic vulnerability curves showing the progression of K_{leaf} with decreasing leaf water potential (ψ_{leaf}), in two forage grass species. The solid black line indicates the ψ_{leaf} at which 50% loss of hydraulic conductivity has occurred (P50), and the dashed red line is where 90% loss of stomatal conductance has occurred. *Lolium multiflorum* Barberia has a 'positive' hydraulic safety margin with stomata closing before P50, whereas *Festuca arundinacea* Flecha has a 'negative' safety margin, closing stomata well after P50 has been transgressed. *Figure adapted from Holloway-Phillips & Brodribb (2011b).*

resistance (both carbon costly) in order to optimize the delivery system with spatial and temporal soil water availability in the given environment. For example, in a glasshouse study that compared the stomatal response of two forage grass species to soil water availability, *Festuca arundinacea* despite being considered more drought tolerant with a larger root system than *Lolium multiflorum*, suffered the most leaf senescence in a restricted root-zone situation due to stomatal closure occurring well after the leaf water potential at which P50 occurred (Holloway-Phillips and Brodribb, 2011a). As such, it was hypothesized that under non-restricted root conditions, where stored soil water is available, prolonged stomatal opening under drying conditions may provide a benefit if the additional carbon gain is directed to root growth or carbohydrate storage, which is an important source of energy during recovery (Volaire, 1995; Volaire et al., 1997). In comparison, more conservative stomatal regulation may be a useful strategy in buffering leaf water status where summer rainfall is intermittent, by preventing leaf senescence during dry spells and rapid utilization of rainfall events through reducing hydraulic recovery times. This is the opposite of the pattern found in grapevines. Cultivar Shiraz originating from a mesic environment displays anisohydric stomatal control compared to the Mediterranean variety Grenache which is isohydric (Schultz, 2003), suggesting that a conservative water strategy is likely to benefit Grenache under prolonged water deficits, whereas Shiraz can be far more 'opportunistic' (*sensu* Jones, 1980) in a mesic environment.

4.3.2 Embolism repair/root pressure

Unlike woody plants, diurnal cavitation in herbaceous species appears to be readily reversed by positive pressures developed in the roots (Stiller et al., 2003). Positive root xylem pressures in woody plants does not tend to exceed 150 kPa (Fisher et al., 1997; Cao et al., 2012) which, in the absence of transpiration, can push water 15 m vertically through the stem, refilling embolized xylem conduits. While this pressure is insufficient to refill embolisms in tall trees, it is adequate to repair embolisms in herbaceous crops. Root pressure and guttation are common among most crop plants, yet the importance of this poorly studied feature of plant function is unknown. Recent studies have shown that cavitation repair (Cao et al., 2012) and the maximum height of bamboo species are constrained by the magnitude of root pressure developed nocturnally. Very little is known about the variation in root pressure among varieties of crop species, but pressure measurements in wheat varieties suggest significant variation exists (Brodribb et al., in prep). Given the extent of cavitation in crop plants, it is likely that the magnitude of root pressure will determine the speed, efficiency and limits of xylem refilling in drying soil. As such this represents a fertile area for future research.

5 CONCLUDING REMARKS

The water transport system in plants represents a new horizon for improving the performance of crop plants. Increased efficiency of transport can lead to crop varieties that do not suffer from midday depression of photosynthesis to increase diurnal carbon gain in environments with sufficient water, while reduced hydraulic conductivity holds promise as a means of generating more efficient water use from crop varieties.

The transfer of knowledge and techniques from the pioneering hydraulic work on woody plants to crops has been slow, but recent years have seen a gradual adoption of hydraulic principles by crop workers, a trend that will no doubt continue into the future. Progress in our understanding of molecular-scale development and function of the vasculature will be an important stepping-stone in the pathway towards manipulating the water transport of crop species.

New knowledge about genetic control and environmental modulation of vascular and stomatal development, understanding of how these traits scale up to crop level (Chapter 1), and the adaptive value of crop traits for specific environments (Chapters 13 and 14) are required. Hence, now is the time to establish, with certainty, the quantitative interactions between plant architecture, anatomy and physiology that regulate the transport of water for carbon in crop plants under realistic field conditions.

References

Aasamaa, K., Niinemets, U., Sober, A., 2005. Leaf hydraulic conductance in relation to anatomical and functional traits during *Populustremula*leaf ontogeny. Tree Physiol. 25, 1409–1418.

Adachi, S., Tsuru, Y., Kondo, M., et al.,2010. Characterization of a rice variety with high hydraulic conductance and identification of the chromosome region responsible using chromosome segment substitution lines. Ann. Bot. 106, 803–811.

Adams, W.W., Watson, A.M., Mueh, K.E., et al.,2007. Photosynthetic acclimation in the context of structural constraints to carbon export from leaves. Photosynth. Res. 94, 455–466.

Ainsworth, E.A., Ort, D.R., 2010. How do we improve crop production in a warming world? Plant Physiol. 154, 526–530.

Alsina, M.M., de Herralde, F., Aranda, X., Save, R., Biel, C., 2007. Water relations and vulnerability to embolism are not related: Experiments with eight grapevine cultivars. Vitis 46, 1–6.

Alsina, M.M., Smart, D.R., Bauerle, T., et al.,2011. Seasonal changes of whole root system conductance by a drought-tolerant grape root system. J. Exp. Bot. 62, 99–109.

Amodeo, G., Door, R., Vallejo, A., Sutka, M., Parisi, M., 1999. Radial and axial water transport in the sugar beet storage root. J. Exp. Bot. 50, 509–516.

Angelopoulos, K., Dichio, B., Xiloyannis, C., 1996. Inhibition of photosynthesis in olive trees (*Oleaeuropaea* L.) during water stress and rewatering. J. Exp. Bot. 47, 1093–1100.

Araus, J.L., Bort, J., Steduto, P., Villegas, D., Royo, C., 2003. Breeding cereals for Mediterranean conditions: ecophysiological clues for biotechnology application. Ann. Appl. Biol. 142, 129–141.

Azaizeh, H., Gunse, B., Steudle, E., 1992. Effects of NaCl and CaCl2 on water transport across root cells of maize (*Zea mays* L.) seedlings. Plant Physiol. 99, 886–894.

Barbe, C., Van Leeuwen, C., Shackel, K.A., Chone, X., 2005. Worldwide use of stem water potential to manage vine water deficit with irrigation in estate vineyards. Am. J. Enol. Viticult. 56, 301A–302A.

Barrowclough, D.E., Peterson, C.A., Steudle, E., 2000. Radial hydraulic conductivity along developing onion roots. J. Exp. Bot. 51, 547–557.

Beaulieu, J.M., Leitch, I.J., Patel, S., Pendharkar, A., Knight, C.A., 2008. Genome size is a strong predictor of cell size and stomatal density in angiosperms. New Phytol. 179, 975–986.

Benabdellah, K., Ruiz-Lozano, J., Aroca, R., 2009. Hydrogen peroxide effects on root hydraulic properties and plasma membrane aquaporin regulation in *Phaseolus vulgaris*. Plant Mol. Biol. 70, 647–661.

Björkman, O., Demmig, B., 1987. Photon yield of O_2 evolution and chlorophyll fluorescence characteristics at 77K among vascular plants of diverse origins. Planta 170, 489–504.

Black, C.C., Tu, Z.-P., Counce, P.A., Yao, P.-F., Angelov, M.N., 1995. An integration of photosynthetic traits and mechanisms that can increase crop photosynthesis and grain production. Photosynth. Res. 46, 169–175.

Black, M.Z., Patterson, K.J., Minchin, P.E.H., Gould, K.S., Clearwater, M.J., 2011. Hydraulic responses of whole vines and individual roots of kiwifruit (*Actinidia chinensis*) following root severance. Tree Physiol. 31, 508–518.

Blackman, C.J., Brodribb, T.J., Jordan, G.J., 2009. Leaf hydraulics and drought stress: response, recovery and survivorship in four woody temperate plant species. Plant Cell Environ. 32, 1584–1595.

Blackman, C.J., Brodribb, T.J., Jordan, G.J., 2012. Leaf hydraulic vulnerability influences species' bioclimatic limits in a diverse group of woody angiosperms. Oecologia 168, 1–10.

Blizzard, W., Boyer, J., 1980. Comparative resistance of the soil and the plant to water transport. Plant Physiol. 66, 809–814.

Blum, A., 2005. Drought resistance, water-use efficiency, and yield potential – are they compatible, dissonant, or mutually exclusive? Aust. J. Agric. Res. 56, 1159–1168.

Boursiac, Y., Chen, S., Luu, D.-T., Sorieul, M., van den Dries, N., Maurel, C., 2005. Early effects of salinity on water transport in Arabidopsis roots. Molecular and cellular features of aquaporin expression. Plant Physiol. 139, 790–805.

Boyce, C.K., Brodribb, T.J., Feild, T.S., Zwieniecki, M.A., 2009. Angiosperm leaf vein evolution was physiologically and environmentally transformative. Proc. R. Soc. Lond. B 276, 1771–1776.

Bramley, H., Turner, D.W., Tyerman, S.D., Turner, N.C., 2007. Water flow in the roots of crop species: The influence of root structure, aquaporin activity, and waterlogging. Adv. Agron. 96, 133–196.

Bramley, H., Turner, N.C., Siddique, K.H.M., 2013. Water use efficiency: governed by an interactive network of physical, biochemical and hydraulic processes. In: Kole, C. (Ed.), Genomics and breeding for climate-resilient crops. Springer-Verlag, Berlin, pp. 225–268.

Bramley, H., Turner, N.C., Turner, D.W., Tyerman, S.D., 2009. Roles of morphology, anatomy, and aquaporins in determining contrasting hydraulic behavior of roots. Plant Physiol. 150, 348–364.

Bramley, H., Turner, N.C., Turner, D.W., Tyerman, S.D., 2010. The contrasting influence of short-term hypoxia on the hydraulic properties of cells and roots of wheat and lupin. Funct. Plant Biol. 37, 183–193.

Brodribb, T., Feild, T., Jordan, G., 2007. Leaf maximum photosynthetic rate and venation are linked by hydraulics. Plant Physiol. 144, 1890–1898.

Brodribb, T.J., Cochard, H., 2009. Hydraulic failure defines the recovery and point of death in water-stressed conifers. Plant Physiol. 149, 575–584.

Brodribb, T.J., Feild, T.S., 2000. Stem hydraulic supply is linked to leaf photosynthetic capacity: evidence from New Caledonian and Tasmanian rainforests. Plant Cell Environ. 23, 1381–1388.

Brodribb, T.J., Feild, T.S., 2010. Leaf hydraulic evolution led a surge in leaf photosynthetic capacity during early angiosperm diversification. Ecol. Lett. 13, 175–183.

Brodribb, T.J., Feild, T.S., Sack, L., 2010. Viewing leaf structure and evolution from a hydraulic perspective. Funct. Plant Biol. 37, 488–498.

Brodribb, T.J., Holbrook, N.M., 2004a. Diurnal depression of leaf hydraulic conductance in a tropical tree species. Plant Cell Environ. 27, 820–827.

Brodribb, T.J., Holbrook, N.M., 2004b. Stomatal protection against hydraulic failure: a comparison of coexisting ferns and angiosperms. New Phytol. 162, 663–670.

Brodribb, T.J., Holbrook, N.M., 2007. Forced depression of leaf hydraulic conductance in situ: effects on the leaf gas exchange of forest trees. Funct. Ecol. 21, 705–712.

Brodribb, T.J., Jordan, G.J., 2008. Internal coordination between hydraulics and stomatal control in leaves. Plant Cell Environ. 31, 1557–1564.

Brodribb, T.J., Jordan, G.J., 2011. Water supply and demand remain balanced during leaf acclimation of Nothofagus cunninghamii trees. New Phytol. 192, 437–448.

Brodribb, T.J., Jordan, G.J., Carpenter, R.J., 2013. Unified changes in cell size permit coordinated leaf evolution. New Phytol. 199, 559–570.

Buchard, C., McCully, M., Canny, M., 1999. Daily embolism and refilling of root xylem vessels in three dicotyledonous crop plants. Agronomie 19, 97–106.

Cao, K.F., Yang, S.J., Zhang, Y.J., Brodribb, T.J., 2012. The maximum height of grasses is determined by roots. Ecol. Lett. 15, 666–672.

Carins Murphy, M.R., Jordan, G.J., Brodribb, T.J., 2012. Differential leaf expansion can enable hydraulic acclimation to sun and shade. Plant, Cell & Environment 35, 1407–1418.

Chen, Y.L., Dunbabin, V.M., Postma, J., et al.,2011. Phenotypic variability and modelling of root structure of wild *Lupinus angustifolius* genotypes. Plant Soil 348, 345–364.

Choat, B., Drayton, W.M., Brodersen, C., et al.,2010. Measurement of vulnerability to water stress-induced cavitation in grapevine: a comparison of four techniques applied to a long-vesseled species. Plant Cell Environ 33, 1502–1512.

Choat, B., Jansen, S., Brodribb, T.J., et al.,2012. Global convergence in the vulnerability of forests to drought. Nature 491, 752–755.

Choudhary, S., Sinclair, T.R., 2013. Hydraulic conductance differences among sorghum genotypes to explain variation in restricted transpiration rates. Funct. Plant Biol. 41, 270–275.

Cochard, H., Badel, E., Herbette, S., Delzon, S., Choat, B., Jansen, S., 2013. Methods for measuring plant vulnerability to cavitation: a critical review. J. Exp. Bot. 64, 4779–4791.

Cochard, H., Ewers, F.W., Tyree, M.T., 1994. Water relations of tropical vine-like bamboo (*Rhipidocladum recemiflorum*): root pressures, vulnerability to cavitation and seasonal changes in embolism. J. Exp. Bot. 45, 1085–1089.

Cochard, H., Venisse, J.S., Barigah, T.S., et al.,2007. Putative role of aquaporins in variable hydraulic conductance of leaves in response to light. Plant Physiol. 143, 122–133.

Comas, L., Becker, S., Cruz, V.M.V., Byrne, P.F., Dierig, D.A., 2013. Root traits contributing to plant productivity under drought. Front. Plant Sci. 4, 1–3.

Condon, A.G., Richards, R.A., Rebetzke, G.J., Farquhar, G.D., 2002. Improving intrinsic water-use efficiency and crop yield. Crop Sci. 42, 122–131.

Condon, A.G., Richards, R.A., Rebetzke, G.J., Farquhar, G.D., 2004. Breeding for high water-use efficiency. J. Exp. Bot. 55, 2447–2460.

Cox, W., Jolliff, G., 1987. Crop-water relations of sunflower and soybean under irrigated and dryland conditions. Crop Sci. 27, 553–557.

Denison, R.F., 2009. Darwinian agriculture: real, imaginary and complex trade-offs as constraints and opportunities. In: Sadras, V., Calderini, D. (Eds.), Crop physiology. Academic Press, San Diego, pp. 214–234.

Devi, M.J., Sadok, W., Sinclair, T.R., 2012. Transpiration response of de-rooted peanut plants to aquaporin inhibitors. Environ. Exp. Bot. 78, 167–172.

deWit, C.T., 1958. Transpiration and crop yields.Verslag. Landbouwk.Onderz. 64.6. Institute of Biological and Chemical Research on Field Crops and Herbage: Wageningen, Netherlands.

Di Pietro, M., Vialaret, J., Li, G.W., et al.,2013. Coordinated post-translational responses of aquaporins to abiotic and nutritional stimuli in Arabidopsis roots. Mol. Cell. Proteom. 12, 3886–3897.

Dixon, H.H., Joly, J., 1895. The path of the transpiration current. Ann. Bot. 9, 416–419.

Domec, J.C., Gartner, B.L., 2003. Relationship between growth rates and xylem hydraulic characteristics in young, mature and old-growth ponderosa pine trees. Plant Cell Environ. 26, 471–483.

Ducrey, M., Huc, R., Ladjal, M., Guehl, J.M., 2008. Variability in growth, carbon isotope composition, leaf gas exchange and hydraulic traits in the eastern Mediterranean cedars *Cedrus libani* and *C. brevifolia*. Tree Physiol. 28, 698–701.

Ennajeh, M., Tounekti, T., Vadel, A.M., Khemira, H., Cochard, H., 2008. Water relations and drought-induced embolism in olive (*Olea europaea*) varieties 'Meski' and 'Chemlali' during servere drought. Tree Physiol. 28, 971–976.

Evans, L.T., Fischer, R.A., 1999. Yield potential: Its definition, measurement, and significance. Crop Sci. 39, 1544–1551.

Fan, Z.X., Zhang, S.B., Hao, G.Y., Ferry Slik, J.W., Cao, K.F., 2012. Hydraulic conductivity traits predict growth rates and adult stature of 40 Asian tropical tree species better than wood density. J. Ecol. 100, 732–741.

Feild, T.S., Brodribb, T.J., 2013. Hydraulic tuning of vein cell microstructure in the evolution of angiosperm venation networks. New Phytol.doi: 10.1111/nph.12311.

Feild, T.S., Brodribb, T.J., Iglesias, A., et al.,2011. Fossil evidence for Cretaceous escalation in angiosperm leaf vein evolution. Proc. Natl. Acad. Sci. USA 108, 8363–8366.

Fischer, R.A., 2007. Understanding the physiological basis of yield potential in wheat. J. Agric. Sci. 145, 99–113.

Fischer, R.A., Rees, D., Sayre, K.D., Lu, Z.-M., Condon, A.G., Larque Savvedra, A., 1998. Wheat yield progress associated with high stomatal conductance and photosynthetic rate, and cooler canopies. Crop Sci. 38, 1467–1475.

Fischer, R.A., Turner, N.C., 1978. Plant productivity in the arid and semiarid zones. Annu. Rev. Plant Physiol. 29, 277–317.

Fisher, J., Angeles, G., Ewers, F., Lopezportillo, J., 1997. Survey of root pressure in tropical vines and woody species. Internl. J. Plant Sci. 158, 44–50.

Flexas, J., Ribas-Carbá, M., Diaz-Espejo, A., Galmés, J., Medrano, H., 2008. Mesophyll conductance to CO2: current knowledge and future prospects. Plant Cell Environ. 31, 602–621.

Flexas, J., Scoffoni, C., Gago, J., Sack, L., 2013. Leaf mesophyll conductance and leaf hydraulic conductance: an introduction to their measurement and coordination. J. Exp. Bot. 64, 3965–3981.

Foulkes, M.J., Snape, J.W., Shearman, V.J., Reynolds, M.P., Gaju, O., Sylvester-Bradley, R., 2007. Genetic progress in yield potential in wheat: recent advances and future prospects. J. Agric. Sci. 145, 17.

Franks, P.J., Beerling, D.J., 2009. Maximum leaf conductance driven by CO2 effects on stomatal size and density over geologic time. Proc. Natl. Acad. Sci. USA 106, 10343–10347.

Frensch, J., Steudle, E., 1989. Axial and radial hydraulic resistance to roots of maize (*Zea mays* L.). Plant Physiol. 91, 719–726.

Gambetta, G.A., Fei, J., Rost, T.L., et al.,2013. Water uptake along the length of grapevine fine roots: developmental anatomy, tissue-specific aquaporin expression, and pathways of water transport. Plant Physiol. 163, 1254–1265.

Gates, D.M., 1964. Leaf temperature and transpiration. Agron. J. 56, 273–277.

Gholipoor, M., Choudhary, S., Sinclair, T.R., Messina, C.D., Cooper, M., 2013. Transpiration response of maize hybrids to atmospheric vapour pressure deficit. J. Agron. Crop Sci. 199, 155–160.

Giordano, M.A., Rijsberman, F.R., Maria Saleth, R., 2006. More crop per drop: revisiting a research paradigm. IWA Publishing, London.

Giraldo, J.P., Wheeler, J.K., Huggett, B.A., Holbrook, N.M., 2013. The role of leaf hydraulic conductance dynamics on the timing of leaf senescence. Funct. Plant Biol. 41, 37–47.

Gloser, V., Zwieniecki, M.A., Orians, C.M., Holbrook, N.M., 2007. Dynamic changes in root hydraulic properties in response to nitrate availability. J. Exp. Bot. 58, 2409–2415.

Gowda, V.R.P., Henry, A., Yamauchi, A., Shashidhar, H.E., Serraj, R., 2011. Root biology and genetic improvement for drought avoidance in rice. Field Crops Res. 122, 1–13.

Hachez, C., Veselov, D., Ye, Q., et al.,2012. Short-term control of maize cell and root water permeability through plasma membrane aquaporin isoforms. Plant Cell Environ. 35, 185–198.

Henry, A., Cal, A.J., Batoto, T.C., Torres, R.O., Serraj, R., 2012. Root attributes affecting water uptake of rice (*Oryza sativa*) under drought. J. Exp. Bot. 63, 4751–4763.

Henzler, T., Waterhouse, R.N., Smyth, A.J., et al.,1999. Diurnal variations in hydraulic conductivity and root pressure can be correlated with the expression of putative aquaporins in the roots of *Lotus japonicus*. Planta 210, 50–60.

Hirasawa, T., Hsiao, T.C., 1999. Some characteristics of reduced leaf photosynthesis at midday in maize growing in the field. Field Crops Res. 62, 53–62.

Holloway-Phillips, M.M., Brodribb, T.J., 2011a. Contrasting hydraulic regulation in closely related forage grasses: implications for plant water use. Funct. Plant Biol. 38, 594–605.

Holloway-Phillips, M.M., Brodribb, T.J., 2011b. Minimum hydraulic safety leads to maximum water-use efficiency in a forage grass. Plant Cell Environ. 34, 302–313.

Hopper, D.W., Ghan, R., Cramer, G.R., 2014. A rapid dehydration leaf assay reveals stomatal response differences in grapevine genotypes. Hortic. Res. 1, 1–8.

Hubbard, R.M., Ryan, M.G., Stiller, V., Sperry, J.S., 2001. Stomatal conductance and photosynthesis vary linearly with plant hydraulic conductance in ponderosa pine. Plant Cell Environ. 24, 113–121.

Idso, S.B., Reginato, R.J., Reicosky, D.C., Hatfield, J.L., 1981. Determining soil induced plant water potential depressions in alfalfa by means of infrared thermometry. Agron. J. 73, 826–830.

Jackson, R.D., Idso, S.B., Reginato, R.J., Pinter, P.J., 1981. Canopy temperature as a crop water stress indicator. Water Resour. Res. 17, 1133–1138.

Javot, H., Maurel, C., 2002. The role of aquaporins in root water uptake. Ann. Bot. 90, 301–313.

Jones, H.G., 1980. Interaction and integration of adaptive responses to water stress: the implications of an unpredictable environment. In: Turner, N.C., Kramer, P.J. (Eds.), Adaptation of plants to water and high temperature stress. Wiley, New York, pp. 353–365.

Kaldenhoff, R., Grote, K., Zhu, J.J., Zimmermann, U., 1998. Significance of plasmalemma aquaporins for water-transport in *Arabidopsis thaliana*. Plant J. 14, 121–128.

Kaufmann, I., Schulze-Till, T., Schneider, H.U., Zimmermann, U., Jakob, P., Wegner, L.H., 2009. Functional repair of embolized vessels in maize roots after temporal drought stress, as demonstrated by magnetic resonance imaging. New Phytol. 184, 245–256.

Koizumi, K., Ookawa, T., Satoh, H., Hirasawa, T., 2007. A wilty mutant of rice has impaired hydraulic conductance. Plant Cell Physiol. 48, 1219–1228.

Lambers, H., Poorter, H., 1992. Inherent variation in growth rate between higher plants: a search for physiological causes and ecological consequences. Advances in Ecological Research. 23, 187–261.

Larese, M.G., Namias, R., Craviotto, R.M., Arango, M.R., Gallo, C., Granitto, P.M., 2014. Automatic classification of legumes using leaf vein image features. Patt. Recog. 47, 158–168.

Lawlor, D.W., 1973. Growth and water absorption of wheat with parts of the roots at different water potentials. New Phytol. 72, 297–305.

Lee, S.H., Chung, G.C., Steudle, E., 2005. Gating of aquaporins by low temperature in roots of chilling-sensitive cucumber and chilling-tolerant figleaf gourd. J. Exp. Bot. 56, 985–995.

Li, G., Santoni, V., Maurel, C., 2013. Plant aquaporins: Roles in plant physiology. Biochim. Biophys. Acta. 1840, 1574–1582.

Li, S., Zhang, Y.L., Sack, L., et al.,2014. The heterogeneity and spatial patterning of structure and physiology across the leaf surface in giant leaves of Alocasia macrorrhiza. Plos One 8, 1–10.

Li, Y.Y., Sperry, J.S., Shao, M.A., 2009. Hydraulic conductance and vulnerability to cavitation in corn (*Zea mays* L.) hybrids of differing drought resistance. Environ. Exp. Bot. 66, 341–346.

Liao, M., Palta, J.A., Fillery, I.R.P., 2006. Root characteristics of vigorous wheat improve early nitrogen uptake. Aust. J. Agric. Res. 57, 1097–1107.

Lobell, D.B., Hammer, G.L., McLean, G., Messina, C., Roberts, M.J., Schlenker, W., 2013. The critical role of extreme heat for maize production in the United States. Nat. Clim. Change 3, 497–501.

Lopes, M.S., Reynolds, M., 2010. Partitioning of assimilates to deeper roots is associated with cooler canopies and increased yield under drought in wheat. Funct. Plant Biol. 37, 147–156.

Maherali, H., Sherrard, M.E., Clifford, M.H., Latta, R.G., 2008. Leaf hydraulic conductivity and photosynthesis are genetically correlated in an annual grass. New Phytol. 180, 240–247.

Manschadi, A.M., Christopher, J., deVoil, P., Hammer, G.L., 2006. The role of root architectural traits in adaptation of wheat to water-limited environments. Funct. Plant Biol. 33, 823–837.

Martre, P., Morillon, R., Barrieu, F., North, G.B., Nobel, P.S., Chrispeels, M.J., 2002. Plasma membrane aquaporins play a significant role during recovery from water deficit. Plant Physiol. 130, 2101–2110.

Matsuo, N., Ozawa, K., Mochizuki, T., 2009. Genotypic differences in root hydraulic conductance of rice (*Oryza sativa* L.) in response to water regimes. Plant Soil 316, 25–34.

Maurel, C., Verdoucq, L., Luu, D.T., Santoni, V., 2008. Plant aquaporins: Membrane channels with multiple integrated functions. Annual Review of Plant Biology. Annual Reviews, Palo Alto, pp. 595–624.

McDowell, N., Pockman, W.T., Allen, C.D., et al.,2008. Mechanisms of plant survival and mortality during drought: why do some plants survive while others succumb to drought? New Phytol. 178, 719–739.

Milburn, J.A., 1973. Cavitation in *Ricinus* by accoustic detection: induction in excised leaves by various factors. Planta 18, 253–265.

Milla, R., de Diego-Vico, N., Martin-Robles, N., 2013. Shifts in stomatal traits following the domestication of plant species. J. Exp. Bot. 64, 3137–3146.

Miralles, D.J., Slafer, G.A., Lynch, V., 1997. Rooting patterns in near-isogenic lines of spring wheat for dwarfism. Plant and Soil. 197, 79–86.

Mullan, D.J., Reynolds, M.P., 2010. Quantifying genetic effects of ground cover on soil water evaporation using digital imaging. Functional Plant Biology 37, 703–712.

Murchie, E.H., Chen, Y.-z., Hubbart, S., Peng, S., Horton, P., 1999. Interactions between senescence and leaf orientation determine in situ patterns of photosynthesis and photoinhibition in field-grown rice. Plant Physiol. 119, 553–563.

Murchie, E.H., Pinto, M., Horton, P., 2009. Agriculture and the new challenge for photosynthesis research. New Phytol. 181, 532–552.

Nardini, A., Salleo, S., 2005. Water stress-induced modifications on leaf hydraulic architecture in sunflower: co-ordination with gas exchange. J. Exp. Bot. 56, 3093–3101.

Nardini, A., Salleo, S., Andri, S., 2005. Circadian regulation of leaf hydraulic conductance in sunflower (*Helianthus annuus* L. cv Margot). Plant Cell Environ. 28, 750–759.

Neufeld, H.S., Grantz, D.A., Meinzer, F.C., Goldstein, G., Crisosto, G.M., Crisosto, C., 1992. Genotypic variability in vulnerability of leaf xylem to cavitation in water-stressed and well-irrigated sugarcane. Plant Physiol. 100, 1020–1028.

Nissanka, S.P., Dixon, A., Tollenaar, M., 1997. Canopy gas exchange response to moisture stress in old and new maize hybrid. Crop Sci. 37 (1), 172–181.

Ocheltree, T.W., Nippert, J.B., Prasad, P.V.V., 2014. Stomatal responses to changes in vapor pressure deficit reflect tissue-specific differences in hydraulic conductance. Plant Cell Environ. 37, 132–139.

Palta, J.A., Chen, X., Milroy, S.P., Rebetzke, G.J., Dreccer, M.F., Watt, M., 2011. Large root systems: are they useful in adapting wheat to dry environments? Funct. Plant Biol. 38, 347–354.

Pantin, F., Simonneau, T., Muller, B., 2012. Coming of leaf age: control of growth by hydraulics and metabolics during leaf ontogeny. New Phytol. 196, 349–366.

Parry, M.A.J., Reynolds, M., Salvucci, M.E., et al.,2011. Raising yield potential of wheat. II. Increasing photosynthetic capacity and efficiency. J. Exp. Bot. 62, 453–467.

Passioura, J.B., 1972. The effect of root geometry on the yield of wheat growing on stored water. Aust. J. Agric. Res. 23, 745–752.

Passioura, J.B., 1977. Grain yield, harvest index, and water use of wheat. J. Aust. Inst. Agric. Res. 43, 117–121.

Passioura, J.B., 1996. Drought and drought tolerance. Plant Growth Reg. 20, 79–83.

Passioura, J.B., Angus, J.F., 2010. Improving productivity of crops in water-limited environments. Adv. Agron. 106, 37–75.

Perrone, I., Gambino, G., Chitarra, W., et al., 2012a. The grapevine root-specific aquaporin VvPIP2; 4N controls root hydraulic conductance and leaf gas exchange under well-watered conditions but not under water stress. Plant Physiol. 160, 965–977.

Perrone, I., Pagliarani, C., Lovisolo, C., Chitarra, W., Roman, F., Schubert, A., 2012b. Recovery from water stress affects grape leaf petiole transcriptome. Planta 235, 1383–1396.

Pittermann, J., Stuart, S.A., Dawson, T.E., Moreau, A., 2012. Cenozoic climate change shaped the evolutionary ecophysiology of the Cupressaceae conifers. Proc. Natl. Acad. Sci. USA 109, 9647–9652.

Pou, A., Medrano, H., Flexas, J., Tyerman, S.D., 2013. A putative role for TIP and PIP aquaporins in dynamics of leaf hydraulic and stomatalconductances in grapevine under water stress and re-watering. Plant Cell Environ. 36, 828–843.

Price, C.A., Knox, S.-J.C., Brodribb, T.J., 2013. The influence of branch order on optimal leaf vein geometries: murray's law and area preserving branching. PLoS One 8, e85420.

Price, C.A., Symonova, O., Mileyko, Y., Hilley, T., Weitz, J.S., 2011. Leaf extraction and analysis framework graphical user interface: segmenting and analyzing the structure of leaf veins and areoles. Plant Physiol. 155, 236–245.

Ranathunge, K., Steudle, E., Lafitte, R., 2003. Control of water uptake by rice (*Oryza sativa* L.): role of the outer part of the root. Planta 217, 193–205.

Raven, J.A., 1977. Evolution of vascular land plants in relation to supracellular transport processes. Adv. Bot. Res. 5, 153–219.

Reynolds, M., Bonnett, D., Chapman, S.C., et al.,2011. Raising yield potential of wheat. I. Overview of a consortium approach and breeding strategies. J. Exp. Bot. 62, 439–452.

Reynolds, M., Tuberosa, R., 2008. Translational research impacting on crop productivity in drought-prone environments. Curr. Opin. Plant Biol. 11, 171–179.

Richards, R.A., Passioura, J.B., 1989a. A breeding program to reduce the diameter of the major xylem vessel in the seminal roots of wheat and its effect on grain-yield in rain-fed environments. Aust. J. Agric. Res. 40, 943–950.

Richards, R.A., Passioura, J.B., 1989b. A breeding program to reduce the diameter of the major xylem vessel in the seminal roots of wheat and its effect on grain yield in rain-fed environments. Aust. J. Agric. Res. 40, 943–950.

Richards, R.A., Rebetzke, G.J., Watt, M., Condon, A.G., Spielmeyer, W., Dolferus, R., 2010. Breeding for improved water productivity in temperate cereals: phenotyping, quantitative trait loci, markers and the selection environment. Funct. Plant Biol. 37, 85–97.

Roberts, M.J., Schlenker, W., Eyer, J., 2012. Agronomic weather measures in econometric models of crop yield with implications for climate change. Am. J. Agric. Econ. 95, 236–243.

Rouschal, E., 1938. Der sommerliche Wasserhaushalt der Macchienpflanzen. Jb. wiss. Bot. 87, 436–523.

Sack, L., Frole, K., 2006. Leaf structural diversity is related to hydraulic capacity in tropical rainforest trees. Ecology 87, 483–491.

Sack, L., Holbrook, N.M., 2006. Leaf hydraulics. Annu. Rev. Plant Physiol. Mol. Biol. 57, 361–381.

Sack, L., Scoffoni, C., 2013. Leaf venation: structure, function, development, evolution, ecology and applications in the past, present and future. New Phytol. 198, 983–1000.

Sack, L., Scoffoni, C., McKown, A.D., et al.,2012. Developmentally based scaling of leaf venation architecture explains global ecological patterns. Nat. Commun. 3, 1–10.

Sadok, W., Sinclair, T.R., 2010. Genetic variability of transpiration response of soybean [*Glycine max* (L.) Merr.] shoots to leaf hydraulic conductance inhibitor $AgNO_3$. Crop Sci. 50, 1423–1430.

Sadok, W., Sinclair, T.R., 2011. Crops yield increase under water-limited conditions: review of recent physiological advances for soybean genetic improvement. Adv. Agron. 113, 325–349.

Sage, R.F., Sage, T.L., Kocacinar, F., 2012. Photorespiration and the evolution of C-4 photosynthesis. Merchant, S.S. (Ed.), Annual Review of Plant Biology, Vol 63, Annual Reviews, Palo Alto, pp. 19–47.

Sage, T.L., Busch, F.A., Johnson, D.C., et al.,2013. Initial events during the evolution of C-4 photosynthesis in C-3 species of Flaveria. Plant Physiol. 163, 1266–1276.

Salekdeh, G.H., Reynolds, M., Bennett, J., Boyer, J., 2009. Conceptual framework for drought phenotyping during molecular breeding. Trends Plant Sci. 14, 488–496.

Schoppach, R., Wauthelet, D., Jeanguenin, L., Sadok, W., 2013. Conservative water use under high evaporative demand associated with smaller root metaxylem and limited trans-membrane water transport in wheat. Funct. Plant Biol. 41, 257–269.

Schultz, H.R., 2003. Differences in hydraulic architecture account for near-isohydric and anisohydric behaviour of two field-grown *Vitis vinifera* L. cultivars during drought. Plant Cell Environ. 26, 1393–1405.

Schulze, E.D., Hall, A.E., 1982. Stomatal responses, water loss and CO_2 assimilation rates of plants in contrasting environment. In: Lange, O.L., Nobel, P.S., Osmond, C.B., Ziegler, H. (Eds.), Physiological plant ecology II. Water relations and carbon assimilation. Springer-Verlag, New York, pp. 181–230.

Shatil-Cohen, A., Attia, Z., Moshelion, M., 2011. Bundle-sheath cell regulation of xylem-mesophyll water transport via aquaporins under drought stress: a target of xylem-borne ABA? Plant J. 67, 72–80.

Siddique, K.H.M., Belford, R.K., Tennant, D., 1990. Root:shoot ratios of old and modern, tall and semi-dwarf wheats in a mediterranean environment. Plant Soil 121, 89–98.

Siefritz, F., Tyree, M.T., Lovisolo, C., Schubert, A., Kaldenhoff, R., 2002. PIP1 plasma membrane aquaporins in tobacco: from cellular effects to function in plants. Plant Cell 14, 869–876.

Sinclair, T.R., 2011. Challenges in breeding for yield increase for drought. Trends Plant Sci. 16, 289–293.

Sinclair, T.R., 2012. Is transpiration efficiency a viable plant trait in breeding for crop improvement? Funct. Plant Biol. 39, 359–365.

Sinclair, T.R., Zwieniecki, M.A., Holbrook, N.M., 2008. Low leaf hydraulic conductance associated with drought tolerance in soybean. Physiol. Plant. 132, 446–451.

Soar, C.J., Collins, M.J., Sadras, V.O., 2009. Irrigated Shiraz vines (*Vitis vinifera*) upregulate gas exchange and maintain berry growth in response to short spells of high maximum temperature in the field. Funct. Plant Biol. 36, 801–814.

Soar, C.J., Speirs, J., Maffei, S.M., Penrose, A.B., McCarthy, M.G., Loveys, B.R., 2006. Grape vine varieties Shiraz and Grenache differ in their stomatal response to VPD: apparent links with ABA physiology and gene expression in leaf tissue. Aust. J. Grape Wine Res. 12, 2–12.

Sperry, J., Ikeda, T., 1997. X ylem cavitation in roots and stems of Douglas fir and white fir. Tree Physiol. 17, 275–280.

Sperry, J.S., 2000. Hydraulic constraints on plant gas exchange. Agric. For. Meteor. 104, 13–23.

Sperry, J.S., Donnelly, J.R., Tyree, M.T., 1988. A method for measuring hydraulic conductivity and embolism in xylem. Plant Cell Environ. 11, 35–40.

Sperry, J.S., Hacke, U.G., 2004. Analysis of circular bordered pit function – I. Angiosperm vessels with homogenous pit membranes. Am. J. Bot. 91, 369–385.

Sperry, J.S., Hacke, U.G., Oren, R., Comstock, J.P., 2002. Water deficits and hydraulic limits to leaf water supply. Plant Cell Environ. 25, 251–263.

Steudle, E., 2000. Water uptake by roots: effects of water deficit. J. Exp. Bot. 51, 1531–1542.

Steduto, P., Hsiao, T.C., Fereres, E., 2007. On the conservative behavior of biomass water productivity. Irrigation Science 25, 189–207.

Stiller, V., Lafitte, H.R., Sperry, J.S., 2003. Hydraulic properties of rice and the response of gas exchange to water stress. Plant Physiol. 132, 1698–1706.

Stiller, V., Sperry, J.S., 2002. Cavitation fatigue and its reversal in sunflower (*Helianthus annus* L.). J. Exp. Bot. 53, 1155–1161.

Sutka, M., Li, G., Boudet, J., Boursiac, Y., Doumas, P., Maurel, C., 2011. Natural variation of root hydraulics in Arabidopsis grown in normal and salt-stressed conditions. Plant Physiol. 155, 1264–1276.

Syvertsen, J.P., Lloyd, J., McConchie, C., Kriedemann, P.E., Farquhar, G.D., 1995. On the relationship between leaf anatomy and CO2 diffusion through the mesophyll of hypostomatous leaves. Plant Cell Environ. 18, 149–157.

Tanaka, Y., Shiraiwa, T., 2009. Stem growth habit affects leaf morphology and gas exchange traits in soybean. Ann. Bot. 104, 1293–1299.

Tanner, C.B., Sinclair, T.R., 1983. Efficient water use in crop production: Research or re-search. In: Taylor, H.M., Jordan, W.R., Sinclair, T.R. (Eds.), Limitations to efficient water use in crop production. ASA, CSSA and SSSA, Madison, WI, pp. 1–27.

Tardieu, F., 2012. Any trait or trait-related allele can confer drought tolerance: just design the right drought scenario. J. Exp. Bot. 63, 25–31.

Taylaran, R.D., Adachi, S., Ookawa, T., Usuda, H., Hirasawa, T., 2011. Hydraulic conductance as well as nitrogen accumulation plays a role in the higher rate of leaf photosynthesis of the most productive variety of rice in Japan. J. Exp. Bot. 62, 4067–4077.

Tcherkez, G.G.B., Farquhar, G.D., Andrews, T.J., 2006. Despite slow catalysis and confused substrate specificity, all ribulose bisphosphate carboxylases may be nearly perfectly optimized. Proc. Natl. Acad. Sci. USA 103, 7246–7251.

Tixier, A., Cochard, H., Badel, E., Dusotoit-Coucaud, A., Jansen, S., Herbette, S., 2013. Arabidopsis thaliana as a model species for xylem hydraulics: does size matter? J. Exp. Bot. 64, 2295–2305.

Törnroth-Horsefield, S., Wang, Y., Hedfalk, K., et al.,2006. Structural mechanism of plant aquaporin gating. Nature 439, 688–694.

Tournaire-Roux, C., Sutka, M., Javot, H., et al.,2003. Cytosolic pH regulates root water transport during anoxic stress through gating of aquaporins. Nature 425, 393–397.

Tsuda, M., Tyree, M.T., 2000. Plant hydraulic conductance measured by the high pressure flow meter in crop plants. J. Exp. Bot. 51, 823–828.

Tyerman, S.D., Niemietz, C.M., Bramley, H., 2002. Plant aquaporins: multifunctional water and solute channels with expanding roles. Plant Cell Environ. 25, 173–194.

Tyree, M.T., Fiscus, E.L., Wullschleger, S.D., Dixon, M.A., 1986. Detection of xylem cavitation in corn under field conditions. Plant Physiol. 82, 597–599.

Tyree, M.T., Zimmermann, M.H., 2002. Xylem structure and the ascent of sap. Springer, Berlin.

Ueno, O., Kawano, Y., Wakayama, M., Takeda, T., 2006. Leaf vascular systems in C3 and C4 grasses: a two-dimensional analysis. Ann. Bot. 97, 611–621.

Uhl, D., Mosbrugger, V., 1999. Leaf venation density as a climate and environmental proxy: a critical review and new data. Palaeogeog. Palaeoclimatol. Palaeoecol. 149, 15–26.

Vandeleur, R.K., Sullivan, W., Athman, A., et al., 2014. Rapid shoot-to-root signalling regulates root hydraulic conductance via aquaporins. Plant Cell Environ. 37, 520–538.

Volaire, F., 1995. Growth, carbohydrate reserves and drought survival strategies of contrasting *Dactylis glomerata* populations in a Mediterranean environment. J. Appl. Ecol. 32, 56–66.

Volaire, F., Leli, F., 1997. Production, persistence, and water-soluble carbohydrate accumulation in 21 contrasting populations of *Dactylis glomerata* L. subjected to severe drought in the south of France. Aust. J. Agric. Res. 48, 933–944.

von Caemmerer, S., Quick, W.P., Furbank, R.T., 2012. The development of C-4 rice: current progress and future challenges. Science 336, 1671–1672.

Waines, J.G., Ehdaie, B., 2007. Domestication and crop physiology: roots of green-revolution wheat. Ann. Bot. 100, 991–998.

Wasson, A.P., Richards, R.A., Chatrath, R., et al.,2012. Traits and selection strategies to improve root systems and water uptake in water-limited wheat crops. J. Exp. Bot. 63, 3485–3498.

Weaver, J.E., Jean, F.C., Crist, J.W., 1922. Development and activities of roots of crop plants: a study in crop ecology, agronomy and horticulture – Faculty Publications. Carnegie Institution of Washington, Washington.

Wherley, B.G., Sinclair, T.R., 2009. Differential sensitivity of C-3 and C-4 turfgrass species to increasing atmospheric vapor pressure deficit. Environ. Exp. Bot. 67, 372–376.

Wiegand, C.L., Namken, L.N., 1966. Influence of plant moisture stress, solar radiation, and air temperature on cotton leaf temperatures. Agron. J. 58, 582–586.

Willson, C.J., Manos, P.S., Jackson, R.B., 2008. Hydraulic traits are influenced by phylogenetic history in the drought-resistant, invasive genus Juniperus (*Cupressaceae*). Am. J. Bot. 95, 299–314.

Wong, S.C., Cowan, I.R., Farquhar, G.D., 1979. Stomatal conductance correlates with photosynthetic capacity. Nature 282, 424–426.

Woodhouse, R., Nobel, P., 1982. Stipe anatomy, water potentials, and xylem conductances in seven species of ferns (*Filicopsida*). Am. J. Bot. 69, 135–140.

Wylie, R.B., 1939. Relations between tissue organization and vein distribution in dicotyledon leaves. Am J. Bot. 26, 219–225.

Zhang, J.-L., Cao, K.-F., 2009. Stem hydraulics mediates leaf water status, carbon gain, nutrient use efficiencies and plant growth rates across dipterocarp species. Funct. Ecol. 23, 658–667.

Zimmermann, M.H., 1983. Xylem structure and the ascent of sap. Springer-Verlag, Berlin.

Zufferey, V., Cochard, H., Ameglio, T., Spring, J.L., Viret, O., 2011. Diurnal cycles of embolism formation and repair in petioles of grapevine (*Vitis vinifera* cv. Chasselas). J. Exp. Bot. 62, 3885–3894.

Zwieniecki, M.A., Boyce, C.K., Holbrook, N.M., 2004. Hydraulic limitations imposed by crown placement determine final size and shape of *Quercus rubra* L. leaves. Plant Cell Environ. 27, 357–365.

PART III

GENETIC IMPROVEMENT AND AGRONOMY

CHAPTER

12

Genetic and environmental effects on crop development determining adaptation and yield

Gustavo A. Slafer[1], *Adriana G. Kantolic*[2], *Maria L. Appendino*[2], *Gabriela Tranquilli*[3], *Daniel J. Miralles*[2,4], *Roxana Savin*[1]

[1]ICREA & Department of Crop and Forest Sciences & AGROTECNIO, University of Lleida, Lleida, Spain

[2]University of Buenos Aires, Buenos Aires, Argentina

[3]Instituto de Recursos Biológicos, INTA, Argentina

[4]IFEVA-CONICET, Buenos Aires, Argentina

1 INTRODUCTION

Crop development is a sequence of phenological events controlled by the genetic background and influenced by external factors, which determines changes in the morphology and/or function of organs (Landsberg, 1977). Although development is a continuous process, the ontogeny of a crop is frequently divided into discrete periods, for instance 'vegetative', 'reproductive' and 'grain-filling' phases (Slafer, 2012).

Patterns of phenological development largely determine the adaptation of a crop to a certain range of environments. For example, genetic improvement in grain yield of wheat has been associated with shorter time from sowing to anthesis in Mediterranean environments of western Australia (Siddique et al., 1989), whereas no consistent trends in phenology were found where drought is present but not necessarily terminal, including environments of Argentina, Canada and the USA (Slafer and Andrade, 1989, 1993; Slafer et al., 1994a) (Fig. 12.1). Even in agricultural lands of the Mediterranean Basin where wheat has been grown for many centuries, breeding during the last century did not clearly change phenological patterns (Acreche et al., 2008).

This chapter focuses on two major morphologically and physiologically contrasting grain crops: wheat and soybean. For both species, we have an advanced understanding of development and

Crop Physiology. DOI: 10.1016/B978-0-12-417104-6.00012-1

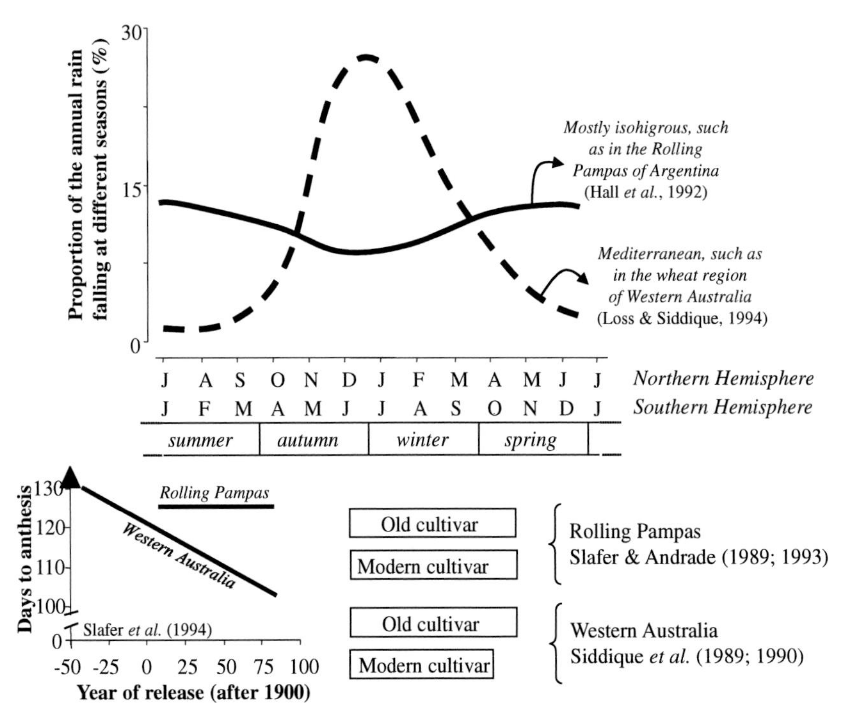

FIG. 12.1 Comparison of the patterns of seasonal rainfall in western Australia and Rolling Pampas of Argentina (top), and the changes in days to anthesis of wheat cultivars released in these regions during the 20th century (bottom). The bars cover, for each type of cultivar and region, common periods from sowing to anthesis. *Souce: Araus et al. (2002).*

physiology in general. Wheat is a determinate, long-day grass of temperate origin, which is responsive to vernalization. Soybean is a typically indeterminate (but with determinate intermediate variants), short-day grain legume of tropical origin, which is insensitive to vernalization. Comparisons with other species are used to highlight the similarities and differences. The aims of this chapter are to outline the developmental characteristics of grain crops and the links between phenology and yield, to revise the mechanisms of environmental and genetic control of development and to explore the possibilities of improving crop adaptation and yield potential through the fine-tuning of developmental patterns.

2 CROP DEVELOPMENT

In this section, we briefly describe the major developmental stages or phases of wheat and soybean separately (as developmental features are in many cases unique), and then discuss the relationships between crop phenology and yield determination.

2.1 Major developmental stages or phases

2.1.1 Wheat

The development of the wheat plant comprises phases defined in terms of microscopic and macroscopic changes that have been integrated into several phenological scales (Miralles and Slafer, 1999). Figure 12.2 shows developmental progress of wheat based on easily recognizable events including microscopic (e.g. double ridges, terminal spikelet initiation) and macroscopic (e.g. crop emergence, heading, anthesis, maturity, harvest) delimiters of phases. In this simple scheme, development involves three major phases:

1. vegetative, when the leaves are initiated
2. reproductive, when first spikelet and then floret development (including floret

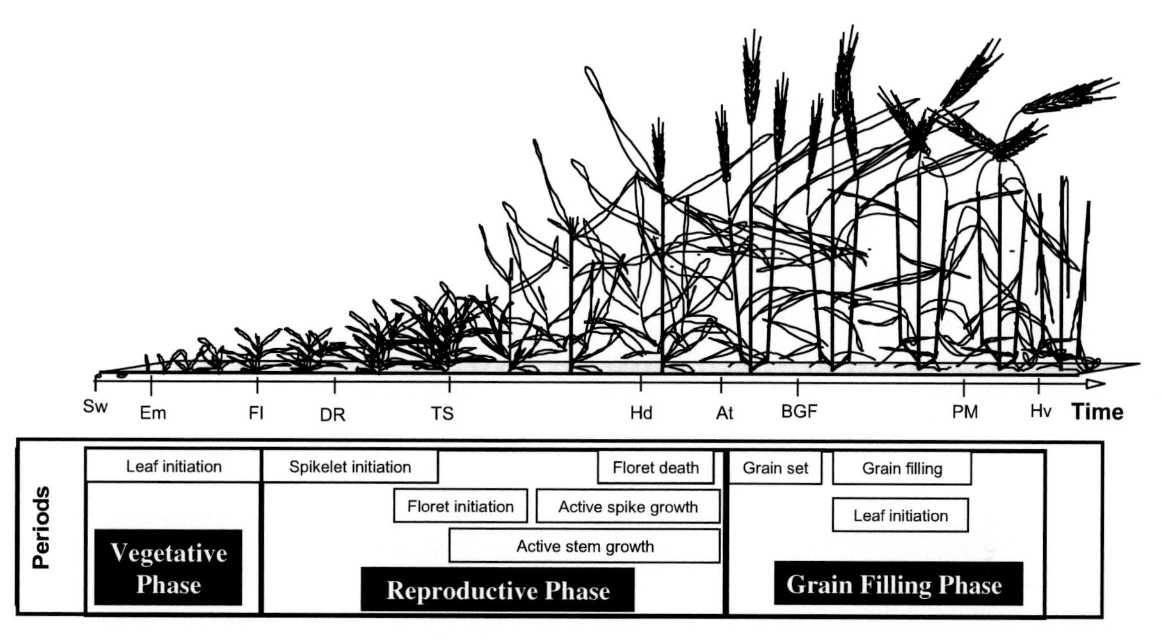

FIG. 12.2 Diagram of wheat growth and development showing the stages: sowing (Sw), seedling emergence (Em), floral initiation (Fl), initiation of the first double ridge (DR), terminal spikelet initiation (TS), heading (Hd), anthesis (At), beginning of the grain-filling period (BGF), physiological maturity (PM), harvest (Hv). Boxes indicate the periods of differentiation or growth of some organs within the vegetative, reproductive, and grain-filling phases. *Source: Slafer and Rawson (1994a).*

mortality) occurs, until the number of fertile florets is determined

3. grain filling, when the grain first develops endosperm cells, and then grows to its final weight.

These phases are delimited by sowing–floral initiation, floral initiation–anthesis and anthesis–maturity (Fig. 12.2). Although they do not delimit major developmental phases, the initiation of both the first double ridge and the terminal spikelet are important early reproductive markers. The former is the first (microscopically) visible sign that the plant is reproductive, while the latter marks the end of the spikelet initiation phase, when the final number of spikelets per spike is determined which, under most field conditions, coincides with the onset of stem elongation.

A mature wheat seed normally contains four leaf primordia (Kirby and Appleyard, 1987; Hay and Kirby, 1991). After sowing, seed imbibition and initiation of leaf primordia is reassumed; a typical seedling has six differentiated leaves at emergence under non-stressful field conditions. Leaf initiation continues until the onset of floral initiation, when the maximum number of leaves in the main shoot is determined. Measured in thermal time, the rate of leaf initiation (or its reciprocal, the plastochron) is relatively constant between different leaf primordia (Kirby et al., 1987; Delécolle et al., 1989), but genetic variation has been reported (e.g. Evans and Blundell, 1994). The timing of floral initiation is therefore a major driver of the length of the crop cycle to anthesis, as all leaf primordia appear at a certain rate (the reciprocal of phyllochron) before the last internode elongates and the crop reaches heading. Phyllochron is approximately constant, although under circumstances of slow development inducing the initiation of a large number of leaf primordia in the main shoot (>10), the phyllochron of later leaves tends to be longer than that of early leaves (Miralles and Slafer, 1999). Although it is frequently assumed that phyllochron is approximately 100°Cd (base temperature 0°C), it is affected by both genetic and environmental factors (e.g. Halloran, 1977; Rawson et al., 1983; Rawson, 1986, 1993; Stapper and Fischer, 1990; Kirby, 1992; Slafer et al., 1994a,b; Slafer and Rawson, 1997).

Cereals develop the capacity to produce a tiller at each phytomer. The process of emergence and growth of tillers, termed tillering, starts when the first tiller bud is mature to grow. The onset of tillering is approximately three phyllochrons after seedling emergence; from then on, the emergence of tillers is closely related to leaf emergence (Masle, 1985; Porter, 1985). Under favorable conditions, the pattern of potential tiller emergence is exponential (Miralles and Slafer, 1999; Alzueta et al., 2012) for a short period up to growth resources become limiting to maintain all tillers. Some tillers die in the reverse order of their emergence, thus contributing to the synchrony and convergence of development in a crop (Hay and Kirby, 1991). This process stabilizes during the period immediately before anthesis, when the number of spikes per unit land area is defined.

In each shoot after floral initiation, the apex starts initiating spikelet primordia; later on, floral initiation starts in the earliest initiated spikelets. Although the double ridge stage has been used as a morphological indication of floral initiation, the first spikelet primordium is normally initiated before double ridge (e.g. Delécolle et al., 1989; Kirby, 1990), so that floral initiation can only be dated *a posteriori* by relating total number of primordia with time (or thermal time), considering the final leaf number (Miralles and Slafer, 1999). The phase of spikelet initiation finishes with the initiation of the terminal spikelet in the apical meristem, when the maximum number of spikelets is fixed. Floral initiation, which had started in the earliest developed spikelets (in the middle third of the spike) before this 'terminal spikelet initiation'stage, continues in all spikelets. Floret development starts in the proximal (to the rachis) positions of each spikelet, and progresses towards the distal positions (e.g. Sibony and Pinthus, 1988). This is why the carpels (at anthesis) and the grains (at maturity) of proximal florets are larger than those in more distal positions (e.g. Rawson and Evans, 1970; Calderini et al., 2001). Floret initiation within each spikelet continues approximately until booting (Kirby, 1988; González et al., 2003b, 2005b; Ferrante et al., 2010, 2013a), reaching a maximum of 6 to 12 floret primordia per spikelet (Sibony and Pinthus, 1988; Youssefian et al., 1992; Miralles et al., 1998), mostly depending on the spikelet position. The development is then arrested in a huge proportion (normally 70–80%) of florets, leading to a large rate of floret mortality coincident with the onset of rapid growth of stems and spikes shortly before anthesis (Kirby, 1988; González et al., 2003b, 2005b). This suggests that competition for assimilates would determine the rate of floret mortality (González et al., 2005b; Ghiglione et al., 2008) as well as the onset of this mortality (González et al., 2011; Ferrante et al., 2013b). Thus, the more the spike can grow at these critical stages, the more florets can reach the stage of fertile florets (and grains afterwards), irrespective of whether this growth is dependent on crop growth or partitioning, or whether it is due to agronomy or genetic improvement (see Slafer (2003); Slafer et al. (2005) for an extended discussion and further references). Just before anthesis, pollination and fertilization occur in fertile florets. Grain set – the proportion of fertile florets producing 'normal' grains – normally ranges between 70 and 90%, likely due to competition for assimilates (Savin and Slafer, 1991; Ferrante et al., 2013a). The period of grain set is characterized by substantial grain development with virtually no grain growth, and is therefore described as the 'lag phase' (Stone and Savin, 1999).

Grain growth and development are normally partitioned into three phases: the above-mentioned lag phase, the effective grain-filling period and the maturation and drying phase (e.g. Bewley and Black, 1985; Savin and Molina-Cano, 2002). Most of the endosperm cells develop during the lag phase, when grains rapidly accumulate water but almost no dry matter (Evers, 1970; Nicolas et al., 1984). The effective grain-filling period involves rapid accumulation of dry matter in the form of seed reserves; water content continues to increase rapidly, and eventually establishes the maximum volume of the seed. During the maturation and drying phases, seeds lose water, reach 'physiological maturity' (maximum dry matter accumulation) and enter a quiescent state (Bewley and Black, 1985). Thus, physiological maturity is the phenological stage that indicates the end of grain growth. The most precise method to determine the timing of physiological maturity is therefore establishing when grain growth ceases (Egli, 1998). However, this laborious method requires frequent consecutive samples to determine that constant grain weight has been reached; hence it only serves to indicate physiological maturity several days after the event (Rondanini et al., 2007). The apparent consistency between the dynamics of water and dry matter in grains of all major crops has led to a reliable, simple method to estimate grain development towards maturity based on the water content of the grains (Box 12.1; see also Fig. 15.4 in Chapter 15).

BOX 12.1

QUANTIFYING GRAIN DEVELOPMENT THROUGH ITS MOISTURE CONTENT

The most common characterization of post-flowering development progress towards maturity has been qualitative, dividing the development into loosely defined grain stages, such as 'aqueous', 'milky', 'dough', and 'hard' grain. As this characterization is based on the proportion of water in grains, it is possible to put forward a quantitative developmental estimate based on the actual grain water content, reflecting the proportion of the time to maturity already elapsed at any time the moisture content of growing grains is measured. For this to be realistic there must be a steady change in this variable during the whole post-flowering period, and for it to be of universal application (a developmental scale applicable to all genotypes of a particular species and to different crop managements), there should be uniform performance across cultivars and environmental conditions.

The scheme indicates that grain growth and grain moisture content dynamics are strongly variable depending on the genotype and the environment, determining large differences in final grain weight. However, the relationship between grain growth and its water content seem much more stable, as there is a positive relationship between the rate of grain growth and the rate of water percentage reduction (the higher the slope of grain dry matter gain the smaller – more negative – the rate of water percentage in grains). If the final grain weight is normalized (by referring in each case the grain weight at any time between anthesis and maturity as a percentage of the final grain weight), there seems to be a universal sharp negative relationship between the grain moisture percentage and grain weight normalized; so that disregarding profound differences in final grain weight and in the dynamics of grain growth, all crops within a particular species reached physiological maturity at a rather similar water content in the grains. Evidences in maize (Saini and Westgate, 2000; Borrás et al., 2003; Borrás and Wesgate, 2006), wheat (Schnyder and Baum, 1992; Calderini et al., 2000), barley (Alvarez Prado et al., 2013), sorghum (Gambín and Borrás, 2005), soybean (Swank et al., 1987) and sunflower (Rondanini et al., 2007) have shown that final grain weight is achieved at, or near to, a particular

BOX 12.1 *(cont.)*

moisture content irrespective of the actual size of the grains (affected by genetic or environmental factors)*, revealing that dry matter accumulation in developing grains and the concurrent loss of water are closely related phenomena.

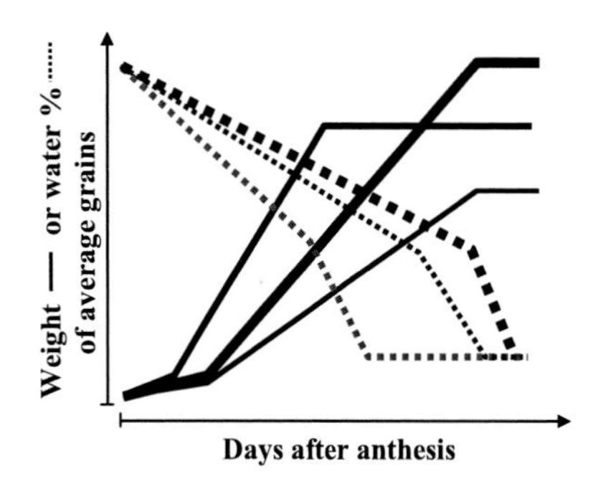

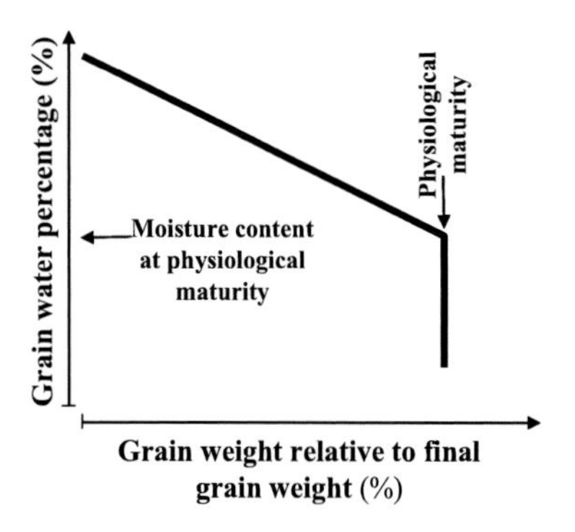

Thus, it seems that duration of grain filling is determined by the interaction between reserve depositions and declining cellular water content, where deposition of reserves such as starch replaces water until critical minimum moisture content is reached. As, for each crop, (1) water percentage at flowering and at maturity are rather constant (for a wide range of grain growing conditions and of final grain weights) and (2) it decreases linearly across the range from flowering to physiological maturity, it can be proposed that the progress of grain development towards maturity may be trustworthily based on the water content of the grains. For instance, if for wheat the limits are ≈80% water content just after anthesis and ≈40% at physiological maturity (Calderini et al., 2000), it can be directly established what proportion of the grain-filling period has elapsed at any time we measure grain moisture content in the field. This quantitative assessment allowing determination of how much of the grain filling has been already completed may be instrumental in management decisions such as when applying a desiccant to the crop to advance harvest without losing yield (e.g. Calviño et al., 2002). Chapter 15 presents the application of the model relating grain growth and grain moisture content dynamics in the analysis of genetic control of grain size in maize.

* In some extreme conditions moisture content at maturity may also be affected within a crop, though assuming a constant value for a particular crop seems quite stable for realistic agronomic conditions.

2.1.2 Soybean

External development in soybean is described in terms of the number of leaves or nodes on the main stem (V-stages) or the presence or growth of reproductive organs in the upper nodes on the main stem, i.e. R-stages (Fig. 12.3). After seedling emergence (VE), the cotyledons open and the two unifoliate leaves unroll (VC); then trifoliate leaves appear and expand at a rate that depends mainly on temperature (Hesketh et al., 1973;

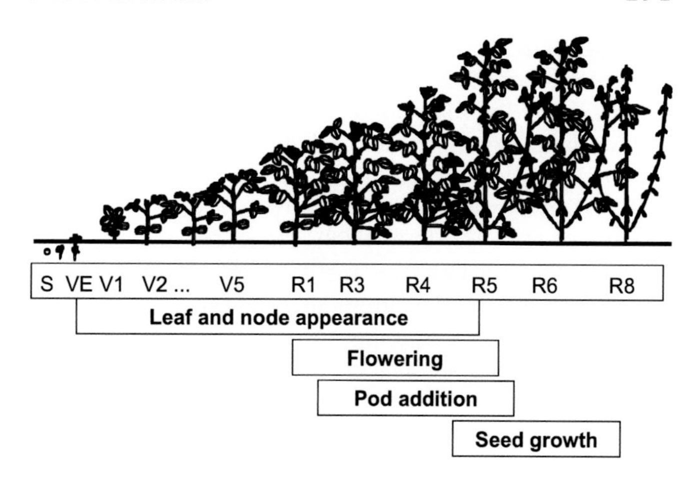

FIG. 12.3 Diagram of soybean growth and development showing the stages: sowing (S), seedling emergence (VE), appearance of different leaves (V1–Vn), until the appearance of the first open flower (R1), followed by different reproductive phases in which flowers become pods, seeds start to grow inside pods (R3–R7) until maturity (R8). Boxes indicate the periods of appearance of leaves, nodes, flowering-pod addition, seed growth.

Thomas and Raper, 1976; Sinclair, 1984a), defining the successive V-stages (V1, V2, Vn), with a phyllochron of approximately 50°Cd (base temperature 11°C; Sinclair et al., 2005). Under many field conditions, approximately one leaf expands every three to four days (Fehr and Caviness, 1977; Bastidas et al., 2008). The opening of the first flower, usually in an axillary raceme on the main stem, defines the flowering stage R1. When flowers open at one of the two uppermost nodes on the main stem, the full bloom stage R2 is defined. Subsequent stages are defined by the presence and size of reproductive organs on one of the four uppermost nodes on the main stem with a fully developed leaf, irrespective of the organs present in other positions of the plant: a pod of 5 mm (R3: beginning pod) or 2 cm (R4: full pod), a seed 3 mm long within the pod (R5: beginning seed) or seeds filling the pod cavity (R6: full seed). Maturity begins when one normal pod on the main stem reaches its mature pod color (R7), and full maturity (R8) is attained when 95% of the pods have reached their mature pod color.

The dormant plumule in the soybean seed has two unifoliate leaves and the first trifoliate leaf primordia initiated during seed development in the mother plant (Lersten and Carlson, 2004). After germination, the pre-formed leaves resume their growth and new foliar primordia are differentiated from the shoot apex; shortly afterwards, branches are differentiated from axillary meristems (Borthwick and Parker, 1938; Sun, 1957; Thomas and Kanchanapoom, 1991). Flower initiation is determined by the appearance of a knob-like primordium in the axil of a bract that precedes the differentiation of floral cycles (Guard, 1931; Carlson and Lersten, 2004). From axillary buds, floral primordia are produced in racemes, while the terminal meristems of the main stem continue differentiating foliar primordia (Borthwick and Parker, 1938). In determinate plants, the terminal apex ceases to differentiate leaf primordia soon after flower initiation and forms a pronounced terminal raceme (Thomas and Kanchanapoom, 1991). In indeterminate plants, the terminal meristem remains vegetative for a longer time and differentiates new foliar primordia, while floral differentiation progresses from the lower to the upper nodes (Saitoh et al., 1999). Eventually, the terminal meristem of indeterminate plants forms a floral primordium and ceases the differentiation of leaves (Caffaro et al., 1988), but does not develop a terminal raceme like determinate genotypes. After floral initiation, the flower primordia progress into complete flowers. Soybean flowers are typically self-pollinated on the day when the corolla opens (Fehr, 1980). From anthesis to pod set, pistil length and weight increase and fertilized ovules develop into embryos (Peterson et al., 1992; Carlson and Lersten, 2004). After a fairly long lag-phase, the seeds begin to accumulate reserves in the cotyledons while water concentration progressively decreases to 55–60%, when physiological maturity is achieved (Swank et al., 1987; Egli, 1998). There are positive correlations between pistil length during early stages of pod development, ovule size and embryo cell numbers, suggesting that changes in external characteristics of flower or pod are correlated with seed development (Peterson et al., 1992). Moreover, seed growth rate during grain filling is highly correlated with the number of cells differentiated in the cotyledons (Egli et al., 1981, 1989; Munier-Jolain and Ney, 1998).

A large proportion of soybean reproductive structures aborts and abscises. Abortion can occur at several stages, including flowers (Huff and Dybing, 1980; Brun and Betts, 1984; Heitholt et al., 1986), immature pods (Hansen and Shibles, 1978; McBlain and Hume, 1981; Egli and Bruening, 2005) or young seeds (Duthion and Pigeaire, 1991; Westgate and Peterson, 1993); when seeds enter the linear phase of growth, the chance of pod abortion decreases (Egli, 1998). Combined flower and pod abscission may range from 32 to 82% (van Schaik and Probst, 1958; Hansen and Shibles, 1978; Wiebold et al., 1981). Reductions in crop photosynthesis increase abortion (Hardman and Brun, 1971; Schou et al., 1978; Egli and Zhen-wen, 1991; Jiang and Egli, 1995); pod age, position and timing of development modify the chances of surviving (Egli, 2005). Abortion and abscission are higher in nodes at the top and the bottom of the canopy than in those in the middle (Heindl and Brun, 1984). Within a node, secondary inflorescences (Piegaire et al., 1988) and distal flowers in the raceme (Peterson et al., 1992) are more likely to abort than primary racemes or proximal flowers, while late developing pods have less chance of surviving than early pods (Huff and Dybing, 1980; Brun and Betts, 1984; Heitholt et al., 1986; Egli and Bruening, 2002, 2006b). Despite its relative long flowering period, the soybean does not have much capacity to recover from flower and pod abortion induced by relatively short-term (≥14 d) reductions in assimilate supply, and final seed number is reduced even though the rest of the period of production of new pods occurs under optimal conditions (Egli, 2010).

Two characteristics in soybean development highly contrast with wheat: (1) the period of leaf appearance and node elongation partially overlaps with the phases of flowering and pod-addition; and (2) reproductive development is highly asynchronous. Both characteristics are related to the difference in growth habit. In determinate soybean, leaf appearance in the main stem ceases soon after R1 (Bernard, 1972). In contrast, indeterminate plants may produce more than two-thirds of main stem nodes after flowering (Heatherly and Smith, 2004); leaf appearance cessation roughly coincides with R5 (Sinclair, 1984b; Bastidas et al., 2008). In semi-determinate types, stem growth ends suddenly after a flowering period, which is almost as long as that of indeterminate plants (Bernard, 1972). Besides these differences, overlapping still exists in determinate soybeans at the plant level, as the production of nodes on the branches is maximum between R1 and R5 in both extreme types (Egli et al., 1985; Board and Settimi, 1986). The degree of overlap and the differences between extreme growth habits are also dependent on environmental conditions. Short photoperiods or very adverse growing conditions reduce the number of nodes appearing on the branches after R1 in indeterminate (Bernard, 1972; Caffaro and Nakayama, 1988) and determinate cultivars (Settimi and Board, 1988; Frederick et al., 2001).

Asynchronous development at the plant level includes inter-nodal and intra-nodal variations. Continued node production on the main stem of the indeterminate cultivars delays flowering of the upper nodes (Saitoh et al., 1999). This in turn leads to high inter-nodal differences in post-flowering development (Munier-Jolain et al., 1993, 1994; Kantolic, 2006; Egli and Bruening, 2006a). In determinate plants, most main stem nodes and the basal nodes of the branches begin to flower near simultaneously (Bernard, 1972; Gai et al., 1984), but flowering at a node level is longer than in indeterminate plants (Gai et al., 1984). At whole-plant level, flowering (defined as the time from R1 until the opening of the last flower) and pod-addition (defined as the period when new pods are formed) phases are generally longer in indeterminate than in determinate plants (Egli and Leggett, 1973; Foley et al., 1986; Egli and Bruening, 2006a). Under normal field conditions, asynchronism declines as the plant approaches maturity (Munier-Jolain et al., 1993; Kantolic, 2006). Seeds developing from flowers opening at later growth stages tend to have shorter seed-fill duration (Egli et al., 1987), and reach maturity only a few days after the first pods lose their green color (Spaeth and Sinclair, 1984).

3 DEVELOPMENTAL RESPONSES TO ENVIRONMENTAL FACTORS

The processes regulating crop development are complex due to interactions between genetic and environmental factors. Water deficit delays phenological development in some species such as sorghum (*Sorghum bicolor*) and quinoa (*Chenopodium quinoa*) (Donatelli et al., 1992; Geerts et al., 2008). Availability of resources, that is water, nutrients, radiation and CO_2, may affect the rate of development (Rawson, 1992, 1993; Evans, 1987; Rodriguez et al., 1994; Arisnabarreta and Miralles, 2004), but these effects are quantitatively minor (Slafer, 1995; Hall et al., 2014). In this section, the analysis of environmental control of development will therefore be restricted to the main environmental drivers of crop development: temperature, including temperature *per se* and vernalization, and photoperiod.

3.1 Temperature *per se*

Of the three major environmental factors, temperature *per se* is the only one that has a universal impact on the rates of development (Aitken, 1974). This means that all crops and all phases of development are sensitive to temperature (Miralles and Slafer, 1999). In general, the higher the temperature, the faster is the rate of development and, consequently, the shorter is the time to complete a particular developmental phase (Slafer and Rawson, 1994a). In all species, developmental responses to temperature start as soon as the seed imbibes (Roberts, 1988), and continue until maturity (Hesketh et al., 1973; Angus et al., 1981; Del Pozzo et al., 1987; Porter et al., 1987; Jones et al., 1991; Slafer and Savin, 1991; Cober et al., 2001; Setiyono et al., 2007). From the various models that have been proposed to predict the timing of development affected by temperature, the most widely accepted is the thermal time (with units of degree days, °Cd; Monteith, 1984). The thermal time model is the calendar time weighted by the thermal conditions; it assumes that the rate of development increases linearly with temperature between the cardinal thresholds of base and optimum temperatures. At temperatures higher than the optimum, there may or may not be a plateau followed by a sharp decrease in rate of development until it becomes zero at the theoretical maximum temperature at which development ceases. Although the general model is universal, the actual cardinal temperatures are generally higher for crops of tropical origin (e.g. soybean, rice, maize, sorghum) than for their temperate counterparts (e.g. wheat, barley, canola). Intra-specific, stage-dependent variation in cardinal temperatures has also been reported (e.g. Angus et al., 1981; Del Pozzo et al., 1987; Porter et al., 1987; Slafer and Savin, 1991; Grimm et al., 1993; Rawson and Richards, 1993; Slafer and Rawson, 1994b, 1995a; Boote et al., 1998; Cober et al., 2001; Setiyono et al., 2007). During the growing cycle, cardinal temperatures increase for wheat (e.g. Slafer and Savin, 1991) but decrease for soybean (Grimm et al., 1994, Setiyono et al., 2007). This reflects the adaptation of wheat to increasing temperatures during its reproductive development, while the opposite occurs in soybeans. Consistent with this proposition, cardinal temperatures decrease with ontogeny in sunflower (Goyne and Hammer, 1982; Chimenti et al., 2001).

3.2 Photoperiod

In comparison to temperature, photoperiod responses are more complex. The degree of variation within species and across stages of development also includes complete insensitivity. Although photoperiodic stimulus is perceived by leaves and transmitted to the apex since the emergence of the crop, many species exhibit a juvenile phase of insensitivity to photoperiod immediately after seedling emergence that imposes a lower limit for the length of the vegetative phase and thus for the final number of leaves. A juvenile phase has been demonstrated for at least some cultivars of soybean (Collinson et al., 1993), maize (Kiniry et al., 1983), barley (Roberts et al., 1988) and sunflower (Villalobos et al., 1996). Other crops such as wheat do not appear to possess a juvenile phase before it becomes sensitive to photoperiod (Hay and Kirby, 1991; Slafer and Rawson, 1995c). These plants perceive the photoperiod stimulus immediately after seedling emergence, and therefore the minimum number of leaves may coincide with the number of leaf primordia initiated by seedling emergence. For example, if photoperiod is sufficiently long after emergence, spring wheat may only have six leaves in the main shoot from seedling emergence to anthesis, which includes the four leaves already present in the embryo plus a couple of leaves initiated from sowing to seedling emergence, when the photoperiod can be perceived (section 2.1.1). Most soybean cultivars have a juvenile phase of variable duration, from 8 to 33 days under optimum temperature (Shanmugasundaram and Tsou, 1978; Board and Settimi, 1988; Ellis et al., 1992; Collinson et al., 1993) but some genotypes have no juvenile phase (Wilkerson et al., 1989; Wang et al., 1998).

Sensitivity to photoperiod is generally quantitative rather than qualitative; that is development is delayed rather than prevented when photoperiod is not optimum (Major, 1980; Summerfield et al., 1993; Slafer and Rawson, 1994a). Most plants are classified by their quantitative photoperiodic response according to the changes in the rate of development and thereby in the length of the phases in response to photoperiod. The two most common categories are 'short-day' and 'long-day' plants. Short-day plants reduce their rates of development (and extend the duration of phases which are sensitive) when photoperiod is lengthened, while long-day plants reduce the duration of their phases when photoperiod is lengthened. Crop species of temperate origin (e.g. wheat, barley, oats, canola, linseed, peas) are long-day plants, while crops of tropical origin (e.g. maize, sorghum, rice, soybean) are short-day plants. Within a particular crop species, genotypes could be classified as follows:

1. Photoperiod insensitive or neutral, if they do not respond to photoperiod in any of its developmental phases; therefore, thermal time to flowering is fairly constant across locations or sowing dates (if insensitive to vernalization), or
2. Photoperiod sensitive, if duration of at least some of its developmental phases increases (short-day species) or decreases (long-day species) in line with photoperiod. Within the sensitive genotypes, there is normally a huge range of genotypic variation.

The optimum photoperiod maximizes the rate of development and, consequently, minimizes the duration of the sensitive phases; photoperiod sensitivity is the delay in duration of a certain stage of development per hour difference between actual and optimum photoperiod. Both optimum photoperiod and photoperiod sensitivity vary among genotypes (e.g. Major, 1980; Davidson et al., 1985; Worland et al., 1994; Slafer and Rawson, 1996; Summerfield et al., 1998; Kantolic and Slafer, 2005). In soybean, cultivars from low maturity groups present a lower sensitivity and a higher photoperiod threshold than genotypes of high maturity groups (Cober et al., 2001; Boote et al., 2003).

Photoperiod sensitivity could be different throughout the crop ontogeny. For instance, while wheat reduces the length of different phases from seedling emergence to flowering as photoperiod is increased, without any sensitivity described for the duration of grain filling, soybean shows photoperiod sensitivity during the whole crop cycle, including grain filling. The variation in sensitivity during different phases has been in fact proposed as a breeding goal to increase the duration of critical phases for yield determination at the expense of the duration of earlier phases (section 5). Highly sensitive cultivars during early phases (e.g. before double ridges in wheat and before R1 in soybean) are usually highly sensitive during later phases (stem elongation in wheat and post-flowering phases in soybean), but the association is not strict (Kantolic and Slafer, 2001); combinations of different sensitivities at different phases might be possible (Slafer et al., 2001).

Both in wheat and in soybean, less-stimulating photoperiods, that is short in wheat and long in soybean, delay both floral initiation and flowering and increase the number of vegetative primordia generated in the apex (Borthwick and Parker, 1938; Rawson, 1971, 1993; Wall and Cartwright, 1974; Halloran, 1977; Thomas and Raper, 1977; Major, 1980; Hadley et al., 1984; Pinthus and Nerson, 1984; Raper and Kramer, 1987; Roberts and Summerfield, 1987; Caffaro and Nakayama, 1988; Sinclair et al., 1991; Evans and Blundell, 1994; Slafer and Rawson, 1994a,b, 1995d, 1996; Upadhyay et al., 1994a,b; Fleming et al., 1997; Kantolic and Slafer, 2001, 2005; Kantolic et al., 2013; Zhang et al., 2001; Miralles et al., 2001, 2003; González et al., 2002). In line with the extended periods, non-stimulating photoperiods modify the number of tillers and the number of leaves per tiller in wheat and the number of branches and the number of nodes in the branches of soybean (Thomas and Raper, 1983; Board and Settimi, 1986; Settimi and Board, 1988; Caffaro and Nakayama, 1988; Miralles and Richards, 2000; Kantolic and Slafer, 2001, 2005, 2007; Kantolic et al., 2013).

Photoperiod also affects developmental rates of soybean after flowering: the duration of the flowering, pod addition phases and the time from R1 to full maturity are increased by direct exposure to long photoperiod (e.g. Johnson et al., 1960; Lawn and Byth, 1973; Major et al., 1975; Thomas and Raper, 1976; Raper and Thomas, 1978; Guiamet and Nakayama, 1984a; Morandi et al., 1988; Summerfield et al., 1998; Kantolic and Slafer, 2001; Han et al., 2006; Kantolic and Slafer, 2007; Kumudini et al., 2007). During seed filling, long photoperiods increase the duration of the lag phase (Zheng et al., 2003; Kantolic, 2006) and delay leaf senescence (Han et al., 2006). It has been proposed that the synchronization in seed maturation, which contrasts with the low synchronism that prevails during the early stages of flowering and pod set, is attributable to photoperiod responses: late developing seeds are generally exposed to short photoperiods that increase their development rate (Gbikpi and Crookston, 1981; Raper and Kramer, 1987). In fact, asynchronous maturity has been described in soybeans exposed to long days under both controlled (Guiamet and Nakayama, 1984b) and field conditions (Mayers et al., 1991). In field experiments that included photoperiod manipulations, asynchronism was quantitatively affected by photoperiod, and the response could be partially reverted by exposure to short photoperiod (Kantolic, 2006).

As photoperiod modulates flowering time and potential plant size, pre-flowering responses to photoperiod have a strong impact on adaptation and potential yield (Roberts et al., 1993). In fact, the classification of soybean cultivars in maturity groups defining broad-sense adaptation is based on pre-flowering developmental response to photoperiod (Summerfield and Roberts, 1985; Boote et al., 1998; Heatherly and Elmore, 2004).

3.3 Vernalization

Vernalization is the acceleration of development by exposing sensitive cultivars to cool temperature during the early stages of crop ontogeny. The plant apex may sense vernalizing temperatures from seed imbibition, throughout the vegetative phase. Vernalization requirements are typical of crops with a temperate origin. In many of these crops, there are 'winter' and 'spring' cultivars. It is frequently assumed that vernalization requirements are characteristic of winter genotypes within a particular species. The difference between spring and winter types may be restricted, however, to the magnitude of vernalization responsiveness (e.g. Levy and Peterson, 1972; Slafer and Rawson, 1994a).

Vernalization may be reversed if the period of low temperature is interrupted, an effect known as 'devernalization'. This was experimentally proven in wheat by Gregory and Purvis (1948) and Purvis and Gregory (1952). Dubert et al. (1992) showed that devernalization may occur at temperatures between 20 and 30°C.

Excluding the effects of temperature *per se* by the calculation of the length of the phases in degree days, photoperiod and vernalization are generally considered to account for most of the differences in development rate among cultivars; any 'residual' difference after the vernalization and photoperiod requirements were satisfied would be the consequence of differences in 'basic development rate' or 'intrinsic earliness' (Major, 1980; Flood and Halloran, 1984; Masle et al., 1989, Slafer, 1996), which are discussed in section 4.1.3.

4 GENETIC CONTROL OF DEVELOPMENT

4.1 Genes affecting development in wheat and related species

The life cycle in wheat is determined by genes that regulate (1) photoperiod response (*Ppd*), (2) vernalization response (*Vrn*) and (3) developmental rates independent of these two environmental factors, called either intrinsic earliness, earliness *per se* or developmental rate genes (*Eps*). The latter are also affected by temperature depending on the gene and allele considered (Slafer and Rawson, 1995b; Appendino and Slafer, 2003). Most of the variation in developmental rates is explained by vernalization and photoperiod response genes, with smaller, more subtle effects of *Eps* alleles (Slafer, 2012; Gomez et al., 2014).

Wheat is an allohexaploid with three genomes, A B and D (Table 12.1): the simultaneous presence of more than one locus implicated in a particular trait (homeologous loci) generates a more complex inheritance pattern than in diploid related species. This polyploid nature prompted an early interest on genetic research in wheat, through a cytogenetic approach, taking advantage of the use of aneuploid genetic stocks. In this way, genetic variability in characters such as vernalization and photoperiod responses are well documented (Worland et al., 1987; Law et al., 1991). Interest in earliness *per se* is more recent, owing to smaller effects and more complex interactions requiring new molecular approaches to understand the *Eps* alleles (Table 12.1).

Understanding genes affecting development in wheat has gained further relevance recently with the development of models using particular gene effects rather than generalized genetic coefficients in crop simulation exercises (Chapter 14). This approach makes models more suitable for developing and testing hypotheses on the genetic improvement value of particular developmental traits based on quantitative predictions of G × E interactions for particular genes (Yin et al. 2000; Hoogenboom and White, 2003; White et al., 2008; Zheng et al., 2013).

4.1.1 Vernalization response genes

Vernalization requirement is the need of fulfillment of a low temperature period in sensitive genotypes to avoid delays in development to reach floral initiation. This requirement prevents the exposure of developing flowers to frost in sensitive cultivars. Most studies about the genetic systems controlling these requirements have been concentrated in the major genes *Vrn1* of *Triticum aestivum*, which explains a large amount of the qualitative variability observed in germplasm and cultivars grown around the world. They are located on homeologous chromosomes 5A (*Vrn-A1*, formerly *Vrn1*); 5B (*Vrn-B1*, formerly *Vrn2*) and 5D (*Vrn-D1*, formerly *Vrn3*) (Flood and Halloran, 1986; Snape et al., 2001); and they map to equivalent position to *Vrn-H1* in *Hordeum vulgare*, *Vrn-R1* in *Secale cereale* and *Vrn-A^{m}1* in diplod wheat *T. monococcum* (Laurie, 1997; Dubcovsky et al., 1998). Additional loci like *Vrn2* (Yan et al., 2004a; Distelfeld et al., 2009a), *Vrn3* (Yan et al., 2006) and *Vrn4* (Yoshida et al., 2010) are also involved in the gene pathway of vernalization (Table 12.1). However, variability on these loci is less known.

TABLE 12.1 *Ppd*, *Vrn* and *Eps* loci of *Triticum aestivum* (T.a.; allohexaploid: A, B and D genomes); *Triticum monococcum* (T.m; diploid: A^m genome) and *Hordeum vulgare* (H.v; diploid: H genome)

	Genomes				
	T.a 2n = 6x			*T.m* 2n = 2x	*H.v* 2n = 2x
Chromosome group	**A**	**B**	**D**	**A^m**	**H**
1					*Ppd-H2*
				Eps-A ^{m} 1	
2	*Ppd-A1*	*Ppd-B1*	*Ppd-D1*		*Ppd-H1*
		Eps2-B			*Eps2-HL* *Eps2-HS*
3	*Eps-AL*				*Eps3-HS*
4					*Eps4-HS*
					Vrn-H2
5	*Eps5*	*Eps5-BL$_{1-2}$*			*Eps5-HL*
	Vrn-A1	*Vrn-B1*	*Vrn-D1*	*Vrn-A ^{m} 1*	*Vrn-H1*
	Vrn-A2	***Vrn.B2***	***Vrn-D2***	*Vrn-A ^{m}2*	
6					*Eps6-HL1* *Eps6-HL2*
7					*Eps7-HS* *Eps7-HL*
		Vrn-B3	*Vrn-D4*		*Vrn-H3*

Numbers from 1 to 7 at the left indicate the chromosome group. Main genes, genetically mapped either as QTL or major genes are in normal type, while not genetically mapped are in bold. Genes are grouped by character. The gene order in the table does not indicate the gene order in the respective chromosome. For gene nomenclature see text in Sections 4.1.1, 4.1.2 and 4.1.3.

The presence of dominance in one or more *Vrn1* loci results in partial or complete elimination of the vernalization requirement, giving rise to spring phenotypes, while winter wheat normally carries recessive alleles. Early studies indicated that the *Vrn-A1* locus has the greatest effect in the elimination of the vernalization requirement (Law et al., 1976; Snape et al., 1976), with some alleles like *Vrn-A1a* conferring insensitivity to vernalization (Appendino and Slafer, 2003; Yan et al., 2004a). Variability in these loci is spread in commercial germplasm but the allelic frequencies may vary depending on the region of cultivation in the world (Stelmak, 1990; Yan et al., 2004a).

A gene primarily detected in the diploid wheat *T. monococcum* is *Vrn-A^{m}2* (mapped on chromosome 5A^m) is also present in *H. vulgare* (*Vrn-H2*). This gene is, in contrast to *Vrn1*, a dominant repressor of flowering and is downregulated by both vernalization and short day (Yan et al., 2004b; Dubcovsky et al., 2006). This genetic factor also regulates flowering by vernalization in polyploid winter wheat. There is no evidence of phenotypic variability in hexaploid wheat but its presence has been demonstrated through RNAi transgenic wheats in winter cultivar Jagger (Yan et al., 2004b) and through chromosome engineering in tetraploid (*T. durum*) wheat (Distelfeld et al., 2009a). Both in barley and diploid wheat, the determination of the vernalization requirement involves an epistatic interaction between *Vrn1* and *Vrn2* (Tranquilli and Dubcovsky, 2000).

Vrn3 is a locus involved in the pathway of vernalization requirement, upregulated by long days. It is located on chromosome 7B (*Vrn-B3*, formerly *Vrn5*), is orthologous to the barley gene *Vrn-H3* located on chromosome 7H, and shows a dominant spring inheritance. The mutation that generates spring genotypes in wheat is not widely spread in commercial germplasm representing a potentially valuable source of genetic diversity (Yan et al., 2006).

Vrn4 is a less characterized gene, and natural variability has been reported only in the D genome (*Vrn-D4*). It promotes flowering and the presence of a dominant *Vrn-D4* allele determines spring growth habit, with a residual response to vernalization. *Vrn-D4* has recently been mapped on chromosome 5D (Yoshida et al., 2010).

Interactions between these genes have been observed, suggesting that all of them integrate the same regulatory pathway of flowering initiation mediated by vernalization. Based on the knowledge acquired from the isolation of *Vrn1*, *Vrn2* and *Vrn3* genes, a model of molecular regulation has been proposed, which integrates also the vernalization and photoperiod pathways (Box 12.2).

4.1.2 Photoperiod response genes

In bread wheat, photoperiod sensitivity is mainly determined by a group of genes located on chromosome group 2 (Table 12.1), namely *Ppd-D1*, formerly *Ppd1* (chromosome 2D), *Ppd-B1*, formerly *Ppd2* (chromosome 2B), and *Ppd-A1*, formerly *Ppd3* (chromosome 2A). Photoperiod insensitivity is of dominant effect, and *Ppd-D1* is the main source of photoperiod insensitivity conferred by the *Ppd-D1a* allele (Worland and Law, 1985; Worland, 1999). *Ppd-B1* is also an important source of photoperiod insensitivity (Scarth and Law, 1984). Although chromosome 2A influences photoperiod sensitivity (Law et al., 1978; Scarth and Law, 1984), to the best of our knowledge *Ppd-A1* has not been genetically mapped. Chromosomes of other groups (1, 3, 4 and 6) may also be involved in photoperiod response, either via modifiers or via major genes (Law, 1987). Law (1987) demonstrated the adaptive roles of *Ppd-D1* and *Ppd-B1* loci using substitution lines with contrasting genotypes in photoperiod response genes.

Two major loci regulating the photoperiod response in barley are *Ppd-H1* and *Ppd-H2* located on chromosomes 2H and 1H, respectively

BOX 12.2

FLOWERING TIME: MODEL OF THE REGULATORY PATHWAY MEDIATED BY VERNALIZATION

Vrn1, *Vrn2* and *Vrn3* genes have been cloned, providing important hints to unravel the regulatory pathway of the spike initiation in response to seasonal cues. These vernalization genes interact and integrate the day-length response to prevent or promote the reproductive stage (reviewed by Trevaskis et al., 2007; Distelfeld et al., 2009b).

Vrn1 encodes a transcriptor factor highly similar to the meristem identity gene *Apetala 1* (*Ap1*) from *Arabidopsis thaliana*, which regulates the transition of the shoot apical meristem from vegetative to reproductive. Sequence changes (insertions or deletions) in regulatory regions have naturally occurred in the three homoeologous *Vrn1* genes giving rise to dominant alleles for spring growth habits.

The *Vrn2* locus includes two tightly linked and related genes (*ZCCT1* and *ZCCT2*), which both repress flowering initiation under long days. In diploid species, like barley or *T. monococcum*, simultaneous deletion or non-functional mutations of both genes were associated with recessive alleles for spring growth habits.

Vrn3 is a functional homolog to the *Flowering Locus T* (*FT*) formerly described as a flowering promoter in *A. thaliana*. The coded protein moves from leaf to shoot apex through the phloem. *Vrn3* accelerates flowering promoting the transcription of *Vrn1* under long days. *Vrn3* (*FT*) is considered a central flowering integrator, since the vernalization and photoperiod signals converge on it. The day-length response is mediated by *Ppd1*, which acts in conjunction with the homologs of the *Arabidopsis* photoperiod gene *CONSTANS* (*CO*). In *Arabidopsis*, long days result in the stabilization of CO proteins, which upregulate *FT* resulting in the acceleration of flowering. This *CO* function would be conserved in temperate cereals.

The model proposes that for a winter, photoperiod-sensitive wheat (ancestral phenotype) sown early in the fall, floral initiation inducted by *Vrn1* is prevented mainly by the action of *Vrn2*, which represses the expression of *Vrn3*, otherwise expected to show a high transcript level under long days. As winter progresses, cold temperatures and short days downregulate *Vrn2*. Low levels of *Vrn2* release *Vrn3* from its repression which, in turn, is induced during the long days of the spring under the regulation mediated by *Ppd1*. *VRN3* protein moves from leaf to apex, where it promotes *Vrn1* transcription leading to the initiation of the reproductive stage. *Vrn1* also acts as a direct or indirect repressor of *Vrn2*, completing the regulatory feedback loop among these genes, which ensures that, once started, the flowering phase progresses steadily. Under this model, any mutation (as those mentioned above) limiting the repressive action of *Vrn2*, either by its own non-functional protein, or by alterations in the repressor recognition sites, will determine a spring wheat.

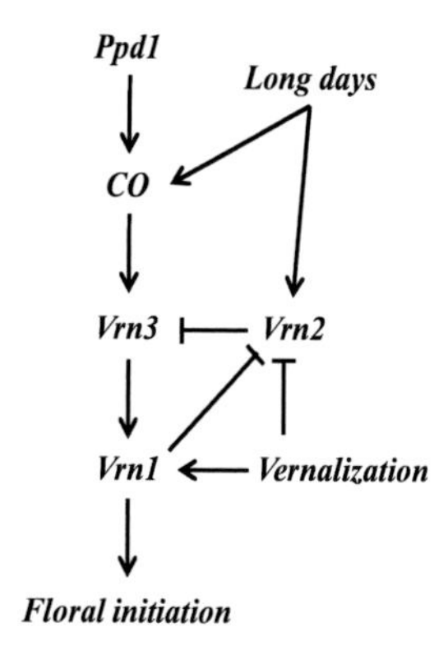

(Table 12.1; Laurie et al., 1995). *Ppd-H1* has been cloned (Turner et al., 2005), and the region of chromosome 2H that contains *Ppd-H1* is collinear with the region of chromosome 2D of wheat where *Ppd-D1* has been mapped (Laurie, 1997; Borner et al., 1998). However, these genes have important differences in the photoperiod response of barley and wheat. In the presence of *Ppd-D1a* (a semi-dominant allele), wheat flowers rapidly on either short or long days, but the recessive genotype delays flowering on short days. In barley, the recessive *Ppd-H1* delays flowering on long days, but has no effect on short days (Laurie et al., 1995; Turner et al., 2005). According to Beales et al. (2007), these wheat and barley *Ppd* genes have contrasting types of mutations, which cause, in wheat, the induction of *Vrn3* (Yan et al., 2006), irrespective of day length, and in barley, the failure to induce this gene correctly on long days (Box 12.2).

Although sowing to heading or anthesis is often considered as a single phase, it is clear that sensitivity to photoperiod changes with developmental phases (e.g. Slafer and Rawson, 1996; Miralles et al., 2000; González et al., 2002). Understanding and manipulating the differential sensitivity to photoperiod at different stages might be useful in increasing yield potential (section 5). However, most attempts to identify genes of photoperiod sensitivity, in particular phases using near-isogenic or recombinant inbred lines for *Ppd* alleles have failed (Scarth et al., 1985; Whitechurch and Slafer, 2001, 2002; Foulkes et al., 2004; González et al., 2005c). This is likely because we only know (and have worked with) a few of the hypothesized genes for photoperiod sensitivity (Table 12.1; Snape et al., 2001). There are other approaches to determine what genes are down- or up-regulated in response to photoperiod (Ghiglione et al., 2008) or to identify genes or quantitative trait loci (QTL) for differences in length of different phases within mapping populations (Borràs-Gelonch et al., 2010) to allow breeding for developmental partitioning (García et al., 2011).

4.1.3 *Earliness* per se *genes*

Time to heading can differ among cultivars by several weeks depending on the interaction between photoperiod, vernalization requirement and the ambient temperature. Where the vernalization and photoperiod requirements have been adequately satisfied, significant differences in time to heading may still persist. The genetic factors underlying these differences have received different names: narrow sense earliness, earliness *per se*, intrinsic earliness or basic development rate genes (e.g. Slafer, 1996, and references quoted therein).

In polyploid species such as *T. aestivum* or diploid species such as *H. vulgare* or *T. monococcum*, this character exhibits a complex genetic base. Earliness *per se* genetic effects have been identified mainly on *T. aestivum* chromosomes of groups 2, 3 and 5 (Table 12.1); a few of them have been mapped as QTL (Scarth and Law, 1984; Miura and Worland, 1994; Worland, 1996; Kato et al., 2002; Toth et al., 2003).

In *T. monococcum*, a QTL for earliness *per se* is located on the distal region of chromosome $1A^mL$ (*Eps-A^m1*) (Bullrich et al., 2002). Allelic variation at this locus modified flowering time with fully vernalized plants grown under long day in a controlled environment. Also, smaller differences under natural conditions, in interaction with photoperiod and vernalization requirements were evidenced. *Eps-A^m1* was then mapped within a 0.8 cM interval using a high-density mapping population and markers generated from the rice collinear region (Valarik et al., 2006). In *H. vulgare*, Laurie et al. (1995) mapped several QTL for *Eps* (Table 12.1), giving evidence of a wide dispersion of candidate genetic factors underlying this character in barley. Future studies will help in the identification of these potential *Eps* genes that, according to Snape et al. (2001), would also be expected to be present in wheat.

4.2 Genes affecting development in soybean

4.2.1 Photoperiod response genes

Genetic control of flowering has been widely used to improve crop adaptation in classical soybean breeding programs (Curtis et al., 2000). The control of time of flowering involves at least eight major loci, each with two alleles: *E1/e1*, *E2/e2* (Bernard, 1971), *E3/e3* (Buzzell, 1971), *E4/e4* (Buzzell and Voldeng, 1980), *E5/e5* (McBlain and Bernard, 1987), *E6/e6* (Bonato and Vello, 1999) and *E7/e7*, closely linked to *E1/e1* (Cober and Voldeng, 2001) and *E8/e8* (Cober et al., 2010). In most cases, the dominant or partially dominant allele lengthens time to flowering in response to photoperiod (McBlain et al., 1987; Saindon et al., 1989; Cober et al., 1996); an exception to this is *E6*, in which early flowering is dominant (Bonato and Vello, 1999). Apparently, the genes of the E series, also known as maturity genes, have no effects on the response of development to temperature (Upadhyay et al., 1994a).

Sensitivity of genotypes to photoperiod depends on the allelic composition of E-genes. Photoperiod sensitivity tends to increase with the number of dominant alleles (Summerfield et al., 1998; Cober et al., 2001; Stewart et al., 2003; Kumudini et al., 2007). The number and type of dominant alleles also seem to modify the photoperiod threshold for response (Messina et al., 2006). Lines with the dominant allele *E1* tended to have a longer juvenile period (Upadhyay et al., 1994b); modeling also supports the role of *E1* in extending the juvenile phase (Messina et al., 2006). Besides their direct effect on photoperiod response, some combinations of alleles seem to have additional advantages in crop adaptation; for instance, the allelic combination of *E1e3e4* is preferable to *e1E3E4* to enhance yield under chilling conditions (Takahashi et al., 2005).

Several QTL associated with flowering time and maturity have been mapped in soybean (Keim et al., 1990; Mansur et al., 1996; Lee et al., 1996; Tasma et al., 2001; Abe et al., 2003; Matsumura et al., 2008); molecular markers are available for marker-assisted breeding for photoperiod sensitivity. Although some association has been found between some loci E and those controlling flowering time in *Arabidopsis thaliana* (Tasma and Shoemaker, 2003), the molecular basis of photoperiodic response in soybean has not been elucidated.

Near-isogenic lines have been developed for the genes of the E-series by back-crosses with different commercial lines (Bernard, 1971; Saindon et al., 1989; Voldeng and Saindon, 1991; Voldeng et al., 1996; Cober and Morrison, 2010). Although the genes that control pre- and post-flowering development are apparently the same, their individual effects may differ depending on the developmental phase. For instance, the allele *E1* alone has a considerable effect on delaying flowering under long days, but has virtually no effects on delaying maturity. In contrast, the effects of *E2, E3, E4, E5* and *E7* on delaying flowering are less marked than those of *E1* but they are effective in delaying maturity under long photoperiods (Bernard, 1971; McBlain and Bernard, 1987; Saindon et al., 1989; Upadhyay et al., 1994a; Cober et al., 1996; Summerfield et al., 1998; Cober and Voldeng, 2001; Xu et al., 2013). The effects of the genes are not purely additive; interactions between them differ depending on developmental stage (Bernard, 1971; Buzzell and Bernard, 1975; Buzzell and Voldeng, 1980; Upadhyay et al., 1994a).

In addition to these major genes, many QTL associated with flowering time and maturity have been mapped in soybean (Keim et al., 1990; Mansur et al., 1996; Lee et al., 1996; Tasma et al., 2001; Abe et al., 2003; Watanabe et al. 2004; Matsumura et al., 2008; Liu and Abe 2010; Cheng et al., 2011). Recent efforts to understand the molecular bases of the major genes and QTL have identified and characterized the soybean orthologs of *Arabidopsis* photoreceptors, clock-associated genes, and flower-identity genes as flowering genes (Watanabe et al., 2012). The functional genes underlying the loci *E3* and *E4* were found to code for phytochrome A3 (Watanabe et al., 2009) and A2 (Liu et al., 2008), respectively. *E2* is a soybean ortholog of the *GIGANTEA* gene (Watanabe et al., 2011). The nuclear protein *GIGANTEA* is involved in the expression of *CONSTANS* (*CO*) and *FLOWERING LOCUS T* (*FT*) in the photoperiodic pathway in *Arabidopsis* (Sawa and Kay 2011); *CONSTANS* (*CO*) is a central regulator of this pathway, triggering the production of the mobile florigen hormone FT that induces flower differentiation (Valverde, 2011). Some studies have suggested that *E1* protein might function as a transcription factor in the phytochrome A signaling pathway, controlling two *GmFTs* genes, orthologs of the *Arabidopsis FT* (Kong 2010; Xia et al., 2012). In fact, soybean not only possesses orthologs for most of the *Arabidopsis* flowering genes but also has multiple copies of most of them; the functions of these multiple orthologs in the control of soybean flowering should be still clarified (Watanabe et al. 2012).

Some QTL have also been identified to control post-flowering photoperiod responses (Watanabe et al., 2004; Cheng et al., 2011; Komatsu et al., 2012). In coincidence with the role of phytochromes mediating some photoperiod responses after flowering (Han et al., 2006), post-flowering photoperiod sensitivity has been associated with *E3* and *E4*, but not with *E1* (Xu et al., 2013). It has been proposed that the *E3* and *E4* alleles inhibit pod development and seed maturation through the activation of a still unknown factor and directly control the persistence of the vegetative activity of the stem apex (Xu et al., 2013).

4.2.2 Long-juvenile genes

Some soybean genotypes delay flowering under short days (Hartwig and Kiihl, 1979). This trait was first identified in a plant introduction designated as P1 159925, and has been subsequently referred to as 'long-juvenile' (Parvez and Gardner, 1987; Hinson, 1989; Wilkerson et al., 1989), although there is no evidence that the trait alters the length of the juvenile period. The long-juvenile trait retards the overall development towards flowering, so that under short days the emergence-to-flowering period is longer in long-juvenile compared with normal genotypes (Sinclair et al., 1991, 2005; Sinclair and Hinson, 1992; Cairo and Morandi, 2006). This trait is useful in tropical and subtropical areas, as flowering can be delayed in spite of the prevailing conditions of high temperature and short photoperiod.

Examining the segregation pattern of this trait in six F2 populations, from crosses between conventional lines and P1 159925, Ray et al. (1995) concluded that the trait is controlled by a single recessive gene (J/j): the conventional phenotype (JJ) has normal photoperiodic response, while the long-juvenile phenotype (jj) delays flowering under short days. Studying the inheritance of the long-juvenile trait under short days for BR80-677 and MG/BR 22 soybeans, Carpentieri-Pípolo et al. (2000, 2002) proposed that the long-juvenile trait may be controlled by three recessive genes; a genotype with a single pair of recessive alleles did not show the long-juvenile characteristic. The recessive genetic combination in three loci causes a longer pre-flowering period than recombination in two loci.

Differences between normal and long-juvenile lines are small under long days (Cregan and Hartwig, 1984; Parvez and Gardner, 1987; Sinclair and Hinson, 1992; Cairo and Morandi, 2006). The genetic background has important effects on the quantitative expression of the trait (Sinclair and Hinson, 1992; Ray et al., 1995; Sinclair et al., 2005). To avoid environmental and genetic effects in the expression of the character, Cairo et al. (2002) have generated molecular markers linked to the juvenile locus in two genetic backgrounds of soybean that can be used for an early discrimination of long-juvenile plants in a segregating population.

4.2.3 *Growth habit genes*

The growth habit or stem termination in soybean is affected by two loci (*Dt1*, *Dt2*) (Bernard, 1972). Determinate growth habit is conditioned by the recessive allele (*dt1dt1*); the dominant gene pair *Dt1Dt1* produces the indeterminate phenotype. A second gene (dominant *Dt2*), independent of the *Dt1* locus, produces a semi-determinate phenotype in the presence of *Dt1*; *dt1* is epistatic to *Dt2-dt2*. However, Bernard (1972) observed that, in some genetic backgrounds, *Dt2* and *dt1dt1* produce indistinguishable phenotypes, and that *Dt1Dt1* may be modified by other genetic factors. A third allele (*dt1-t*) was reported at the *dt1* locus (Thompson et al., 1997); plants with this allele present a tall determinate phenotype. Although *dt1-tdt1-t* and *Dt2Dt2* phenotypes are similar in plant height, *dt1-tdt1-t* is more similar to *dt1dt1* when considering leaf and stem traits at the top of the plant.

Isogenic lines for indeterminate/determinate growth habit alleles at the *Dt1* locus in combination with different photoperiod sensitive/insensitive alleles at loci *E1*, *E2*, *E3*, *E4*, and *E7* have been developed to allow comparison of differences between the two growth habits in a wide range of maturity (Cober and Morrison, 2010). When comparing isogenic lines of the same maturity group, determinate lines were always shorter than indeterminate ones but determinate and indeterminate isogenic lines had similar seed yields (Cober and Morrison, 2010).

The *Dt1* gene is an ortholog of *Arabidopsis TERMINAL FLOWER1* (*TFL1*), *GmTFL1b* (Liu et al., 2010; Tian et al., 2010). The *Dt1* expression is under the control of the two phyA-genes, *E3* and *E4*. When photoperiod-sensitive plants having *E1* and a dominant allele at either the *E3* or *E4* locus (or both) are exposed to non-inductive long days, the vegetative activity at the stem apex meristem is retained to produce more nodes due to the enhanced expression of *Dt1* (Xu et al., 2013).

5 CAN WE IMPROVE CROP ADAPTATION AND YIELD POTENTIAL THROUGH FINE-TUNING DEVELOPMENTAL RATES?

Although yield components are being formed all the time from sowing to maturity (e.g. Slafer et al., 1996; Slafer, 2003), there are particular phases that are more relevant for yield. This means that there is scope for improving yield through manipulation of phenology. The prerequisites for this to be effective is that we must (1) recognize the phases which are actually critical, (2) be able to manipulate development to avoid stressful conditions (adaptation) or take advantage of resource availability (yield potential) in these critical phases and (3) evaluate trade-offs between yield components when the developmental phases are modified. In wheat and soybean, yield relates with number of grains rather than average grain size (Slafer and Andrade, 1993; Magrin et al., 1993; Egli, 1998). Evolutionary principles explain the dominant role of grain number and the secondary influence of grain size (Sadras, 2007; Sadras and Slafer 2012; Slafer et al. 2014).

Number of grains per unit area is largely determined by the events during the stem elongation phase in wheat, while in soybean, the critical phase for seed number goes from soon after flowering to early seed filling. These species-specific periods are known as 'critical period' for yield determination. Thus, crop ontogeny should be tailored to avoid stress during the most critical stages (Lawn and Imrie, 1994) and to capture the environmental conditions that favor grain number; environmental characterizations quantifying likelihood of stress in critical period are therefore important (Chapter 13). With good supply of nutrients and water, the number of grains per unit area is proportional to the amount of solar radiation affecting growth and negatively related to mean temperature affecting development. Thus, their combined effect can be described by the photothermal quotient defined as the ratio of radiation and temperature (Nix, 1976). Fischer (1984, 1985) demonstrated a strong correlation between the number of grains per unit area and the photothermal quotient during the critical period of wheat. In soybean, capture of solar radiation during the pod-setting stage is closely associated with the number of grains per unit area (Kantolic and Slafer, 2001; Calviño et al., 2003).

5.1 Crop development and adaptation

An important objective of crop adaptation is to match crop development phases with optimum environmental conditions, particularly, the timing of flowering is critical. If flowering is too early, plant growth may be insufficient to produce a minimum amount of biomass compatible with reasonable yields (Mayers et al., 1991). This is why early vigor of the crop is much more important for 'short-season' crops (e.g. spring cereals grown at high latitudes) than for 'full season' crops. Chapter 4 discusses in detail the challenges of growing temperate crops in northern Europe, including the role of phenology as the key adaptive trait in these extreme environments. On the other hand, if flowering is too late, the period available for grain growth may be too short and/or too stressful (Lawn et al., 1995). Therefore, the above-mentioned extremes (early and late flowering) define the length of the growing season, and the pre-flowering development may be manipulated to improve adaptation by balancing the optimum time of flowering (Rhone et al., 2010) and the consequent duration of pre- and post-flowering phases.

Phenological adaptation is particularly critical in stressful environments. When water or nutrients are scarce, vegetative growth may become limiting, increasing the length of the preflowering phase, which may increase the size of both canopy and root system. Cultivars with a longer vegetative period may have deeper root systems and better capacity of extracting water from deeper soil layers than early flowering ones (e.g. Giménez and Fereres, 1986; Dardanelli et al., 2004), which may be useful provided the soil holds water deep in the profile. In contrast, long-cycle cultivars may deplete more water before the critical periods (Edwards and Purcell, 2005), risking more severe stress when crop yield is most sensitive; under these circumstances, early-flowering cultivars may produce larger yields when moisture stress develops late in the season (Fig. 12.1; Kane and Grabau, 1992).

Crop development may also improve cultivar adaptation by reducing the risk of biotic stresses. Early-maturing wheat had been useful for escaping rust damage in Australia (Park et al. 2009), while management techniques based primarily on early-maturing cultivars of soybean effectively reduced the impact of stem rot (*Sclerotinia sclerotiorum*) in Argentina (Ploper, 2004).

In soybean, the stem-termination habit, which modifies the length of the phase of node production independently of the duration of pre-flowering stages, is not directly associated with potential yield, but confers some characteristics that may modify cultivar adaptation. Determinate growth habit is useful in reducing plant height and lodging but can result in excessive dwarfing in early-maturing soybean (Cober and Tanner, 1995). Indeterminate cultivars tend to yield better than determinate ones in yield-restricted environments and late plantings (Beaver and Johnson, 1981; Robinson and Wilcox, 1998; Kilgore-Norquest and Sneller, 2002). However, the adaptive value of stem termination types to particular environments may also depend on genetic background (Heatherly and Elmore, 2004).

5.2 Crop development and yield potential

A large body of evidence indicates that reduction in canopy photosynthesis before the onset of stem elongation in wheat or before

R1 in soybean seldom reduces the final number of seeds, while crop growth reduction after stem elongation in wheat or from R2 to R5 in soybean is directly related with the number of seeds or pods set (Fischer, 1985, 2011; Egli and Zhen-wen, 1991; Savin and Slafer, 1991; Jiang and Egli, 1993, 1995; Board et al., 1995; Board and Tan, 1995; Abbate et al., 1997; Demotes-Mainard and Jeuffroy, 2004; González et al., 2005a; Miralles and Slafer, 2007; Ferrante et al., 2012). In wheat, where grain filling is mostly sink limited (Slafer and Savin, 1994; Borrás et al., 2004), reduction in grain number cannot be compensated by an increase of grain weight, except for small compensations that might occur if the decrease in grain number brings about increases in carpel size and concomitantly greater grain weight potential (Calderini and Reynolds, 2000; Calderini et al., 2001; Ugarte et al., 2007). In soybeans, where seed filling is usually more limited by the source (Egli, 1999, 2004; Borrás et al., 2004; Egli and Bruening, 2006b), reductions in seed production by stresses during R1–R3 may be partially compensated by favorable conditions during seed filling.

Within this context, from the developmental point of view, it has been proposed that the length of the critical phase might be extended at the expense of the duration of earlier phases as a means of increasing yield potential both in wheat (Slafer et al., 2001, 2005; Miralles and Slafer, 2007) and soybean (Kantolic et al., 2007). Briefly, the length of the critical phase in both crops is (1) highly relevant in the determination of seed number per unit land area (Fig. 12.4a) and (2) sensitive to photoperiod (Fig. 12.4b). In experiments where crops were exposed to different day lengths, the number of grains metre^{-2} increased with increasing duration of the critical phase (Fig. 12.4c). In soybean, the increase in seed number in response to long photoperiod is related to both more nodes per plant and more seeds per node, which actively accumulate during the critical period; this is strongly supported by experiments both in controlled environments (Guiamet and Nakayama, 1984a,b; Morandi et al., 1988) and in the field (Kantolic and Slafer, 2001, 2005, 2007; Kantolic et al., 2013). In wheat, the increase in seed number in response to short photoperiod is mainly related to the fate of floret primordia during the 'floret mortality' period: with longer photoperiod and shorter phase, the proportion of floret primordia that develops towards fertile florets is consistently reduced (González et al., 2003b, 2005b; Ghiglione et al. 2008; Serrago et al., 2008). Early studies where the duration of the stem elongation phase was modified demonstrated that increasing assimilate allocation to the spike can improve survival of florets in the middle of the spikelet, mostly in the third to fifth position from the rachis within the spikelet (González et al., 2005a). Furthermore, recent evidence supports that both the onset and rate of floret death are strongly linked to resource availability for the developing (and growing) florets (González et al., 2011; Ferrante et al., 2013b). More detailed studies suggest that mortality is linked to autophagy (Box 12.3). From the association between seed set, duration of critical phases and the photoperiodic control of these phases, it has been proposed that selecting for photoperiodic sensitivity could increase grain set (Fig. 12.4d).

6 CONCLUDING REMARKS

In this chapter, the particularities of development of wheat and soybean have been discussed to highlight the importance of identifying the genetic and environmental controls of the phenological pattern. This knowledge is a prerequisite to understand, predict and manipulate the association between crop cycle, the resources and the environmental constraints to favor the coincidence of the critical period with the most favorable conditions. Although the cycle to match crops and environmental factors has been determined in most production systems, further improvement is feasible by manipulation of critical periods.

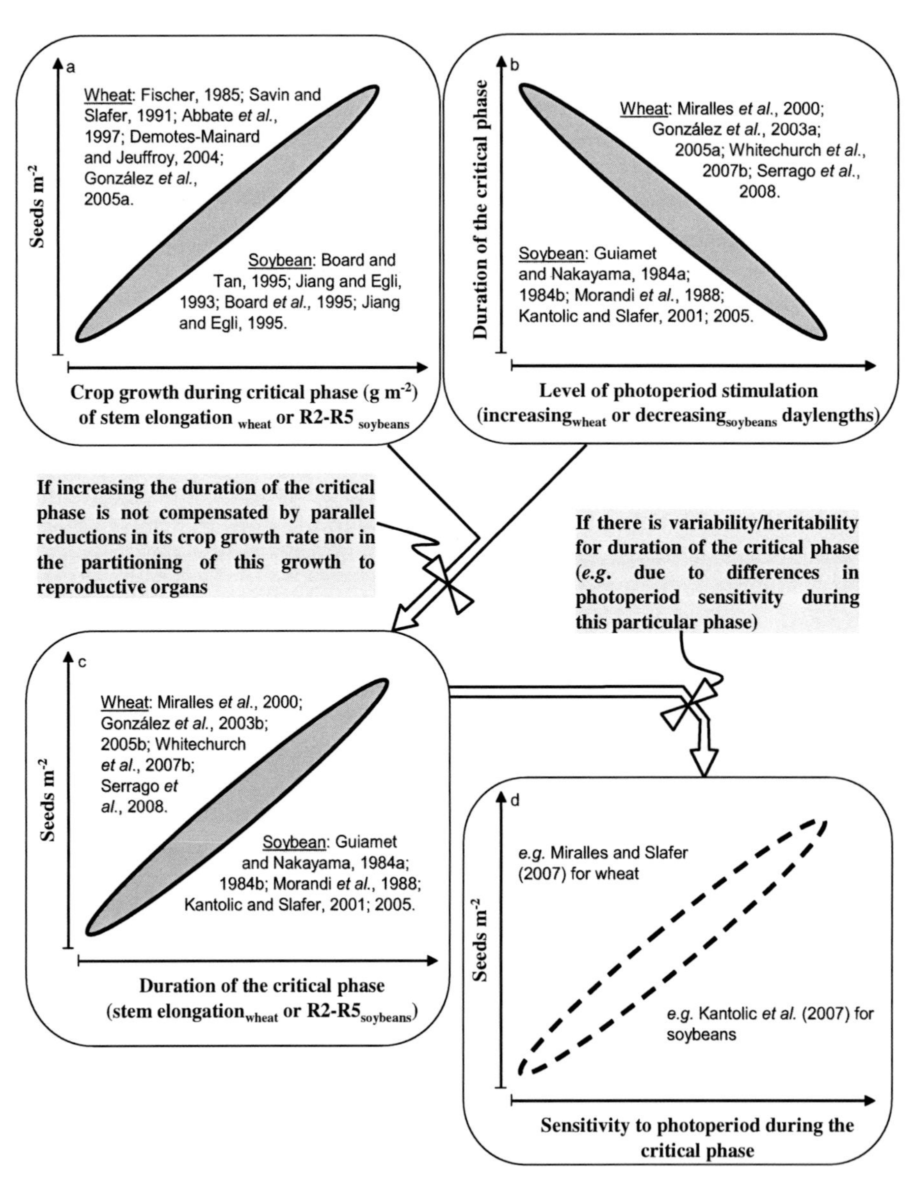

FIG. 12.4 The top two panels summarize two principles: seed number is strongly responsive to crop growth during a critical phase for seed number determination ((a) with some of the many references evidencing this relationship), and these phases in wheat and soybean are sensitive to photoperiod so that the less inductive the photoperiod the longer the phase (b). Lack of or incomplete compensation between crop duration and growth rate during the critical phase, means that longer duration favor higher seed set, as shown when the duration of the critical phase is altered by manipulating photoperiod only during this phase (c). Thus, increasing sensitivity to photoperiod during this particular phase would increase seed number m^{-2} through increasing growth, allowing in turn the set of more inflorescences, more seeds per inflorescence or both (d). A more comprehensive treatment can be found in Slafer et al. (2001), Miralles and Slafer (2007) and Slafer et al. (2014) for wheat, and Kantolic et al. (2007) for soybeans.

BOX 12.3

LIVE AND LET DIE: LIKELY MECHANISMS OF FLORET MORTALITY IN WHEAT

As in many other species, the number of grains per unit land area is the main yield component explaining the variations in wheat yield. Although each spikelet within the wheat spike has the capacity to produce a very large number of floret primordia (≈10–12 in central spikelets), many of these developing florets fail to reach the stage of fertile floret at anthesis. As (1) the number of floret primordia developed in each spikelet is enormously higher than the number of fertile florets at anthesis and (2) the proportion of grain set (grains per fertile floret) is normally quite high, the capacity of floret primordia to survive and continue developing all its floret organs constitutes a major bottleneck process for reaching a high number of fertile florets, and thereby grains, in wheat.

Understanding the mechanisms of floret mortality from a combined approach including (1) anatomy changes of the primordia at cellular level, (2) physiological development and (3) gene expression (transcriptomic patterns) during the active spike growth period would be a way to gain insight into the process for increasing floret survival in wheat. In a study where a combination of microarray, biochemical and anatomical approaches were used to investigate the origin of floret mortality, Ghiglione et al. (2008) modified the duration of the spike growth period by modifying the photoperiod only during the phase of spike growth (until then all plants were in the same condition). They found that the acceleration of floret primordia mortality induced by extending photoperiod (e.g. González et al., 2003b; 2005b), which is illustrated below, was associated with the expression of genes involved in photosynthesis, photo-protection and carbohydrates metabolism. The expression of marker genes associated with floral development cell proliferation and programmed cell death were activated with long photoperiod, i.e. the signal that determined the mortality of distal floret within the spikelets.

Anatomy and morphological changes in the primordia were found and cells of the ovaries of aborting florets revealed the formation of the vacuoles that become dense and increased in size, the development of dense globular bodies (authophagosomes), chromatin condensation of the nucleus and vanishing of nucleolus in the

Plants grown under field conditions (i.e. relatively short days)

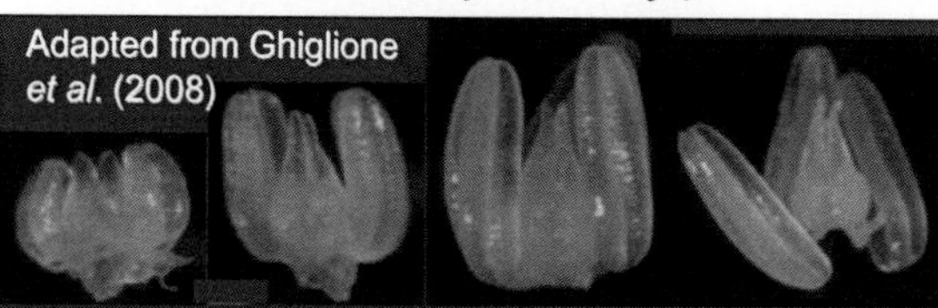

Floret development in a plant growing under normal field conditions through stages 6, 7, 8 and 8.5 of the Waddington *et al.* (1983) scale

Plants grown under extended photoperiod during the spike growth period

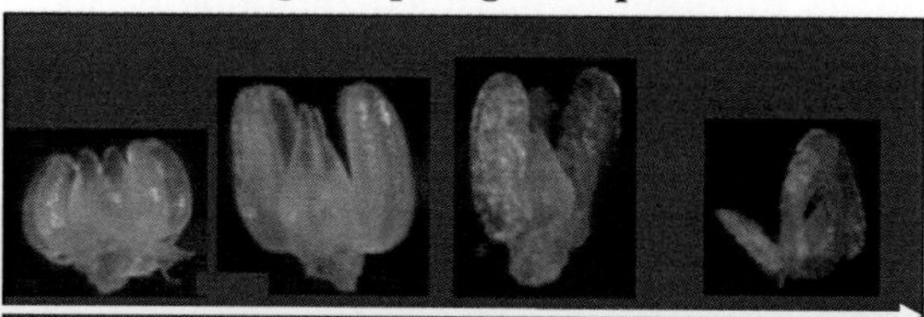

Floret mortality in a similar floret that reached the stage 7 normally, illustrating the decay of floret organs and floret abortion, due to accelerated development

BOX 12.3 *(cont.)*

cells of the ovaries of aborting florets suggesting that floret abortion was a programmed cell death by authophagy rather than passive death or necrosis. Finally, another mechanism involved in floret abortion is the level of soluble carbohydrates as the decrease of soluble sugars (glucose and fructose) during the spike growth period enhanced the floret abortion of the primordia previously initiated. This agrees with previous studies showing that the fate of florets was largely determined by the acquisition of assimilates by the growing spike when affected by either genetic (Miralles et al., 1998; González et al., 2005c) or environmental (González et al., 2005a; Ferrante et al., 2010) factors.

The study of floret initiation and mortality in wheat from a multidisciplinary approach allows the exploration of different mechanisms involved in the process under study improving the understanding of the process. Thus, the identification of candidate genes controlling the process of survival of the floret primordia previously initiated might constitute a major step to improve further the number of grains per unit land area and yield in cereals and in other crops.

The critical period may occur before (e.g. in wheat) or after (e.g. in soybean) flowering, but it is clear that, in both species, in spite of their large morphological and physiological differences, the growth during this period defines crop yield in most environments. Improving our knowledge of genetic and environmental drivers of the expression of genes that control flowering time should improve our precision in positioning the critical period when the highest level of resources is expected, and stresses are less likely (Chapter 13).

In both species, the length of the critical phase is positively related with the number of seeds, and its duration is modified by photoperiod. Manipulative experiments described in this chapter showed that increasing sensitivity to photoperiod during the critical phase for grain number determination may actually raise yield potential. The longer the critical phase, the more the crop may grow, supporting more grains to be set. So yield could be improved if this phase is lengthened without modifying the whole crop cycle duration. In wheat and barley, there is a large variation in duration of stem elongation independently of the total duration to anthesis (Kernich et al., 1997; Whitechurch et al., 2007a) which is partially due to variation in photoperiod sensitivity during this phase (Whitechurch et al., 2007b). In fact, in exceptional cases, empirical breeding may have made use of this variability (Abeledo et al., 2001). In soybean, simulation studies have shown that shortening the pre-flowering period, without changing the duration of the whole cycle, could increase yields in a broad range of latitudes and environmental conditions (Kantolic et al., 2007). The main problem to manipulate this trait is to identify clearly the genetic basis of photoperiod sensitivity of the critical phase. This is not simple. However, even though no single major allele has been particularly linked with photoperiod sensitivity during the critical phase for yield determination, different sources of evidence reinforce the idea that photoperiod sensitivity of individual phases may be independent of each other. This would allow exploiting this trait to change the length of the critical period without altering the duration of the whole crop cycle. Moreover, identifying the genes involved in floret survival (e.g. Ghiglione et al., 2008) is the first step to understanding their environmental modulation.

References

Abbate, P.E., Andrade, F.H., Culot, J.P., Bindraban, P.S., 1997. Grain yield in wheat: effects of radiation during spike growth period. Field Crops Res. 54, 245–257.

Abe, J., Xu, D., Miyano, A., Komatsu, K., Kanazawa, A., Shimamoto, Y., 2003. Photoperiod-insensitive Japanese soybean landraces differ at two maturity loci. Crop Sci. 43, 1300–1304.

Abeledo, L.G., Calderini, D.F., Slafer, G.A., 2001. Genetic improvement of barley yield potential and its physiological determinants in Argentina (1944-1998). Euphytica 130, 325–334.

Acreche, M., Briceño-Félix, G., Martín Sánchez, J.A., Slafer, G.A., 2008. Physiological bases of genetic gains in Mediterranean bread wheat yield in Spain. Eur. J. Agron. 28, 162–170.

Aitken, Y., 1974. Flowering time, climate and genotype. Melbourne University Press, Melbourne.

Alvarez Prado, S., Gallardo, J.M., Serrago, R.A., Kruk, B.C., Miralles, D.J., 2013. Comparative behavior of wheat and barley associated with field release and grain weight determination. Field Crops Res. 144, 28–33.

Alzueta, I., Abeledo, G.L., Mignone, C.M., Miralles, D.J., 2012. Differences between wheat and barley in leaf and tillering coordination under contrasting nitrogen and sulfur conditions. Eur. J. Agron. 41, 92–102.

Angus, J.F., Mackenzie, D.H., Morton, R., Schafer, C.A., 1981. Phasic development in field crops. II. Thermal and photoperiodic responses of spring wheat. Field Crops Res. 4, 269–283.

Appendino, M.L., Slafer, G.A., 2003. Earliness per se and its dependence upon temperature in diploid wheat lines differing in the allelic constitution of a major gene (*Eps-A^{m}1*). J. Agric. Sci. (Camb.) 141, 149–154.

Araus, J.L., Slafer, G.A., Reynolds, M.P., Royo, C., 2002. Plant breeding and water relations in C3 cereals: what should we breed for? Ann. Bot. 89, 925–940.

Arisnabarreta, S., Miralles, D.J., 2004. The influence of fertilizer nitrogen application on development and number of reproductive primordia in field grown two- and six-rowed barleys. Aust. J. Agric. Res. 55, 357–366.

Bastidas, A.M., Setiyono, T.D., Dobermann, A., et al., 2008. Soybean sowing date: the vegetative, reproductive and agronomic impacts. Crop Sci. 48, 727–740.

Beales, J., Turner, A., Griffiths, S., Snape, J., Laurie, D., 2007. A pseudo response regulator is misexpressed in the photoperiod insensitive *Ppd-D1a* mutant of wheat (*Triticum aestivum*). Theor. Appl. Genet. 115, 721–735.

Beaver, J.S., Johnson, R.R., 1981. Yield stability of determinate and indeterminate soybeans adapted to the Northern United States. Crop Sci. 21, 449–454.

Bernard, R.L., 1971. Two major genes for time of flowering and maturity in soybeans. Crop Sci. 11, 242–244.

Bernard, R.L., 1972. Two genes affecting stem termination in soybeans. Crop Sci. 12, 235–239.

Bewley, J.D., Black, M., 1985. Seeds: physiology of development and germination. Plenum, New York.

Board, J.E., Settimi, J.R., 1986. Photoperiod effect before and after flowering on branch development in determinate soybean. Agron. J. 78, 995–1002.

Board, J.E., Settimi, J.R., 1988. Photoperiod requirements for flowering and flower production in soybean. Agron. J. 80, 518–525.

Board, J.E., Tan, Q., 1995. Assimilatory capacity effects on soybean yield components and pod number. Crop Sci. 35, 846–851.

Board, J.E., Wier, A.T., Boethel, D.J., 1995. Source strength influence on soybean yield formation during early and late reproductive development. Crop Sci. 35, 1104–1110.

Bonato, E.R., Vello, N.A., 1999. E6, a dominant gene conditioning early flowering and maturity in soybeans. Gen. Mol. Biol. 22, 229–232.

Boote, K.J., Jones, J.W., Hoogenboom, G., 1998. Simulation of crop growth: CROPGRO model. In: Peart, R.M., Curtis, R.B. (Eds.), Agricultural systems modeling and simulation. Marcel Dekker, New York, pp. 651–692.

Boote, K.J., Jones, J.W., Batchelor, W.D., Nafziger, E.D., Myers, O., 2003. Genetic coefficients in the CROPGRO–soybean model: links to field performance and genomics. Agron. J. 95, 32–51.

Borner, A., Korzum, V., Worland, A.J., 1998. Comparative genetic mapping of loci affecting plant height and development in cereals. Euphytica 100, 245–248.

Borrás, L., Slafer, G.A., Otegui, M.E., 2004. Seed dry weight response to source-sink manipulations in wheat, maize and soybean. A quantitative reappraisal. Field Crops Res. 86, 131–146.

Borrás, L., Wesgate, M.E., 2006. Predicting maize kernel sink capacity early in development. Field Crops Res. 95, 223–233.

Borrás, L., Westgate, M.E., Otegui, M.E., 2003. Control of kernel weight and kernel water relations by post-flowering source-sink ratio in maize. Ann. Bot. 91, 857–867.

Borrás-Gelonch, G., Slafer, G.A., Casas, A.M., van Eeuwijk, F., Romagosa, I., 2010. Genetic control of pre-heading phases and other traits related to development in a double haploid barley (*Hordeum vulgare* L.) population. Field Crops Res. 119, 36–47.

Borthwick, H.A., Parker, M.W., 1938. Influence of photoperiods upon the differentiation of meristems and the blossoming of Biloxi soy beans. Bot. Gaz. 9, 825–839.

Brun, W.A., Betts, K.J., 1984. Source/sink relations of abscising and nonabscising soybean flowers. Plant Physiol. 75, 187–191.

Bullrich, L., Appendino, M.L., Tranquilli, G., Lewis, S., Dubcovsky, J., 2002. Mapping of a thermo-sensitive earliness *per se* gene on *Triticum monococcum* chromosome 1. Am. Theor. Appl. Genet. 105, 585–593.

Buzzell, R.I., 1971. Inheritance of a soybean flowering response to fluorescent-daylength conditions. Can. J. Genet. Cytol. 13, 703–707.

Buzzell, R.I., Bernard, R.L., 1975. E2 and E3 maturity gene tests. Soybean Genet. Newsl. 2, 47–49.

Buzzell, R.I., Voldeng, H.D., 1980. Inheritance of insensitivity to long daylength. Soybean Genet. Newsl. 7, 26–29.

Caffaro, S.V., Martignone, R.A., Torres, R., Nakayama, F., 1988. Photoperiod regulation of vegetative growth and meristem behavior toward flower initiation of an indeterminate soybean. Bot. Gaz. 149, 311–316.

Caffaro, S.V., Nakayama, F., 1988. Vegetative activity of the main stem terminal bud under photoperiod and flower removal treatments in soybean. Aust. J. Plant Physiol. 15, 475–480.

Cairo, C.A., Morandi, E.N., 2006. Modificación de la respuesta fotoperiódica para la floración en la soja inducida por el gen juvenil. In: Proceedings of the III MERCOSUR Soybean Conference. Rosario, Argentina. pp. 95–98.

Cairo, C.A., Stein, J., Delgado, L., et al., 2002. Tagging the juvenile locus in soybean [*Glycine max* (L.) Merr.] with molecular markers. Euphytica 124, 387–395.

Calderini, D.F., Abeledo, L.G., Slafer, G.A., 2000. Physiological maturity in wheat based on kernel water and dry matter. Agron. J. 92, 895–901.

Calderini, D.F., Reynolds, M.P., 2000. Changes in grain weight as a consequence of de-graining treatments at pre- and post-anthesis in synthetic hexaploid lines of wheat (*Triticum durum* × *T. tauschii*). Aust. J. Plant Physiol. 27, 183–191.

Calderini, D.F., Savin, R., Abeledo, L.G., Reynolds, M.P., Slafer, G.A., 2001. The importance of the immediately preceding anthesis period for grain weight determination in wheat. Euphytica 119, 199–204.

Calviño, P.A., Sadras, V.O., Andrade, F.H., 2003. Development, growth and yield of late-sown soybeans in the southern Pampas. Eur. J. Agron. 19, 265–275.

Calviño, P.A., Studdert, G.A., Abbate, P.E., Andrade, F.H., Redolatti, M., 2002. Use of non-selective herbicides for wheat physiological and harvest maturity acceleration. Field Crops Res. 77, 191–199.

Carlson, J.B., Lersten, N.R., 2004. Reproductive morphology. In: Boerma, H.R., Specht, J.E., (eds), Soybeans: improvement, production, and uses, 3rd edn. Agronomy Series No. 16. ASA CSSA SSSA: Madison, WI, pp. 59–95.

Carpentieri-Pípolo, V., Almeida, L.A., Kiihl, R.A.S., 2002. Inheritance of a long juvenile period under short-day conditions in soybean. Gen. Mol. Biol. 25, 463–469.

Carpentieri-Pipolo, V., Almeida, L.A., Kiihl, R.A.S., Rosolem, C.A., 2000. Inheritance of long juvenile period under short day conditions for the BR80-6778 soybean (*Glycine max* (L.) Merrill) line. Euphytica 112, 203–209.

Cheng, L., Wang, Y., Zhang, C., et al., 2011. Genetic analysis and QTL detection of reproductive period and post-flowering photoperiod responses in soybean. Theor. Appl. Genet. 123, 421–429.

Chimenti, C.A., Hall, A.J., López, M.S., 2001. Embryo growth rate and duration in sunflower as affected by temperature. Field Crops Res. 69, 81–88.

Cober, E.R., Stewart, D.W., Voldeng, H.D., 2001. Photoperiod and temperature responses in early-maturing, near-isogenic soybean lines. Crop Sci. 41, 721–727.

Cober, E.R., Tanner, J.W., 1995. Performance of related indeterminate and tall determinate soybean lines in short-season areas. Crop Sci. 35, 361–364.

Cober, E.R., Tanner, J.W., Voldeng, H.D., 1996. Genetic control of photoperiod response in early-maturing, near isogenic soybean lines. Crop Sci. 36, 601–605.

Cober, E.R., Voldeng, H.D., 2001. A new soybean maturity and photoperiod-sensitivity locus linked to E1 and T. Crop Sci. 41, 698–701.

Cober, E.R., Molnar, S.J., Charette, M., Voldeng, H.D., 2010. A new locus for early maturity in soybean. Crop Sci. 50, 524–527.

Cober, E.R., Morrison, M.J., 2010. Regulation of seed yield and agronomic characters by photoperiod sensitivity and growth habit genes in soybean. Theor. Appl Genet. 120, 1005–1012.

Collinson, S.T., Summerfield, R.J., Ellis, R.H., Roberts, E.H., 1993. Durations of the photoperiod-sensitive and photoperiodinsensitive phases of development to flowering in four cultivars of soybean [*Glycine max* (L) Merrill]. Ann. Bot. 71, 389–394.

Cregan, P.B., Hartwig, E.E., 1984. Characterization of flowering response to photoperiod in diverse soybean genotypes. Crop Sci. 24, 659–662.

Curtis, D.F., Tanner, J.W., Luzzi, B.M., Hume, D.J., 2000. Agronomic and phenological differences of soybean isolines differing in maturity and growth habit. Crop Sci. 40, 1624–1629.

Dardanelli, J.L., Ritchie, J.T., Calmon, M., Andriani, J.M., Collino, D.J., 2004. An empirical model for root water uptake. Field Crops Res. 87, 59–71.

Davidson, J.L., Christian, K.R., Jones, D.B., Bremner, P.M., 1985. Responses of wheat to vernalization and photoperiod. Aust. J. Agric. Res. 36, 347–359.

Delécolle, R., Hay, R.K.M., Guerif, M., Pluchard, P., Varlet-Grancher, C., 1989. A method of describing the progress of apical development in wheat based on the time course of organogenesis. Field Crops Res. 21, 147–160.

Del Pozzo, A.H., Garcia-Huidobro, J., Novoa, R., Villaseca, S., 1987. Relationship of base temperature to development of spring wheat. Exp. Agric. 23, 21–30.

Demotes-Mainard, S., Jeuffroy, M.H., 2004. Effects of nitrogen and radiation on dry matter and nitrogen accumulation in the spike of winter wheat. Field Crops Res. 87, 221–233.

Distelfeld, A., Li, C., Dubcovsky, J., 2009b. Regulation of flowering in temperate cereals. Curr. Opin. Plant Biol. 12, 178–184.

Distelfeld, A., Tranquilli, G., Li, C., Yan, L., Dubcovsky, J., 2009a. Genetic and molecular characterization of the *VRN2* loci in tetraploid wheat. Plant Physiol. 149, 245–257.

Donatelli, M., Hammer, G.L., Vanderlip, R.L., 1992. Genotype and water limitation effects on phenology, growth, and transpiration efficiency in grain sorghum. Crop Sci. 32, 781–786.

Dubcovsky, J., Lijavetzky, D., Appendino, M.L., Tranquilli, G., 1998. Comparative RFLP mapping of *Triticum monococcum* genes controlling vernalization requirement. Theor. Appl. Genet. 97, 968–975.

Dubcovsky, J., Loukoianv, A., Fu, D., Valarik, M., Sanchez, A., Yan, L., 2006. Effect of photoperiod on the regulation of wheat vernalization genes *VRN1* and *VRN2*. Plant Mol. Biol. 60, 469–480.

Dubert, F., Filek, M., Marcinska, I., Skoczowski, A., 1992. Influence of warm intervals on the effect of vernalization and the composition of phospholipid fatty acids in seedlings of winter wheat. J. Agron. Crop Sci. 168, 133–141.

Duthion, C., Pigeaire, A., 1991. Seed lengths corresponding to the final stage in seed abortion in three grain legumes. Crop Sci. 31, 1579–1583.

Edwards, J.T., Purcell, L.C., 2005. Soybean yield and biomass responses to increasing plant population among diverse maturity groups: I. Agronomic characteristics. Crop Sci. 45, 1770–1777.

Egli, D.B., 1998. Seed biology and the yield of grain crops. CAB International, Wallingford.

Egli, D.B., 1999. Variation in leaf starch and sink limitations during seed filling in soybean. Crop Sci. 36, 1361–1368.

Egli, D.B., 2004. Seed-fill duration and yield of grain crops. Adv. Agron. 83, 243–279.

Egli, D.B., 2005. Flowering, pod set and reproductive success in soya bean. J. Agron. Crop Sci. 191, 283–291.

Egli, D.B., 2010. Soybean reproductive sink size and short-term reductions in photosynthesis during flowering and pod set. Crop Sci. 50, 1971–1977.

Egli, D.B., Bruening, W.P., 2002. Flowering and fruit set dynamics during synchronous flowering at phloem-isolated nodes in soybean. Field Crops Res. 79, 9–19.

Egli, D.B., Bruening, W.P., 2005. Shade and temporal distribution of pod production and pod set in soybean. Crop Sci. 45, 1764–1769.

Egli, D.B., Bruening, W.P., 2006a. Temporal profiles of pod production and pod set in soybean. Eur. J. Agron. 24, 11–18.

Egli, D.B., Bruening, W.P., 2006b. Fruit development and reproductive survival in soybean: position and age effects. Field Crops Res. 98, 195–202.

Egli, D.B., Fraser, J., Leggett, J.E., Poneleit, C.G., 1981. Control of seed growth in soybean [*Glycine max* (L.) Merrill]. Ann. Bot. 48, 171–176.

Egli, D.B., Guffy, R.D., Leggett, J.L., 1985. Partitioning of assimilates between vegetative and reproductive growth in soybean. Agron. J. 77, 917–922.

Egli, D.B., Leggett, J.L., 1973. Dry matter accumulation patterns in determinate and indeterminate soybeans. Crop Sci. 13, 220–222.

Egli, D.B., Ramseur, E.L., Zhen-wen, Y., Sullivan, C.H., 1989. Source-sink alterations affect the number of cells in soybean cotyledons. Crop Sci. 29, 732–735.

Egli, D.B., Wiralaga, R.A., Bustamam, T., Zhen-Wen, Y., TeKrony, D.M., 1987. Time of flower opening and seed mass in soybean. Agron. J. 79, 697–700.

Egli, D.B., Zhen-wen, Y., 1991. Crop growth rate and seeds per unit area in soybeans. Crop Sci. 31, 439–442.

Ellis, R.H., Collinson, S.T., Hudson, D., Patefield, W.M., 1992. The analysis of reciprocal transfer experiments to estimate the durations of the photoperiod-sensitive and photoperiod-insensitive phases of plant development: an example in soya bean. Ann. Bot. 70, 87–92.

Evans, L.T., 1987. Short day induction of inflorescence initiation in some winter wheat varieties. Aust. J. Plant Physiol. 14, 277–286.

Evans, L.T., Blundell, C., 1994. Some aspects of photoperiodism in wheat and its wild relatives. Aust. J. Plant Physiol. 21, 551–562.

Evers, A.D., 1970. Development of the endosperm of wheat. Ann. Bot. (Lond.) 34, 547–555.

Fehr, W.R., 1980. Soybean. In: Fehr, W.R., Handley, H.H. (Eds.), Hybridization of crop plants. ASA CSSSA, Madison, pp. 589–599.

Fehr, W.R., Caviness, C.E., 1977. Stages of soybean development. Special Report No. 80. Iowa State University: Ames, IA, p. 11.

Ferrante, A., Savin, R., Slafer, G.A., 2010. Floret development of durum wheat in response to nitrogen availabilities. J. Exp. Bot. 61, 4351–4359.

Ferrante, A., Savin, R., Slafer, G.A., 2012. Differences in yield physiology between modern, well adapted durum wheat cultivars grown under contrasting conditions. Field Crops Research 136, 52–64.

Ferrante, A., Savin, R., Slafer, G.A., 2013a. Floret development and grain setting differences between modern durum wheats under contrasting nitrogen availability. J. Exp. Bot. 64, 169–184.

Ferrante, A., Savin, R., Slafer, G.A., 2013b. Is floret primordia death triggered by floret development in durum wheat? J. Exp. Bot. 64, 2859–2869.

Fischer, R.A., 1984. Wheat. In: Smith, W.H., Banta, J.J. (Eds.), Potential productivity of field crops under different environments. IRRI, Los Baños, Philippines, pp. 129–154.

Fischer, R.A., 1985. Number of kernels in wheat crops and the influence of solar radiation and temperature. J. Agric. Sci. 105, 447–461.

Fischer, R.A., 2011. Wheat physiology: a review of recent developments. Crop Past. Sci. 62, 95–114.

Fleming, J.E., Ellis, R.H., John, P., Summerfield, R.J., Roberts, E.H., 1997. Developmental implications of photoperiod sensitivity in soyabean (*Glycine max* (L.) Merr.). Int. J. Plant Sci. 158, 142–151.

Flood, R.G., Halloran, G.M., 1984. Basic development rate in spring wheat. Agron. J. 76, 260–264.

Flood, R.G., Halloran, G.M., 1986. Genetics and physiology of vernalization response in wheat. Adv. Agron. 39, 87–123.

Foley, T.C., Orf, J.H., Lambert, J.W., 1986. Performance of related determinate and indeterminate soybean lines. Crop Sci. 26, 5–8.

Foulkes, M.J., Sylvester-Bradley, R., Worland, A.J., Snape, J.W., 2004. Effects of a photoperiod-response gene *Ppd-D1* on yield potential and drought resistance in UK winter wheat. Euphytica 135, 63–73.

Frederick, J.R., Camp, C.R., Bauer, P.J., 2001. Drought-stress effects on branch and mainstem seed yield and yield components of determinate soybean. Crop Sci. 41, 759–763.

Gai, J., Palmer, R.G., Fehr, W.R., 1984. Bloom and pod set in determinate and indeterminate soybeans grown in China. Agron. J. 76, 979–984.

García, G.A., Serrago, R.A., Appendino, M.L., et al., 2011. Variability of duration of pre-anthesis phases as a strategy for increasing wheat grain yield. Field Crops Res. 124, 408–416.

Gambín, B.L., Borrás, L., 2005. Sorghum kernel weight: growth patterns from different positions within the panicle. Crop Sci. 45, 553–561.

Gbikpi, P.J., Crookston, R.K., 1981. Effect of flowering date on accumulation of dry matter and protein in soybean seeds. Crop Sci. 21, 652–655.

Geerts, S., Raes, D., Garcia, M., Mendoza, J., Huanca, R., 2008. Crop water use indicators to quantify the flexible phenology of quinoa (*Chenopodium quinoa* Willd.) in response to drought stress. Field Crops Res. 108, 150–156.

Ghiglione, H.O., González, F.G., Serrago, R., et al., 2008. Autophagy regulated by day length determines the number of fertile florets in wheat. Plant J. 55, 1010–1024.

Giménez, C., Fereres, E., 1986. Genetic variability in sunflower cultivars under drought. II. Growth and water relations. Aust. J. Agric. Res. 37, 583–597.

Gomez, D., Vanzetti, L., Helguera, M., Lombardo, L., Fraschina, J., Miralles, D.J., 2014. Effect of *Vrn-1*, *Ppd-1* genes and earliness *per se* on heading time in Argentinean Bread wheat cultivars. Field Crops Res. 158, 73–81.

González, F.G., Slafer, G.A., Miralles, D.J., 2002. Vernalization and photoperiod responses in wheat reproductive phases. Field Crops Res. 74, 183–195.

González, F.G., Slafer, G.A., Miralles, D.J., 2003a. Grain and floret number in response to photoperiod during stem elongation in fully and slightly vernalized wheats. Field Crops Res. 81, 17–27.

González, F.G., Slafer, G.A., Miralles, D.J., 2003b. Floret development and spike growth as affected by photoperiod during stem elongation in wheat. Field Crops Res. 81, 29–38.

González, F.G., Slafer, G.A., Miralles, D.J., 2005a. Photoperiod during stem elongation in wheat: is its impact on fertile floret and grain number determination similar to that of radiation? Funct. Plant Biol. 32, 181–188.

González, F.G., Slafer, G.A., Miralles, D.J., 2005b. Floret development and survival in wheat plants exposed to contrasting photoperiod and radiation environments during stem elongation. Funct. Plant Biol. 32, 189–197.

González, F.G., Slafer, G.A., Miralles, D.J., 2005c. Pre-anthesis development and number of fertile florets in wheat as affected by photoperiod sensitivity genes *Ppd-D1* and *Ppd-B1*. Euphytica 146, 253–269.

González, F.G., Miralles, D.J., Slafer, G.A., 2011. Wheat floret survival as related to pre-anthesis spike growth. J. Exp. Bot. 62, 4889–4901.

Goyne, P.J., Hammer, G.L., 1982. Phenology of sunflower cultivars. II. Controlled environment studies of temperature and photoperiod effects. Aust. J. Agric. Res. 33, 251–261.

Gregory, F.G., Purvis, O.N., 1948. Reversal of vernalization by high temperature. Nature 161, 859–860.

Grimm, S.S., Jones, J.W., Boote, K.J., Herzog, D.C., 1994. Modeling the occurrence of reproductive stages after flowering for four soybean cultivars. Agron. J. 86, 31–38.

Grimm, S.S., Jones, J.W., Boote, K.J., Hesketh, J.D., 1993. Parameter estimation for predicting flowering date of soybean cultivars. Crop Sci. 33, 137–144.

Guard, A.T., 1931. Development of floral organs of the soy bean. Bot. Gaz. 91, 97–102.

Guiamet, J.J., Nakayama, F., 1984a. The effects of long days upon reproductive growth in soybeans (*Glycine max* (L.) Merr.) cv. Williams. Jpn. J. Crop Sci. 53, 35–40.

Guiamet, J.J., Nakayama, F., 1984b. Varietal responses of soybeans (*Glycine max* (L.) Merr.) to long days during reproductive growth. Jpn. J. Crop Sci. 53, 299–306.

Hadley, P., Roberts, E.H., Summerfield, R.J., Minchin, F.R., 1984. Effects of temperature and photoperiod on flowering in soya bean [*Glycine max* (L.) Merrill]: a quantitative model. Ann. Bot. 53, 669–681.

Hall, A.J., Savin, R., Slafer, G.A., 2014. Is time to flowering in wheat and barley influenced by nitrogen? A critical appraisal of recent published reports. Eur. J. Agron. 54, 40–46.

Halloran, G.M., 1977. Developmental basis of maturity differences in spring wheat. Agron. J. 69, 899–902.

Han, T., Wu, C., Tong, Z., Mentreddy, R.S., Tan, K., Gai, J., 2006. Postflowering photoperiod regulates vegetative growth and reproductive development of soybean. Environ. Exp. Bot. 55, 120–129.

Hansen, W.R., Shibles, R., 1978. Seasonal log of the flowering and podding activity of field-grown soybeans. Agron. J. 70, 47–50.

Hardman, L.L., Brun, W.A., 1971. Effects of atmospheric carbon dioxide enrichment at different development stages on growth and yield components of soybeans. Crop Sci. 11, 886–888.

Hartwig, E.E., Kiihl, R.A.S., 1979. Identification and utilization of a delayed flowering character in soybean for short-day conditions. Field Crops Res. 2, 145–151.

Hay, R.K.M., Kirby, E.J.M., 1991. Convergence and synchrony a review of the coordination of development in wheat. Aust. J. Agric. Res. 42, 661–700.

Heatherly, L.G., Elmore, R.W., 2004. Managing inputs for peak production. In: Boerma, H.R., Specht, J.E., (eds), Soybeans: improvement, production, and uses, 3rd edn. Agronomy Monograph 16. ASA, CSSA, SSSA: Madison, pp. 451–536.

Heatherly, L.G., Smith, J.R., 2004. Effect of soybean stem growth habit on height and node number after beginning bloom in the midsouthern USA. Crop Sci. 44, 1855–1858.

Heindl, J.C., Brun, W.A., 1984. Patterns of reproductive abscission, seed yield, and yield components in soybean. Crop Sci. 24, 542–545.

Heitholt, J.J., Egli, D.B., Leggett, J.E., 1986. Characteristics of reproductive abortion in soybean. Crop Sci. 26, 589–595.

Hesketh, J.D., Myhre, D.L., Willey, C.R., 1973. Temperature control of time intervals between vegetative and reproductive events in soybeans. Crop Sci. 13, 250–254.

Hinson, K., 1989. Use of a long juvenile trait in cultivar development. In: Pascale, A.J. (Ed.), In: Proceedings of the World Soybean Research Conference. Buenos Aires, Argentina. pp. 983–987.

Hoogenboom, G., White, J.W., 2003. Improving physiological assumptions of simulation models by using gene-based approaches. Agron. J. 95, 82–89.

Huff, A., Dybing, C.D., 1980. Factors affecting shedding of fl owers in soybean (*Glycine max* (L.) Merrill). J. Exp. Bot. 31, 751–762.

Jiang, H., Egli, D.B., 1993. Shade induced changes in flower and pod number and flower and fruit abscission in soybean. Agron. J. 85, 221–225.

Jiang, H., Egli, D.B., 1995. Soybean seed number and crop growth rate during flowering. Agron. J. 87, 264–267.

Johnson, H.A., Borthwick, H.W., Leffel, R.C., 1960. Effect of photoperiod and time of planting on rates of development on the soybean in various stages of life cycle. Bot. Gaz. 122, 77–95.

Jones, J.W., Boote, K.J., Jagtap, S.S., Mishoe, J.W., 1991. Soybean development. In: Hanks, R.J., Ritchie, J.T. (Eds.), Modelling plant and soil systems. ASA, Madison, pp. 71–90.

Kane, M.V., Grabau, L.J., 1992. Early planted and early maturing soybean cropping system: growth, development, and yield. Agron. J. 84, 769–773.

Kantolic, A.G., 2006. Duración del período crítico y defini--ción del número de granos en soja: Cambios asociados a la respuesta fotoperiódica en postfloración de genotipos indeterminados de los grupos cuatro y cinco. PhD Thesis, Graduate School Alberto Soriano, Faculty of Agronomy, Buenos Aires University.

Kantolic, A.G., Mercau, J.L., Slafer, G.A., Sadras, V.O., 2007. Simulated yield advantages of extending post-flowering development at the expense of a shorter pre-flowering development in soybean. Field Crops Res. 101, 321–330.

Kantolic, A.G., Slafer, G.A., 2001. Photoperiod sensitivity after flowering and seed number determination in indeterminate soybean cultivars. Field Crops Res. 72, 109–118.

Kantolic, A.G., Slafer, G.A., 2005. Reproductive development and yield components in indeterminate soybean as affected by post-flowering photoperiod. Field Crops Res. 93, 212–222.

Kantolic, A.G., Slafer, G.A., 2007. Development and seed number in indeterminate soybean as affected by timing and duration of exposure to long photoperiod after flowering. Ann. Bot. 99, 925–933.

Kantolic, A.G., Peralta, G.E., Slafer, G.A., 2013. Seed number responses to extended photoperiod and shading during reproductive stages in indeterminate soybean. Eur. J. Agron. 51, 91–100.

Kato, K., Miura, H., Sawada, S., 2002. Characterization of *QEet.ocs-5A-1*, a quantitative trait locus for ear emergence time on wheat chromosome 5AL. Plant Breed. 121, 389–393.

Keim, P., Diers, B.W., Olson, T.C., Shoemaker, R.G., 1990. RFLP mapping in the soybean: association between marker loci and variation in quantitative traits. Genetics 126, 735–742.

Kernich, G.C., Halloran, G.M., Flood, R.G., 1997. Variation in duration of pre-anthesis phases of development in barley (*Hordeum vulgare*). Aust. J. Agric. Res. 48, 59–66.

Kilgore-Norquest, L., Sneller, C.H., 2002. Effect of stem termination on soybean traits in southern U.S. production systems. Crop Sci. 40, 83–90.

Kiniry, J.R., Ritchie, J.T., Musser, R.L., 1983. Dynamic nature of photoperiod response in maize. Agron. J. 75, 700–703.

Kirby, E.J.M., 1988. Analysis of leaf, stem and ear growth in wheat from terminal spikelet stage to anthesis. Field Crops Res. 18, 127–140.

Kirby, E.J.M., 1990. Co-ordination of leaf emergence and leaf and spikelet primordium initiation in wheat. Field Crops Res. 25, 253–264.

Kirby, E.J.M., 1992. A field study of the number of main shoot leaves in wheat in relation to vernalization and photoperiod. J. Agric. Sci. 118, 271–278.

Kirby, E.J.M., Appleyard, M., 1987. Cereal development guide. NAC Cereal Unit, Stoneleigh, p. 95.

Kirby, E.J.M., Porter, J.R., Day, W., et al., 1987. An analysis of primordium initiation in Avalon winter wheat crops with different sowing dates and at nine sites in England, Scotland. J. Agric. Sci. 109, 107–121.

Komatsu, K., Hwang, T.Y., Takahashi, M., et al., 2012. Identification of QTL controlling post-flowering period in soybean. Breed Sci. 61, 646–652.

Kong, F., Liu, B., Xia, Z., et al., 2010. Two coordinately regulated homologs of *FLOWERING LOCUS T* are involved in the control of photoperiodic flowering in soybean. Plant Physiol. 154, 1220–1231.

Kumudini, S.V., Pallikonda, P.K., Steele, C., 2007. Photoperiod and E-genes influence the duration of the reproductive phase in soybean. Crop Sci. 47, 1510–1517.

Landsberg, J.J., 1977. Effects of weather on plant development. In: Lansdberg, J.J., Cutting, C.V. (Eds.), Environmental effects on crop physiology. Academic Press, London, pp. 289–307.

Laurie, D.A., 1997. Comparative genetics of flowering time in cereals. Plant Mol. Biol. 35, 167–177.

Laurie, D.A., Pratchet, N., Bezant, J., Snape, J.W., 1995. RFLP mapping of 13 loci controlling flowering time in a winter–spring barley (*Hordeum vulgare* L.) cross. Genome 38, 575–585.

Law, C.N., 1987. The genetic control of day length response in wheat. In: Atherton, J.G. (Ed.), Manipulation of flowering. Butterworths, London, pp. 225–240.

Law, C.N., Dean, C., Coupland, G., 1991. Genes controlling flowering and strategies for their isolation and characterization. In: Jordan, B.R. (Ed.), The molecular biology of flowering. CAB International, Wallingford, pp. 47–68.

Law, C.N., Sutka, J., Worland, A.J., 1978. A genetic study of day length response in wheat. Heredity 41, 185–191.

Law, C.N., Worland, A.J., Giorgi, B., 1976. The genetic control of ear emergence by chromosomes 5A and 5D of wheat. Heredity 36, 49–58.

Lawn, R.J., Byth, D.E., 1973. Response of soybeans to planting date in South-Eastern Queensland. Aust. J. Agric. Res. 24, 67–80.

Lawn, R.J., Imrie, B.C., 1994. Exploiting phenology in crop improvement: matching genotypes to the environment. Crop Physiol. Abstr. 20, 467–476.

Lawn, R.J., Summerfield, R.J., Ellis, R.H., et al., 1995. Towards the reliable prediction of time to flowering in six annual crops. VI. Applications in crop improvement. Exp. Agric. 31, 89–108.

Lee, S.H., Bailey, M.A., Mian, M.A.R., et al., 1996. Molecular markers associated with soybean plant height, lodging, and maturity across locations. Crop Sci. 36, 728–735.

Lersten, N.R., Carlson, J.B., 2004. Vegetative morphology. In: Boerma, H.R., Specht, J.E., (eds.), Soybeans: improvement, production, and uses, 3rd edn. Agronomy Series No. 16. ASA CSSA SSSA: Madison, pp. 15–57.

Levy, J., Peterson, M.L., 1972. Responses of spring wheats to vernalization and photoperiod. Crop Sci. 12, 487–490.

Liu, B., Abe, J., 2010. QTL mapping for photoperiod insensitivity of a Japanese soybean landrace Sakamotowase. J. Hered. 101, 251–256.

Liu, B., Kanazawa, A., Matsumura, H., Takahashi, R., Harada, K., Abe, J., 2008. Genetic redundancy in soybean photoresponses associated with duplication of the phytochrome A gene. Genetics 180, 995–1007.

Liu, B., Watanabe, S., Uchiyama, T., et al., 2010. The soybean stem growth habit gene *Dt1* is an ortholog of *Arabidopsis TERMINAL FLOWER 1*. Plant Physiol. 153, 198–210.

Magrin, G.O., Hall, A.J., Baldy, C., Grondona, M.O., 1993. Spatial and interannual variations in the photothermal quotient: implications for the potential kernel number of wheat crops in Argentina. Agric. Forest Meteorol. 67, 29–41.

Major, D.J., 1980. Photoperiod response characteristics controlling flowering of nine crop species. Can. J. Plant Sci. 60, 777–784.

Major, D., Johnson, D., Tanner, J., Anderson, I., 1975. Effects of daylength and temperature on soybean development. Crop Sci. 15, 174–179.

Mansur, L.M., Orf, J.H., Chase, K., Jarvik, T., Cregan, P.B., Lark, K.G., 1996. Genetic mapping of agronomic traits using recombinant inbred lines of soybean. Crop Sci. 36, 1327–1336.

Masle, J., 1985. Competition among tillers in winter wheat: consequences for growth and development of the crop. In: Day, W., Atkin, R.K. (Eds.), Wheat growth and modeling. Plenum Press, New York, pp. 33–54.

Masle, J., Doussinault, G., Sun, B., 1989. Response of wheat genotypes to temperature and photoperiod in natural conditions. Crop Sci. 29, 712–721.

Matsumura, H., Liu, B., Abe, J., Takahashi, R., 2008. AFLP mapping of soybean maturity gene *E4*. J. Hered. 99, 193–197.

Mayers, J.D., Lawn, R.J., Byth, D.E., 1991. Agronomic studies on soybean (*Glycine max* L. Merrill) in the dry season of the tropics. I. Limits to yield imposed by phenology. Aust. J. Agric. Res. 42, 1075–1092.

McBlain, B.A., Bernard, R.L., 1987. A new gene affecting the time of flowering and maturity in soybeans. J. Hered. 78, 160–162.

McBlain, B.A., Hesketh, J.D., Bernard, R.L., 1987. Genetic effects on reproductive phenology in soybean isolines differing in maturity genes. Can. J. Plant Sci. 67, 105–116.

McBlain, B.A., Hume, D.J., 1981. Reproductive abortion, yield components, and nitrogen content in three early soybean cultivars. Can. J. Plant Sci. 61, 499–505.

Messina, C.D., Jones, J.W., Boote, K.J., Vallejos, C.E., 2006. A gene-based model to simulate soybean development and yield responses to environment. Crop Sci. 46, 456–466.

Miralles, D.J., Ferro, B.C., Slafer, G.A., 2001. Developmental responses to sowing date in wheat, barley and rapeseed. Field Crops Res. 71, 211–223.

Miralles, D.F., Katz, S.D., Colloca, A., Slafer, G.A., 1998. Floret development in near isogenic wheat lines differing in plant height. Field Crops Res. 59, 21–30.

Miralles, D.J., Richards, R.A., 2000. Response of leaf and tiller emergence and primordium initiation in wheat and barley to interchanged photoperiod. Ann. Bot. 85, 655–663.

Miralles, D.J., Richards, R.A., Slafer, G.A., 2000. Duration of stem elongation period influences the number of fertile florets in wheat and barley. Aust. J. Plant Physiol. 27, 931–940.

Miralles, D.J., Slafer, G.A., 1999. Wheat development. In: Satorre, E.H., Slafer, G.A. (Eds.), Wheat: ecology and physiology of yield determination. Food Product Press, New York, pp. 13–43.

Miralles, D.J., Slafer, G.A., 2007. Sink limitations to yield in wheat: how could it be reduced? J. Agric. Sci. 145, 139–149.

Miralles, D.J., Slafer, G.A., Richards, R.A., Rawson, H.M., 2003. Quantitative developmental response to the length of exposure to long photoperiod in wheat and barley. J. Agric. Sci. 141, 159–167.

Miura, H., Worland, A.J., 1994. Genetic control of vernalization, day-length response and earliness *per se* by homeologous group-3 chromosomes in wheat. Plant Breed. 113, 160–169.

Monteith, J.L., 1984. Consistency and convenience in the choice of units for agricultural science. Exp. Agric. 20, 105–117.

Morandi, E.N., Casano, L.M., Reggiardo, L.M., 1988. Post-flowering photoperiodic effect on reproductive efficiency and seed growth in soybean. Field Crops Res. 18, 227–241.

Munier-Jolain, N.G., Ney, B., 1998. Seed growth rate in grain legumes. II. Seed growth rate depends on cotyledon cell number. J. Exp. Bot. 49, 1974–1976.

Munier-Jolain, N.G., Ney, B., Duthion, C., 1993. Sequential development of flowers and seeds on the mainstem of an indeterminate soybean. Crop Sci. 33, 768–771.

Munier-Jolain, N.G., Ney, B., Duthion, C., 1994. Reproductive development of an indeterminate soybean as affected by morphological position. Crop Sci. 34, 1009–1013.

Nicolas, M.E., Gleadow, R.M., Dalling, M.J., 1984. Effects of drought and high temperature on grain growth in wheat. Aust. J. Plant Physiol. 11, 553–566.

Nix, H.A., 1976. Climate and crop productivity in Australia. In: Yoshida, S. (Ed.), Climate and rice. International Rice Research Institute, Los Baños, Philippines, pp. 495–507.

Park, R., Ayliffe, M., Burdon, J., Guest, D., 2009. Dynamics of crop-pathogen interactions: from gene to continental scale. In: Sadras, V.O., Calderini, D.F. (Eds.), Crop physiology: applications for genetic improvement and agronomy. Academic Press, San Diego, pp. 423–447.

Parvez, A.Q., Gardner, F.P., 1987. Daylength and sowing date responses of soybean lines with "juvenile" trait. Crop Sci. 27, 305–310.

Peterson, C.M., Mosjidis, C.O'H., Dute, R.R., Westgate, M.E., 1992. A flower and pod staging system for soybean. Ann. Bot. 69, 59–67.

Piegaire, A., Sebillote, M., Blanchet, R., 1988. Water stress in indeterminate soybeans: no critical stage in fruit development. Agronomie 8, 881–888.

Pinthus, M.J., Nerson, H., 1984. Effects of photoperiod at different growth stages on the initiation of spikelet primordial in wheat. Aust. J. Plant Physiol. 11, 17–22.

Ploper, L.D., 2004. Economic importance of and control strategies for the major soybean diseases in Argentina. In: Moscardi, F., Hoffmann-Campo, D.C., Ferreira Saravia, O., Galerani, P.R., Krzyzanowski, F.C., Carrao-Panizzi, M.C. (Eds.), In: Proceedings of the VII World Soybean Research Conference. Londrina, Brazil, pp. 606–614.

Porter, J.R., 1985. Approaches to modelling canopy development in wheat. In: Day, W., Atkin, R.K. (Eds.), Wheat growth and modeling. Plenum Press, New York, pp. 69–81.

Porter, J.R., Kirby, E.J.M., Day, W., et al., 1987. An analysis of morphological development stages in Avalon winter wheat crops with different sowing dates and at ten sites in England, Scotland. J. Agric. Sci. 109, 107–121.

Purvis, O.N., Gregory, F.G., 1952. Studies in vernalization in cereals. XII. The reversibility by high temperature of the vernalized condition in Petkus winter rye. Ann. Bot. 1, 1–21.

Raper, Jr., C.D., Thomas, J.F., 1978. Photoperiodic alteration of dry matter partitioning and seed yield in soybeans. Crop Sci. 18, 654–656.

Raper, C.D., Jr, Kramer, P.J., 1987. Stress physiology. In: Wilcox, J.R., (ed.), Soybeans: improvement, production, and uses, 2nd edn. Agronomy Monograph No. 16. ASA CSSA SSA: Madison, pp. 589–641.

Rawson, H.M., 1971. An upper limit for spikelet number per ear in wheat as controlled by photoperiod. Aust. J. Agric. Res. 22, 537–546.

Rawson, H.M., 1986. High temperature tolerant wheat: a description of variation and a search for some limitations to productivity. Field Crops Res. 14, 197–212.

Rawson, H.M., 1992. Plant responses to temperature under conditions of elevated CO2. Aust. J. Bot. 40, 473–490.

Rawson, H.M., 1993. Radiation effects on development rate in a spring wheat grown under different photoperiods and high and low temperatures. Aust. J. Plant Physiol. 20, 719–727.

Rawson, H.M., Evans, L.T., 1970. The pattern of grain growth within the ear of wheat. Aust. J. Biol. Sci. 23, 753–764.

Rawson, H.M., Hindmarsh, J.H., Fischer, R.A., Stockman, Y.M., 1983. Changes in leaf photosynthesis with plant ontogeny and relationships with yield per ear in wheat cultivars and 120 progeny. Aust. J. Plant Physiol. 10, 503–514.

Rawson, H.M., Richards, R.A., 1993. Effects of high temperature and photoperiod on floral development in wheat isolines differing in vernalization and photoperiod genes. Field Crops Res. 32, 181–192.

Ray, J.D., Hinson, K., Mankono, E.B., Malo, F.M., 1995. Genetic control of a long-juvenile trait in soybean. Crop Sci. 35, 1001–1006.

Rhone, B., Vitalis, R., Goldringer, I., Bonnin, I., 2010. Evolution of flowering time in experimental wheat populations: A comprehensive approach to detect genetic signatures of natural selection. Evolution 64, 2110–2125.

Roberts, E.H., 1988. Temperature and seed germination. In: Long, S.P., Woodward, F.I. (Eds.), Plants and temperature. Society of Experimental Biology, Company of Biologists, Cambridge, pp. 109–132.

Roberts, E.H., Summerfield, R.J., 1987. Measurement and prediction of flowering in annual crops. In: Atherton, J.G., (ed.), Manipulation of flowering. Proceedings of 45th Easter School in Agricultural Science, University of Nottingham, pp. 17–50.

Roberts, E.H., Summerfield, R.J., Cooper, J.P., Ellis, R.H., 1988. Environmental control of flowering in barley (*Hordeum vulgare* L.). I. Photoperiod limits to long day responses and photoperiod insensitive phases and effects of low temperature and short day vernalization. Ann. Bot. 62, 127–144.

Roberts, E., Summerfield, R., Ellis, R., Qi, A., 1993. Adaptation of flowering in crops to climate. Outlook Agric. 22, 105–110.

Robinson, S.L., Wilcox, J.R., 1998. Comparison of determinate and indeterminate soybean near-isolines and their response to row-spacing and planting date. Crop Sci. 38, 1554–1557.

Rodriguez, D., Santa Maria, G.E., Pomar, M.C., 1994. Phosphorus deficiency affects the early development of wheat plants. J. Agron. Crop Sci. 173, 69–72.

Rondanini, D.P., Savin, R., Hall, A.J., 2007. Estimation of physiological maturity in sunflower as a function of fruit water concentration. Eur. J. Agron. 26, 295–309.

Sadras, V.O., 2007. Evolutionary aspects of the trade-off between seed size and number in crops. Field Crops Res. 100, 125–138.

Sadras, V.O., Slafer, G.A., 2012. Environmental modulation of yield components in cereals: Heritabilities reveal a hierarchy of phenotypic plasticities. Field Crops Res. 127, 215–224.

Saindon, G., Voldeng, H.D., Beversdorf, W.D., Buzzell, R.I., 1989. Genetic control of long daylength response in soybean. Crop Sci. 29, 1436–1439.

Saini, H.S., Westgate, M.E., 2000. Reproductive development in grain crops during drought. Adv. Agron. 68, 59–96.

Saitoh, K., Wakui, N., Mahmood, T., Kuroda, T., 1999. Differentiation and development of floral organs at each node and raceme order in an indeterminate type soybean. Plant Prod. Sci. 2, 47–50.

Savin, R., Molina-Cano, J.L., 2002. Changes in malting quality and its determinants in response to abiotic stresses. In: Slafer, G.A., Molina-Cano, J.L., Savin, R., Araus, J.L., Romagosa, I. (Eds.), Barley science. Recent advances from molecular biology to agronomy of yield and quality. Food Product Press, New York, pp. 523–550.

Savin, R., Slafer, G.A., 1991. Shading effects on the yield of an Argentinian wheat cultivar. J. Agric. Sci. (Camb.) 116, 1–7.

Sawa, M., Kay, S.A., 2011. *GIGANTEA* directly activates *Flowering Locus T* in *Arabidopsis thaliana*. Proc. Natl. Acad. Sci. USA 108, 11698–11703.

Scarth, R., Kirby, E.J.M., Law, C.N., 1985. Effects of the photoperiod genes *Ppd1* and *Ppd2* on growth and development of the shoot apex in wheat. Ann. Bot. 55, 351–359.

Scarth, R., Law, C.N., 1984. The control of day length response in wheat by the group 2 chromosome. Z. Pflanzensuchtung 92, 140–150.

Schnyder, H., Baum, U., 1992. Growth of the grain of wheat (*Triticum aestivum* L.). The relationship between water content and dry matter accumulation. Eur. J. Agron. 1, 51–57.

Schou, J.B., Jeffers, D.L., Streeter, J.G., 1978. Effects of reflectors, black boards, or shades applied at different stages of plant development on yield of soybeans. Crop Sci. 18, 29–34.

Serrago, R.A., Miralles, D.J., Slafer, G.A., 2008. Floret fertility in wheat as affected by photoperiod during stem elongation and removal of spikelets at booting. Eur. J. Agron. 28, 301–308.

Setiyono, T.D., Weiss, A., Specht, J., Bastidas, A.M., Cassman, K.G., Dobermann, A., 2007. Understanding and modelling the effect of temperature and daylength on soybean phenology under high-yield conditions. Field Crops Res. 100, 257–271.

Settimi, J.R., Board, J.E., 1988. Photoperiod and planting date effects on the spatial distribution of branch development in soybean. Crop Sci. 28, 259–263.

Shanmugasundaram, S., Tsou, S.C.S., 1978. Photoperiod and critical duration for flower induction in soybean. Crop Sci. 18, 598–601.

Sibony, M., Pinthus, M.J., 1988. Floret initiation and development in spring wheat (*Triticum aestivum* L.). Ann. Bot. 62, 473–479.

Siddique, K.H.M., Belford, R.K., Perry, M.W., Tennan, D., 1989. Growth, development and light interception of old and modern wheat cultivars in a Mediterranean environment. Aust. J. Agric. Res. 40, 473–487.

Siddique, K.H.M., Tennant, D., Perry, M.W., Belford, R.K., 1990. Water use and water use efficiency of old and modern wheat cultivars in a Mediterranean environment. Aust. J. Agric. Res. 41, 431–447.

Sinclair, T.R., 1984a. Leaf area development in field-grown soybeans. Agron. J. 76, 141–146.

Sinclair, T.R., 1984b. Cessation of leaf emergence in indeterminate soybeans. Crop Sci. 24, 483–486.

Sinclair, T.R., Hinson, K., 1992. Soybean flowering in response to the long-juvenile trait. Crop Sci. 32, 1242–1248.

Sinclair, T.R., Kitani, S., Hinson, K., Bruniard, J., Horie, T., 1991. Soybean flowering date: linear and logistic models based on temperature and photoperiod. Crop Sci. 31, 786–790.

Sinclair, T.R., Neumaier, N., Farias, J.R.B., Nepomuceno, A.L., 2005. Comparison of vegetative development in soybean cultivars for low-latitude environments. Field Crops Res. 92, 53–59.

Slafer, G.A., 1995. Wheat development as affected by radiation at two temperatures. J. Agron. Crop Sci. 175, 249–263.

Slafer, G.A., 1996. Differences in phasic development rate amongst wheat cultivars independent of responses to photoperiod and vernalization. A viewpoint of the intrinsic earliness hypothesis. J. Agric. Sci. (Camb.) 126, 403–419.

Slafer, G.A., 2003. Genetic basis of yield as viewed from a crop physiologist's perspective. Ann. Appl. Biol. 142, 117–128.

Slafer, G.A., 2012. Wheat development: its role in phenotyping and improving crop adaptation. In: Reynolds, M.P., Pask, A.J.D., Mullan, D.M. (Eds.), Physiological breeding I: Interdisciplinary approaches to improve crop adaptation. CIMMYT, Mexico DF, pp. 107–121.

Slafer, G.A., Abeledo, L.G., Miralles, D.J., González, F.G., Whitechurch, E.M., 2001. Photoperiod sensitivity during stem-elongation phase as an avenue to rise potential yields in wheat. Euphytica 119, 191–197.

Slafer, G.A., Andrade, F.H., 1989. Genetic improvement in bread wheat (*Triticum aestivum*) yield in Argentina. Field Crops Res. 21, 289–296.

Slafer, G.A., Andrade, F.H., 1993. Physiological attributes related to the generation of grain yield in bread wheat cultivars released at different eras. Field Crops Res. 31, 351–367.

Slafer, G.A., Araus, J.L., Royo, C., García del Moral, L.F., 2005. Promising ecophysiological traits for genetic improvement of cereal yields in Mediterranean environments. Ann. Appl. Biol. 146, 61–70.

Slafer, G.A., Calderini, D.F., Miralles, D.J., 1996. Yield components and compensation in wheat: opportunities for further increasing yield potential. In: Reynolds, M.P., Ortiz-Monasterio, J.I., McNab, A. (Eds.), Application of physiology in wheat breeding. CIMMYT, Mexico, DF, pp. 101–133.

Slafer, G.A., Halloran, G.M., Connor, D.J., 1994b. Development rate in wheat as affected by duration and rate of change of photoperiod. Ann. Bot. 73, 671–677.

Slafer, G.A., Rawson, H.M., 1994a. Sensitivity of wheat phasic development to major environmental factors: a re-examination of some assumptions made by physiologists and modellers. Aust. J. Plant Physiol. 21, 393–426.

Slafer, G.A., Rawson, H.M., 1994b. Does temperature affect final numbers of primordia in wheat? Field Crops Res. 39, 111–117.

Slafer, G.A., Rawson, H.M., 1995a. Base and optimum temperatures vary with genotype and stage of development in wheat. Plant Cell Environ. 18, 671–679.

Slafer, G.A., Rawson, H.M., 1995b. Intrinsic earliness and basic development rate assessed for their response to temperature in wheat. Euphytica 83, 175–183.

Slafer, G.A., Rawson, H.M., 1995c. Development in wheat as affected by timing and length of exposure to long photoperiod. J. Exp. Bot. 46, 1877–1886.

Slafer, G.A., Rawson, H.M., 1995d. Photoperiod - temperature interactions in contrasting wheat genotypes: time to heading and final leaf number. Field Crops Res. 44, 73–83.

Slafer, G.A., Rawson, H.M., 1996. Responses to photoperiod change with phenophase and temperature during wheat development. Field Crops Res. 46, 1–13.

Slafer, G.A., Rawson, H.M., 1997. Phyllochron in wheat as affected by photoperiod under two temperature regimes. Aust. J. Plant Physiol. 24, 151–158.

Slafer, G.A., Satorre, E.H., Andrade, F.H., 1994a. Increases in grain yield in bread wheat from breeding and associated physiological changes. In: Slafer, G.A. (Ed.), Genetic improvement of field crops. Marcel Dekker, New York, pp. 1–68.

Slafer, G.A., Savin, R., 1991. Developmental base temperature in different phenological phases of wheat (*Triticum aestivum*). J. Exp. Bot. 42, 1077–1082.

Slafer, G.A., Savin, R., 1994. Sink-source relationships and grain mass at different positions within the spike in wheat. Field Crops Res. 37, 39–49.

Snape, J.W., Butterworth, K., Whitechurch, E.M., Worland, A.J., 2001. Waiting for fine times: genetics of flowering time in wheat. Euphytica 119, 185–190.

Snape, J., Law, C.N., Worland, A.J., 1976. Chromosome variation for loci controlling ear emergence time on chromosome 5 A of wheat. Heredity 37, 335–340.

Spaeth, S.C., Sinclair, T.R., 1984. Soybean seed growth I. Timing of growth of individual seeds. Agron. J. 76, 123–127.

Stapper, M., Fischer, R.A., 1990. Genotype, sowing date and planting spacing influence on high-yielding irrigated wheat in southern New South Wales. I. Phasic development, canopy growth and spike production. Aust. J. Agric. Res. 41, 997–1019.

Stelmak, A.F., 1990. Geographic distribution of *Vrn* genes in landraces and improved varieties of spring bread wheat. Euphytica 45, 113–118.

Stewart, D.W., Cober, E.R., Bernard, R.L., 2003. Modeling genetic effects on the photothermal response of soybean phonological development. Agron. J. 95, 65–70.

Stone, P.J., Savin, R., 1999. Grain quality and its physiological determinants. In: Satorre, E.H., Slafer, G.A. (Eds.), Wheat: ecology and physiology of yield determination. Food Product Press, New York, pp. 85–120.

Summerfield, R.J., Asumadu, H., Ellis, R.H., Qi, A., 1998. Characterization of the photoperiodic response of post-flowering development in maturity isolines of soyabean [*Glycine max* (L.) Merrill] 'Clark'. Ann. Bot. 82, 765–771.

Summerfield, R.J., Lawn, R.J., Qi, A., et al., 1993. Towards a reliable prediction of time to flowering in six annual crops. II. Soyabean (*Glycine max*). Exp. Agric. 29, 253–289.

Summerfield, R.J., Roberts, E.H., 1985. *Glycine max*. Halevy, A.H. (Ed.), Handbook of flowering, Vol. 1, CRC Press, Boca Raton, pp. 100–117.

Sun, C.N., 1957. Histogenesis of the leaf and structure of the shoot apex in *Glycine max* (L:) Merrill. Bull. Torrey Bot. Club 94, 163–174.

Swank, J.C., Egli, D.B., Pfeiffer, T.W., 1987. Seed growth characteristics of soybean genotypes differing in duration of seed fill. Crop Sci. 27, 85–89.

Takahashi, R., Benitez, E.R., Funatsuki, H., Ohnishi, S., 2005. Soybean maturity and pubescence color genes improve chilling tolerance. Crop Sci. 45, 1387–1393.

Tasma, I.M., Lorenzen, L.L., Green, D.E., Shoemaker, R.C., 2001. Mapping genetic loci for flowering time, maturity and photoperiod insensitivity in soybean. Mol. Breed. 8, 25–35.

Tasma, I.M., Shoemaker, R.C., 2003. Mapping flowering time gene homologous in soybean and their association with maturity (E) loci. Crop Sci. 43, 319–328.

Thomas, J.F., Kanchanapoom, M.L., 1991. Shoot meristem activity during floral transition in *Glycine max* (L.). Merr. Bot. Gaz. 152, 139–147.

Thomas, J.F., Raper, Jr., C.D., 1976. Photoperiodic control of seed filling for soybeans. Crop Sci. 16, 667–672.

Thomas, J.F., Raper, Jr., C.D., 1977. Morphological response of soybeans as governed by photoperiod, temperature and age at treatment. Bot. Gaz. 138, 321–328.

Thomas, J.F., Raper, Jr., C.D., 1983. Photoperiod and temperature regulation of floral initiation and anthesis in soya bean. Ann. Bot. 51, 481–489.

Thompson, J.A., Bernard, R.L., Nelson, R.L., 1997. A third allele at the soybean *dt1* locus. Crop Sci. 37, 757–762.

Tian, Z., Wang, X., Lee, R., et al., 2010. Artificial selection for determinate growth habit in soybean. Proceedings of the National Academy of Sciences 107, 8563–8568.

Toth, B., Galiba, G., Feher, E., Sutka, J., Snape, J., 2003. Mapping genes affecting flowering time and frost resistance on chromosome 5B of wheat. Theor. Appl. Genet. 107, 509–514.

Tranquilli, G., Dubcovsky, J., 2000. Epistatic interaction between vernalization genes *Vrn-A^{m}1* and *Vrn-A^{m}2* in diploid wheat. J. Hered. 91, 304–306.

Trevaskis, B., Hemming, M.N., Dennis, E.S., Peacock, W.J., 2007. The molecular basis of vernalization-induced flowering in cereals. Trends Plant Sci. 12, 352–357.

Turner, A., Beales, J., Faure, S., Dunford, R.P., Laurie, D.A., 2005. The pseudo response regulator *Ppd-H1* provides adaptation to photoperiod in barley. Science 310, 1031–1034.

Ugarte, C., Calderini, D.F., Slafer, G.A., 2007. Grain weight and grain number responsiveness to pre-anthesis temperature in wheat, barley, and triticale. Field Crops Res. 100, 240–248.

Upadhyay, A.P., Ellis, R.H., Roberts, E.H., Qi, A., 1994a. Characterization of photothermal flowering responses in maturity isolines of soyabean [*Glycine max* (L.) Merrill] cv. Clark. Ann. Bot. 74, 87–96.

Upadhyay, A.P., Summerfield, R.J., Ellis, R.H., Roberts, E.H., Qi, A., 1994b. Variation in the duration of the photoperiodsensitive and photoperiod insensitive phases of development to flowering among eight maturity isolines of soya bean [*Glycine max* (L.) Merrill]. Ann. Bot. 74, 97–101.

Valarik, M., Linkiewicz, A., Dubcovsky, J., 2006. A microcolinearity study at the earliness *per se* gene *Eps-A^{m}1* region reveals an ancient duplication that preceded the wheatrice divergence. Theor. Appl. Genet. 112, 945–995.

Valverde, F., 2011. CONSTANS and the evolutionary origin of photoperiodic timing of flowering. J. Exp. Bot. 62, 2453–2463.

van Schaik, P.H., Probst, A.H., 1958. Effects of some environmental factors on flower production and reproductive efficiency in soybeans. Agron. J. 50, 192–197.

Villalobos, F.J., Hall, A.J., Ritchie, F.T., Orgaz, F., 1996. OILCROP-SUN: a development, growth and yield model of the sunflower crop. Agron. J. 88, 403–415.

Voldeng, H.D., Cober, E.R., Saindon, G., Morrison, M.J., 1996. Registration of seven early-maturing Harosoy nearisogenic soybean lines. Crop Sci. 36, 478.

Voldeng, H.D., Saindon, G., 1991. Registration of seven long daylength insensitive soybean genetic stocks. Crop Sci. 31, 1398–1399.

Waddington, S.R., Cartwright, P.M., Wall, P.C., 1983. A quantitative scale of spike initial and pistil development in barley and wheat. Ann. Bot. 51, 119–130.

Wall, P.C., Cartwright, P.M., 1974. Effects of photoperiod, temperature and vernalization on the phenology and spikelet number of spring wheats. Ann. Appl. Biol. 72, 299–309.

Wang, Z., Reddy, V.R., Acock, M.C., 1998. Testing for early photoperiod insensitivity in soybean. Agron. J. 90, 389–392.

Watanabe, S., Hideshima, R., Xia, Z., et al., 2009. Map-based cloning of the gene associated with the soybean maturity locus E3. Genetics 182, 1251–1262.

Watanabe, S., Harada, S.K., Abe, J., 2012. Genetic and molecular bases of photoperiod responses of flowering in soybean. Breed Sci. 61, 531–543.

Watanabe, S., Tajuddin, T., Yamanaka, N., Hayashi, M., Harada, K., 2004. Analysis of QTLs for reproductive development and seed quality traits in soybean using recombinant inbred lines. Breed Sci. 54, 399–407.

Watanabe, S., Xia, Z., Hideshima, R., et al., 2011. A map-based cloning strategy employing a residual heterozygous line reveals that the *GIGANTEA* gene is involved in soybean maturity and flowering. Genetics 188, 395–407.

Westgate, M.E., Peterson, C.M., 1993. Flower and pod development in water defi cient soybeans (*Glycine max* (L.) Merr.). J. Exp. Bot. 44, 109–117.

White, J.W., Herndl, M., Hunt, L.A., Payne, T.S., Hoogenboom, G., 2008. Simulation-based analysis of effects of *Vrn* and *Ppd* loci on flowering in wheat. Crop Sci. 48, 678–687.

Whitechurch, E.M., Slafer, G.A., 2001. Responses to photoperiod before and after jointing in wheat substitution lines. Euphytica 118, 47–51.

Whitechurch, E.M., Slafer, G.A., 2002. Contrasting *Ppd* genes in wheat affect sensitivity to photoperiod in different phases. Field Crops Res. 73, 95–105.

Whitechurch, E.M., Slafer, G.A., Miralles, D.J., 2007a. Variability in the duration of stem elongation in wheat and barley. J. Agron. Crop Sci. 193, 138–145.

Whitechurch, E.M., Slafer, G.A., Miralles, D.J., 2007b. Variability in the duration of stem elongation in wheat genotypes, sensitivity to photoperiod, and vernalization. J. Agron. Crop Sci. 193, 131–137.

Wiebold, W.J., Ashley, D.A., Boerma, H.R., 1981. Reproductive abscission levels and patterns for eleven determinate soybean cultivars. Agron. J. 73, 43–46.

Wilkerson, G.G., Jones, K.W., Boote, K.J., Buol, G.S., 1989. Photoperiodically sensitive interval in time to flower of soybean. Crop Sci. 29, 721–726.

Worland, A.J., 1996. The influence of flowering time genes on environmental adaptability in European wheats. Euphytica 89, 49–57.

Worland, A.J., 1999. The importance of Italian wheats to worldwide varietal improvement. J. Gen. Breed. 53, 165–173.

Worland, A.J., Appendino, M.L., Sayers, L., 1994. The distribution, in European winter wheats, of genes that influence ecoclimatic adaptability whilst determining photoperiodic insensitivity and plant height. Euphytica 80, 219–228.

Worland, A.J., Gale, M.D., Law, C.N., 1987. Wheat genetics. In: Lupton, F.G.H. (Ed.), Wheat breeding. Chapman Hall, New York, pp. 129–171.

Worland, A.J., Law, C.N., 1985. Genetic analysis of chromosome 2D of wheat. I. The location of genes affecting height, day length insensitivity, and yellow rust resistance. Z. Pflanzensuchtung 96, 331–345.

Xia, Z., Watanabe, S., Yamada, T., et al., 2012. Positional cloning and characterization reveal the molecular basis for soybean maturity locus *E1* that regulates photoperiodic flowering. Proceedings of the National Academy of Sciences 109, E2155–E2164, doi:10.1073/pnas.1117982109.

Xu, M., Xu, Z., Liu, B., et al., 2013. Genetic variation in four maturity genes affects photoperiod insensitivity and PHYA-regulated post-flowering responses of soybean. BMC Plant Biol. 13, 91.

Yan, L., Fu, D., Li, C., et al., 2006. The wheat and barley vernalization gene *Vrn3* is an orthologue of *FT*. Proc. Natl. Acad. Sci. USA 103, 19581–19586.

Yan, L., Helguera, M., Kato, K., Fukuyama, S., Sherman, J., Dubcovsk, J., 2004a. Allelic variation at the *VRN-1* promoter region in polyploid wheat. Theor. Appl. Genet. 109, 1677–1686.

Yan, L., Loukoianov, A., Blechl, A., et al., 2004b. The wheat *VRN2* gene is a flowering repressor down-regulated by vernalization. Science 303, 1340–1644.

Yin, X.Y., Chasalow, S.D., Dourleijn, C.J., Stam, P., Kropff, M.J., 2000. Coupling estimated effects of QTLs for physiological traits to a crop growth model: predicting yield variation among recombinant inbred lines in barley. Heredity 85, 539–549.

Yoshida, T., Nishida, H., Zhu, J., et al., 2010. *Vrn-D4* is a vernalization gene located on the centromeric region of chromosome 5D in hexaploid wheat. Theor. Appl. Genet. 120, 543–552.

Youssefian, S., Kirby, E.J.M., Gale, M.D., 1992. Pleiotropic effects of the G.A. insensitive *Rht* dwarfing gene in wheat. 2. Effects on leaf, stem, and ear growth. Field Crops Res. 28, 191–210.

Zhang, L., Wang, R., Hesketh, J.D., 2001. Effects of photoperiod on growth and development of soybean floral bud in different maturity. Agron. J. 93, 944–948.

Zheng, B., Biddulph, B., Li, D., Kuchel, H., Chapman, S., 2013. Quantification of the effects of *VRN1* and *Ppd-D1* to predict spring wheat (*Triticum aestivum*) heading time across diverse environments. J. Exp. Bot. 64, 3747–3761.

Zheng, S.H., Maeda, A., Fukuyama, M., 2003. Genotypic and environmental variation of lag period of pod growth in soybean. Plant Prod. Sci. 6, 243–246.

CHAPTER

13

Characterizing the crop environment – nature, significance and applications

Karine Chenu

The University of Queensland, Queensland Alliance for Agriculture and Food Innovation (QAAFI), Toowoomba, Queensland, Australia

1 INTRODUCTION

Progress in plant improvement for yield relies on the identification of genotypes better adapted to their production environment. However, the complexity of genotype × environment (G × E) interactions typically reduces heritability of yield, resulting in slow breeding progress, especially in production environments where complex abiotic stresses, such as drought, are frequent. Over the last decades, productivity improvement in crops like wheat has been limited, especially in rain-fed regions (Graybosch and Peterson, 2010; Richards et al., 2010). A possible factor inhibiting progress may be that breeders tend to focus on selection for disease resistance, and then yield and quality (Richards, 1996), but they generally lack reliable methods for selecting tolerance to abiotic stresses, such as drought.

A range of abiotic stresses affect to varying degrees different genes and physiological processes depending on the timing, intensity, duration and history (acclimation) of the stress in regards to the crop cycle (Slafer, 2003; Hammer et al., 2006; Fischer, 2011). For drought, G × E interactions may originate from environmental variations in the timing and severity of the water deficit (Cooper et al., 1999a; Chenu et al., 2011), from genetic variations, e.g. in flowering time or in rate of water uptake (Pantuwan et al., 2002; Chenu et al., 2009a), and from interactions with nutrient supply and biotic stresses (Cooper et al., 1999b; Bänziger and Cooper, 2001; Chapter 7). As a result, the pertinence of gene, trait and germplasm evaluation is highly dependent on how relevant the growing environment is compared to the target population of environments (TPE; Comstock, 1977).

In breeding trials, the source of yield variation arises primarily from the environment (E), secondly from the (G ×E) interactions and, finally, from the genotype (G), i.e. E > G × E > G.

Crop Physiology. DOI: 10.1016/B978-0-12-417104-6.00013-3

For instance, for a series of crops grown in different environments including field pea in Canada (Yang et al., 2005), sunflower in Argentina (de la Vega et al., 2007), wheat in Queensland (Cooper et al., 1995), sugar beet in Europe (Hoffmann et al., 2009), and maize in Midwestern USA (Alwala et al., 2010), the yield variation explained by the environment (E) was between 48 and 93%, compared to 4–35% for G × E interactions and 1–16% for the genotype (G). However, the resources invested to deal with these sources of variation rank in the opposite order: G > G × E > E (Bernardo, 2001; Sadras et al., 2009; Varshney et al., 2011) and refined methods used to characterize genotypes often contrast with the coarse methods applied to characterize breeding-trial environments (Varshney et al., 2011). Different methods have been developed and deployed to deal with environmental variability and G × E in breeding and genetic studies. They include increasingly sophisticated statistical tools (Cooper et al., 1995; Basford and Cooper, 1998; Lacaze et al., 2009; van Eeuwijk et al., 2010), and environmental characterization as described here.

This chapter focuses on characterization of the crop environment in a breeding context, with an emphasis on drought stresses. The first section examines characterization of the target population of environments (TPE) to understand better the prevalent types of crop production environments over long-term periods. The second section focuses on characterization of local environments (e.g. trials) and how these can be used to extrapolate locally observed phenotypes to overall performances expected in the TPE. In the third section, managed environment facilities are discussed as another way to deal with environmental variability, as management interventions can be used to target environments that best reflect the TPE or that maximize relevant germplasm discrimination. The final section (4) illustrates how environment characterization can assist the identification of adapted traits for the TPE.

2 CHARACTERIZATION OF THE TARGET POPULATION OF ENVIRONMENTS (TPE) – A BETTER UNDERSTANDING OF THE NATURE, DISTRIBUTION AND FREQUENCY OF THE MAIN ENVIRONMENT CLASSES

2.1 Importance of characterizing the TPE

Multi-environment trials (MET) of small breeding programs typically involve trials over a few years at several locations, and are often considered to reflect the overall TPE. While screening for genotype performance in MET allows germplasm improvement for the tested environments, MET may misrepresent the TPE and thus lead to the release of germplasm that is poorly adapted to the true production environment. Hence, breeder's elite pool of germplasm can oscillate between adapted and non-adapted depending, for example, on how seasonal conditions affect the MET. For instance, selecting for yield during a few consecutive wet years in a generally drought-prone TPE can bias the selection towards slow-maturing genotypes, as these can accumulate more resources over their longer lifespan than quick-maturing genotypes, and thus better perform during *good* years. However in *bad* years, slow-maturing genotypes are more likely to suffer severe terminal drought. Selection can thus potentially alternate between late and early maturity or between low and high values of other adaptive traits, as crop environments vary across seasons. To avoid such bias, breeders should consider how their MET compare to the overall TPE. This requires understanding of the TPE.

2.2 Different types of environment classification

The characterization of a TPE is typically conducted to identify key environment classes affecting genotype performance. These environment classes (e.g. 'mega-environments', 'macro-environments', 'environment types') are groupings of environments that are relatively homogeneous within the TPE, in which genotypes are expected to perform similarly. Environments may be grouped according to geographical locations (country, region), physical characteristics (e.g. altitude, photoperiod, temperature, rainfall), stress factors (e.g. drought, heat) or directly by their effect on genotype performance. These criteria have been used in a variety of combinations and, more recently, with additional input from crop modeling. Ultimately, environment classes are typically described as either

1. *'groups of locations'*, also called *'mega-environments'* when they refer to a broad spatial area. These classes generally encompass various types of stresses, and year-to-year variations. The international maize and wheat improvement center (CIMMYT) defines a mega-environment as a 'broad, not necessarily contiguous area, occurring in more than one country and frequently transcontinental, defined by similar biotic and abiotic stresses, cropping system requirements, consumer preferences and, for convenience, by a volume of production' (e.g. Braun et al., 1996)
2. *'environment types'* when they refer to specific environments experienced by crops. These are not location specific but are defined by *explicit ranges of climatic factors or stress patterns which affect the crops*. Thus, environment types are based on specific site–year–management–genotype combinations rather than on locations. Accordingly, various environment types can be found at a specific location or within a given mega-environment.

Overall, methods used to characterize a TPE are mostly based either on (1) yield, which is the *end-product* trait broadly integrating all environmental factors that influence crops over their lifespan (Chapman et al., 1997; Chauhan et al., 2008; Hernandez-Segundo et al., 2009), (2) multifactor pedo-climatic analysis (Runge, 1968; Pollak and Corbett, 1993; Chapman and Barreto, 1996; Hodson and White, 2007), or (3) stress index reflecting the intensity and timing of key stress(es) to which crops are exposed (e.g. Lacaze and Roumet, 2004; Chapman, 2008).

In all cases, the quality of the TPE characterization depends on the relevance of the database (e.g. representativeness of the sites, years and management practices) and on the criteria and method used to define the environment classes.

2.2.1 *Yield-based characterization*

TRIAL-BASED CHARACTERIZATION – CAPTURING G $\times$ E INTERACTIONS

The advantage of using yield to characterize the TPE is that this trait is usually the main target, and reflects the integration of environmental factors that influenced crop growth (e.g. temperature, drought, nutrient availability). Accordingly, breeders typically define environment groups based on similarities in germplasm yield response rather than using environmental variations *per se* and use performance in multi-year, multilocation trials to characterize their TPE. Hence, environment classes are generally defined by homogeneous subsets of trials or sites identified using clustering techniques applied to genotype performance data (Horner and Frey, 1957; Abou-El-Fittouh et al., 1969; Ghaderi et al., 1980; Brown et al., 1983; Collaku et al., 2002). Such groupings allow G $\times$ E interactions to be minimized within groups, while maximized among groups (Malhotra and Singh, 1991; Russell et al., 2003). This approach has been in use for decades. For example, Horner and Frey (1957) reduced G $\times$ E variances of oat trials by 11, 21, 30 and 40% by dividing the state of Iowa (USA) into 2, 3, 4 and 5 subregions, respectively. More sophisticated grouping methods have been developed since (DeLacy and Cooper, 1990; Crossa et al., 1993; Trethowan et al., 2001; Yang et al., 2005).

A major limitation of yield-based approaches is that, in many breeding programs, they are applied to only a limited set of years and sites that are likely to misrepresent the TPE, as longer-term weather patterns of the TPE are not considered. Yang et al. (2005) illustrated how characterization of an extreme case, using single-year data, led to site grouping that was highly varied from year to year. Their results highlight the importance of working with a relevant set of trials, preferably over a long period, when characterizing a TPE. While germplasm pools are constantly evolving in breeding programs with frequent turn-over of genotypes, long-term consistency can be achieved by comparing germplasm performance to a check variety that is grown consistently over the years and is used as a sort of 'environmental probe' (Cooper and Fox, 1996; Brancourt-Hulmel, 1999; Mathews et al., 2011; Forkman, 2013).

Long-term analyses of field trials have been used by major breeding programs to define large TPEs. For instance, the ICARDA–CIMMYT barley breeding program used yield data of 25–50 genotypes from 750 trials (235 locations in 75 countries over 27 years with 10–50 trials per year) to identify three mega-environments across the world (Hernandez-Segundo et al., 2009). For convenience, the mega-environments were also coarsely described in terms of environmental factors (temperature and precipitation). Numerous similar studies have been performed for various crops, usually at smaller scales, for fewer years and within specific production regions or countries (e.g. de la Vega et al., 2001; Yang et al., 2005).

As both biotic and abiotic factors can influence the determination of environment classes (e.g. Hernandez-Segundo et al., 2009), trials affected by pest and diseases are often excluded from the TPE characterization, while tolerance to biotic factors is screened separately (e.g. Mathews et al., 2011). In practice, disease occurrence is often predominately associated with specific environment classes (e.g. Bänziger et al., 2006; Hernandez-Segundo et al., 2009).

MODELLING-BASED CHARACTERIZATION – SIMULATION OF A WIDE RANGE OF ENVIRONMENTS

Modeling tools offer another avenue to characterize TPEs as they allow more comprehensive environmental sampling than is possible with experiments (e.g. Löffler et al., 2005; Hammer and Jordan, 2007). While the relevance of sites (soil and climate) and management practices chosen to represent the cropping production system of the region remain criteria of prime importance, crop productivity can be simulated over decades or more using crop modeling, provided soil and historical climatic data are available. Characterizations based on simulated yield have been done for maize and mungbean in Australia (Fig. 13.1; Chauhan et al., 2013, 2014) and for chickpea in India and Australia (Chauhan et al., 2008). In addition to yield, these authors also considered flowering time and maturity to at least partly account for phenology impact on yield. While simulation approaches help to explore spatial and temporal variability for yield (or other traits), these characterizations remain constrained by the limits of the models used (e.g. generally do not account for extreme temperatures and toxicities). In addition, simulation-based characterizations are currently restricted to a limited number of genotypes with studies generally conducted for a standard well-characterized genotype. However, such TPE characterizations can be complemented by experimental data with multiple genotypes to ensure their relevance (e.g. Löffler et al., 2005; Chenu et al., 2011). To be useful, environment classes are expected to be such that $G \times E$ (or the ratio $G \times E/G$) is substantially lower in trials within, compared to across, environment classes.

Yield-based TPE can be characterized by environment classes that are either spatially contiguous (e.g. Bänziger et al., 2006; Chauhan et al., 2013) or not (e.g. Yang et al., 2005; Hernandez-Segundo et al., 2009). In some cases, the identification of continuous or discontinuous

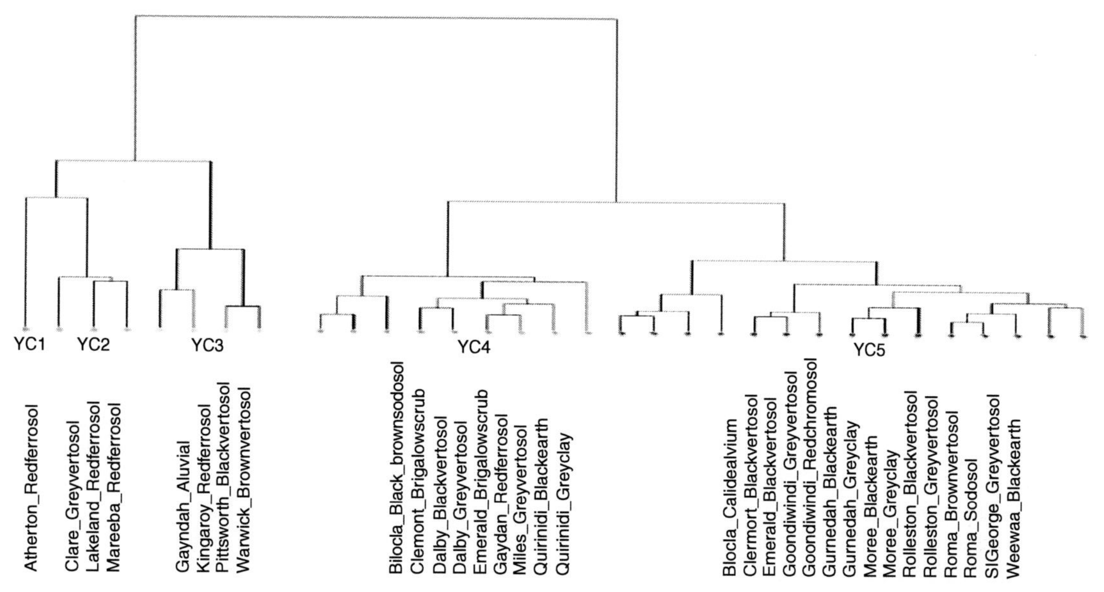

FIG. 13.1 Maize environment characterization based on yield variation. Sites of the maize breeding program from north-eastern Australia were classified based on long-term simulation, using hierarchical clustering on cumulative probability of simulated yield. *Source: Chauhan et al. (2013).*

classes may only depend on the method used for the characterization (Chauhan et al., 2014), highlighting the sensitivity and thus importance of the method used to characterize a TPE.

2.2.2 Pedo-climatic characterization

Physical variables (soil and climate characteristics) can be considered logical bases for TPE characterization (e.g. Pollak and Corbett, 1993; Berger and Turner, 2007), given their long-established influence on crop performance (e.g. Yates and Cochran, 1938; Grafius and Kiesling, 1960) and on G × E interactions (e.g. Nor and Cady, 1979; Malosetti et al., 2004; Dreccer et al., 2008; van Eeuwijk et al., 2010). TPE characterization based on site pedo-climatic conditions offers substantial advantages, as they allow wide spatial and temporal characterizations (provided data are available) without experimentation. Accordingly, new evaluation sites can also be potentially classified and integrated into an existing pedo-climatic characterization, without any need for new multiple-year trials beforehand.

Facing the major challenge to breed wheat and maize for the benefit of the developing world, CIMMYT began defining mega-environments in the late 1980s to develop breeding programs targeting each mega-environment. In the case of bread and durum wheat, CIMMYT is interested in a crop which occupies an estimated 200 million ha globally, ranging from sea level to over 3500 m of altitude, and from the equator to Canada, Europe, and Asia in the northern hemisphere and to South America and Australia in the southern hemisphere (FAOSTAT, 2005). The CIMMYT wheat mega-environments were first defined generically on the bases of key components such as rainfall and temperature, e.g. 'high rainfall' vs 'low rainfall' or 'moderate cold' vs 'severe cold' (Braun et al., 1996). Since, advances in geographic information systems (GIS), agro-climatic datasets (e.g. Hijmans et al., 2005) and irrigated area databases (Siebert et al., 2005) have permitted global mapping of wheat mega-environments based on more quantitative climate, soil, and management data that are now used for breeding (Hodson and White, 2007).

A slightly different approach has been used to characterize the production environment of maize in southern Africa, where the crop is grown on over 12 million hectares (FAOSTAT, 2003). Here, eight mega-environments were identified from a cluster analysis applied to the most prominent G × E grouped trial sites, using data from 290 genotypes over 3 years at 94 sites (Setimela et al., 2002; Bänziger et al., 2006). As the mega-environments could be distinguished by environmental factors such as season rainfall, maximum temperature and subsoil pH, a map was produced to delineate the geographic boundaries of the mega-environments using GIS (Fig. 13.2a; Hodson et al., 2002).

Private breeding companies have also defined and used TPE characterization. Pioneer Hi-Bred International Inc. identified five mega-environments for maize in the USA based on variation in photoperiod, maximum temperature and average radiation (Fig. 13.2b; Löffler et al., 2005). In their case, the environment classes were defined for different periods of the crop cycle and calculated through crop modeling.

While both pedo-climatic and yield-based characterizations may improve our understanding of production environments, these are typically used to group locations into mega-environments and are thus *de facto* considering environmental conditions as 'static'. Methods have been developed to integrate the year-to-year variability in the characterization of sites. However, the classification of a site into a single mega-environment infers that all trials at this site will be classified as belonging to the same mega-environment, irrespective of specific weather and management conditions for each season. In reality, sites may belong to different mega-environments depending on the season (e.g. Trethowan et al., 2005a; Yang et al., 2005). This highlights the need for a more 'dynamic' approach that relates to the environment that crops are experiencing during part of or the whole crop cycle. In contrast to the 'site' approaches, dynamic approaches apply to specific trials or even to specific genotypes in those trials (e.g. an early-maturing genotype may avoid a terminal stress). As for the 'site' approaches, characterizations can be based on yield or pedo-climatic data as long as environments are classified based on conditions that apply to the crop (i.e. in terms of the specific genotype–site–year–management combination) and not the site alone (i.e. with all environments at a site classified into a unique mega-environment).

2.2.3 Specific-stress characterization

To focus on the environment experienced by the crops *per se*, TPE characterization can be based on stress patterns or environmental factors occurring over the crop cycle or at specific phenological stages. Modeling approaches have been developed to estimate phenological stages (e.g. Löffler et al., 2005; Zheng et al., 2013), quantify the local plant/crop environment (e.g. Chelle, 2005; Chenu et al., 2008b), and capture the interactions between the plants and their environment (e.g. Chenu et al., 2008a). For instance in the case of drought, crop models capture feedbacks between plant growth and soil water depletion, e.g. genotypes with rapidly developing canopies will generally deplete soil water quicker than less vigorous genotypes. Accordingly, crop models have been shown to characterize water-limited environments better than standard climatic indices (Muchow et al., 1996). As a result, they have been used to characterize drought environments (see below), in a similar way as previously done with weather indices (e.g. Palmer, 1965; Chapman and Barreto, 1996).

TPE characterization has applied to different stresses (e.g. drought, heat). In the case of drought, long-term simulations have been carried out for various sites, soils and

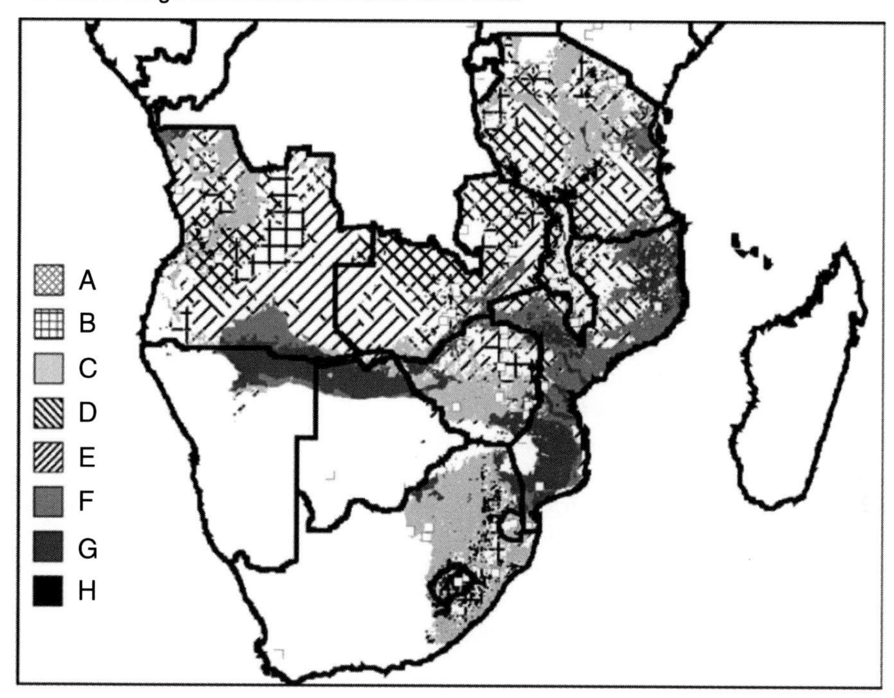

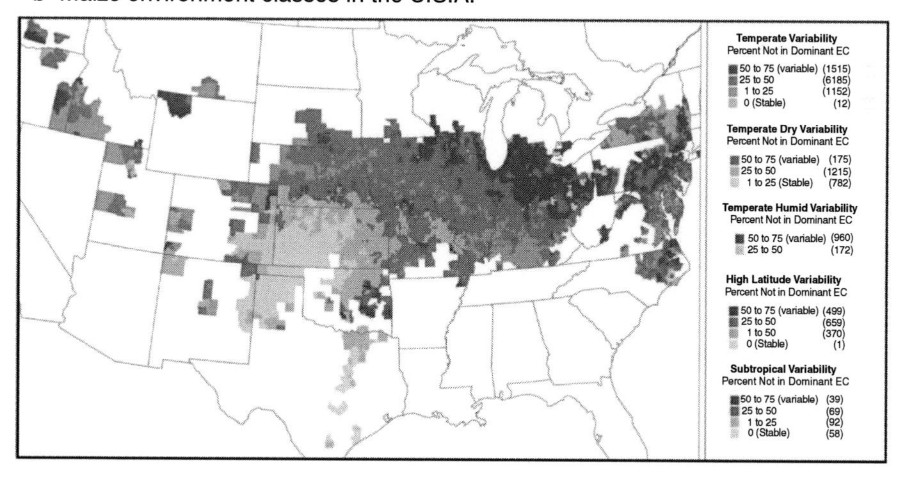

FIG. 13.2 Environment characterization based on pedo-climatic factors. (a) Public breeding organization CIMMYT identified eight maize mega-environments (A–H) in southern Africa that were defined by combination of maximum temperature, seasonal rainfall and subsoil pH. Figure from Bänziger et al. (2006). (b) Private breeding company Pioneer Hi-Bred International Inc. identified five major abiotic environment classes (EC) for the US Corn Belt, for which geographic distribution and variability are presented here. The major environment classes ('Subtropical', 'High Latitude', 'Temperate Dry', 'Temperate Humid' and 'Temperate') were defined in regards to photoperiod, maximum temperature and average solar radiation. *Source: Löffler et al. (2005).* Please see color plate at the back of the book.

management strategies to determine the timing, intensity and frequency of drought that crops experience in a TPE (Fig. 13.3). Such studies have been carried out for various crops and regions, such as sorghum, wheat, barley, field pea, maize and mungbean in Australia (Chapman et al., 2000a; Chenu et al., 2009b, 2011, 2013a; Sadras et al., 2012; Chauhan et al., 2013, 2014); rice and maize in part of Brazil (Heinemann et al., 2008; Chenu et al., 2009a); millet in India (Kholová et al., 2013) and maize in Europe (Harrison et al., 2014). These drought characterizations are all based on the approach first proposed by Muchow et al. (1996) and Chapman et al. (2000a), where environments are clustered based on the drought seasonal pattern experienced by the crop, and usually centered at flowering. The drought pattern in this case corresponds to the fluctuation over time of a water-stress index calculated as the ratio between modeled water supply and demand (Chapman et al., 1993; Chenu et al., 2013a), which indicates the degree to which the soil water extractable by the roots ('water supply') is able to match the potential transpiration ('water demand'). Better to represent TPEs, Chenu et al. (2011) have proposed applying a weight to the simulations based on the degree to which they represent the production environments. Chenu et al. (2013a) also stressed the importance of the choice of soil, site (climate) and management (including the cultivar) given their substantial influence on the environment characterization (Fig. 13.4). However, the nature of the drought patterns can be quite consistent across regions, despite large variation in soil characteristics and rainfall patterns (Chenu et al., 2013a). This might be partly explained by the method applied (e.g. smoothing of the drought pattern by averaging stress indexes every 100°Cd) and by producer practices that restrain certain types of stresses (e.g. early vegetative stress). While the nature of the main drought patterns can be quite stable across regions, their frequency of occurrence varies greatly both spatially and temporally (Fig. 13.4; e.g. Chapman et al., 2000b, 2002a; Chenu et al., 2011, 2013a). Interestingly, similarities in drought patterns can also be observed at a broader level, across crops and TPEs (Fig. 13.3). Most TPEs seem to include an environment type defined by little or no stress, while other environment types typically involve water stresses that begin before or at flowering (Fig. 13.3). The stresses are then generally released during grain filling in one environment type, while they are maintained till maturity in another environment type. In none of the previously mentioned studies was an early vegetative stress identified as a characteristic of a major environment type, partly due to the fact that stress around germination was not examined, as producers rarely sow in conditions where this type of risk is high. However, these early-season stresses may indeed occur in some TPEs (e.g. sowing of subsistence crops in developing countries, Chapter 5) and could become more frequent with the increasing adoption of dry sowing in places like Australia. From the studies mentioned above (Fig. 13.3), it appears that the earliest vegetative stresses of the environment types tend to begin 300–600°Cd before flowering in wheat, maize and field pea in Australia, and millet in India, when they are expected to greatly impact grain number and thus yield (e.g. Chenu et al., 2013a).

Minimum and maximum temperature for either critical stages or the whole crop cycle have also been used to characterize environments (e.g. Löffler et al., 2005; Sadras et al., 2012; Zheng et al., 2012). Correlations exist among climatic/stress variables, as they often co-vary in space and time (Rodriguez and Sadras, 2007). For instance, negative associations exist between water supply/demand ratio and maximum temperature (e.g. Sadras et al., 2012). However, despite the statistically significant association between these variables, severe water stress can occur under mild temperatures and, conversely, thermal stress can occur with mild water stress.

Different stresses and/or climatic variables can be combined either with a unique or with several environment characterizations, although

a- Australian rainfed wheat

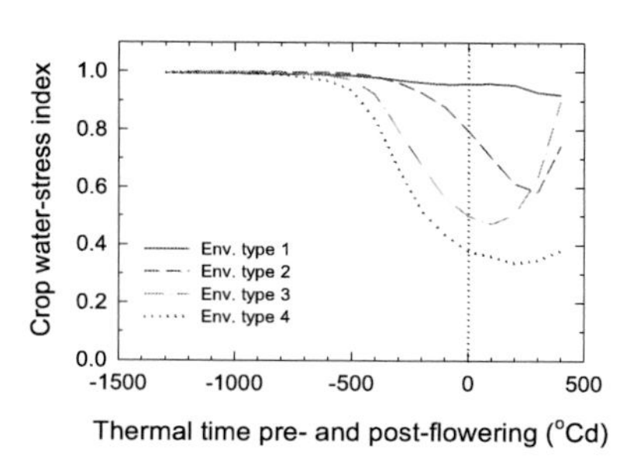

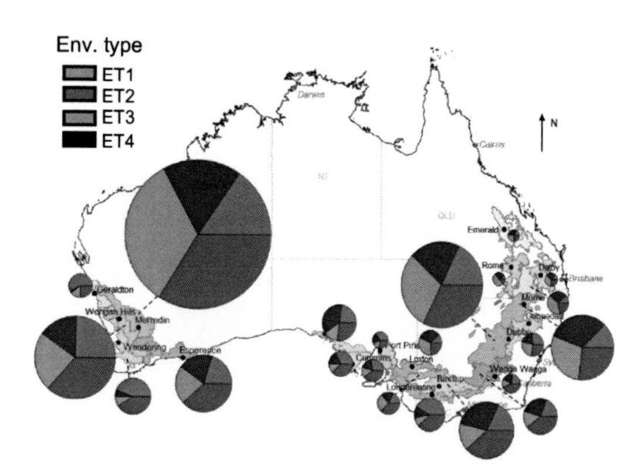

b- European rainfed maize

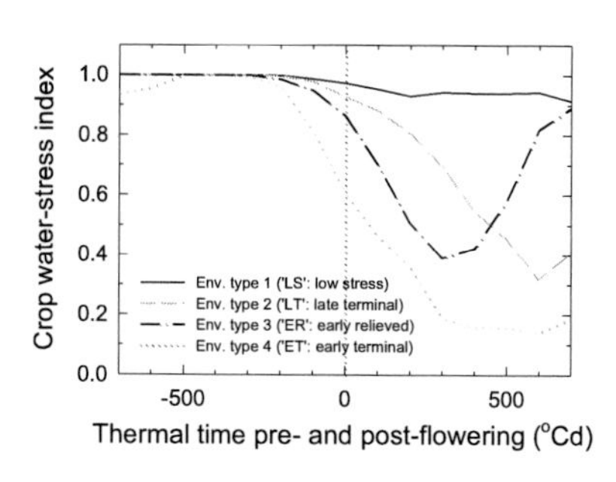

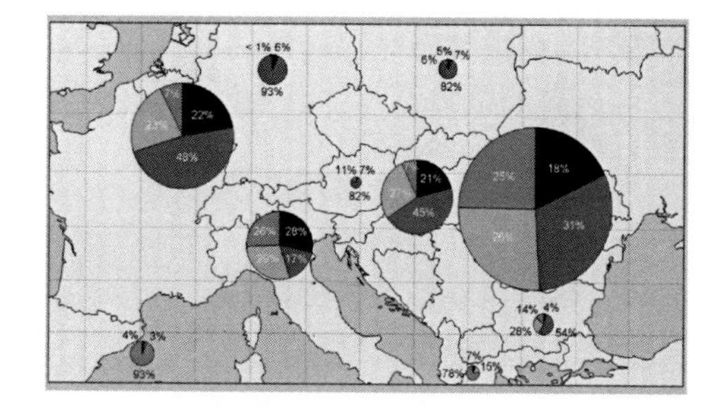

c- Australian rainfed field pea

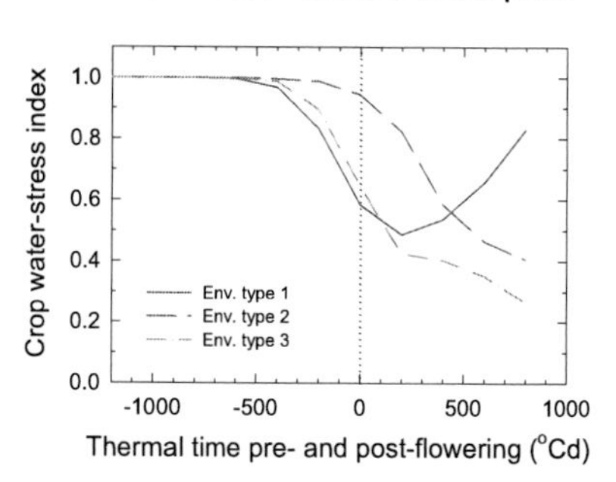

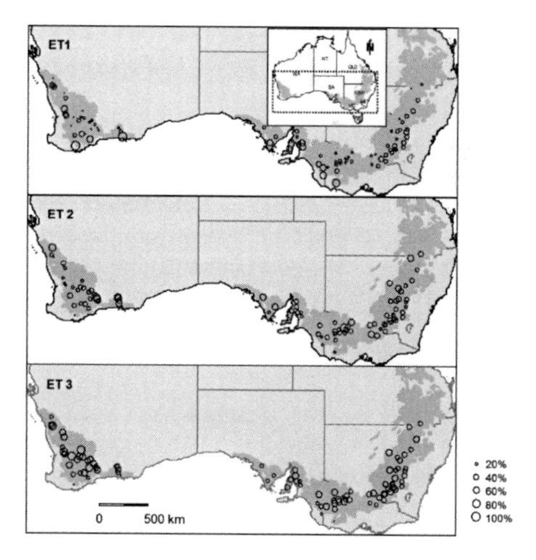

FIG. 13.3 Drought environment characterization for (a) rain-fed wheat in Australia, (b) maize in Europe and (c) field pea in Australia. Dominant water-stress index patterns expressed as thermal time before or after anthesis are presented on the left of the figure, while the distribution of their frequency is displayed on the right. In (a–b), the pie-chart size is proportional to the regional (for wheat) or national (for maize) average cropped area (wheat) or harvest (maize), while in (c), the size of the circle corresponds to the frequency of environment types (ET) at various locations. The crop water-stress index (or 'water supply/demand ratio') indicates the degree to which the potential water supply that depends on the volume and wetness of soil explored by roots ('water supply') is able to match the 'water demand' of the canopy, which is influenced by radiation and temperature and air humidity conditions. See Chapman et al. (1993) or Chenu et al. (2013a) for details. Please see color plate at the back of the book.

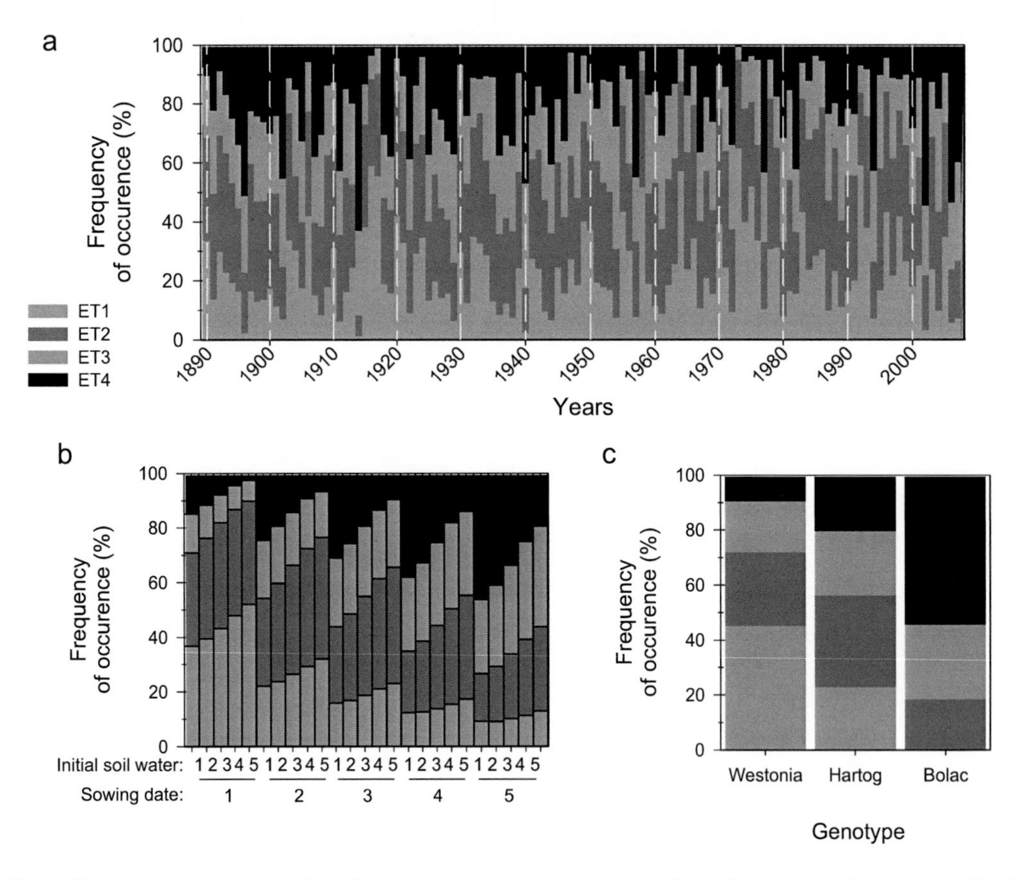

FIG. 13.4 Change in occurrence of drought-environment types for rain-fed Australian wheat (a) over time, (b) with different management practices and (c) for different genotypes. The environment types (ET1–4) correspond to those described in Figure 13.3a. Sowing dates are presented from the earliest (1) to the latest (5), with each sowing date representing 20% of the simulated sowing opportunities. Initial soil water increased from the lowest (1; most severe conditions) to the highest (5; less severe) values, each representing 20% of the simulated initial soil water availability. Simulations were performed for a standard, medium-maturing variety ('Hartog') in (a–b) as well as for an early-maturing ('Westonia') and a late-maturing ('Bolac') variety in (c). For details and information concerning spatial variability, see Chenu et al. (2013a). *Source: Chenu et al. (2013a).* Please see color plate at the back of the book.

correlations between variables and interactions of their impact on crops complicate the task. In Australian rain-fed maize and mungbean, Chauhan et al. (2013, 2014) showed that environment classes for yield distribution were related to the environment types for drought, thus confirming the relevance of drought characterization in the search for crop adaptation in these TPEs.

How to characterize a TPE depends on the region, crop and management practices, as well as the aim, e.g. broad vs specific adaptation. However, to be directly useful for breeders, the characterization should have a substantial influence on G × E, and assist in better understanding of the genetic correlations for genotype performance among testing sites. Hence, most relevant characterizations are likely to concern major limiting stresses (e.g. drought in Australia) or combine a series of important environmental variables (e.g. Löffler et al., 2005).

2.3 Comparison of environments (regions), genotypes and management practices to assist site and management selection, germplasm exchange, global data analysis, and adaptation to future climates

2.3.1 Breeding-trial locations and management practices

Brown et al. (1983) defined an *optimum* selection environment as one in which (1) the targeted trait is expressed, (2) genetic variance is maximized, (3) environmental and G × E variances are minimized, (4) the growing region of the germplasm tested is accurately represented, (5) the environment is accessible for efficient and inexpensive screening and (6) conditions 1–5 are consistent over years. An environment characterization of the TPE provides a basis for choosing representative sites and management practices, and may also possibly allow the scaling down of testing programs by identifying smaller sets of representative test sites. For instance, based on groupings derived from correlations for traits between sites, Guitard (1960) found that the number of barley test locations in Alberta could be reduced from 10 to 5 without appreciable loss of information. Using similar analyses in sunflower, de la Vega and Chapman (2006) revealed that hybrid performance in remote northern Argentina could be well predicted by late-sown trials in central Argentina (the main breeding zone), which has since facilitated the early identification of candidate northern hybrids and allowed substantial savings of breeding resources.

The selection of most evaluation sites typically relies on the postulate that a genotype developed at one location (or with a managed stress) is expected to perform well at other locations within the same mega-environment (e.g. Malhotra and Singh, 1991). Although environments vary from one year to another due to climatic variability, developing germplasm in locations where major types of stress are predominant is expected to improve rates of yield gain over selection cycles (e.g. Chapman et al., 2003; Hammer et al., 2005). Where year-to-year variations are high, evaluation trials should thus be located at sites known to experience high frequency of targeted environment types, and/or use managed environments (section 4).

2.3.2 Exchange of germplasm

In addition to rationalizing testing-trial locations and management, TPE characterization may be useful for improving the efficiency of national and international germplasm exchanges, and introducing high-yielding cultivars into new regions based on environmental similarities. Using such an approach, CIMMYT is running selection trials in worldwide mega-environments and developing widely-adapted germplasm (e.g. Mathews et al., 2007, 2011; Hernandez-Segundo et al., 2009). Large importations of CIMMYT wheat germplasm have occurred in Australia given the environmental similarities between certain CIMMYT testing sites and the drought-prone Australian production area (e.g. Cooper et al., 1997; Brennan and Fox, 1998; Brennan and Quade, 2006). In particular, wheat grown in Australia's north-eastern region experiences drought stresses similar to those of CIMMYT mega-environments ME1, ME4c and ME5, which are also generated via specific management practices in Mexico (e.g. Braun et al., 1996; Trethowan et al., 2005b; Rattey and Shorter, 2010). Accordingly, strong genetic correlations have been observed between Australian environments and those at CIMMYT stations in Mexico (Mathews et al., 2007). The value of germplasm exchange across breeding programs that share similar mega-environments is such that Australia now systematically introduces wheat germplasm from CIMMYT (CAIGE program; http://gwis.lafs.uq.edu.au/index.php/cagesections).

Other tools have been developed to mine germplasm of potential interest for particular environment classes. For instance, based on the premise that adaptive traits should reflect the selection pressures of the environment from which accessions were originally sampled (e.g. Mackay et al., 2005), the Focused Identification of Germplasm Strategy (FIGS) combines environmental and plant characteristics to facilitate the identification of germplasm with traits of potential interest for particular environments (e.g. Khazaei et al., 2013).

2.3.3 *Adaptation to future climates*

Using climatic forecasts in crop modeling is a way to simulate future expected changes within TPEs. Using this approach, a recent study revealed that no major change in the main range of drought patterns is foreseen for European maize, however, their occurrence is expected to shift toward more frequent incidence of the more severe drought patterns (Harrison et al., 2014). Similar results were found previously for wheat in Australia (Chenu and Chapman, 2012).

Simulations of adaptation to future climates are still in the early stages and need to be interpreted carefully, as (1) current studies typically do not account for adaptation in management practices (e.g. Zheng et al., 2012), (2) daily prediction of future climates (as required for crop modeling) remain uncertain with predictive climate models under constant evolution (e.g. Semenov et al., 2010), and (3) crop models are currently missing the capability to simulate extreme events (e.g. Lobell et al., 2012). Chapter 20 further discusses the application of modeling tools to characterize cropping systems under future climate scenarios.

3 TRIAL CHARACTERIZATION – ADDING VALUE TO FIELD DATA THROUGH IMPROVED UNDERSTANDING OF THE GENOTYPIC VARIABILITY

3.1 Environment characterization at the trial level

Breeders typically characterize the environment of selection trials indirectly, based on genotype discrimination for crop performance as described in section 2.2.1. As for the TPE characterization, alternatives have been deployed, with trial characterization based on pedo-climatic or stress variables. The relevance of these alternative approaches depends on how variable the considered factors are (i.e. factors stable across trials are not helpful), and how these factors influence yield G × E interactions, especially cross-over interactions (i.e. variation in genotypes ranking among trials). Advantages of pedo-climatic and stress-based methods include: (1) independence from the performance of genotypes used in trials; and (2) capacity for comparison of trials from a breeding program (MET) with the targeted environments (TPE), thus indicating how well the TPE is represented by the trials.

To enable MET–TPE comparison, trial characterization has to be performed in ways that relate to the TPE characterization. When the TPE is defined by continuous geographical area (Fig. 13.2a; Bänziger et al., 2006) or by climatic characteristics (e.g. Hodson and White, 2007), the trial characterization is generally based on the location or the long-term climatic records at the site (i.e. with no account of the particular seasonal environment specific to the trial). For characterizations using pedo-climatic variables or stress indexes, which account for the conditions that crops experience at a particular trial (e.g. Chenu et al., 2011), data specific to the trial are required. This trial information can come from measurements for certain pedo-climatic factors, or from modeling in the case of complex stresses such as drought, where interactions between the crops and their environment have to be accounted for (e.g. Chapman et al., 2000a; Chenu et al., 2011). An example of drought characterization for a MET is presented in Figure 13.5, including steps for soil and climate characterization, trial modeling, testing/validation of the model for those trials using a check genotype, classification of simulated drought patterns from the trials into their respective environment types, and finally G × E analysis. Other types of detailed

a- Weather and soil measurements

b- Multiple Environment Trials (MET)

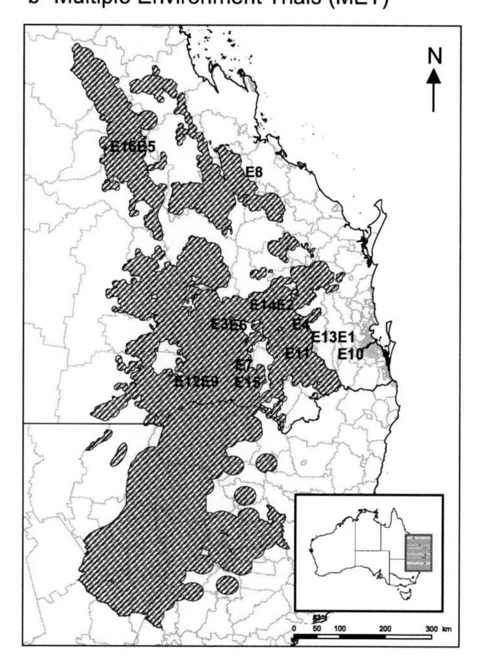

c- Trial simulation

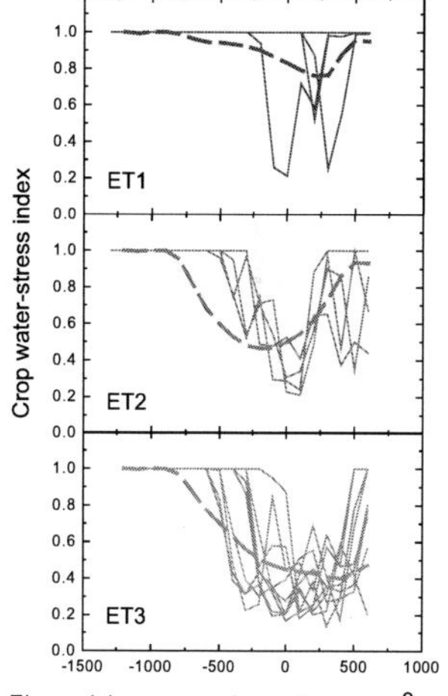

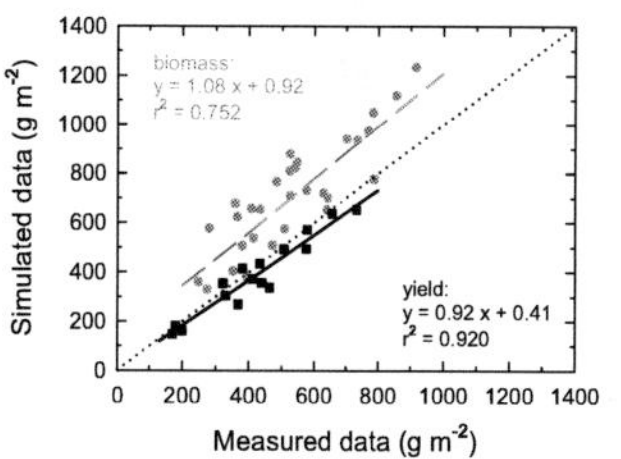

d- Genotype x Environment analysis

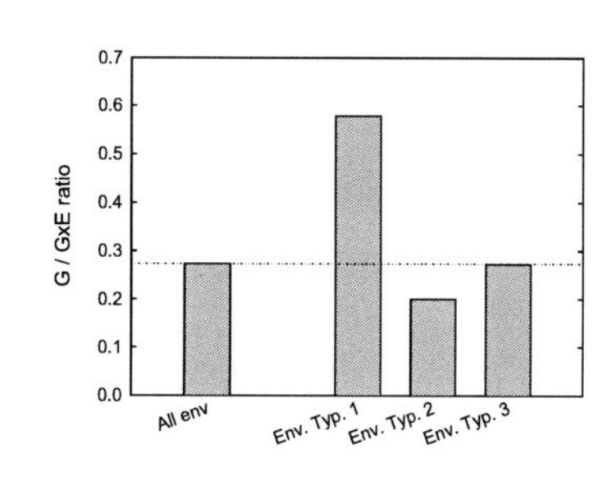

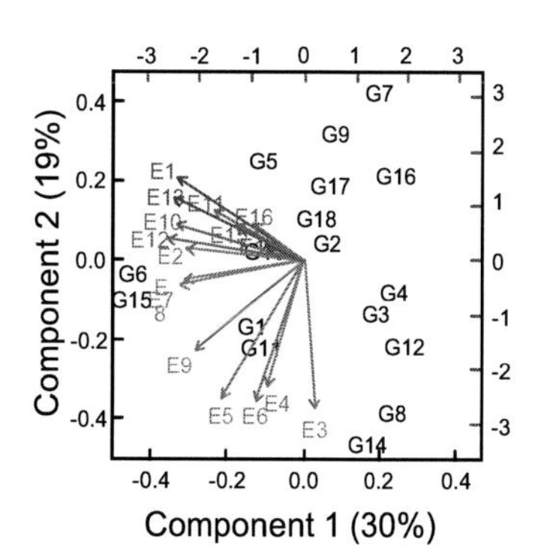

FIG. 13.5 Environment characterization for wheat breeding trials in north-eastern Australia. Measurements of weather and soil data, in particular estimation of the drained upper limit and the crop lower limit (a) were required to simulate crops at specific trials. The example presented here involves 16 wheat trials (E1–E16) of a breeding program in the Australian north-eastern production area (red hashed area) (b), for which drought patterns were characterized using crop modeling (c). Simulations were done for a standard cultivar with the APSIM crop model (Keating et al., 2003). The main drought-environment types (ET1–3) of the studied TPE are presented with dashed lines in (c). Out of the 16 trials, three were classified as environment type 1 (ET1), four trials ET2, and nine trials ET3. Overall, compared to the TPE, this 16-trial MET slightly over-represented ET1 (22% vs 16% in the TPE) and slightly under-represented ET2 (28% vs 34%), while both the TPE and the MET had the same proportion of ET3 (50%). Including the drought characterization in the trial analysis assisted the interpretation of the observed genotype-environment interactions (d). Figure adapted from Chenu *et al.* (2011), with permission from Oxford University Press and the Queensland Government © 2011. Please see color plate at the back of the book.

trial characterizations have been proposed for studies on plant physiology in both field and controlled environments (e.g. Chelle, 2005; Louarn et al., 2008; Chenu et al., 2008b).

While pests and diseases are usually excluded from the TPE characterization, they might strongly impact genotype performance in MET, and have to be considered appropriately. Affected trials might need to be analyzed separately (e.g. Mathews et al., 2011). Alternatively, the importance of a disease might justify its direct integration into the TPE and trial characterizations (e.g. Löffler et al., 2005).

Variations in genotype maturity are also to be considered and accounted for in the trial characterization or data analysis, as they strongly influence G × E interactions for yield (e.g. Bänziger and Cooper, 2001). Genotypes with contrasting phenology (e.g. short vs long crop cycle) are likely to experience different environments even within the same trial (e.g. stress occurring after vs before flowering). The environments of particular genotypes can be characterized separately, e.g. using crop modeling or, alternatively, yield in each trial can be adjusted for flowering date or other traits (e.g. Bänziger et al., 2006).

3.2 Genotype–environment interpretation

The main objective of grouping trials is to reduce the G × E within subgroups of trials (e.g. grouping by environment types) and identify adapted genotypes that perform consistently better in a group of environments. Thus, the aim is to minimize the G × E variance component while maximizing the G component.

Grouping trials from a MET based on genotype performance is common but may, as discussed previously, result in environment groups that are unstable over time (e.g. Malhotra and Singh, 1991). Alternatively, grouping trials based on previously defined environment classes (preferably from an independent dataset reflecting the broad TPE) is more challenging, but allows a more comprehensive analysis in respect to adaptation in the TPE. An example is given in Figure 13.5, where drought patterns of wheat MET are compared to the TPE main drought-environment types. Here, grouping trials (18 genotypes, 16 trials from 11 sites over 3 years) into the TPE environment types allowed a better interpretation of the G × E interactions, in particular, by explaining a greater proportion of the genotypic variability in the generally non-stressed environment type (doubling in the G/G × E ratio in Figure 13.5d; Chenu et al., 2011). Another interesting example is given for maize in the USA, where grouping trials (18 genotypes grown in 266 environments from 90 sites over 3 years) by the TPE environment classes explained up to 30% of the G × E variance (Löffler et al., 2005). Here, genetic correlations between trials grouped within each environment class were generally greater than those between trials across environment classes. As expected, however, in both examples, a substantial portion of the G × E remained unexplained by the environment classification which focused on a restricted part of the environmental variation.

3.3 Trial representativeness and weighted selection

While increasing MET size may assist breeding programs better to sample the TPE, resources limit the number of trials. In practice, trials that diverge from the TPE are likely to bias genetic analysis and to divert selection from the optimum for the TPE. Cooper et al. (1996) demonstrated theoretically how mismatches between the frequencies of environment types sampled in a MET and their true frequencies in the TPE can reduce genetic improvement for the TPE.

'Weighted selection' is a breeding strategy that uses environment characterization to anticipate and adjust for deviations from the TPE. In this approach, trial data from a MET are weighted according to the frequency of the trial's environment class within the TPE in order to 'balance' the MET. Such weighting of trial data based on the similarity of trials to the target environments (or 'to a hypothetical most frequently encountered environment') was proposed in the 1980s (Fox and Rosielle, 1982; Brennan and Sheppard, 1985). Simulation analyses have since assessed the value of weighting genotype performance with respect to trial representativeness in the TPE (Fig. 13.6). The advantage of such weighted selection has been demonstrated *in silico* for variable environments, especially for MET of limited size or for germplasm pools with high cross-over interactions (Fig. 13.6a; Podlich et al., 1999). Chapman et al. (2000b) also showed that the expected yield of a genotype when weighting MET data was quite stable over the years and was similar to the expected average yield of this genotype in the TPE, whereas the unweighted average yield fluctuated greatly over the years (Fig. 13.6b).

Given the limited resources available to breeding programs, weighting germplasm performance is an effective way to increase the value of collected data and to improve selection towards elite germplasm with better performance in the TPE.

4 MANAGED ENVIRONMENTS – INCREASING THE RELEVANCE OF PHENOTYPING ENVIRONMENTS

While different tools exist to increase the relevance and exploitation of MET, the identification of superior genotypes may still be limited by the nature of the selection environments (MET), which only represent a limited sample of the multitude of complex environments that constitute the TPE. To deal with the variability and unpredictability of field environments, managed environments are being used by both public and private organizations, as they permit selection under more controlled conditions (e.g. Campos et al., 2004; Kirigwi et al., 2004; Trethowan et al., 2005b; Bänziger et al., 2006; Rebetzke et al., 2013).

4.1 Managed environments to target specific stresses in the field

While breeders commonly inoculate dedicated trials to assess crop tolerance to specific diseases, some breeding programs are now, in a similar way, managing abiotic stress at key sites. For instance, CIMMYT undertakes managed drought trials in Ciudad Obregón (Mexico), where the dry desert climate allows various kinds of drought stresses to be reliably imposed via irrigation scheduling (e.g. Edmeades et al., 1989, 1999). Byrne et al. (1995) demonstrated that significantly greater yield stability and higher rate of yield gain could be achieved when selecting under managed levels of drought in one location compared to conventional MET at a range of international locations. Using managed environments over several cycles of selection has contributed to improving yield simultaneously for drought and irrigated conditions (e.g. Chapman et al., 1997). Other types of managed environments have been designed, as in southern Africa, where selecting in environments managed for nitrogen and drought resulted in a consistent yield advantage over maize hybrids selected more conventionally, especially in low to medium yielding environments (Bänziger et al., 2006). In this case, the authors attributed their success partly to (1) selection in managed environments where high-priority stresses were imposed to keep heritability high and to maximize genotype-by-stress interactions (even if those stresses were more severe than typical stresses from the TPE), and (2) selection for stress tolerance at early breeding stages when genetic variance is large (which contrasts with classical breeding approaches where genotypes are usually exposed to abiotic stresses at later breeding stages, in MET).

Managed environments have also been created to target representative environments. For instance, in Australia, a web-based modeling application ('StressMaster') is used at representative sites to define irrigation strategies that target the predominant drought patterns of the TPE (Fig. 13.7; Chenu et al., 2013b). In this case, managed environments are used to assess the value of drought tolerant germplasm, traits and genes in representative environments

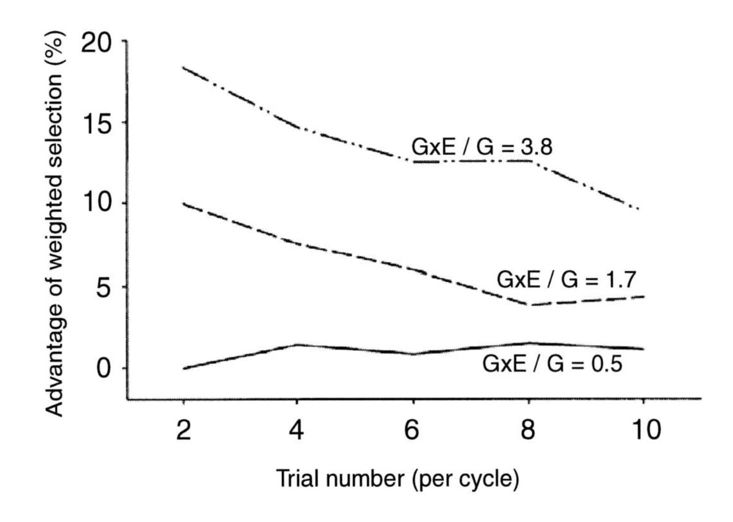

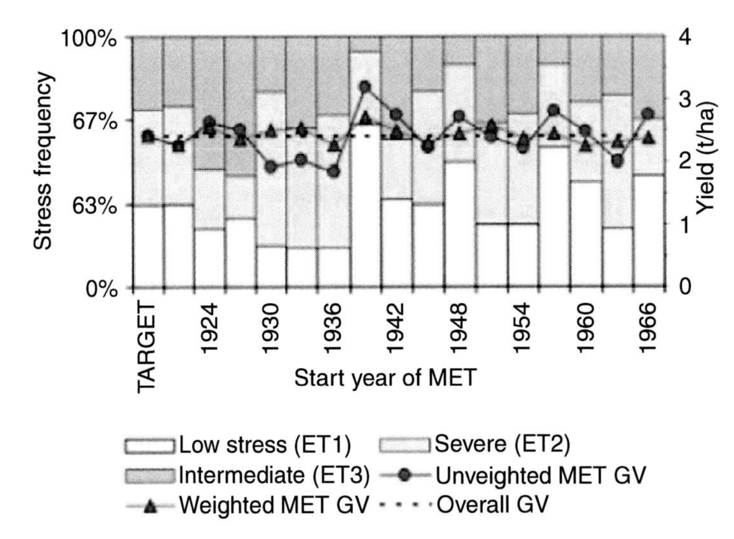

FIG. 13.6 Advantage of weighted selection illustrated in simulation studies for (a) contrasting levels of cross-over G × E interactions and different numbers of trials per testing cycle, in a wheat breeding program; and for (b) yield evaluation in weighted and unweighted multi-environment trials in sorghum. In (a), virtual genotypes were tested *in silico* in 2–10 trials over 2 years during 1 to 10 breeding cycles (the average over all cycles is presented here). The advantage of weighting the trial data according to the trial-environment expected frequency of occurrence in the TPE increases with the level of cross-over G × E, while decreasing with the size of the MET. Source: Podlich et al. (1999). In (b), the genotype value (GV) was either unweighted (i.e. average yield, circle) or weighted (triangle) according to the frequency of occurrence of the specific drought-environment types (ET) in trials compared to the frequency in the TPE. The data are for simulations for 1 hybrid in 16 consecutive multi-environment trials (each of 6 locations over 3 years). Weighting trial data (triangle) resulted in a better prediction of the *overall* value of the genotype (dashed horizontal line) in the TPE ('Target' bar). *Source: Chapman et al. (2000b).*

a- Australian managed-environment facilities

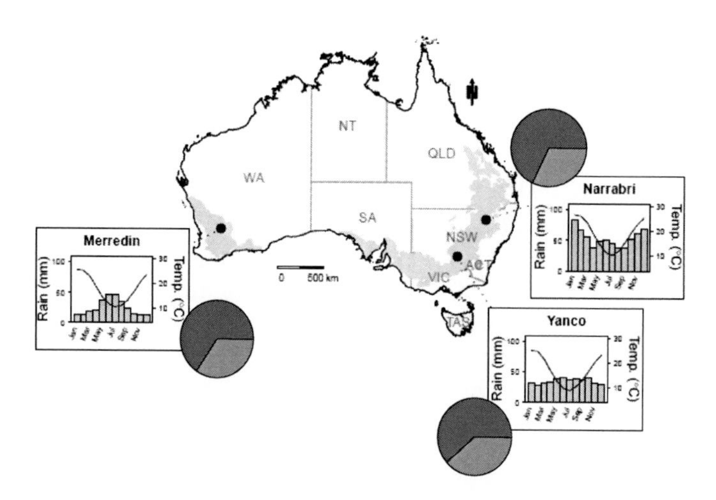

c- Targeted genetic analysis

Traits	No. genetic backgrounds
Awn presence	Five
Canopy stay-green	Two
Canopy temperature	Two
Carbon isotope discrimination	Six
Early vigour	Six
Grain fertility	Three
Leaf glaucousness	Two
Plant development	Two
Reduced-tillering	Six
Root vigour	Two
Stem carbohydrates	Two

b- Modelled-assisted management of drought patterns (StressMaster application)

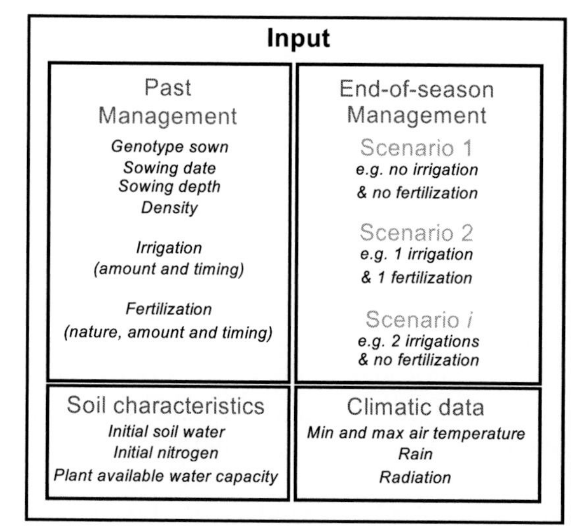

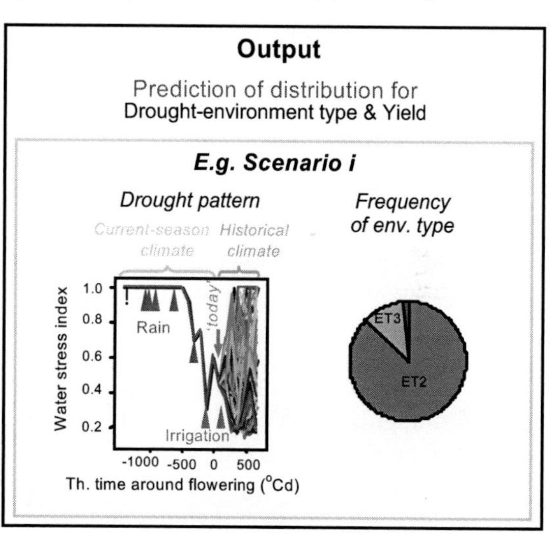

FIG. 13.7 Managed-environment facilities targeting representative drought patterns. Three locations have been chosen across Australia to represent the variability in soil and weather conditions observed across the wheat-belt (a). To adapt for the year-to-year variability that affects the occurrence of drought-environment types (pies in (a); the environment types are defined in Fig. 13.3a), irrigation is applied to target specific drought types. Irrigation scheduling is assisted by a model-based application (StressMaster) that keeps track of the crop water-stress index as the season progresses, and that allows testing of various future end-of-the-season management scenarios (b). The managed-environment facilities are used to test the value of traits with potential for drought adaptation in diverse genetic backgrounds (c). *Sources: Rebetzke et al. (2013) and Chenu et al. (2013b).* Please see color plate at the back of the book.

at relatively large scale (Fig. 13.7; Rebetzke et al., 2013).

4.2 Phenotyping platforms in artificial environments

Following the trends in genomics, phenomics platforms are now being developed worldwide for screening purposes as well as to provide new insights into gene function and environmental responses (e.g. Granier et al., 2006; Berger et al., 2010; Furbank and Tester, 2011). The advantage of such platforms is their ability to phenotype quickly large numbers of genotypes, and to impose different levels of stress in a controlled manner. They are typically used to identify genotypes, traits and genes of potential interest for crop-adaptation improvement, and to develop and test hypotheses concerning processes and their involvement in plant response. The value of these platforms to aid improvement for the industry depends on how the results transfer to the TPE (e.g. Passioura, 2012). The effect of genes and physiological processes involved in adaptation to complex stress such as drought vary greatly depending on the timing, severity, duration and history of stress (e.g. Slafer, 2003), e.g. plant survival and crop performance under drought are regulated by different processes and genetic controls (e.g. Skirycz et al., 2011). Typically, genes and traits beneficial to yield in some environments can have negative effects in others (e.g. Chenu et al., 2009a; Tardieu, 2012). Overall, given the complexity of the interactions across traits and the interactions with the environment (e.g. competition for light, water, nutrients), discoveries about complex traits made in growth chambers or glasshouses need to be validated in field environments and in genetic backgrounds of interest to test their relevance for breeding (e.g. Tardieu, 2012; Passioura, 2012; Rebetzke et al., 2014). Chapter 15 presents a case study in maize where physiologically-meaningful traits are quantified in breeding-relevant field trials.

5 CROP PLASTICITY AND ENVIRONMENT TYPES – IDENTIFICATION OF KEY TRAITS FOR POTENTIAL ADAPTATION

5.1 Which traits for which environment types?

Given the complex G × E interactions that occur over the crop cycle, the relative importance of traits (e.g. maturity, tillering, leaf and root growth) may vary depending on the environment (e.g. Hammer and Vanderlip, 1989; Van Oosterom et al., 2003; Chenu et al., 2009a; Veyradier et al., 2013). For instance, when a drought stress occurs early in development (e.g. ET3–4 for Australian wheat and ET2–3 for Australian field pea; Fig. 13.3a and c), improvement of vigor and maintenance of crop growth rate around flowering can contribute to seed set and yield. However, high biomass at flowering might also be detrimental, as excessive leaf area can prematurely exhaust soil water, thus limiting photosynthesis, reducing seed growth and decreasing yield if the stress is not relieved during the seed-filling period (e.g. ET4 for wheat, ET3 for field pea). Such trade-offs between advantageous/detrimental traits or genetic controls depending on the environment types are further illustrated in Chenu et al. (2009a).

Grouping environments into classes is expected to reveal groups of genotypes with common traits that contribute to adaptation to specific classes of environments. Accordingly, phenotypic selection in breeding has been applied either directly for yield, for other (secondary) traits, or a combination of both, depending on the environment targeted (e.g. Araus et al., 2008; Cattivelli et al., 2008). Selection indexes have also been created to weight the importance of traits, with traits and weights chosen based on the type of environments considered (e.g. Bänziger et al., 2006). Such trait-selection approaches are particularly interesting in conditions where yield has a low heritability while secondary traits have high heritability and genetic variance.

For traits to be useful in selection, they must be associated with some interesting genetic variation (e.g. have the potential to improve yield), and be less subjected to G × E than yield in at least some relevant types of environments (e.g. Chapman et al., 2002b; Araus et al., 2008). In maize, this is the case for traits like increased seed number, synchronized male–female flowering, and adjusted crop maturity (Bolaños and Edmeades, 1996; Duvick, 2005; Cooper et al., 2009; Chapter 15). Harrison et al. (2014) have tested the value of these traits in their European TPE (Fig. 13.8) and found that traits having positive effects on yield in crops subjected to one type of drought-stress pattern did not necessarily have positive effects on yield in other drought patterns. This sort of result could explain the lack of consistent progress achieved in breeding when selecting for genotypes with high expression of key traits (Bolaños and Edmeades, 1996), even though such traits are highly correlated with yield (Campos et al., 2004).

While germplasm may share common traits within environment classes, each environmental class typically groups a wide range of environments, and different adaptation strategies may still be required to improve yield within environment classes. For instance, wheat under severe drought (e.g. ET3–4 in Fig. 13.3a) may benefit from deep roots in deep clay soils that store substantial soil moisture (Manschadi et al., 2006), but this may not be so valuable in light sandy soils, even where crops are experiencing the same type of drought (ET3–4). Environment classes encompass a lot of environmental variability and the effect of traits of interest may differ even within environment classes.

To aid identification of potential traits of interest, modeling approaches have been designed to predict potential consequences of trait × management combinations in the TPE and generate information to help define ideotypes (e.g. Chapter 14). Numerous studies have explored the putative value of potential trait variation in a range of species (e.g. Spitters and Schapendonk 1990; Muchow et al., 1991; Aggarwal et al., 1997; Boote et al., 2001; Asseng and van Herwaarden 2003; Sinclair et al., 2005), some looking more specifically at response to environment types (Fig. 13.8; Hammer et al., 2005; Chenu et al., 2009a; Harrison et al., 2014). While such approaches have merit, they nevertheless require confidence in the adequacy of the model to simulate the effects of trait variations, and may require additional experimental evidence to confirm the findings.

Direct evaluation of traits is taking place in numerous experimental studies, ranging from growth chambers to field, and from phenotyping on small numbers of genotypes to screening of large populations (e.g. Fig. 13.7). Ultimately, examples where the influence of traits is evaluated in representative environments of the TPE (via MET or managed environments) and in genetic backgrounds of relevance for breeding are important for bridging the gap between research and industry (Rebetzke et al., 2013).

5.2 Design and evaluation of breeding strategies

The value of traits in breeding, over several cycles of selection, can be evaluated when combining breeding-system and crop models. Such an approach has highlighted the importance to consider appropriately both (1) the traits to select for and (2) the selection environments (Fig. 13.9; Cooper et al., 2002; Chapman et al., 2003). Gene/trait impact and fixation rate over breeding cycles vary greatly with the type of environment. Hence, depending on the trait, alleles targeted when considering the whole TPE might differ from those targeted in a specific environment type (Fig. 13.9a). For instance, in Australian sorghum, a phenology–ideotype under terminal drought is genetically distant from the phenology–ideotype for the TPE (Fig. 13.9a(iii)). Accordingly, when selecting for increased yield,

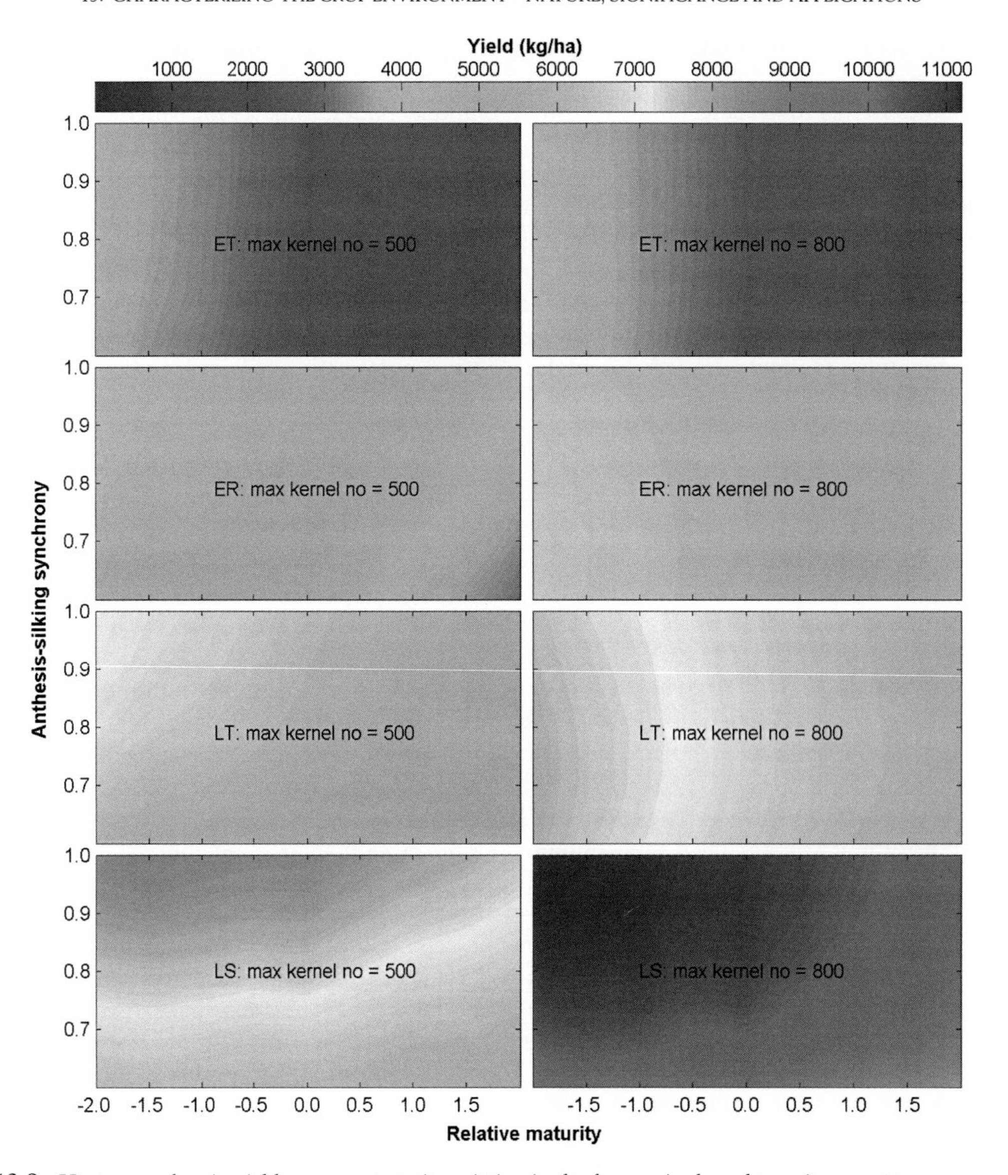

FIG. 13.8 Heat map of grain yield response to trait variation in the four main drought-environment types occurring in the European rain-fed maize cropping area. Each panel presents yield for variation in anthesis-silking synchrony (standard genotype = 0.8; higher values improve synchrony) and in relative maturity (expressed as number of leaves difference from the standard genotype). Panels in the left and right columns show yields for maximum grain number per plant of 500 and 800, respectively. Drought-environment types (ET = ET4, ER = ET3, LT = ET2 and LS = ET1) are as presented in Figure 13.3b. *Source: Harrison et al. (2014).*

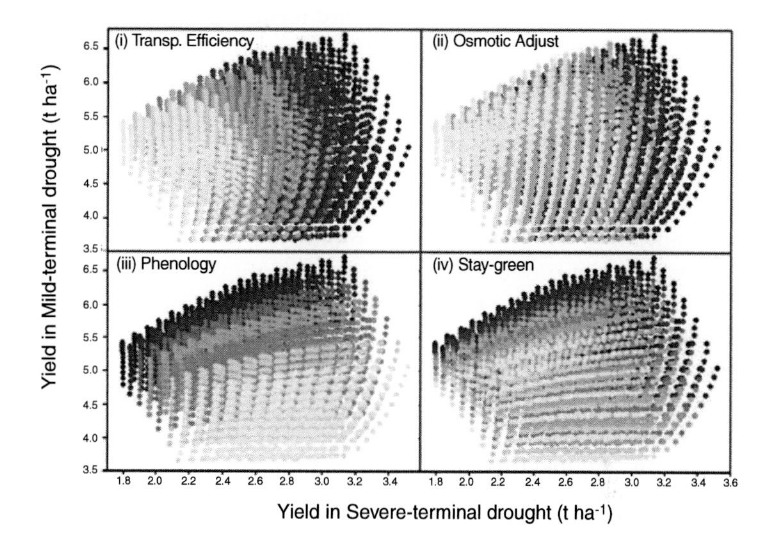

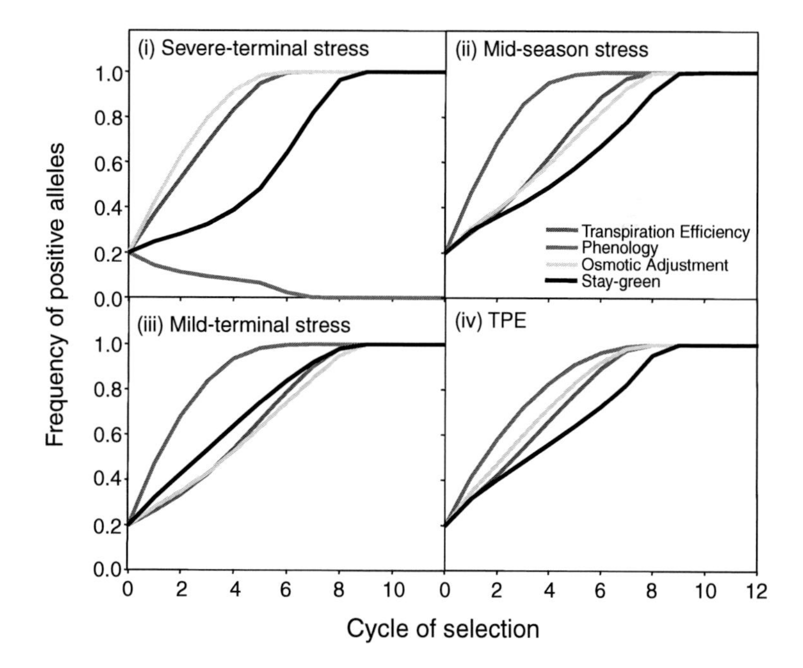

FIG. 13.9 Effect of drought-environment types on (a) simulated yield for trait ideotypes and on (b) the change in allele frequency while selecting for simulated yield. Sorghum simulations generated for 15 genes associated with four adaptive traits identified by genetic and physiological studies as important for drought tolerance: transpiration efficiency (five genes), phenology (three genes), osmotic adjustment (two genes), and stay-green (five genes). In (a), the yield of virtual genotypes in the 'mild terminal stress' and the 'severe terminal stress' environment types is presented with a color code that indicates the genetic distance from the trait ideotype of the TPE (blue: no allele different from the ideotype, yellow: all alleles different). Colors are presented for ideotypes for (i) transpiration efficiency, (ii) osmotic adjustment, (iii) phenology and (iv) stay-green. In (b), change in frequency of alleles with a positive effect on the four traits over cycles of selection, when selection is conducted in the different environment types. Simulations for the sorghum region of north-eastern Australia. *Figures adapted from Cooper et al. (2002) reprinted with permission from IOS Press, and from Chapman et al. (2003).* Please see color plate at the back of the book.

the frequency of late-flowering alleles decreased over breeding cycles in severe terminal stress (Fig. 13.9b(i)), while rapidly increasing when selecting in the other environment types (Fig. 13.9Bb(ii–iii)). While these results may appear trivial when considering flowering, the method applies to various traits of different complexity (Fig. 13.9).

In addition, combining crop models with breeding-system models also provides quantitative information on how the traits would likely be selected over time, in particular in different sampling of the environments (Chapman et al., 2003).

Integrated modeling gives a framework to explore the implications of interactions between the genetic architecture of traits, the selection environments, and breeding strategies (e.g. Cooper et al., 2002). It also provides hypotheses for further experimentation to test our current understanding, thus offering a foundation for defining priorities in research (physiology, genetics and modeling) and for assisting the design of efficient breeding strategies.

6 CONCLUDING REMARKS – PERSPECTIVE

The complexity of the phenotypic landscape arises from multiple and challenging G × E interactions, that impede progress in crop productivity. Environment characterization plays an important role in improving the efficiency of breeding programs, in particular in allowing (1) better understanding of the TPE, (2) informed choice for locations of selection trials, (3) global analysis of trials from similar environments, where germplasm may express interesting adaptation strategies, (4) weighing of germplasm performance (yield and/or secondary traits) based on the degree to which the trial environment represents the TPE, and also in assisting the identification of (5) relevant environmental indices for selection in the TPE, (6) accurate environmental targets for managed environments, and (7) pertinent adaptive physiological traits and underlying genes of relevance for environment classes in the TPE.

Commonly, while assessing selection strategies, breeders may be too concerned with keeping high heritability for yield while ignoring the need to represent adequately the TPE. The effectiveness of genotype evaluation largely depends on the genetic correlation between their performance in multi-environment trials and in the TPE. Weighting trial data is a way to reduce bias in MET data and obtain a more reliable assessment of the genotype value for the TPE. In parallel, managed environments might be considered to target specific environments of value for the TPE, either because they are representative of TPE environment types or because they allow good germplasm discrimination for stresses of importance in the TPE. Secondary traits which affect yield while maintaining a high heritability under stress can also be useful to improve crop adaptation. And while it was not the focus of this chapter, the importance of improved statistical design and analysis techniques (e.g. Gilmour et al., 1997) should not be neglected.

Experimental and *in-silico* research are continuously developing methods, tools and knowledge to increase understanding of the processes involved in plant adaptation. Improved productivity of cropping systems is coming from breeding (genotype), agronomy (management), and economics (feasibility, market). Beyond crop improvement, environment characterization offers benefits such as managing financial risks associated with crop production, which can benefit not only growers but also financial institutions, input suppliers, and end users (Avey et al., 2011). Finally, in a context of seasonal climate variability and long-term climate change (Chapter 20), modeling frameworks are opening promising avenues to simulate and improve understanding of crop performance and adaptation where climate (E), farmer practices (M) and genotype responses (G) are constantly changing.

Acknowledgment

Thanks to Scott Chapman, Graeme Hammer, Victor Sadras, Jack Christopher, Yash Chauhan and Greg Rebetzke for meaningful discussions and comments.

References

Abou-El-Fittouh, H.A., Rawlings, J.O., Miller, P.A., 1969. Classification of environments to control genotype by environment interactions with an application to cotton. Crop Sci. 9, 135–140.

Aggarwal, P.K., Kropff, M.J., Cassman, K.G., ten Berge, H.F.M., 1997. Simulating genotypic strategies for increasing rice yield potential in irrigated, tropical environments. Field Crops Res. 51, 5–17.

Alwala, S., Kwolek, T., McPherson, M., Pellow, J., Meyer, D., 2010. A comprehensive comparison between Eberhart and Russell joint regression and GGE biplot analyses to identify stable and high yielding maize hybrids. Field Crops Res. 119, 225–230.

Araus, J.L., Slafer, G.A., Royo, C., Serret, M.D., 2008. Breeding for yield potential and stress adaptation in cereals. Crit. Rev. Plant Sci. 27, 377–412.

Asseng, S., van Herwaarden, A.F., 2003. Analysis of the benefits to wheat yield from assimilates stored prior to grain filling in a range of environments. Plant Soil 256, 217–229.

Avey, D.P., Bax, P.L., Brooke, R.G., et al., 2011. Method for using environmental classification for making crop production decisions, involves determining lending terms for lender to finance producer for crop production using agricultural input based on profile of producer. Pioneer Hi-Bred Int Inc.US Patent 8046280 B2.

Bänziger, M., Cooper, M., 2001. Breeding for low input conditions and consequences for participatory plant breeding: Examples from tropical maize and wheat. Euphytica 122, 503–519.

Bänziger, M., Setimela, P.S., Hodson, D., Vivek, B., 2006. Breeding for improved abiotic stress tolerance in maize adapted to southern Africa. Agric. Water Manag. 80, 212–224.

Basford, K.E., Cooper, M., 1998. Genotype × environment interactions and some considerations of their implications for wheat breeding in Australia. Aust. J. Agric. Res. 49, 153–174.

Berger, J.D., Turner, N.C., 2007. The ecology of chickpea: evolution, distribution, stresses and adaptation from an agro-climatic perspective. In: Yadav, S., Redden, R., Chen, W., Sharma, B. (Eds.), Chickpea breeding and management. CAB International, Wallingford, pp. 47–71.

Berger, B., Parent, B., Tester, M., 2010. High-throughput shoot imaging to study drought responses. J. Exp. Bot. 61, 3519–3528.

Bernardo, R., 2001. What if we knew all the genes for a quantitative trait in hybrid crops? Crop Sci. 41, 1–4.

Bolaños, J., Edmeades, G.O., 1996. The importance of the anthesis-silking interval in breeding for drought tolerance in tropical maize. Field Crops Res. 48, 65–80.

Boote, K.J., Kropff, M.J., Bindraban, P.S., 2001. Physiology and modelling of traits in crop plants: implications for genetic improvement. Agric. Syst. 70, 395–420.

Brancourt-Hulmel, M., 1999. Crop diagnosis and probe genotypes for interpreting genotype environment interaction in winter wheat trials. Theor. Appl. Genet. 99, 1018–1030.

Braun, H.J., Rajaram, S., van Ginkel, M., 1996. CIMMYT's approach to breeding for wide adaptation. Euphytica 92, 175–183.

Brennan, P.S., Sheppard, J.A., 1985. Retrospective assessment of environments in the determination of an objective strategy for the evaluation of the relative yield of wheat cultivars. Euphytica 34, 397–408.

Brennan, J.P., Fox, P.N., 1998. Impact of CIMMYT varieties on the genetic diversity of wheat in Australia, 1973-1993. Aust. J. Agric. Res. 49, 175–178.

Brennan, J.P., Quade, K.J., 2006. Evolving usage of materials from CIMMYT in developing Australian wheat varieties. Aust. J. Agric. Res. 57, 947–952.

Brown, K.D., Sorrells, M.E., Coffman, W.R., 1983. A method for classification and evaluation of testing environments. Crop Sci. 23, 889–893.

Byrne, P.F., Bolanos, J., Edmeades, G.O., Eaton, D.L., 1995. Gains from selection under drought versus multilocation testing in related tropical maize populations. Crop Sci. 35, 63–69.

Campos, H., Cooper, M., Habben, J.E., Edmeades, G.O., Schussler, J.R., 2004. Improving drought tolerance in maize: a view from industry. Field Crops Res. 90, 19–34.

Cattivelli, L., Rizza, F., Badeck, F.-W., et al., 2008. Drought tolerance improvement in crop plants: An integrated view from breeding to genomics. Field Crops Res. 105, 1–14.

Chapman, S.C., Hammer, G.L., Meinke, H., 1993. A sunflower simulation model: I. Model development. Agron. J. 85, 725–735.

Chapman, S.C., Barreto, H.J., 1996. Using simulation models and spatial databases to improve the efficiency of plant breeding programs. *In* : Cooper, M., Hammer, G.L., (eds), Plant adaptation and crop improvement. pp. 563–587. CAB international, Wallington, UK.

Chapman, S.C., Crossa, J., Edmeades, G.O., 1997. Genotype by environment effects and selection for drought tolerance in tropical maize. 1. Two mode pattern analysis of yield. Euphytica 95, 1–9.

Chapman, S.C., Cooper, M., Hammer, G.L., Butler, D.G., 2000a. Genotype by environment interactions affecting grain sorghum. II. Frequencies of different seasonal patterns of drought stress are related to location effects on hybrid yields. Aust. J. Agric. Res. 51, 209–221.

Chapman, S.C., Hammer, G.L., Butler, D.G., Cooper, M., 2000b. Genotype by environment interactions affecting grain sorghum. III. Temporal sequences and spatial patterns in the target population of environments. Aust. J. Agric. Res. 51, 223–233, http://www.publish.csiro.au/nid/40/paper/AR99022.htm.

Chapman, S.C., Cooper, M., Hammer, G.L., 2002a. Using crop simulation to generate genotype by environment interaction effects for sorghum in water-limited environments. Aust. J. Agric. Res. 53, 379–389.

Chapman, S.C., Hammer, G.L., Podlich, D.W., Cooper, M., 2002b. Linking biophysical and genetic models to integrate physiology, molecular biology and plant breeding. In: Kang, M.S. (Ed.), Quantitative genetics, genomics and plant breeding. CAB International, Wallingford, pp. 167–187.

Chapman, S., Cooper, M., Podlich, D., Hammer, G., 2003. Evaluating plant breeding strategies by simulating gene action and dryland environment effects. Agron. J. 95, 99–113.

Chapman, S.C., 2008. Use of crop models to understand genotype by environment interactions for drought in real-world and simulated plant breeding trials. Euphytica 161, 195–208.

Chauhan, Y.S., Rachaputi, R.C.N., 2014. Defining agro-ecological regions for field crops in variable target production environments: A case study on mungbean in the northern grains region of Australia. Agric. Forest Meteorol. 194, 207–221.

Chauhan, Y., Wright, G., Rachaputi, N., McCosker, K., 2008. Identifying chickpea homoclimes using the APSIM chickpea model. Aust. J. Agric. Res. 59, 260–269.

Chauhan, Y.S., Solomon, K.F., Rodriguez, D., 2013. Characterization of north-eastern Australian environments using APSIM for increasing rainfed maize production. Field Crops Res. 144, 245–255.

Chelle, M., 2005. Phylloclimate or the climate perceived by individual plant organs: What is it? How to model it? What for? New Phytol. 166, 781–790.

Chenu, K., Chapman, S.C., Hammer, G.L., McLean, G., Ben-Haj-Salah, H., Tardieu, F., 2008a. Short-term responses of leaf growth rate to water deficit scale up to whole-plant and crop levels: an integrated modelling approach in maize. Plant Cell Environ. 31, 378–391.

Chenu, K., Rey, H., Dauzat, J., Lydie, G., Lecoeur, J., 2008b. Estimation of light interception in research environments: a joint approach using directional light sensors and 3D virtual plants applied to sunflower (*Helianthus annuus*) and *Arabidopsis thaliana* in natural and artificial conditions. Funct. Plant Biol. 35, 850–866.

Chenu, K., Chapman, S.C., Tardieu, F., McLean, G., Welcker, C., Hammer, G.L., 2009a. Simulating the yield impacts of organ-level quantitative trait loci associated with drought response in maize: A "gene-to-phenotype" modeling approach. Genetics 183, 1507–1523.

Chenu, K., McIntyre, K., Chapman, S.C., 2009b. Environment characterisation as an aid to improve barley adaptation in water-limited environments. Australian Barley Technical Symposium. 13-16 September 2009, Twin Waters, Australia. p. 9.

Chenu, K., Cooper, M., Hammer, G.L., Mathews, K.L., Dreccer, M.F., Chapman, S.C., 2011. Environment characterization as an aid to wheat improvement: interpreting genotype-environment interactions by modelling water-deficit patterns in North-Eastern Australia. J. Exp. Bot. 62, 1743–1755.

Chenu, K., Chapman, S.C., 2012. Drought experienced by Australian wheat: current and future trends. 16th Australian Agronomy Conference. Armidale, NSW, Australia. p. 7.

Chenu, K., Deihimfard, R., Chapman, S.C., 2013a. Large-scale characterization of drought pattern: a continent-wide modelling approach applied to the Australian wheat belt spatial and temporal trends. New Phytol. 198, 801–820.

Chenu, K., Doherty, A., Rebetzke, G.J., Chapman, S.C., 2013b. StressMaster: a web application for dynamic modelling of the environment to assist in crop improvement for drought adaptation. In: Sievänen, R., Nikinmaa, E., Godin, C., Lintunen, A., Nygren, P. (Eds.), 7th International Conference on Functional-Structural Plant Models. Saariselkä, Finland. pp. 317–319.

Collaku, A., Harrison, S.A., Finney, P.L., Van Sanford, D.A., 2002. Clustering of environments of southern soft red winter wheat region for milling and baking quality attributes. Crop Sci. 42, 58–63.

Comstock, R.E., 1977. Quantitative genetics and the design of breeding programs. In: Proceedings of the international conference on quantitative genetics. Iowa State University Press, Ames, USA, pp. 705–718.

Cooper, M., Woodruff, D.R., Eisemann, R.L., Brennan, P.S., Delacy, I.H., 1995. A selection strategy to accommodate genotype-by-environment interaction for grain yield of wheat: Managed environments for selection among genotypes. Theor. Appl. Genet. 90, 492–502.

Cooper, M., DeLacy, I.H., Basford, K., 1996. Relationships among analytical methods used to analyse genotypic adaptation in multi-environment trials. In: Cooper, M., Hammer, G.L. (Eds.), Plant adaptation and crop improvement. CAB International, Wallingford, pp. 193–224.

Cooper, M., Fox, P.N., 1996. Environmental characterization based on probe and reference genotypes. *In* : Cooper, M., Hammer, G.L., (eds), Plant adaptation and crop improvement. pp. 529–549. CAB International, Wallington, UK.

Cooper, M., Stucker, R.E., DeLacy, I.H., Harch, B.D., 1997. Wheat breeding nurseries, target environments, and indirect selection for grain yield. Crop Sci. 37, 1168–1176.

Cooper, M., Rajatasereekul, S., Immark, S., Fukai, S., Basnayake, J., 1999a. Rainfed lowland rice breeding strategies for Northeast Thailand. I. Genotypic variation and genotype × environment interactions for grain yield. Field Crops Res. 64, 131–151.

Cooper, M.E., Podlich, D.W., Fukai, S., 1999b. Combining information from multi-environment trials and molecular markers to select adaptive traits for yield improvement of rice in water-limited environments. In: Ito, O., O'Toole, J., Hardy, B. (Eds.), Genetic improvement of rice for water-limited environments. International Rice Research Institute, Makati City, The Philippines, pp. 13–33.

Cooper, M., Chapman, S.C., Podlich, D.W., Hammer, G.L., 2002. The GP problem: Quantifying gene-to-phenotype relationships. In Silico Biol. 2, 151–164.

Cooper, M., van Eeuwijk, F.A., Hammer, G.L., Podlich, D.W., Messina, C., 2009. Modeling QTL for complex traits: detection and context for plant breeding. Curr. Opin. Plant Biol. 12, 231–240.

Crossa, J., Cornelius, P.L., Seyedsadr, M., Byrne, P., 1993. A shifted multiplicative model cluster analysis for grouping environments without genotypic rank change. Theor. Appl. Genet. 85, 577–586.

de la Vega, A.J., Chapman, S.C., Hall, A.J., 2001. Genotype by environment interaction and indirect selection for yield in sunflower I. Two-mode pattern analysis of oil and biomass yield across environments in Argentina. Field Crops Res. 72, 17–38.

de la Vega, A.J., Chapman, S.C., 2006. Defining sunflower selection strategies for a highly heterogeneous target population of environments. Crop Sci. 46, 136–144.

de la Vega, A.J., DeLacy, I.H., Chapman, S.C., 2007. Changes in agronomic traits of sunflower hybrids over 20 years of breeding in central Argentina. Field Crops Res. 100, 73–81.

DeLacy, I.H., Cooper, M., 1990. Pattern analysis for the analysis of regional variety trials. In: Kang, M.S. (Ed.), Genotype-by-environment interaction and plant breeding. Louisiana State University, Baton Rouge, LA, pp. 301–334.

Dreccer, M.F., Chapman, S.C., Ogbonnaya, F.C., Borgognone, M.G., Trethowan, R.M., 2008. Crop and environmental attributes underpinning genotype by environment interaction in synthetic-derived bread wheat evaluated in Mexico and Australia. Aust. J. Agric. Res. 59, 447–460.

Duvick, D.N., 2005. The contribution of breeding to yield advances in maize (*Zea mays* L.). In: Sparks, D.L., (ed.), Advances in agronomy, 86. pp. 83-145. Academic Press, Elsevier. https://www.elsevier.com/books/advances-in-agronomy/sparks/978-0-12-000784-4

Edmeades, G.O., Bolaños, .J., Lafitte, H.R., Pfeiffer, W., Rajaram, S., Fischer, R.A., 1989. Traditional approaches in breeding for drought resistance in cereals. In: Baker, F.W.G. (Ed.), Drought resistance in cereals. ICSU Press/CABI, Paris/Wallingford, pp. 27–52.

Edmeades, G.O., Bolaños, J., Chapman, S.C., Lafitte, H.R., Bänziger, M., 1999. Selection improves drought tolerance in tropical maize populations: I. Gains in biomass, grain yield, and harvest index. Crop Sci. 39, 1306–1315.

FAOSTAT 2003. Statistical Database of the Food and Agriculture Organization of the United Nations. http://www.fao.org/waicent/portal/statistics_en.asp

FAOSTAT 2005. FAO Statistical Databases. Available online at http://faostat.fao.org/

Fischer, R.A., 2011. Wheat physiology: a review of recent developments. Crop Past. Sci. 62, 95–114.

Forkman, J., 2013. The use of a reference variety for comparisons in incomplete series of crop variety trials. J. Appl. Stat. 40, 2681–2698.

Fox, P.N., Rosielle, A.A., 1982. Reference sets of genotypes and selection for yield in unpredictable environments. Crop Sci. 22, 1171–1175.

Furbank, R.T., Tester, M., 2011. Phenomics – technologies to relieve the phenotyping bottleneck. Trends Plant Sci. 16, 635–644.

Ghaderi, A., Everson, E.H., Cress, C.E., 1980. Classification of environments and genotypes in wheat. Crop Sci. 20, 707–710.

Gilmour, A.R., Cullis, B.R., Verbyla, A.P., Gleeson, A.C., 1997. Accounting for natural and extraneous variation in the analysis of field experiments. J. Agric. Biol. Environ. Stat. 2, 269–293.

Grafius, J.E., Kiesling, R.L., 1960. Prediction of relative yields of different oat varieties based on known environmental variables. Agron. J. 52, 396–399.

Granier, C., Aguirrezabal, L., Chenu, K., et al., 2006. PHENOPSIS, an automated platform for reproducible phenotyping of plant responses to soil water deficit in *Arabidopsis thaliana* permitted the identification of an accession with low sensitivity to soil water deficit. New Phytol. 169, 623–635.

Graybosch, R.A., Peterson, C.J., 2010. Genetic improvement in winter wheat yields in the Great Plains of North America, 1959-2008. Crop Sci. 50, 1882–1890.

Guitard, A.A., 1960. The use of diallel correlations for determining the relative locational performance of varieties of barley. Can. J. Plant Sci. 40, 645–651.

Hammer, G.L., Vanderlip, R.L., 1989. Genotype-by-environment interaction in grain-sorghum. 3. Modeling the impact in field environments. Crop Sci. 29, 385–391.

Hammer, G.L., Chapman, S., van Oosterom, E., Podlich, D.W., 2005. Trait physiology and crop modelling as a framework to link phenotypic complexity to underlying genetic systems. Aust. J. Agric. Res. 56, 947–960.

Hammer, G., Cooper, M., Tardieu, F., et al., 2006. Models for navigating biological complexity in breeding improved crop plants. Trends Plant Sci. 11, 587–593.

Hammer, G.L., Jordan, D.R., 2007. An integrated systems approach to crop improvement. In : Spiertz, J.H.J., Struikc, P.C., van Laar, H.H., (eds), Scale and complexity in plant systems research: gene-plant-crop relations. pp. 45–61. Springer. Printed in the Netherlands.

Harrison, M.T., Hammer, G.L., Messina, C.D., Dong, Z., Tardieu, F., 2014. Characterizing drought stress and trait influence on maize yield under current and future conditions. Glob. Change Biol. 20, 867–878.

Heinemann, A.B., Dingkuhn, M., Luquet, D., Combres, J.C., Chapman, S., 2008. Characterization of drought stress environments for upland rice and maize in central Brazil. Euphytica 162, 395–410.

Hernandez-Segundo, E., Capettini, F., Trethowan, R., et al., 2009. Mega-environment identification for barley based on twenty-seven years of global grain yield data. Crop Sci. 49, 1705–1718.

Hijmans, R.J., Cameron, S.E., Parra, J.L., Jones, P.G., Jarvis, A., 2005. Very high resolution interpolated climate surfaces for global land areas. Internatl. J. Climatol. 25, 1965–1978.

Hodson, D.P., Martinez-Romero, E., White, J.W., Corbett, J.D., Bänziger, M., 2002. Africa maize research atlas (v. 3.0). CD-ROM Publication CIMMYT, Mexico, DF, Mexico.

Hodson, D.P., White, J.W., 2007. Use of spatial analyses for global characterization of wheat-based production systems. J. Agric. Sci. 145, 115–125.

Hoffmann, C.M., Huijbregts, T., van Swaaij, N., Jansen, R., 2009. Impact of different environments in Europe on yield and quality of sugar beet genotypes.. Eur. J. Agronomy 30, 17–26.

Horner, T.W., Frey, K.J., 1957. Methods for determining natural areas for oat varietal recommendations. Agron. J. 49, 313–315.

Keating, B.A., Carberry, P.S., Hammer, G.L., et al., 2003. An overview of APSIM, a model designed for farming systems simulation. Eur. J. Agron. 18, 267–288.

Khazaei, H., Street, K., Bari, A., Mackay, M., Stoddard, F.L., 2013. The FIGS (Focused Identification of Germplasm Strategy) approach identifies traits related to drought adaptation in *Vicia faba* genetic resources. Plos One 8, e63107–e163107.

Kholova, J., McLean, G., Vadez, V., Craufurd, P., Hammer, G.L., 2013. Drought stress characterization of post-rainy season (rabi) sorghum in India. Field Crops Res. 141, 38–46.

Kirigwi, F.M., van Ginkel, M., Trethowan, R., Sears, R.G., Rajaram, S., Paulsen, G.M., 2004. Evaluation of selection strategies for wheat adaptation across water regimes. Euphytica 135, 361–371.

Lacaze, X., Roumet, P., 2004. Environment characterisation for the interpretation of environmental effect and genotype × environment interaction. Theor. App. Genet. 109, 1632–1640.

Lacaze, X., Hayes, P.M., Korol, A., 2009. Genetics of phenotypic plasticity: QTL analysis in barley, *Hordeum vulgare*. Heredity 102, 163–173.

Lobell, D.B., Sibley, A., Ortiz-Monasterio, I., 2012. Extreme heat effects on wheat senescence in India. Nat. Climate Change 2, 186–189.

Löffler, C.M., Wei, J., Fast, T., et al., 2005. Classification of maize environments using crop simulation and geographic information systems. Crop Sci. 45, 1708–1716.

Louarn, G., Chenu, K., Fournier, C., Andrieu, B., Giauffret, C., 2008. Relative contributions of light interception and radiation use efficiency to the reduction of maize productivity under cold temperatures. Funct. Plant Biol. 35, 885–899.

Mackay, M., von Bothmer, R., Skovmand, B., 2005. Conservation and utilization of plant genetic resources – future directions. Czech J. Genet. Plant Breed. 41, 335–344.

Malhotra, R.S., Singh, K.B., 1991. Classification of chickpea growing environments to control genotype by environment interaction. Euphytica 58, 5–12.

Malosetti, M., Voltas, J., Romagosa, I., Ullrich, S.E., van Eeuwijk, F.A., 2004. Mixed models including environmental covariables for studying QTL by environment interaction. Euphytica 137, 139–145.

Manschadi, A.M., Christopher, J., Devoil, P., Hammer, G.L., 2006. The role of root architectural traits in adaptation of wheat to water-limited environments. Funct. Plant Biol. 33, 823–837.

Mathews, K.L., Chapman, S.C., Trethowan, R., et al., 2007. Global adaptation patterns of Australian and CIMMYT spring bread wheat. Theor. Appl. Genet. 115, 819–835.

Mathews, K.L., Trethowan, R., Milgate, A.W., et al., 2011. Indirect selection using reference and probe genotype performance in multi-environment trials. Crop Past. Sci. 62, 313–327.

Muchow, R.C., Hammer, G.L., Carberry, P.S., 1991. Optimising crop and cultivar selection in response to climatic risk. In: Muchow, R.C., Bellamy, J.A. (Eds.), Climatic risk in crop production: models and management for the semiarid tropics and subtropics. CAB International, Wallingford, UK, pp. 235–262.

Muchow, R.C., Cooper, M., Hammer, G.L., 1996. Characterizing environmental challenges using models. In: Cooper, M., Hammer, G.L. (Eds.), Plant adaptation and crop improvement. CAB International, Wallingford, pp. 349–364.

Nor, K.M., Cady, F.B., 1979. Methodology for identifying wide adaptability in crops. Agron. J. 71, 556–559.

Palmer, W., 1965. Meteorological drought. US Department of Commerce Weather Bureau Research paper no. 45: 58 pp.

Pantuwan, G., Fukai, S., Cooper, M., Rajatasereekul, S., O'Toole, J.C., 2002. Yield response of rice (*Oryza sativa* L.) genotypes to different types of drought under rainfed lowlands – Part 1. Grain yield and yield components. Field Crops Res. 73, 153–168.

Passioura, J.B., 2012. Phenotyping for drought tolerance in grain crops: when is it useful to breeders? Funct. Plant Biol. 39, 851–859.

Podlich, D.W., Cooper, M., Basford, K.E., 1999. Computer simulation of a selection strategy to accommodate genotype-by-environment interaction in a wheat recurrent selection program. Plant Breed. 118, 17–28.

Pollak, L.M., Corbett, J.D., 1993. Using GIS datasets to classify maize-growing regions in Mexico and Central America. Agron. J. 85, 1133–1139.

Rattey, A., Shorter, R., 2010. Evaluation of CIMMYT conventional and synthetic spring wheat germplasm in rainfed sub-tropical environments. I. Grain yield. Field Crops Res. 118, 273–281.

Rebetzke, G.J., Chenu, K., Biddulph, B., et al., 2013. A multisite managed environment facility for targeted trait and germplasm phenotyping. Funct. Plant Biol. 40, 1–13, http://www.publish.csiro.au/nid/102/paper/FP12180.htm.

Rebetzke, G.J., Fischer, R.A., van Herwaarden, A.F., et al., 2014. Plot size matters: interference from intergenotypic competition in plant phenotyping studies. Funct. Plant Biol. 41, 107–118.

Richards, R., 1996. Increasing the yield potential of wheat: manipulating sources and sinks. In: Rajaram, S., Reynolds, M. (Eds.), Increasing yield potential in wheat: breaking barriers. CIMMYT, Mexico, pp. 134–149.

Richards, R.A., Rebetzke, G.J., Watt, M., Condon, A.G., Spielmeyer, W., Dolferus, R., 2010. Breeding for improved water productivity in temperate cereals: phenotyping, quantitative trait loci, markers and the selection environment. Funct. Plant Biol. 37, 85–97.

Rodriguez, D., Sadras, V.O., 2007. The limit to wheat water-use efficiency in eastern Australia. I. Gradients in the radiation environment and atmospheric demand. Aust. J. Agric. Res. 58, 287–302.

Runge, E.C.A., 1968. Effects of rainfall and temperature interactions during growing season on corn yield. Agron. J. 60, 503–507.

Russell, W.K., Eskridge, K.M., Travnicek, D.A., Guillen-Portal, F.R., 2003. Clustering environments to minimize change in rank of cultivars. Crop Sci. 43, 858–864.

Sadras, V.O., Denison, R.F., 2009. Do plant parts compete for resources? An evolutionary viewpoint. New Phytol. 183, 565–574.

Sadras, V.O., Lake, L., Chenu, K., McMurray, L.S., Leonforte, A., 2012. Water and thermal regimes for field pea in Australia and their implications for breeding. Crop Past. Sci. 63, 33–44, http://www.publish.csiro.au/nid/40/paper/CP11321.htm.

Semenov, M.A., Donatelli, M., Stratonovitch, P., Chatzidaki, E., Baruth, B., 2010. ELPIS: a dataset of local-scale daily climate scenarios for Europe. Climate Res. 44, 3–15.

Setimela, P., Chitalu, Z., Jonazi, J., Mambo, A., Hodson, D., Bänziger, M., 2002. Revision of maize megaenvironments in the Southern African Development Community (SADC) region. Arnel R Hallauer International Symposium on Plant Breeding. CIMMYT Mexico, Mexico, pp. 246–247.

Siebert, S., Feick, S., Döll, P., Hoogeveen, J., 2005. Global Map of Irrigation Areas Version 3.0. University of Frankfurt (Main) and FAO, Frankfurt and Rome.

Sinclair, T.R., Hammer, G.L., van Oosterom, E.J., 2005. Potential yield and water-use efficiency benefits in sorghum from limited maximum transpiration rate. Funct. Plant Biol. 32, 945–952.

Skirycz, A., Vandenbroucke, K., Clauw, P., et al., 2011. Survival and growth of *Arabidopsis* plants given limited water are not equal. Nat. Biotechnol. 29, 212–214.

Slafer, G.A., 2003. Genetic basis of yield as viewed from a crop physiologist's perspective. Ann. Appl. Biol. 142, 117–128.

Spitters, C.J.T., Schapendonk, A., 1990. Evaluation of breeding strategies for drought tolerance in potato by means of crop growth simulation. Plant Soil 123, 193–203.

Tardieu, F., 2012. Any trait or trait-related allele can confer drought tolerance: just design the right drought scenario. J. Exp. Bot. 63, 25–31.

Trethowan, R.M., Crossa, J., van Ginkel, M., Rajaram, S., 2001. Relationships among bread wheat international yield testing locations in dry areas. Crop Sci. 41, 1461–1469.

Trethowan, R., Hodson, D.P., Braun, H.J., Pfeiffer, W., van Ginkel, M., 2005a. Wheat breeding environments. In: Lantican, M.A., Dubin, H.J., Morris, M.L. (Eds.), Impacts of international wheat breeding research in the developing world, 1988-2002. DF CIMMYT, Mexico, pp. 4–11.

Trethowan, R.M., Reynolds, M., Sayre, K., Ortiz-Monasterio, I., 2005b. Adapting wheat cultivars to resource conserving farming practices and human nutritional needs. Ann. Appl. Biol. 146, 405–413.

van Eeuwijk, F.A., Bink, M.C.A.M., Chenu, K., Chapman, S.C., 2010. Detection and use of QTL for complex traits in multiple environments. Curr. Opin. Plant Biol. 13, 193–205.

Van Oosterom, E.J., Bidinger, F.R., Weltzien, E.R., 2003. A yield architecture framework to explain adaptation of pearl millet to environmental stress. Field Crops Res. 80, 33–56.

Varshney, R.K., Bansal, K.C., Aggarwal, P.K., Datta, S.K., Craufurd, P.Q., 2011. Agricultural biotechnology for crop improvement in a variable climate: hope or hype? Trends Plant Sci. 16, 363–371.

Veyradier, M., Christopher, J., Chenu, K., 2013. Quantifying the potential yield benefit of root traits. In: Sievänen, R., Nikinmaa, E., Godin, C., Lintunen, A., Nygre, P. (Eds.), 7th International Conference on Functional-Structural Plant Models. Saariselkä, Finland. pp. 317–319.

Yang, R.-C., Blade, S.F., Crossa, J., Stanton, D., Bandara, M.S., 2005. Identifying isoyield environments for field pea production. Crop Sci. 45, 106–113.

Yates, F., Cochran, W.G., 1938. The analysis of groups of experiments. J. Agric. Sci. 28, 556–580.

Zheng, B., Chenu, K., Dreccer, M.F., Chapman, S.C., 2012. Breeding for the future: what are the potential impacts of future frost and heat events on sowing and flowering time requirements for Australian bread wheat (*Triticum aestivium*) varieties? Glob. Change Biol. 18, 2899–2914.

Zheng, B., Biddulph, B., Li, D., Kuchel, H., Chapman, S., 2013. Quantification of the effects of VRN1 and Ppd-D1 to predict spring wheat (*Triticum aestivum*) heading time across diverse environments. J. Exp. Bot. doi:10.1093/jxb/ert209. http://jxb.oxfordjournals.org/content/early/2013/07/17/jxb.ert209.full.pdf+html

CHAPTER

14

Model-assisted phenotyping and ideotype design

Pierre Martre[1,2], Bénédicte Quilot-Turion[3], Delphine Luquet[4], Mohammed-Mahmoud Ould-Sidi Memmah[5], Karine Chenu[6], Philippe Debaeke[7,8]

[1]INRA, UMR1095 Genetics, Diversity and Ecophysiology of Cereals, Clermont-Ferrand, France

[2]Blaise Pascal University, UMR1095 Genetics, Diversity and Ecophysiology of Cereals, Aubière, France

[3]INRA, UR1052 Genetics and Improvement of Fruit and Vegetables, Avignon, France

[4]CIRAD, Department of Biological Systems, UMR1334 Genetic Improvement and Adaptation of Mediterranean and Tropical Plants, Montpellier, France

[5]INRA, UR1115, Plantes et Systèmes de Culture Horticoles, Avignon, France

[6]The University of Queensland, Queensland Alliance for Agriculture and Food Innovation, Toowoomba, Queensland, Australia

[7]INRA, UMR1248 Agroécologie, Innovations, Territoires, Castanet-Tolosan, France

[8]INPT, Université Toulouse, UMR1248 Agroécologie, Innovations, Territoires, Toulouse, France

1 INTRODUCTION

The whole plant is the central scale of analysis and integration to improve plant population performance (Hammer et al., 2010; Keurentjes et al., 2011; Pedró et al., 2012). This is true from both agronomic and ecological points of view, i.e. regarding production or survival (Dingkuhn et al., 2007). Plant growth results from multiple interactions and trade-offs among processes of various nature (e.g. morphological, physiological, biochemical) acting at different scales (Chapter 1) that can compete for the same resources internally to the plant. These processes can be characterized by traits (Box 14.1) that are potentially linked, both physiologically

Crop Physiology. DOI: 10.1016/B978-0-12-417104-6.00014-5

BOX 14.1

GLOSSARY

Emergent property. In system theory, qualifies a higher-level property which is or can be deduced from the properties of the lower level entities.

Genotype. The inherited instructions an organism carried within its genetic code. Not all organisms with the same genotype have the same phenotype because their morphology and physiology are modified by environmental and developmental conditions. Likewise, not all organisms with a similar phenotype have the same genotype. Genotype may refer to the specific allelic composition of the entire genome, and by extension to set of genes, or a specific gene.

G × E × M interaction. Indicates that the relative performance of genotypes (G) varies with environmental (E) conditions and with crop management (M). Sometimes M is included in E. G × E (× M) interaction is attributed to the dependence of expression of underlying genes or QTL on environments (QTL × E interaction). This interaction has often been conceptualized by the following relationship: G + E + G × E + error (e) → Phenotype (P).

High-throughput phenotyping. Plant phenotyping is the experimental assessment of individual quantitative traits (e.g. growth, development, tolerance, resistance, architecture, physiology) that forms the basis for more complex traits. High-throughput phenotyping involves comprehensive and fast measurements of phenotypes in the lab, the greenhouse or the field. See Chapter 15 for an example.

Ideotype. Combination of morphological and/or physiological traits, or their genetic bases, optimizing crop performance to a particular biophysical environment, crop management, and end-use.

Multicriteria optimization. Also known as multiobjective or multiattribute optimization. An area of multiple-criteria decision making, which deals with mathematical optimization problems involving simultaneous optimization of multiple objective functions.

Phenotypic plasticity. The ability of a cell, tissue, organ, organism, or species to change its phenotype in response to environmental signals. Induced changes may be morphological or physiological and may or may not be permanent during the lifespan of the considered entity.

Pleiotropy. The phenomenon of a single trait or gene (loci) modifying multiple phenotypic traits that are apparently unrelated.

Quantitative trait loci (QTL). Chromosomal segments (loci) at which the allelic variability is statistically linked to a quantitative trait. QTL may vary depending on the population and the environment studied.

Phenotype. The expression in a particular environment of a specific genotype through its morphology, development, cellular, biochemical or physiological properties.

Phenome. The set of all phenotypes expressed by a cell, tissue, organ, organism, or species.

Phenomics. The study of the phenome and how it is determined by the genotype and the environment, particularly when studied in relation to the set of genes and the non-coding sequences (genomics), transcripts (transcriptomics), proteins (proteomics) or metabolites (metabolomics).

Trait. A distinct variant of a phenotypic property of an organism that may be inherited, be environmentally determined or be a combination of both.

(Rebolledo et al., 2013) and genetically (ter Steege et al., 2005). Such linkages could result either from human or natural selection (Rebolledo et al., 2013) and are particularly challenging regarding the improvement of the plant system. Key traits for improving performances at the crop level may be viewed as regulatory hubs with pleiotropic (Box 14.1) actions. These linkages and trade-offs make it difficult to decipher phenotype (Box 14.1) construction and improve crop performance in various agro-climatic conditions. This is becoming even more challenging with the increasing complexity of the plant characteristics sought in breeding programs aiming at combining yield and quality (Chapter 17), disease tolerance and agronomic adaptation in current and future climates (Chapter 20), and in multiple purpose crops, such as dual-purpose sorghum (Gutjahr et al., 2013) or wheat (Harrison et al., 2011). There is thus the need to better understand interactions and trade-offs between traits or processes contributing to crop performance and their genetic bases.

By formalizing traits as the result of genotypic and environmental effects and the relations among traits, ecophysiological models provide a platform for integrative analyses of the impact of a combination of traits on whole-plant and crop phenotype (e.g. Hammer et al., 2009; Messina et al., 2009; Bertin et al., 2010). The application of such models in the context of phenotype analyses and ideotyping strongly relies on the use of mathematical tools to quantify the effect of individual traits within a trait network from plant measurements made on various genotypes in a range of environments (Farnsworth and Niklas, 1995; Dingkuhn et al. 2007). Model-assisted phenotyping, where a genotype is characterized by a set of traits, provides a phenotypic fingerprint that can be used to explore trait correlations in a population and ultimately to connect model parameters to genetic information.

Ecophysiological models also provide a platform for quantifying the impact of 'simple' traits (individually, or in interaction with other traits in a trait network) on more integrated traits such as yield, in a range of agro-climatic conditions. While breeders have traditionally favored broad adaptation (i.e. development of genotypes with improved performance across *all* environments), modeling opens new avenues to develop genotypes specifically adapted to a set of conditions of particular interest such as hostile soils, new cultivation techniques, and future climates. Robust statistical methods for quantitative analyses of model parameter influence (sensitivity) on plant performance have been developed (Saltelli et al., 2000) and were recently used to study ecophysiological models (e.g. Makowski et al., 2006; He et al., 2010). Such methods allow quantifying the influence of 'simple' traits, individually and in combinations, on more complex traits such as yield in different environments. This constitutes a first step toward multicriteria optimization (Box 14.1) of yield, quality, disease tolerance, and resource uses.

The use of ecophysiological models to assist in plant and crop phenotyping, allows quantifying meaningful traits that can hardly be estimated experimentally on a large number of genotypes (e.g. cold requirement, plant state variables such as labile C/N concentration), using measurements of plant response variables that can more easily be obtained experimentally. Simulations with the estimated genotypic parameters also allow analyzing the behavior of state variables (e.g. internal pools of carbon) that cannot be quantified experimentally (e.g. Luquet et al., 2012a).

One of the main challenges to progress in this direction is to build ecophysiological models that integrate genetic information associated to specific process(es) and simulate interactions among genetic, physiological and environmental controls to estimate the value of integrated traits (i.e. emergent properties, Box 14.1) in various conditions (e.g. Bertin et al., 2010; Hammer et al., 2010). In this chapter, we argue that ecophysiological models can help breeders and geneticists 'to transition from statistical approaches in analyzing genotype-by-environment interactions to a knowledge-based view that emphasizes crop responses to specific environmental factor' as called by Edmeades et al. (2004). After defining the ideotype concept in the framework of ecophysiological modeling, we discuss how ecophysiological models can help identify influential traits in given environments and cropping systems and predicting genotypic variation in different environments. We then review recent studies applying ecophysiological models to design varietal types or virtual genotypes better adapted to given environment–management combinations.

2 THE IDEOTYPE CONCEPT: ITS USEFULNESS AND LIMITATIONS FOR BREEDING AND VARIETAL CHOICE

The ideotype approach (also called analytical or physiological trait-based approach) was proposed by Donald (1968) to overcome the limitations of the methods used by breeders, namely 'selection for yield (empirical method)' and 'defect/default elimination'. Although these two empirical methods had been effective for improving disease resistance and grain yield, Donald proposed as an alternative first to define an efficient plant type theoretically, based on our knowledge of crop physiology and then breed for it. He defined an ideotype as 'a biological model which is expected to perform or behave in a predictable manner within a defined environment' (Donald, 1968). This conceptual plant model was supposed 'to yield a greater quantity or quality of grain, oil or other useful product when developed as a cultivar'.

According to this approach, breeders should select directly for the plant ideotype, rather than empirically for grain yield. As several target traits besides final grain yield were provided, concrete guidelines have since been made to conduct the breeding process (Sedgley, 1991; Rasmusson, 1991; Reynolds et al., 2009). For instance, guided by the idea of improving light capture and assimilate partitioning in cereals, Donald identified short stature, strong stem, few small erect leaves, large erect awned ears, low tillering capacity (oligoculm), disease resistance, local adaptation and low plant competitive ability as important target traits for wheat crops sown at high density, under non-limiting conditions. Breeding for such an ideotype has resulted in improved lodging resistance and higher harvest index (Hamblin, 1993).

In the literature, the ideotype concept generally refers to the breeding process, but it can also be extended to the seeking of the best crop phenotype to grow in given environments, with defined cropping systems and for targeted end uses. Commercial varieties can be far from an ideotype viewed as a theoretical objective, the variety choice can be optimized even with a limited range of traits opportunities. Therefore, we suggest broadening the ideotype definition, to the combination of morphological and physiological traits (or their genetic bases) conferring to a crop a satisfying adaptation to a particular biophysical environment, crop management, and end use.

Ideotype breeding was initially developed for annual crops, mainly cereals (Mock and Pearce, 1975; Rasmusson, 1987; Peng et al., 2008; Hanocq et al., 2009) and was later applied to forest and fruit tree species (Dickmann et al., 1994; Socias et al., 1998; Lauri and Costes, 2005; Cilas et al., 2006), emphasizing the generic value of the concept. Since then, the ideotype approach was expanded and refined to include other concerns such as market, new outlets, climate change adaptation (Semenov and Stratonovitch, 2013; Semenov et al., 2014), emerging pests and diseases, and changes in farming system (Jeuffroy et al., 2013). Attempts have also been made to define traits at the biochemical and molecular levels (e.g. Reynolds et al., 1996). Overall, the level of knowledge to support the ideotype approach depends on the crop and final trait (e.g. disease resistant, end-use quality) considered.

The initial ideotype of Donald was built for low- or non-stress environments where light capture was the major limitation to grain yield. For water-limited environments, the difficulty to design an ideotype is that water deficit affects crop growth and development to a different extent depending on the timing, severity and duration of stress episodes, the history of stresses during the growing season and the interactions between water deficit and other factors such as temperature and nutrient availability. Therefore, it is unlikely that a single trait will improve plant performance in all scenarios of water deficit (Tardieu, 2012). Therefore, specific ideotypes should be conceived for targeted environments. For environments with large inter-annual weather variability, stability of performance is often considered an important varietal characteristic (Braun et al., 1992), but selecting for more specific adaptation could be an alternative (Ceccarelli, 1989). Chapter 13 deals in detail with quantitative environmental characterization in a context of crop adaptation.

While productivity improvement has slowed down in crops like wheat in recent years (e.g. Brisson et al., 2010; Richards et al., 2010), the major benefits of ideotype design are seen as being conceptual and analytical rather than in direct yield improvements (Hamblin, 1993). Progress from using ideotype breeding has often not been as fast as hoped. However, it stimulated interactions among breeders about the utility of the concept and among physiologists, agronomists, breeders and, more recently, modelers about what traits might be important and in which conditions (Marshall, 1991; Hamblin, 1993; Chapman et al., 2003; Foulkes et al., 2011).

Recently, Andrivon et al. (2013) considered three views of ideotypes: (1) the historical, 'genetic' view, as described above; (2) the 'agronomic' view, where new genotypes are designed for specific cropping systems; (3) the 'modeling' view, where the best combinations of traits (usually represented by model parameters) are identified from formal or simulation experiments. They concluded that these views of the ideotype should lead to different breeding strategies. The emergence of new objectives (e.g. low input systems, double purpose crops) and new constraints (e.g. increasing risk of extreme weather events, price volatility) is now arguing for both new breeding objectives and new design methods. Designing crop ideotypes for these new targets is a burning point and no review has recently addressed this subject.

The ideotype design process (ideotyping) could be split into three steps (Fig. 14.1):

1. Definition of the main goal (target) for the breeding process (e.g. breeding for improved water-deficit tolerance)
2. Identification of morpho-physiological traits to reach the defined goal and the way to assemble them within an ideotype (e.g. developing early maturing cultivars or cultivars maintaining photosynthesis under stress or both)
3. Multicriteria assessment of the suggested ideotypes to prove the agronomic relevance of trait integration in target environments (through simulations or field experiments).

Generally, the ideotype is thought of in terms of crop improvement via breeding, but crop management (e.g. sowing density, row width, nitrogen fertilization, irrigation) may also produce the desired ideotypes by exploiting phenotypic plasticity (Box 14.1). For instance, this is the case when dealing with plant architecture traits and crop canopies to limit the epidemic development of pests (Andrivon et al., 2013; Desanlis et al., 2013). So ideotyping may result from breeding and varietal choice but also from crop management and cropping system strategies. While crop management is often added in a second step as an effective driver to complement genetic gains, greatest productivity improvement may arise when combining together both breeding and agronomic practices (e.g. Duvick et al., 2004) (Fig. 14.1). Part 1 of this book presents further examples of the synergy between breeding and agronomy in contrasting cropping systems.

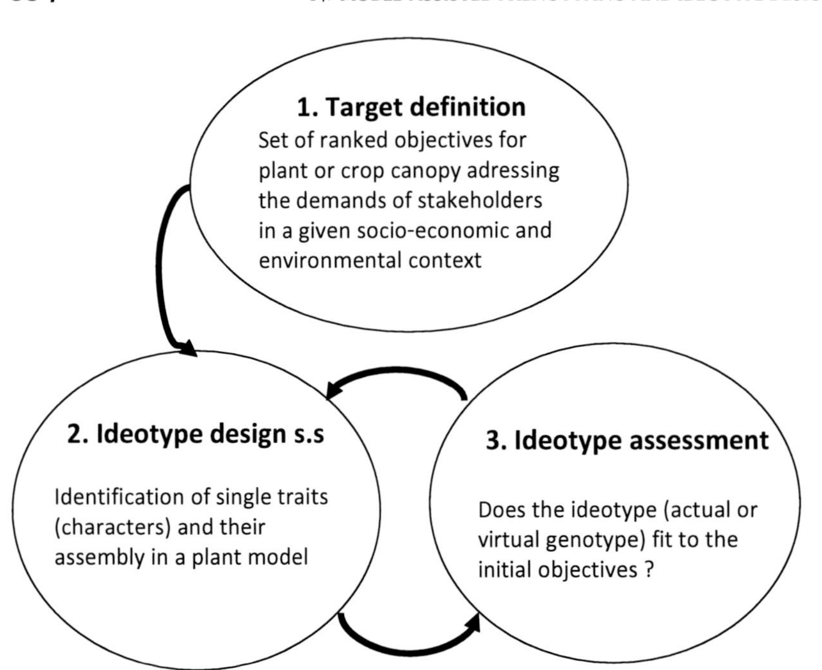

FIG. 14.1 Scheme of the three main steps for ideotype design (ideotyping).

3 HOW TO DEAL WITH GENETIC CONTROL IN ECOPHYSIOLOGICAL MODELS?

To support ideotyping, ecophysiological models need to integrate traits that can be reliably estimated based on genetic information. In this section, we present (1) the different levels of complexity found in models concerning the integration of genetic details, with examples on models dealing with quantitative trait loci (QTL)/genes effects; (2) applications of modeling to support trait assessment in multienvironments and to simulate genotype × environment interactions; (3) the potential of new technologies in this framework and in particular to decipher QTL × environment interactions; and, finally, (4) the use of integrated models in breeding.

3.1 Required level of complexity

White and Hoogenboom (2003) identified six classes of models in relation to genetic detail:

1. Generic model with no reference to species
2. Species-specific model with no reference to cultivars
3. Genetic differences represented by cultivar-specific parameters
4. Genetic differences represented by gene actions modeled through their effects on model parameters
5. Genetic differences represented by genotypes, with gene action explicitly simulated based on knowledge of regulation of gene expression and effects of gene products
6. Genetic differences represented by genotypes, with the gene action simulated at the level of interactions of regulators, gene products, and other metabolites.

Historically, 'generic' (class 1) and 'species-specific' (class 2) models were developed first, in the 1970s. Progressively, basic genotypic information has been included, so that most ecophysiological models are now 'cultivar-specific' (class 3). Researchers are currently developing 'gene-specific' models (class 4), which include information from major genes associated to 'simple' traits or from quantitative trait loci (Box 14.1) associated to more complex traits. These models have been proposed for hypothetical genes when key genetic controls were unknown (e.g. Chapman et al., 2003; Hammer et al., 2005) and for genetic controls identified experimentally (Fig. 14.2; Chenu et al., 2009). To date, models based on the 'regulation of gene expression' (class 5) are rare as the understanding of gene action is restricted to particular physiological processes mostly in model species (e.g. Welch et al., 2005; Chew et al., 2012). Despite incomplete information, models of class 5 can be proposed for crop species. For instance, to test physiological assumptions and to improve simulations with the common bean model GeneCro, Hoogenboom and White (2003) introduced an unknown gene (White and Hoogenboom, 1996) which affects the expression of other genes. Lastly, models that simulate gene action based on interactions of regulators, gene products, and other metabolites (class 6) have only been developed for unicellular organisms (Karr et al., 2012; Sanghvi et al., 2013).

Overall, while models of classes 5 and 6 can be relevant for 'simple' traits associated to a restricted number of genes (e.g. flowering time, resistance to a pathogen), more complex traits are usually modeled via QTL (class 4) and/or as emergent properties. The level of detail and complexity of the model (the 'required' complexity) are also defined depending on the targeted applications of the model. Increasing the complexity of a model does not necessary lead to better predictions (e.g. Challinor et al., 2014). Typically, a trade-off between complexity and accuracy has to be addressed to avoid over parameterization (e.g. Reynolds and Acock, 1985) and to limit uncertainty. However, well-known processes that affect traits of interest or key variables of the system might be worth including with further detail in a model. Hence, models typically contain different levels of detail and include submodels of different classes. They evolve with time and may be enhanced by new knowledge focused on central points for predictions.

3.2 Integration of QTL/genes in ecophysiological models

Few ecophysiological models include genetic controls (class 4). There, genes or QTL are associated with parameters of the model, and genotypes are defined by a set of parameters which depend on their allelic combination. Robust modeling requires parameters to be constant under a wide range of environmental conditions (Boote et al., 2001; Tardieu, 2003). They often display quantitative and continuous variations in populations, in the same way as variables classically measured (e.g. yield or biomass). However, the QTL associated to these parameters do not systematically co-localize with the QTL for the more integrated variables, thus highlighting the complexity of the system and genetic independence of the trait plasticity and the trait *per se*. For instance, lack of co-localization was found between QTL for final leaf length of maize under water deficit, and QTL for the parameters associated to leaf expansion response to water deficit (Reymond et al., 2004). Co-localization of QTL for different traits or parameters can nevertheless help better understanding of the processes involved, and thus assist model improvement. For instance, co-localizations between QTL for leaf elongation and anthesis–silking interval in maize suggest that these traits might be regulated by the same underlying process (e.g. tissue elongation for either the leaves or the silks; Fig. 14.2a) (Welcker et al., 2007).

While early work on QTL-based modeling at the crop level has highlighted the need to integrate physiologically-based processes in crops models (e.g. Yin et al., 2000), promising results have since been obtained using physiological components for traits such as leaf elongation (Reymond et al., 2003), plant development (Yin

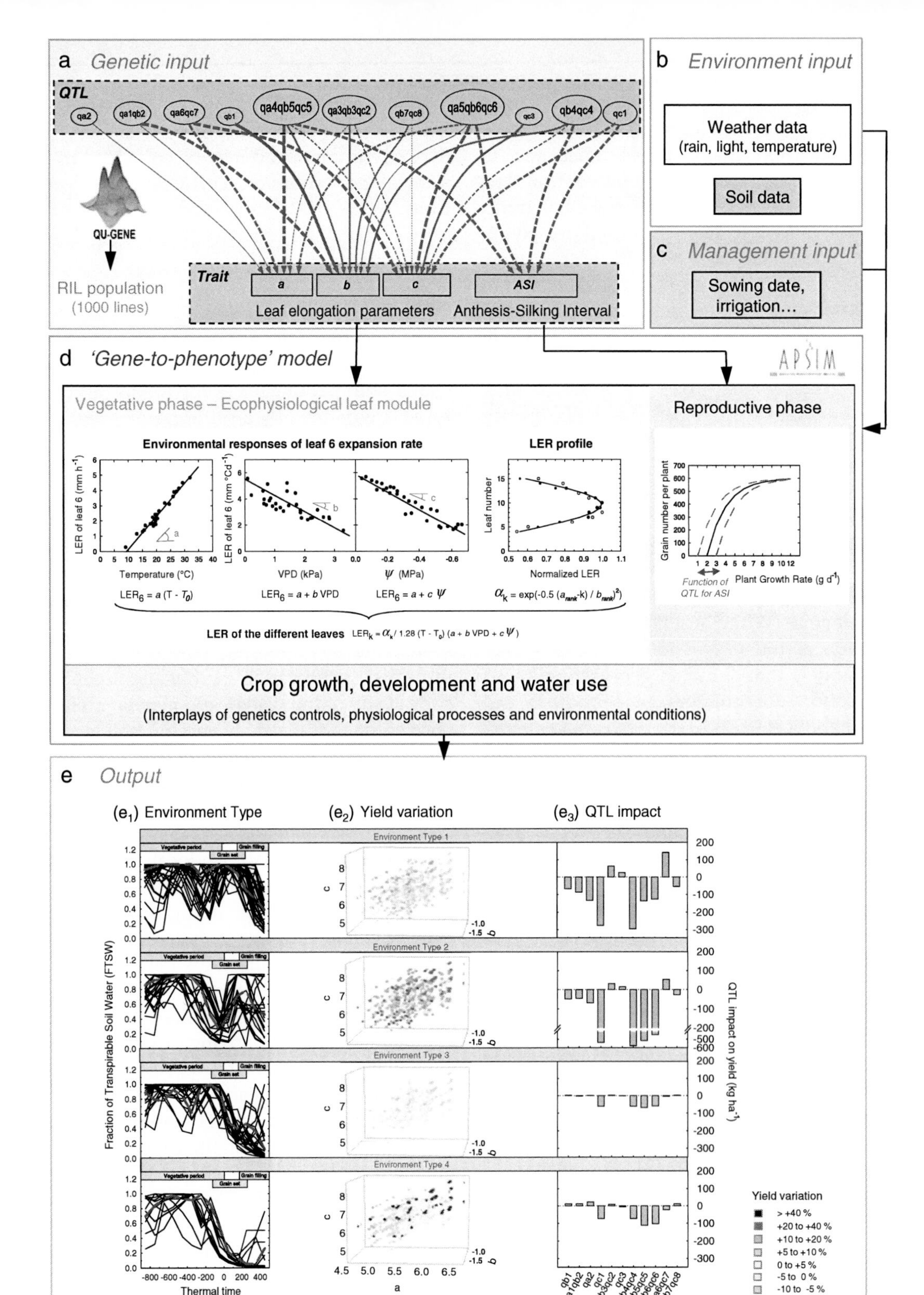

a Genetic input
QTL
qa2
qa1qb2
qa6qc7
qb1
qa4qb5qc5
qa3qb3qc2
qb7qc8
qa5qb6qc6
qc3
qb4qc4
qc1
QU-GENE
RIL population
(1000 lines)
Trait
a
b
c
ASI
Leaf elongation parameters
Anthesis-Silking Interval
b Environment input
Weather data
(rain, light, temperature)
Soil data
c Management input
Sowing date, irrigation...
d 'Gene-to-phenotype' model
APSIM
Vegetative phase – Ecophysiological leaf module
Environmental responses of leaf 6 expansion rate
LER profile
Reproductive phase
Temperature (°C)
VPD (kPa)
Ψ (MPa)
Normalized LER
Leaf number
Grain number per plant
Plant Growth Rate (g d-1)
Function of QTL for ASI
LER6 = a (T - T0)
LER6 = a + b VPD
LER6 = a + c Ψ
αk = exp(-0.5 (arank-k) / brank)2)
LER of the different leaves LERk = αk / 1.28 (T - T0) (a + b VPD + c Ψ)
Crop growth, development and water use
(Interplays of genetics controls, physiological processes and environmental conditions)
e Output
(e1) Environment Type
(e2) Yield variation
(e3) QTL impact
Environment Type 1
Environment Type 2
Environment Type 3
Environment Type 4
Vegetative period
Grain set
Grain filling
Fraction of Transpirable Soil Water (FTSW)
Thermal time before/after flowering (°Cd)
Parameter
QTL impact on yield (kg ha-1)
QTL
Yield variation
> +40 %
+20 to +40 %
+10 to +20 %
+5 to +10 %
0 to +5 %
-5 to 0 %
-10 to -5 %
-20 to -10 %
-40 to -20 %
< -40 %

◀ **FIG. 14.2** 'Gene-to-phenotype' modeling to capture QTL (quantitative trait loci) effects and gene/QTL × environment interaction from organ to crop levels. Genetic knowledge of 'simple' component traits was used to parameterize the model and to infer the impact of single QTL or QTL combinations on complex traits. In this example, organ-level QTL for leaf and silk elongation of maize were inputs to a modified version of the APSIM ecophysiological model. The impact of the environmentally stable QTL was tested in different environments for 1000 recombinant lines (RILs) simulated with the quantitative genetics model QU-GENE (Podlich and Cooper, 1998) (a). The QTL were associated first with the additive effects (red dashed line, negative; blue solid line, positive) affecting leaf elongation rate (LER) response to temperature (parameter *a*), evaporative demand (parameter *b*) and soil water deficit (parameter *c*) (Reymond et al., 2003), and secondly, with assumed pleiotropic effects on silk elongation and anthesis-silking interval (ASI) under drought (Welcker et al., 2007). These responses were integrated in a leaf module of APSIM, and the response of QTL for ASI was integrated in a reproductive module of APSIM (d). Overall, the model integrated genetic (a), environmental (b) and management (c) information to account for the complex interplay of genetic, physiological and environmental controls throughout the crop cycle (d) (Chenu et al., 2008). After characterizing the drought environment types based on the FTSW (fraction of transpirable soil water) in Sete Laogas, Brazil (e_1), simulations were undertaken for four representative drought patterns (red dashed line). Genotype × environment interaction was generated for simulated yield (e_2) and the impact of the organ-level QTL highly varied depending on the environment considered (e_3). For instance, many positive-effect QTL in low/mild stress environments (Environment Types 1 and 2) had a negative impact in a severe reproductive stress environment (Environment Types 3 and 4) and vice versa. Two QTL (qa6qc7 and qa4qb5qc5) with similar effects on the LER response to temperature (parameter *a*) had contrasting effects on simulated yield (e_3) (Chenu et al., 2009). *Adapted from Chenu et al. (2008, 2009).* Please see color plate at the back of the book.

et al., 2005; Messina et al., 2006), early plant growth (Brunel et al., 2009), nitrogen uptake and root growth and architecture (Laperche et al., 2006) and peach fruit growth and sweetness (Quilot et al., 2005). In each of these studies, QTL associated with the considered traits and processes were identified. Tests of the model against independent data (new genotypes and environmental conditions) were also promising (e.g. Reymond et al., 2003). More details on the approach can be found in recent reviews (e.g. Hammer et al., 2006; Yin and Struik, 2008; Messina et al., 2009; Bertin et al., 2010).

One challenge for ecophysiological modeling is to extend the approach to other component traits. *Ideally*, complex traits simulated by a model (e.g. yield, fruit quality) *should* be modeled as emergent properties from component traits, and these component traits *should* be modeled based on input parameters that are stable across environments and genotype specific.

3.3 Ecophysiological modeling to support trait assessment in multienvironments

Estimating parameter values of populations presents a major advantage, as by construction, the parameters are supposed to be independent from the environment and have thus a greater heritability than associated traits. However, a major drawback concerns the observations needed to parameterize the models for a large number of genotypes. First, whole-plant ecophysiological models usually comprise a large number of parameters, typically from 50 to 200. Second, some parameters are not accessible for measurements, but could represent key traits.

To overcome these difficulties, one possible approach is to identify parameters that have larger impact on the targeted traits (e.g. via a sensitivity analysis of model outputs) or are strongly correlated with such parameters (or their associated traits; Box 14.2). Note that to be of interest, those parameters have to vary among genotypes and be quantifiable with relevant accuracy either experimentally or through numerical optimization. This strategy was applied in the 'Virtual Fruit' model (Génard et al., 2007) in peach and resulted, by successive steps, in reducing the initial set of 39 parameters (Quilot et al., 2005) to 25 parameters inducing significant output variations, from which 16 could be measured in the population studied, and only 10 parameters displayed significant genetic variation.

BOX 14.2

IDENTIFYING INFLUENTIAL TRAITS THROUGH GLOBAL SENSITIVITY ANALYSES

Ecophysiological models can help develop hypotheses starting near the top of the trait hierarchy leading to integrated character such as grain yield and to identify putative influential physiological traits in target environments (Sinclair et al., 2004). Before addressing the question of how to translate the information of model simulations to knowledge that can be used by physiologists or geneticists, we need a better understanding of the model properties and behavior. One of the best ways to do that is to conduct a global uncertainty and sensitivity analysis of the model to investigate its behavior in response to variations in inputs (Cariboni et al., 2007). By perturbing model parameters associated with simple physiological traits, uncertainty and sensitivity analyses allow investigation of crop responses and can help identify those traits that lead to high and stable grain yields in the target environments.

In wheat, and more generally in cereals, maintaining grain protein concentration while increasing grain yield represents a challenge for plant breeders because of the genetic and physiological antagonism between these two characters (e.g. Cooper et al., 2001; Oury et al., 2003; Aguirrezábal et al., 2009). A global sensitivity analysis of the wheat ecophysiological model *SirusQuality*2 was conducted with the aim of identifying candidate traits to increase both grain yield and grain protein concentration (He et al., 2010). *SiriusQuality*2 (http://www1.clermont.inra.fr/siriusquality) accounts for canopy development, and capture and allocation of C and N at the organ and phytomer levels (Martre et al., 2006; Ferrise et al., 2010). Chapter 17 presents further detail of these applications of ecophysiological models to questions of grain quality, and quality-yield trade-offs.

Three contrasting European sites were considered and simulations were performed using long-term weather data and two nitrogen treatments to quantify the effect of parameter uncertainty on grain yield and protein concentration under variable environments. The overall influence of all of the 75 crop parameters of *Sirius-Quality*2 on grain yield and protein concentration was first analyzed using the semiquantitative Morris method (Morris et al., 1991). Forty-one influential parameters with respect to grain yield and protein concentration were identified and their individual (first-order) and total effects on the model outputs were investigated using the extended Fourier amplitude sensitivity test (E-FAST; Saltelli et al., 1999).

The Morris analysis showed that most influential parameters are also involved in curvature or interaction effects (He et al., 2010). In other words, the 'overall' importance of a model parameter is primarily determined by the non-linear response of the model or by its interactions with other model parameters. This result was confirmed by the E-FAST analysis, where 8 to 56% of the variance for grain yield and protein concentration was accounted for by the interactions between the parameters and the overall effect of most individual parameters was dominated by their interactions with other parameters (Fig. 14.3). Interestingly, the contribution of the interactions was twofold higher under low N (averaging 34%) than under high N (averaging 16%) supply. This result indicates that the expression of the effect of a trait at the crop level depends on the value of the other traits, but also on crop management. The direct implication of this result for plant breeding is that we need to select for combinations of traits (ideotypes), rather than for a single trait.

BOX 14.2 (*cont.*)

When the total sensitivity index (first-order effect plus interaction effects) of the parameters is considered, for each site/N supply/model output, 90% of the sum the total sensitivity index was accounted for by only 6 to 17 parameters (Fig. 14.3). The low number of influential parameters with respect to grain yield and protein concentration may reflect the many compensatory effects between traits (e.g. grain size vs grain number, or light saturated photosynthesis vs leaf surface area) and the fact that complex characters such as grain yield and protein concentration are inherently determined at the population level rather than at the organ or plant level (Sinclair et al., 2004).

Under non-limiting N supply a few influential parameters with respect to grain yield could be identified (e.g. radiation-use efficiency, potential duration of grain filling or phyllochron); however, under limiting N more than 10 parameters showed equivalent and small effects. All the parameters had opposite effects on grain yield and protein concentration, but leaf and stem N storage capacity appeared as good candidates to shift the negative relationship between grain yield and protein concentration. These results are consistent with both a previous sensitivity analysis of *SiriusQuality*1 (Martre et al., 2007), and experimental results (Shearman et al., 2005; Gaju et al., 2014). Therefore, grain yield and protein concentration are influenced by several traits and processes and the ranking of the influential traits with respect to grain yield and grain protein concentration depends both on the environment and N supply.

Another approach consists of estimating the parameters of interest by a global calibration of the model. For example, in rice, the four parameters of the *EcoMeristem* model that had the most effect on simulation outputs were calibrated in more than 200 genotypes using a hybrid optimization approach (Box 14.3; Luquet et al., 2012b). The other model parameters were fixed at rice-specific values, either because they are known to have low variation or because their variation has low impact on simulated growth.

3.4 Robust simulation of genotype × environment interactions

The foremost expectation of ecophysiological models is the simulation of genotypic variation in different environments and to capture G × E × M interactions (Box 14.1) (Asseng et al., 2002; Boote et al., 2003; Chapman et al., 2003; Hoogenboom and White, 2003). However, despite promising results from the work mentioned above, three main shortcomings remain.

First, shortcomings related to the proper characterization of the environment (Chapter 13). The record of climatic inputs and soil parameters is crucial to (1) classify the environments experienced by crops and (2) identify and quantify the main environmental factors affecting genotypic expression of traits of interest. To study the response of specific traits in detail, high-throughput phenotyping platforms have recently been established in greenhouses or in the field, allowing fine environmental monitoring (e.g. Granier et al., 2006; Furbank and Tester, 2011). Furthermore, progress is being made in the phenotyping of physiologically meaningful, yield-related traits that are relevant to breeding (Chapter 15).

Second, shortcomings concerning ecophysiological models themselves and the prediction of G × E interactions under a wide range of conditions. Typically, models must be improved to

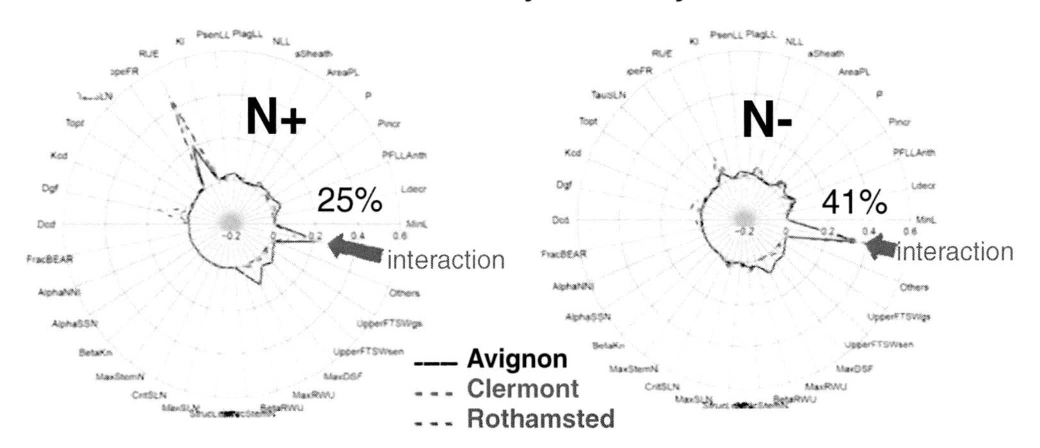
1st order sensitivity index for yield
N+
25%
interaction
N-
41%
interaction
Avignon
Clermont
Rothamsted

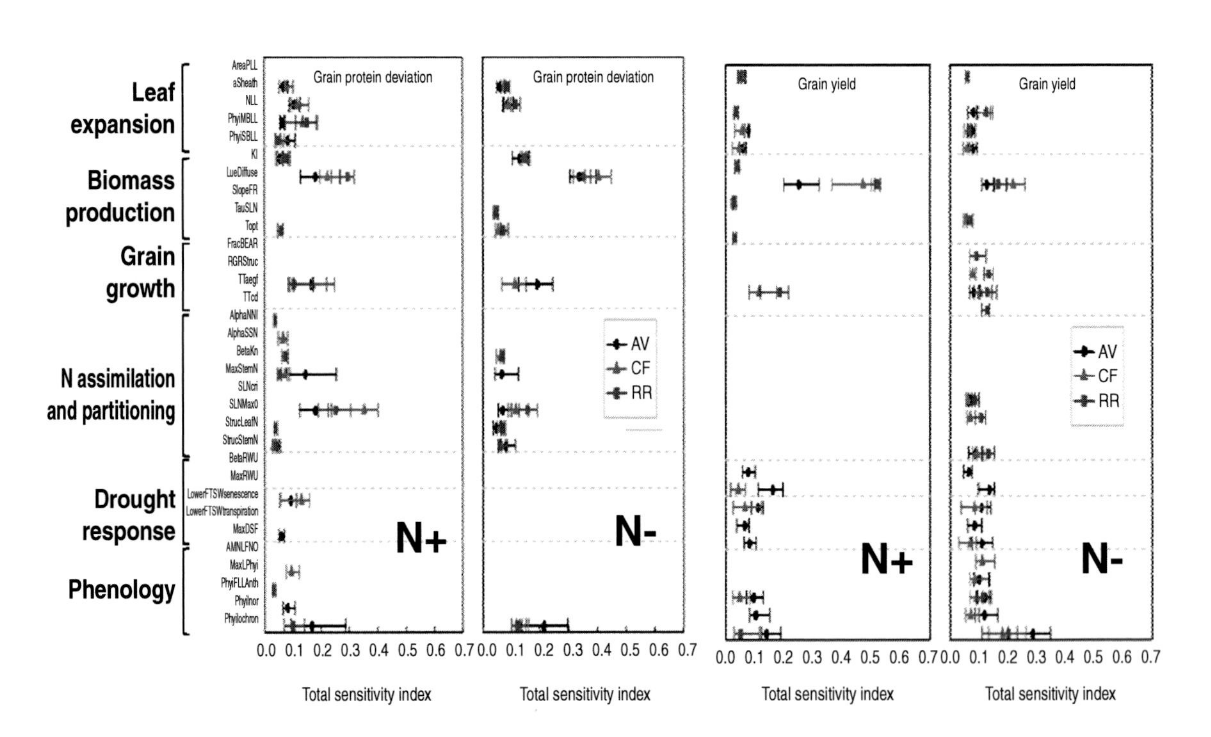
Total sensitivity index for grain protein
Total sensitivity index for yield
Leaf expansion
Biomass production
Grain growth
N assimilation and partitioning
Drought response
Phenology
Grain protein deviation
Grain yield
AV
CF
RR
N+
N-
Total sensitivity index

FIG. 14.3 Global sensitivity analysis of the 41 most influential parameters of the wheat ecophysiological *SiriusQuality2* calculated at three sites in France and the UK for 40 years of weather data (1970–2009) and with high (N+) and low (N−) nitrogen supply. A parameter is considered as influential with respect to a given trait if its sensitivity index is >0.1 for that trait (i.e. if it explains more than 10% of the variance for that trait due the global parameter perturbation). *Top graphs*, radar plots of the median values of the extended Fourier amplitude sensitivity test (E-FAST) first-order sensitivity index (S_i). *Bottom graphs*, medians of the E-FAST total sensitivity index (S_{Ti}). Parameters are grouped according to the submodels (processes) to which they belong (indicated on the left of the graphs). The error bars represent the 25% and 75% percentiles. AV, Avignon, France; CF, Clermont-Ferrand, France; RR, Harpenden, UK. For each site, N treatment, and response variable, only the parameters contributing to 90% of the sum of the total sensitivity index in at least 50% of the years are shown. In the top graphs, 'Others' indicates S_i contributed by the rest of 9 parameters not represented, and 'Interactions' indicates the total sensitivity index contributed by interactions involving the 41 parameters. Parameters are: **AreaPLL** (cm^2 lamina^{-1}), maximum potential surface area of the penultimate leaf lamina; **aSheath** (dimensionless), constant of the quadratic function relating the surface area of leaf sheath between two successive ligules and leaf rank after floral initiation; **NLL** (number of leaf), number of leaves produced after floral initiation; **PhyllMBLL** (dimensionless), potential phyllochronic duration between end of expansion and beginning of senescence for the leaves produced after floral initiation; **PhylSBLL** (dimensionless), potential phyllochronic duration between end of expansion and beginning of senescence for the leaves produced before floral initiation; **Kl** (m^2 (ground) m^{-2} (leaf)), light extinction coefficient; **LueDiffuse** (g (DM) MJ^{-1}), potential radiation-use efficiency (RUE) under overcast conditions; **SlopeFR** (dimensionless), slope of the relationship between RUE and the ratio of diffuse to total solar radiation; **TauSLN** (m^2 (leaf) g^{-1} (N)), relative rate of increase of RUE with specific leaf nitrogen; **Topt** (°C), optimal temperature for RUE; **FracBEAR** (grain g^{-1} (DM)), ratio of grain number to ear dry matter at anthesis; **RGRStruc** ((°Cday) $^{-1}$), relative rate of accumulation of grain structural dry mass; **TTaegf** (°Cday), grain-filling duration (from anthesis to physiological maturity); **TTcd** (°Cday), duration of the endosperm cell division phase; **AlphaNNI** (g (N) g^{-1} (DM)), scaling coefficient of the N dilution curve; **AlphaSSN** (g (N) m^{-2}), scaling coefficient of the allometric relation between area-based:lamina and sheath N content; **BetaKn** (dimensionless), scaling exponent of the relationship between the ratio of nitrogen to light extinction coefficients and the nitrogen nutrition index; **MaxStemN** (g (N) m^{-2}), maximum potential stem N concentration; **SLNcri** (g (N) m^{-2} (leaf)), critical area-based nitrogen content for leaf expansion; **SLNmax0** (g (N) m^{-2} (leaf)), maximum potential specific leaf N of the top leaf layer; **StrucLeafN** (g (N) g^{-1} (DM)), structural N concentration of the leaves; **StrucStemN** (m (N) g^{-1} (DM)), structural N concentration of the true stem; **BetaRWU** (dimensionless), efficiency of the root system to extract water through the vertical soil profile; **MaxRWU** (d^{-1}), maximum relative rate of root water uptake from the top soil layer; **LowerFTSWsenescence** (dimensionless), fraction of transpirable soil water value for which DSFmax is reached; **LowerFTSWtranspiration** (dimensionless), fraction of transpirable soil water for which the stomatal conductance equals zero; **MaxDSF** (dimensionless), maximum rate of acceleration of leaf senescence in response to soil water deficit; **AMNLFNO** (number of leaf), absolute minimum possible leaf number; **MaxLPhyll** (number of leaf), leaf number above which P is increased by PhyllIncr; **PhyllFLLAnth** (dimensionless), phyllochronic duration of the period between flag leaf ligule appearance and anthesis; **PhyllIncr** (dimensionless), factor increasing the phyllochron for leaf number higher than PhyllIncr; **Phyllochron** (°Cday leaf^{-1}), phyllochron.

(1) simulate phenotypic plasticity for key traits and trait combinations (Dingkuhn et al., 2005; Hammer et al., 2005) in a wide range of environments (Rötter et al., 2011; Lobell et al., 2012; Asseng et al., 2013); and (2) give a representative description of physiological processes and be mechanistic enough to bridge the gap between complex traits and genes. This implies that models must account for interconnections and feedback regulations among subsystem components (e.g. organs or tissues), biological processes (e.g. photosynthesis or protein synthesis) and with the environment (e.g. see review by Bertin et al., 2010). While progress is made in this direction, we are still far from understanding all these interactions.

Third, shortcomings related to the characterization of the genetic controls: (1) parameters are not always directly measureable (e.g. root characteristics in a 'natural' environment); (2) quantifying the genetic parameters for a large number of genotypes remains limiting; (3) conventional QTL analyses are typically performed in biparental populations, which tremendously restricts the genetic variation explored, and the analyses rely on the recombination events taking

BOX 14.3

MODEL-ASSISTED PHENOTYPING

The ecophysiological model *EcoMeristem* (Luquet et al., 2006) aims at simulating plant morphogenesis in crop stands and its phenotypic plasticity, in terms of leaf size, appearance, growth and senescence rates and tillering, depending on genotypic and environmental characteristics. Genotypes are defined by sets of parameters of equations formalizing morphogenetic (organ appearance rate, dimensioning at initiation time, expansion, and tillering), physiological processes (light interception and conversion efficiencies) and their regulation by plant state variables defined as the ratio between plant supply and demand for water and carbohydrates that result from genotypic characteristics and environmental conditions. The model was validated in its capacity to represent rice phenotypic plasticity and diversity for the above mentioned morphogenetic processes underling rice early vigor. To evaluate the genetic variability of parameters controlling rice early vigor in *EcoMeristem* and their responses to drought, the model was calibrated for 200 accessions of an *Oryza sativa* L. ssp. *Japonica* diversity panel grown in a greenhouse (Luquet et al., 2012).

Nine parameters (Fig. 14.4) were calibrated using the R package Genoud (Sekhon and Mebane, 2011) to minimize model error for shoot dry mass, surface area of the youngest expanded leaf, leaf and tiller number at the onset and end of a dry down period applied at leaf 6 stage on the main stem until the fraction of transpirable soil water (FTSW) reached a value of 0.2. Inputs were daily weather (incident global radiation, air temperature, potential evapotranspiration), soil volume, and soil drained upper limit and crop lower limit. Parameters were calibrated for three independent experiments.

Most of the calibrated parameters strongly discriminated among genotypes in terms of early vigor and response to drought. However, drought response parameters showed much less heritability compared to morphogenetic, constitutive parameters. Based on correlation and PCA (Fig. 14.4), parameters clustered the collection of accessions in three groups: vigorous but with low drought tolerance; low vigor and good drought tolerance, and a group of intermediate genotypes. This clustering was consistent with that performed based on morphogenetic and metabolic measurements on a subpanel of 50 genotypes of the same collection (Rebolledo et al., 2012). Therefore, this approach confirmed that early vigor and drought tolerance are physiologically linked, but also possibly genetically (negatively), which may have implications when selecting for both traits.

The ranges of parameter values explored accordingly can be considered as representative of existing genetic diversity for the studied species. This information can be used for genome-wide association studies, to explore trait combinations maximizing plant performance, in terms of early vigor as in the present study, but also in terms of early vigor impact on final grain yield, in target environments. In the case of rice presented here, preliminary results suggest that higher biomass accumulation would be possible by overcoming the antagonism between early vigor and drought tolerance.

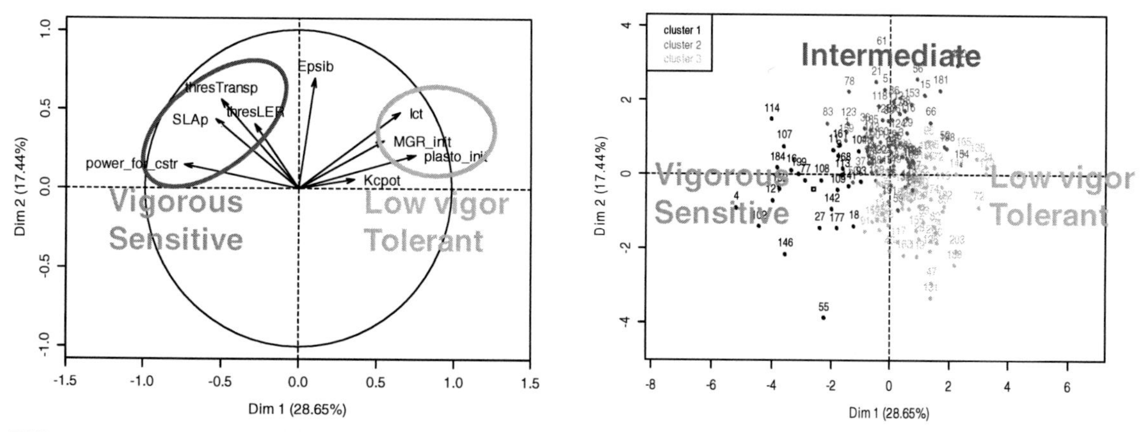

FIG. 14.4 Principal component analysis (PCA) of nine parameters of the ecophysiological model *EcoMeristem* estimated for 200 genotypes of a rice (*Oryza sativa* L. ssp. *Japonica*) diversity panel. Left, projection of the model parameters on the first two axes of the PCA. Right, projection of the genotypes on the first two axes of the PCA. A hierarchical cluster analysis on principal components revealed three groups of genotypes with contrasted early vigor and sensitivity to water deficit. **Epsib**, light-use efficiency; **SLAp**, slope of the logarithmic relation between specific leaf area and its rank on a given tiller; **power_for_cstr**, parameter for reducing Epsib in response to water deficit estimated by f the fraction of transpirable soil water (FTSW); **thresTransp** and **thresLER**, thresholdFTSW at which leaf transpiration and expansion rates, respectively, start decreasing linearly with drying soil; **Ict**, plant C supply to demand ratio enabling tillering; **MGR**, meristem growth rate; **Kc-pot**, additive parameter defining the potential length of leaf n depending on the length of leaf $n-1$. **Plasto_init**, phyllochron (°Cd). *Adapted from Luquet et al. (2012a).*

place in the F1 generation; (4) most genetic analyses performed do not adequately describe the epistatic and pleiotropic effects of the loci.

3.5 New technologies and their potential for gene-to-phenotype modeling

Technological progresses may offer solutions to some of the drawbacks mentioned above. In terms of molecular genetics, Chapter 18 outlines the recent progress in technologies. Briefly, with progress in DNA marker assays and sequencing technologies, it is now feasible to genotype thousands of plants with high densities of markers. In addition, approaches have been developed to analyze the genetic control of quantitative traits, such as association mapping, nested association mapping (e.g. Yu et al., 2006; Brachi et al., 2010) and multiparent advanced generation intercross (MAGIC) populations (Cavanagh et al., 2008). They potentially address the major limitations of available mapping resources. At the same time, new statistical methods have been developed to detect QTL involved in response curves ('functional mapping'). For example, Ma et al. (2002) combined logistic growth curves and QTL mapping within a mixed model approach, which proved to be powerful to estimate accurately QTL effects and positions (Wu et al., 2002, 2003). Using a similar framework, Malosetti et al. (2006) proposed a non-linear extension of classical mixed models. van Eeuwijk et al. (2010) reviewed advanced statistical methods, e.g. to perform multienvironment and/or multitrait QTL mapping.

In terms of phenotyping, field and controlled-conditions platforms (Pieruschka, and Poorter, 2012; White et al., 2012; Fiorani and Schurr, 2013) and networks of field experiments (Hammer et al., 2006; Tardieu and Tuberosa, 2010; Messina et al., 2011; Rebetzke et al., 2013) partly answer the need to characterize precisely relevant traits in contrasting environments. Models are also used to identify relevant traits to phenotype (e.g. Reymond et al., 2004; Martre et al., 2007) and to choose field locations or environmental conditions to impose (e.g. Chenu et al., 2013; Chapter 13).

Concerning model development, the strategy to search for physiological processes that are stable across environments has already turned out promising. Incorporating such processes into ecophysiological models that account for interactions at the crop level has been successfully achieved in the APSIM-Maize model with functions describing leaf expansion rate (Fig. 14.2d) that can be linked to QTL effects.

3.6 Deciphering QTL × environment interactions

The understanding of gene/QTL × E interactions is central to improve fine construction of adapted genotypes. Ecophysiological models have a place of choice to highlight QTL × E interactions, as they facilitate the interpretation of the G × E interactions through environmental characterization (Chapman, 2008; Chenu et al., 2011; Chapter 13) and are useful to dissect complex traits into component traits with higher heritability (Tardieu and Tuberosa, 2010; van Eeuwijk et al., 2010). In addition, 'gene-to-phenotype' models simulate QTL × E interactions, and allow interpretation within a genotype–environment–management framework for the target agricultural system. In this direction, Chenu et al. (2009) develop a 'gene-to-phenotype' modeling approach (Fig. 14.2), introducing organ-growth QTL involved in leaf elongation into the APSIM-Maize model. Their simulations highlighted the importance of the genetic architecture and G × E interactions when assessing the QTL impact on yield (Chenu et al., 2009) as, for example, QTL could have positive or negative impact on yield depending on the pattern of drought.

3.7 Link with breeding

The enormous number of combinations that breeders would ideally analyze to identify best-adapted genotypes highlights a major interest for predictive approaches. Over the last decades, top-down approaches from whole-plant phenotypes to the molecular genomic level (Hammer et al., 2004) have been developed to simulate gene-to-phenotype associations for traits such as plant phenology (e.g. Hoogenboom et al., 2004; Messina et al., 2006) or responses of plant growth and architecture to environment (Tardieu, 2003; Yin et al., 2000; Hammer et al., 2006). While limitations remain, gene-to-phenotype models are being used in breeding (Messina et al., 2011). In addition, robust gene-to-phenotype models linked to quantitative genetics models like QU-GENE (Podlich and Cooper, 1998), give new opportunities to explore alternative selection methods and assist breeding for complex traits in broad or specific environments (Chapman et al., 2003).

4 TOOLS FOR OPTIMIZING TRAIT COMBINATIONS AND MODEL-BASED IDEOTYPING

To meet the demand for multiobjective attributes, the critical question is how to design best combinations of genetic resources and cultural practices adapted to, and respectful of specific environments. In other terms, the question is 'How to optimize the strong G × E × M interactions to design plant ideotypes that meet multicriteria objectives?' This requires integrating two old coexisting visions: the breeder's view of optimizing G × E interactions, and the agronomist's view doing the same for G × M interactions (Messina et al., 2009). The approach relies on the potentialities of integrating genetic information into ecophysiological models to capture G × E × M interactions. The combination of genetic parameters (fingerprint of the genotype), and cultural practices can then be optimized to design new genotypes coupled with adequate management, adapted to target environments (e.g. Hammer et al., 2006; Letort et al., 2008).

The design of ideotypes is usually based on antagonistic criteria with respect to strong constraints (biological, economical, ecological, or environmental). The resulting fitness landscapes are often very complex, and researchers initially applied techniques such as trial and error (Haverkort and Grashoff, 2004; Herndl et al., 2007) or sensitivity analysis (Habekotte, 1997) to identify potential ideotypes. The high number of combinations to identify best-adapted genotypes highlights the impossibility of exploring exhaustively the whole G × E × M space this way (Messina et al., 2009). Overall, the model-based design of ideotypes is a very difficult non-linear multiobjective optimization problem that resists the classical simulation and optimization methods.

Effective optimization methods have been recently proposed to resolve this problem and bio-inspired optimization algorithms (e.g. genetic algorithms, particle swarm optimization algorithms) are increasingly used for model-based design of ideotypes (Letort et al., 2008; Qi et al., 2010; Kadrani et al., 2012; Quilot-Turion et al., 2012; Semenov and Stratonovitch, 2013; Semenov et al., 2014), or optimization of management scenarios (Grechi et al., 2012). Multi-objective evolutionary optimization algorithms allow exploration of high-dimension solution spaces in a reasonable computation time. These methods do not require any derivative information and can address the complex multi-objective optimization problems (e.g. very large search spaces, uncertainty, noise, disjoint Pareto curves, etc.) that resist traditional optimization methods. These methods provide the decision maker with a set of diversified solutions with reduced, but sufficient, cardinality. The decision maker will thus have the final choice of the best suited trade-off between criteria and will be provided with the corresponding optimal ideotypes or management practices.

Despite recent advances in the use of optimization methods applied to model-based approaches, their application to plant and crop design and to management is still in its infancy. Some of the studies focused on the optimization of the management scenarios for designing sustainable and integrated production systems, while others dealt with the optimization of morphological or physiological traits taking advantage of complex interactions between traits in different environments. Others stepped further to the optimization of the allelic variations using detailed genetic information allele effects and interactions. Finally, a step further is also to optimize selection strategies to help the creation of optimized genotypes.

4.1 Optimization of cultural practices

Ould-Sidi and Lescourret (2011) reviewed model-based design of integrated production systems. They discussed many examples using simulation and optimization techniques and recommended the use of integrative modeling platforms, process-oriented modeling and object/component-oriented techniques to improve the genericity, modularity, and re-use of ecophysiological models and data sharing. They also suggested using ecophysiological models that are spatially explicit and employing multi-objective optimization algorithms based on the Pareto dominance concept.

Two recent studies using a model-based approach coupled with optimization algorithms have to be mentioned because of their relevance to management decisions. Grechi et al. (2012) interfaced a process-based model with a multi-objective optimization algorithm to design new management scenarios in line with integrated production systems objectives. The model describes the interacting peach–aphid dynamics, fruit production and fruit quality, and their control by (1) cultural practices, (2) release of the biological control agent *Harmonia axyridis Pallas* (Coleptera: Coccinellidae), and (3) insecticide applications. The simulations were performed for three virtual farmer profiles differing in the relative importance given to each of the performance criteria. Four pest management strategies namely 'no-treatment', 'conventional' (insecticide-based), 'organic', and 'integrated' were simulated. Simulations showed that agronomic performances were largely explained by cultural practices, while aphid pressure was largely explained by pest control strategies. The scenarios using the 'conventional' pest control strategy were the best, regardless of the farmer profile. However, under the 'no-treatment' strategy (only cultural practices), the resulting scenarios have very good performance except for aphid infestation criterion. Optimal values of cultural management variables displayed a high variability between the farmer production profiles under the 'no-treatment' and 'integrated' strategies, while they were independent of the farmer profile under the 'organic' and 'conventional' strategies.

Wu et al. (2012) proposed an optimal control method for solving a water supply problem for optimal sunflower fruit filling. Using a sunflower ecophysiological model, they compared the numerical solutions, obtained through an iterative optimization gradient-based procedure, and those obtained using genetic algorithms in previous studies (Wu et al., 2005). The authors stated that further improvements in sunflower yield have been found using the optimal control method.

4.2 Optimization of genetic parameters

This approach lies in finding combinations of values of the genetic parameters that best satisfy the fixed objectives. Such a set of values would form the 'parametric phenotype' or genotype 'fingerprint' of the ideotype. Only a few studies presented a proof of concept of the approach linking ecophysiological models and optimization methods (Letort et al., 2008; Qi et al., 2010; Quilot-Turion et al., 2012; Semenov and Stratonovitch, 2013; Semenov et al., 2014).

In a theoretical study, Letort et al. (2008) used a particle swarm optimization algorithm to optimize the cob sink strength and the coefficients of the cob sink variation function of the GreenLab-Maize to maximize either the cob weight (single objective problem) or both the cob weight and leaf and stem weight (multiobjective problem) for multiusage maize. Their results illustrated the power of such a combined approach to improve the design of ideotypes, especially in case of trade-off between traits. However, a limitation to the use of particle swarm algorithms is their weak exploitation ability and diversity (local search) keeping within the swarm (population of solutions). The consequence is usually a poor choice for the decision maker.

Quilot-Turion et al. (2012) used the Virtual Fruit model (Génard et al., 2010) to design peach genotypes with enhanced fruit quality and resistance to brown rot for given cultural scenarios. They focused on six genetic parameters combined to create the genotypes. Three traits simulated by the model, of major importance for fruit quality and sensitivity to brown rot, were taken into account to evaluate the genotypes. Simulations were performed for four cultural scenarios (two levels of crop load and two water regimens) to analyze the putative impact of cultural practices on the optimized solutions. The authors used NSGA-II (non-dominated sorting genetic algorithm II) with a modified stopping criterion based on the crowding distance (mechanism of diversity control used by the NSGA-II) to speed up the computation (Ould-Sidi et al., 2012). The modified algorithm obtained solutions similar in quality to those of the original version but after significantly fewer generations. The resulting reduction in computational time for the optimization provides opportunities for further studies (Kadrani et al., 2012, 2013).

The results confirmed the strong antagonism between the criteria considered. Large fruits had a weak sweetness and high crack density and for a given mass, those with improved sweetness had higher crack density. In a current breeding scheme, fruit mass would be the only criterion considered but alternative schemes could be considered in the future, favoring organoleptic quality or environment-friendly practices. In those cases, some interesting optimized solutions were identified (Quilot-Turion et al., 2012).

A main weakness of this approach lies in the lack of quantitative relationships between genes and model parameters. It only provides a picture of the optimized space of solutions considering the functioning of the system driven by biophysical constraints only. The suggested solutions represent ideal genotypes the breeder is not sure to be able to create. To produce more realistic genotypes, genetic constraints (e.g. pleiotropic and epistatic effects, gene × environment interactions) may be integrated to the optimization scheme.

4.3 Optimization of allelic combinations

The difficult issue that remains to be solved for a complete optimization of G × E × M interactions is the integration in the optimization scheme of the complex genetic architecture controlling the model parameters. Two options can be considered. First, known genetic constraints may be included in the definition of the space of parameter variation to be explored during the optimization. Discontinuous space and links between parameters could be added as constraints of optimization at this step. Then, allelic combinations may be inferred from the parameter values. Letort et al. (2008) tested this option for a virtual maize mapping population built from a simple genetic model (virtual genes and virtual chromosomes). They implemented a genetic algorithm to optimize the allelic combination that maximized cob weight. The use of ecophysiological models to help breeding by optimization has not been stepped over since this theoretical study.

The second option is the direct optimization of allelic combinations. This option requires that genetic information is combined to ecophysiological models and may allow consideration of complex genetic models. Some steps have already been made towards this integration and the resulting models are referred to as gene-to-phenotype or QTL-based models (section 3). Chenu et al. (2009) used such a gene-to-phenotype model to explore G × E interactions and the complexity of the results highlighted the importance of genetic architecture in the generation of phenotypes (Fig. 14.2). Considerable work is needed to introduce complex genetic architecture in gene-to-phenotype models and to evaluate their predictive capacity against empirical data. However, more than targeting an exact prediction of the phenotypic value of specific genes, these approaches are designed to help understand the dynamic nature of the gene-to-phenotype problem and explore genotype–environment systems for defining priorities in model development, trait to focus on, etc.

4.4 Towards virtual breeding

To conduct the approach towards its final goal, the next step is to develop a method to optimize the selection schemes necessary to obtain plants as analogous as possible to the ideotypes. The idea is to adjust the selection strategies (e.g. frequency of different alleles in the population, number of selection cycles) in contrasting environments and cultural practices, while considering the genetic diversity and available germplasm. A significant step towards this goal was made by Messina et al. (2011), who developed an integrated method based on round-trip between modeling and experiment. They used the APSIM-Maize model (Keating et al., 2003) coupled to the breeding model QU-GENE (Podlich and Cooper, 1998; Podlich et al., 2004) to reveal interesting trajectories in maize breeding. They included in the model genetic variation for five adaptive traits using an additive genetic model based on three genes and two alleles per locus. Maize phenotypes were simulated for 16 contrasting environments and for 50 years of weather.

For each environment type and management, reciprocal recurrent selection was simulated using QU-GENE. This study demonstrated the validity of the method to explore the G × E × M space, to integrate and apply plant physiology concepts to plant breeding and the value of leveraging this knowledge to develop improved crops. With progress in crop physiology, genomics, genetics, model development and optimization, such integrated approaches are evolving to increase breeding in efficiency.

5 FUTURE PROSPECTS

Ecophysiological models are powerful tools to understand G × E × M interactions (e.g. Hammer et al., 2010), identify key traits of interest for target environments (e.g. Manschadi et al., 2006; Martre et al., 2007; He et al., 2010; Messina et al., 2011; Veyradier et al., 2013), ease ideotype design (e.g. Quilot-Turion et al., 2012; Semenov and Stratonovitch, 2013) and develop adaptive strategies to cope with climate change (e.g. Asseng et al., 2011; Singh et al., 2012; Zheng et al., 2012; Semenov et al., 2014; Chapter 20).

Preliminary studies reveal the potential of model-based approaches in optimizing allelic combinations of genotypes and cultural practices. This would lead to defining an ideotype adapted to a given environment. Then, based on the available genetic material, a second optimization step (virtual breeding) could help identify better selection strategies leading to new varieties resembling the identified ideotype. This modeling framework has potential to integrate information to predict the potential behavior of a genotype in a given environment under the effect of management practices. Linking such models to multiobjective optimization methods can then be suitable to design innovative ideotypes that would optimize the genotypes and the cultural practices in a given environment (Jeuffroy et al., 2013).

Nevertheless, several scientific and technical challenges need to be overcome, and resulting ideotypes require experimental test (Andrivon et al., 2013). It appears essential to solve the question of the structure (e.g. interactions between processes in relation to the action of environmental factor on the processes) of ecophysiological models needed to link phenotype and genotype. The domain of validity and realism of the models in terms of modeled processes and their interactions also need further analyses (Rötter et al., 2011; Boote et al., 2013). Compensation of error for processes or interactions between processes (e.g. feedbacks) or parameters that are not taken into account in the model but contribute to the observed phenotype in given environments also need to be solved (Challinor et al., 2014). Moreover considerable efforts are needed to develop robust links between genetic and physiological determinants and the variation of traits relevant to breeders, but we believe that this framework will soon support agronomists and breeders in the definition and creation of improved varieties and sustainable systems that will answer future needs.

References

Aguirrezábal, L., Martre, P., Pereyra-Irujo, G., Izquierdo, N., Allard, V., 2009. Management and breeding strategies for the improvement of grain and oil quality. In: Sadras, V.O., Calderini, D.F. (Eds.), Crop physiology. Applications for genetic improvement and agronomy. Academic Press, San Diego, pp. 387–421.

Andrivon, D., Giorgetti, C., Baranger, A., et al., 2013. Defining and designing plant architectural ideotypes to control epidemics? Eur. J. Plant Pathol. 135, 611–617.

Asseng, S., Ewert, F., Rosenzweig, C., et al., 2013. Uncertainty in simulating wheat yields under climate change. Nature Clim. Change 3, 827–832.

Asseng, S., Foster, I.A.N., Turner, N.C., 2011. The impact of temperature variability on wheat yields. Glob. Change Biol. 17, 997–1012.

Asseng, S., Turner, N.C., Ray, J.D., Keating, B.A., 2002. A simulation analysis that predicts the influence of physiological traits on the potential yield of wheat. Eur. J. Agron. 17, 123–141.

Bertin, N., Martre, P., Genard, M., Quilot, B., Salon, C., 2010. Under what circumstances can process-based simulation models link genotype to phenotype for complex traits? Case study of fruit and grain quality traits. J. Exp. Bot. 61, 955–967.

Boote, K.J., Jones, J.W., Batchelor, W.D., Nafziger, E.D., Myers, O., 2003. Genetic coefficients in the CROPGRO–Soybean model: Links to field performance and genomics. Agron. J. 95, 32–51.

Boote, K.J., Jones, J.W., White, J.W., Asseng, S., Lizaso, J.I., 2013. Putting mechanisms into crop production models. Plant Cell Environ. 36, 1658–1672.

Boote, K.J., Kropff, M.J., Bindraban, P.S., 2001. Physiology and modelling of traits in crop plants: implications for genetic improvement. Agric. Syst. 70, 395–420.

Brachi, B., Faure, N., Horton, M., et al., 2010. Linkage and association mapping of *Arabidopsis thaliana* flowering time in nature. Plos Genet. 6, e1000940.

Braun, H.J., Pfeiffer, W.H., Pollmer, W.G., 1992. Environments for selecting widely adapted spring wheats. Crop Sci. 22, 1420–1427.

Brisson, N., Gate, P., Gouache, D., Charmet, G., Oury, F.-X., Huard, F., 2010. Why are wheat yields stagnating in Europe? A comprehensive data analysis for France. Field Crops Res. 119, 201–212.

Brunel, S., Teulat-Merah, B., Wagner, M.H., Huguet, T., Prosperi, J.M., Dürr, C., 2009. Using a model-based framework for analysing genetic diversity during germination and heterotrophic growth of *Medicago truncatula*. Ann Bot. 103, 1103–1117.

Cariboni, J., Gatelli, D., Liska, R., Saltelli, A., 2007. The role of sensitivity analysis in ecological modelling. Ecol. Model. 203, 167–182.

Cavanagh, C., Morell, M., Mackay, I., Powell, W., 2008. From mutations to MAGIC: resources for gene discovery, validation and delivery in crop plants. Curr. Opin. Plant Biol. 11, 215–221.

Ceccarelli, S., 1989. Wide adaptation: how wide? Euphytica 40, 197–205.

Challinor, A., Martre, P., Asseng, A., Ewert, F., Thornton, P.K., 2014. Making the most of climate impacts ensembles. Nat. Clim. Change 4, 77–80.

Chapman, S.C., 2008. Use of crop models to understand genotype by environment interactions for drought in real-world and simulated plant breeding trials. Euphytica 161, 195–208.

Chapman, S., Cooper, M., Podlich, D., Hammer, G., 2003. Evaluating plant breeding strategies by simulating gene action and dryland environment effects. Agron. J. 95, 99–113.

Chenu, K., Chapman, S.C., Hammer, G.L., McLean, G., Ben-Haj-Salah, H., Tardieu, F., 2008. Short-term responses of leaf growth rate to water deficit scale up to whole-plant and crop levels: an integrated modelling approach in maize. Plant Cell Environ. 31, 378–391.

Chenu, K., Chapman, S., Tardieu, F., McLean, G., Welcker, C., Hammer, G., 2009. Simulating the yield impacts of organ-level quantitative trait loci associated with drought response in maize: a "gene-to-phenotype" modeling approach. Genetics 183, 1507–1523.

Chenu, K., Cooper, M., Hammer, G.L., Mathews, K.L., Dreccer, M.F., Chapman, S.C., 2011. Environment characterization as an aid to wheat improvement: interpreting genotype–environment interactions by modelling water-deficit patterns in North-Eastern Australia. J. Exp. Bot. 62, 1743–1755.

Chenu, K., Deihimfard, R., Chapman, S.C., 2013. Large-scale characterization of drought pattern: a continent-wide modelling approach applied to the Australian wheatbelt spatial and temporal trends. New Phytol. 198, 801–820.

Chew, Y.H., Wilczek, A.M., Williams, M., Welch, S.M., Schmitt, J., Halliday, K.J., 2012. An augmented Arabidopsis phenology model reveals seasonal temperature control of flowering time. New Phytol. 194, 654–665.

Cilas, C., Bar-Hen, A., Montagnon, C., Godin, C., 2006. Definition of architectural ideotypes for good yield capacity in *Coffea canephora*. Ann. Bot. 97, 405–411.

Cooper, M., Woodruff, D.R., Phillips, I.G., Basford, K.E., Gilmour, A.R., 2001. Genotype-by-management interactions for grain yield and grain protein concentration of wheat. Field Crops Res. 69, 47–67.

Desanlis, M., Aubertot, J.N., Mestries, E., Debaeke, P., 2013. Analysis of the influence of a sunflower canopy on *Phomopsis helianthi* epidemics as a function of cropping practices. Field Crops Res. 149, 63–75.

Dingkuhn, M., Luquet, D., Clément-Vidal, A., Tambour, L., Kim, H.K., Song, Y.H., 2007. Is plant growth driven by sink regulation? Implications for crop models, phenotyping approaches and ideotypes. In: Spiertz, J.H.J., Struik, P.C., van Laar, H.H. (Eds.), Scale and complexity in plant systems research: gene-plant-crop relations. Springer, Dordrecht, The Netherlands, pp. 157–170.

Dingkuhn, M., Luquet, D., Quilot, B., de Reffye, P., 2005. Environmental and genetic control of morphogenesis in crops: towards models simulating phenotypic plasticity. Aust. J. Agric. Res. 56, 1289–1302.

Dickmann, D.I., Gold, M.A., Flore, J.A., 1994. The ideotype concept and the genetic improvement of tree crops. Plant Breed. Rev. 12, 163–193.

Donald, C.M., 1968. The breeding of crop ideotype. Euphytica 17, 385–403.

Duvick, D.N., Smith, J.S.C., Cooper, M., 2004. Long-term selection in a commercial hybrid maize breeding program. Plant Breed. Rev. 24, 109–151.

Edmeades, G.O., McMaster, G.S., White, J.W., Campos, H., 2004. Genomics and the physiologist: bridging the gap between genes and crop response. Field Crops Res. 90, 5–18.

Farnsworth, K.D., Niklas, K.J., 1995. Theories of optimization, form and function in branching architecture in plants. Funct. Ecol. 9, 355–363.

Ferrise, R., Triossi, A., Stratonovitch, P., Bindi, M., Martre, P., 2010. Sowing date and nitrogen fertilisation effects on dry matter and nitrogen dynamics for durum wheat: An experimental and simulation study. Field Crops Res. 117, 245–257.

Fiorani, F., Schurr, U., 2013. Future scenarios for plant phenotyping. Annu. Rev. Plant Biol. 64, 267–291.

Foulkes, M.J., Slafer, G.A., Davies, W.J., et al., 2011. Raising yield potential of wheat. III. Optimizing partitioning to grain while maintaining lodging resistance. J. Exp. Bot. 62, 469–486.

Furbank, R.T., Tester, M., 2011. Phenomics – technologies to relieve the phenotyping bottleneck. Trends Plant Sci. 16, 635–644.

Gaju, O., Allard, V., Martre, P., et al., 2014. Nitrogen partitioning and remobilization in relation to leaf senescence, grain yield and grain nitrogen concentration in wheat cultivars. Field Crops Res. 155, 213–223.

Génard, M., Bertin, N., Borel, C., et al., 2007. Towards a virtual fruit focusing on quality: modelling features and potential uses. J. Exp. Bot. 58, 917–928.

Génard, M., Bertin, N., Gautier, H., Lescourret, F., Quilot, B., 2010. Virtual profiling: a new way to analyse phenotypes. Plant J. 62, 344–355.

Granier, C., Aguirrezabal, L., Chenu, K., et al., 2006. PHENOPSIS, an automated platform for reproducible phenotyping of plant responses to soil water deficit in Arabidopsis thaliana permitted the identification of an accession with low sensitivity to soil water deficit. New Phytol. 169, 623–635.

Grechi, I., Ould-Sidi, M.M., Hilgert, N., Senoussi, R., Sauphanor, B., Lescourret, F., 2012. Designing integrated management scenarios using simulation-based and multi-objective optimization: Application to the peach tree–*Myzus persicae* aphid system. Ecol. Model. 246, 47–59.

Gutjahr, S., Vaksmann, M., Dingkuhn, M., et al., 2013. Grain, sugar and biomass accumulation in tropical sorghums. I. Trade-offs and effects of phenological plasticity. Funct. Plant Biol. 40, 342–354.

Habekotte, B., 1997. Options for increasing seed yield of winter oilseed rape (*Brassica napus* L.): a simulation study. Field Crops Res. 54, 109–126.

Hamblin, J., 1993. The ideotype concept: useful or outdated? In: Buxton, D.R., Shibles, R., Forsberg, A. et al. (Eds.), International crop science I. CSSA, Madison, pp. 589–597.

Hammer, G., Cooper, M., Tardieu, F., et al., 2006. Models for navigating biological complexity in breeding improved crop plants. Trends Plant Sci. 11, 587–593.

Hammer, G.L., Dong, Z., McLean, G., et al., 2009. Can changes in canopy and/or root system architecture explain historical maize yield trends in the U.S. corn belt? Crop Sci. 49, 299–312.

Hammer, G.L., Chapman, S., van Oosterom, E., Podlich, D.W., 2005. Trait physiology and crop modelling as a framework to link phenotypic complexity to underlying genetic systems. Aust. J. Agric. Res. 56, 947–960.

Hammer, G.L., Sinclair, T.R., Chapman, S.C., van Oosterom, E., 2004. On systems thinking, systems biology, and the in silico plant. Plant Physiol. 134, 909–911.

Hammer, G.L., van Oosterom, E.J., McLean, G., et al., 2010. Adapting APSIM to model the physiology and genetics of complex adaptive traits in field crops. J. Exp. Bot. 61, 2185–2202.

Hanocq, E., Jeuffroy, M.H., Lejeune-Hénaut, I., Munier-Jolain, N., 2009. Construire des idéotypes pour des systèmes de culture variés en pois d'hiver. Innovat. Agron. 7, 14–28.

Harrison, M.T., Evans, J.R., Dove, H., Moore, A.D., 2011. Dual-purpose cereals: can the relative influences of management and environment on crop recovery and grain yield be dissected? Crop Past. Sci. 62, 930–946.

Haverkort, A.J., Grashoff, C., 2004. IDEOTYPING-POTATO: a modelling approach to genotype performance. In: MacKerron, D.K.L., Haverkort, A.J. (Eds.), Decision support systems in potato production: bringing models to practice. Wageningen Academic Publishers, Wageningen, pp. 198–211.

He, J., Stratonovitch, P., Allard, V., Semenov, M.A., Martre, P., 2010. Global sensitivity analysis of the process-based wheat simulation model *SiriusQuality*1 identifies key genotypic parameters and unravels parameters interactions. Proc. Soc. Behav. Sci. 2, 7676–7677.

Herndl, M., Shan, C., Wang, P., Graeff, S., Claupein, W., 2007. A model based ideotyping approach for wheat under different environmental conditions in North China plain. Agric. Sci. China 6, 1426–1436.

Hoogenboom, G., White, J.W., 2003. Improving physiological assumptions of simulation models by using gene-based approaches. Agron. J. 95, 82–89.

Hoogenboom, G., White, J.W., Messina, R., 2004. From genome to crop: integration through simulation modeling. Field Crops Res. 90, 145–163.

Jeuffroy, M.H., Casadebaig, P., Debaeke, P., Loyce, C., Meynard, J.M., 2013. Agronomic model uses to predict cultivar performance in various environments and cropping systems. A review. Agron. Sustain. Dev. doi: 10.1007/s13593-013-0170-9.

Kadrani, A., Sidi, M.M.O., Quilot-Turion, B., Génard, M., Lescourret, F., 2012. Particle swarm optimization to design ideotypes for sustainable fruit production systems. Internatl. J. Swarm Intell. Res. 3, 1–19.

Kadrani, A., Ould-Sidi, M.-M., Quilot-Turion, B., Génard, M., Lescourret, F., 2013. Comparison of evolutionary and swarm intelligence-based approaches in the improvement of fruit quality. International Symposium on Operational Research and Applications (ISORAP 2013). Marrakesh, Morocco, May 08-10, 2013.

Karr, J.R., Sanghvi, J.C., Macklin, D.N., et al., 2012. A whole-cell computational model predicts phenotype from genotype. Cell 150, 389–401.

Keurentjes, J.J.B., Angenent, G.C., Dicke, M., et al., 2011. Redefining plant systems biology: from cell to ecosystem. Trends Plant Sci. 16, 183–190.

Keating, B.A., Carberry, P.S., Hammer, G.L., et al., 2003. An overview of APSIM, a model designed for farming systems simulation. Eur. J. Agron. 18, 267–288.

Laperche, A., Devienne-Baret, F., Maury, O., Le Gouis, J., Ney, B., 2006. A simplified conceptual model of carbon/nitrogen functioning for QTL analysis of winter wheat adaptation to nitrogen deficiency. Theor. Appl. Genet. 113, 1131–1146.

Lauri, P.E., Costes, E., 2005. Progress in whole-tree architectural studies for apple cultivar characterization at INRA, France – Contribution to the ideotype approach. Acta Hort. 663, 357–362.

Letort, V., Mahe, P., Cournède, P.H., De Reffye, P., Courtois, B., 2008. Quantitative genetics and functional-structural plant growth models: simulation of quantitative trait loci detection for model parameters and application to potential yield optimization. Ann. Bot. 101, 1243–1254.

Lobell, D.B., Sibley, A., Ortiz-Monasterio, I., 2012. Extreme heat effects on wheat senescence in India. Nat. Clim. Change 2, 186–189.

Luquet, D., Dingkuhn, M., Kim, H.K., Tambour, L., Clément-Vidal, A., 2006. EcoMeristem, a model of morphogenesis and competition among sinks in rice. 1. Concept, validation and sensitivity analysis. Funct. Plant Biol. 33, 309–323.

Luquet, D., Rebolledo, M.C., Soulié, J.C., 2012a. Functional-structural plant modeling to support complex trait phenotyping: Case of rice early vigour and drought tolerance using EcoMeristem model. IEEE International Symposium. 4 (PMA'12). In: Kang, M., Dumont, Y., Guo, Y. (Eds.), Plant growth modeling, simulation, visualization and applications (PMA), 2012 IEEE Fourth International Symposium, Shangai, China, pp. 270–277.

Luquet, D., Soulié, J.C., Rebolledo, M.C., Rouan, L., Clément-Vidal, A., Dingkuhn, M., 2012b. Developmental dynamics and early growth vigour in rice 2. Modelling genetic diversity using EcoMeristem. J. Agron. Crop Sci. 198, 385–398.

Ma, C.X., Casella, G., Wu, R., 2002. Functional mapping of quantitative trait loci underlying the character process: a theoretical framework. Genetics 161, 1751–1762.

Makowski, D., Naud, C., Jeuffroy, M.H., Barbottin, A., Monod, H., 2006. Global sensitivity analysis for calculating the contribution of genetic parameters to the variance of crop model prediction. Reliab. Eng. Syst. Safe. 91, 1142–1147.

Malosetti, M., Visser, R.G.F., Celis-Gamboa, C., Eeuwijk, F.A., 2006. QTL methodology for response curves on the basis of non-linear mixed models, with an illustration to senescence in potato. Theor. Appl. Genet. 113, 288–300.

Manschadi, A.M., Christopher, J., deVoil, P., Hammer, G.L., 2006. The role of root architectural traits in adaptation of wheat to water-limited environments. Funct. Plant Biol. 33, 823–837.

Marshall, D.R., 1991. Alternative approach and perspectives in breeding for higher yields. Field Crops Res. 26, 171–190.

Martre, P., Jamieson, P.D., Semenov, M.A., Zyskowski, R.F., Porter, J.R., Triboi, E., 2006. Modelling protein content and composition in relation to crop nitrogen dynamics for wheat. Eur. J. Agron. 25, 138–154.

Martre, P., Semenov, M.A., Jamieson, P.D., 2007. Simulation analysis of physiological traits to improve yield, nitrogen use efficiency and grain protein concentration in wheat. In: Spiertz, J.H.J., Struik, P.C., Van Laar, H.H. (Eds.), Scale and complexity in plant systems research, gene-plant-crop relations. Springer, Dordrecht, The Netherlands, pp. 181–201.

Messina, C., Hammer, G.L., Dong, Z., Podlich, D., Cooper, M., 2009. Modelling crop improvement in a G × E × M framework via gene-trait-phenotype relationships. In: Sadras, V.O., Calderini, D.F. (Eds.), Crop physiology. Applications for genetic improvement and agronomy. Academic Press, San Diego, pp. 235–265.

Messina, C.D., Jones, J.W., Boote, K.J., Vallejos, C.E., 2006. A gene-based model to simulate soybean development and yield responses to environment. Crop Sci. 46, 456–466.

Messina, C.D., Podlich, D., Dong, Z., Samples, M., Cooper, M., 2011. Yield-trait performance landscapes: from theory to application in breeding maize for drought tolerance. J. Exp. Bot. 62, 855–868.

Mock, J.J., Pearce, R.B., 1975. An ideotype of maize. Euphytica 24, 613–623.

Morris, M.D., 1991. Factorial sampling plans for preliminary computational experiments. Technometrics 33, 161–174.

Ould-Sidi, M.M., Lescourret, F., 2011. Model-based design of integrated production systems: a review. Agron. Sustain. Dev. 31, 571–588.

Ould-Sidi, M.M., Kadrani, A., Quilot-Turion, B., Lescourret, F., Génard, M., 2012. Compromising NSGA-II performances and stopping criteria: case of virtual peach design. The 4th International Conference on Metaheuristics and Nature Inspired Computing (META'2012). October 27-31, 2012, Port Kintaoui, Tunisia.

Oury, F.X., Berard, P., Brancourt-Hulmel, M., et al., 2003. Yield and grain protein concentration in bread wheat: a review and a study of multi-annual data from a French breeding program. J. Genet. Breed. 57, 59–68.

Pedró, A., Savin, R., Slafer, G.A., 2012. Crop productivity as related to single-plant traits at key phenological stages in durum wheat. Field Crops Res. 138, 42–51.

Peng, S., Khush, G.S., Virk, P., Tang, Q., Zou, Y., 2008. Progress in ideotype breeding to increase rice yield potential. Field Crop Res. 108, 32–38.

Pieruschka, R., Poorter, H., 2012. Phenotyping plants: genes, phenes and machines Introduction. Funct. Plant Biol. 39, 813–820.

Podlich, D., Cooper, M., 1998. QU-GENE: a simulation platform for quantitative analysis of genetic models. Bioinformatics 14, 632–653.

Podlich, D.W., Winkler, C.R., Cooper, M., 2004. Mapping as you go: an effective approach for marker-assisted selection of complex traits. Crop Sci. 44, 1560–1571.

Qi, R., Ma, Y., Hu, B., de Reffye, P., Cournede, P.H., 2010. Optimization of source-sink dynamics in plant growth for ideotype breeding: A case study on maize. Comput. Electron. Agr. 71, 96–105.

Quilot, B., Kervella, J., Génard, M., Lescourret, F., 2005. Analysing the genetic control of peach fruit quality through an ecophysiological model combined with a QTL approach. J. Exp. Bot. 56, 3083–3092.

Quilot-Turion, B., Ould-Sidi, M.M., Kadrani, A., Hilgert, N., Génard, M., Lescourret, F., 2012. Optimization of parameters of the 'Virtual Fruit' model to design peach genotype for sustainable production systems. Eur. J. Agron. 42, 34–48.

Rasmusson, D.C., 1987. An evaluation of ideotype breeding. Crop Sci. 27, 1140–1146.

Rasmusson, D.C., 1991. A plant breeder's experience with ideotype breeding. Field Crops Res. 26, 191–200.

Rebetzke, G.J., Chenu, K., Biddulph, B., et al., 2013. A multisite managed environment facility for targeted trait and germplasm phenotyping. Funct. Plant Biol. 40, 1–13.

Rebolledo, M.C., Dingkuhn, M., Clément-Vidal, A., Rouan, L., Dingkuhn, M., 2012. Phenomics of rice early vigour and drought response: Are sugar related and morphogenetic traits relevant? Rice J. 5, 22.

Rebolledo, M.C., Luquet, D., Courtois, B., et al., 2013. Can early vigour occur in combination with drought tolerance and efficient water use in rice genotypes? Funct. Plant Biol. 40, 582–594.

Reymond, M., Muller, B., Leonardi, A., Charcosset, A., Tardieu, F., 2003. Combining quantitative trait loci analysis and an ecophysiological model to analyze the genetic variability of the responses of maize leaf growth to temperature and water deficit. Plant Physiol. 131, 664–675.

Reymond, M., Muller, B., Tardieu, F., 2004. Dealing with the genotype x environment interaction via a modelling approach: a comparison of QTLs of maize leaf length or width with QTL of model parameters. J. Exp. Bot. 55, 2461–2472.

Reynolds, J.F., Acock, B., 1985. Predicting the response of plants to increasing carbon dioxide: A critique of plant growth models. Ecol. Model. 29, 107–129.

Reynolds, M., Manes, Y., Izanloo, A., Langridge, P., 2009. Phenotyping approaches for physiological breeding and gene discovery in wheat. Ann. Appl. Biol. 155, 309–320.

Reynolds, M.P., Rajaram, S., McNab, A. (Eds.), 1996. Increasing yield potential in wheat: breaking the barriers. Proceedings of a Workshop Held in Ciudad Obregón, Sonora, Mexico, 28-30 Mar. Mexico, DF, CIMMYT.

Richards, R.A., Rebetzke, G.J., Watt, M., Condon, A.G., Spielmeyer, W., Dolferus, R., 2010. Breeding for improved water productivity in temperate cereals: phenotyping, quantitative trait loci, markers and the selection environment. Funct. Plant Biol. 37, 85–97.

Rötter, R.P., Carter, T.R., Olesen, J.E., Porter, J.R., 2011. Crop-climate models need an overhaul. Nat. Clim. Change 1, 175–177.

Sanghvi, J.C., Regot, S., Carrasco, S., et al., 2013. Accelerated discovery via a whole-cell model. Nat. Meth. 10, 1192–1195.

Saltelli, A., Chan, K., Scott, E.M., 2000. Sensitivity analysis. John Wiley and Sons, New York.

Saltelli, A., Tarantola, S., Chan, K.P.-S., 1999. A quantitative model-independent method for global sensitivity analysis of model output. Technometrics 41, 39–56.

Sekhon, S., Mebane, W.R., 2011. Genetic optimization using derivatives: The genoud Package for R. J Stat. Softw. 42, 1–26.

Sedgley, R.H., 1991. An appraisal of the Donald ideotype after 21 years. Field Crops Res. 26, 93–112.

Semenov, M.A., Stratonovitch, P., 2013. Designing high-yielding wheat ideotypes for a changing climate. Food Energ. Secur. 2, 185–196.

Semenov, M.A., Stratonovitch, P., Alghabari, F., Gooding, M.J., 2014. Adapting wheat in Europe for climate change. J. Cer. Sci., http://dx.doi.org/10.1016/j.jcs.2014.01.006.

Shearman, V.J., Sylvester-Bradley, R., Scott, R.K., Foulkes, M.J., 2005. Physiological processes associated with wheat yield progress in the UK. Crop Sci. 45, 175–185.

Sinclair, T.R., Purcell, L.C., Sneller, C.H., 2004. Crop transformation and the challenge to increase yield potential. Trends Plant. Sci. 9, 70–75.

Singh, P., Boote, K.J., Kumar, U., Srinivas, K., Nigam, S.N., Jones, J.W., 2012. Evaluation of genetic traits for improving productivity and adaptation of groundnut to climate change in India. J. Agron. Crop Sci. 198, 399–413.

Socias, R., Felipe, A.J., Gómez Aparisi, J., García, J.E., Dicenta, F., 1998. The ideotype concept in almond. Acta Hort. 470, 51–56.

Tardieu, F., 2003. Virtual plants: modelling as a tool for the genomics of tolerance to water deficit. Trends Plant Sci. 8, 9–14.

Tardieu, F., Tuberosa, R., 2010. Dissection and modelling of abiotic stress tolerance in plants. Curr. Opin. Plant Biol. 13, 206–212.

Tardieu, F., 2012. Any trait or trait-related allele can confer drought tolerance: just design the right drought scenario. J. Exp. Bot. 63, 25–31.

ter Steege, M.W., den Ouden, F.M., Lambers, H., Stam, P., Peeters, A.J., 2005. Genetic and physiological architecture of early vigor in *Aegilops tauschii*, the D-genome donor of hexaploid wheat. A quantitative trait loci analysis. Plant Physiol. 139, 1078–1094.

van Eeuwijk, F., Bink, M., Chenu, K., Chapman, S., 2010. Detection and use of QTL for complex traits in multiple environments. Curr. Opin. Plant Biol. 13, 193–205.

Veyradier, M., Christopher, J., Chenu, K., 2013. Quantifying the potential yield benefit of root traits. In: Sievänen, R., Nikinmaa, E., Godin, C., Lintunen, A., Nygren, P. (Eds.), Proceedings of the 7th International Conference on Functional-Structural Plant Models. Saariselkä, Finland, 9-14 June. pp. 317–319.

Welch, S.M., Dong, Z.S., Roe, J.L., Das, S., 2005. Flowering time control: gene network modelling and the link to quantitative genetics. Aust. J. Agric. Res. 56, 919–936.

White, J.W., Andrade-Sanchez, P., Gore, M.A., et al., 2012. Field-based phenomics for plant genetics research. Field Crops Res. 133, 101–112.

Welcker, C., Boussuge, B., Bencivenni, C., Ribaut, J.M., Tardieu, F., 2007. Are source and sink strengths genetically linked in maize plants subjected to water deficit? A QTL study of the responses of leaf growth and of anthesis-silking interval to water deficit. J. Exp. Bot. 58, 339–349.

White, J.W., Hoogenboom, G., 1996. Simulating effects of genes for physiological traits in a process-oriented crop model. Agron. J. 88, 416–442.

White, J.W., Hoogenboom, G., 2003. Gene-based approaches to crop simulation: Past experiences and future opportunities. Agron. J. 95, 52–64.

Wu, L., de Reffye, P., Hu, B.G., Le Dimet, F.X., Cournède, P.H., 2005. A water supply optimization problem for plant growth based on GreenLab model. In: Kamgnia, E., Philippe, B., Slimani, Y. (Eds.), 7ème Colloque Africain sur la Recherche en Informatique. Hammamet, Tunisie, novembre 2004. pp. 194–207.

Wu, L., Le Dimet, F.X., de Reffye, P., Hu, B.G., Cournède, P.H., Kang, M.Z., 2012. An optimal control methodology for plant growth – Case study of a water supply problem of sunflower. Math. Comput. Simulat. 82, 909–923.

Wu, R., Ma, C.X., Zhao, W., Casella, G., 2003. Functional mapping for quantitative trait loci governing growth rates: a parametric model. Physiol. Genomics 14, 241–249.

Wu, W., Zhou, Y., Li, W., Mao, D., Chen, Q., 2002. Mapping of quantitative trait loci based on growth models. Theor. Appl. Genet. 105, 1043–1049.

Yin, X., Struik, P.C., 2008. Applying modelling experiences from the past to shape crop systems biology: the need to converge crop physiology and functional genomics. New Phytol. 179, 629–642.

Yin, X., Chasalow, S.C., Dourleijn, C.J., Stam, P., Kropff, M.J., 2000. Coupling estimated effects of QTL for physiological traits to a crop growth model: predicting yield variation among recombinant inbred lines in barley. Heredity 85, 539–549.

Yin, X., Struik, P.C., Tang, J., Qi, C., Liu, T., 2005. Model analysis of flowering phenology in recombinant inbred lines of barley. J. Exp. Bot. 56, 959–965.

Yu, J.M., Pressoir, G., Briggs, W.H., et al., 2006. A unified mixed-model method for association mapping that accounts for multiple levels of relatedness. Nat. Genet. 38, 203–208.

Zheng, B., Chenu, K., Dreccer, M.F., Chapman, S.C., 2012. Breeding for the future: what are the potential impacts of future frost and heat events on sowing and flowering time requirements for Australian bread wheat (*Triticum aestivium*) varieties? Glob. Change Biol. 18, 2899–2914.

CHAPTER

15

Crop phenotyping for physiological breeding in grain crops: A case study for maize

Maria E. Otegui[1], *Lucas Borrás*[2], *Gustavo A. Maddonni*[1]

[1]IFEVA-CONICET & University of Buenos Aires, Buenos Aries, Argentina

[2]CONICET & University of Rosario, Rosario, Santa Fe, Argentina

1 INTRODUCTION

From the time of Mendel's laws rediscovery to present, grain yield (GY) of cereals has increased steadily. In wheat, this trend was mostly related to the combination of higher nitrogen use and shorter plants, predominantly by means of *Rht* genes (Bell et al., 1995). In maize (*Zea mays* L.), massive adoption of hybrids has been critical, starting with double crosses in the 1930s up to the complete exploitation of heterosis benefits with single crosses in the 1960s (Cooper et al., 2004); Chapter 18 presents an updated view of the genetic basis of heterosis including the new concepts of genetic diversity accounting for copy number variation (CNV) and presence/absence variants (PAVs). Independent of the species, breeding was almost exclusively driven by the selection of yield *per se* plus defensive traits (resistance to diseases, reduced lodging) and seed quality. This approach has been extraordinarily successful up to the present.

So-called 'traditional breeding' has been able to hold genetic gains high enough to satisfy the rise in demand associated with increasing population, and this was more evident for maize than for any other cereal crop (Fischer and Edmeades, 2010; Ray et al., 2013). In their approach to yield improvement, the inclusion of secondary traits has been limited to highly stress-prone environments, where selection for yield *per se* is usually inefficient due to reduced heritability (Rosielle and Hamblin, 1981; Ziyomo and Bernardo, 2013). In the case of maize, public programs like those of CIMMYT (Bänziger et al., 1997), but not commercial breeding, have targeted stressful environments. Bänziger and

Crop Physiology. DOI: 10.1016/B978-0-12-417104-6.00015-7

Lafitte (1997) listed four secondary traits related to stress adaptation in maize: the anthesis–silking interval (ASI), number of ears per plant, chlorophyll concentration index, and leaf senescence index; this short list reflects the restrictive requirements for a secondary trait to be useful. These requirements include high heritability in poor environments, a high correlation with yield, and to be of fast assessment and low cost (Richards, 2006). This original list may also reflect the lack of adequate phenotyping tools and skills at the time it was proposed (Passioura, 2012), which set a limit to the breadth of selected traits and of subsequent comparisons between indirect selection methods based on them (Ziyomo and Bernardo, 2013).

The advent of molecular tools in the 1980s (Chapter 18) did not produce a drastic change in breeding methods until very recently. This includes the era of genetically modified crops that for maize started in 1996 with the release of *Bt* corn in the USA, as well as the broad expansion of marker-assisted selection during the last decade. In spite of early forecasts, the focus of these techniques has remained chiefly on biotic constraints and quality traits (Slafer, 2003; Shi et al., 2013). The inclusion of molecular markers in crop simulation models as actual genetic coefficients for the prediction of the physiological determinants of seed yield (Messina et al., 2006; Chenu et al., 2009; Chapter 14) remains a major challenge. The reasons for this difficult task can be traced to more than a decade ago, when Miflin (2000) alerted on the huge gap between the large amount of plant genetic information (in his own words 'a genocentric view of priorities') that found no correspondence in our understanding of the phenotype. There is a 'phenotype gap' (Brown and Peters, 1996) that must be bridged for the full exploitation of marker-assisted technologies, particularly when breeding for abiotic stress tolerance (Collins et al., 2008; Passioura, 2012; Chapter 19). This requires a shift towards a more 'phenocentric approach' (Miflin, 2000).

Until recently, breeders, molecular geneticists, agronomists and crop physiologists were unaware of the mentioned gap, and relied on their independent capacities for sustaining current increases in cereal production. Gloomy reports, however, warned us of a possibly large imbalance between cereal supply and demand in the near future (Ray et al., 2013). This imbalance will be related not only to climate change predictions, which are not very conclusive (IPCC, 2007), but also to the combination of population growth, improved living standards in developing countries and use of grains for biofuels. Based on this scenario, required yield gains should at least double current values to meet demand estimates by 2050 (Ray et al., 2013); this seems *a priori* an impossible task for wheat and rice (Hall and Richards, 2013). Suddenly, the phenotype gap and the need for effective interdisciplinary studies caught our immediate attention (Trachsel et al., 2011; Masuka et al., 2012; Tuberosa, 2012; White et al., 2012).

In this chapter, we briefly review the physiological determinants of maize yield. We also analyze gains in yield and breeding effects on this trait and its physiological determinants. We discuss field-based phenomics for secondary traits, and the approach to detection of candidate quantitative trait loci (QTL) behind their genetic control. For this purpose, we use maize as a model crop and a general physiological framework for the analysis of yield determination, based on resource capture, resource-use efficiency for biomass production and biomass partitioning to reproductive structures (Gifford et al., 1984; Passioura, 1996). We dissect these major traits into minor components (Lee and Tollenaar, 2007), most of them included in current crop simulation models. Focus is set on trait relationships and the parameters of these relationships, emphasizing the relative importance of multiyear trials. In this process, we move along the organ–plant–crop levels, including the analysis of phenotypic plasticity, genotypic variation and trait heritability (Sadras and Slafer, 2012; D'Andrea et al., 2013).

2 TRAIT DISSECTION OF THE GENERAL PHYSIOLOGICAL MODEL OF GRAIN YIELD DETERMINATION IN MAIZE CROPS

A critical aspect of marker-assisted selection is the correct choice of traits to be included in the analysis (Chapter 19). Trait inclusion must be based on a thorough understanding of the effect of each trait on yield, which depends upon selection of the correct conceptual framework (Salekdeh et al., 2009) and clear knowledge of the relationships among traits. This is particularly important for avoiding commonly found biases produced when scaling up from the organ to the plant and further to the crop level, and from controlled conditions in growth chambers or greenhouses to the actual field environment (Passioura, 2010, 2012; Chapter 1). It is also important for the distinction of possible trade-offs among traits, as detected for N-use related traits (discussed in section 4) and between seed number and seed weight across (Sadras, 2007; Gambín and Borrás, 2010) and within (Uribelarrea et al., 2008) species.

There is a general consensus among those interested in physiological breeding on the importance of a resource-focused approach (Eqs 15.1 and 15.2) for yield dissection in minor but relevant traits for yield definition (Richards et al., 2002, 2010; Reynolds et al., 2007; Salekdeh et al., 2009).

$$\text{Biomass} = \text{Resource offer} \times \text{Resource capture} \times \text{Resource use efficiency} \quad (15.1)$$

$$\text{GY} = \text{Biomass} \times \text{Harvest Index} \quad (15.2)$$

In this approach, biomass production and harvest index (i.e. the proportion of total biomass that is partitioned to grains) are the major traits that determine final GY. These traits can be dissected into minor components within a model of physiological processes that take place along the crop growth cycle (Fig. 15.1). Traits related to biomass production are referred to as the source components of the model, whereas those related to its allocation into the grains are referred to as sink components (Lee and Tollenaar, 2007). Biomass production depends upon the amount of photosynthetically active radiation (PAR) that is captured by the canopy along its cycle and the photosynthetic capacity of the canopy (Long et al., 2006). PAR capture depends upon the interception efficiency of the crop (e_i: quotient between intercepted and incident PAR), which is related to the rate of canopy development, its maximum size and persistence, and its architecture given by vertical and azimuthal leaf angles (Pepper et al., 1977; Girardin and Tollenaar, 1994). Changes in canopy architecture affect e_i (Maddonni and Otegui, 1996; Maddonni et al., 2001). This led Duncan (1971) to define an ideal maize plant (ideotype; Chapter 14) with a gradient of vertical leaf angles from erect at the top of the canopy to more planophyle at the bottom for high leaf area index, LAI >4. Canopy photosynthetic capacity is represented by the conversion efficiency or radiation-use efficiency (RUE; Fig. 15.1d), which can be expressed in terms of PAR or solar radiation × 0.45 (Monteith, 1965). RUE depends upon the balance between the photosynthetic rate of green tissues and respiratory losses (Long et al., 2006). This balance varies markedly among species and growing conditions, and maize leads the RUE recorded in grain crops (Sinclair and Muchow, 1999).

Harvest index is primarily linked to the number of grains (D'Andrea et al., 2006) which, in maize, is set during a relatively short critical period (≈30 days) that is centered at silking (Fischer and Palmer, 1984; Kiniry and Ritchie, 1985) and matches the stages of ear elongation (Otegui and Bonhomme, 1998), silk extrusion

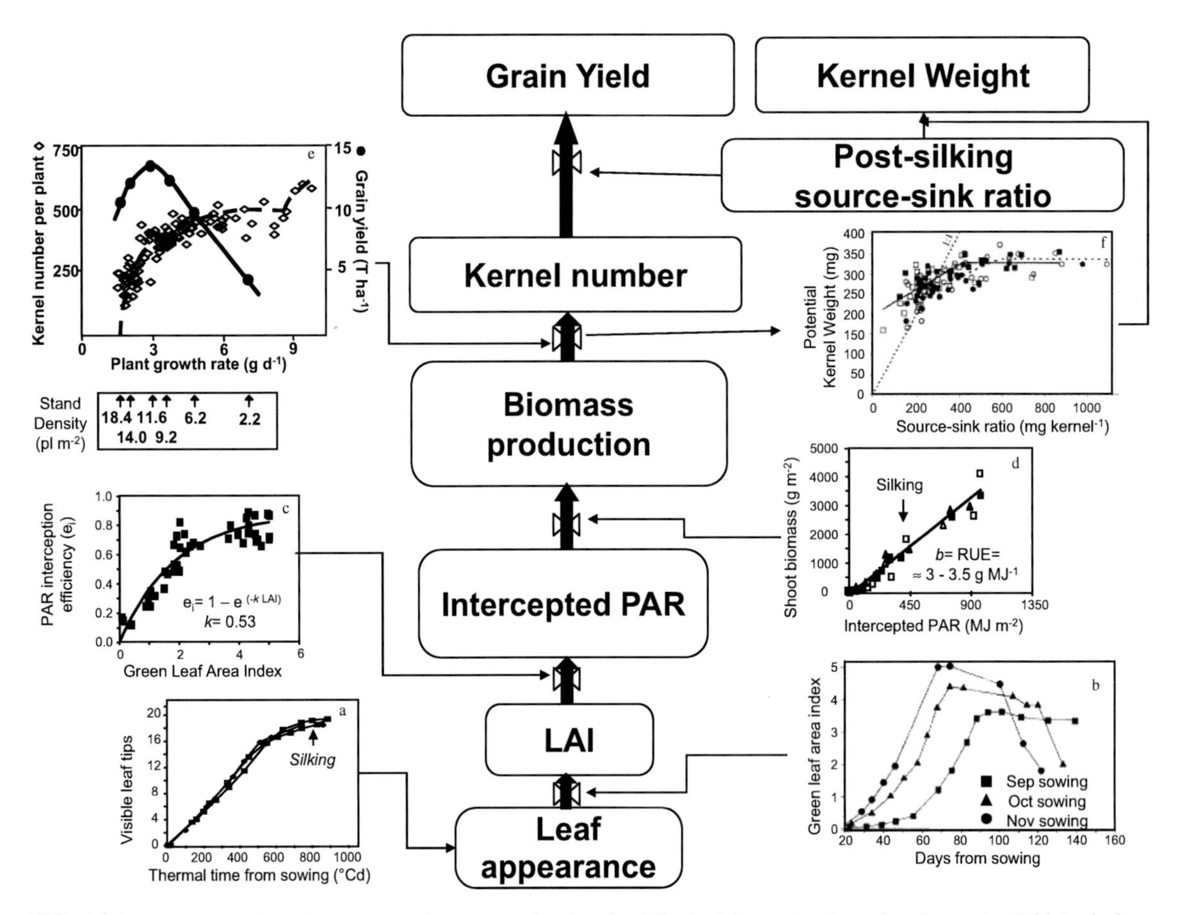

FIG. 15.1 Diagram linking key traits and processes for the physiological determination of maize grain yield, including (a) visible leaf tips evolution, (b) green leaf area index (LAI) evolution, (c) the response of PAR interception efficiency to green LAI and the exponential model fitted to data, (d) the response of cumulative shoot biomass to cumulative PAR and the linear model fitted to data, (e) the response of kernel number per plant to plant growth rate during the critical period for kernel set and the related response of grain yield ha^{-1} (stand densities for each grain yield level are indicated in the bottom box), and (f) response of individual final kernel weight to plant growth per kernel during active grain filling (i.e. the source–sink ratio). *Adapted from Cárcova et al. (2003b).* RUE: radiation-use efficiency; e_i: light interception efficiency; k: attenuation coefficient.

(Bassetti and Westgate, 1993a; Cárcova et al., 2003a), pollination (Bassetti and Westgate, 1993b), and ovary fertilization (Westgate and Boyer, 1986). Plant (Tollenaar et al., 1992; Andrade et al., 1999) and ear (Pagano and Maddonni, 2007) growth rates during this period control final kernel number per plant, whereas kernel desiccation rate (Borrás et al., 2003b; Gambín et al., 2007) and the source–sink ratio (Jones and Simmons, 1983; Gambín et al., 2006) control final kernel weight. The environment and management practices exert their control on all model components, modulating the source–sink balance along the cycle (Andrade et al., 2002; Borrás and Otegui, 2001; Borrás et al., 2004) and consequently yield (Fig. 15.1).

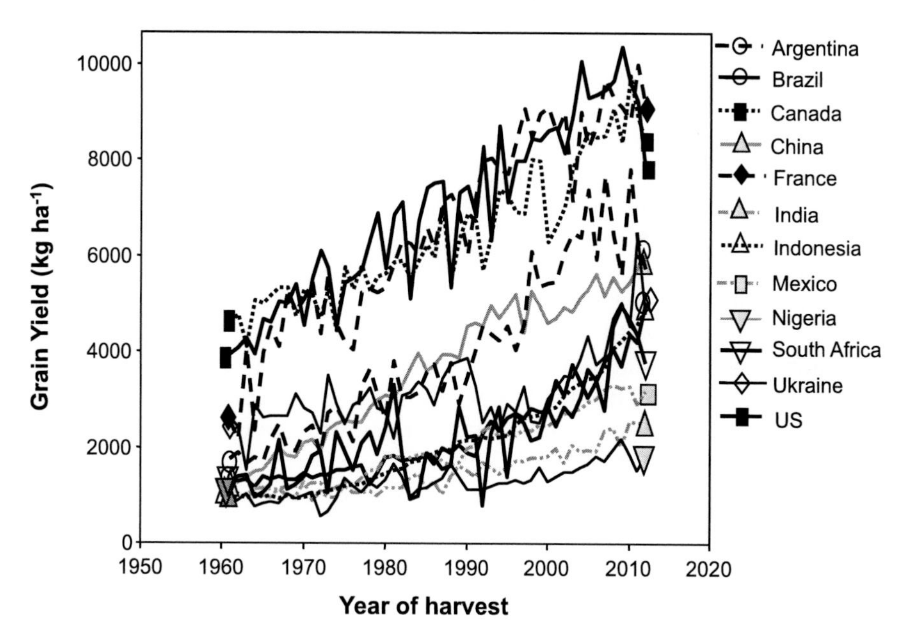

FIG. 15.2 Evolution of country mean grain yield between 1960 and 2012. Data correspond to the 2012 top 12 maize-producing countries (FAOSTAT, 2014).

Most elements of this conceptual framework are currently incorporated in broadly used crop simulation models, such as CERES-Maize (Jones and Kiniry, 1986). The evaluation of yield determination with these models has been utilized for understanding management and environmental effects and their interactions (Savin et al., 1995; Löffler et al., 2005; Mercau et al., 2007; Grassini et al., 2009, 2011), the interpretation of past breeding effects (Bell and Fischer, 1994; Hammer et al., 2009) and expected benefits of trait modification (Hammer et al., 2010; Sinclair et al., 2010). An additional advantage of these models is that genetic controls could be easily coupled to each trait of interest (Hoogenboom et al., 2004; Messina et al., 2006; Chenu et al., 2009), transforming their current generic (most traits) or cultivar dependent (a few traits covered by phenotype descriptors that are erroneously named 'genetic coefficients') values into actual genetic coefficients. Chapter 14 expands on this issue.

3 GAINS IN MAIZE YIELD

Gains in yield have been documented extensively for most crops (references in Slafer, 1994). In this book, time trends in yield are analyzed for major crops in production systems of the USA (Chapter 2), China (Chapter 3), Finland (Chapter 4) and Chile (Chapter 5). Three cereals (rice, wheat and maize) deserve special attention due to their exceptional importance in human nutrition (Fischer and Edmeades, 2010). In maize, as in all grain crops, these gains have a genetic as well as an agronomic origin (Cardwell, 1982; Russell, 1984; Duvick, 2005), and probably most yield improvement in this species is due to the interaction between both sources of variation (Lee and Tollenaar, 2007). Interestingly, maize holds the largest global rate of yield increase since the 1960s (1.6% $year^{-1}$) when these three cereals are compared (1% $year^{-1}$ for rice and 0.9% $year^{-1}$ for wheat; Ray et al., 2013), and there is no clear evidence of plateau in the last decade as detected for rice and wheat (Hall and Richards, 2013). Current high and steady gains in yield cannot be explained by heterosis, which reached its maximum exploitation with the single-cross hybrids that met massive adoption before the end of the 20th century among most leading producing countries (Cooper et al., 2004). No significant increase in heterosis has been registered since the introduction of these hybrids, because similar yield gains have been obtained among elite parental inbreds and their F_1 progeny (Duvick, 2005). The steady genetic gains in yield among single-cross hybrids have been attributed (Tollenaar and Lee, 2002) to their enhanced stress tolerance, obtained through selection for yield stability across target environments (i.e. wide adaptation). The improved performance of newer transgenic hybrids in era studies (Cooper et al., 2004) is related to higher yield potential and insect control, but they do not out-yield their non-transgenic counterparts when pest damage is minimized (Ma and Subedi, 2005).

Despite the high global gains in maize yield, the rates vary broadly among top producing countries for the 1990–2012 period when steady yield increases could be observed for most of these countries (Fig. 15.2, Table 15.1). High gains ($\geq$94 kg ha^{-1} year$-^{-1}$) were obtained for most countries with mean GY $\geq$2900 kg ha^{-1} (averaged for the same period) independently of other considerations (economic development, soils, climate, production system, farm size). Argentina, South Africa, Canada, Indonesia, Ukraine, Brazil, the USA, and France belong to this group (Table 15.1). On the contrary, very low yield gains ($\leq$57 kg ha^{-1} year^{-1}) were achieved for Nigeria, India, and Mexico with mean GY $\leq$2700 kg ha^{-1} (Table 15.1). China represents an exception, with high yields that are not accompanied by high rates. Computation of yield gains in percent (Table 15.1) gave a somewhat

TABLE 15.1 Main features of top maize-producing countries

	Production[a]	Harvested area[a]	Grain yield[b]	Grain yield gain[c]	
	Million T	Million ha	kg ha^{-1}	kg ha^{-1} y^{-1}	% y^{-1}
USA	274.0	35.4	8542	109	1.28
China	208.0	35.0	5054	50	0.99
Brazil	71.1	14.2	3140	114	3.63
Mexico	22.1	6.9	2658	57	2.14
Argentina	21.2	3.7	5560	144	2.59
India	21.1	8.4	1906	47	2.47
Ukraine	21.0	4.4	3623	114	3.15
Indonesia	19.4	4.0	3114	123	3.95
France	15.6	1.7	8492	94	1.11
South Africa	11.8	3.1	2928	125	4.27
Canada	11.7	1.4	7655	124	1.62
Nigeria	9.4	5.2	1496	38	2.54

FAOSTAT, 2014. [a]2012 records; [b]average of the 1990–2012 period; [c]computed for the 1990–2012 period. Percent gain was computed as the quotient between gain in kg ha^{-1} y^{-1} and mean grain yield ($\times$ 100).

different perspective. On one hand, the average gain is high (2.5% year^{-1}) and in agreement with other recent computations (Ray et al., 2013). On the other hand, the variation across top producing countries is very large (0.99–4.27% year^{-1}), with largest values (>3% year^{-1}) among countries with mean GY <5000 kg ha^{-1}, and smallest values (<2% year^{-1}) among countries with mean GY >7000 kg ha^{-1}. Again, China is an exception with a relatively high yield and the lowest percent rate. An optimum response function fits the described trends in yield gain (Fig. 15.3a), provided China was excluded from the analysis. The response, however, changes markedly when the dependent variable is percent gain (Fig. 15.3b). There is a decline in percent gain with increasing yield, but the general fit discriminates two groups of countries. Groups share the rate of change in percent gain (slope of the relationship) but differ markedly in its potential value (ordinate of the relationship). Developed countries (USA, France and Canada) belong to the high potential group, together with Argentina, Ukraine, Brazil, Indonesia and South Africa. Assuming linearity over the whole range, the potential percent yield gain of this group would be 5.32 % year^{-1} and the extrapolated x-intercept suggests zero gain at a mean yield of 11561 kg ha^{-1}. China, Mexico, India and Nigeria belong to the low potential group, for which maximum estimated percent yield gain would be 3.28 % year^{-1} and zero gain would be attained at a mean yield of 7304 kg ha^{-1}.

Several considerations that exceed the scope of this chapter may explain the described contrast seen in Figure 15.3b. From a breeding perspective, a common feature of the low yield potential group is the still extensive use of open-pollination varieties, whereas hybrids use predominates in countries of the high yield potential group. Of interest, differences among countries may compare to those that can be found among regions of a country (i.e. target environments in a breeding program). For instance, USA gain in yield estimated by Fischer and Edmeades (2010) for the 1960–2007 period (114 kg ha^{-1} year^{-1}) was very similar to our computation for the last 22 years for this country (109 kg ha^{-1} year^{-1}), but their estimate for the state of Iowa after 1993 almost doubled these figures (214 kg ha^{-1} year^{-1}). We did not detect such a difference for top-yielding areas of other countries, like the Ontario province (149.5 kg ha^{-1} year^{-1}, based on data by Lee and Tollenaar, 2007) in Canada (estimated country gain of 124 kg ha^{-1} year^{-1}) or the Santa Fe province (149.6 kg ha^{-1} year^{-1}; based

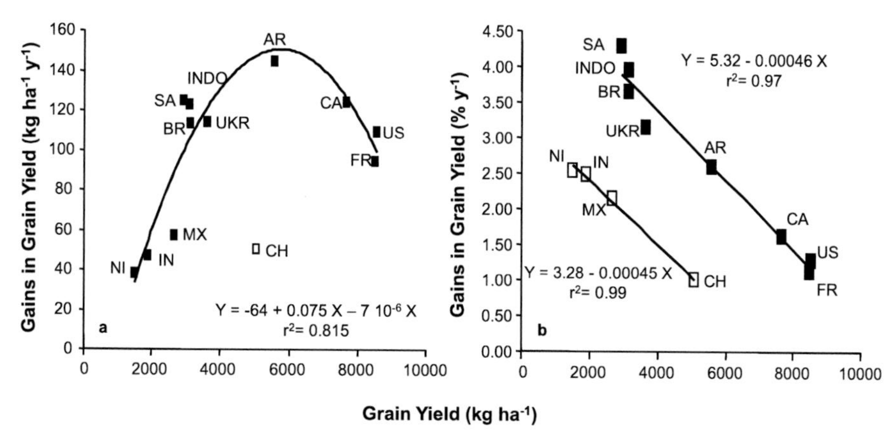

FIG. 15.3 Response of grain yield gain to mean grain yield for different countries in (a) kg ha^{-1} year^{-1}, and (b) % year^{-1}. AR: Argentina; BR: Brazil; CA: Canada; CH: China; FR: France; IN: India; INDO: Indonesia; MX: Mexico, NI: Nigeria; SA: South Africa, UKR: Ukraine; US: USA. Lines represent fitted models. Model in (a) does not include China.

on national statistics from MINAGRI, 2013) in Argentina (estimated country gain of 144 kg ha^{-1} $year^{-1}$). Discrepancies of this nature may probably arise from the relationship between the target population of environments of main breeding programs and the country area cropped to a given species (Cooper et al., 2005). These two aspects are usually tightly related in countries with a relatively small area cropped to maize, but the relationship tends to relax with increasing area. In our analysis, gains in yield were always high (≥94 kg ha^{-1} $year^{-1}$) among the countries that harvested less than 4.4 million ha of maize during 2012 (Ukraine, Indonesia, Argentina, South Africa, France, and Canada). On the contrary, the yield gains ranged between 38 (Nigeria) and 114 kg ha^{-1} $year^{-1}$ (Brazil) among the countries that harvested more than 5.2 million ha of maize during the same year (Table 15.1). Described trends highlight the possible impact on maize production of an apparently easy to reach increase in yield gain (e.g. 2% $year^{-1}$) in the low potential group, particularly in China due to its large influence in global maize production.

4 BREEDING EFFECTS ON THE PHYSIOLOGICAL DETERMINANTS OF MAIZE YIELD

Up to the 1990s, the analysis of breeding effects on phenotypic attributes other than yield (secondary traits) had been limited to a few defensive traits, such as resistance to biotic stress, premature senescence and lodging (Russell, 1991; Duvick, 1992, 2005), and to traits of easy and fast assessment related to reproductive success (Russell, 1985; Duvick, 1992). From these studies we learned that, after selection for an adequate fit of the crop cycle to the available season for maize growth (Tollenaar, 1989; Russell, 1991), breeding caused (1) improved tolerance to low temperature that allowed for earlier sowing, which in most cool-temperate environments allowed crops to explore higher solar radiation conditions during grain filling (Chapter 2 discusses the agronomic implications of this change), (2) a reduction in tassel size (and consequently of apical dominance) and in the anthesis–silking interval (indicative of increased biomass partitioning to the ear), which were accompanied by an increase in the number of grained-ears per plant (prolificacy) and the concomitant increase in kernel number (i.e. enhanced sink), (3) a more erectophyle leaf growth habit (i.e. more upright leaf angle), which allowed for increasing stand density with no penalty in prolificacy, (4) an increase in the proportion of the cycle represented by the post-silking period as compared to the pre-silking period (Russell, 1985; Tollenaar, 1989) together with an enhanced post-silking stay-green, which collectively allowed for an adequate assimilate supply to the augmented number of grains to be filled (i.e. enhanced source), and (5) a decline in grain quality measured as percent protein content.

It was not until the end of the 20th century that breeding effects on the physiological determinants of yield (Fig. 15.1) received our thorough attention, including actual measurements of the traits of interest in a single set of experiments and not only indirect evidence from many proxy traits as in ERA-hybrid studies of the USA (Duvick et al., 2004; Duvick, 2005; Mansfield and Mumm, 2014). The first research was performed with a set of nine short-season hybrids (relative maturity <110) bred for the cool high-latitude (>40°N) environment of Canada (Tollenaar, 1991; Tollenaar et al., 1992), and the subsequent analysis was developed with a set of seven hybrids bred for the extended season (relative maturity >118) of the mid-latitude (<35°S) environment of the Argentine Pampas (Luque et al., 2006). Both studies combined the era analysis with the response to stand density, which expanded the breadth of the results. The main findings of the Canadian research were supportive of ERA-hybrid studies for the USA (e.g. more erect leaves, prolonged stay-green), but added evidence of (1) enhanced dry matter accumulation at late stages of the cycle among newer hybrids (Tollenaar, 1991), and (2) reduced barrenness of these hybrids at high stand densities due to their improved plant growth rate around flowering and not to a decrease in threshold plant growth rate for ear formation (Tollenaar et al., 1992). These trends were accompanied by a somewhat larger LAI and slightly higher stay-green (Valentinuz and Tollenaar, 2004). Results from the Argentine study also demonstrated that breeding effects on physiological traits started at the critical period of ear elongation, and were evident as improved RUE, plant growth rate and biomass partitioning to the ear around silking. The final number of kernels responded to these trends, and improved biomass production after silking allowed for an almost constant source–sink ratio during active grain filling. For this reason, there was no trade-off between kernel number and kernel weight, and increases in the former translated into equivalent increases in yield. In agreement with the Canadian study, harvest index increased and plant growth rate threshold for plant barrenness did not seem to be reduced among newer hybrids. The points of disagreement were (1) yield gains at no competition for resources (very low stand density), for which a positive trend was detected in Argentina but not in Canada or the USA (Duvick, 1997), (2) leaf angle, for which there were differences among evaluated Argentine hybrids but no clear trend of the type detected for Canadian and USA hybrids, and (3) no trend detected for the relative duration of the post-silking period in Argentina. A distinctive aspect of the Argentine study was (1) the computation of genetic gains for all evaluated traits, including the parameters of trait responses to stand density, and (2) the demonstration of variations in these gains due to environmental effects. For instance, genetic gains were computed for the three parameters of the quadratic response of yield to stand density, which were always significant ($P < 0.001$) for the ordinate value (representative of yield when there is no competition for resources or potential yield) but were significant ($P < 0.001$) only under good growing conditions for the linear coefficient (representative of the response of yield to increased stand density until maximum yield is reached) and the quadratic coefficient (representative of the decrease in yield at stand densities above the optimum level). A similar comment applies to the genetic gain in kernel number per unit plant growth rate around silking, which was significant ($P < 0.05$) only in the high yielding environment. On the contrary, the post-silking source–sink ratio declined in response to increased stand density, but no breeding effect was detected for the parameters of fitted models, confirming the stability of this trait across breeding eras.

Other studies addressed in more detail the effect of breeding on the physiological determinants of canopy architecture, ei, and RUE, as well as in N-use related traits. It is important to recall that leaf area varies markedly across maize hybrids (Dwyer et al., 1992), and that post-silking leaf senescence and N distribution are strongly linked to sink activity (Wolfe et al., 1988a,b) combined with genotype differences in the stay-green trait (Pommel et al., 2006; Ning et al., 2013). Consequently, the largest values of associated secondary traits (leaf mass, photosynthesis, chlorophyll content) are usually registered in leaves near to the grained ears (Sadras et al., 2000). Sangoi et al. (2002) and Lee and Tollenaar (2007) demonstrated that maize breeding produced a more 'compact' plant type, with shorter internodes and more erect (and less curved) leaves among newer hybrids. These changes are known to affect the quantity and quality (red:far red ratio) of light distribution across leaf strata (Borrás et al., 2003a; Maddonni et al., 2006), with associated effects on maize yield in response to changes in common agronomic practices such as stand density and row spacing (Ottman and Welch, 1989; Maddonni et al., 2006; Robles et al., 2012). Compact hybrids, however, may not necessarily represent an advantage for biomass production, because there is also evidence of reductions in post-silking RUE within this type of canopy (Maddonni et al., 2006). The decrease in RUE may be linked to an imbalance between sunlit and shaded leaves (Boote and Pickering, 1994) that may cause (1) poor photosynthetic activity at the bottom of the canopy, and consequently reduced root activity for the uptake of soil resources, and (2) other photomorphogenic reactions, like differential capacity for internode elongation or leaf reorientation in response to crowding stress (Maddonni et al., 2001, 2002). Modern maize hybrids exhibit less internode elongation than the old ones in response to increased stand density (Sangoi et al., 2002), with the concomitant benefit of a reduced stalk breakage. It is still unknown if this response is due to differences in the pattern of light distribution or to contrasting sensitivity to the increased proportion of far-red light among hybrids released at different breeding eras that vary in plant height.

Breeding effects on traits related to maize N use are also poorly understood, and have been receiving increased attention in recent years. We learned that yield gains are expected to be reduced in low-N environments (Castleberry et al., 1984), and that seed companies usually avoid these environments because it is not profitable to develop genotypes for poor-yielding areas. We also learned that breeding may have reduced the genotypic differences related to N-uptake capacity due to the practice of applying high N fertilizer rates in commercial programs (Lafitte et al., 1997). Finally, the apparent negative effects of these trends seemed to have no negative effect on current maize productivity because of a trade-off with increased N-use efficiency (NUE). However, a recent analysis of ERA-hybrids showed that the rate of yield gain in unfertilized crops was 56 kg grain ha^{-1} $year^{-1}$, compared with a rate of 86 kg grain ha^{-1} $year^{-1}$ for crops with 252 kg N ha^{-1} (Haegele et al., 2014). This result indicates that the response at low N almost doubled the response rate at high N (only 30 kg grain ha^{-1} $year^{-1}$). The study included 21 conventional, non-transgenic hybrids released in the USA between 1967 and 2006, and confirmed the enhanced NUE produced by breeding but also an improved N uptake among newer hybrids. Consistent with this, physiological studies demonstrated that modern hybrids have higher photosynthetic capacity at low N levels because they have reduced non-stomatal limitations to maintenance of PEPCase activity, as well as of chlorophyll and soluble protein content (Ding et al., 2005). Similarly, a study of recent breeding effects on Chinese maize hybrids detected a reduction in root length density in the topsoil but no change in this trait in the deep soil, which had no consequence in N-uptake efficiency and plant productivity (Chen et al., 2014). Moreover, it seems that root growth angles become steeper under reduced soil N (Trachsel et al., 2013), probably contributing to the positioning of roots in soil layers with enhanced N availability. In summary, all available evidence supports the notion of positive breeding effects on maize NUE, and recent evidence indicates no clear negative breeding effects on N-uptake capacity. This lack of negative effects on N use due to trade-offs among root-related traits is probably linked to the permanent effort of maize breeding programs for improving crop performance in water-limited environments (Bänziger et al., 2002).

5 FIELD-BASED PHENOTYPING OF PHYSIOLOGICAL TRAITS

In spite of our improved knowledge about breeding effects on the physiological determinants of yield in maize (previous section) and other important grain crop species (Slafer and Andrade, 1993; López Pereira et al., 1999a,b, 2000; Specht et al., 1999; Kumudini et al., 2002, 2008), there is little evidence of the inclusion of secondary traits in breeding programs. Richards (2006) reviewed seven successful examples (i.e. those that were able to deliver cultivars used by farmers), all oriented to drought-prone environments. Most of them correspond to wheat breeding in Australia (crop cycle duration, axial resistance to water flow, transpiration efficiency based on carbon isotope discrimination, and osmotic adjustment), and the rest corresponds to maize in Mexico by CIMMYT (anthesis–silking interval), soybean by the USDA (nitrogen fixation in dry soils), and sorghum again in Australia (enhanced stay-green). As pointed out by the author, common features of most of these traits are their lack of a direct relation with plant–water status, their quantitative genetic control, and their influence on growth and development during long periods of the crop cycle. Their effect is exerted through one or more of the components in the resource-driven Eq. 15.1 and/or through harvest index (Eq. 15.2).

There are well-known restrictions a secondary trait must overcome to be useful in traditional breeding. From the strictly genetic point of view, its heritability must be high enough to produce a robust selection index for improving the result of traditional selection based on yield *per se* (Bänziger and Lafitte, 1997), it must not result in yield penalties under favorable growing conditions, and it must not have negative effects on other important attributes (pleiotropic effects). From an operative perspective, it must be an easy, fast, and inexpensive measurement; i.e. the cost of including the secondary trait must be over-compensated by the benefit of its inclusion (Brennan and Martin, 2007).

The advent of molecular tools combined with adequate mapping populations strengthen the focus on the operative problems at the crop level. This is evident in the large number of opinion papers on this topic (Furbank and Tester, 2011; Tuberosa, 2012; Masuka et al., 2012), and reviews on new technologies for field-based high-throughput phenotyping based on remote sensing (White et al., 2012; Araus and Cairns, 2014) published in recent years. Meanwhile, the imbalance between genomics and phenomics still remains, delivering studies that lack the necessary understanding of the ecophysiological processes that support yield determination (Sadras et al., 2013).

There is no question that our ability to measure crop physiological traits on large breeding populations is still unresolved if breeding for yield is to be helped with secondary traits. The wheat Green Revolution was based on a differential biomass partitioning between stem and spike during late stem elongation, a trait that is still of great importance for yield improvement today (Dreccer et al., 2009). However, we currently have not developed a single tool to phenotype large breeding populations for canopy biomass production and partitioning. We still rely on a direct process of sampling–processing–drying of biomass for its correct assessment. The only partial improvement towards a simplification of this process has been the application of a morphometric approach (i.e. allometry) developed by Andrade et al. (1999) and Vega et al. (2000, 2001) for a few grain crops (maize, sunflower and soybean), which seems reasonably robust for maize. This method was necessary for the correct analysis of the response of kernel number per plant to plant growth rate during the period of kernel set. The general response for maize is represented in Figure 15.1e, and the correct estimation of the curvilinear function parameters was inexact when it was based on mean values obtained at the plot level (Andrade et al., 2002) rather than at the individual plant level (Vega et al., 2000, 2001). A non-destructive approach was necessary for surveying always the same plants during the critical period and solving this restriction.

In maize, the morphometric approach for biomass estimation was expanded to the whole cycle (Maddonni and Otegui, 2004), and also to ear biomass determination for the evaluation of biomass partitioning to the ear and the response of kernel number per plant to ear growth rate (Pagano et al., 2007; Pagano and Maddonni, 2007). Indirect biomass estimation requires establishing models (independent for the ear:shoot and the rest of the plant) between actual individual plant biomass and its dimensions for a set of plants. This set must be large enough to capture all the expected variation in plant size of a given treatment (e.g. genotypes). Established models are subsequently used for the non-destructive biomass estimation of plants in the field for each entry of interest. Usual plant dimensions assessed for maize are (1) stem volume, based on plant height to the uppermost collar and mean stem diameter at the base of the stem (average of maximum and minimum values), for the estimation of vegetative biomass (i.e. excluding the ear:shoot), and (2) maximum ear diameter (ideally combined with maximum ear length for the computation of ear volume), for the estimation of ear:shoot biomass.

Allometric assessment of maize biomass has been predominantly used in the analysis of differences among a few genotypes (up to 12) exposed to abiotic stresses, including water (Echarte and Tollenaar, 2006) and nitrogen deficiencies (D'Andrea et al., 2006, 2008; Rossini et al., 2012; Ciampitti et al., 2012), and above-optimum temperatures (Cicchino et al., 2010b; RattalinoEdreira and Otegui, 2012, 2013). It has also been used for a detailed evaluation of the phenotypic variation among versions of transgenic and non-transgenic iso-hybrids (Laserna et al., 2012). These experiments never exceeded 80 plots. Nevertheless, allometry was also used successfully for the evaluation of more genotypes in studies of parent–progeny relationships in complete or incomplete dialelic experiments grown in different combinations of N availability (D'Andrea et al., 2009; Munaro et al., 2011a, 2011b), where the number of plots increased up to 216. More recently, it has been used in field phenotyping of families of recombinant inbred lines (RILs) aimed to subsequent QTL analysis (Palmieri et al., 2010; Mirabilio et al., 2010; Amelong et al., 2012). In RILs experiments, the number of evaluated plots rose from 400 to 735. Most of these studies included other non-destructive or destructive evaluations needed for the correct interpretation of resource-driven effects on maize yield determination. Other non-destructive analyses of plant growth include leaf area and plant architecture. The former is based on the sum of individual leaf areas, which is well represented by the product of leaf length × leaf maximum width × 0.75 (Montgomery, 1911). This trend seems very stable across genotypes and time (Prévot et al., 1991; Stewart and Dwyer, 1999). Vertical and horizontal leaf angles at stem insertion can be easily measured with a protractor (Maddonni et al., 2001), and for the former it may be important to compute leaf length up to the flagging point (Pepper et al., 1977). The general estimation of maximum leaf area index, vertical leaf angle and stay-green has been part of traditional maize breeding (Duvick, 2005; Fischer and Edmeades, 2010), but its assessment using methods described above has been limited to a recent study of breeding effects in the USA (Mansfield and Mumm, 2014). Additional examples of non-destructive surveys are (1) the anthesis and silking dates of individual plants (Uribelarrea et al., 2002), necessary for ASI computation and correct allometric computation of ear biomass, (2) chlorophyll status by SPAD, as an indirect estimation of N limitations to photosynthesis (Echarte et al., 2008), (3) the light interception efficiency (Maddonni and Otegui, 1996), necessary for computing the amount of daily intercepted PAR that will be associated to shoot biomass for estimating RUE (Fig. 15.1d), and (4) canopy infrared thermometry, for the estimation of leaf conductance limitations to transpiration under water deficit and direct heat effects (Cárcova et al., 1998; Reynolds et al., 2007; Cicchino et al., 2010a; RattalinoEdreira and Otegui, 2012). Examples of destructive surveys are all traits related to kernel growth (rate and duration) and its water relationships (Alvarez Prado et al., 2013; Box 12.1 in Chapter 12), as well as final yield and kernel number.

Mentioned phenotyping methods are simple enough to be conducted by most people with minimum training and supervision, and the necessary equipment is (ridiculously) inexpensive as compared to most equipment used for genomics analyses (Chapter 18). There are, however, many issues still pending for the application of crop physiology in a commercial breeding program, where thousands of plots must be surveyed. The most obvious is that most measurements are time consuming and/or may be limited in the opportunity to perform them. Moreover, there are traits for which phenotyping tools are still poorly developed even for studies aimed exclusively at physiological analysis, as all those related to the root system.

An additional, yet not too obvious, constraint for developing phenotyping methods is the need for multidisciplinary teams. These teams must include breeders, physiologists and molecular biologists, but also statisticians, mathematicians, and engineers. Most important: as scientists we usually receive no training for interacting in such a type of team; Chapter 1 touches on epistemological barriers for effective cross-disciplinary research. Furthermore, funding agencies using publication records as selection criteria are not prone to support projects aimed at developing technical methods. The direct consequence of all these constraints is that not too many scientists are eager to spend the necessary time in field-based phenotyping. The indirect one is that this lack of engagement still sets a serious limit to the application of methods aimed at the evaluation of physiological traits in actual breeding programs.

6 GENETIC STRUCTURE OF MAIZE PHYSIOLOGICAL TRAITS

The use of molecular markers in current breeding programs follows a traditional black box approach (Eathington et al., 2007), where markers correlated to yield are selected and combined when developing new elite germplasm. Their physiological effect is rarely tested. Breeding based on physiological traits must investigate the existence of genotypic variability for non-traditional yield-related traits and their predominant genetic control. This is not routinely done in any breeding program. The former is a prerequisite for their inclusion in the breeding pipe-line (Austin, 1993), the latter refers to additive (represented by general combining ability of inbreds, GCA) and non-additive (represented by specific combining ability of inbreds, SCA) gene-action effects. Information on these topics is abundant for maize yield and some secondary traits descriptive of plant (e.g. plant height), ear (ear length, kernel-row number, number of ears, etc.) or kernel shape (Hallauer et al., 2010). On the contrary, research on these topics for the physiological processes underlying yield determination (Fig. 15.1) is limited to a single study (Lee et al., 2005) that covered few traits (biomass at silking and maturity, LAI, stay-green and harvest index). The objective of Lee et al. (2005) was to examine the influence of GCA and SCA on the relationships between yield and its physiological determinants and, for this purpose, they used 12 F1 short-season hybrids derived from the mating of three male and four female inbreds. Unfortunately, they gave no information on parent–progeny relationships for each evaluated trait (Falconer and Mackay, 1996), and failed in dissecting yield in terms of its physiological determinants with a quantitative genetic model. They attributed this failure to the low contribution of genetic effects to differences between hybrids and the relatively high error terms linked to year-related interactions. They also alerted on the expected difficulty of QTL analysis for these types of traits.

Based on a thorough dissection of yield (Table 15.2), D'Andrea et al. (2013) demonstrated that the huge difference between inbreds and hybrids in yield and its physiological determinants caused by heterotic effects (Munaro et al., 2011a, 2011b) did not cause equivalent differences in phenotypic plasticity (Sadras and Slafer, 2012) between them (Table 15.2).

TABLE 15.2 Phenotypic plasticity and parent–progeny relationship (PPR) quantified by D'Andrea et al. (2013) in a set of irrigated experiments that included 18 genotypes (6 inbreds and 12 derived hybrids), 2 contrasting N levels and 3 growing seasons

Trait group	Trait description and units	Phenotypic plasticity[a]		Parent–progeny relationship[b]
		Inbreds	Hybrids	
Phenology	Thermal time to anthesis (°Cd)	0.14	0.11	0.822 ***
	Thermal time to silking (°Cd)	0.15	0.14	0.846 ***
	Anthesis–silking interval (days)	1.35	1.60	0.851 ***
Tissue expansion and light capture	Maximum leaf area index (green leaf m^2 soil m^{-2})	0.47	0.47	0.552 ***
	Green leaf area index at physiological maturity (green leaf m^2 soil m^{-2})	1.32	1.23	0.424 ***
	Maximum proportion of incident PAR intercepted	0.52	0.32	0.382 ***
	Proportion of incident PAR intercepted at physiological maturity	0.89	0.83	0.466 ***
	Amount of incident PAR intercepted daily during the critical period (MJ m^{-2} d^{-1})	0.44	0.44	0.195 ns
	Amount of incident PAR intercepted during grain filling (MJ m^{-2})	0.69	0.79	0.642 ***
	Amount of incident PAR intercepted up to physiological maturity (MJ m^{-2})	0.45	0.46	0.636 ***
Photosynthetic capacity and biomass production	Plant growth rate during the critical period (g d^{-1})	0.60	0.72	−0.014 ns
	Plant biomass at silking (g)	0.55	0.60	−0.006 ns
	Plant biomass at physiological maturity (g)	0.52	0.70	0.294 *
	Radiation-use efficiency during the critical period (g MJ^{-1})	0.50	0.48	0.292 *
	Radiation-use efficiency during grain filling (g MJ^{-1})	0.87	0.70	0.045 ns
	Radiation-use efficiency to physiological maturity (g MJ^{-1})	0.34	0.46	0.207 ns
Biomass partitioning and reproductive efficiency	Ear growth rate during the critical period (g d^{-1})	0.80	0.82	0.521 ***
	Biomass partitioning to the ear during the critical period	0.52	0.52	0.561 ***
	Plant biomass reproductive efficiency (kernels d g^{-1})	0.58	0.38	0.278 *
	Apical ear biomass reproductive efficiency (kernels d g^{-1})	0.51	0.46	0.204 ns
	Harvest index	0.54	0.26	0.684 ***

(continued)

TABLE 15.2 Phenotypic plasticity and parent–progeny relationship (PPR) quantified by D'Andrea et al. (2013) in a set of irrigated experiments that included 18 genotypes (6 inbreds and 12 derived hybrids), 2 contrasting N levels and 3 growing seasons *(cont.)*

Trait group	Trait description and units	Phenotypic plasticity[a]		Parent–progeny relationship[b]
		Inbreds	Hybrids	
Grain yield and its components	Kernel number per plant	0.85	0.61	0.117 ns
	Individual kernel weight (mg)	0.26	0.31	0.537 ***
	Plant grain yield (g)	0.89	0.87	0.201 ns
N metabolism	Total N uptake per plant at physiological maturity (g)	0.92	1.30	0.497 ***
	Percent grain protein	0.39	0.59	0.250 *
	N harvest index (grain N per unit shoot N at physiological maturity)	0.60	0.28	0.272 *
	N proportion in plant biomass at physiological maturity	0.64	0.77	0.031 ns
	N-use efficiency for grain production (g grain g absorbed N^{-1})	1.03	0.92	0.592 ***

*,**,*** Significant at $P < 0.05$, 0.01 and 0.001, respectively. ns: not signficant. [a]Difference between normalized values corresponding to the 10th and 90th percentile of trait distribution; [b]slope of the relationship between standardized hybrid value and standardized mid-parent value (i.e. average between parental inbreds value).

Moreover, using a parent–progeny approach D'Andrea et al. (2013) demonstrated that an inbred's phenotype gave a good prediction of a hybrid's phenotype for most traits. Observed trends, however, varied with the environment, as also reported by Mansfield and Mumm (2014) for broad and narrow sense heritabilities for yield. Data standardization allowed D'Andrea et al. (2013) to make a distinction between true and fake parent–progeny relationships (Table 15.2). The latter were driven exclusively by the environment, and this group included yield and its main determinant (kernel number), as well as plant growth rate during the critical period for kernel set and some RUE values. Additionally, they identified 12 inbred traits that may allow yield improvement at the hybrid level across environments. Ear growth rate during the mentioned critical period and NUE stand out in this important list, because they exerted both direct and indirect effects on yield. The indirect effect was mediated by traits such as harvest index, the ASI and biomass partitioning to the ear during the flowering period (i.e. quotient between ear and plant growth rates). Provided this level of phenotyping is available for mapping populations, QTL studies are expected to improve our insight on these effects and yield responses.

The genetic structure of many traits in terms of QTL detection has received considerable attention in the last decade (Cattivelli et al., 2008; Collins et al., 2008; Chapter 19). However, the genetic basis of physiological traits related to the definition of yield are rare, and examples equivalent to that of trait dissection of leaf expansion made by Reymond et al. (2003) for an individual leaf do not exist at the crop level. A relevant problem of QTL detection for traits under complex quantitative control is the common and usually large QTL × environment interaction, which urges multienvironment analysis and quantitative environmental characterization for drawing robust conclusions (Chapter 13). A large part of this problem derives from the primary interest on traits of high phenotypic plasticity (i.e. largely influenced by the environment), as yield or its main determinant, kernel number. But it is also due to the lack of a correct physiological background for defining the right process-driven (Fig. 15.1) or resource-driven relationships for analysis (Eq. 15.1), with focus on model parameters rather than on the traits linked by the model (Chapter 14). A good example is the response of kernel number per plant to plant growth rate during the critical period (Fig. 15.1e). This is a genotype specific relationship (Echarte et al., 2004; D'Andrea et al., 2008), but relatively stable across environments (Andrade et al., 2002) because most part of environmental effects are captured by the independent variable. Consequently, QTL analysis should focus on the main parameters of the relationship (i.e. the threshold plant growth rate for avoiding plant barrenness, the slope at this threshold rate, and the plateau of maximum kernel number) rather than on the evaluated variables (Amelong et al., 2012). An example that illustrates this procedure at the crop level is a recent QTL analysis of traits that control the process of grain filling (Fig. 15.4) and final kernel weight in maize (Alvarez Prado et al., 2013). Previous research on this topic (Liu et al., 2011; Li et al., 2012) reported several QTL for kernel-filling rate, but the analysis based on calendar days instead of degree units, for example, turned the results inconsistent across

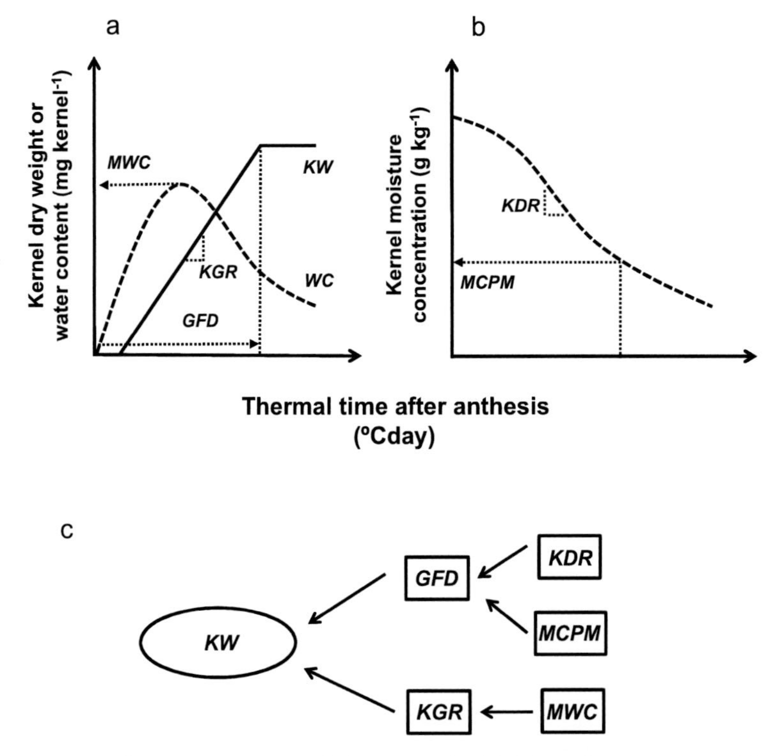

FIG. 15.4 Phenotypic grain-filling traits of interest: (a) kernel weight (KW), kernel water content (WC), kernel growth rate (KGR), grain-filling duration (GFD), maximum water content (MWC); (b) moisture concentration at physiological maturity (MCPM) and kernel desiccation rate (KDR) during the effective grain-filling period; and (c) conceptual representation of trait correlations based on physiological mechanisms. Based on Gambín and Borrás (2011).

genotypes and environments. Improvement of the analytical and phenotyping protocol, the addition of a few more relevant traits (kernel–water relations), and a multitrait multienvironment analysis revealed QTL for kernel growth rate and maximum water content frequently associated with an emergent QTL for kernel weight (Alvarez Prado et al., 2013).

7 CONCLUSIONS

The genomic era unraveled many mysteries linked to genes and their control of several traits in crops of interest for humankind. It also revealed a huge gap between plant genetics and our understanding of plant (and even more crop) phenotypic variation across genotypes and environments. As crop physiologists, we have been usually prompted to deliver information related to a reduced number of commercial hybrids grown at a narrow range of environments, but never for the large numbers of genotypes and sites included in a breeding program. At present, our traditional tools for addressing crop physiology questions are suitable for dealing with no more than 1000–2000 plots per growing season and, consequently, adequate for genetic studies that can be seized with dialelic designs or single RIL populations. There is a need for interdisciplinary teams focused on the development of phenotyping protocols for addressing questions relevant for breeding programs. Recent maize hybrids delivered by the seed industry in the USA (Cooper et al., 2014) are an example of mentioned restrictions and alternative solutions. Selection of drought-tolerant genotypes was not based on the complete phenotyping of the physiological determinants of crop yield in the entire commercial breeding program. It was supported by the use of targeted phenotyping and strong statistical tools combined with managed environments and a crop model approach, giving evidence of the feasibility of an interdisciplinary solution to a complex problem.

References

Alvarez Prado, S., López, C.G., Gambín, B.L., Abertondo, V.J., Borrás, L., 2013. Dissecting the genetic basis of physiological processes determining maize kernel weight using the IBM (B73 × Mo17) Syn4 population (2013). Field Crops Res. 145, 33–43.

Amelong, A., Gambín, B.L., Borrás, L., 2012. QTLs asociados a procesos fisiológicos relacionados con la determinación de número de granos en maíz. Book of Abstracts.XXIX Reunión Argentina de Fisiología Vegetal: Mar del Plata, Argentina.

Andrade, F.H., Echarte, L., Rizzalli, R., Della Maggiora, A., Casanovas, M., 2002. Kernel number prediction in maize under nitrogen or water stress. Crop Sci. 42, 1173–1179.

Andrade, F.H., Vega, C., Uhart, S., Cirilo, A., Cantarero, M., Valentinuz, O., 1999. Kernel number determination in maize. Crop Sci. 39, 453–459.

Araus, J.L., Cairns, J.E., 2014. Field high-throughput phenotyping: The new crop breeding frontier. Trends Plant Sci. 19, 52–61.

Austin, R.B., 1993. Augmenting yield-based selection. In: Hayward, M.D., Bosemark, N.O., Romagosa, I. (Eds.), Plant breeding: principles and prospects. Chapman and Hall, London, pp. 391–405.

Bänziger, M., Edmeades, G.O., Lafitte, R., 2002. Physiological mechanisms contributing to the increased N stress tolerance of tropical maize selected for drought tolerance. Field Crops Res. 75, 223–233.

Bänziger, M., Betrán, F.J., Lafitte, H.R., 1997. Efficiency of high-nitrogen selection environments for improving maize for low-nitrogen target environments. Crop Sci. 37, 1103–1109.

Bänziger, M., Lafitte, H.R., 1997. Efficiency of secondary traits for improving maize for low-nitrogen target environments. Crop Sci. 37, 1110–1117.

Bassetti, P., Westgate, M.E., 1993a. Emergence, elongation, and senescence of maize silks. Crop Sci. 33, 271–275.

Bassetti, P., Westgate, M.E., 1993b. Senescence and receptivity of maize silks. Crop Sci. 33, 275–278.

Bell, M.A., Fischer, R.A., 1994. Using yield prediction models to assess yield gains: A case study for wheat. Field Crops Res. 36, 161–166.

Bell, M.A., Fischer, R.A., Byerlee, D., Sayre, K., 1995. Genetic and agronomic contributions to yield gains: A case study for wheat. Field Crops Res. 44, 55–65.

Boote, K.J., Pickering, N.B., 1994. Modeling photosynthesis of row canopies. HortScience 29, 1423–1434.

Borrás, L., Maddonni, G.A., Otegui, M.E., 2003a. Leaf senescence in maize hybrids: stand density, row spacing, and source-sink ratio effects. Field Crops Res. 82, 13–26.

Borrás, L., Otegui, M.E., 2001. Maize kernel weight response to postflowering source–sink ratio. Crop Sci. 41, 1816–1822.

Borrás, L., Slafer, G.A., Otegui, M.E., 2004. Seed dry weight response to source–sink manipulations in wheat, maize and soybean: A quantitative reappraisal. Field Crops Res. 86, 131–146.

Borrás, L., Westgate, M.E., Otegui, M.E., 2003b. Control of kernel weight and kernel water relations by post-flowering source–sink ratio in maize. Ann. Bot. 91, 857–867.

Brennan, J.P., Martin, P.J., 2007. Returns to investment in new breeding technologies. Euphytica 157, 337–349.

Brown, S.D., Peters, J., 1996. Combining mutagenesis and genomics in the mouse – closing the phenotype gap. Trends Genet. 12, 433–435.

Cárcova, J., Andrieu, B., Otegui, M.E., 2003a. Silk elongation in maize: Relationship with flower development and pollination. Crop Sci. 43, 914–920.

Cárcova, J., Borrás, L., Otegui, M.E., 2003b. Cicloontogénico, dinámica del des arrollo y generación del rendimiento y la calidad en maíz. In: Satorre, E., Benech-Arnold, R., Slafer, G.A. et al., (Eds.), Producción de cultivos de granos. Bases Funcionales para su Manejo, Editorial Facultad de Agronomía, Argentina, pp. 135–166.

Cárcova, J., Maddonni, G.A., Ghersa, C.M., 1998. Crop water stress index of three maize hybrids grown in soil with different quality. Field Crops Res. 55, 165–174.

Cardwell, V.B., 1982. Fifty years of Minnesota corn production: sources of yield increase. Agron. J. 74, 984–990.

Castleberry, R.M., Crum, C.W., Krull, C.F., 1984. Genetic yield improvement of U.S. maize cultivars under varying fertility and climatic environments. Crop Sci. 24, 33–36.

Cattivelli, L., Rizza, F., Badeck, F.W., et al., 2008. Drought tolerance improvement in crop plants: an integrated view from breeding to genomics. Field Crop Res. 105, 1–14.

Chen, X., Zhang, J., Chen, Y., et al., 2014. Changes in root size and distribution in relation to nitrogen accumulation during maize breeding in China. Plant Soil 374, 121–130.

Chenu, K., Chapman, S.C., Tardieu, F., McLean, G., Welcker, C., Hammer, G.L., 2009. Simulating the yield impacts of organ-level quantitative trait loci associated with drought response in maize: A "gene-to-phenotype" modeling approach. Genetics 183, 1507–1523.

Ciampitti, I.A., Zhang, H., Friedemann, P., Vyn, T.J., 2012. Potential physiological frameworks for mid-season field phenotyping of final plant nitrogen uptake, nitrogen use efficiency, and grain yield in maize. Crop Sci. 52, 2728–2742.

Cicchino, M., RattalinoEdreira, J.I., Otegui, M.E., 2010a. Heat stress during late vegetative growth of maize: Effects on phenology and assessment of optimum temperature. Crop Sci. 50, 1432–1436.

Cicchino, M., RattalinoEdreira, J.I., Uribelarrea, M., Otegui, M.E., 2010b. Heat stress in field-grown maize: Response of physiological determinants of grain yield. Crop Sci. 50, 1438–1448.

Collins, N.C., Tardieu, F., Tuberosa, R., 2008. Quantitative trait loci and crop performance under abiotic stress: Where do we stand? Plant Physiol. 147, 469–486.

Cooper, M., Gho, C., Leafgren, R., Tang, T., Messina, C., 2014. Breeding drought-tolerant maize hybrids for the US corn-belt: discovery to product. J. Exp. Bot. (in press), doi:10.1093/jxb/eru064.

Cooper, M., Podlich, D.W., Smith, O.S., 2005. Gene-to-phenotype models and complex trait genetics. Aust. J. Agric. Res. 56, 895–918.

Cooper, M., Smith, O.S., Graham, G., Arthur, L., Feng, L., Podlich, D.W., 2004. Genomics, genetics, and plant breeding: A private sector perspective. Crop Sci. 44, 1907–1913.

D'Andrea, K.E., Otegui, M.E., Cirilo, A.G., 2008. Kernel number determination differs among maize hybrids in response to nitrogen. Field Crops Res. 105, 228–239.

D'Andrea, K.E., Otegui, M.E., Cirilo, A.G., Eyhérabide, G., 2006. Genotypic variability in morphological and physiological traits among maize inbred lines – Nitrogen responses. Crop Sci. 46, 1266–1276.

D'Andrea, K.E., Otegui, M.E., Cirilo, A.G., Eyhérabide, G.H., 2009. Ecophysiological traits in maize hybrids and their parental inbred lines: Phenotyping of responses to contrasting nitrogen supply levels. Field Crops Res. 114, 147–158.

D'Andrea, K.E., Otegui, M.E., Cirilo, A.G., Eyhérabide, G.H., 2013. Parent-progeny relationships between maize inbreds and hybrids: Analysis of grain yield and its determinants for contrasting soil nitrogen conditions. Crop Sci. 53, 2147–2161.

Ding, L., Wang, K.J., Iang, G.M., et al., 2005. Effects of nitrogen deficiency on photosynthetic traits of maize hybrids released in different years. Ann. Bot. 96, 925–930.

Dreccer, M.F., van Herwaarden, A.F., Chapman, S.C., 2009. Grain number and grain weight in wheat lines contrasting for stem water soluble carbohydrate concentration. Field Crops Res. 112, 43–54.

Duncan, W.G., 1971. Leaf angles, leaf area, and crop photosynthesis. Crop Sci. 11, 482–485.

Duvick, D.N., 1992. Genetic contributions to advances in yield of U.S. maize. Maydica 37, 69–79.

Duvick, D.N., 1997. What is yield? In: Edmeades, G.O., Bänziger, M., Mickelson, H.R., Peña-Valdivia, C.B. (Eds.), Developing drought- and low N-tolerant maize. Symposium Proceedings. CIMMYT, El Batan, Mexico, pp. 332–335.

Duvick, D.N., 2005. The contribution of breeding to yield advances in maize (*Zea mays* L.). Adv. Agron. 86, 83–145.

Duvick, D.N., Smith, J.S.C., Cooper, M., 2004. Long-term selection in a commercial hybrid maize breeding program. Janick, J. (Ed.), Plant breeding reviews, vol. 24, Wiley, New York, pp. 109–151.

Dwyer, L.M., Steward, D.W., Hamilton, R.I., Howing, L., 1992. Ear position and vertical distribution of leaf area in corn. Agron. J. 84, 430–438.

Eathington, S.R., Crosbie, T.M., Edwards, M.D., Reiter, R.S., Bull, J.K., 2007. Molecular markers in a commercial breeding program. Crop Sci. 47, S154–S163.

Echarte, L., Andrade, F.H., Vega, C.R.C., Tollenaar, M., 2004. Kernel number determination in Argentinean maize hybrids released between 1965 and 1993. Crop Sci. 44, 1654–1661.

Echarte, L., Rothstein, S., Tollenaar, M., 2008. The response of leaf photosynthesis and dry matter accumulation to nitrogen supply in an older and a newer maize hybrid. Crop Sci. 48, 656–665.

Echarte, L., Tollenaar, M., 2006. Kernel set in maize hybrids and their inbred lines exposed to stress. Crop Sci. 46, 870–878.

Falconer, D.S., Mackay, T.F.C., 1996. Introduction to quantitative genetics, 4th edn. Pearson Education Limited. Prentice Hall: Essex, pp. 480.

FAOSTAT, 2014. Webpage of the Food and Agriculture Organization of the United Nations. faostat.fao.org.

Fischer, K.S., Palmer, F.E., 1984. Tropical maize. In: Goldsworthy, P.R., Fischer, N.M. (Eds.), The physiology of tropical field crops. John Wiley & Sons: Chichester, pp. 213–248.

Fischer, R.A., Edmeades, G.O., 2010. Breeding and cereal yield progress. Crop Sci. 50, S-85–S-98.

Furbank, R.T., Tester, M., 2011. Phenomics – technologies to relieve the phenotyping bottleneck. Trends Plant Sci. 16, 635–644.

Gambín, B.L., Borrás, L., 2011. Genotypic diversity in sorghum inbred lines for grain-filling patterns and other related agronomic traits. Crop Past. Sci. 62, 1026–1036.

Gambín, B.L., Borrás, L., Otegui, M.E., 2006. Source-sink relations and kernel weight differences in maize temperate hybrids. Field Crops Res. 95, 316–326.

Gambín, B.L., Borrás, L., Otegui, M.E., 2007. Kernel water relations and duration of grain filling in maize temperate hybrids. Field Crops Res. 101, 1–9.

Gambín, B.L., Borrás, L., 2010. Resource distribution and the trade-off between seed number and seed weight: a comparison across crop species. Ann. Appl. Biol. 156, 91–102.

Gifford, R.M., Thorne, J.H., Hitz, W.D., Giaquinta, R.T., 1984. Crop productivity and photoassimilate partitioning. Science 225, 801–808.

Girardin, P., Tollenaar, M., 1994. Effects of intraspecific interference on maize leaf azimuth. Crop Sci. 34, 151–155.

Grassini, P., Thorburn, J., Burr, C., Cassman, K.G., 2011. High-yield irrigated maize in the Western U.S. Corn Belt: I. On-farm yield, yield potential, and impact of agronomic practices. Field Crops Res. 120, 142–150.

Grassini, P., Yang, H., Cassman, K.G., 2009. Limits to maize productivity in Western Corn-Belt: A simulation analysis for fully irrigated and rainfed conditions. Agric. For. Meteorol. 149, 1254–1265.

Haegele, J.W., Cook, K.A., Nichols, D.M., Below, F.E., 2014. Changes in nitrogen use traits associated with genetic improvement for grain yield of maize hybrids released in different decades. Crop Sci. 53, 1256–1268.

Hall, A.J., Richards, R.A., 2013. Prognosis for genetic improvement of yield potential and water-limited yield of major grain crops. Field Crops Res. 143, 18–33.

Hallauer, A.R., Carena, M.J., Miranda Filho, J.B., 2010. Quantitative genetics in maize breeding, 3rd edn. Springer Science: New York, USA.

Hammer, G.L., Van Oosterom, E., McLean, G., et al., 2010. Adapting APSIM to model the physiology and genetics of complex adaptive traits in field crops. J. Exp. Bot. 61, 2185–2202.

Hammer, G.L., Dong, Z., McLean, G., et al., 2009. Can changes in canopy and/or root system architecture explain historical maize yield trends in the U.S. corn belt? Crop Sci. 49, 299–312.

Hoogenboom, G., White, J.W., Messina, C.D., 2004. From genome to crop: Integration through simulation modeling. Field Crops Res. 90, 145–163.

IPCC, 2007. Climate change 2007: synthesis report. Contribution of working groups I, II and III to the fourth assessment report of the intergovernmental panel on climate change core writing team. In: Pachauri, R.K., Reisinger, A., (Eds.) IPCC: Geneva, Switzerland, p. 104.

Jones, C.A., Kiniry, J.R., 1986. CERES-Maize: A simulation model of maize growth and development. Texas A&M University Press: College Station.

Jones, R.J., Simmons, S.R., 1983. Effect of altered source-sink ratio on growth of maize kernels. Crop Sci. 23, 129–134.

Kiniry, J.R., Ritchie, J.T., 1985. Shade-sensitive interval of kernel number in maize. Agron. J. 77, 711–715.

Kumudini, S., Hume, D.J., Chu, G., 2002. Genetic improvement in short-season soybeans: II. Nitrogen accumulation, remobilization, and partitioning. Crop Sci. 42, 141–145.

Kumudini, S., Omielan, J., Hume, D.J., 2008. Soybean genetic improvement in yield and the effect of late-season shading and nitrogen source and supply. Agron. J. 100, 400–405.

Lafitte, H.R., Edmeades, G.O., Taba, S., 1997. Adaptive strategies identified among tropical maize landraces for nitrogen-limited environments. Field Crops Res. 49, 187–204.

Laserna, M.P., Maddonni, G.A., López, C.G., 2012. Phenotypic variations between non-transgenic and transgenic maize hybrids. Field Crops Res. 134, 175–184.

Lee, E.A., Ahmadzadeh, A., Tollenaar, M., 2005. Quantitative genetic analysis of the physiological processes underlying maize grain yield. Crop Sci. 45, 981–987.

Lee, E.A., Tollenaar, M., 2007. Physiological basis of successful breeding strategies for maize grain yield. Crop Sci. 47, S202–S215.

Li, Y., Yang, M., Dong, Y., et al., 2012. Three main genetic regions for grain development revealed through QTL detection and meta-analysis in maize. Mol. Breed. 30, 195–211.

Liu, Z.H., Ji, H.Q., Cui, Z.T., et al., 2011. QTL detected for grain filling rate in maize using a RIL population. Mol. Breed. 27, 25–36.

Löffler, C.M., Wei, J., Fast, T., et al., 2005. Classification of maize environments using crop simulation and geographic information systems. Crop Sci. 45, 1708–1716.

Long, S.P., Zhu, X.-G., Naidu, S.L., Ort, D.R., 2006. Can improvement in photosynthesis increase crop yields? Plant Cell Environ. 29, 315–330.

López Pereira, M., Sadras, V.O., Trápani, N., 1999a. Genetic improvement of sunflower in Argentina between 1930 and 1995. I. Yield and its components. Field Crops Res. 62, 157–166.

Lopez Pereira, M., Trápani, N., Sadras, V.O., 1999b. Genetic improvement of sunflower in Argentina between 1930 and 1995. II. Phenological development, growth and source-sink relationship. Field Crops Res. 63, 247–254.

Lopez Pereira, M., Trápani, N., Sadras, V.O., 2000. Genetic improvement of sunflower in Argentina between 1930 and 1995: Part III. Dry matter partitioning and grain composition. Field Crops Res. 67, 215–221.

Luque, S.F., Cirilo, A.G., Otegui, M.E., 2006. Genetic gains in grain yield and related physiological attributes in Argentine maize hybrids. Field Crops Res. 95, 383–397.

Ma, B.L., Subedi, K.D., 2005. Development, yield, grain moisture and nitrogen uptake of Bt corn hybrids and their conventional near-isolines. Field Crops Res. 93, 199–211.

Maddonni, G.A., Cirilo, A.G., Otegui, M.E., 2006. Row width and maize grain yield. Agron. J. 98, 1532–1543.

Maddonni, G.A., Otegui, M.E., 1996. Leaf area, light interception, and crop development in maize. Field Crops Res. 48, 81–87.

Maddonni, G.A., Otegui, M.E., 2004. Intra-specific competition in maize: Early establishment of hierarchies among plants affects final kernel set. Field Crops Res. 85, 1–13.

Maddonni, G.A., Otegui, M.E., Andrieu, B., Chelle, M., Casal, J.J., 2002. Maize leaves turn away from neighbors. Plant Physiol. 130, 1181–1189.

Maddonni, G.A., Otegui, M.E., Cirilo, A.G., 2001. Plant population density, row spacing and hybrid effects on maize canopy architecture and light attenuation. Field Crops Res. 71, 183–193.

Mansfield, B.D., Mumm, R.H., 2014. Survey of plant density tolerance in U.S. maize germplasm. Crop Sci. 54, 157–173.

Masuka, B., Araus, J.L., Das, B., Sonder, K., Cairns, J.E., 2012. Phenotyping for abiotic stress tolerance in maize. J. Integr. Plant Biol. 54, 238–249.

Mercau, J.L., Dardanelli, J.L., Collino, D.J., Andriani, J.M., Irigoyen, A., Satorre, E.H., 2007. Predicting on-farm soybean yields in the pampas using CROPGRO-soybean. Field Crops Res. 100, 200–209.

Messina, C.D., Jones, J.W., Boote, K.J., Vallejos, C.E., 2006. A gene-based model to simulate soybean development and yield responses to environment. Crop Sci. 46, 456–466.

Miflin, B., 2000. Crop improvement in the 21st century. J. Exp. Bot. 51, 1–8.

MINAGRI, 2013. Sistema integrado de información agropecuaria. Ministerio de Agricultura, Ganadería y Pesca de la Nación: Argentina. In: www.siia.gov.ar.

Mirabilio, V., D'Andrea, K.E., Otegui, M.E., Cirilo, A.G., Eyhérabide, G.H., 2010. Variabilidad genotípica en líneas endocriadas recombinantes de maíz: I. Estudio de l heredabiliad para los determinante seco fisiológicos del rendimiento. In: Proceedings of the IXth National Maize Congress. Rosario, Argentina. pp. 353–355.

Monteith, J.L., 1965. Radiation and crops. Exp. Agric. 1, 241–251.

Montgomery, E.C., 1911. Correlation studies in corn. In: Nebraska Agric. Exp. Station Annual report 24th. Lincoln, NE, pp. 108-159.

Munaro, E.M., D'Andrea, K.E., Otegui, M.E., Cirilo, A.G., Eyhérabide, G.H., 2011a. Heterotic response for grain yield and ecophysiological related traits to nitrogen availability in maize Crop Sci. 51, 1172–1187.

Munaro, E.M., Eyhérabide, G.H., D'Andrea, K.E., Cirilo, A.G., Otegui, M.E., 2011b. Heterosis × environment interaction in maize: What drives heterosis for grain yield? Field Crops Res. 124, 441–449.

Ning, P., Li, S., Yu, P., Zhang, Y., Li, C., 2013. Post-silking accumulation and partitioning of dry matter, nitrogen, phosphorus and potassium in maize varieties differing in leaf longevity. Field Crops Res. 144, 19–27.

Otegui, M.E., Bonhomme, R., 1998. Grain yield components in maize I. Ear growth and kernel set. Field Crops Res. 56, 247–256.

Ottman, M.J., Welch, L.F., 1989. Planting patterns and radiation interception, plant nutrient concentration, and yield in corn. Agron. J. 81, 167–174.

Pagano, E., Cela, S., Maddonni, G.A., Otegui, M.E., 2007. Intra-specific competition in maize: Ear development, flowering dynamics and kernel set of early-established plant hierarchies. Field Crops Res. 102, 198–209.

Pagano, E., Maddonni, G.A., 2007. Intra-specific competition in maize: Early established hierarchies differ in plant growth and biomass partitioning to the ear around silking. Field Crops Res. 101, 306–320.

Palmieri, E., D'Andrea, K.E., Otegui, M.E., Cirilo, A.G., Eyhérabide, G.H., 2010. Variabilidad genotípica en líneas endocriadas recombinantes de maíz: II. Estudio de la heredabilidad para los determinantes numéricos del rendimiento. Proceedings of the IXth National Maize Congress. Rosario, Argentina. pp. 358–359.

Passioura, J., 1996. Drought and drought tolerance. Plant Growth Regul. 20, 79–83.

Passioura, J., 2010. Scaling up: The essence of effective agricultural research. Funct. Plant Biol. 37, 585–591.

Passioura, J., 2012. Phenotyping for drought tolerance in grain crops: When is it useful to breeders? Funct. Plant Biol. 39, 851–859.

Pepper, G., Pearce, R.B., Mock, J.J., 1977. Leaf orientation and yield of maize. Crop Sci. 17, 883–886.

Pommel, B., Gallais, A., Coque, M., et al., 2006. Carbon and nitrogen allocation and grain filling in three maize hybrids differing in leaf senescence. Eur. J. Agron. 24, 203–211.

Prévot, L., Aries, F., Monestiez, P., 1991. Modélisation de la structure géométrique du maïs. Agronomie 11, 491–503.

RattalinoEdreira, J.I., Otegui, M.E., 2012. Heat stress in temperate and tropical maize hybrids: Differences in crop growth, biomass partitioning and reserves use. Field Crops Res. 130, 87–98.

RattalinoEdreira, J.I., Otegui, M.E., 2013. Heat stress in temperate and tropical maize hybrids: A novel approach for assessing sources of kernel loss in field conditions. Field Crops Res. 142, 58–67.

Ray, D.K., Mueller, N.D., West, P.C., Foley, J.A., 2013. Yield trends are insufficient to double global crop production by 2050. PloS One 8, 1–8.

Reymond, M., Muller, B., Leonardi, A., Charcosset, A., Tardieu, F., 2003. Combining quantitative trait loci analysis and an ecophysiological model to analyze the genetic variability of the response of maize leaf growth to temperature and water deficit. Plant Physiol. 131, 664–675.

Reynolds, M.P., Saint Pierre, C., Saad, A.S.I., Vargas, M., Condon, A.G., 2007. Evaluating potential genetic gains in wheat associated with stress-adaptive trait expression in elite genetic resources under drought and heat stress. Crop Sci. 47, S172–S189.

Richards, R.A., 2006. Physiological traits used in the breeding of new cultivars for water-scarce environments. Agric. Water Manag. 80, 197–211.

Richards, R.A., Rebetzke, G.J., Condon, A.G., van Herwaarden, A.F., 2002. Breeding opportunities for increasing the efficiency of water use and crop yield in temperate cereals. Crop Sci. 42, 111–121.

Richards, R.A., Rebetzke, G.J., Watt, M., Condon, A.G., Spielmeyer, W., Dolferus, R., 2010. Breeding for improved water productivity in temperate cereals: Phenotyping, quantitative trait loci, markers and the selection environment. Funct. Plant Biol. 37, 85–97.

Robles, M., Ciampitti, I.A., Vyn, T.J., 2012. Responses of maize hybrids to twin-row spatial arrangement at multiple plant densities. Agron. J. 104, 1747–1756.

Rosielle, A.A., Hamblin, J., 1981. Theoretical aspects of selection for yield in stress and non-stress environments. Crop Sci. 21, 943–946.

Rossini, M.A., Maddonni, G.A., Otegui, M.E., 2012. Interplant variability in maize crops grown under contrasting N × stand density combinations: Links between development, growth and kernel set. Field Crops Res. 133, 90–100.

Russell, W.A., 1985. Evaluations for plant, ear, and grain traits of maize cultivars representing different eras of breeding. Maydica 30, 85–96.

Russell, W.A., 1984. Agronomic performance of maize cultivars representing different eras of breeding. Maydica 29, 375–390.

Russell, W.A., 1991. Genetic improvement of maize yields. Adv. Agron. 46, 245–298.

Sadras, V.O., Echarte, L., Andrade, F., 2000. Profile of leaf senescence during reproductive growth of sunflower and maize. Ann. Bot. 85, 187–195.

Sadras, V.O., 2007. Evolutionary aspects of the trade-off between seed size and number in crops. Field Crops Res. 100, 125–138.

Sadras, V.O., Rebetzke, G.J., Edmeades, G.O., 2013. The phenotype and the components of phenotypic variance of crop traits. Field Crops Res. 154, 255–259.

Sadras, V.O., Slafer, G.A., 2012. Environmental modulation of yield components in cereals: Heritabilities reveal a hierarchy of phenotypic plasticities. Field Crops Res. 127, 215–224.

Salekdeh, G.H., Reynolds, M., Bennett, J., Boyer, J., 2009. Conceptual framework for drought phenotyping during molecular breeding. Trends Plant Sci. 14, 488–496.

Sangoi, L., Gracietti, M.A., Rampazzo, C., Bianchetti, P., 2002. Response of Brazilian maize hybrids from different eras to changes in plant density. Field Crops Res. 79, 39–51.

Savin, R., Satorre, E.H., Hall, A.J., Slafer, G.A., 1995. Assessing strategies for wheat cropping in the monsoonal climate of the Pampas using the CERES-wheat simulation model. Field Crops Res. 42, 81–91.

Shi, G., Chavas, J.-P., Lauer, J., 2013. Commercialized transgenic traits, maize productivity and yield risk. Nat. Biotechnol. 2, 111–114.

Sinclair, T.R., Messina, C.D., Beatty, A., Samples, M., 2010. Assessment across the United States of the benefits of altered soybean drought traits. Agron. J. 102, 475–482.

Sinclair, T.R., Muchow, R.C., 1999. Radiation use efficiency. Adv. Agron. 65, 215–265.

Slafer, G.A., 1994. Genetic improvement of field crops. Marcel Dekker, Inc.: New York.

Slafer, G.A., 2003. Genetic basis of yield as viewed from a crop physiologist's perspective. Ann. Appl. Biol. 143, 117–128.

Slafer, G.A., Andrade, F.H., 1993. Physiological attributes related to the generation of grain yield in bread wheat cultivars released at different eras. Field Crops Res. 31, 351–367.

Specht, J.E., Hume, D.J., Kumudini, S.V., 1999. Soybean yield potential – A genetic and physiological perspective. Crop Sci. 39, 1560–1570.

Stewart, D.W., Dwyer, L.M., 1999. Mathematical characterization of leaf shape and area of maize hybrids. Crop Sci. 39, 422–427.

Tollenaar, M., 1989. Genetic improvement in grain yield of commercial maize hybrids grown in Ontario from 1959 to 1980. Crop Sci. 29, 1365–1371.

Tollenaar, M., 1991. Physiological basis of genetic improvement of maize hybrids in Ontario from 1959 to 1988. Crop Sci. 31, 119–124.

Tollenaar, M., Lee, E.A., 2002. Yield potential, yield stability and stress tolerance in maize. Field Crops Res. 75, 161–169.

Tollenaar, M., Dwyer, L.M., Stewart, D.W., 1992. Ear and kernel formation in maize hybrids representing three decades of grain yield improvement in Ontario. Crop Sci. 32, 432–438.

Trachsel, S., Kaeppler, S.M., Brown, K.M., Lynch, J.P., 2013. Maize root growth angles become steeper under low N conditions. Field Crops Res. 140, 18–31.

Trachsel, S., Kaeppler, S.M., Brown, K.M., Lynch, J.P., 2011. Shovelomics: High throughput phenotyping of maize (*Zea mays* L.) root architecture in the field. Plant Soil 341, 75–87.

Tuberosa, R., 2012. Phenotyping for drought tolerance of crops in the genomics era. Front. Physiol., Article 347.

Uribelarrea, M., Cárcova, J., Borrás, L., Otegui, M.E., 2008. Enhanced kernel set promoted by synchronous pollination determines a tradeoff between kernel number and kernel weight in maize. Field Crops Res. 105, 172–181.

Uribelarrea, M., Cárcova, J., Otegui, M.E., Westgate, M.E., 2002. Pollen production, pollination dynamics, and kernel set in maize. Crop Sci. 42, 1910–1918.

Valentinuz, O.R., Tollenaar, M., 2004. Vertical profile of leaf senescence during the grain-filling period in older and newer maize hybrids. Crop Sci. 44, 827–834.

Vega, C.R.C., Andrade, F.H., Sadras, V.O., Uhart, S.A., Valentinuz, O.R., 2001. Seed number as a function of growth. A comparative study in soybean, sunflower and maize. Crop Sci. 41, 748–754.

Vega, C.R.C., Sadras, V.O., Andrade, F.H., Uhart, S.A., 2000. Reproductive allometry in soybean, maize and sunflower. Ann. Bot. 85, 461–468.

Westgate, M.E., Boyer, J.S., 1986. Reproduction at low silk and pollen water potentials in maize. Crop Sci. 26, 951–956.

White, J.W., Andrade-Sanchez, P., Gore, M.A., et al., 2012. Field-based phenomics for plant genetics research. Field Crops Res. 133, 101–112.

Wolfe, D.W., Henderson, D.W., Hsiao, T.C., Alvino, A., 1988a. Interactive water and nitrogen effects on senescence of maize. I. Leaf area duration, nitrogen distribution, and yield. Agron. J. 80, 859–864.

Wolfe, D.W., Henderson, D.W., Hsiao, T.C., Alvino, A., 1988b. Interactive water and nitrogen effects on senescence of maize. II. Photosynthetic decline and longevity of individual leaves. Agron. J. 80, 865–870.

Ziyomo, C., Bernardo, R., 2013. Drought tolerance in maize: Indirect selection through secondary traits versus genome-wide selection. Crop Sci. 53, 1269–1275.

CHAPTER

16

Breeding challenge: improving yield potential

M.J. Foulkes[1], M.P. Reynolds[2]

[1]University of Nottingham, School of Biosciences, Nottingham, UK

[2]CIMMYT Wheat Program, Mexico

1 RATIONALE FOR RAISING YIELD POTENTIAL

It is estimated that 1.16–1.31% $year^{-1}$ compound rates of increase in grain yield are needed to satisfy projected demand in cereals for food, feed and biofuels for 2050 (Bruinsma, 2009; Fischer, 2009). Over the 20-year period to 2012 chosen to estimate current rates of progress, the linear rates of yield change for the world (Fig. 16.1) have been 33 kg ha^{-1} $year^{-1}$ (wheat), 43 kg ha^{-1} $year^{-1}$ (rice), and 73 kg ha^{-1} $year^{-1}$ (maize). Relative rates of yield increase are declining and, expressed relative to predicted yield in 2009, are 1.1% $year^{-1}$ for wheat, 1.0% $year^{-1}$ for rice, and 1.4% $year^{-1}$ for maize. With the exception of maize in some regions (Fischer and Edmeades, 2010; Chapter 2), there is no evidence for exponential growth in yield and, in countries like France and the UK, there is evidence of a yield plateau for wheat (Brisson et al. 2010; Mackay et al., 2011). Furthermore, there is evidence of abrupt decreases in rate of yield gain, including rice in eastern Asia and wheat in northwest Europe, which account for 31% of total global rice, wheat and maize production (Grassini et al., 2014). The leveling off of yield may occur because: (1) farmers cannot achieve the crop and soil management required to reach attainable yield; and/or (2) crop response to additional inputs exhibits a diminishing marginal yield benefit as yield approaches the ceiling (Cassman et al., 2010). Average regional and national yields can be predicted to plateau when they reach 70–90% of yield potential (Cassman, 1999; Cassman et al., 2003; Grassini et al., 2009).

Crop yield potential (YP) is defined as the maximum attainable yield per unit land area that can be achieved by a particular crop cultivar in an environment to which it is adapted when pests and diseases are effectively controlled, and nutrients and water are non-limiting (water stress being eliminated by full irrigation or

Crop Physiology. DOI: 10.1016/B978-0-12-417104-6.00016-9

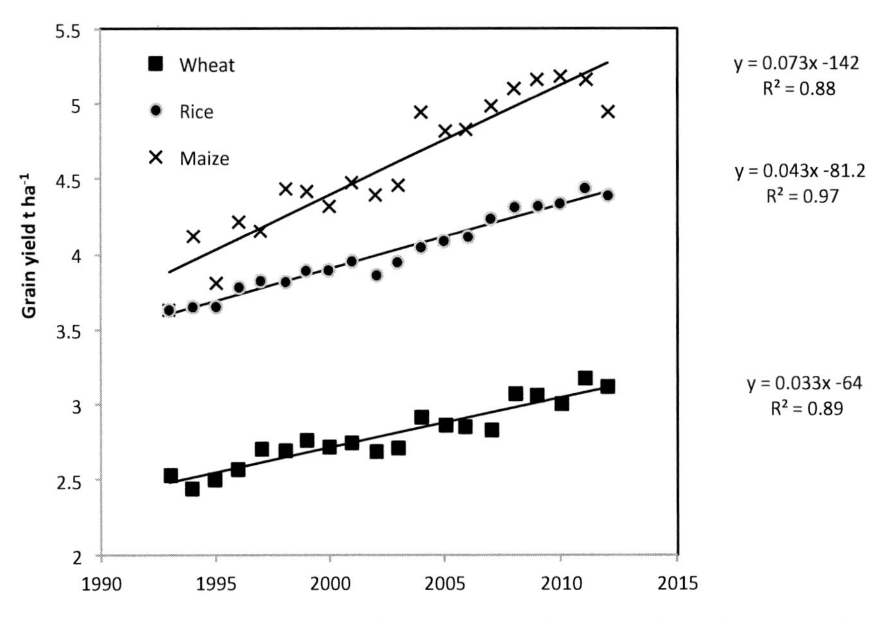

FIG. 16.1 Global average yield for wheat, mazie and rice 1993–2012. The present chapter focuses mainly on strategies to raise yield potential in wheat (*Triticum aestivum* L.), maize (*Zea mays*) and rice (*Oriza sativa* L.). The three species dominate cereals globally and provide approximately 50% of human food calories (Tweeten and Thompson, 2008). The physiological traits associated with current rates of yield gains are examined and then the major breeding challenges for raising future yield potential are considered.

ample rainfall) (Evans and Fischer, 1983). Attainable yield (AY) may be defined as the yield a skilful farmer should reach when taking judicious account of economics and risk (van Ittersum and Rabbinge, 1997). The exploitable yield gap (i.e. gap between farm yield and attainable yield) was recently estimated at 30% for winter wheat in the UK, 50% for spring wheat in Mexico, 70% for japonica rice in Japan, 100% or 58 % for indica rice in the Phillipinnes in wet and dry seasons, respectively, and >200% for maize in sub-Saharan Africa (Fischer and Edmeades, 2010). On a national scale, yield gaps for maize in the USA are around 50% (Fischer and Edmeades, 2010), but average irrigated maize yield in high-yielding districts in Nebraska has not increased since 2001, suggesting that leading growers are approaching an upper biophysical limit for crop productivity (Chapter 2). Therefore, in some developing countries, especially in Africa, there is scope for raising yields through application of existing technologies such as fertilization (Tittonell et al., 2008) but this will require infrastructure investment and availability of inputs (Hall and Richards, 2013).

Given the extent of yield gaps mentioned above, at first sight it may not appear cost effective to invest in increasing genetic yield potential. However, while improved agronomic practices are necessary to realize the full potential of a crop's genome, their implementation is much less straightforward – both practically and economically – for farmers than changing cultivars (Maydup et al., 2012). Furthermore, increasing yield potential is the only avenue to improve productivity where growers have fully closed the exploitable gap. Economic analysis has typically found that yield gains result equally from both breeding and crop management interventions (e.g. Fischer, 2009) and, in the last decades, there is evidence that plant breeding is becoming a proportionally larger component of yield growth (Fischer and Edmeades, 2010; Mackay et al., 2011). Arguments in favor of investment in genetic improvement, other than the ready adoption of new cultivars, include: (1) theoretical considerations which support the idea that radiation-use efficiency (above-ground dry matter per unit radiation interception; RUE) can be increased by 50% at least in C_3 species (Zhu et al., 2010), implying significant scope for raising yield potential, albeit it is recognized that RUE does not scale linearly with grain yield (e.g. Sinclair et al., 2004); (2) strong precedents for yield improvement through breeding starting with and extending well beyond the Green Revolution (e.g. Sharma et al. 2012); (3) the fast-growing fields of both genetics with availability of genome sequences for several crops, and high-throughput field phenotyping platforms (Araus and Cairns, 2014) that offer considerable promise for screening genetic resources, parental characterization and progeny selection (Chapter 18); and (4) the existence of well established national and international crop improvement networks, such as those coordinated by CGIAR centers, that enable new genotypes to be rapidly and extensively tested in and delivered to representative target regions (Braun et al., 2010).

2 RELATIONSHIP BETWEEN YIELD POTENTIAL AND YIELD UNDER ABIOTIC STRESS

Selection for greater yield potential has frequently resulted in higher production in environments subject to abiotic stress (usually water and heat), e.g. in wheat (Calderini and Slafer, 1999; Slafer et al., 1999; Richards et al., 2002; Araus et al., 2002; Reynolds and Borlaug, 2006). The point at which there is a crossover interaction implying that selection for yield potential will result in cultivars which in absolute terms perform less well than those selected under drought for wheat is estimated to be in environments with yields <2.0 t ha^{-1} (Araus et al., 2002), i.e. in severely stressed environments. CIMMYT's experience with maize shows that genetic correlations between yield in stressed and unstressed environments remain positive but tend to non-significance when stress reduces yields by about 50% probably due to radical changes in relative allocation of assimilates to male vs female reproductive structures (Edmeades et al., 2000). Therefore, an important outcome of breeding for yield potential is higher attainable yields under moderate abiotic stresses. In many farms in drought environments because of the overriding impact of low rainfall, growers make most of their profit in a few 'good' seasons. Hayman (2007), for example, estimated that in a series of 18 seasons the business profit for grain growers in western Australia ranged from slight loss (about $10K) to over $200k profit. The top five seasons accounted for 52% of the series profit. Therefore, in addition to the biological considerations, there is a strong economic justification to maintain the capacity to capture the benefits of good seasons, even in the dry environments.

3 CURRENT RATES OF PROGRESS IN YIELD POTENTIAL AND ASSOCIATED TRAITS

3.1 Current rates of yield progress

3.1.1 Maize

Continued rises in maize farmer yield (Fischer and Edmeades, 2010), especially from 1961 through 1990, were largely due to increases in fertilizer use, chemical weed control, and higher plant densities, coupled with the use of hybrids that could respond to fertilizers and tolerate crowding (Cardwell, 1982). The rapid increase in yields in maize in the last decades was associated with a positive interaction between improved hybrids and higher plant density, which now averages around 80000 plants ha^{-1} in the USA

(Chapter 2). Furthermore, genetic improvements in tolerance to cold and waterlogged soils have played an important part in allowing the earlier planting of maize and the expansion of conservation tillage practices that themselves favor earlier planting. Studies of Pioneer hybrids released in Iowa from 1930 to 2002 at their optimum density showed linear yield growth of 0.8% year^{-1}of 2002 hybrid yields (79 kg ha^{-1} year^{-1}) (Cooper et al., 2004) and 1% (116 kg ha^{-1} year^{-1}) to 2007 (Hammer et al., 2009). Messina et al. (2009) provide data that suggest that under full irrigation the yield of maize in a target population of environments was increasing at 0.79% year^{-1} in 2007. On the other hand, Grassini et al. (2011) in irrigated high yield environments in Nebraska suggested farm yields are approaching or have reached their economic upper limit and Cassman et al. (2010) suggest a yield plateau may be developing for irrigated maize in the USA. Overall, evidence suggests relative rates of potential yield progress fall below the necessary exponential rate required to meet projected demand for 2050 with evidence of yield plateaus in some countries.

3.1.2 Rice

Japonica rice has generally shown slow progress in YP in recent years, e.g. 0.4% year^{-1} from 1961 to 2008 relative to yields in 1968 (Fischer and Edmeades 2010). A focus on better rice quality, requiring reduced N fertilization, appears to have taken up much of the breeding effort. Progress in the YP of inbred indica rice since the release of 'IR72' at IRRI in 1988 seems also to have been slow (Peng et al., 1999, 2000, 2010), with no progress reported in YP, although YP was maintained with improved disease and insect resistance and grain quality. Since the early 1990s, IRRI breeders have made a major effort to boost rice YP by ideotype breeding first for the new plant type (NPT) and then a second generation of NPT2 cultivars, which perform only slightly better than the best contemporaneous cultivars from the conventional inbred breeding program (Yang et al., 2007). The best hybrid 'super' rices in China appear to have NPT2 traits of higher biomass and light-saturated photosynthetic rate (A_{max}) around heading, higher specific leaf weight, and higher leaf chlorophyll than older hybrid check cultivars, which they out-yielded by 10 to 20% (Peng et al., 2008). The yield advantage of hybrid rice is typically 10–15% more than inbred cultivars (e.g. Peng et al., 1999; Yang et al., 2007; Bueno and Lafarge, 2009; Lafarge and Bueno, 2009; Li et al., 2009; Bueno et al., 2010). Hybrid rice now covers close to 60% (mean for 2004–2008) of the area sown to rice in China (Li et al., 2009), and the area is also expanding in south-east and south Asia (Janaiah and Xie, 2010). Li et al. (2009) show that commercial hybrid rice yields have been growing at a steady 46.4 kg ha^{-1} year^{-1} over the last three decades (0.64% year^{-1}). Nevertheless, there is evidence of stagnation in the rate of gain in average rice yields in major rice-producing countries (Cassman et al., 2003). Grassini et al. (2014), in an analysis of yield trends from 1965 to 2010 in countries and regions worldwide, reported statistically significant upper yield plateaus for rice grown in China, Korea and California representing 33% of global rice.

3.1.3 Wheat

Several studies have examined yield progress in wheat in field side-by-side experiments on sets of historic CIMMYT-derived spring wheat cultivars and/or advanced lines. Waddington et al. (1986) observed genetic progress for seven spring wheat CIMMYT cultivars of 59 kg ha^{-1} (1.1 %) year^{-1} from 1950 to 1982 in NW Mexico. Sayre et al. (1997) reported progress of YP in NW Mexico for eight CIMMYT spring wheat cultivars from 1962 to 1988 at 67 kg ha^{-1} year^{-1} (0.88% year^{-1}) in high YP irrigated environments. The genetic yield progress of 26 spring wheat advanced lines released by CIMMYT from 1977 to 2008 was evaluated by Lopes et al. (2012) in irrigated conditions; grain yield progress was linear and about 0.7% year^{-1}. Generally, these investigations indicated that genetic progress of CIMMYT spring wheat has continued in recent decades but at a slightly slower rate. Shearman et al. (2005) in a study on cultivars released in the UK from 1972 to 1995 showed a relative rate of genetic gain of 0.59% $year^{-1}$ in UK winter wheat. Clarke et al. (2012) reported on genetic gains in winter wheat cultivars (11 feed and 9 bread-making cultivars) released in the UK between 1953 and 2008. Results showed that breeders continued to increase yield potential linearly by 0.067 t ha^{-1} $year^{-1}$ (relative rate 0.6% $year^{-1}$ expressed relative to yield in 2007) for bread-making cultivars and at 0.049 t ha^{-1} $year^{-1}$ (relative rate 0.5% $year^{-1}$ expressed relative to 2007) for feed cultivars. Mackay et al. (2011), from an analysis of genetic gain in winter wheat in the UK since 1947 which involved 3590 site–year combinations, found a linear rate of yield increase of 69 kg ha^{-1} $year^{-1}$ between 1948 and 2007 (relative rate of 0.76% $year^{-1}$ in 2007). Several other studies worldwide do show recent genetic gains for wheat yield potential, e.g. Zhou et al. (2007) and Zheng et al. (2011) in China, Peltonen-Sainio et al. (2009) in Finland and Sadras and Lawson (2013) in Australia. However, there are also cases in which the genetic progress appears to be showing a plateau, e.g. winter wheat in the great plains of North America between 1984 and 2008 (Graybosch and Peterson, 2010), yield improvement in the last years in Spain (Acreche et al., 2008) and no clear genetic gain in yield potential could be found in Chile after 1990 (Matus et al. 2012). In addition, Brisson et al. (2010) found genetic progress has not intrinsically declined in France but has been counteracted from 1990 onward by climate change (particularly by heat stress). In an analysis of independent data sets reported in Argentina, Australia, China, Italy, France, Serbia, the UK and the USA, the rates of yield improvement with year of cultivar release in wheat were shown to align positively with the environmental mean yield, i.e. the rate of yield gain was proportional to the yield of the background environment (Sadras and Lawson, 2013).

Overall, there is evidence that relative rates of yield are falling below the required increase of ca. 1.2% $year^{-1}$, e.g. in wheat (Brisson et al., 2010; Mackay et al., 2011), rice (Peng et al., 1999) and in maize (Cassman et al., 2010; Fischer and Edmeades, 2010). In at least some countries (regions) of the globe, it appears yield is plateauing; an extended discussion of the possible basis of the yield plateau in the major grain crops was recently provided by Hall and Richards (2013).

3.2 Traits associated with yield progress

3.2.1 Maize

Luque (2000) compared seven maize hybrids released from 1965 and 1997 in Argentina under irrigation and reported grain yield increased steadily with year of release from 7.4 to 14.8 t ha^{-1} associated with later anthesis, greater stay-green, greater anthesis RUE, improved partitioning to the ear around flowering, shortened anthesis–silking interval, and better grain set. Maize yield progress in southern Canada from 1959 and 1988 was associated with greater biomass and an extended period of photosynthesis after flowering (Tollenaar, 1991). Changes in yield and associated traits in Pioneer maize hybrids released between 1930 and 2002 were reported by Duvick (2005). Major traits underlying yield gains included: the yield at optimum density (optimum density increased to >80 000 plants ha^{-1} from 30 000 plants ha^{-1}); increased biomass; longer grain-filling duration; a more erect canopy (although green leaf area index (LAI) was little changed); increased stay-green; reduced tassel size; and anthesis–silking interval reduced to near zero. Tollenaar and Lee (2006) examining a set of early-maturing temperate hybrids in Canada reported similar findings for traits underlying yield gains to those of Duvick (2005) and also found that selection has increased both crop growth rate during grain filling, and leaf area index (ratio of leaf lamina surface area (one side) to ground area).

The improvement in maize grain yield through the last decades has indirectly been accompanied by a decline in grain N concentration (Duvick, 1997; Ciampitti and Vyn, 2012). Therefore, better understanding of the sources of grain N uptake in maize and especially the trade-off between N remobilization and whole-plant N uptake during the reproductive stage is needed to help guide future improvements in yield and N-use efficiency (grain yield dry matter/N supply from fertilizer and/or soil N; NUE). A literature review was performed by Ciampitti and Vyn (2012) to investigate changes over time in grain N sources and on N partitioning to the grain and stover plant fractions at maturity. The data set analysis was based on 100 reports, divided into: (1) research conducted from 1940 to 1990 – 'Old Era' – and (2) research conducted from 1991 to 2011 – 'New Era'. The main findings were that (1) reproductive (post-silk emergence) whole-plant N uptake contributed proportionally more to grain N for the New Era while reproductive N and remobilized N contributed equally to grain N for the Old Era, (2) remobilized N was primarily associated with vegetative (pre-silk emergence) whole-plant N uptake, which was constant across eras, (3) complex plant regulation processes (source–sink) appeared to influence whole plant N uptake, and (4) plant N uptake increased at maturity in both eras. Whether continued lowering of grain N concentration is a sustainable approach to pursue further gains in grain yield will depend to some extent on alternative destinations for grain maize as a final product (food, feed, fiber, and fuel).

3.2.2 *Rice*

The high yield potential of japonica rice cultivar, Takanari, relative to its lower yielding predecessors was associated with high filled spikelet number m^{-2} and high crop growth rate and RUE during the late jointing period (just before heading), higher RUE and non-structural carbohydrate content at heading and more soil nitrogen uptake (Takai et al., 2006; Katsura et al., 2007). The increased crop growth rate of Takanari was associated with greater canopy photosynthesis (Katsura et al., 2009). Ohsumi et al. (2007) confirmed a higher light-saturated leaf photosynthetic rate in Takanari compared with other high-yielding cultivars and showed it to be associated with higher stomatal conductance rather than higher specific leaf N content (N content per unit leaf lamina area; SLN).

The development of 'super' rice in China (Cheng et al., 1998) with F1 hybrid cultivars using a combination of the ideotype approach and interspecific heterosis produced yield gains 8–15% higher than the hybrid check cultivars associated with more biomass than ordinary hybrid and inbred varieties. The yield advantage of super hybrid rice cultivars developed in China was attributed to a substantial increase of biomass rather than harvest index (HI) (Cheng et al., 2007; Yang et al., 2007). Higher biomass was associated with a long growth duration and high accumulated incident radiation and significantly larger panicle size (spikelets per panicle), which resulted in larger sink size (spikelets m^{-2}) (Cheng et al., 1998). Higher biomass with a long growth duration may imply a greater demand for nutrients. However, in a study of 'super' hybrid, ordinary hybrid, and inbred varieties of cultivars, Zhang et al. (2009) concluded that the 'super' hybrid group cultivars did not necessarily require more N fertilizer to produce high grain yield. Improvement in plant type design was achieved in China's 'super' hybrid rice by emphasizing the top three leaves and panicle position within a canopy that meet the demand of heavy panicles for a large source supply during the post-anthesis period (Cheng et al., 2007). The success of China's 'super' hybrid rice was partially the result of assembling the good components of IRRI's NPT design, including large panicle size, reduced tillering capacity, and improved lodging resistance. However, crop RUE did not explain the yield superiority of 'super' hybrid rice. As with wheat, breeding of higher yielding rice cultivars may have been associated with progress from sink limitation towards a closer balance with source limitation (Dingkuhn and Kropff, 1996).

3.3.3 Wheat

Clarke et al. (2012) reported that, for UK winter wheat varieties (11 feed and 9 bread-making cultivars) released from 1953 to 2008, YP increased linearly by 0.067 t ha^{-1} $year^{-1}$ for bread-making cultivars and by 0.049 t ha^{-1} $year^{-1}$ for feed cultivars. For feed varieties, above-ground biomass increased linearly by 0.031 t ha^{-1} $year^{-1}$; and for both groups HI increased with year of release but progress was slowing with a quadratic curve fitting data better than a linear curve. There was a linear increase in grains m^{-2} for the feed varieties. For the bread-making varieties, grains m^{-2} did not increase although there was a linear increase in grains ear^{-1}. The only component that did not change was grain weight. For both the feed and bread-making varieties, there was no change in anthesis date but days from emergence to GS31 was positively correlated with grain number ($P < 0.001$). For UK wheat cultivars introduced from 1972 to 1995, Shearman et al. (2005) observed yield gains associated with biomass increase, higher pre-anthesis RUE and stem water-soluble carbohydrate (WSC). For spring wheat grown in NW Mexico under fully irrigated conditions using six key semi-dwarf bread wheat cultivars spanning 1962 to 1988 and grown with irrigation and full foliar disease protection, Sayre et al. (1997) showed YP progress of 0.8% $year^{-1}$ associated with increased grains m^{-2} ($r^2 = 0.71$, $P < 0.001$) and HI ($r^2 = 0.66$, $P < 0.05$). For the same cultivars, Fischer et al. (1998) reported higher stomatal conductance pre-anthesis with year of release and found a positive association between grains m^{-2} and grain yield and crop growth rate and RUE in the 3-week period before flowering in one year out of the three. Greater light-saturated photosynthetic rate was found in the more recent cultivars. Changes in the flag leaf with breeding progress at CIMMYT (e.g. smaller, more erect flag leaves, higher N and chlorophyll per unit area) were similar to those reported for winter wheat progress in the UK (Shearman et al., 2005). In contrast, Lopes et al. (2012) reported that yield progress of CIMMYT advanced lines from 1977 to 2008 was associated with increased grain weight and greater stay-green and Aisawi (2011) found that yield progress at CIMMYT from 1966 to 2009 in semi-dwarf cultivars was associated with changes in grain weight and not grain number.

Xiao et al. (2012) reported that genetic gain in Shandong Province, China, over the past four decades was 62 kg ha^{-1} $year^{-1}$, largely associated with increased grains m^{-2} and biomass together with increased HI and reduced plant height. Significant genetic changes were also observed for leaf area index, chlorophyll content and stem water-soluble carbohydrate content at anthesis and photosynthesis rate during grain filling. Similar to the findings of Shearman et al. (2005), increases in grains m^{-2} and biomass were apparently achieved through improving crop photosynthesis at and after heading, and increased stem WSC in stems at anthesis may have contributed to grain filling. An overview of candidate physiological traits for yield potential and the associations between them in maize, rice and wheat is shown in Figure 16.2.

Changes in N uptake are also important as they have implications for management (N fertilizer) and grain protein. Various studies worldwide have identified genetic associations between grain yield and NUE components under contrasting conditions of high and low N input supply. Genetic gains in NUE with breeding under low N supply have been related mainly to improvements in N-uptake efficiency (above-ground N uptake/N supply from fertilizer and/or soil N; NUpE) in spring wheat in Mexico (Ortiz-Monasterio et al., 1997) and Finland (Muurinen et al., 2006) and to N-utilization efficiency (grain DM/above-ground N uptake; NUtE) in winter wheat in France (Brancourt-Hulmel

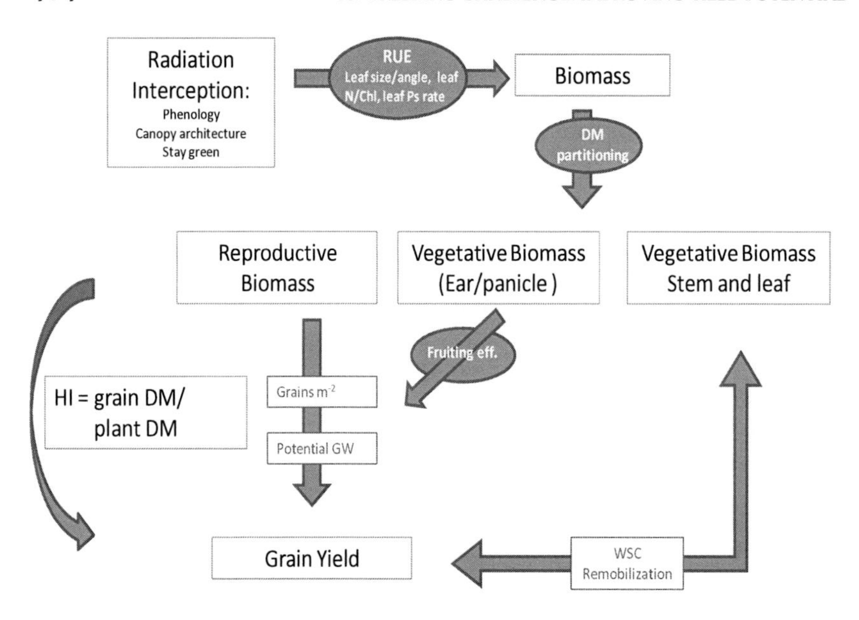

FIG. 16.2 Links between physiological traits associated with yield potential in maize, rice and wheat: radiation-use efficiency (RUE), potential grain weight (potential GW), fruiting efficiency (fruiting eff.), above-ground DM accumulated pre-anthesis (vegetative biomass), above-ground DM accumulation post-anthesis (reproductive biomass), water-soluble carbohydrate (WSC).

et al., 2003) and the UK (Foulkes et al., 1998). Under high N supply, wheat breeding improved mainly NUtE in Mexico (Fischer and Wall, 1976), Argentina (Calderini et al., 1995), France (Brancourt-Hulmel et al., 2003) and in various countries (Paccaud et al., 1985; Feil, 1992). In contrast, studies in the UK (Foulkes et al., 1998), Mexico (Ortiz-Monasterio et al., 1997) and Finland (Muurinen et al., 2006) found that increases in NUE with year of cultivar release were explained approximately equally by N uptake and N utilization efficiency. Sadras and Lawson (2013) examining a collection of Australian wheats released between 1958 and 2007 found increased NUE was mainly associated with nitrogen uptake, although there was a secondary contribution of reduced grain N concentration particularly under environments that favored high grain N%. Sylvester-Bradley and Kindred (2009), examining yields of 34 winter wheat cultivars released from the late 1970s to the 2000s, found yield increases were associated with increases in the economic optimum N amount, but there was no significant change in NUE with the optimum amount of applied N for wheat. As with maize, the optimization of traits to pursue further gains in grain yield will depend to some extent on alternative destinations for grain wheat as a final product (bread-making, feed, biscuit-making, biofuel, etc); for example, the stay-green trait may not be beneficial if it is associated with lowering of grain N concentration in bread-making varieties (Foulkes et al., 2009).

4 OPPORTUNITIES FOR FUTURE GAINS IN YIELD POTENTIAL

Prospects for further progress in yield potential must build on the physiological understanding set out above, including recent reviews on wheat (Sylvester-Bradley et al., 2005; Reynolds et al., 2012) and on all three crops (Fischer and Edmeades, 2010; Hall and Richards, 2013). Overall, there may be scope for raising HI in the short to medium term above current values of ≈0.50–0.55 in maize and wheat (Fischer and Edmeades 2010; Foulkes et al., 2011) and ≈0.45–0.50 in rice (Fischer and Edmeades, 2010), e.g. through further decreasing assimilate partitioning to structural stem DM component in the pre-anthesis period (Foulkes et al., 2011).

However, it may be difficult to increase HI significantly over ≈0.60 if the risk of lodging is to be kept at a reasonable level (Berry et al., 2007). Biomass production, therefore, must be the main medium-to-long-term avenue for yield potential progress. This implies greater radiation-use efficiency, since most crops grown under high yield potential conditions intercept >90% of the incident solar radiation apart from short phases prior to stem extension and during the latter stages of grain filling, and there may be restricted scope to extend crop duration. Under such conditions, four key breeding challenges to enhance YP progress will be to: (1) optimize rooting traits determining water and nutrient uptake (underpinning RUE and biomass gains); (2) increase radiation-use efficiency; (3) increase spike/panicle fertility through increased spike/panicle DM partitioning and/or fruiting efficiency (grains per unit spike/panicle DM at anthesis); and (4) increase potential grain size (which may upregulate RUE during the post-anthesis phase). The prospects for advances in these four physiological trait areas are now examined.

4.1 Optimize rooting traits

Crop improvement for enhanced biomass seems likely to be increasingly dependent on breeding for deeper or denser root systems which may promote enhanced soil water and nutrient capture in the absence of improvements in above-ground water- and nutrient- utilization efficiencies. Raising RUE may also partly depend on an increased capacity for water and nutrient uptake particularly during the latter stages of grain filling. For example, the principal restriction to future genetic gains in biomass for UK rain-fed winter wheat was estimated to be water availability (Sylvester-Bradley et al., 2005). A deeper relative distribution of roots while maintaining root DM ratio (root DM/total DM) could comprise part of an ideotype to maximize grain yield in future breeding programs (Carvalho and Foulkes, 2011) and further improvements in root architecture could focus on root proliferation at depth in wheat (Foulkes et al. 2009). Modeling studies suggest that deeper roots and more effective root systems have contributed to increased maize yields at higher planting densities in Iowa (e.g. Hammer et al., 2009). Lynch (2007) has successfully selected for multigenic root traits to increase phosphorus acquisition in maize. Nevertheless, breeding for root characteristics has seldom been implemented to date, principally because of the difficulties of scoring root phenotypes directly and the absence of suitable proxy measurements. In wheat, there is evidence that indirect selection for increased leaf vigor has enlarged the root system through increases in root biomass and length and root length density (Palta et al., 2011). This has contributed to increasing the capture of water and nitrogen early in the season, and facilitates the capture of additional water for grain filling under water-limited environments. Lower canopy temperatures might be taken as an indirect indication of a greater root water-uptake capacity, but higher stomatal conductance would produce a similar signal (Reynolds et al., 2009).

Genetic variation in root system size has been widely reported in grain crops (e.g. O'Toole and Bland, 1987; Hoad et al., 2001; Ehdaie and Waines, 2003), but root distribution varies strongly with soil characteristics such as water and nutrient availability and mechanical impedance. The root DM ratio of wheat is typically ≈30% during early vegetative growth decreasing to ≈10% by anthesis (Gregory et al., 1978, 1989; Miralles et al., 1997). Effects of increasing plant height on root DM partitioning have been studied using isolines and are generally either neutral or negative (e.g. Gregory et al., 1989; Miralles et al., 1997). The existence of significant genetic variation for rooting traits has not resulted to date, except for a few exceptions as mentioned above, in the incorporation of rooting traits in conventional breeding. In a recent review, Hall and Richards (2013) concluded that root architecture and root function are likely to be multigenic and hence much more difficult to select for and less amenable to marker-assisted selection. For example, xylem vessel diameter in the seminal roots of wheat is a trait which has a moderate heritability and can be selected for, but it is controlled by multiple genes each with a small effect. Future genetic progress could potentially be accelerated by the application of rapidly emerging genetic resources facilitating the fine mapping of root quantitative trait loci (QTL) and the development of markers for marker-assisted selection. There is a need for high precision root phenotyping because the genetic differences may be small, and detailed physiological measurements (e.g. of root length density at depth) are difficult when large numbers of genotypes are involved.

Field phenotyping methods for roots in cereals were reviewed by Manske et al. (2001) and Polomski and Kuhn (2002), including the use of rhizotrons and assessments of root parameters from soil cores (root washing and root counts/image analysis). The use of root-observation chambers and a non-destructive digital imaging technique offers some promise (Manschadi et al., 2006, 2010), but may be less suitable for screening of root traits that are expressed at later stages of crop development. Laboratory screens have focused mainly on seedlings, with traits correlating to field performance in only some cases (Wojciechowski et al., 2009; Bai et al., 2013). Although several screening tests have been designed to generate accurate and robust data from seedlings grown under artificial conditions, these phenotypes can rarely be extrapolated to field conditions because of the pronounced plasticity of root growth and development processes. The development of methods that measure changes in the root DNA concentration in soil, to allow comparison of soil DNA concentrations from different wheat genotypes, could eliminate the need for separation of roots from soil and permits large-scale phenotyping of root genotypes and responses to environmental stresses in the field (Chun et al., 2013).

4.2 Increase radiation-use efficiency

Yield potential can be expressed very simply as a function of the light intercepted (LI) and RUE, whose product is biomass, and the partitioning of biomass to grain yield, i.e. harvest index (Chapter 15). Increasing RUE is one avenue to increase the crop growth rate (above-ground dry matter per unit area per day; CGR) in the pre-anthesis period. This may in turn increase assimilate supply to the spike/panicle pre-anthesis and increase grains m^{-2}. An increase in grains m^{-2} may then act to enhance post-anthesis RUE through alleviation of feedback inhibition of photosynthesis. There is evidence that grain growth in maize, rice and wheat is limited by grain sink or co-limited by grain source and sink size. For example, grain yield improvement was highly associated with grain number per unit area in wheat (Canevara et al., 1994; Sayre et al., 1997; Brancourt-Hulmel et al., 2003; Shearman et al., 2005; Peltonen-Saino et al., 2009), rice (Horie et al., 1997; Peng et al., 2000) and maize (Andrade et al., 1999; Echarte et al., 2000; Tollenaar et al., 2000). In wheat, current evidence suggests grain sink strength remains a critical yield-limiting factor (Borrás et al., 2004; Miralles and Slafer, 2007; Fischer, 2009; Serrago et al., 2013) and that improving the balance between source and sink as outlined above is a promising approach for raising yield potential (Reynolds et al., 2001, 2005).

Genetic gains in ear or panicle biomass at anthesis have been achieved historically through enhanced partitioning of dry matter to these organs in the critical period during stem extension in small grain cereals (Chapter 12). It is generally accepted that the critical period covers the phase from late stem elongation to anthesis in wheat (Fischer, 1985), of the 14 days prior to full heading in rice (Takai et al., 2006) and from −230 to 100°Cd from silking (the active stem elongation period) in maize (Otegui and Bonhomme, 1998). There is some evidence that RUE may now be directly contributing to gains in ear or panicle DM accumulation as well during the critical period (Horie et al., 2003; Shearman et al., 2005; Takai et al., 2006). In this situation, increased RUE during the critical pre-anthesis period may directly cause increases in grain number which, in turn, may enhance RUE in the post-anthesis period as a result of improved grain sink strength. In rice, Horie et al. (2003) observed a significant positive relationship between crop growth rate during the period from 2 weeks before heading to final heading and grain yield, and stressed the importance of higher crop growth rate (CGR; above-ground dry matter m^{-2} day^{-1}) on the basis of spikelet formation. Takai et al. (2006) examining eight genotypes also found a positive association between CGR from 14 days before heading to heading and grain yield in rice; genotypes having higher CGR produced more spikelets m^{-2}. Interestingly, the large genotypic variability in CGR during the 14-day period was mainly derived from RUE. Likewise a positive association between RUE during the period from the onset of stem elongation to flowering and each of grains m^{-2} and grain yield was observed in winter wheat among a set of cultivars released from 1972 to 1995 in the UK (Shearman et al., 2005).

One of the challenges of understanding and improving RUE is that it is a moving target (Reynolds et al., 2005). It is affected by light intensity that varies on an hourly, daily and seasonal basis, as well as the age of the photosynthetic tissue (Murchie and Reynolds, 2012). It has also been shown that RUE is not always expressed at its maximum capacity, responding significantly to sink strength. For example, post-anthesis RUE in wheat is strongly influenced by partitioning to reproductive structures. Older cultivars with smaller reproductive sinks show significantly reduced RUE during grain filling (Calderini et al. 1997; Acreche and Slafer 2009), and larger sink strength increased crop growth and RUE during grain filling for both *Rht* (Miralles et al., 1997) and *Lr19* introgression lines (Reynolds et al., 2001, 2009). These results show the importance of optimizing source: sink, especially in the context of crops in which investment is made to improve genetically for RUE.

Consistent with theoretical predictions for C_3 species (Zhu et al., 2010), increases in above-ground biomass of wheat have been reported in recent years (Shearman et al., 2005; Bustos et al., 2013; Sadras et al., 2012) and some physiological and genetic bases identified (Reynolds et al., 2009). Key to achieving progress in this complex area will be obtaining a better understanding of the genetic and molecular control of how partitioning of assimilates at key developmental stages affects ear/panicle fertility and hence the determination of grain number and sink strength. At the metabolic level, strategies include modifying specificity, catalytic rate and regulation of Rubisco, upregulating Calvin cycle enzymes, introducing chloroplast CO_2 concentrating mechanisms, or introducing all of the genes necessary to express a full C_4 photosynthesis (Zhu et al., 2010; Parry et al., 2011).

Current wheat, rice and maize are well below the theoretical photosynthetic limit, which is commonly taken from the response of photosynthesis to radiation at low levels of radiation; according to Zhu et al. (2008), for C_3 crops, this amounts to the capture (net of respiration) of 4.6% of intercepted total solar radiant energy as carbohydrate energy or about 2.7 g DM MJ^{-1} intercepted total solar radiation. The highest recorded conversion rates for the full crop life cycle are around half of this; for shorter periods 2 g DM MJ^{-1} have been recorded. The corresponding theoretical limit for C_4 crops is 6.0% or 3.5 g DM MJ^{-1} of total incident solar energy (Zhu et al., 2008). Actual rates found in maize fall well short of this. Here, we explore further the possible physiological mechanisms by which breeding may enhance RUE.

4.2.1 Radiation-use efficiency at the canopy level

At the canopy level, modification of leaf architecture is expected to improve RUE by permitting a light distribution profile that reduces the number of leaves experiencing wasteful and potentially destructive supersaturated light levels, while increasing light penetration to canopy levels where photosynthesis responds linearly to light. Effects of canopy architecture on RUE have been observed with RUE being higher for large wheat canopies with more erect leaves, associated with reduced light saturation of the upper leaves (Evans, 1973; Araus et al., 1993). The erectophile leaf trait was introduced into wheat from sources such as *T. sphaerococcum* and modern wheat frequently shows more erectophile canopies (Fischer, 2007a). Nonetheless, modeling suggests that there may still be scope for further optimization of both light and N distribution in the canopy (Mussgnug et al., 2007).

It has been suggested that to maximize carbon gain by a canopy, N should be optimally distributed so that leaves receiving the greatest photon flux densities have the largest specific leaf N content (Field and Mooney, 1983; Grindlay, 1997). Theoretically, canopy photosynthesis is maximized when each leaf in the canopy receives irradiance in proportion to the associated photosynthetic capacity (Farquhar, 1989). A vertical N distribution that follows the light gradient would allow higher photosynthesis compared with that expected from a uniform N distribution (Mooney and Gulmon, 1979). The 'optimization' theory proposed by Hirose and Werger (1987) suggests that lamina N distribution within a vegetative canopy optimizes whole canopy photosynthesis. It proposes that, within a dense canopy, leaf lamina N distribution is driven by the light gradient such that SLN follows an exponential function of the downward cumulative leaf area index with an extinction coefficient for N (K_N) equal to that for light (K_L). In wheat, observed N gradients are generally less steep than predicted with the optimization theory, but do demonstrate that SLN follows an exponential gradient with vertical depth in the canopy (Critchley, 2001; Pask, 2009). Alternatively, Chen et al. (1993) proposed the co-ordination theory to explain N distribution in a canopy. This theory explicitly takes photosynthetic processes into account at the leaf scale to explain the relationship between light and N vertical distribution in vegetative canopies. Specific leaf N is computed to maintain a balance between the Rubisco-limited rate of carboxylation and the electron transport-limited rate of carboxylation, which depends on the amount of intercepted light.

There is relatively little information on genetic diversity in the vertical distribution of N in the canopy (Chapter 8), despite the large amount of work published on light interception and attenuation by crop canopies. In wheat, small differences were observed in the distribution of N in the top four leaves at anthesis between two UK winter wheat cultivars, Soissons and Spark but, overall, the distributions were close to that predicted by the optimization theory in both cultivars (Critchley, 2001). Similarly, the vertical distribution of N at anthesis was close to the optimum, as defined in the optimization theory, and did not differ significantly for two French winter wheat cultivars, Apache and Isengrain, until almost the end of grain filling (Bertheloot et al., 2008). The role of N dynamics on canopy photosynthesis and crop productivity will likely become even more important in the future because of the increase of atmospheric CO_2 concentration (Kim et al., 2001; Anten et al., 2004). Overall, the reported investigations indicate that actual plants tend to distribute N more uniformly than the optimal distribution. The difference between actual and optimal distribution implies that optimal N distribution leads to leaf N per unit area that is too low at the bottom and too high at the top to be realized. Hirose (2005) suggested that the reason for this may be that some N may not be capable of translocation, and a certain amount of N is necessary to utilize sunflecks that leaves receive in lower layers in the canopy (Pons et al., 1993). Sadras et al. (2013) investigated crop and leaf photosynthetic traits in a set of Australian wheat cultivars released between 1958 and 2007. Radiation-use efficiency increased linearly with year of cultivar release at a rate of 0.012 g MJ^{-1} $year^{-1}$. There was a sharp extinction of nitrogen concentration with canopy depth relative to the extinction of radiation in older varieties that shifted to a flatter nitrogen–radiation extinction in newer varieties. Increased RUE in newer varieties was associated with the relaxation in the coupling between the extinction of nitrogen and radiation in the profile which was in turn partially related to the improved N status of modern varieties, as quantified with the N nutrition index (Chapter 8).

Measurement of the relative contribution of spike photosynthesis in wheat (Tambussi et al., 2007) to overall canopy photosynthesis has never been considered as a selection criterion despite the large proportion of light that spikes intercept during grain filling. However, recent comparative studies of the integrated contribution of spike photosynthesis to grain weight showed highly significant genetic effects (Molero and Reynolds, unpublished data). At the level of plant growth and development, a more optimal balance between source and sink is expected to improve overall RUE (Reynolds et al., 2012; Serrago et al., 2013).

4.2.2 *Increase leaf photosynthetic rate*

In cereals, RUE only increases at a low rate as A_{max} increases above values of about 20–30 μmol CO_2 m^{-2} s^{-1} (Monteith, 1977; Sinclair and Horie, 1989). This is because individual leaves may operate well below light saturation in the canopy and because of the need to account for dark respiration (Reynolds et al., 2000). Selection for photosynthetic parameters other than A_{max} may introduce pleiotropic trade-offs with greater SLN being associated with reduced leaf size and light interception (Austin et al., 1982). Since leaf size and thickness are associated with anatomical structure, there may be merit in investigating the cellular basis of differences in leaf specific weight including the number of chloroplasts per cell and mesophyll cell size. Changes in mesophyll cell size and area per leaf induced by ploidy have been linked to photosynthetic rate in wheat (Austin et al., 1982). However, sacrificing leaf size to gain a higher rate of leaf photosynthesis per unit leaf area in future breeding strategies must be considered carefully.

4.2.3 *Decrease respiration*

Respiration in crops has received less attention than photosynthesis due to difficulties in measurement and the fact that it is rather heterogeneous in the plant depending on tissue type and substrate. Respiration nonetheless is critical in determining yield, and therefore potentially influences RUE. It is highly responsive to temperature and may become an issue as global temperatures rise (Peng et al., 2004). Respiration is commonly divided into growth and maintenance components, with each exerting differing effects. An important early observation was that LAI does not increase proportionally with respiration rate, meaning that canopy assimilation can continue to respond positively to irradiance. Respiration therefore may be positively but nonlinearly related to photosynthesis.

Is it possible to reduce 'wasteful' respiration in crop canopies? The ratio of respiration to photosynthesis is thought to be close to the optimum (Amthor, 2000), although it is suggested that the data for respiration and regulation in different tissue types during development need refinement, and that improvements may come from reduction of maintenance respiration (Amthor, 2000). There has been recent work showing that genetic manipulation of respiration in tomato has a knock-on effect on both biomass and yield (Nunes-Nesi, 2005).

Respiration may consume 30–80% of the carbon fixed (Atkin et al., 2005), increasing with temperature and depending on phenological stage (McCullough and Hunt, 1993). High respiration rates (especially at night) can increase reactive oxygen species (ROS), leading to cell damage and affecting pollen viability (Prasad et al., 1999). Recent work highlighting the importance of increased night-time temperature on productivity in rice (Sheehy et al., 2006) and wheat (Tester and Langridge, 2010; Lizana and Calderini, 2013) and the high sensitivity of respiration to temperature in general, suggests that the environmental responses of crop respiration to temperature is an important area on which to focus.

4.3 Increase ear/panicle DM partitioning and fruiting efficiency

The proportion of shoot biomass partitioned to the ear or panicle at anthesis is typically in the range 0.15–0.20% in modern wheat and rice cultivars (Peng et al., 2000; Spink et al., 2000; Shearman et al., 2005; Gaju, 2007). To increase partitioning of assimilate to the ear or panicle further, future avenues would include reducing partitioning of assimilate to competing sinks: stems, roots (structural and non-structural carbohydrate components) and leaves. Reducing partitioning to the leaf lamina may be incompatible with raising RUE (although it may be possible to modify leaf traits, e.g. specific leaf area, to boost RUE while maintaining leaf partitioning). Potential avenues for increasing ear/panicle partitioning therefore seem mainly to rely on reducing partitioning to the stems or roots. Reducing the DM partitioning to the structural stem DM during stem extension would potentially favor ear/panicle DM partitioning. The study of Beed et al. (2007) on winter wheat in the UK demonstrated shading from flag-leaf emergence to heading reduced non-structural (water-soluble carbohydrate) stem DM proportionally more than the structural stem DM, implying that stem WSC DM does not compete strongly with the ear for assimilate. Selecting for lower structural stem DM (or higher stem WSC DM) may therefore be a favored mechanism to raise ear/panicle partitioning index (ear/panicle DM/above-ground DM at anthesis; EPPI) and HI. Cultivars allocating more dry matter to stem carbohydrate reserves should be better positioned to buffer grain yield and maximize HI. In wheat, optimized partitioning for enhancing ear DM partitioning and yield potential was recently reviewed by Foulkes et al. (2011). Those authors suggested strategies for: (1) optimizing phenological pattern and spike growth rate to maximize EPPI and grain number; and (2) desensitizing floret abortion to environmental cues to maximize EPPI and grain number. In maize, increased grain number of modern hybrids was related to greater dry matter accumulation around silking and to increased DM partitioning to the ear (Tollenaar and Lee, 2006; Echarte and Tollenaar, 2006; Chapter 15).

Fischer and Stockman (1986), examining tall and dwarf isolines of wheat, observed that competition between the ear and stem growth began at a relatively early stage during stem elongation and that allometric relationships may be established shortly after onset of stem elongation affecting ear growth. Siddique et al. (1989) also found that the difference in the ear to stem ratio between tall and semi-dwarf Australian isolines was evident soon after the terminal spikelet stage. A constant ratio was observed between relative growth rates of the ear and the stem in tall and semi-dwarf lines. A greater ear:stem ratio at anthesis for the semi-dwarfs was attributed to a bigger intercept of the regression of ln ear DM vs ln stem DM; and semi-dwarf isolines had a lower predicted stem DM per shoot at ear DM (= 1 mg).

The complement to ear/panicle DM partitioning at anthesis that may be improved to enhance further grain number is the fruiting efficiency (FE; grains g^{-1} ear/panicle DM at anthesis) (Abbate et al., 1997; Fischer, 2011; Foulkes et al., 2011; Lázaro and Abatte, 2011). There is often a negative relationship between EPPI and FE in wheat (Gaju et al., 2009; Gonzalez et al.,

2011; Lázaro and Abbate, 2011), although other studies did not find a negative association (Ferrante et al., 2012; Bustos et al., 2013). The cause of the frequently observed negative relationship between EPPI and FE is still unclear. In a study of two bread wheat cultivars, EPPI was apparently associated with the proportion of some of the ear morphological components, but none of these associations was significant (Abbate et al., 1995, 1997). Abbate et al. (1998) studied Argentinean wheat cultivars released after 1980 and did not find an association between EPPI and the rachis proportion. Fischer (2007b) similarly observed no association between EPPI and the weight of glumes or awns. On the other hand, Slafer and Andrade (1993) observed that higher grains m^{-2} among bread wheat genotypes was associated with allocating a higher proportion of ear DM to reproductive (developing florets) rather than structural (rachis, glumes and paleas) organs within the ears. Improvement of vascular connections within the rachilla may be one avenue to minimize the trade-off between EPPI and FE in wheat breeding, since floret fertility of more distal florets still may be limited by resistance according to the vascular connections within the rachilla (Hanif and Langer, 1972; Minchin et al., 1993; Bancal and Soltani, 2002). In maize, increased grain number of modern hybrids was related to more grains per unit ear dry weight (Tollenaar and Lee, 2006; Echarte and Tollenaar, 2006). More grains, amounting to a greater grain-filling sink, may also be increasing photosynthesis and dry weight accumulation during this grain-filling period through feedback mechanisms.

4.4 Strategies to optimize potential grain size

Strategies to increase potential grain size were reviewed in the original version of this chapter (Foulkes et al., 2009). In brief, the switch between determination of grain number and individual grain weight at anthesis is not absolute; individual grain weight can be affected by assimilate supply for a period before anthesis (Calderini et al., 1999), and grain number can be affected by assimilate supply for a period after anthesis (Beed et al., 2007). It would clearly be disadvantageous for a total potential grain weight (grain number $\times$ potential weight $grain^{-1}$, PWG) to be set before anthesis which differs substantially from the likely availability of assimilate for grain storage after anthesis. This could partly explain why grain number is typically well related to grain yield in cereals (Sinclair and Jamieson, 2006, 2008; Fischer, 2008). The last stage in setting the balance between sink and subsequent source capacities of the inflorescence is in the determination of potential weight $grain^{-1}$ (PWG). More endosperm cells are initiated than eventually survive, so final cell number depends on cell loss at the end of the cell division phase (in wheat ≈15 days after pollination; Gao et al., 1992). Cell expansion then continues and, although final cell volume is less important in determining PGW (Brocklehurst et al., 1978), this also can influence PWG. Thus PGW is not fully determined until about one-third of grain DM has accumulated (Borrás et al., 2004). In general, source–sink manipulation experiments (involving de-graining or defoliation during early grain development) show PWG to be responsive to assimilate supply (Brocklehusrt, 1977; Singh and Jenner, 1984; Fischer and HilleRisLambers, 1978) up to, but not beyond, a certain point (Borrás et al., 2004). Possibly more subtle signals, that correlate with assimilate supply, are controlling cell division.

The challenge for future yield improvement is to devise strategies that will enhance the scope for PWG to respond to availability of assimilates without unduly enhancing the risk of incomplete grain filling. Possible physiological objectives are to reduce the DM requirements of cell division and expansion, to understand and change the signals to which cell division is responding, or to increase the concurrence of cell expansion and grain DM deposition. This last might be achieved for each grain or, more controversially (because of increased variation in maturity), by extending the disparity between the development of grains within the inflorescence.

There is also evidence that final grain size and the floret cavity volume may be influenced by mechanical constraints (Millet, 1986; Foulkes et al., 2011). This could explain the strong maternal influence on final grain weight reported for several species (Jones et al., 1996; Millet and Pinthus, 1980). The importance of the pericarp on early grain development is well established (Rijven and Banbury, 1960), but the physiological processes of the pericarp which affect the final size of the grain are only recently being investigated (e.g. Schruff et al., 2006; Song et al., 2007). Recent studies indicate expansion of the pericarp may be controlled by the rheological properties of the cell wall through the action of expansins (Lizana et al., 2010), proteins allowing the loosening of cell walls (McQueen-Mason et al., 1992). Hormones also play a major role in the coordination of grain tissue expansion. In rice, the endosperm concentration in cytokinins during the period of active endosperm cell division was related to the endosperm cell number and the final grain weight at different positions within the panicle for genotypes of contrasting potential grain size (Yang et al., 2002). Also in rice the ABA to ethylene ratio was positively associated with endosperm cell division and grain-filling rates (Yang et al., 2006). The effect of hormones on grain development is closely related to sugar metabolism and signaling (Cheng and Chourey, 1999).

It will be important for plant breeding for enhanced potential grain size to be underpinned by genetical analysis, and it is encouraging in this regard that, for a range of different wheat crosses, some QTL controlling grain yield have been found to work primarily through individual grain weight without pleiotropic effects on grain number (Snape et al., 2007). Recent results in rice have demonstrated the important role of several transcription factors and E3 ubiquitin ligases in sugar and hormone signaling networks controlling the early stages of grain development (e.g. Song et al., 2007).

5 TRAIT-BASED BREEDING

Trait selection has made continual progress in breeding through incorporating agronomic traits such as height and flowering time, resistance to a spectrum of prevalent pests and diseases, quality parameters determined by end use, and yield based on multilocation trials (Braun et al., 2010). However, to accelerate genetic gains, physiological trait (PTs) must now be considered in crossing strategies.

The PTs related to light interception (LI), such as stand establishment, ground cover, canopy architecture, and nitrogen partitioning within the canopy, show significant genetic variation in conventional gene pools and are relatively straightforward to phenotype on a routine basis with the possible exception of canopy architecture as it affects light extinction at different canopy levels. They are also highly amenable to visual selection, suggesting that they are probably not currently major bottlenecks for improving yield potential, a picture which could change as bottlenecks in RUE and partitioning are removed.

Although physiological traits related to RUE are generally more challenging to measure, the potential to increase RUE is supported by theory as well as observations of increased biomass in recent cultivars, as already outlined. Initially, the focus should be on integrative traits such as growth rate at key phenological stages (such as when grain number and grain weight potential are being deterimined), as well as expression of final biomass. Later, these traits can be dissected into processes at the canopy, plant, organ, tissue, cellular, and subcellular levels. Strong candidates for non-transgenic targets include optimal light and pigment distribution in canopies to increase canopy level RUE, and reduce mesophyll resistance to CO_2 (Murchie and Reynolds, 2012). As mentioned earlier, transgenic targets include enzymes of the Calvin cycle as well as C_4-type traits.

While HI has increased steadily, with the exception of the deployment of major-effect alleles at the *Rht*, *Ppd* and *Vrn* loci (Mathews et al., 2006), optimal expression of HI is still achieved empirically and is subject to seasonal effects (Ugarte et al., 2007). In wheat, for example, genetic variation in expression of HI can be found typically in the range of 0.4–0.55 in elite cultivars worldwide (Sayre et al., 1997; Shearman et al., 2005; Zheng et al. 2011). Maximum expression will require a clear understanding of how to optimize dry matter partitioning to reproductive structures without under-investing in roots, stems and leaves on which both grain yield and lodging resistance are also dependent. Crop phenology must be conducive to spike fertility as well as being tailored to different photoperiod and temperature regimens (Chapter 12).

Hybridization schemes should be designed to combine PTs (Box 16.1) such that cumulative gene action is probable. Given limited

BOX 16.1

A CONCEPTUAL PLATFORM FOR DESIGNING CROSSES THAT COMBINE COMPLEMENTARY YIELD POTENTIAL TRAITS IN WHEAT (BASED ON TRAITS REVIEWED IN REYNOLDS ET AL., 2009).

The simple model is based on the assumption that since light interception (LI), radiation-use efficiency (RUE) and harvest index (HI) are the three main drivers of yield, combining traits from different drivers is likely to result in cumulative genetic gains.

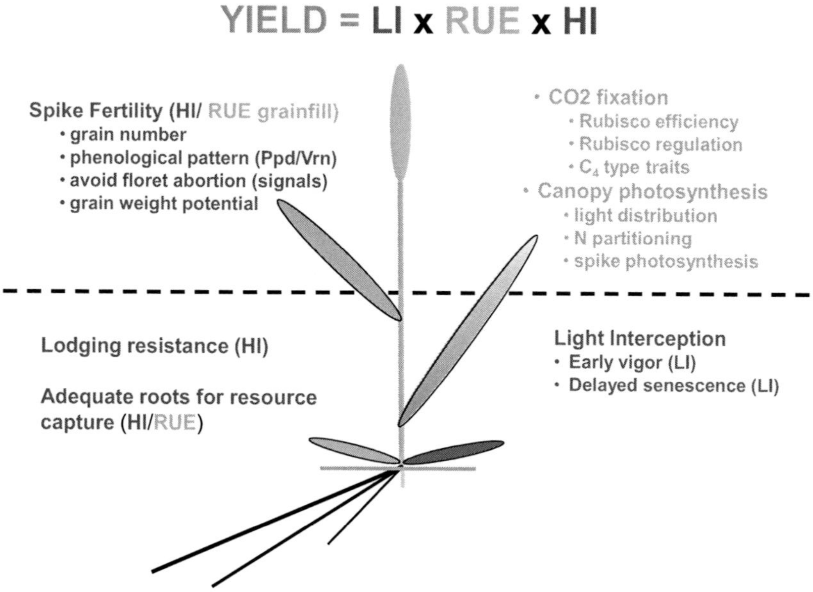

understanding of the genetic basis of traits that contribute to yield, it is impossible to predict the outcome of combining theoretically complementary characteristics. Nonetheless, selecting for PTs is a practical means of increasing the probability of achieving cumulative gene action, as demonstrated by recent impacts in breeding for drought adaptation (Rebetzke et al., 2009; Reynolds et al. 2009). This approach could be complemented by marker-assisted crossing and selection as more information from genetic dissection of complex physiological traits becomes available as well as diagnostic markers such as *Ppd-1*, *Vrn-1*, *Vrn-3*, and *Rht-1*

Diagnostic markers are not yet available for the majority of PT, precluding marker-assisted selection. Notwithstanding the potential value of genomic selection in this context (Chapter 18), a few high-throughput phenotyping approaches can be applied in progeny screening. One is canopy temperature (CT), that under appropriate conditions is well correlated with stomatal conductance (Amani et al., 2006), and therefore a proxy for photosynthetic rate and high sink demand. Using airborne remote sensing platforms, a range of spectral indices can be measured at high throughput (Araus and Cairns, 2014), including the normalized vegetative difference index (NDVI) and the water index, both of which can estimate relative differences in biomass among plots at similar growth stages (Babar et al., 2006).

6 CONCLUDING REMARKS

Rates of genetic gain in yield potential may be slowing in wheat, rice and (albeit to a lesser extent) maize, but it appears that breeding progress has not stopped. This may be associated with harvest index having approached values close to theoretical maxima and biomass gains being harder to achieve. In wheat and rice, at least, there is evidence that modern cultivars may be in a closer source:sink balance than their predecessors. Evidence now points to the contribution of increased photosynthetic activity before and around flowering for recent genetic increases in yield potential. Therefore, more attention is recommended to photosynthetic activity before and around anthesis in in-depth studies. In addition, studies on the N status of modern varieties are justified including associations between RUE in newer varieties and the coupling between the extinction of nitrogen and radiation in the profile. Furthermore, increasing RUE may require a more detailed focus on root traits with increased water and nutrient uptake required to support high biomass, particularly in the latter stages of grain filling to maintain RUE. A relatively deeper distribution of roots is a prospective characteristic which deserves further study along with the development of high-throughput root phenotyping with application in breeding. Improved crop designs may also relate to: (1) optimizing structural stem DM partitioning during stem elongation, to increase ear/panicle DM partitioning at anthesis; (2) increasing fruiting efficiency to enhance grains per unit area; and (3) increasing potential grain size by reducing the dry matter requirements of cell division and expansion and improved understanding of the signals to which cell division is responding. Hybridization schemes should be designed to combine physiological traits such that cumulative gene action is probable. This approach could be complemented by marker-assisted crossing and selection as more information from genetic dissection of complex physiological traits becomes available. Genetic studies should focus on finding QTL that affect biomass without affecting harvest index, and vice-versa, or grain number without affecting grain weight. Increased understanding of the physiological processes underlying yield potential at the crop level of organization is required, to exploit key traits either directly in breeding or through contributing to the development and use of molecular markers for these quantitative complex traits.

References

Abbate, P.E., Andrade, F.H., Culot, J.P., 1995. The effects of radiation and nitrogen on number of grains in wheat. J. Agric. Sci., Camb. 124, 351–360.

Abbate, P.E., Andrade, F.H., Culot, J.P., Bindraban, P.S., 1997. Grain yield in wheat: effects of radiation during spike growth period. Field Crops Res. 54, 245–257.

Abbate, P.E., Andrade, D.H., Lazaro, L., Baraitti, J.H., Beradocco, H.G., Inza, V.H., Marturano, F., 1998. Grain yield in recent Argentine wheat cultivars. Crop Sci. 38, 1203–1209.

Acreche, M.M., Slafer, G.A., 2009. Grain weight, radiation interception and use efficiency as affected by sink-strength in Mediterranean wheats released from 1940 to 2005. Field Crops Res. 110, 98–105.

Acreche, M., Briceno-Felix, G., Martın Sanchez, J.A., Slafer, G.A., 2008. Physiological bases of genetic gains in Mediterranean bread wheat yield in Spain. Eur. J. Agron. 28, 162–170.

Aisawi, K.A.B., 2011. Physolgical processes associated with genetic progress in wheat (*Triticum aestivum* L.). PhD Thesis, University of Nottingham, 239 pp.

Aisawi, K.A.B., 2012. Physiological processes associated with genetic progress in yield potential of wheat (*Triticum aestivum* L.). PhD Thesis, University of Nottingham, 266 pp.

Amani, I., Fischer, R.A., Reynolds, M.P., 1996. Evaluation of canopy temperature as a screening tool for heat tolerance in spring wheat. J. Agron. Crop Sci. 176, 119–129.

Amthor, J.S., 2000. The McCree-de Wit-Penning de Vries-Thornley respiration paradigms: 30 years later. Ann. Bot. 86, 1–20.

Andrade, F.H., Vega, C., Uhart, S., Cirilo, A., Cantarero, M., Valentinuz, V., 1999. Kernel number determination in maize. Crop Sci. 39, 453–459.

Anten, N.P.R., Hirose, T., Onoda, Y., et al., 2004. Elevated CO2 and nitrogen availability have interactive effects on canopy carbon gain in rice. New Phytol. 161, 459–471.

Araus, J.L., Reynolds, M.P., Acedevo, E., 1993. Leaf structure, leaf posture, growth, grain yield and carbon isotope discrimination in wheat. Crop Sci. 33, 1273–1279.

Araus, J.L., Slafer, G.A., Reynolds, M.P., Royo, C., 2002. Plant breeding and drought in C3 cereals: what should we breed for? Ann. Bot. 89, 925–940.

Araus, J.L., Cairns, J., 2014. Field high-throughput phenotyping: the new crop breeding frontier. Trends Plant Sci. 19, 52–61.

Atkin, O.K., Bruhn, D., Hurry, V.M., Tjoelker, M.G., 2005. Evans Review No. 2: The hot and the cold: unravelling the variable response of plant respiration to temperature. Funct. Plant Biol. 32, 87–105.

Austin, R.B., Morgan, C.L., Ford, M.A., Bhagwat, S.G., 1982. Flag leaf photosynthesis of *Triticum aestivum* and related diploid and tetraploid species. Ann. Bot. 49, 177–189.

Babar, M.A., Reynolds, M.P., Van Ginkel, M., Klatt, A.R., Raun, W.R., Stone, M.L., 2006. Spectral reflectance to estimate genetic variation for in-season biomass, leaf chlorophyll and canopy temperature in wheat. Crop Sci. 46, 1046–1057.

Bai, C., Liang, Y., Hawkesford, M.J., 2013. Identification of QTLs associated with seedling root traits and their correlation with plant height in wheat. J. Exp. Bot. 64, 1745–1753.

Bancal, P., Soltani, F., 2002. Source–sink partitioning. Do we need Münch? J Exp. Bot. 53, 1919–1928.

Beed, F.D., Paveley, N.D., Sylvester-Bradley, R., 2007. Predictability of wheat growth and yield in light-limited conditions. J. Agric. Sci. (Camb.) 145, 63–79.

Berry, P.M., Sylvester-Bradley, R., Berry, S., 2007. Ideotype for lodging resistant wheat. Euphytica 154, 165–179.

Bertheloot, J., Martre, P., Andrieu, B., 2008. Dynamics of light and nitrogen distribution during grain filling within wheat canopy. Plant Physiol. 148, 1707–1720.

Borrás, L., Slafer, G.A., Otegui, M.E., 2004. Seed dry weight response to source–sink manipulations in wheat, maize and soybean: a quantitative reappraisal. Field Crops Res. 86, 131–146.

Brisson, N., Gate, P., Gouache, D., Charmet, G., Oury, F.-X., Huard, F., 2010. Why are wheat yields stagnating in Europe? A comprehensive data analysis for France. Field Crops Res. 119, 201–212.

Brocklehusrt, P.A., 1977. Factors controlling grain weight in wheat. Nature 266, 348–349.

Bruinsma, J., 2009. The resource outlook to 2050. By how much do land, water use and crop yields need to increase by 2050? In: Proc. FAO Expert Meeting on How to Feed the World in 2050, 24–26 June, 2009. FAO, Rome (available at http://www.fao.org/wsfs/forum2050/background-documents/expert-papers/en/).

Brancourt-Hulmel, M., Doussinault, G., Lecomte, C., Bérard, P., Le Buanec, B., Trottet, M., 2003. Genetic improvement of agronomic traits of winter wheat cultivars released in France from 1946 to 1992. Crop Sci. 43, 37–45.

Braun, H.J., Atlin, G., Payne, T., 2010. Multi-location testing as a tool to identify plant response to global climate change. In: Reynolds, M.P. (Ed.), Climate change and crop production. CABI Climate Change Series, pp. 115–138.

Brocklehurst, P.A., Moss, J.P., Williams, W., 1978. Effects of irradiance and water supply on grain development in wheat. Ann. Appl. Biol. 90, 265–276.

Bueno, C.S., Lafarge, T., 2009. Higher crop performance of rice hybrids than of elite inbreds in the tropics: 1. Hybrids accumulate more biomass during each phenological phase. Field Crops Res. 112, 229–237.

Bueno, C.S., Pasuquin, E., Tubaña, B., Lafarge, T., 2010. Improving sink regulation, and searching for promising traits associated with hybrids, as a key avenue to increase yield potential of the rice crop in the tropics. Field Crops Res. 118, 199–207.

Bustos, D.V., Hasan, A.K., Reynolds, M.P., Calderini, D.F., 2013. Combining high grain number and weight through a DH-population to improve grain yield potential of wheat in high-yielding environments. Field Crops Res. 145, 106–115.

Calderini, D.F., Slafer, G.A., 1999. Has yield stability changed with genetic improvement of wheat yield? Euphytica 107, 51–59.

Calderini, D.F., Torres-Leon, S., Slafer, G.A., 1995. Consequences of wheat breeding on nitrogen and phosphorus yield, grain nitrogen and phosphorus concentration and associated traits. Ann. Bot. 76, 315–322.

Calderini, D.F., Dreccer, M.F., Slafer, G.A., 1997. Consequences of plant breeding on biomass growth, radiation interception and radiation use efficiency in wheat. Field Crops Res. 52, 271–281.

Calderini, D.F., Abeledo, L.G., Savin, R., Slafer, G.A., 1999. Effects of temperature and carpel size during pre-anthesis on potential grain weight in wheat. J. Agric. Sci. 132, 453–459.

Canevara, M.G., Romani, M., Corbellini, M., Perenzin, M., Borghi, B., 1994. Evolutionary trends in morphological, physiological, agronomical and qualitative traits of *Triticum aestivum* L. cultivars bred in Italy since 1900. Eur. J. Agron., 175–185.

Cardwell, V.B., 1982. Fifty years of Minnesota corn production: Sources of yield increase. Agron. J. 74, 984–990.

Carvalho, P., Foulkes, M.J., 2011. Roots and the uptake of water and nutrients. In: Meyers, R.A. (Ed.), Encyclopedia of sustainability science and technology. Springer, Heidelberg, pp. 1390–1404.

Cassman, K.G., 1999. Ecological intensification of cereal production systems: Yield potential, soil quality, and precision agriculture. Proc. Natl. Acad. Sci. USA 96, 5952–5959.

Cassman, K.G., Dobermann, A., Walters, D.T., Yang, H., 2003. Meeting cereal demand while protecting natural resources and improving environmental quality. Annu. Rev. Environ. Resour. 28, 315–358.

Cassman, K.G., Grassini, P., van Wart, J., 2010. Crop yield potential, yield trends and global food security in a changing climate. In: Hillel, D., Rosenzweig, C. (Eds.), Handbook of climate change and agroecosystems: impacts, adaptation, and mitigation. World Scientific, pp. 37–51.

Chen, J.L., Reynolds, J.F., Harley, P.C., Tenhunen, J.D., 1993. Coordination theory of leaf nitrogen distribution in a canopy. Oecologia 93, 63–69.

Cheng, W.H., Chourey, P.S., 1999. Genetic evidence that invertase mediated release of hexoses is critical for appropriate carbon partitioning and normal seed development in maize. Theor. Appl. Genet. 98, 485–495.

Cheng, S., Liao, X., Min, S., 1998. China's "super" rice research: background, goals and issues. China Rice 1, 3–5, (in Chinese).

Cheng, S., Cao, L.-Y., Zhuang, J.-Y., et al., 2007. Super hybrid rice breeding in China: achievements and prospects. J. Integr. Plant Biol. 49, 805–810.

Chun, Y., Huang, L., Kuchel, H., et al., 2013. A DNA-based method for studying root responses to drought in field-grown wheat genotypes. Sci. Rep. 12, 3194.

Ciampitti, I.A., Vyn, T.J., 2012. Grain nitrogen source changes over time in maize: a review. Crop Sci. 53, 366–377.

Clarke, S., Sylvester-Bradley, R., Foulkes, J., et al., 2012. Adapting wheat to global warming or 'ERYCC' earliness and resilience for yield in a changing climate. HGCA Research and Development Report 496. Home-Grown Cereals Authority: London.

Cooper, M., Smith, O.S., Graham, G., Arthur, L., Feng, L., Podlich, D.W., 2004. Genomics, genetics, and plant breeding: A private perspective. Crop Sci. 44, 1907–1913.

Critchley, C.S., 2001. A physiological explanation for the canopy nitrogen requirement of wheat. PhD Thesis, University of Nottingham.

Dingkuhn, M., Kropff, M., 1996. Rice. In: Zamski, E., Schaffer, A.A. (Eds.), Photo-assimilate distribution in plants and crops: source-sink relationships. Marcel Dekker Inc, New York, pp. 519–547.

Duvick, D.N., 1997. What is yield? *In*: Edmeades, G.O., Banziger, M., Mickelson, H.R., Pena-Valdivia, C.B. (Eds.), Developing drought- and low N-tolerant maize. Proceedings of a Symposium, CIMMYT, El Batan, Mexico, 25–29 March 1996. CIMMYT: Mexico, D.F., pp. 332–335.

Duvick, D.N., 2005. The contribution of breeding to yield advances in maize (*Zea mays* L.). Adv. Agron. 86, 83–145.

Echarte, L., Luque, S., Andrade, F.H., et al., 2000. Response of maize kernel number to plant density in Argentinean hybrids released between 1965 and 1995. Field Crops Res. 68, 1–8.

Echarte, L., Tollenaar, Y., 2006. Kernel set in maize hybrids and their inbred lines exposed to stress. Crop Sci. 46, 870–878.

Edmeades, G.O., Banzinger, M., Ribaut, J-M., 2000. Maize improvement for drought-limited environments. In: Otegui, M.A., Slafer, G.A. (Eds.), Physiological bases for maize improvement. Food Products Press, New York, pp. 75-112.

Ehdaie, B., Waines, J.G., 2003. 1RS translocation increases root biomass in Veery-type wheat isogenic lines and associates with grain yield. In: Pogna, N.E., Romano, M., Pogna, E.A., Galterio, G. (Eds.), In: Proceedings of the 10th International Wheat Genetics Symposium. Rome. ISC: Paestum. pp. 693–695.

Evans, L.T., 1973. The effect of light on plant growth, development and yield. In: Slatyer, R.O. (Ed.), Plant response to climatic factors. UNESCO, Paris, pp. 21–35.

Evans, L.T., 1993. Feeding the ten billion. Cambridge University Press, Cambridge.

Evans, L.T., Fischer, R.A., 1983. Yield potential: its definition, measurement, and significance. Crop Sci. 39, 1544–1551.

Farquhar, G.D., 1989. Models of integrated photosynthesis of cells and leaves. Philosoph. Transact. R. Soc. Lond. Ser. B 323, 357–367.

Feil, B., 1992. Breeding progress in small grain cereals, a comparison of old and modern cultivars. Plant Breed. 108, 1–11.

Ferrante, A., Savin, R., Slafer, G.A., 2012. Differences in yield physiology between modern, well adapted durum wheat cultivars grown under contrasting conditions. Field Crops Res. 136, 52–64.

Field, C., Mooney, H.A., 1983. Leaf age and seasonal effects on light, water and nitrogen use efficiency in a California shrub. Oecologia 56, 348–355.

Fischer, R.A., 1985. Number of kernels in wheat crops and the influence of solar radiation and temperature. J. Agric. Sci. (Camb.) 105, 447–461.

Fischer, R.A., 2007a. Understanding the physiological basis of yield potential in wheat. J. Agric. Sci. 145, 99–113.

Fischer, R.A., 2007b. The importance of grain or kernel number in wheat: A reply to Sinclair and Jamieson. Field Crops Res. 105, 15–21.

Fischer, R.A., 2008. Improvements in wheat yield: Farrer, physiology and functional genomics. Agric. Sci., 6–18, NS 1/08.

Fischer, R.A., 2009. Farming systems of Australia: Exploiting the synergy between genetic improvement and agronomy. In: Sadras, V., Calderini, D. (Eds.), Crop physiology: Applications for genetic improvements and agronomy. Elsevier, Amsterdam, pp. 23–54.

Fischer, R.A., 2011. Wheat physiology: a review of recent developments. Crop & Pasture Science 62, 95–114.

Fischer, R.A., Edmeades, G.O., 2010. Breeding and cereal yield progress. Crop Sci. 50, 585–598.

Fischer, R.A., HillRisLambers, D., 1978. Effect of environment and cultivar on source limitation to grain weight in wheat. Aust. J. Agric. Res. 29, 433–458.

Fischer, R.A., Rees, D., Sayre, K.D., Lu, Z.-M., Condon, A.G., Larque-Saavedra, A., 1998. Wheat yield progress associated with higher stomatal conductance and photosynthetic rate, and cooler canopies. Crop Sci. 38, 1467–1475.

Fischer, R.A., Stockman, Y.M., 1986. Increase kernel number in Norin 10-derived dwarf wheat: evaluation of the cause. Australian Journal of Plant Physiol. 13, 767–784.

Fischer, R.A., Wall, P.C., 1976. Wheat breeding in Mexico and yield increases. J. Aust. Inst. Agric. Sci. 42, 139–148.

Foulkes, M.J., Reynolds, M.P., Sylvester-Bradley, R., 2009. Genetic improvement of grain crops: yield potential. In: Sadras, V., Calderini, D. (Eds.), Crop physiology: Applications for genetic improvements and agronomy. Elsevier, Amsterdam, pp. 355–385.

Foulkes, M.J., Slafer, G.A., Davies, W.J., Berry, P.M., Sylvester-Bradley, R., Martre, P., Calderini, D.F., Giffiths, S., Reynolds, M.P., 2011. Raising yield potential of wheat. III. Optimizing partitioning to grain while maintaining lodging resistance. J. Exp. Bot. 62, 469–486.

Foulkes, M.J., Sylvester-Bradley, R., Scott, R.K., 1998. Evidence for differences between winter wheat cultivars in acquisition of soil mineral nitrogen and uptake and utilization of applied fertiliser nitrogen. J. Agric. Sci. (Camb.) 130, 29–44.

Gaju, O., 2007. Identifying physiological processes limiting genetic improvement of ear fertility in wheat. PhD thesis, University of Nottingham, UK. 230 pp.

Gaju, O., Reynolds, M.P., Sparkes, D.L., Foulkes, M.J., 2009. Relationships between large-spike phenotype, grain number, and yield potential in spring wheat. Crop Sci. 49, 961–973.

González, F.G., Terrile, I.I., Falcón, M.O., 2011. Spike Fertility and Duration of Stem Elongation as Promising Traits to Improve Potential Grain Number (and Yield): Variation in Modern Argentinean Wheats. Crop Sci. 51, 1693–1702.

Gao, X., Francis, D., Ormorod, J.C., Bennett, M.D., 1992. Changes in cell number and cell division activity during endosperm development in allohexaploid wheat, *Triticum aestivum* L. J. Exp. Bot. 43, 1602–1606.

Grassini, P., Yang, H., Cassman, K.G., 2009. Limits to maize productivity in western Corn Belt: A simulation analysis for fully irrigated and rainfed conditions. Agric. For. Meteorol. 149, 1254–1265.

Grassini, P., Thorburn, J., Burr, C., Cassman, K.G., 2011. High-yield irrigated maize in the Western U.S. Corn Belt: I. On-farm yield, yield potential, and impact of agronomic practices. Field Crops Res. 120, 142–150.

Grassini, P., Eskridge, K., Cassman, K.G., 2014. Distinguishing between yield advances and yield plateaus in historical crop production trends. Nat. Commun. 4, 2918.

Graybosch, R.A., Peterson, C.J., 2010. Genetic improvement in winter wheat yields in the Great Plains of North America, 1959–2008. Crop Sci. 50, 1882–1890.

Gregory, P.J., McGowan, M., Biscoe, P.V., Hunter, B., 1978. Water relations of winter wheat .1. Growth of root system. J. Agric. Sci. 91, 91–102.

Gregory, P.J., Brown, S.C., 1989. Root growth, water use and yield of crops in dry environments: what characteristics are desirable? Aspects Appl. Biol. 22, 235–243.

Grindlay, D.J.C., 1997. Towards an explanation of crop nitrogen demand based on the optimisation of leaf nitrogen per unit leaf area. J. Agric. Sci. (Camb.) 128, 377–396.

Hall, A.J., Richards, R.A., 2013. Prognosis for genetic improvement of yield potential and water-limited yield of major grain crops. Field Crops Res. 143, 18–33.

Hammer, G.L., Dong, Z., McLean, G., et al., 2009. Can changes in canopy and/or root system architecture explain historical maize yield trends in the U.S. Corn Belt? Crop Sci. 49, 299–312.

Hanif, M., Langer, R.H.M., 1872. The vascular system of the spikelet in wheat (*Triticum aestivum*). Ann. Bot. 36, 721–727.

Hayman, P.H., 2007. The impact of drought on grain farms – best opportunities for improving farmers livelihoods. In: Australian winter cereals pre-breeding alliance. workshop on pre-breeding for better performance under drought, Canberra, 5-6 September, 2007.

Hirose, T., 2005. Development of the Monsi-Saeki theory on canopy structure and function. Ann. Bot. 94, 483–495.

Hirose, T., Werger, M.J.A., 1987. Maximizing daily canopy photosynthesis with respect to the leaf nitrogen allocation pattern in the canopy. Oceologia 72, 520–526.

Hoad, S.P., Russell, G., Lucas, M.E., Bingham, I.J., 2001. The management of wheat, barley and oat root systems. In: Sparks, D.L. (Ed.), Advances in Agronomy, Vol 74, Elsevier Academic Press Inc, San Diego, pp. 193–246.

Horie, T., Ohnishi, M., Angus, J.F., Lewin, L.G., Tsukaguchi, T., Matano, T., 1997. Physiological characteristics of high-yielding rice inferred from cross-location experiments. Field Crops Res. 52, 55–67.

Horie, T., Lubis, I., Takai, T., et al., 2003. Physiological traits associated with high yield potential in rice. In: Mew, T.W., Brar, D.S., Peng, S., Dawe, D., Hardy, B. (Eds.), Rice science: innovations and impact for livelihood. IRRI, Los Baños, Philippines, pp. 117–145.

Janaiah, A., Xie, F., 2010. Hybrid rice adoption in India: farm level impacts and challenges. IRRI Tech. Bull. no. 14. IRRI: Los Baños, Philippines.

Jones, R.J., Schreiber, B.M., Roessler, J.A., 1996. Kernel sink capacity in maize: genotypic and maternal regulation. Crop Sci. 36, 301–306.

Katsura, K., Maeda, S., Horie, T., Shiraiwa, T., 2007. Analysis of yield attributes and crop physiological traits of Liangyoupeiju, a hybrid rice recently bred in China. Field Crops Res. 103, 170–177.

Katsura, K., Maeda, S., Horie, T., Shiraiwa, T., 2009. Estimation of respiratory parameters for rice based on long-term and intermittent measurement of canopy CO_2 exchange rates in S-98 WWW.CROPS.ORG CROP SCIENCE, VOL. 50, MARCH–APRIL 2010 the field. Field Crops Res. 111, 85-91.

Kim, H.Y., Lieffering, M., Miura, S., Kobayashi, K., Okada, M., 2001. Growth and nitrogen uptake of CO2-enriched rice under field conditions. New Phytol. 150, 223–230.

Lafarge, T., Bueno, C.S., 2009. Higher crop performance of rice hybrids than elite inbreds in the tropics. 2. Does sink regulation, rather than sink size, play a major role? Field Crops Res. 114, 434–440.

Lázaro, L., Abbate, P.E., 2011. Cultivar effects on relationship between grain number and photothermal quotient or spike weight in wheat. J. Agric. Sci. 150, 447–459.

Li, J., Xin,Y., Yuan, L., 2009. Hybrid rice technology development: ensuring China's food security. IPFRI discussion paper 0918. IPFRI: Washington, DC.

Lizana, X.C., Riegel, R., Gomez, L.D., et al., 2010. Expansins expression is associated with grain size dynamics in wheat (*Triticum aestivum* L.). J. Exp. Bot. 61, 1147–1157.

Lizana, X.C., Calderini, D.F., 2013. Yield and grain quality of wheat in response to increased temperatures at key periods for grain number and grain weight determination: considerations for the climatic change scenarios of Chile. J. Agric. Sci. 151, 209–221.

Lopes, M.S., Reynolds, M.P., Manes, Y., Singh, R.P., Crossa, J., Braun, H.J., 2012. Genetic yield gains and changes in associated traits of CIMMYT spring bread wheat in a 'Historic' set representing 30 years of breeding. Crop Sci. 52, 1123–1131.

Luque, S.L., 2000. Bases ecofisiologicas de la ganancia genetica en el rendimiento del maiz en la Argentina en los ultimos 30 ãnos. MSc Thesis. University of Buenos Aires, 96 p.

Lynch, J.P., 2007. Roots of the second green revolution. Aust. J. Bot. 55, 493–512.

Mackay, I., Horwell, A., Garner, J., White, J., McKee, J., Philpott, H., 2011. Reanalyses of the historic series of UK variety trials to quantify the contributions of genetic and environmental factors to trends and variability in yield over time. Theor. Appl. Genet. 22, 225–238.

Manschadi, A.M., Christopher, J., de Voil, P., Hammer, G.L., 2006. The role of root architectural traits in adaptation of wheat to water-limited environments. Funct. Plant Biol. 33, 823–837.

Manschadi, A.M., Christopher, J.T., Hammer, G.L., deVoil, P., 2010. Experimental and modelling studies of drought-adaptive root architectural traits in wheat (*Triticum aestivum* L.). Plant Biosyst. 144, 458–462.

Manske, G.G.B, Ortiz-Monasterio, J.I., Vlek, P.L.G., 2001. Techniques for measuring genetic diversity in roots. In: Reynolds, M.P., Ortiz-Monasterio, J.I., McNab, A. (Eds.), Application of Physiology in Wheat Breeding. CIMMIYT, Mexico.

Matus, I., Mellado, M., Pinares, M., Madariaga, R., del Pozo, A., 2012. A. Genetic progress in winter wheat cultivars released in Chile from 1920 to 2000. Chil. J. Agic. Res. 72, 303–308.

Mathews, K.L., Chapman, S.C., Trethowan, R., et al., 2006. Global adaptation of spring bread and durum wheat lines near-isogenic for major reduced height genes. Crop Sci. 46, 603–613.

Maydup, M.L., Antonietta, M., Guiamet, J.J., Tambussi, E.A., 2012. The contribution of green parts of the ear to grain filling in old and modern cultivars of bread wheat (*Triticum aestivum* L.): Evidence for genetic gains over the past century. Field Crops Res. 134, 208–215.

McCullough, D.E., Hunt, L.A., 1993. Mature tissue and crop canopy respiratory characteristics of rye, triticale and wheat. Ann. Bot. (Lond.) 72, 269–282.

McQueen-Mason, S.J., Durachko, D.M., Cosgrove, D.J., 1992. Two endogenous proteins that induce cell wall extension in plants. Plant Cell 4, 1425–1433.

Messina, C.D., Hammer, G., Dong, Z., Podlich, D., Cooper, M., 2009. Modeling crop improvement in a GxExM framework via gene-trait-phenotype relationships. In: Sadras, V.O., Calderini, D.F. (Eds.), Crop Physiology: Applications for Genetic Improvement and Agronomy. Elsevier, Amsterdam, pp. 235–265.

Millet, E., 1986. Relationships between grain weight and the size of floret cavity in the wheat spike. Ann. Bot. 58, 417–423.

Millet, E., Pinthus, M.J., 1980. Genotypic effects of the material tissues of wheat on its grain weight. Theoret. Appl. Genet. 58, 247–252.

Minchin, P.E.H., Thorpe, M.R., Farrar, J.F., 1993. A simple mechanistic model of phloem transport which explains sink priority. Exp. Bot. 44 (5), 947–955.

Miralles, D.J., Slafer, G.A., 2007. Sink limitations to yield in wheat: how could it be reduced? J. Agric. Sci. (Camb.) 145, 139–150.

Miralles, D., Slafer, J., Lynch, V., 1997. Rooting patterns in near-isogenic lines of spring wheat for dwarfism. Plant Soil 197, 79–86.

Monteith, J.L., 1977. Climate and the efficiency of crop production in Britain. Philosophical Transactions of the Royal Society, London B. 281, 277–294.

Mooney, H.A., Gulmon, S.L., 1979. Environmental and evolutionary constraints on the photosynthetic characteristics of higher plants. In: Solbrig, O.T., Jain, S., Johnson, G.B., Raven, P.H. (Eds.), Topics in plant population biology. Columbia University Press, New York, pp. 316–337.

Murchie, E., Reynolds, M.P., 2012. Crop radiation capture and use efficiency. In: Meyers, R.A. (Ed.), Encyclopedia of sustainability science and technology. Springer.

Muurinen, S., Slafer, G.A., Peltonen-Sainio, P., 2006. Breeding effects on nitrogen use efficiency of spring cereals under northern conditions. Crop Sci. 46, 561–568.

Mussgnug, J.H., Thomas-Hall, S., Rupprecht, J., et al., 2007. Engineering photosynthetic light capture: impacts on improved solar energy to biomass conversion. Plant Biotechnol. J. 5, 802–814.

Nunes-Nesi, A., Carrari, F., Lytovchenko, A., et al., 2005. Enhanced photosynthetic performance and growth as a consequence of decreasing mitochondrial malate dehydrogenase activity in transgenic tomato plants. Plant Physiol. 137, 611–622.

Ohsumi, A., Hamasaki, A., Nakagawa, H., Yoshida, H., Shiraiwa, T., Horie, T., 2007. A model explaining genotypic and ontogenetic variation of leaf photosynthetic rate in rice (*Oryza sativa*) based on leaf nitrogen content and stomatal conductance. Ann. Bot. 99, 265–723.

Ortiz-Monasterio, J.I., Lobell, D.B., 2007. Remote sensing assessment of yield losses due to sub-optimal planting dates and fallow period weed management. Field Crops Res. 101, 80–87.

Otegui, M.E., Bonhomme, R., 1998. Grain yield components in maize I. Ear growth and kernel set. Field Crops Res. 56, 247–256.

O'Toole, J.C., Bland, W.L., 1987. Genotypic variation in crop plant-root systems. Adv. Agron. 41, 91–145.

Paccaud, F.X., Fossanti, A., Cao, H.S., 1985. Breeding for quality and yield in winter wheat: consequences for nitrogen uptake and nitrogen partitioning efficiency. Z. Pflanzenzucht. 94, 89–100.

Palta, J.A., Chen, X., Milroy, S.P., Rebetzke, G.J., Dreccer, M.F., Watt, M., 2011. Large root systems: are they useful in adapting wheat to dry environments? Funct. Plant Biol. 38, 347–354.

Parry, M.A.J., Reynolds, M.P., Salvucci, M.E., et al., 2011. Raising yield potential of wheat. II. Increasing photosynthetic capacity and efficiency. J. Exp. Bot. 62, 453–467.

Pask, A., 2009. Optimising nitrogen storage in wheat canopies for genetic reduction in fertiliser nitrogen inputs. PhD thesis, University of Nottingham, UK.

Peltonen-Sainio, P., Jauhiainen, L., Laurila, I.P., 2009. Cereal yield trends in northern European conditions: Changes in yield potential and its realisation. Field Crops Res. 110, 85–90.

Peng, S., Cassman, K.G., Virmani, S.S., Sheehy, J., Khush, G.S., 1999. Yield potential trends of tropical rice since the release ofIR8 and the challenge of increasing rice yield potential. Crop Sci. 39, 1552–1559.

Peng, S., Huang, J., Cassman, K.G., Laza, R.C., Visperas, R., Khush, G.S., 2010. The importance of maintenance breeding: a case study of the first miracle rice variety-IR8. Field Crops Res. 119, 342–347.

Peng, S.B., Huang, J.L., Sheehy, J.E., Laza, R.C., Visperas, R.M., Zhong, X.H., Centeno, G.S., Khush, G.S., Cassman, K.G., 2004. Rice Yields Decline with Higher Night Temperature from Global Warming. Proceedings of the National Academy of Sciences of the United States of America, 101, 9971–9975.

Peng, S., Khush, G.S., Virk, P., Tang, Q., Zou, Y., 2008. Progressin ideotype breeding to increase rice yield potential. Field Crops Res. 108, 32–38.

Peng, S., Laza, R.C., Visperas, R.M., Sanico, A.L., Cassman, K.G., Khush, G.S., 2000. Grain yield of rice cultivars and lines developed in the Philippines since 1966. Crop Sci. 40, 307–314.

Polomski, J., Kuhn, N., 2002. Root research methods. In: Waisel, Y., Eshel, A., Kafkafi, U. (Eds.), Plant roots: the hidden half. 3rd edn Marcel Dekker, Inc, New York, pp. 300–306.

Pons, T.L., Van Rijnbeek, H., Scheurwater, I., Van der Werf, A., 1993. Importance of the gradient in photosynthetically active radiation in a vegetation stand for leaf nitrogen allocation in two monocotyledons. Oecologia 95, 416–424.

Prasad, P.V.V., Craufurd, P.Q., Summerfield, R.J., 1999. Fruit number in relation to pollen production and viability in groundnut exposed to short episodes of heat stress. Ann. Bot. (Lond.) 84, 381–386.

Rebetzke, G.J., Chapma, S.C., MacIntryre, I., McIntryre, R., Condon, A., Van Herwaarden, A., 2009. Grain yield improvement in water-limited environments. In: Carver, B.F. (Ed.), Wheat: science and trade. Wiley–Blackwell, Ames, Iowa, pp. 215–249.

Reynolds, M.P., Van Ginkel, M., Ribaut, J.M., 2000. Avenues for genetic modification of radiation use efficiency in wheat. J. Exp. Bot. 51, 459–473.

Reynolds, M.P., Calderini, D.F., Condon, A.G., Rajaram, R., 2001. Physiological basis of yield gains in wheat associated with the LR19 translocation from Agropyron elongatum. Euphytica 119, 137–141.

Reynolds, M.P., Pellegrineschi, A., Skovmand, B., 2005. Sink-limitation to yield and biomass: a summary of some investigations in spring wheat. Ann. Appl. Biol. 146, 39–49.

Reynolds, M.P., Borlaug, N.E., 2006. Impacts of breeding on international collaborative wheat improvement. J. Agric. Sci. (Camb.) 144, 3–17.

Reynolds, M.P., Foulkes, J., Furbank, R., et al., 2012. Achieving yield gains in wheat. Plant Cell Environ. 35, 1799–1823.

Reynolds, M.P., Manes, Y., Izanloo, A., Langridge, P., 2009. Phenotyping for physiological breeding and gene discovery in wheat. Ann. App. Biol. 155, 309–320.

Richards, R.A., Rebetzke, G.A., Condon, A.G., van Herwaarden, A.F., 2002. Breeding opportunities for increasing the efficiency of water use and crop yield in temperate cereals. Crop Sci. 42, 111–121.

Rijven A.H.G.C., Banbury, C.A., 1960. Role of the grain coat in wheat grain development. Nature 188, 546–547.

Sadras, V.O., Lawson, C., Montoro, A., 2012. Photosynthetic traits of Australian wheat varieties released between 1958 and 2007. Field Crops Res. 134, 19–29.

Sadras, V.O., Lawson, C., 2013. Nitrogen and water-use efficiency of Australian wheat varieties released between 1958 and 2007. Eur. J. Agron. 46, 36–41.

Sayre, K.D., Rajaram, S., Fischer, R.A., 1997. Yield potential progress in short wheats in northwest Mexico. Crop Sci. 37, 36–42.

Schruff, M.C., Spielman, M., Tiwari, S., Adams, S., Fenby, N., Scott, R.J., 2006. The AUXIN RESPONSE FACTOR 2 gene of Arabidopsis links auxin signalling, cell division, and the size of seeds and other organs. Development 133, 251–261.

Serrago, R.A., Alzueta, I., Savin, R., Slafer, G.A., 2013. Understanding grain yield responses to source–sink ratios during grain filling in wheat and barley under contrasting environments. Field Crops Res. 150, 42–51.

Sharma, R.C., Crossa, J., Velu, G., et al., 2012. Genetic gains for grain yield in CIMMYT spring bread wheat across international environments. Crop Sci. 52, 1522–1533.

Shearman, V.J., Sylvester-Bradley, R., Scott, R.K., Foulkes, M.J., 2005. Physiological processes associated with wheat yield progress in the UK. Crop Sci. 45, 175–185.

Sheehy, J.E., Mitchell, P.L., Ferrer, A.B., 2006. Decline in rice grain yields with temperature: Models and correlations can give different estimates. Field Crops Res. 98, 151–156.

Slafer, G.A., Andrade, F.H., 1993. Physiological attributes related to the generation of grain yield in bread wheat cultivars released at different eras. Field Crops Res. 31, 351–367.

Slafer, G.A., Araus, J.L., Richards, R.A., 1999. Physiological traits that increase the yield potential of wheat. In: Satorre, E.H., Slafer, G.A. (Eds.), Wheat: ecology and physiology of yield determination. Food Products Press, New York, pp. 379–415.

Siddique, K.H.M., Kirby, E.J.M., Perry, M.W., 1989. Ear: stem ratio in old and modern wheat varieties; relationship with improvement in number of grains per ear and yield. Field Crops Res. 21, 59–78.

Sinclair, T.R., Horie, T., 1989. Leaf nitrogen, photosynthesis, and crop radiation use efficiency: a review. Crop Sci. 29, 90–98.

Sinclair, T.R., Jamieson, P.D., 2006. Grain number, wheat yield, and bottling beer: an analysis. Field Crops Res. 98, 60–67.

Sinclair, T.R., Jamieson, P.D., 2008. Yield and grain number of wheat: A correlation or causal relationship? Authors' response to The importance of grain or kernel number in wheat: A reply to Sinclair and Jamieson by R.A. Fischer. Field Crops Res. 105, 22–26.

Sinclair, T.R., Purcell, L.C., Sneller, C.H., 2004. Crop transformation and the challenge to increase yield potential. Trends Plant Sci. 9, 1360–1385.

Singh, B.K., Jenner, C.F., 1984. Factors controlling endosperm cell number and grain dry weight in wheat effects of shading on intact plants and of variation in nutritional supply to detached cultured ears. Aust. J. Plant Physiol. 11, 151–164.

Snape, J.W., Foulkes, M.J., Simmonds, J., et al., 2007. Dissecting gene × environmental effects on wheat yields via QTL and physiological analysis. Euphytica 154, 401–408.

Song, X.-J., Huang, W., Shi, M., Zhu, M.-Z., Lin, H.- X., 2007. A QTL for rice grain width and weight encodes a previously unknown RING-type E3 ubiquitin ligase. Nat. Genet. 39, 623–630.

Spink, J., Semere, T., Sparkes, D.L., Whaley, J.M., Foulkes, M.J., Clare, R.W., Scott, R.K., 2000. Effect of sowing date on the optimum plant density of winter wheat. Ann. Appl. Biol. 137, 179–188.

Sylvester-Bradley, R., Foulkes, J., Reynolds, M., 2005. Future wheat yields: Evidence, theory and conjecture. In: Sylvester-Bradley, R., Wiseman, J. (Eds.), Yields of farmed species. Nottingham University Press, Nottingham, pp. 233–260.

Sylvester-Bradley, R., Kindred, D.R., 2009. Analysing nitrogen responses of cereals to prioritize routes to the improvement of nitrogen use efficiency. J. Exp. Bot. 60, 1939–1951.

Takai, T., Matsuura, S., Nishio, T., Ohsumi, A., Shiraiwa, T., Horie, T., 2006. Rice yield potential is closely related to crop growth rate during late reproductive period. Field Crops Res. 96, 328–335.

Tambussi, E.A., Bort, J., Guiamet, J.J., Nogues, S., Araus, J.L., 2007. The photosynthetic role of ears in C3 cereals: metabolism, water use efficiency and contribution to grain yield. Crit. Rev. Plant Sci. 26, 1–16.

Tester, M., Langridge, P., 2010. Breeding Technologies to Increase Crop Production in a Changing World. Science 327, 818–822.

Tittonell, P., Shepherd, K.D., Vanlauwe, B., Giller, K.E., 2008. Unravelling the effects of soil and crop management on maize productivity in smallholder agricultural systems of western Kenya – an application of classification and regression tree analysis. Agric. Ecosyst. Environ. 123, 137–150.

Tollenaar, M., 1991. Physiological basis of genetic improvement of maize hybrids in Ontario from 1959 to 1988. Crop Sci. 31, 119–124.

Tollenaar, M., Lee, E.A., 2006. Dissection of physiological processes underlying grain yield in maize by examining genetic improvement and heterosis. Maydica 51, 399–408.

Tollenaar, M., Ying, J. Duvick, D.N. 2000. Genetic gain in corn hybrids from the Northern and Central Corn Belt. In: Proceedings of 55th Corn Sorghum Research Conference 5–8 December. Chicago, IL. ASTA Washington, DC, pp. 53–62.

Tweeten, L., Thompson, S.R., 2008. Long-term agricultural output supply-demand balance and real farm and food prices. Working Paper AEDE-WP 0044-08. The Ohio State University.

van Ittersum, M.K., Rabbinge, R., 1997. Concepts in production ecology for analysis and quantification of agricultural input-output combinations. Field Crops Res. 52, 197–208.

Waddington, S.R., Ransom, J.K., Osmanzai, M., Saunders, D.A., 1986. Improvement in the yield potential of bread wheat adapted to northwest Mexico. Crop Sci. 26, 698–703.

Wojciechowski, T., Gooding, M.J., Ramsay, L., Gregory, P.J., 2009. The effects of dwarfing genes on seedling root growth of wheat. J. Exp. Bot. 60, 2565–2573.

Ugarte, C., Calderini, D.F., Slafer, G.A., 2007. Grain weight and grain number responsiveness to pre-anthesis temperature in wheat, barley and triticale. Field Crops Res. 100, 240–248.

Xiao, Y.G., Qian, Z.G., Wu, K., et al., 2012. Genetic gains in grain yield and physiological traits of winter wheat in Shandong Province, China, from 1969 to 2006. Crop Sci. 52, 44–56.

Yang, J.C., Zhang, J.H., Huang, Z.L., Wang, Z.Q., Zhu, Q.S., Liu, L.J., 2002. Correlation of cytokinin levels in the endosperms and roots with cell number and cell division activity during endosperm development in rice. Ann. Bot. 90, 369–377.

Yang, J., Zhang, J., Wang, Z., Liu, K., Wang, P., 2006. Post-anthesis development of inferior and superior spikelets in rice in relation to abscisic acid and ethylene. J. Exp. Bot. 57, 149–160.

Yang, W., Peng, S., Laza, R.C., Visperas, R.M., Dionisio-Sese, M.L., 2007. Grain yield and yield attributes of new plant type and hybrid rice. Crop Sci. 47, 1393–1400.

Zhang, Y., Tang, Q., Zou, Y., et al., 2009. Yield potential and radiation use efficiency of "super" hybrid rice grown under subtropical conditions. Field Crops Res. 114, 91–98.

Zheng, T.C., Zhang, X.K., Yin, G.H., et al., 2011. Genetic gains in grain yield, net photosynthesis and stomatal conductance achieved in Henan Province of China between 1981 and 2008. Field Crops Res. 122, 225–233.

Zhou, Y., He, Z.H., Sui, X.X., Xia, X.C., Zhang, X.K., Zhang, G.S., 2007. Genetic improvement of grain yield and associated traits in the northern China winter wheat region from 1960 to 2000. Crop Sci. 47, 245–253.

Zhu, X.-G., Long, S.P., Ort, D.R., 2008. What is the maximum efficiency with which photosynthesis can convert solar energy into biomass? Curr. Opin. Biotechnol. 19, 153–159.

Zhu, X.-G., Long, S.P., Ort, D.R., 2010. Improving photosynthetic efficiency for greater yield. Annu. Rev. Plant Biol. 61, 235–261.

CHAPTER

17

Improving grain quality: ecophysiological and modeling tools to develop management and breeding strategies

Luis Aguirrezábal[1], *Pierre Martre*[2,3], *Gustavo Pereyra-Irujo*[1], *María Mercedes Echarte*[1], *Natalia Izquierdo*[1]

[1]Unidad Integrada Balcarce (Facultad de Ciencias Agrarias, Universidad Nacional de Mar del Plata – Instituto Nacional de Tecnología Agropecuaria, Estación Experimental Balcarce, Balcarce, Argentina

[2]INRA, UMR1095 Genetic, Diversity and Ecophysiology of Cereals, Clermont-Ferrand, France

[3]Blaise Pascal University, UMR1095 Genetic, Diversity and Ecophysiology of Cereals, Aubière, France

1 INTRODUCTION

The quality of a harvested organ can most simply be defined as its suitability for the intended market or processing and product manufacture. The term quality may therefore encompass many criteria, as the compositional and textural requirements may vary from one product to another. Sunflower and bread wheat are crops traditionally grown for production of edible oil, starch or proteins for human consumption. Thus, oilseed sunflower quality includes the potential industrial yield, the nutritional value of the oil and its stability (Box 17.1). For bread wheat, quality depends on the milling performance, the dough rheology, the baking quality, the nutritional value, and its suitability for storage. Moreover, the relative importance of quality attributes of a given end product changes along the market chain from grower to consumer. For

Crop Physiology. DOI: 10.1016/B978-0-12-417104-6.00017-0

BOX 17.1

SELECTED PROPERTIES OF SUNFLOWER OILS AND ITS COMPONENTS

Vegetable oils are composed mainly of triglycerides (or triacylglycerol, TAG), which consist of three fatty acids bound to a glycerol backbone. Fatty acids are usually classified based on the length of their carbon chain and the number and position of double bonds. The fatty acid composition defines the nutritional, industrial and organoleptic quality of the oil.

Saturated fatty acids, with the exception of stearic acid, increase the levels of cholesterol in humans (Velasco and Fernández-Martínez, 2002). Polyunsaturated acids, such as linoleic acid, are essential to mammals and have a potent hypocholesterolemic effect and reduce the risk of cardiovascular diseases (Kris-Etherton and Yu, 1997). They can also be used as feed to dairy cattle yielding milk with a high level of conjugated linoleic acid (CLA; Kelly et al., 1998). On the other hand, saturated fatty acids provide the oil with a higher oxidative stability than unsaturated fatty acids. Therefore, for the cooking and food industry, sunflower oils with oleic acid concentration near those of mid-oleic cultivars (oleic acid concentration between 60 and 79%) are often preferred (Binkoski et al., 2005). Oils with high concentration of oleic acid also allow for margarine with less proportion of undesirable *trans* fatty acids.

Although vegetable oils are currently used primarily for edible applications (e.g. cooking oils and margarine), they hold considerable potential for a wide range of uses depending upon the physicochemical properties conferred by their constituent fatty acids. As crude oil supplies decline, vegetable oils are gaining increasing interest as substitutes for petroleum-derived materials in fuels, lubricants, and specialty chemicals (Metzger and Bornsheuer, 2006). For example, conventional vegetable oils, such as those derived from soybean seeds, are currently used for biodiesel production, while oxidatively stable vegetable oils that have high and mid-oleic acid contents are finding increasing use as high temperature, biodegradable lubricants (Sharma et al., 2005; Hill et al., 2006). Through metabolic engineering of their fatty acid compositions, it is possible to expand the potential uses of vegetable oils as renewable substitutes for petroleum-derived chemical feedstocks. Recent efforts to produce unusual fatty acids in engineered seeds for novel industrial oils have focused largely on divergent forms of the D12-oleic acid desaturase (FAD2; Cahoon and Kinney, 2005). Divergent forms of FAD2 have been identified in seeds from several non-agronomic species that catalyze a remarkably wide range of fatty acid modifications, including hydroxylation, epoxygenation, and double bond conjugation (Cahoon and Kinney, 2005). Many of these modified fatty acids have potential industrial significance. The production of vegetable oils that contain economically relevant amounts of these fatty acids hinges on an increased understanding of factors that mediate fatty acid biosynthesis. For biodiesel, oils containing fatty acids with a low degree of unsaturation are preferable, because they decrease the iodine index (inversely related to stability), and increase biodiesel viscosity and the cetane number (a measure of combustion quality; Clements, 1996).

Tocopherols (α, β, γ and δ isomers) are natural antioxidants that inhibit lipid oxidation in biological systems by stabilizing hydroperoxyl and other free radicals (Bramley et al., 2000). The antioxidant activity of tocopherols increases oil stability (Martínez de la Cuesta et al., 1995; Bramley

BOX 17.1 *(cont'd)*

et al., 2000). Tocopherols are essential for humans and they have been associated with delayed cellular aging (White and Xing, 1997), reduced risks of cardiovascular diseases and regression of several cancers in cell culture. At high temperatures, β- and γ-tocopherol present a higher antioxidant activity than α-tocopherol and are thus preferred for cooking oil. On the other hand, α-tocopherol has the highest vitamin E activity compared to the other three isomers (Mullor, 1968).

example, elevated oil or protein concentration is of economic importance for farmers since a premium is often paid for these attributes, while high baking quality in wheat or oxidative stability in oilseeds are of economic importance to food manufacturers. Quality criteria of a given crop species also depends on the product end use (Box 17.1). For example, high-protein flours are required for leavened bread or pasta, while low-protein counterparts are desirable for biscuits, crackers, cake, or oriental noodles. An efficient agro-industrial production system, therefore, needs to know the year-to-year variation in composition of raw materials that can be obtained in different regions.

The identification of novel genes or loci with major effects on quality traits has prompted the development of new cultivars with improved quality (Velasco and Fernández-Martínez, 2002; DePauw et al., 2007; Chapter 18). Nonetheless, dealing with genotype (G) by environment (E) interactions, and with pleiotropic effects (i.e. trait by trait interactions) remains a major difficulty in plant breeding, especially for grain quality (de la Vega and Chapman, 2001). A physiological perspective provides useful insights into G × E interactions, as shown in this and other chapters of this book (Chapters 13, 14 and 19). In this chapter, we show that grain oil and protein concentration and composition is primarily determined by processes at the crop level, and therefore could not be correctly understood or predicted by extrapolating from the individual plant to the population. We will show how physiological concepts classically applied to yield analysis (e.g. identification of critical periods, kinetics of biomass accumulation and partitioning) can be used to investigate and model the genetic and environmental determinants of grain oil and protein concentration and composition.

In this chapter, quality is restricted to the biochemical composition of grains and the focus is on sunflower as an oilseed model and bread wheat as a cereal model. Comparisons with other species are included to emphasize similarities and differences with these model crops. The rationale for the use of these model species is twofold. First, grain oil and protein, major storage compounds in sunflower and wheat, are important in human and animal diets and for food processing, and increasingly important for non-food uses. Second, our knowledge of quality aspects in these two species is sufficient to allow for meaningful quantitative models that capture major genetic, environmental and G × E interaction effects (Martre et al., 2011).

The chapter comprises three main parts. First, we review the effects of environmental and genetic factors, and their interactions, on grain oil and protein concentration and composition. Second, we outline process-based crop models accounting for grain yield in both species, and for concentration and composition of oil (sunflower) and protein (wheat). Third, we use a combination of modeling and experiments to derive relationships between quality traits and yield, and outline management and breeding strategies for the improvement of grain quality.

2 ENVIRONMENTAL AND GENETIC EFFECTS ON GRAIN COMPOSITION

2.1 Oil concentration

Grain oil accumulates mainly in the endosperm (e.g. castor bean and oat), embryo cotyledons and axis (e.g. rapeseed and sunflower), embryo scutellum and aleurone layer (e.g. maize) or mesocarp (e.g. olive and palm oil) depending on the species (Murphy, 2001). For oilseed crops, grain oil concentration, usually expressed as percentage of grain dry mass, determines the industrial yield of the grains. As a consequence, in many countries, sunflower grains with an oil concentration above a threshold are paid a premium.

Grain oil concentration is genetically determined, and plant breeding has improved it in many crops. Modern high-oil sunflower hybrids (47–53% oil; de la Vega et al., 2007; Izquierdo et al., 2008) have replaced low-oil varieties and hybrids (38–47% oil). For maize and oat, high-oil hybrids (up to 20%) have been developed, although most current hybrids have 3.9–5.8% oil (Frey and Holland, 1999; Dudley and Lambert, 2004). Genetic increase of grain oil concentration has been mainly achieved by increasing the proportion of tissue in which oil is stored (e.g. Doehlert and Lambert, 1991, in maize; Tang et al., 2006, in sunflower) rather than by increasing the oil concentration in this tissue. Furthermore, the difference in oil concentration between low- and high-oil sunflower hybrids is due to a longer grain-filling period rather than to a higher rate of oil accumulation (Mantese et al., 2006; Izquierdo et al., 2008).

High oil concentration is associated with a high sensitivity of oil concentration to environmental conditions, as opposed to the almost stable oil concentration of low-oil hybrids (Dosio et al., 2000; Izquierdo et al., 2008), even when genotypes differ only in a single genetic region (8.1 cM) conferring the high-oil trait (León et al., 1996). Environmentally-induced variation of sunflower grain oil concentration has been largely related to variation in embryo oil concentration (Santalla et al., 2002, Izquierdo et al., 2008). However, variation in grain oil concentration has been associated with changes in the embryo-to-pericarp ratio when growing temperatures exceed 30°C (Rondanini et al., 2003).

Grain oils are synthesized from carbohydrates either from current photosynthesis or reserves. For sunflowers grown under very different conditions, the contribution of carbon fixed during the pre-flowering period to the grain carbon content can range from 15 to 27% (Hall et al., 1990). Therefore, grain oil accumulation greatly depends on the carbon economy of the crop during grain filling (Andrade and Ferreiro, 1996; Dosio et al., 2000). Since grain oil is mainly derived from post-flowering carbon assimilation, final grain oil concentration is mainly determined by the amount of intercepted solar radiation (ISR) per plant during the grain-filling period (Andrade and Ferreiro, 1996; Dosio et al., 2000; Izquierdo et al., 2008). Likewise, Izquierdo (2007) found reductions in oil concentration of rapeseed grains when incident radiation was severely reduced. In contrast, Andrade and Ferreiro (1996) did not detect any effect of ISR on grain oil concentration in soybean and maize. Other authors have attempted to use the source–sink ratio as the explanatory variable, finding either no relationship (Ruiz and Maddonni, 2006) or a relationship similar to the one obtained by considering just ISR (Izquierdo et al., 2008, Echarte et al., 2012). Ruiz and Maddonni (2006) concluded that oil concentration did not change with varying source–sink ratio by performing a meta-analysis of published data for different hybrids. It is possible, however, that genetic variability of underlying physiological processes (e.g. assimilate partitioning) mask the effect of source–sink ratio. Later, Echarte et al. (2012) showed that carbohydrate equivalents appropriately explained differences of grain oil concentration in sunflower plants with similar

ISR during the grain-filling period when the source is manipulated by directly injecting sucrose into the receptacles. These results suggest that ISR effects on oil concentration might be driven by changes in carbon fixation and allocation to the grains.

In sunflower, final grain dry mass and oil concentration are most sensitive to the amount of ISR between 250 and 450°C d after flowering (base temperature 6°C; Aguirrezábal et al. 2003). A linear-plateau relationship between final grain oil concentration and ISR during this period was found, with a maximum (OC_{max}) at approximately 26.3 MJ $plant^{-1}$ for high-oil hybrids (Fig. 17.1a; Dosio et al., 2000; Izquierdo et al., 2008). Recently, Rondanini et al. (2013) found similar results

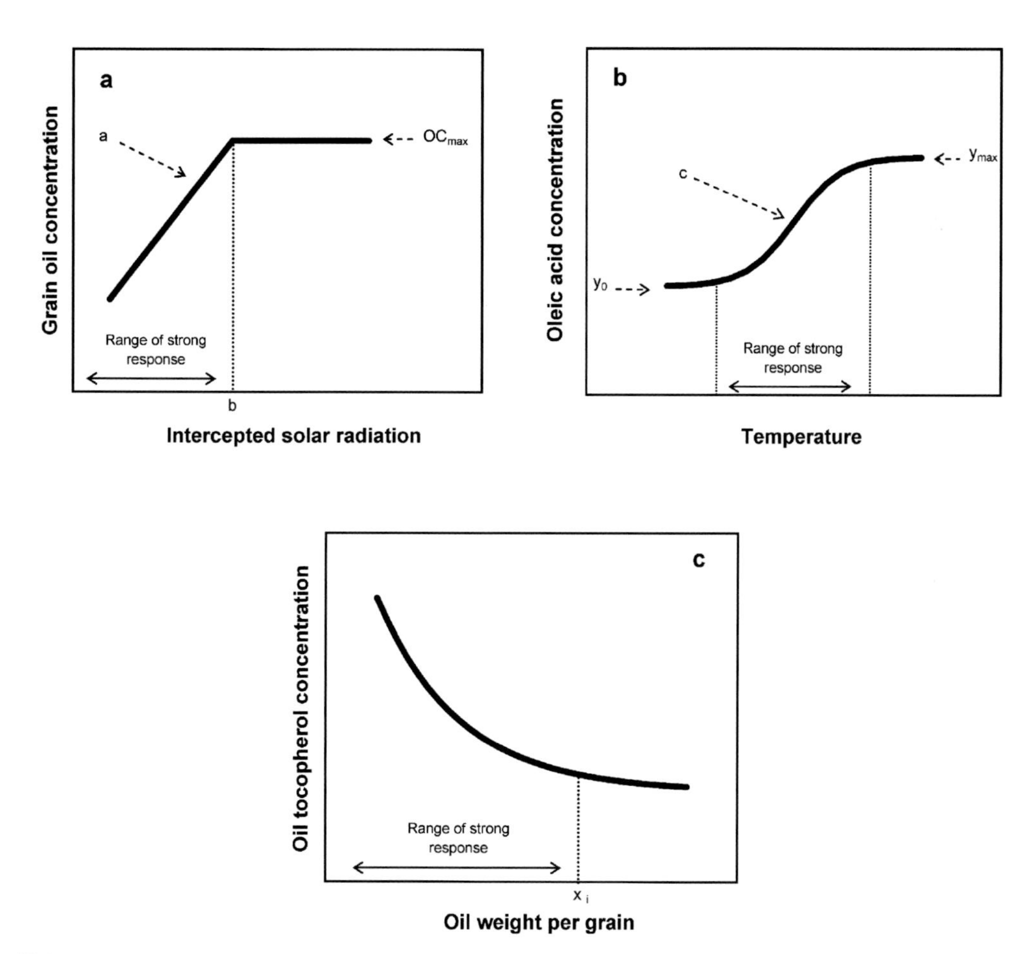

FIG. 17.1 Modeling quality traits in sunflower. (a) Linear-plateau response of grain oil concentration to intercepted solar radiation during the period 250–450°Cday after flowering; parameters are the slope of the linear phase (a), the minimum intercepted radiation to obtain the maximum oil content (b), and the maximum oil concentration (OC_{max}). (b) Sigmoid function describing the relationship between oleic acid concentration and minimum night temperature during the period 100–300°Cday after flowering; parameters are the maximum slope (c), the minimum (y_0) and the maximum oleic acid concentration (y_{max}). (c) Negative exponential relationship between oil tocopherol concentration and oil weight per grain; X_i is the oil weight per grain above which oil tocopherol concentration shows a stabilization-like phase.

for spring canola, but with a broader critical period from flowering to 730°C d after flowering (base temperature 0°C).

The response of grain oil concentration to temperature during grain filling depends on the species. For flax, both the quantity of oil per grain and oil concentration decreases linearly with daily average temperature between 13 and 25°C (Green, 1986). For soybean, grain oil concentration increases with daily average temperature up to ≈28°C, and decreases with higher temperatures (Piper and Boote, 1999; Thomas et al., 2003). In sunflower, grain oil concentration shows a biphasic response to daily mean temperature, with no response up to 17–22°C depending on the hybrid, and a steep decrease at higher temperature (Cantagallo et al., 2004; Angeloni et al., 2012). The magnitude of oil concentration decrease depends on the stage of grain filling when high temperature events occur (Rondanini et al., 2003, 2006). Although most of this effect is related to the lower accumulation of ISR during the critical period (Aguirrezábal et al., 2003) due to the shortened grain-filling period under higher temperature (Ploschuk and Hall, 1995; Villalobos et al., 1996), a direct effect of temperature on oil synthesis should be considered (Angeloni et al., unpublished). For instance, assuming that grain oil concentration is governed by carbon allocation to the grains, effects of temperature on carbon fixation and photorespiration rate could be reflected in grain oil concentration.

Water deficit during grain filling can reduce grain oil concentration (Hall et al., 1985; Roche et al., 2006; Alahdadi et al., 2011; Sezen et al., 2011). Long term, slow developing water deficit decreases growth by slowing rates of cell division and expansion (Lawlor and Cornic, 2002). This effect could significantly affect the amount of ISR, and thus grain oil concentration. More severe water deficits can also reduce leaf photosynthesis or accelerate leaf senescence (Chapter 10), thereby reducing the availability of carbohydrate for oil synthesis (Tezara et al., 1999; Lawlor, 2002). Water deficit effects are closely related to the phenological stage when they occur (Karam et al., 2007).

Crop biomass increases with nitrogen supply according to the law of diminishing returns (Chapter 8). High nitrogen supply promoting both leaf area and ISR could thus be expected to increase grain oil concentration. However, excess of nitrogen can reduce sunflower oil concentration, mainly through an increase in protein concentration with no increase in biomass (Steer et al., 1984, and references therein). Zheljazkov et al. (2008, 2009) observed that grain oil concentration of different sunflower hybrids decreased with increased nitrogen supply, but grain oil yield was unaffected due to higher total grain dry mass yield. The same response of grain oil concentration to N rate has been recently reported in canola (Zheljazkov et al., 2013).

2.2 Oil composition

Fatty acid composition is the main determinant of oil quality, as outlined in Box 17.1. Fatty acid chains are built through a cycle of biochemical reactions involving the multienzyme fatty acid synthase (FAS) and the low molecular acyl-carrier protein (ACP). They are converted to acyl-CoA and then incorporated into triacylglycerol (TAG). Desaturation of fatty acids is mediated mainly by the enzymes stearoyl-ACP desaturase, oleoil-ACP desaturase, and linoleoil-ACP desaturase. Although our knowledge of the enzymes involved in the synthesis and desaturation of oil fatty acids, the genes that code for these enzymes, and their environmental regulation has increased over the last years, our understanding of the regulation of oil synthesis and storage is still far from complete.

The fatty acid composition of the oil of different species has been modified by breeding and mutagenesis (Fig. 17.2; Lagravere et al., 1998; Fernández-Martínez et al., 1989; Lacombe and Bervillé, 2000, Haddadi et al., 2011). Sunflower high oleic cultivars carry a knockdown

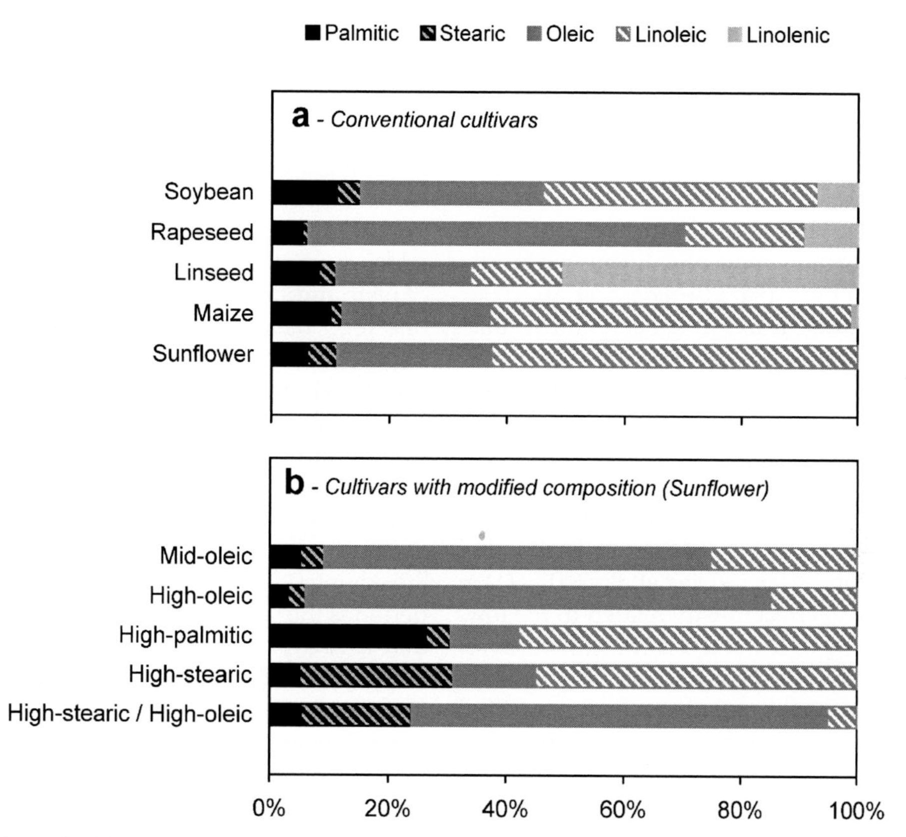

FIG. 17.2 Typical oil fatty acid composition for (a) conventional cultivars of soybean, rapeseed, linseed, maize and sunflower and (b) sunflower cultivars with modified fatty acid composition: mid oleic, high oleic, high palmitic, high stearic–high linoleic, and high stearic–high oleic. *Data from Velasco and Fernández-Martínez (2002), Aguirrezábal and Pereyra (1998), and Izquierdo (unpublished).*

mutation (named Pervenets) on an oleate desaturase gene resulting in a reduced oleoil-ACP desaturase activity (Garcés and Mancha, 1991; Lacombe et al., 2009). The causal mutation for this shift in oil composition has been identified as a tandem duplication of the oleate desaturase FAD2-1 gene (Schuppert et al., 2006). This mutation is present in most of the current high oleic sunflower hybrids. A new mutation of the same gene, comprising an 800 base pair nucleotide insertion and encoding a truncated oleoil-ACP desaturase protein has been recently reported (León et al., 2013). Mid oleic sunflowers are produced by crossing high oleic and traditional lines. High stearic sunflowers result from low activity of stearoyl-ACP desaturase (Cantisán et al., 2000) and high activity of the enzymes FAT-A and FAT-B, which increase the accumulation of stearic acid in the endoplasmic reticulum and thus reduce its desaturation (Salas et al., 2008). High palmitic cultivars, in turn, have been related to a significant decrease of the last step of intraplastidial elongation catalyzed by the FASII enzyme complex (Salas et al., 2004). Inbred lines combining the high stearic with the high oleic trait (high stearic–high oleic) have also been obtained (Serrano-Vega et al., 2005). High oleic, high linoleic and high stearic cultivars of soybean have been developed (Schnebly et al., 1996; Stojšin et al., 1998; Primomo et al., 2002; Serrano-Vega et al., 2005), but they have not been released to the seed market yet. Low-erucic rapeseed cultivars for human consumption are currently available, as well as high-erucic cultivars for industrial uses (Velasco et al., 1999).

Oil fatty acid composition varies according to the environmental conditions during grain filling (Strecker et al., 1997; Pritchard et al., 2000; Roche et al., 2006). It has long been known that an increase in temperature increases the oleic-to-linoleic acid ratio in the oil of several oilseed crops (Canvin, 1965), by affecting the total activity of the desaturases enzymes (e.g. oleoil-ACP desaturase in sunflower; Garcés et al., 1992; Kabbaj et al., 1996). In sunflower, temperature regulates oleoil-ACP desaturase activity by two mechanisms (García-Díaz et al., 2002): (1) a long-term direct effect mostly related to the low thermal stability of the enzyme (Martínez-Rivas et al., 2003); and (2) a short-term indirect effect of temperature, which determines the availability of oxygen and, in turn, regulates the activity of oleoil-ACP desaturase (Rolletschek et al., 2007).

Oleic acid concentration was better related to minimum night temperature (MNT) than to other temperature descriptors (Izquierdo et al., 2006). The circadian rhythm of the oleoil-ACP desaturase activity seems to be associated with this effect of temperature during the dark period (Pleite et al., 2008). Daytime temperature above 30°C also affects fatty acid composition in sunflower (Rondanini et al., 2003), but probably through different, and possibly interacting, mechanisms involved in responses to moderately high temperatures. In soybean, oleic acid concentration of oil was unrelated to night-time or minimum temperatures, while daily average temperature better explained the fatty acid response (Izquierdo, 2007). This might be related to the light-dependent biosynthesis of some fatty acid in green seeds, compared to the small contribution of synthesis in darkness (12–20%) (Browse et al., 1981; Ohlrogge and Jaworski, 1997).

The total activity of oleoil-ACP desaturase is maximum early during grain filling (Garcés et al. 1992; Kabbaj et al., 1996), and therefore the effect of temperature during this period could be stronger than during the rest of grain filling. Correspondingly, MNT between 100 and 300°Cd after flowering (base temperature 6°C) accounted for most of the variability in the concentration of oleic and linoleic acids in two traditional and one high oleic hybrids (Izquierdo et al., 2006; Izquierdo and Aguirrezábal, 2008). The critical period for the effect of temperature on fatty acid composition of sunflower oil has been confirmed for five genotypes grown in 11 locations under field conditions (Pepper, unpublished). This supports the idea (usually assumed in crop simulation models; e.g. Villalobos et al., 1996; Stöckle et al., 2003) that the timing of a critical period is the same for different genotypes, given that it is expressed in relation to developmental events and in thermal time (or normalized to a standard temperature). Inspired by results in sunflower, Baux et al. (2013) found a critical period for the effect of temperature on linolenic acid concentration in winter rapeseed grains. The period defined in this work was shorter than the one observed by Rondanini et al. (2013) for the effect of ISR on grain dry mass and oil content and corresponded to the second half period of dry matter and oil accumulation.

The concentration of oleic acid shows a sigmoidal response to MNT (Fig. 17.1b; Izquierdo and Aguirrezábal, 2008). Differences in the parameters of this function highlight the genetic variability of this response. Differences among traditional hybrids were observed for the minimum and maximum oleic acid concentration attainable, for the maximum slope and the range of temperatures for which oleic acid concentration changes with temperature; differences were especially high for the minimum concentration of this fatty acid (from 15.4 to 32.5% of total fatty acids in oil). The high-oleic hybrid tested showed lower sensitivity to temperature as evidenced by smaller difference between the parameters for minimum and maximum oleic acid concentration. Since the same function characterized the response of several hybrids, it is possible to incorporate the estimation of the oil fatty acid composition easily into crop simulation models using hybrid-specific parameters. The analysis of the variability of these parameters gives important information on the nature of the observed genotype by environment interactions and provides new traits (parameters) that are independent of the environment. This can be used (1) to obtain new genotypes with specific responses to environmental variables, as proposed by Tardieu (2003), and discussed further in section 4.4, and (2) to define working conditions for breeding for a certain trait (e.g. low temperature locations for improving minimum oleic acid attainable).

A positive correlation between oleic acid percentage and the amount of ISR has been reported for sunflower (Seiler, 1986) and soybean oil (Kane et al., 1997). In sunflower, differences in oleic acid driven by ISR could be higher than 10% (Izquierdo et al., 2009). Variations in oleic acid concentration through changes in intercepted radiation in soybean and maize (Izquierdo et al., 2009) were similar to those driven by latitude, extreme sowing dates and temperatures (Muratorio et al., 2001, Izquierdo and Aguirrezábal, 2005; Izquierdo et al., 2009). In sunflower, oleic acid percentage increased with ISR per plant up to a maximum value at high radiation levels (Echarte et al., 2010).

Izquierdo et al. (2009) observed differences in oleic acid percentage response to seasonal variation in ISR. Later, Echarte et al. (2012) showed that fatty acid composition of sunflower oil does not depend directly on ISR but on the assimilate availability for the grains. These authors proved their hypothesis by supplying large amounts of sucrose to shaded plants. Effects of reduced photosynthesis were overcome by sucrose injection. In this context, carbohydrate equivalents come out as a robust variable and provided a framework that allowed analyzing together data from field and greenhouse experiments performed under very different conditions.

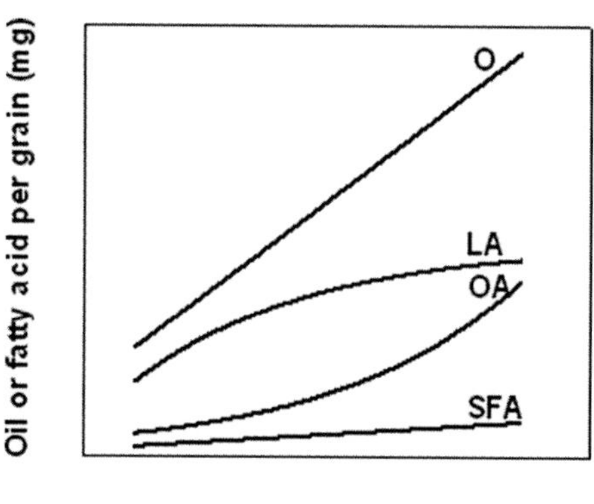

FIG. 17.3 Oil and fatty acids content dependence on carbohydrate equivalents. O: oil weight per grain, LA: linoleic acid per grain, OA: oleic acid per grain, SFA: saturated fatty acids per grain. Redrawn with permission from Echarte et al. (2012).

The amount of saturated fatty acids increase linearly with carbon allocated to the grains, while oleic acid increases exponentially and linoleic acid increases up to a maximum. The analysis of oil and fatty acids content as a function of carbohydrate equivalents (Fig. 17.3) resulted in a following conceptual model. In this model, at very low assimilate allocated to the grains, both first steps of oil synthesis and oleic acid desaturation, a key controlling step in fatty acids biosynthesis pathway (Garcés et al., 1992), occur at submaximal substrate concentration. In these conditions, oil accumulates and most oleic acid can be transformed into linoleic acid. When assimilate allocated to the grain increases further, the capacity of oleic acid oxidation step reaches a maximum and oleic acid begins to accumulate. Thus, oleic acid and oil content increase in the grain at the same rate. Linear responses of oil and saturated fatty acids content to carbohydrate equivalent might represent that carbon allocated to the grains can be mostly turned into oleic acid by first steps of oil synthesis for a wide range of substrate available.

Under natural conditions, temperature and incident solar radiation (one of the components of ISR per plant) are often correlated. By combining field experiments at different latitude and *in situ* temperature × radiation experiments, Echarte et al. (2010) showed that MNT and ISR per plant have independent and additive effects on fatty acid composition of sunflower oil. In addition, ISR and daily mean temperature during grain filling independently affected oleic acid percentage of soybean and maize genotypes with increased oleic acid percentage (Zuil et al., 2012). Recently, Echarte et al. (2013) found that the critical period for the effect of ISR on fatty acid composition is different from the one reported for MNT (Izquierdo et al., 2006).

Environmental factors that exert their effects additively act upon different causal sequences leading to the response (Salisbury and Ross, 1992). Temperature affects total activity of oleate desaturase, the enzyme that catalyzes the conversion of oleic to linoleic acid. Total activity of this enzyme peaks at early grain filling (Garcés and Mancha, 1991; Gray and Kekwick, 1996; Kabbaj et al., 1996). Considering that the effect of temperature on fatty acid composition is more important during early stages of grain filling, when the quantity of oil accumulated per grain is small, Izquierdo et al. (2002) suggested a 'memory effect' of an early temperature treatment on the fatty acid desaturation mechanism. This work showed that once oleate desaturase activity has been defined by temperature between 100 and 300°Cd after flowering (Izquierdo et al., 2006), fatty acid composition could still vary depending on the quantity of carbon allocated to the grains between 350 and 450°Cd after flowering. At higher irradiances oleate desaturase (FAD2) would be substrate-saturated and thus increased carbon availability would lead to a relative accumulation of oleic acid.

Environmental factors other than temperature and solar radiation can also affect fatty acid composition. Water availability (Baldini et al., 2002; Anastasi et al., 2010; Sezen et al., 2011), nitrogen availability (Steer and Seiler, 1990; Zheljazkov et al., 2008, 2009), soil salinity (Irving et al., 1988; Di Caterina et al., 2007), and pathogens (Zimmer and Zimmerman, 1972) affect fatty acid composition of sunflower oil. Although physiological mechanisms underlying these responses are still unknown, part of these effects could be mediated by intercepted radiation effects.

Since sunflower was initially domesticated for its edible seeds (Putt, 1997), as opposed to explicit selection for oil content and/or composition, selective pressures may have affected palatability or germinability (Chapman and Burke, 2012). More recently, the cultivated sunflower gene pool has been subjected to intense selection for high oil content and specific oil composition (Burke et al., 2005; Putt, 1997). The best-known case of oil composition selection is high oleic acid lines. In these genotypes, a trade-off between sunflower yield and oil quality has been reported (e.g. Pereyra-Irujo and Aguirrezábal, 2007). Since selection at a single locus can result in sweeps spanning relatively large genomic regions, it is possible that selection at a nearby locus has produced the observed pattern.

Genotypes carrying the high stearic mutation also modify their oil fatty acid composition depending on environmental conditions. Increasing intercepted solar radiation curvilinearly increased oleic acid percentage reaching a maximum concentration at high levels of radiation (Martínez et al., 2012), with a correlative decrease in linoleic acid concentration and minor variations in saturated fatty acids. Oil fatty acid composition of genotypes with the high stearic mutation also depends on temperature during grain filling. Like in traditional genotypes, increasing temperature increases the oleic-to-linoleic acid ratio and reduces the concentration of saturated fatty acids. The magnitude of these changes depends on the presence of the high oleic mutation, since high stearic–high linoleic genotypes varied their fatty acid composition more than high stearic–high oleic ones (Izquierdo et al., 2013).

2.3 Oil tocopherol and phytosterol concentration

Further than acyl-lipids, the dietary quality of vegetable oils depends on the concentration and composition of microconstituents of the non-glyceride portion termed unsaponifiable matter. In sunflower oil, this portion generally ranges from 0.5 to 2% and contains molecules of paramount interest for human nutrition and health: tocopherols and phytosterols.

Tocopherols (Box 17.1) are synthesized from specific precursors, via the isoprenoid pathway or the homogentisic acid pathway (Bramley et al., 2000). Their synthesis is regulated by the availability of these precursors. In plants, four isomers of tocopherol have been identified (α-, β-, γ- and δ- tocopherol), which differ in the position of the methyl group in the molecule, and also in their *in vitro* and *in vivo* antioxidant activities (Kamal-Eldin and Appelqvist, 1996). The enzymes catalyzing the methylations thus control the relative quantities of these isomers. In sunflower, α-tocopherol accounts for 91–97% of total tocopherols, with little genetic variability (Nolasco et al., 2006).

Tocopherol concentration in sunflower grain is very sensitive to environmental conditions (Kandil et al., 1990; Marquard, 1990; Velasco et al., 2002). The amount of ISR during grain filling was negatively correlated to tocopherol concentration in the oil of sunflower (Nolasco et al., 2004), soybean, maize and rape (Izquierdo et al., 2011). Although intercepted radiation increases the synthesis of tocopherol and oil, the former would be more affected and therefore diluted in the oil (Izquierdo et al., 2011). The effect of temperature on tocopherol concentration, however, has not been clearly identified, with many contradictory reports (Dolde et al., 1999; Almonor et al., 1998; Izquierdo et al., 2007).

Grain tocopherols are present in a small quantity, typically 500–1200 μg g oil^{-1} (Marquard, 1990; Nolasco et al., 2004), and their concentration depends on the amount of both tocopherols and oil. Velasco et al. (2002) did not find correlations between seed oil and seed tocopherol content, but conversely, Nolasco et al. (2004) found that oil weight per grain accounted for 73% of the variation in total tocopherol concentration in sunflower oil. The relationship between oil tocopherol concentration and oil weight per grain showed a dilution-like shape (Fig. 17.1c), independently of the location, the hybrid and its maximum oil concentration, or the intercepted radiation during grain filling (Nolasco et al., 2004). Oil weight per grain also accounted for much of the variation in oil tocopherol concentration in traditional varieties of soybean and rapeseed but not in maize, a species with a low and stable grain oil concentration (Izquierdo et al., 2011). The simple relationship between tocopherol concentration in oil and oil weight per grain can be easily incorporated into crop simulation models, and used to identify management practices useful to obtain grains with a larger quantity of tocopherols. These relationships could also be useful for commercialization and grain processing, since the tocopherol concentration of a grain lot could be estimated from weight per grain and oil concentration, which are simpler, faster, and less expensive to measure than tocopherol concentration.

Phytosterols are naturally present in plants, playing important roles in membrane fluidity and permeability (Schaller, 2003), embryogenesis (Clouse, 2000), and as precursors of brassinosteroid hormones involved in plant growth and development (Lindsey et al., 2003). They have a proved role in reducing low-density lipoprotein cholesterol (LDL; Ostlund, 2007) and present other interesting properties, such as anti-cancer and anti-oxidation activities (Awad et al., 2003).

Among oilseed species, sunflower is one of the most significant sources of phytosterols, which can exceed 7000 mg kg^{-1} of crude oil with predominance of 4-desmethyl sterols class, represented specially by Δ7-sitosterol, stigmasterol, campesterol and 7-stigmastenol (Piironen et al., 2000). The different sterols accumulate at genotype-dependent rates suggesting different regulation related to genetic background. The absence of a direct correlation between fatty acid and sterols biosynthesis suggests that they are independently regulated, despite a common substrate (acetylCoA) in the metabolic pathways of both components. Environmental factors modify phytosterols content and composition in the oil (Roche et al., 2006, 2010). Both high temperature and water deficit increase concentration of phytosterols in sunflower grain (Anastasi et al., 2010). Recently, Nolasco et al. (unpublished) found a negative correlation between phytosterols and oil content, similar to the one observed for tocopherols. Thus, environmental effects on phytosterol concentration could be explained, as was considered for tocopherols, by direct effects on oil content that result in phytosterol dilution in the oil.

2.4 Protein concentration

Grain protein concentration, usually expressed in percent of grain dry mass, is the main determinant of the end-use value of most cereal and grain legume species. This is particularly true for wheat, which is mostly consumed by humans after processing. Cereal grains contain a relatively small concentration of protein, typically between 8 and 18%, but the large dependence on maize, wheat, and rice as main sources of carbohydrate and energy means these species account for 85% of dietary proteins for humans (Shewry, 2007).

In the field, variations in grain protein concentration induced by weather, water and nitrogen availability, especially during the grain-filling period, are much larger than variations due to genotype (Cooper et al., 2001). For wheat, variations in grain protein concentration in response to temperature, solar radiation, CO_2 or soil water availability are mainly related to variation in the quantity of carbon compounds (i.e. starch and oil) per grain, while the quantity of nitrogen compounds (i.e. proteins) per grain is relatively stable (Panozzo and Eagles, 1999; Triboi and Triboi-Blondel, 2002; Triboi et al., 2006). In oilcrops, an increase in oil concentration is generally associated with a decrease in protein concentration (López Pereira et al., 2000; Morrison et al., 2000; Uribelarrea et al., 2004), as a result of a dilution effect (Connor and Sadras, 1992).

The interplay between carbon and nitrogen compounds leading to a final concentration of protein in grain can most simply be explained by the effects of environmental factors during the grain-filling period on the rate and duration of accumulation of starch, oil and protein. Starch, oil and protein depositions in the grain are relatively independent from each other and are controlled differently (Jenner et al., 1991). The synthesis of grain starch and oil mostly relies on current photosynthesis. Therefore, the quantity of starch and oil per grain is mainly determined by the duration in days of the grain-filling period. The rate of accumulation of carbon compounds per day is little modified by post-anthesis environmental factors, while the duration of grain filling in thermal time shows little variation (Triboi et al., 2003; DuPont et al., 2006b). This explains the close correlation between grain dry mass and the rate of accumulation of grain dry matter degree-day^{-1} or with the duration of grain filling in days. In contrast, under most conditions, the synthesis of grain protein relies mostly on nitrogen remobilization from the vegetative organs. Chapter 8 discusses further the relative contribution of current assimilation and reserves to the carbon and nitrogen economy of grains, and emphasizes differences between wheat and rice, for which 60–95% of the grain N at harvest comes from remobilization of stored N, and maize where this proportion is only 45–65%.

Temperature and water supply have small effects on the rate of grain nitrogen accumulation degree-day^{-1} (Triboi et al., 2003; DuPont et al., 2006b), which means that any temperature-driven decrease in the duration in days of grain filling is compensated for by an increase in the rate of accumulation of grain nitrogen day^{-1}. Under water deficit or high temperature, the higher daily rate of nitrogen remobilization from vegetative organs to grain accelerates canopy senescence and reduces remobilization of stored pre-anthesis carbon (Palta et al., 1994; Triboi and Triboi-Blondel, 2002). Under very high temperature (maximum daily temperature higher than 30–35°C, depending on the species), this compensation decreases and the quantity of nitrogen per grain may decrease. Environmental constraints during the early phase of grain development (i.e. endosperm or cotyledon cell division) may reduce the rate of accumulation of carbon, which accentuates the increases of grain protein concentration (Gooding et al., 2003). The effects of these environmental factors before anthesis on grain protein concentration depend mainly on their effect on the sink-to-source ratio, but the processes described above still drives grain protein concentration (Chapter 8).

In contrast with the effects of post-flowering temperature, radiation, or water supply, the effects of nitrogen supply on grain protein concentration are mostly due to changes in the amount of nitrogen per grain which, for most crop species, is regulated by nitrogen availability in the vegetative organs and in the soil during the grain-filling period (e.g. Lhuillier-Soundele et al., 1999; Martre et al., 2003). The number of grains set usually matches the growth capacity of the canopy during the grain-filling period (Sinclair and Jamieson, 2006) and, as a consequence, average grain dry mass is little modified or increases slightly under pre- and/or post-flowering soil nitrogen shortage (Triboi et al., 2003; Uribelarrea et al., 2004).

To model the effects of the environment on protein concentration, it is necessary to take into account the dynamics of carbon and nitrogen accumulation in the grain. This has been successfully achieved in the wheat model *SiriusQuality*1 (Martre et al., 2006), which is described in section 3.2. In this model, the accumulation of structural proteins and carbon, which occurs during the stage of endosperm cell division and DNA endoreduplication, is assumed to be sink regulated and is driven by temperature. In contrast, the accumulation of storage proteins and starch, which occurs after the period of endosperm cell division, is assumed to be source regulated (i.e. independent of the number of grains per unit ground area), and is set daily to be proportional to the current amount of vegetative non-structural nitrogen.

2.5 Protein composition

Storage proteins account for 70–80% of the total quantity of reduced nitrogen in mature grains of cereals and grain legumes, and their composition greatly influences their processing and nutritional quality (Wieser and Zimmermann, 2000; Gras et al., 2001; Khatkar et al., 2002). It is important to note that grain protein concentration has decreased linearly with the year of cultivar release but flour functionality has increased significantly (Ortiz-Monasterio et al., 1997; Fufa et al., 2005). This trend is partly associated with the selection of favorable storage protein alleles after the mid-1980s (Branlard et al., 2001). Although the qualitative protein composition of grain depends on the genotype, its quantitative composition is largely determined by environmental factors including nitrogen and sulfur availability (Graybosch et al., 1996; Huebner et al., 1997; Panozzo and Eagles, 2000; Zhu and Khan, 2001).

Wheat storage proteins are controlled by over one hundred genes located at different loci (Shewry and Halford, 2003), coding for high-molecular-weight-glutenin subunits (HMW-GS), low-molecular-weight-glutenin subunits (LMW-GS), α/β-gliadins, γ-gliadins, and ω-gliadins. Gliadins and glutenins are collectively referred to as prolamins. Glutenins are polymeric proteins that form very large macropolymers with viscoelastic properties (responsible for the unique rheological properties of wheat flour; Don et al., 2003), while gliadins are important for dough extensibility (Branlard et al., 2001). Grain storage proteins accumulate mainly during the linear phase of grain filling. For wheat during the desiccation phase after physiological maturity (Box 12.1 in Chapter 12), glutenin proteins form very large polymers (Carceller and Aussenac, 2001), whose size distribution is critical for wheat flour functionality (Don et al., 2003).

Environmentally-induced changes in grain protein composition are associated with the altered expression of gene encoding storage proteins, in response to signals that indicate the relative availability of nitrogen and sulfur (Bevan et al., 1993; Peak et al., 1997; Chiaiese et al., 2004; Hernandez-Sebastia et al., 2005). These signals trigger transduction pathways in developing grains that, in general, balance the storage of nitrogen and sulfur to maintain homeostasis of the total amount of protein per grain (Tabe et al., 2002; Islam et al., 2005). Emergent properties of these regulation networks, which are still poorly understood at the molecular level, are allometric relations between the amount of nitrogen per grain and the amount of the different storage protein fractions (Sexton et al., 1998b for soybean; Landry, 2002 for maize; Triboi et al., 2003 for wheat). These relations are independent of the causes of variations in the quantity of nitrogen per grain and are similar for developing and mature grains (Fig. 17.4a–c, compare main panels and insets; Daniel and Triboi, 2001; Triboi et al., 2003). This means that environmental factors, including supply of water and nitrogen, do not directly influence grain protein accumulation, but only indirectly through their effect on grain nitrogen accumulation. It thus appears that the gene regulatory network involved in the control of the synthesis of storage proteins is coordinated in such a way that the grain reacts in a predictable manner to nitrogen availability, yielding a meta-mechanism at the grain level (Martre et al., 2003).

A consequence of this meta-mechanism is that grain protein composition is closely related to the total quantity of proteins per grain, independently of the cause of its variation. For wheat, the proportion of total glutenin in grain protein, as well as the proportion of each HMW-GS in total HMW-GS (Wieser and Zimmermann, 2000; DuPont et al., 2007), appear to be independent of the quantity of nitrogen per grain, whereas the proportion of gliadin in grain protein shows significant environmental variation (Fig. 17.4c). Changes in the proportion of gliadin are directly related to the variations in the quantity of nitrogen per grain, and are compensated by a proportional decrease of non-prolamin protein. These variations in grain protein composition are accompanied by changes in amino acid composition with the total quantity of nitrogen per grain (Eppendorfer, 1978; Mossé et al., 1985). A practical implication of these relations is that grain protein and amino acid composition can be calculated directly from the quantity of nitrogen per grain independently of the growing conditions.

In a recent study, Plessis et al. (2013) analyzed the genetic and environmental variability of the allometric parameters of grain nitrogen allocation for wheat. They showed that the scaling exponents for the allocation of total grain nitrogen to gliadins and glutenins are similar across environments, but have a large genetic variability. None of the scaling coefficients and exponents was associated with the same loci, suggesting that their genetic control is different. The scaling exponents were not correlated with any other variable and were associated with loci independently of other variables. This means that using the parameters of a model that are independent of the environment allowed identification of a level of genetic regulation of the synthesis of grain storage protein that could not be detected through basic compositional data. Only three loci were found to be associated with this parameter, two of which were in strong linkage disequilibrium with NAC transcription factors.

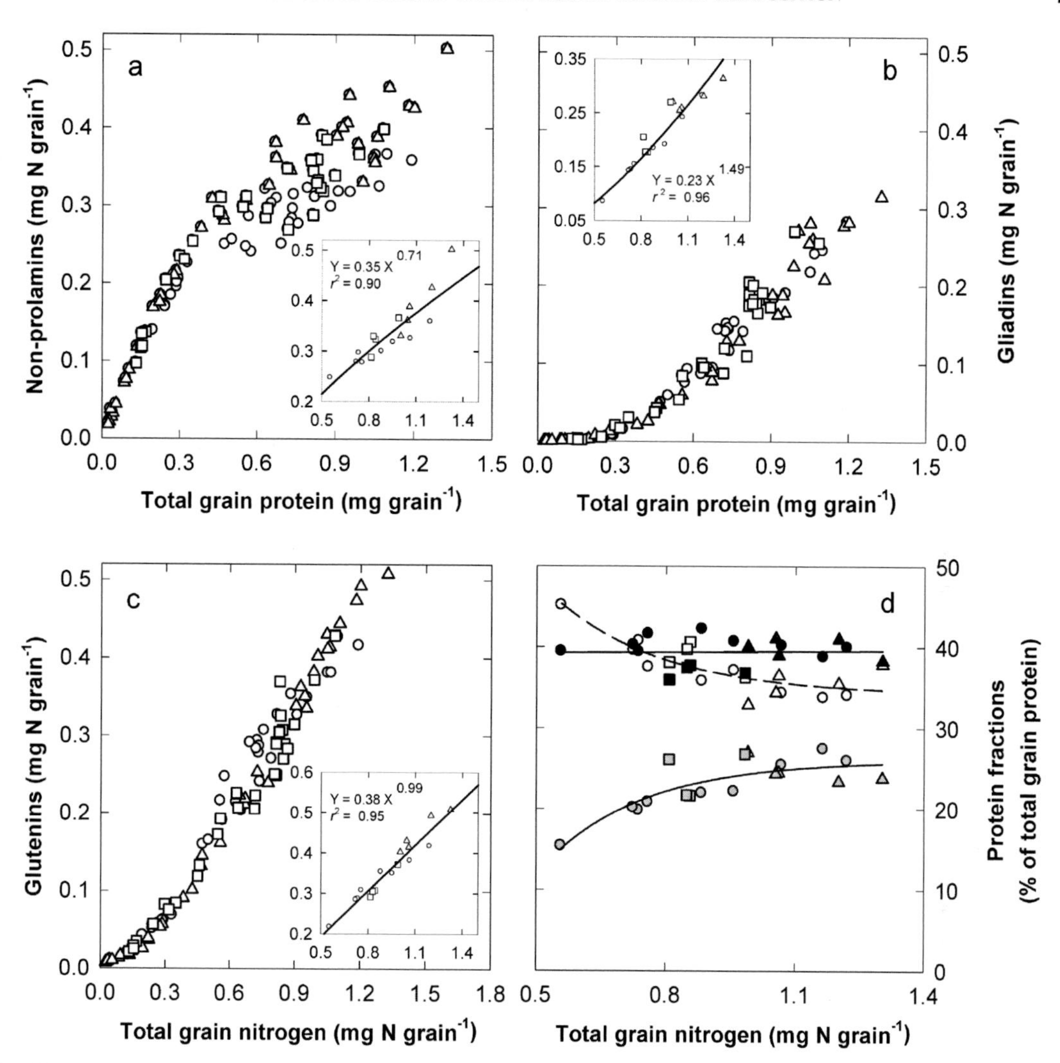

FIG. 17.4 Quantities of (a) non-prolamin proteins, (b) gliadins, (c) and glutenins per grain versus the total quantity of nitrogen per grain for developing and mature grains of bread wheat. Insets show the allometric relations between the quantity of each protein fraction and the total nitrogen content of mature grains. (d) Percentages of non-prolamin proteins (open symbols), gliadins (gray symbols), and glutenins (closed symbols) vs the total quantity of nitrogen per grain for mature grains. Crops were grown in semi-controlled environments with different post-anthesis temperature (triangles) and water supplies (rectangles), and in the field with different rates and times of nitrogen fertilization (circles). Redrawn with permission from Triboi et al. (2003).

Interestingly, transcription factors known to control expression of grain storage protein in cereals by binding grain storage protein gene promoters were associated with grain storage protein composition, while transcription factors with unknown functions were associated with the grain nitrogen allocation parameters, indicative of both direct and indirect transcriptional regulation for grain storage protein composition and allocation. However, in a previous study,

Ravel et al. (2009) found that the nucleotide polymorphism within the transcription factor Storage Protein Activator (*Spa-A*), known to bind conserved cis-motifs in the promoter of cereal grain storage proteins, were associated with differences in the scaling coefficient of the allometric relationship between total grain N and gliadins and, consequently, of the gliadin-to-glutenin ratio. Genetic mapping analysis showed that the *Spa-A* locus is also significantly associated with dough viscoelasticity (Ravel et al., 2009). Both studies indicate that the allometric relationships of grain nitrogen allocation underlie the dynamics structuring these regulatory networks in which transcription factors are key regulators.

Nitrogen supply determines the level of protein accumulation in the grain and its gross allocation between storage protein fractions but, at a given level of nitrogen supply, sulfur supply fine-tunes the composition of the protein fractions by regulating the expression of individual storage protein genes (Hagan et al., 2003; Chiaiese et al., 2004). Interestingly, DuPont et al. (2006a) noted that any factors that increase the amount of nitrogen per grain also increase the proportion of sulfur-poor (ω-gliadins and HMW-GS) storage protein at the expense of sulfur-rich (α/β-gliadins, γ-gliadins, and LMW-GS) storage protein. These authors proposed a putative molecular mechanism by which nitrogen activates the synthesis of glutamine and proline, thus favoring the synthesis of sulfur-poor proteins rich in these amino acids. An alternative hypothesis is that the accumulation of sulfur-rich storage proteins, determined by sulfur availability, generates a signal of sulfur deficiency, which increases the expression of sulfur-poor storage proteins to the extent of nitrogen available to it. The latter hypothesis is substantiated by reported differences in temporal appearance and spatial distribution of sulfur-rich and sulfur-poor storage proteins in developing grain of maize (Lending and Larkins, 1989) and wheat (Panozzo et al., 2001).

Very high temperature, above a threshold of daily average temperature of about 30°C, can have marked effects on wheat dough strength which are largely independent of grain protein concentration (Randall and Moss, 1990; Blumenthal et al., 1991; Wardlaw et al., 2002). Relatively small variations in the proportions of the different types of gliadin and glutenin subunits have been reported in response to chronic or short periods of very high temperature and it is difficult to relate convincingly such variations to changes in flour functionality. Weaker dough from grains that experience one or several days of very high temperature have been related to a marked decrease in the proportion of large molecular size glutenin polymers (Ciaffi et al., 1996; Corbellini et al., 1998; Wardlaw et al., 2002; Don et al., 2005). The aggregation of glutenin proteins, which occurs mainly during grain desiccation after physiological maturity (Carceller and Aussenac, 2001; Ferreira et al., 2012), is likely the major process responsible for heat-shock-related dough weakening.

Extended periods of high temperature are common in many cereal-growing areas of the world, and above-optimal temperature is one of the major factors affecting small grain cereal yield and composition (Randall and Moss, 1990; Borghi et al., 1995; Graybosch et al., 1995). Some studies suggest that very high temperature around mid-grain filling has positive effects on wheat dough strength (Stone et al., 1997; Panozzo and Eagles, 2000), whereas very high temperature around physiological maturity has a negative effect on dough strength (Randall and Moss, 1990; Blumenthal et al., 1991; Stone and Nicolas, 1996). This difference in the response of dough properties according to the timing of heat-shock are related to different effects on the gliadin-to-glutenin ratio and the size of glutenin polymers at each developmental stage (Ferreira et al., 2012). Further experiments to identify the relative sensitivity of different development stages in terms of gliadin and glutenin accumulation and glutenin polymer formation are clearly required. Important genetic variability in the relative response of grain protein composition and flour functionality to very high temperature has been reported (Blumenthal et al., 1995; Stone et al., 1997; Spiertz et al., 2006), but the genetic and physiological bases of these differences are still largely unknown.

3 INTEGRATION OF QUALITY TRAITS INTO CROP SIMULATION MODELS

Models can range from detailed mechanistic descriptions to simple response curves to environmental variables, which are 'meta-mechanisms' at the plant or crop level (Tardieu, 2003). If models are robust enough, one set of parameters represents one genotype (Hammer et al., 2006), and thus they can be used to analyze complex traits with G × E and pleitropic effects. This underlies model-assisted phenotyping and ideotype design, as explained in Chapter 14.

Crop simulation models for grain yield have been published for all the major crop species. Some sunflower models also simulate oil yield (Villalobos et al., 1996; Casadebaig et al., 2011) and oil quality (Pereyra-Irujo and Aguirrezábal, 2007). Most wheat models simulate crop nitrogen accumulation and partitioning, primarily because crop nitrogen status greatly affects crop biomass and grain yield; however, few models have extended the simulation of nitrogen dynamics to grain nitrogen concentration (Sexton et al., 1998a; Asseng et al., 2002; Martre et al., 2006). In this section, we describe simulation models for sunflower and wheat, which have integrated grain oil or protein quality modules, based on the relationships described in section 2. In the following section, these models will be used to design strategies to improve quality traits through crop management or breeding.

3.1 Modeling oil concentration and composition in sunflower

The model of Pereyra-Irujo and Aguirrezábal (2007) integrates empirical relationships (section 2) as shown in Figure 17.5a. The model includes a temperature-driven phenology module that allows establishment of the critical periods for determination of yield and quality components. Grain oil concentration and yield components require an estimate of intercepted radiation per plant, which is determined at the population level. Leaf growth is predicted through its relationship with temperature and plant density. Radiation interception depends on leaf area index (determined by plant density and individual plant leaf area) and incident solar radiation at the top of the canopy. Fatty acid composition was initially predicted through its relationship with temperature during its critical period (Fig. 17.1b). The recent incorporation of the effect of radiation during its own critical period improved the estimation of fatty acid composition (Echarte et al., 2013). In field experiments, where radiation was artificially modified by shading, oleic acid estimation improved by 14% in terms of relative root mean square error (RRMSE) when radiation effect was taken into account while, for plants grown under full sunlight, the estimation improved by just 5% (Jaimes F., unpublished). This low apparent improvement of oleic acid estimation under current growing conditions could gain importance under global change scenarios, where a dimming effect and a shorter period for radiation accumulation due to higher temperatures are expected to occur (Chapter 20). Then, oil tocopherol concentration is calculated from oil weight per grain (Fig. 17.1c) which, in turn, depends on grain weight and grain oil concentration.

This model provided estimations of grain yield similar to those of a more complex model, and good estimates of oil quality and intermediate variables such as phenology and intercepted

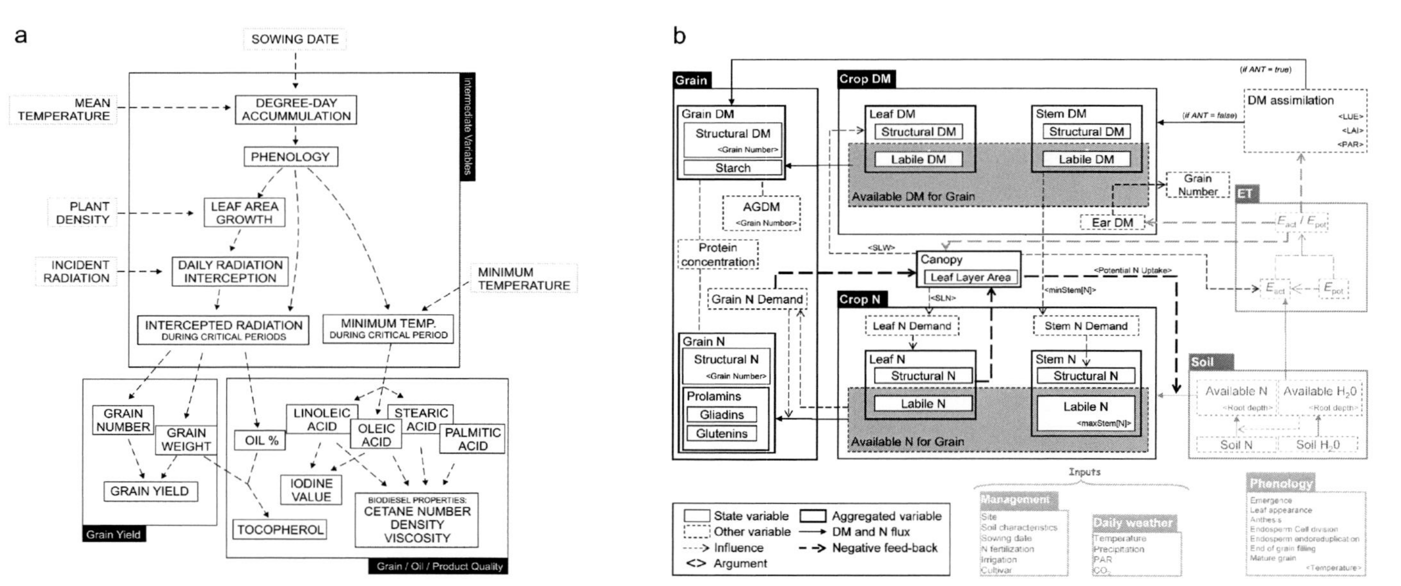

FIG. 17.5 (a) Schematic representation of the sunflower simulation model from Pereyra-Irujo and Aguirrezábal (2007) and Pereyra-Irujo et al. (2009) showing the relationships between input variables (outside boxes), intermediate variables, and output variables (yield and quality). Adapted from Pereyra-Irujo and Aguirrezábal (2007). (b) Schematic representation of the wheat simulation model *SiriusQuality*1 (Martre et al., 2006) showing the main variables, influences, and feedbacks. AGDM: average grain dry mass; ANT: anthesis; DM: dry matter; N: nitrogen; E_{act}: actual evapotranspiration; E_{pot}: potential evapotranspiration; ET: evapotranspiration; LAI: leaf area index; LUE: light use efficiency; maxStem[N]: maximum stem nitrogen concentration; minStem[N]: minimum stem nitrogen concentration; PAR: photosynthetically active radiation; SLN: leaf nitrogen mass per unit of leaf surface area; SLW: leaf dry mass per unit of leaf surface area.

radiation over a wide range of environmental conditions (Pereyra-Irujo and Aguirrezábal, 2007; Pereyra-Irujo et al., 2009). The original model was expanded to simulate quality traits for biodiesel (i.e. density, kinematic viscosity, heating value, cetane number and iodine value) using validated empirical relationships with oil fatty acid composition (Pereyra-Irujo et al., 2009). It was also useful for estimating possible effects of warming on oil concentration and quality of different sunflower hybrids at different locations in Argentina (Echarte et al., 2009).

3.2 Modeling grain protein concentration and composition in wheat

Simulation of grain nitrogen content requires first simulating the uptake of nitrogen into the vegetative tissues of the plant, followed by a step where nitrogen is transferred to the grains from vegetative tissues. A mechanistic approach to simulate crop nitrogen accumulation and partitioning has been implemented in the wheat simulation model Sirius (Jamieson and Semenov, 2000). Nitrogen is distributed to leaf and stem tissues separately, with simplifying assumptions that nitrogen per unit leaf area is constant at the canopy level (Grindlay, 1997), but that the stem could store nitrogen. The effect of nitrogen shortage is first to reduce stem nitrogen concentration and then leaf expansion to maintain specific leaf nitrogen concentration of growing leaves. One advantage of this approach is that it reduces the number of parameters and the need to define stress factors, compared with the demand-driven approaches based on nitrogen dilution (e.g. van Keulen and Seligman, 1987; Stöckle and Debaeke, 1997; Brisson et al., 1998). This approach also provides more plasticity in the response of the crop to nitrogen availability. The latest version of Sirius (*SiriusQuality*, http://www1.clermont.inra.fr/siriusquality/), considers the vertical distribution of leaf lamina, leaf sheath and internode dry matter and nitrogen down the canopy. The model uses the approach of Bertheloot et al. (2008) and the scaling relationships between canopy size and nitrogen vertical distribution described by Moreau et al. (2012). A schematic representation of *SiriusQuality* is shown in Figure 17.5b.

SiriusQuality also attempts to model nitrogen transfer to the grain more mechanistically than previous models (Sinclair and Amir, 1992; Sinclair and Muchow, 1995; Gabrielle et al., 1998; Bouman and van Laar, 2006). Accumulation of structural proteins and carbon, during endosperm cell division and DNA endoreduplication, and of storage proteins and starch, after the period of endosperm cell division, is explicitly considered (Martre et al., 2006). The accumulation of structural proteins and carbon per grain is assumed to be sink regulated and is driven by temperature. On the other hand, the accumulation of storage proteins and starch is assumed to be source regulated (i.e. independent of the number of grains per unit ground area), and is set daily to be proportional to the current amount of vegetative non-structural nitrogen. Sirius and *SiriusQuality*1 successfully simulate yield and protein concentration for contrasting pre- and post-flowering nitrogen supplies, post-anthesis temperature, and water supply in France (Martre et al., 2006) and Italy (Ferrise et al., 2010), under the dryer climate of Arizona and Australia across variation in air CO_2 concentration, water and nitrogen supplies, and cultivars (Jamieson et al., 2000, 2008; Ewert et al., 2002), and under the cooler climate of New Zealand for a range of rates and patterns of water and nitrogen availabilities, locations and cultivars (Jamieson and Semenov, 2000; Armour et al., 2004).

The allometric relations between the total amount of nitrogen per grain and the amount of the storage protein fractions presented in section 2.5 have been used to model gliadin and glutenin accumulation in developing wheat grain (Martre et al., 2003). This model has been implemented in *SiriusQuality* and has been evaluated against a wide range of nitrogen supply and post-anthesis temperature and water supply (Martre et al., 2006). The existence of environment-independent relations of nitrogen allocation for different cereals species (but with different parameter values), suggest that the model developed for wheat can be used to analyze and simulate the allocation of storage proteins for other cereals. The next step would be to model the effect of sulfur availability on the allocation of storage proteins and its interaction with nitrogen availability.

4 APPLYING CROP PHYSIOLOGY TO OBTAIN A SPECIFIC QUALITY AND HIGH YIELDS

In many agronomically relevant conditions, the trade-off between yield and quality is an obstacle to achieve the dual objective of high yield and high quality. This section thus presents a physiological viewpoint of the relationships between yield and oil attributes in sunflower and between yield and protein in wheat, and outlines physiological concepts of potential value for management and breeding aimed at improving grain quality.

4.1 Sunflower yield and oil composition

The interactions between yield and quality in sunflower were analyzed for a traditional and a high-oleic sunflower hybrid using the model presented in section 3.1 (Fig. 17.5a). Quality attributes were estimated for a wide range of sowing dates and three plant densities, using 35 years of weather data from three locations with contrasting radiation and temperature regimens in Argentina. The same interactions were analyzed experimentally in a trial network where five hybrids were sown in 11 locations between 26.7°S and 38.6°S (Peper et al., 2007). Large variation in observed and simulated yield and grain and oil quality traits were obtained both among and within locations.

The direction of the correlation between yield and quality traits evidently depends on the nature of the quality trait. For most of the situations, grain oil concentration and linoleic acid concentration were positively correlated with yield, while the concentration of oleic acid (Fig. 17.6a) and oil tocopherol concentration were negatively correlated with yield. Experimental results for the traditional hybrids in the trial network showed a similar pattern of negative correlation between oleic acid concentration and yield (Peper et al., 2007). Current knowledge suggests that the trade-off between yield and oil quality is mainly driven by temperature.

Temperature differences among locations were the main cause of the negative correlations between simulated yield and oleic acid. Oleic acid concentration increases with the temperature between 100 and 300°Cd after flowering (Izquierdo et al., 2006), whereas grain yield decreases with temperature around flowering (Cantagallo et al., 1997) and during early grain filling, i.e. between 250 and 450°Cd after flowering (Aguirrezábal et al., 2003), when the grain number and grain dry mass are determined, respectively. Because of this overlap of the critical periods for oleic acid concentration and yield, it is not easy to avoid this negative correlation when sowing a traditional sunflower hybrid. However, because this overlap was only partial, in a given year and location, temperature fluctuations over the reproductive period can reverse the general tendencies. So, despite the average negative correlation, it is possible to calculate, from model outputs, the probability of obtaining relatively high product quality (in this case oleic acid concentration) with a high yield at a given location. For instance, the probability of obtaining oil with more than 35% oleic acid in Paraná is 50% when yield potential is lower than 2.5 t ha^{-1}, and decreases to 27% for higher yields (Pereyra-Irujo and Aguirrezábal, unpublished results).

In some cases, genetic improvement can break negative correlations between quality traits and

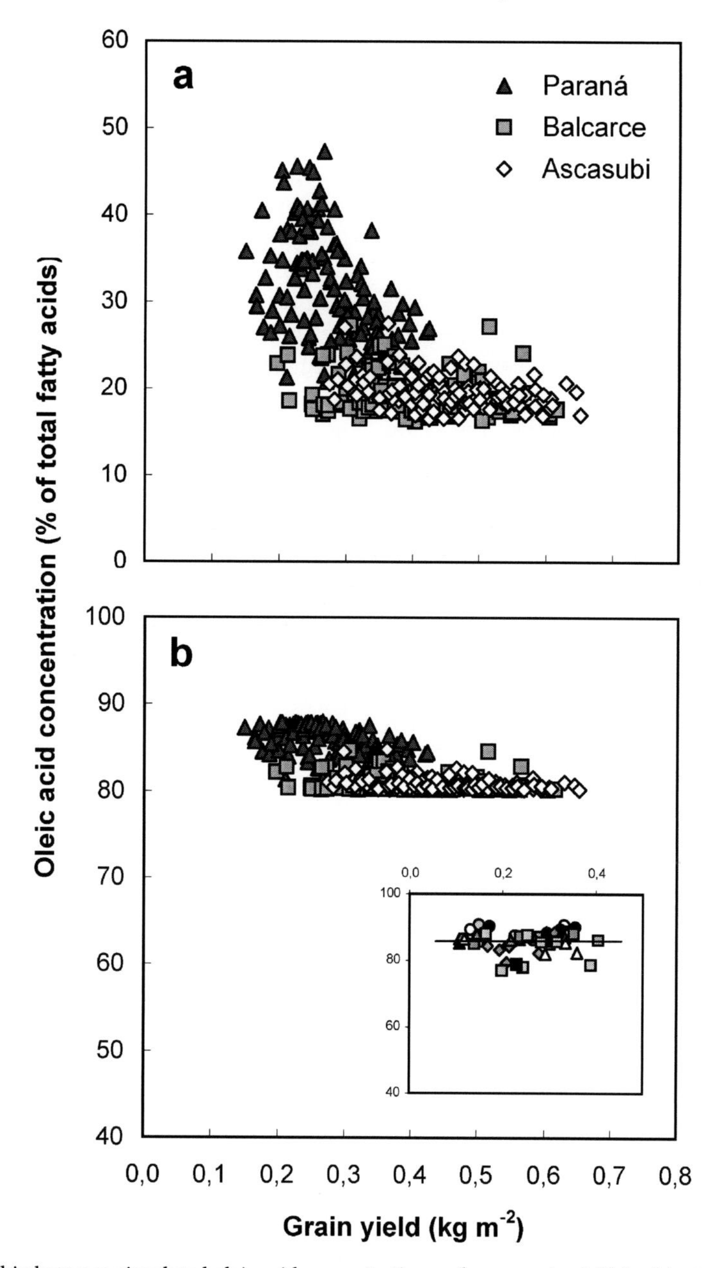

FIG. 17.6 Relationship between simulated oleic acid concentration and crop grain yield for (a) a traditional hybrid and (b) a high-oleic hybrid, for three contrasting locations in Argentina. Inset: relationship between measured oleic acid concentration and grain yield for 15 high-oleic sunflower hybrids sown in different locations in Argentina (Pereyra-Irujo and Aguirrezábal, unpublished results).

yield. This is easiest when quality traits depend on relatively few genes, as for high-oleic trait in sunflower (section 2.2). For instance, simulated and experimental data showed that oleic acid concentration for sunflower cultivars with high-oleic genes was independent of yield (Fig. 17.6b; Peper et al., 2007).

In sunflower, oil tocopherol concentration was negatively correlated with yield (Pereyra-Irujo and Aguirrezábal, 2007). In this case, the response was the same across locations, but depended strongly on the sowing density. This was because at higher densities, the same yield can be obtained with a higher number of smaller grains, for which the response curve of tocopherol concentration to variations in oil concentration is steepest. The negative correlation between oil tocopherol concentration and yield results from the relationship between oil tocopherol concentration and oil weight per grain. This relationship follows a dilution-like curve, with a steeper slope for grains with less than 15 μg oil (Fig. 17.1c). Within this range, oil tocopherol is therefore inversely correlated with weight per grain, which is the main driver of oil weight per grain, and also one of the two components of yield. However, it is well known that yield is mainly driven by the number of grains m^{-1} and that the correlation between grain number and weight per grain is weak for sunflower (Cantagallo et al., 1997) and annual crops in general (Chapters 4, 6, 7 and 16). Therefore, high oil tocopherol concentration could be obtained together with high yield, provided that this yield results mainly from many small grains. This deduction, based on simulations and application of physiological principles, was experimentally confirmed in experiments where oil tocopherol concentration was negatively related to oil weight per grain, but no clear correlation was found with yield (Peper et al., 2007). For hybrids genetically producing small grains, oil tocopherol concentration above 800 μg g oil^{-1} and yield above 3.1 t ha^{-1} were obtained when grain number was higher than 6700 grains m^{-2} (A. Peper, personal communication), i.e. close to the maximum number of grains m^{-2} observed under normal/regular field conditions (Lopez Pereira et al., 1999). Although environmental effects on phytosterol concentration have been less explored and the negative correlation between tocopherols and oil has not been incorporated into the model yet, it is likely that improvement of phytosterol concentration will be achieved by the same strategy as the improvement of tocopherol concentration.

4.2 Wheat yield and protein concentration

Negative relationships between grain protein concentration and yield have been known in wheat for more than 70 years (Waldron, 1933; Metzger, 1935; Neatby and McCalla, 1938; Grant and McCalla, 1949), and crop physiologists have analyzed the effect of environmental factors on this relation for over 40 years (Terman et al., 1969). Since the pioneer works on wheat, this negative grain yield–protein concentration relation has been reported for all the major crops, including maize (Dubley et al., 1977), sunflower (López Pereira et al., 2000), soybean (Wilcox and Cavins, 1995), and cowpea (Olusola Bayo, 1997). This correlation is observed when comparing different genotypes and also for a given genotype in response to environmental conditions or management practices.

Several putative physiological causes of the relationship between protein concentration and yield have been proposed, but the picture is still incomplete (reviewed in Feil, 1997). It has been hypothesized that energetic cost of protein synthesis limits crop nitrogen assimilation and protein synthesis in the grain (Bhatia and Rabson, 1976; Munier-Jolain and Salon, 2005). However, nitrogen uptake during the pre- and/or post-flowering periods and/or the efficiency of nitrogen translocation from vegetative organs to grains are more likely to explain this correlation (Kade et al., 2005). Consistently with this hypothesis, late (between heading and flowering) application of nitrogen fertilizer often increases grain protein concentration without reducing yield (Triboi and Triboi-Blondel, 2002). It also has been suggested that the observed genetic correlation between grain yield and protein concentration is at least in part due to the increase in dry matter harvest index (i.e. straw-to-grain dry mass ratio), which has accompanied most of the genetic progress in yield over the last 50 years, resulting in a reduction of the storage capacity of the crop for nitrogen (Kramer and Kozlowski, 1979; Herzog and Stamp, 1983). There is contradicting evidence concerning this issue (Cox et al., 1985; Slafer et al., 1990; Bänziger et al., 1992; Calderini et al., 1995; Uribelarrea et al., 2004; Abeledo et al., 2008), which possibly results from different patterns of spatial and temporal availability of soil nitrogen (taking into account soil moisture) throughout the growing season. Alternative hypotheses include the lower protein concentration of grain in distal florets in the spike (Simmons and Moss, 1978; Herzog and Stamp, 1983). Modern wheat cultivars setting more grains in distal positions of the spikelets could affect the balance between N and C of grains.

Some authors (e.g. Feil, 1997) argued that the seemingly universal genetic negative correlation between grain yield and protein concentration is an artifact of inadequate availability of soil nitrogen in most experiments. However, several reports showed significant variations of crop nitrogen yield independently of yield suggesting that a significant part of the negative relation is under genetic control (e.g. Monaghan et al. 2001; Laperche et al. 2007).

The effects of weather and specific environmental factors on the relationship between grain yield and protein concentration were studied using *SiriusQuality*1 (Martre et al., 2006). The model was run for a wide range of nitrogen availabilities, using 100 years of synthetic daily weather data representing typical conditions in France, as well as scenarios where temperature or radiation are increased or decreased.

At medium-nitrogen supply, a unique relationship was observed, independently of the causes of yield and grain protein concentration variation (Fig. 17.7a). The simulated effect of post-flowering temperature, radiation and water supply were very similar to that observed experimentally (Terman et al., 1969; Pushman and Bingham, 1976; Terman, 1979; Triboi et al., 2006). Interestingly, the large range of simulated yield shows that the grain yield–protein concentration is not linear. The analysis of the few experimental results with significant ranges of grain yield also indicates that the relation deviates from linearity (Metzger, 1935; Simmonds, 1995; Rharrabti et al., 2001). The non-linearity is due to the dilution of a roughly constant amount of nitrogen in an increasing amount of carbon components. The negative relation also tends to deviate slightly from the grain nitrogen yield isopleths showing decreases of the simulated crop grain nitrogen yield for the higher yielding conditions. This was associated with a decrease of the total crop N yield rather than with a systematic effect on the nitrogen harvest index (data not shown). These high yielding conditions were associated with long growth cycles (when simulation was performed considering that average daily temperature was increased by 3°C) with late grain filling occurring under water-limited conditions thus limiting crop nitrogen accumulation. This phenomenon possibly relied more on an interaction between the simulated treatment and the climate structure of the study site rather than an absolute outcome of the model. More generally, it is interesting to note that in *SiriusQuality*1, the dynamic of nitrogen and dry matter are mostly independent from each other. This relation is therefore an emergent property of the model. The model was thus able to simulate the observed effects of temperature, radiation, and water and nitrogen supplies on the negative relation, although it was not part of the model assumptions. These results give confidence for using *SiriusQuality*1 to analyze environmentally-induced variations of grain yield and protein

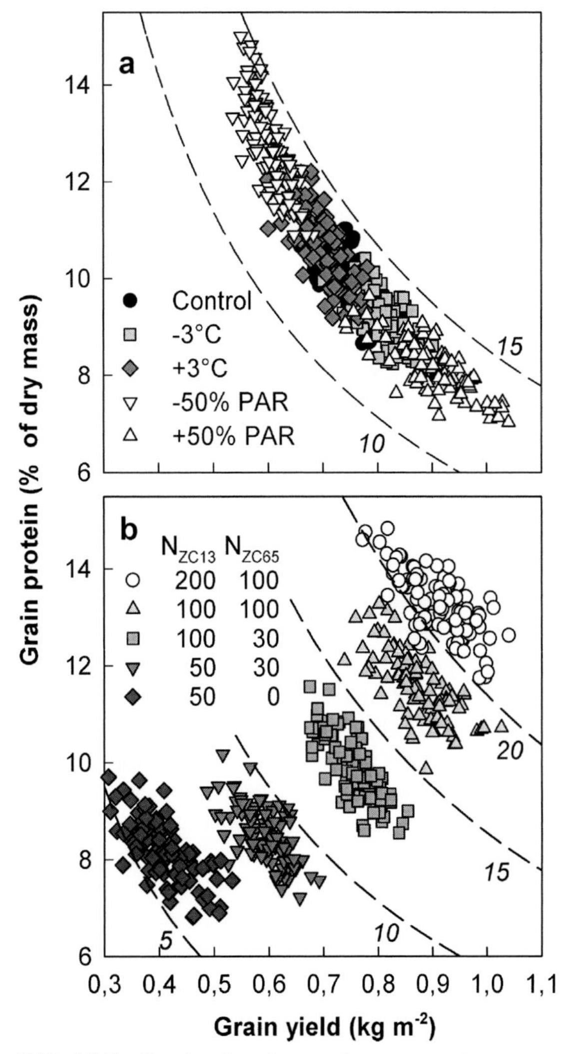

FIG. 17.7 Simulated grain protein concentration versus grain yield of wheat crops in response to variations (a) of the post-flowering temperature (average daily temperature increased or decreased by 3°C) or solar radiation (daily cumulated radiation increased or decreased by 50%), and (b) of nitrogen fertilization (indicated rates of nitrogen were applied when the crops had 3 leaves [N_{ZC13}] or at anthesis [N_{ZC65}]). In (a) the crops received 10 g N m^{-2} at 3 leaves and 10 g N m^{-2} at anthesis. Simulations were performed with the wheat simulation model *SiriusQuality*1 (Martre et al., 2006) for 100 years daily synthetic weather generated with the LARS-WG stochastic weather generator (Semenov and Brooks, 1999) for Clermont-Ferrand, France. Dashed lines are grain nitrogen yield isopleths (g N m^{-2}). Details of the soil and cultivar characteristics are given in Martre et al. (2007).

concentration, and the interactions between carbon and nitrogen metabolisms at the crop level.

At low nitrogen availability, grain yield increases linearly with the amount of nitrogen available, and a constant grain protein concentration is expected. For higher availability of nitrogen, the rate of increase of grain yield with soil nitrogen availability decreases (nitrogen-use efficiency decreases), grain protein concentration increases faster than grain yield, and a positive correlation between grain yield and grain protein concentration is expected (Fowler et al., 1990). Similarly, the simulated effect of nitrogen fertilizer on the grain yield–protein concentration relation (Fig. 17.7b) agrees with the literature (Pushman and Bingham, 1976; Oury et al., 2003; Triboi et al., 2006). Some experimental data suggest possible weaker correlation with very high or low nitrogen supplies (Kramer, 1979; Fowler et al., 1990). Such effects were not observed in the present simulations but extreme treatments would probably have more drastic effect on the yield–grain protein concentration relationship where inter-annual variability for water stress during the growth cycle is more marked.

A summary of the effects of environmental variables on grain yield and protein concentration is shown in Figure 17.8. At a given nitrogen supply and sink–source ratio, environmental factors modify grain yield and protein concentration symmetrically and a close negative correlation between grain yield and protein concentration is observed (Terman et al., 1969; Terman, 1979; Triboi et al., 2006). If environmental conditions reduce the sink–source ratio, then a partial compensation mechanism may occur, and the grain yield–protein concentration relation is then shifted. If the sink capacity (number of grains) is limited by early drought, high temperature, or genetically, then the duration of grain filling can be shortened. In this case, yield decreases despite an increase of single grain dry mass. At the grain level, the sink–source ratio increases more for nitrogen than for carbon,

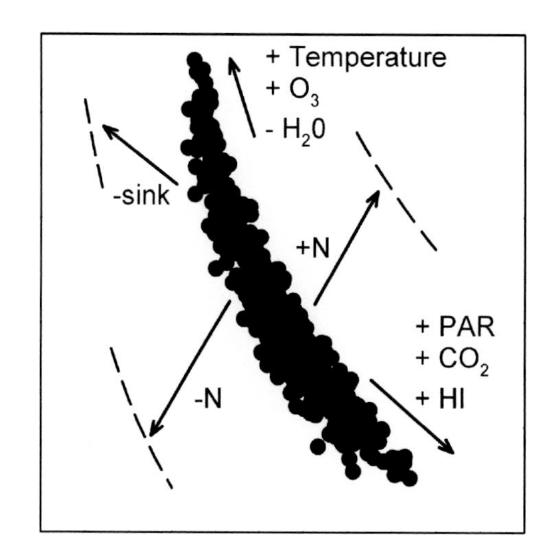

FIG. 17.8 Summary of the effects of temperature, CO_2, O_3, radiation, and supply of nitrogen and water on the relationship between protein concentration and grain yield. Genetic progress has continuously increased grain yield while grain N concentration has decreased linearly with yield (closed symbols). The environmental factors have contrasted effects on this relation, depending if they modify or not the sink:source ratio, and on their effects on grain yield. HI: harvest index; PAR: photosynthetically active radiation. *Adapted with permission from Triboi et al. (2006).*

and grain protein concentration can be higher than under non-limiting sink. Under limiting nitrogen, the slope of the genetic yield–protein concentration relationship decreased (Triboi et al., 2006), and thus grain protein concentration becomes slightly more sensitive to yield variations than under non-limiting nitrogen (Fig. 17.8).

The yield–grain protein concentration relationship is often hidden by environmental and management effects and, on average, can often be non-existent (Simmonds, 1995; Oury et al., 2003; Munier-Jolain and Salon, 2005). As shown here, this is explained by the different effects of environmental factors and management practices on crop dry mass and nitrogen dynamics, but these effects can be successfully analyzed through simulation models.

4.3 Management strategies for obtaining a target grain and oil composition

4.3.1 *Grain oil concentration*

Grain oil concentration in sunflower depends on intercepted radiation per plant during a critical period (Fig. 17.1a). High oil concentration would therefore require adjustment of management practices (e.g. sowing date, choice of cultivar, sowing density) to enhance intercepted radiation during such a period. For instance, late sowing decreases not only grain yield but also grain oil concentration in high-oil sunflower hybrids. For three locations in Argentina, simulated grain oil concentration decreased when sowing was delayed (Pereyra-Irujo and Aguirrezábal, 2007). This was related to the faster decline of radiation in relation to temperature after the summer solstice causing a steady decline in the amount of radiation intercepted per plant. Delaying sowing decreased grain oil concentration at 1–2% month^{-1}, but with a different magnitude according to the location. When a low intercepted radiation during grain filling is expected (e.g. late sowing), oil yield of low-oil hybrids is similar to that of high-oil hybrids (Izquierdo et al., 2008), and therefore cultivar choice can be based on other traits.

A fine-tuning of nitrogen supply and demand is necessary to obtain high oil concentration. Excessive fertilization can reduce grain oil concentration and therefore decrease the commercial quality of the product in sunflower and rape (e.g. Steer et al., 1984; Jeuffroy et al., 2006). This situation can be frequent in sunflower, a crop with low nitrogen requirements with respect to other species (Andrade et al., 1996). Models can assist in a more precise prevision of nitrogen requirements for specific yield and grain quality targets (Jeuffroy et al., 2006).

4.3.2 *Oil fatty acid composition*

Hybrid selection, sowing date and location are the three main management practices that can be used to obtain oils with different

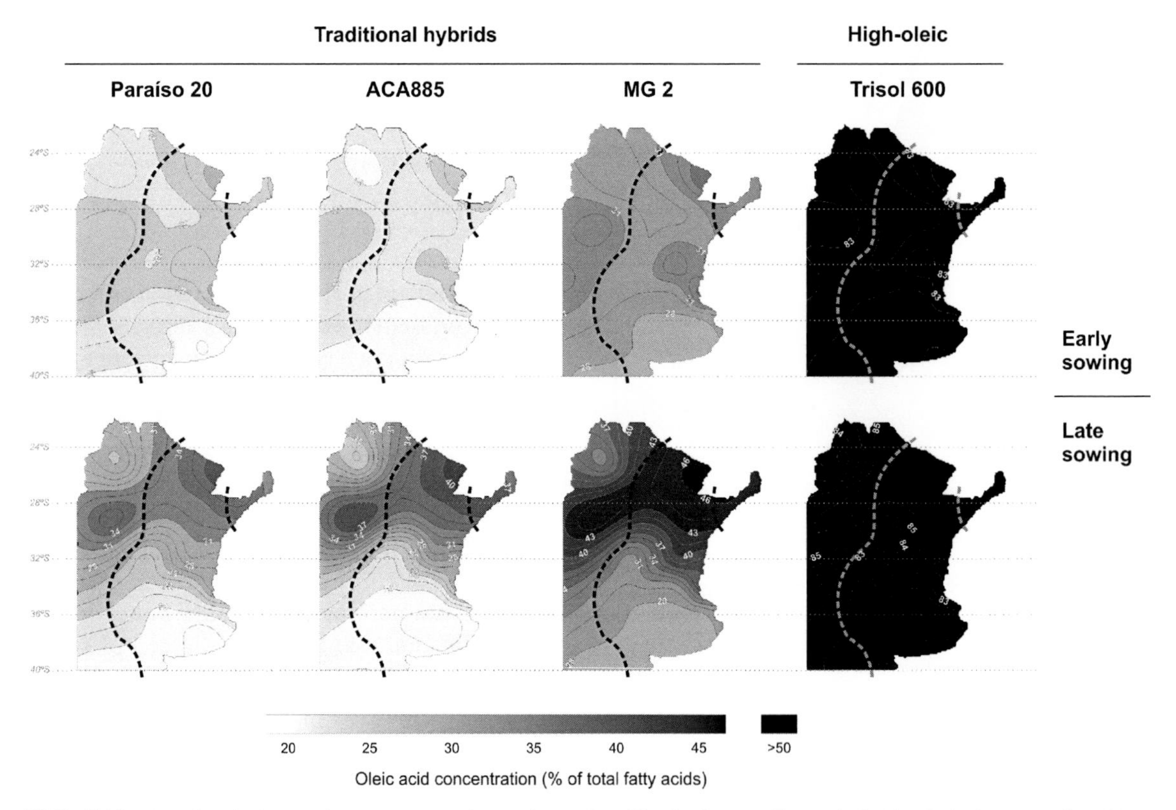

FIG. 17.9 Simulated oleic acid concentration for traditional and high-oleic sunflower hybrids. Early (top maps) and late (bottom maps) sowings were used in simulations. Simulations using the optimum sowing date (not shown) yielded results intermediate between those of early and late sowings. The main sunflower growing region of Argentina is between the two dotted lines. Details of the simulations are given in Pereyra-Irujo et al. (2009).

properties (Box 17.1), as illustrated in Figure 17.9. To obtain oil with a high proportion of oleic acid using traditional hybrids, crops should be grown in warm regions, sowing date should be adjusted so that minimum temperatures and radiation are high during grain filling, and hybrids with a high maximum concentration of oleic acid should be used (about 50%, close to a mid-oleic hybrid). On the other hand, the analysis of fatty acid composition of high-oleic hybrids in comparative yield trials showed these genotypes would yield high mono-unsaturated oil (80% oleic acid or higher) independently of location or sowing date (Echarte et al., 2008). To obtain oils with high proportion of linoleic acid, hybrids with a low minimum oleic acid concentration (high maximum linoleic acid concentration) should be used, and they should be sown early at high latitudes.

Several quality parameters of biodiesel are highly dependent on fatty acid composition (Clements, 1996). Simulated density, kinematic viscosity, and heating value produced from sunflower oil were very stable, whereas simulated iodine value and cetane number were highly variable between hybrids, regions and sowing dates (Pereyra-Irujo et al., 2009). For the high-oleic sunflower cultivar, all the analyzed parameters fell within the limits of the two main biodiesel standards (ASTM D6751 from the USA, and EN 14214 from Europe). For traditional cultivars, the US standard was met in almost all cases, whereas the European specifications were met only by the hybrid with the highest maximum oleic acid concentration. For a given cultivar, biodiesel quality tended to be higher at lower latitudes and in late sowings, following the degree of oil saturation (Fig. 17.9).

Climate models predict an increase in global mean surface temperature of about 1.0–3.5°C in the next century (Christensen et al., 2007; Chapter 20). Regional climate change projections allowed an estimation of the possible effects of warming on fatty acid composition of sunflower oil grown at different latitudes (Echarte et al., 2009). Simulations predict an increase in oleic acid concentration for every location considered, with stronger effect in warmer regions. Although, in general, warming would have a positive effect on oil quality, it would probably have the opposite effect on crop yield (Pereyra-Irujo and Aguirrezábal, 2007).

4.3.3 Grain protein concentration

Management of nitrogen fertilizer is critical to obtain a targeted grain yield and protein concentration in diverse cropping systems (Chapters 2, 3 5, 6 and 7). This practice has an important economical cost for the farmer, e.g. in France it represents about 60% of the cost of growing a wheat crop, and excess nitrogen has a potentially deleterious environmental effect (Chapter 3). Chapter 8 details approaches for diagnostic of nitrogen deficiency from a physiological perspective. Until recently, considerations of fertilizer needs have been mostly driven by yield rather than protein targets, except in crops such as barley where narrower bands of grain protein have a more marked influence on grain price. Crop simulation models offer opportunities to manage nitrogen fertilization to optimize both grain yield and protein concentration, and also cultivar choice (Jeuffroy et al., 2013).

4.4 Strategies for genetic improvement of quality traits

4.4.1 Grain oil concentration

As mentioned earlier, genetic analyses of crosses between high-oil sunflower hybrids have identified alleles which could potentially increase grain oil concentration (Mestries et al., 1998; Mokrani et al., 2002; Bert et al., 2003), although these studies did not link the differences in oil concentration to anatomical or physiological characteristics. In maize, Yang et al. (2012) found that QTL for oil content co-localized with QTL of physical characteristics of the grain (e.g. endosperm to embryo weight ratio). This kind of information could be of importance to determine, for instance, if future increases in grain oil concentration through decreased pericarp weight (and therefore decreased grain weight) could be negatively correlated to yield, if grain number does not compensate for reduced grain weight (Mantese et al., 2006).

As genetic differences become smaller among modern, high-oil, sunflower hybrids, and given the relative environmental instability of their oil concentration (as compared to low-oil hybrids), the effect of the environment becomes increasingly important. However, the effect of environmental conditions and their interactions with genotype on oil concentration have received less attention than the effect of genotype. The well-known effect of radiation on grain oil concentration (Fig. 17.1a) is probably one of the causes of the frequent co-localization of QTL for flowering date and grain oil concentration since solar radiation intensity during grain filling is lower when flowering is delayed (León et al., 2000). In sunflower, a critical period during grain filling has been identified (Aguirrezábal et al., 2003), during which grain oil concentration is closely associated with intercepted radiation and temperature, following a predictable response curve. Based on this relationship, two complementary strategies for

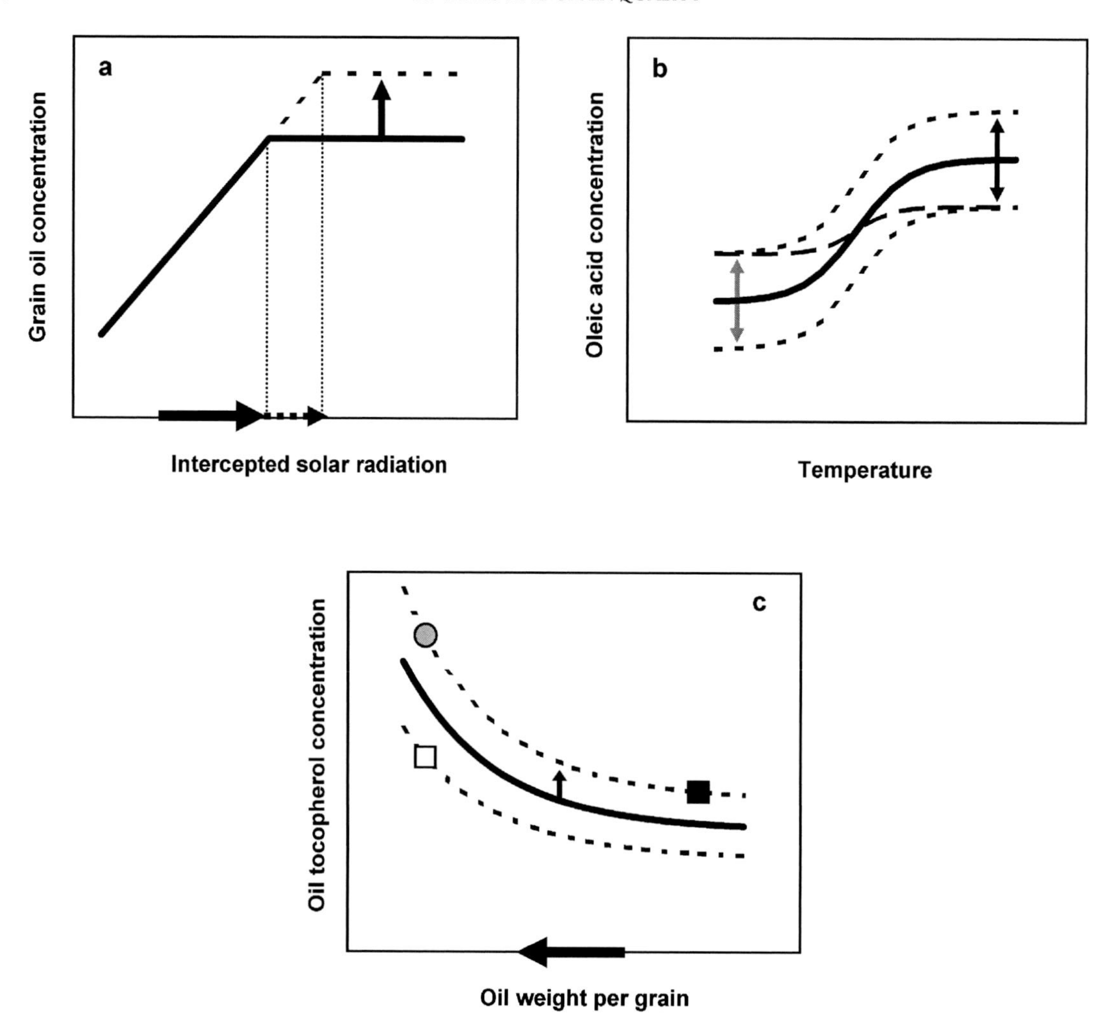

FIG. 17.10 Main avenues proposed for the genetic modification of sunflower quality traits: (a) grain oil concentration: increased radiation interception during the critical period (arrows along the x-axis), and higher maximum oil concentration (dotted line); (b) oleic acid concentration: changes in the plateau values for minimum (gray arrows) and maximum (black arrows) oleic acid concentration, either to obtain higher or lower average concentration (dotted lines), or increased stability (dashed line); and (c) oil tocopherol concentration: increased tocopherol concentration relative to that expected through the dilution curve (vertical arrow), and decreased oil weight per grain (arrow along the x-axis), combined to obtain high oil tocopherol concentration (gray circle); black and white squares: see text.

improving oil concentration can be proposed (Fig. 17.10a):

1. Selecting for traits that increase radiation interception during this period such as stay-green or resistance to leaf diseases should increase oil concentration (Fig. 17.10a) and maybe also increase yield. Also, care should be taken that increases in 'greenness' effectively result in increased photosynthesis, which is not always the case with stay-green

mutants (Hall, 2001). Likewise, visual disease symptoms are not always correlated to reductions in photosynthesis (e.g. Sadras et al., 2000).

2. Another strategy could be to select for higher maximum oil concentration (dotted line in Fig. 17.10a). A simple way to measure this trait would be post-flowering thinning of crops sown at a normal density. This would increase radiation interception per plant, without the confounding effects of high grain number per plant or high grain weight potential (Aguirrezábal et al., 2003).

4.4.2 *Oil fatty acid composition*

Crop breeding has achieved numerous modifications of the fatty acid profile of oil in many species, mainly through the incorporation of genes with large effects (for a review see Velasco and Fernández-Martínez, 2002). The effect of the environment on these genotypes is relatively small (e.g. up to 6.9% for high-oleic sunflower; Izquierdo and Aguirrezábal, 2008), but it can still be a concern, especially with strict market standards. The stability of improved fatty acid compositions and the achievement of a desired quality under specific environments are important breeding objectives.

The parameters of the model in section 2.2 showed a wide variation between traditional, high-linoleic sunflower hybrids (Izquierdo and Aguirrezábal, 2008). The stability of minimum and maximum oleic acid concentrations for temperatures outside the range of strong response (Fig. 17.1b) provide a simple way to screen for high or low values of individual parameters (y_0 and y_{max}, dotted lines in Fig. 17.10b), by sowing at locations or dates of known temperature, or under semi-controlled conditions (e.g. greenhouse). Of these two parameters, y_0 showed the highest genetic variability. Such screening could be simpler and cheaper than screening for the trait itself (oleic acid percentage). Another approach could be to develop cultivars for specific environments. For instance, a genotype yielding very-high-linoleic oil under low temperature could be obtained by selecting for a low y_0, irrespective of other parameters.

This model is also valid for high-oleic hybrids. Breeders have improved high-oleic sunflower genotypes mostly by indirectly selecting for a high y_0. There are, however, high-oleic genotypes that differ in their response to temperature, although this genetic variability has not been quantified using the approach described above. An interesting breeding goal for these genotypes would be to increase the stability of fatty acid composition by selecting a low difference between minimum and maximum concentration of oleic acid (dashed line in Fig. 17.10b).

For other species, the lack of a robust model, and the need to take into account the effect of radiation (and the interaction between temperature and radiation), could make these approaches more difficult.

4.4.3 *Oil tocopherol concentration*

In sunflower, an inverse relationship between tocopherol concentration and oil weight per grain (Fig. 17.1c) accounts for most of the environmental effect and, therefore, this relationship can theoretically be used to quantify the genetic variability of oil tocopherol concentration. Nolasco et al. (2006) analyzed the genetic variability of tocopherol concentration among sunflower hybrids, finding the effect of the environment to be larger than the genotypic effect. A re-analysis of data from Nolasco et al. (2004) through a two-way analysis of variation (ANOVA), using the ratio between the measured and expected (according to the dilution-like curve) values (as opposed to the raw tocopherol concentration data), showed that environmental and G $\times$ E effects were reduced, the effect of the genotype was increased, and differences between genotypes were detected.

Based on this relationship, two strategies for improving oil tocopherol concentration are proposed:

1. An ideotype for high tocopherol concentration in the oil could be defined as a plant with a small amount of oil in each grain. Reducing oil weight per grain (solid arrow along the x-axis in Fig. 17.10c) should preferably be achieved through smaller grains, and not through reduced grain oil concentration (which is an important quality trait *per se*). Therefore, to avoid negative consequences on grain yield, a smaller grain weight should be compensated for by an increased grain number.
2. Using the relationship between tocopherol concentration and oil weight per grain, screening for genotypes with an improved tocopherol concentration, independently of genetic or environmental variations in oil concentration or grain weight. Such a genotype would be one with a positive deviation from the expected 'dilution' curve (e.g. black square in Fig. 17.10c). A genotype with a higher absolute tocopherol concentration, but with a negative value relative to the curve (e.g. white squares in Fig. 17.10c) would be expected to have a negative direct effect on tocopherol concentration. This latter genotype should have, however, a positive indirect effect through a decreased oil weight per grain, according to strategy (1). Provided the independence of these two effects, the two strategies presented here could hypothetically be applied simultaneously (gray circle in Fig. 17.10c) and would be equally useful to improve concentration of phytosterols.

4.4.4 Grain protein concentration

The yield–protein concentration correlation and the large genotype by management and genotype by environment interaction components of variance relative to the genotypic component (Cooper et al., 2001) have significantly restrained genetic improvements of grain protein concentration. Breaking this negative relationship remains a major challenge for cereal breeders (DePauw et al., 2007; Oury and Godin, 2007).

Both grain yield and protein concentration are genetically determined by a large number of independent loci. On this basis, some authors have concluded that genetic restrictions caused by linkage or pleiotropy are not of sufficient magnitude to hinder the improvement of both traits simultaneously (Kibite and Evans, 1984; Monaghan et al., 2001). Indeed, several breeding programs have been successful in breaking or shifting this correlation, thus demonstrating that there are no physiological or genetic barriers to breeding high-yielding, high-protein genotypes. Cober and Voldeng (2000) developed high-protein soybean populations exhibiting very low or no association between grain yield and protein concentration. Similarly, high-protein and low-protein maize strains resulting from long-term divergent recurrent selection allowed translation of the negative yield–grain protein concentration relation (Uribelarrea et al., 2004). More recently, the transfer of a chromosomic region from an accession of emmer wheat (*Triticum turgidum* L. var. *dicoccoides*) associated with high grain protein concentration into high-yielding bread wheat (DePauw et al., 2007) and durum wheat (Chee et al., 2001) also shifted the correlation.

In an analysis of 27 wheat genotypes grown in multienvironment field trials in France, the genetic deviation from the negative yield–grain protein was related to post-anthesis nitrogen uptake, independently of anthesis date or shoot nitrogen (Bogard et al., 2010). This strongly supports the hypothesis that, where water deficit does not limit post-anthesis soil nitrogen uptake, increasing the capacity of a genotype to assimilate soil nitrogen after anthesis will shift the grain yield–grain protein concentration genetic antagonism (Monaghan et al., 2001; Triboi and Triboi-Blondel, 2002). Where water deficit reduces post-anthesis soil nitrogen uptake, increasing the remobilization of vegetative nitrogen to grains is a promising strategy (e.g. Avni et al., 2014).

Models can be used to assess the link between physiological traits (model parameter) and grain yield or protein concentration (Boote et al., 2001; Hammer et al., 2006). The wheat simulation models Sirius and *SiriusQuality*1 (Martre et al., 2007; Semenov et al., 2007) and APSIM-Nwheat (Asseng and Milroy, 2006) have been used to analyze the effect of single plant or crop traits on both grain yield and protein concentration. The effect of climate was assessed by running the models for 32 to 100 years of weather data. These simulations showed that variations in weather and nitrogen treatments induced larger variations in grain yield and protein concentration than most of the physiological traits considered, and revealed strong trait-by-nitrogen and trait-by-water interactions. Chapter 14 illustrates a more systematic analysis for the wheat simulation model *SiriusQualiy* where the effect of all parameters was quantified simultaneously. This analysis showed that the effects of parameters on grain yield and protein concentration are mainly non-linear with strong interactions among parameters and strong influence of climate and crop management on the first order effect of most traits

Monaghan et al. (2001) and Bogard et al. (2010) showed that, under western European conditions, positive departures from the grain yield–protein concentration relation were associated with the amount of nitrogen accumulated by the crop after flowering. The sensitivity analysis of *SiriusQuality*1 was consistent with this finding: for high nitrogen supply, increasing the maximum rate of nitrogen uptake during grain filling allowed a shift of the negative correlation between protein concentration and grain yield (Fig. 17.11b).

To analyze the effect of the rate of nitrogen remobilization from leaves and stems to grains on crop dry matter and nitrogen dynamics, a scaling parameter (β) changing proportionally the daily rate of accumulation of grain nitrogen was introduced in *SiriusQuality*1 (Martre et al., 2007). For both low and high nitrogen supply, simulated grain yield increased significantly when the rate of grain nitrogen accumulation decreased, whereas grain nitrogen yield, post-flowering nitrogen uptake, and nitrogen harvest index were very similar, even for high nitrogen supply (Fig. 17.11c, d). Therefore, increasing the daily rate of grain nitrogen accumulation leads to a dilution of simulated grain nitrogen. Surprisingly, about 75% of the increase in grain yield was due to the increase in dry matter harvest index, which increased by about 6% when β decreased from 1 to 0.7. In good agreement with this result, pot-grown stay-green mutants of durum wheat in a greenhouse had a higher grain yield per plant than the wild type, but the grain nitrogen yield was the same for the mutants and the wild type; therefore the grain protein concentration was lower for the stay-green mutants than for the wild type (Spano et al. 2003). In contrast, Uauy et al. (2006a,b) reported that acceleration of canopy senescence in transgenic wheat plants, under-expressing an NAC transcription factor located at a QTL for high grain protein concentration, paralleled acceleration of nutrient remobilization from leaves to grains, leading to higher grain protein concentration. Similarly Avni et al. (2014) reported that 'loss-of-function' mutants for this transcription factor have a delayed monocarpic leaf senescence, leading to lower grain protein concentration for the mutants compared with the wild type, but grain yield was similar for both the mutants and the wild type. Chapter 10 presents an extensive account of environmental modulation and genetic control of leaf senescence.

A survey of UK winter wheat cultivars revealed a positive association between grain yield and stem nitrogen concentration at flowering (Shearman et al., 2005). In the sensitivity analysis of *SiriusQuality*1, the nitrogen storage capacity of both leaves and stem was considered. Under limiting nitrogen supply, increasing the storage

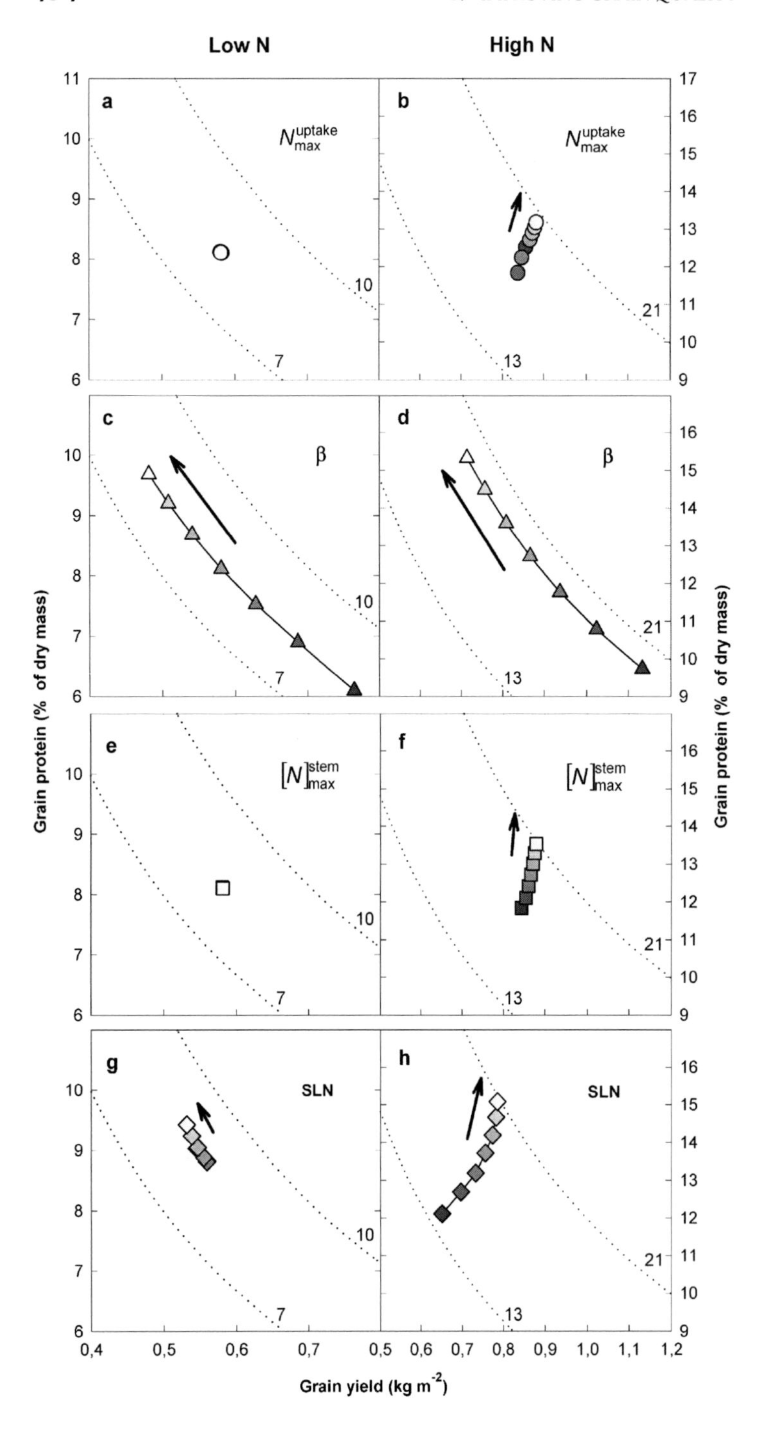

FIG. 17.11 Simulated grain protein concentration versus grain yield for wheat crops in response to (a, b) variations of the maximum rate of nitrogen uptake at anthesis (N_{max}^{uptake}, default value 4 g N m^{-2} ground d^{-1}), (c, d) a scaling parameter modifying the daily rate of grain nitrogen accumulation (β, default value 1), (e, f) the nitrogen storage capacity of the stem ($[N]_{max}^{system}$, default value 10 mg N g^{-1} DM), and (g, h) leaf nitrogen mass per unit of leaf surface area (SLN, default value 1.5 g N m^{-2} leaf). Simulations were performed with the wheat simulation model *SiriusQuality*1 (Martre et al., 2006). Data are means for 32 years at Clermont-Ferrand at low (8 g N m^{-2} applied in two splits; a, c, e, g) and high (25 g N m^{-2} applied in four splits; b, d, f, h) nitrogen supplies. The gray intensity of the symbols decreases as the parameters increase by 10% increments from −30% to +30% of their default value. Arrows indicate the way of the increase of the parameter values. Dotted lines are grain nitrogen yield (g N m^{-2}) isopleths. Details of the soil and cultivar characteristics are given in Martre et al. (2007). *Adapted with permission from Martre et al. (2007).*

capacity of the leaves or stems did not shift the negative correlation between grain yield and protein concentration (Fig. 17.11e,g). The increase of leaf nitrogen mass per unit leaf surface area (i.e. specific leaf nitrogen) was associated with a reduction in grain yield because it reduced leaf expansion due to nitrogen shortage. In contrast, under high nitrogen supply, increasing maximum stem nitrogen concentration, or leaf nitrogen mass per unit of leaf surface area resulted in a significant increase in both grain yield and grain protein concentration (Fig. 17.11f, h). Similar trends in grain yield and protein concentration were observed when the efficiency of nitrogen remobilization was modified, but changes in simulated grain yield and protein concentration were limited. This simulation analysis shows that, under high nitrogen inputs, increasing the nitrogen storage capacity of the leaves and stem, and/or the efficiency of nitrogen remobilization, may significantly shift the negative grain yield–protein concentration. Moreover, it may also reduce the risk of nitrogen losses by leaching and volatilization.

5 CONCLUDING REMARKS

Diverse industries require grains with high and reliable quality for specific uses, and breeders have responded by tailoring new cultivars to these demands (DePauw et al., 2007; Velasco and Fernández-Martínez, 2002). There is, therefore, an increasing need for knowledge of the physiology of quality traits, in support of both breeding efforts and crop management aimed at the dual objective of high yield and high quality. While breeding for crop yield has been always performed at the population (crop) level, the improvement of quality traits has mostly focused on individual plants or plant parts (i.e. grain). Combining experiments and simulations, we demonstrated that most quality traits, and certainly quality–yield interactions, cannot be correctly understood or predicted by extrapolating from individual plants to the population. The determination of quality traits – like that of yield components – should be analyzed at the crop level, and crop physiology concepts and methods classically applied to yield analysis are powerful tools for this analysis.

Much of our current knowledge about the physiology of grain quality in oilseed crops and cereals has been obtained in sunflower and bread wheat, hence the value of these species as models for quality studies illustrated in this chapter. Examples were presented where relationships (and underlying processes) between a given quality trait and its predictor first identified in model crops were then found to be common to other crops. This research strategy seems promising for further insight into the physiology of quality traits in several crops.

Mathematical crop models have incorporated quality modules that allow simultaneous simulation of grain yield and quality (section 3). Several cases were presented to illustrate how, by means of these models, crop physiology can be used for designing better management and breeding strategies for improving quality in oil crops and cereals. Interestingly, these models are more than a means to gather physiological knowledge; they also help create new knowledge by highlighting emergent properties that arise from the combination of different processes. For example, crop models realistically reproduced relationships between yield and quality traits (section 4), which are assumed to be independent processes in the models.

Fine-tuning of resource with production is necessary to improve efficiency in agriculture. Experimental determination of input requirements for optimum production has been successful but is limited in space and time and, therefore, transferable only in general terms (Lawlor et al., 2002). Better understanding of biochemical and physiological processes underlying plant response to the environment will help to obtain a desired product while improving resource-use efficiency.

Dealing with genotype by environment interactions is still one of the major challenges for plant breeders. As a complement to classical approaches, quantitative relationships between quality traits and environmental drivers in this chapter are a more robust means of quantifying genotype by environment interactions, and of generating new 'traits' (model parameters) that are largely independent of the environment. These relationships represent meta-mechanisms at the plant or crop level, and therefore could help to fill the current gap between genotype and phenotype, especially for complex traits. Coupling physiological and genetic approaches to improve quality traits, we argue, represents one of the next challenges for crop physiology and breeding.

Acknowledgments

Luis Aguirrezábal, María Mercedes Echarte, Gustavo Pereyra Irujo and Natalia Izquierdo are members of CONICET (Consejo Nacional de Investigaciones Científicas y Técnicas, Argentina). Some unpublished results presented in this chapter were funded by CONICET (PIP 0362) and ANPCyT (PICT 08 0941).

References

Abeledo, L.G., Calderini, D.F., Slafer, G.A., 2008. Nitrogen economy in old and modern malting barleys. Field Crops Res. 106, 171–178.

Aguirrezábal, L.A.N., Lavaud, Y., Dosio, G.A.A., Izquierdo, N.G., Andrade, F.H., González, L.M., 2003. Intercepted solar radiation during seed filling determines sunflower weight per seed and oil concentration. Crop Sci. 43, 152–161.

Aguirrezábal, L.A.N., Pereyra, V.R., 1998. Girasol. In: Aguirrezábal, L.A.N., Andrade, F.H. (Eds.), Calidad de productos agrícolas. Bases ecofisiológicas, genéticas y de manejo agronómico. Editorial Unidad Integrada Balcarce–Ediciones técnicas Morgan–Publicaciones Nidera, Buenos Aires (Argentina), pp. 140–185.

Alahdadi, I., Oraki, H., Khajani, F.P., 2011. Effect of water stress on yield and yield components of sunflower hybrids. Afr. J. Biotechnol. 10, 6504–6509.

Almonor, G.O., Fenner, G.P., Wilson, R.F., 1998. Temperature effects on tocopherol composition in soybean with genetically improved oil quality. J. Am. Oil Chem. Soc. 75, 591–596.

Anastasi, U., Santonoceto, C., Giuffrè, A.M., Sortino, O., Gresta, F., Abbate, V., 2010. Yield performance and grain lipid composition of standard and oleic sunflower as affected by water supply. Field Crops Res. 119, 145–153.

Andrade, F., Cirilo, A., Uhart, S., Otegui, M., 1996. Ecofísiología del cultivo de maíz. Editorial La Barrosa, Dekalb Press, INTA, FCA-UNMdP.

Andrade, F.H., Ferreiro, M.A., 1996. Reproductive growth of maize, sunflower and soybean at different source levels during grain filling. Field Crops Res. 48, 155–165.

Angeloni, P.N., Echarte, M.M., Aguirrezábal, L.A.N., 2012. Efecto de la temperatura sobre el peso y concentración de aceite de los granos de genotipos de girasol tradicional y alto oleico. 18th Internacional Sunflower Conference, Mar del Plata, Argentina.

Armour, T., Jamieson, P.D., Zyskowski, R.F., 2004. Using the sirius wheat calculator to manage wheat quality – the Canterbury experience. Agron. New Zealand 34, 171–176.

Asseng, S., Milroy, S.P., 2006. Simulation of environmental and genetic effects on grain protein concentration in wheat. Eur. J. Agron. 25, 119–128.

Asseng, S., Bar-Tal, A., Bowden, J.W., et al., 2002. Simulation of grain protein content with APSIM-Nwheat. Eur. J. Agron. 16, 25–42.

Avni, R., Zhao, R., Pearce, S., et al., 2014. Functional characterization of GPC-1 genes in hexaploid wheat. Planta 239, 313–324.

Awad, A.B., Roy, R., Fink, C.S., 2003. β-sitosterol, a plant sterol, induces apoptosis and activates key caspases in MDA-MB-231 human breast cancer cells. Oncol. Rep. 10, 497–500.

Baldini, M., Giovanardi, R., Tahmasebi-enferadi, S., Vannozzi, G.P., 2002. Effects of water regime on fatty acid accumulation and final fatty acid composition in the oil of standard and high oleic sunflower hybrids. Ital. J. Agron. 6, 119–126.

Bänziger, M., Feil, B., Schmid, J.E., Stamp, P., 1992. Genotypic variation in grain nitrogen content of wheat as affected by mineral nitrogen supply in the soil. Eur. J. Agron. 1, 155–162.

Baux, A., Colbach, N., Allirand, J.M., Jullien, A., Ney, B., Pellet, D., 2013. Insights into temperature effects on the fatty acid composition of oilseed rape varieties. Eur. J. Agron. 49, 12–19.

Bert, P.F., Jouan, I., Tourvieille De Labrouhe, D., et al., 2003. Comparative genetic analysis of quantitative traits in sunflower (*Helianthus annuus* L.). 2. Characterisation of QTL involved in developmental and agronomic traits. Theor. Appl. Genet. 107, 181–189.

Bertheloot, J., Andrieu, B., Fournier, C., Martre, P., 2008. A process-based model to simulate nitrogen distribution in wheat (*Triticum aestivum*) during grain-filling. Funct. Plant Biol. 35, 781–796.

Bevan, M., Colot, V., Hammondkosack, M., et al., 1993. Transcriptional control of plant storage protein genes. Philosoph. Transact. R. Soc. Lond. Ser. B Biol. Sci. 342, 209–215.

Bhatia, C.R., Rabson, R., 1976. Bioenergetic considerations in cereal breeding for protein improvement. Science 194, 1419–1421.

Binkoski, A., Kris-Etherton, P., Wilson, T., Mountain, M., Nicolosi, R., 2005. Balance of unsaturated fatty acids is important to a cholesterol-lowering diet: comparison of mid-oleic sunflower oil and olive oil on cardiovascular disease risk factors. J. Am. Dietet. Assoc. 105, 1080–1086.

Blumenthal, C., Bekes, F., Gras, P.W., Barlow, E.W.R., Wrigley, C.W., 1995. Identification of wheat genotypes tolerant to the effects of heat stress on grain quality. Cereal Chem. 72, 539–544.

Blumenthal, C.S., Batey, I.L., Bekes, F., Wrigley, C.W., Barlow, E.W.R., 1991. Seasonal changes in wheat-grain quality associated with high temperatures during grain filling. Aust. J. Agric. Res. 42, 21.

Bogard, M., Allard, V., Brancourt-Hulmel, M., et al., 2010. Deviations from the grain protein concentration – grain yield negative relationship are highly correlated to post-anthesis N uptake in winter wheat. J. Exp. Bot. 61, 4303–4312.

Boote, K.J., Kropff, M.J., Bindraban, P.S., 2001. Physiology and modelling of traits in crop plants: implications for genetic improvement. Agric. Sys. 70, 395–420.

Borghi, B., Corbellini, M., Ciaffi, M., et al., 1995. Effect of heat shock during grain filling on grain quality of bread and durum wheats. Aust. J. Agric. Res. 46, 1365–1380.

Bouman, B.A.M., van Laar, H.H., 2006. Description and evaluation of the rice growth model ORYZA2000 under nitrogen-limited conditions. Agric. Sys. 87, 249–273.

Bramley, P., Elmadfa, I., Kafatos, A., et al., 2000. Vitamin E. J. Sci. Food Agric. 80, 913–938.

Branlard, G., Dardevet, M., Saccomano, R., Lagoutte, F., Gourdon, J., 2001. Genetic diversity of wheat storage proteins and bread wheat quality. Euphytica 119, 59–67.

Brisson, N., Mary, B., Ripoche, D., et al., 1998. STICS: a generic model for the simulation of crops and their water and nitrogen balances. I. Theory and parameterization applied to wheat and corn. Agronomie 18, 311–346.

Browse, J., Roughan, P.G., Slack, C.R., 1981. Light control of fatty acid synthesis and diurnal fluctuations of fatty acid composition in leaves. Biochemical Journal 196 (1), 347–354.

Burke, J.M., Knapp, S.J., Rieseberg, L.H., 2005. Genetic consequences of selection during the evolution of cultivated sunflower. Genetics 171, 1933–1940.

Cahoon, E.B., Kinney, A.J., 2005. Production of vegetable oils with novel properties: using genomic tools to probe and manipulate fatty acid metabolism. Eur. J. Lipid Sci. Technol. 107, 239–243.

Calderini, D.F., Torroes-León, S., Slafer, G.A., 1995. Consequences of wheat breeding on nitrogen and phosphorus yield, grain nitrogen and phosphorus concentration and associated traits. Ann. Bot. 76, 315–322.

Cantagallo, J., Chimenti, C., Hall, A., 1997. Number of seeds per unit area in sunflower correlates well with a photothermal quotient. Crop Sci. 37, 1780–1786.

Cantagallo, J.E., Medan, D., Hall, A.J., 2004. Grain number in sunflower as affected by shading during floret growth, anthesis and grain setting. Field Crops Res. 85, 191–202.

Cantisán, S., Martínez-Force, E., Garcés, R., 2000. Enzymatic studies of high stearic acid sunflower seed mutants. Plant Physiol. Biochem. 38, 377–382.

Canvin, D., 1965. The effect of temperature on the oil content and fatty acid composition of the oils from several seed crops. Can. J. Bot. 43, 63–69.

Carceller, J.L., Aussenac, T., 2001. SDS-insoluble glutenin polymer formation in developing grains of hexaploid wheat: the role of the ratio of high to low molecular weight glutenin subunits and drying rate during ripening. Aust. J. Plant Physiol. 28, 193–201.

Casadebaig, P., Guilioni, L., Lecoeur, J., Christophe, A., Champolivier, L., Debaeke, P., 2011. SUNFLO, a model to simulate genotype-specific performance of the sunflower crop in contrasting environments. Agric. For. Meteorol. 151, 163–178.

Chapman, M.A., Burke, J.M., 2012. Evidence of selection on fatty acid biosynthetic genes during the evolution of cultivated sunflower. Theor. Appl. Genet. 125, 897–907.

Chee, P.W., Elias, E.M., Anderson, J.A., Kianian, S.F., 2001. Evaluation of a high grain protein QTL from *Triticum turgidum* L. var. dicoccoides in an adapted durum wheat background. Crop Sci. 41, 295–301.

Chiaiese, P., Ohkama-Ohtsu, N., Molvig, L., et al., 2004. Sulphur and nitrogen nutrition influence the response of chickpea seeds to an added, transgenic sink for organic sulphur. J. Exp. Bot. 55, 1889–1901.

Christensen, J.H., Carter, T.R., Rummukainen, M., Amanatidis, G., 2007. Evaluating the performance and utility of regional climate models: The PRUDENCE project. Clim. Change 81, 1–6.

Ciaffi, M., Tozzi, L., Borghi, B., Corbellini, M., Lafiandra, D., 1996. Effect of heat shock during grain filling on the gluten protein composition of bread wheat. J. Cereal Sci. 24, 91–100.

Clements, L., 1996. Blending rules for formulating biodiesel fuel. The National Biodiesel Board.

Clouse, S.D., 2000. Plant development: A role for sterols in embryogenesis. Curr. Biol. 10, 601–604.

Cober, E.R., Voldeng, H.D., 2000. Developing high-protein, high-yield soybean populations and lines. Crop Sci. 40, 39–42.

Connor, D., Sadras, V., 1992. Physiology of yield expression in sunflower. Field Crops Res. 30, 333–389.

Cooper, M., Woodruff, D.R., Phillips, I.G., Basford, K.E., Gilmour, A.R., 2001. Genotype-by-management interactions for grain yield and grain protein concentration of wheat. Field Crops Res. 69, 47–67.

Corbellini, M., Mazza, L., Ciaffi, M., Lafiandra, D., Borghi, B., 1998. Effect of heat shock during grain filling on protein composition and technological quality of wheats. Euphytica 100, 147–154.

Cox, M.C., Qualset, C.O., Rains, D.W., 1985. Genetic variation for nitrogen assimilation and translocation in wheat. I. Dry matter and nitoren accumulation. Crop Sci. 25, 430–435.

Daniel, C., Triboi, E., 2001. Effects of temperature and nitrogen nutrition on the accumulation of gliadins analysed by RP-HPLC. Aust. J. Plant Physiol. 28, 1197–1205.

de la Vega, A., Chapman, S., 2001. Genotype by environment interaction and indirect selection for yield in sunflower: II. Three-mode principal component analysis of oil and biomass yield across environments in Argentina. Field Crops Res. 72, 39–50.

de la Vega, A.J., DeLacy, I.H., Chapman, S.C., 2007. Changes in agronomic traits of sunflower hybrids over 20 years of breeding in central Argentina. Field Crops Res. 100, 73–81.

DePauw, R., Knox, R., Clarke, F., et al., 2007. Shifting undesirable correlations. Euphytica 157, 409–415.

Di Caterina, R., Giuliani, M.M., Rotunno, T., De Caro, A., Flagella, Z., 2007. Influence of salt stress on seed yield and oil quality of two sunflower hybrids. Ann. Appl. Biol. 151, 145–154.

Doehlert, D.C., Lambert, R.J., 1991. Metabolic characteristics associated with starch, protein, and oil deposition in developing maize kernels. Crop Sci. 31, 151–157.

Dolde, D., Vlahakis, C., Hazebroek, J., 1999. Tocopherols in breeding lines and effects of planting location, fatty acid composition, and temperature during development. J. Am. Oil Chem. Soc. 76, 349–355.

Don, C., Lichtendonk, W.J., Plijter, J.J., Hamer, R.J., 2003. Understanding the link between GMP and dough: from glutenin particles in flour towards developed dough. J. Cereal Sci. 38, 157–165.

Don, C., Lookhart, G., Naeem, H., MacRitchie, F., Hamer, R.J., 2005. Heat stress and genotype affect the glutenin particles of the glutenin macropolymer-gel fraction. J. Cereal Sci. 42, 69–80.

Dosio, G.A.A., Aguirrezábal, L.A.N., Andrade, F.H., Pereyra, V.R., 2000. Solar radiation intercepted during seed filling and oil production in two sunflower hybrids. Crop Sci. 40, 1637–1644.

Dubley, J.W., Lambert, R.J., de la Roche, L.A., 1977. Genetic analysis of crosses among corn strains divergently selected for percent oil and protein. Crop Sci. 17, 111–117.

Dudley, J.W., Lambert, R.J., 2004. 100 generations of selection for oil and protein in corn. Plant Breeding Reviews 24 (1), 79–110.

DuPont, F.M., Chan, R., Lopez, R., 2007. Molar fractions of high-molecular-weight glutenin subunits are stable when wheat is grown under various mineral nutrition and temperature regimens. J. Cereal Sci. 45, 134–139.

DuPont, F.M., Hurkman, W.J., Vensel, W.H., et al., 2006a. Differential accumulation of sulfur-rich and sulfur-poor wheat flour proteins is affected by temperature and mineral nutrition during grain development. J. Cereal Sci. 44, 101–112.

DuPont, F.M., Hurkman, W.J., Vensel, W.H., et al., 2006b. Protein accumulation and composition in wheat grains: effects of mineral nutrients and high temperature. Eur. J. Agron. 25, 96–107.

Echarte, M.M., Quiroz, F.J., Izquierdo, N.G., Quillehauquy, V., Aguirrezábal, L.A.N., 2008. Evaluación de la estabilidad de híbridos alto oleico de girasol. XXXVII Congreso Argentino de Genética. Tandil, Buenos Aires, Argentina.

Echarte, M.M., Pereyra Irujo, G.A., Covi, M., Izquierdo, N.G., Aguirrezábal, L.A.N., 2009. Producing better sunflower oils in a changing environment. In: Advances in fats and oils research. Editorial Research Signpost, Kerala (India), pp. 1–23.

Echarte, M.M., Angeloni, P., Jaimes, F., et al., 2010. Night temperature and intercepted solar radiation additively contribute to oleic acid percentage in sunflower oil. Field Crops Res. 119, 27–35.

Echarte, M.M., Alberdi, I., Aguirrezábal, L.A.N., 2012. Post-flowering assimilate availability regulates oil fatty acid composition in sunflower grains. Crop Sci. 52, 818–829.

Echarte, M.M., Puntel, L., Aguirrezábal, L.A.N., 2013. Assessment of the critical period for the effect of intercepted solar radiation on sunflower oil fatty acid composition. Field Crops Res. 149, 213–222.

Eppendorfer, W.H., 1978. Effects of nitrogen, phosphorus and potassium on amino acid composition and on relationships between nitrogen and amino acids in wheat and oat grain. J. Sci. Food Agric. 29, 995–1001.

Ewert, F., Rodriguez, D., Jamieson, P., et al., 2002. Effects of elevated CO_2 and drought on wheat: testing crop simulation models for different experimental and climatic conditions. Agric. Ecosyst. Environ. 93, 249–266.

Feil, B., 1997. The inverse yield-protein relationship in cereals: possibilities and limitations for genetically improving the grain protein yield. Trends Agron. 1, 103–119.

Fernández-Martínez, J., Jimenez, A., Dominguez, J., Garcia, J., Garces, R., Mancha, M., 1989. Genetic analysis of the high oleic acid content in cultivated sunflower (*Helianthus annuus* L.). Euphytica 41, 39–51.

Ferreira, M.S.L., Martre, P., Mangavel, C., et al., 2012. Physicochemical control of durum wheat grain filling and glutenin polymer assembly under different temperature regimes. J. Cereal Sci. 56, 58–66.

Ferrise, R., Triossi, A., Stratonovitch, P., Bindi, M., Martre, P., 2010. Sowing date and nitrogen fertilisation effects on dry matter and nitrogen dynamics for durum wheat: An experimental and simulation study. Field Crops Res. 117, 245–257.

Fowler, D.B., Brydon, J., Darroch, B.A., Entz, M.H., Johnston, A.M., 1990. Environment and genotype influence on grain protein concentration of wheat and rye. Agron. J. 82, 655–664.

Frey, K.J., Holland, J.B., 1999. Nine cycles of recurrent selection for increased groat-oil content in oat. Crop Sci. 39, 1636–1641.

Fufa, H., Baenziger, P., Beecher, B., Graybosch, R., Eskridge, K., Nelson, L., 2005. Genetic improvement trends in agronomic performances and end-use quality characteristics among hard red winter wheat cultivars in Nebraska. Euphytica 144, 187–198.

Gabrielle, B., Denoroy, P., Gosse, G., Justes, E., Andersen, M.N., 1998. Development and evaluation of a CERES-type model for winter oilseed rape. Field Crops Res. 57, 95–111.

Garcés, R., Mancha, M., 1991. *In vitro* oleate desaturase in developing sunflower seeds. Phytochemistry 30, 2127–2130.

Garcés, R., Sarmiento, C., Mancha, M., 1992. Temperature regulation of oleate desaturase in sunflower (*Helianthus annuus* L.) seeds. Planta 186, 461–465.

García-Díaz, M.T., Martínez-Rivas, J.M., Mancha, M., 2002. Temperature and oxygen regulation of oleate desaturation in developing sunflower (*Helianthus annuus*) seeds. Physiol. Plant. 114, 13–20.

Gooding, M.J., Ellis, R.H., Shewry, P.R., Schofield, J.D., 2003. Effects of restricted water availability and increased temperature on the grain filling, drying and quality of winter wheat. J. Cereal Sci. 37, 295–309.

Grant, M.N., McCalla, A.G., 1949. Yield and protein content of wheat and barley. I. Interrelation of yield and protein content of random selections from single crosses. Can. J. Res. 27, 230–240.

Gras, P.W., Anderssen, R.S., Keentok, M., Békés, F., Appels, R., 2001. Gluten protein functionality in wheat flour processing: a review. Aust. J. Agric. Res. 52, 1311–1323.

Gray, D.A., Kekwick, R.G.O., 1996. Oleate desaturase activity in sunflower (*Helianthus annuus*) seeds and its relation to associated constituents during seed development. Plant Sci. 115, 39–47.

Graybosch, R.A., Peterson, C.J., Baenziger, P.S., Shelton, D.R., 1995. Environmental modification of hard red winter wheat flour protein composition. J. Cereal Sci. 22, 45–51.

Graybosch, R.A., Peterson, C.J., Shelton, D.R., Baenziger, P.S., 1996. Genotypic and environmental modification of wheat flour protein composition in relation to end-use quality. Crop Sci. 36, 296–300.

Green, A.G., 1986. Effect of temperature during seed maturation on the oil composition of low-linolenic genotypes of flax. Crop Sci. 26, 961–965.

Grindlay, D.J.C., 1997. Towards an explanation of crop nitrogen demand based on the optimization of leaf nitrogen per unit leaf area. J. Agric. Sci. 128, 377–396.

Haddadi, P., Yazdi-samadi, B., Berger, M., Naghavi, M.R., Calmon, A., Sarrafi, A., 2011. Genetic variability of seed-quality traits in gamma-induced mutants of sunflower (*Helianthus annuus* L.) under water-stressed condition. Euphytica 178, 247–259.

Hagan, N.D., Upadhyaya, N., Tabe, L.M., Higgins, T.J.V., 2003. The redistribution of protein sulfur in transgenic rice expressing a gene for a foreign, sulfur-rich protein. Plant J. 34, 1–11.

Hall, A., Chimenti, C., Vilella, F., Freier, G., 1985. Timing of water stress effects on yield components in sunflower. Proceedings of the 11th International Sunflower Conference. Mar del Plata, Argentina. pp. 131–136.

Hall, A., Whitfield, D., Connor, D., 1990. Contribution of pre-anthesis assimilates to grain-filling in irrigated and water-stressed sunflower crops II. Estimates from a carbon budget. Field Crops Res. 24, 273–294.

Hall, A.J., 2001. Sunflower ecophysiology: Some unresolved issues. Oleagin. Corps Grass Lipid. 8, 15–21.

Hammer, G., Cooper, M., Tardieu, F., et al., 2006. Models for navigating biological complexity in breeding improved crop plants. Trends Plant. Sci. 11, 587–593.

Hernandez-Sebastia, C., Marsolais, F., Saravitz, C., Israel, D., Dewey, R.E., Huber, S.C., 2005. Free amino acid profiles suggest a possible role for asparagine in the control of storage-product accumulation in developing seeds of low- and high-protein soybean lines. J. Exp. Bot. 56, 1951–1963.

Herzog, H., Stamp, P., 1983. Dry matter and nitrogen accumulation in grains at different ear positions in 'gigas', semidwarf and normal spring wheats. Euphytica 32, 511–520.

Hill, J., Nelson, E., Tilman, D., Polasky, S., Tiffany, D., 2006. Environmental, economic, and energetic costs and benefits of biodiesel and ethanol biofuels. Proc. Natl. Acad. Sci. USA 103, 11206–11210.

Huebner, F.R., Nelsen, T.C., Chung, O.K., Bietz, J.A., 1997. Protein distributions among hard red winter wheat varieties as related to environment and baking quality. Cereal Chem. 74, 123–128.

Irving, D., Shannon, M., Breda, V., Mackey, B., 1988. Salinity effects on yield and oil quality of high-linoleate and high-oleate cultivars of safflower. J. Agric. Food. Chem. 36, 37–42.

Islam, N., Upadhyaya, N.M., Campbell, P.M., Akhurst, R., Hagan, N., Higgins, T.J.V., 2005. Decreased accumulation of glutelin types in rice grains constitutively expressing a sunflower seed albumin gene. Phytochemistry 66, 2534–2539.

Izquierdo, N., 2007. Factores determinantes de la calidad de aceite en diversas especies. Tesis Doctoral. Universidad Nacional de Mar del Plata, Balcarce, Argentina.

Izquierdo, N., Aguirrezábal, L., 2005. Composición en ácidos grasos del aceite de híbridos de girasol cultivados en Argentina. Caracterización y modelado. Aceit. Gras XV, 338–343.

Izquierdo, N., Aguirrezábal, L., Andrade, F., Pereyra, V., 2002. Night temperature affects fatty acid composition in sunflower oil depending on the hybrid and the phenological stage. Field Crops Res. 77, 115–126.

Izquierdo, N.G., Aguirrezábal, L.A.N., 2008. Genetic variability in the response of fatty acid composition to minimum night temperature during grain filling in sunflower. Field Crops Res. 106, 116–125.

Izquierdo, N.G., Aguirrezábal, L.A.N., Andrade, F.H., Cantarero, M.G., 2006. Modeling the response of fatty acid composition to temperature in a traditional sunflower hybrid. Agron. J. 98, 451–461.

Izquierdo, N.G., Dosio, G.A.A., Cantarero, M., Lujan, J., Aguirrezábal, L.A.N., 2008. Weight per grain, oil concentration, and solar radiation intercepted during grain filling in black hull and striped hull sunflower hybrids. Crop Sci. 48, 688–699.

Izquierdo, N.G., Mascioli, S., Aguirrezábal, L.A.N., Nolasco, S.M., 2007. Temperature influence during seed filling on tocopherol concentration in a traditional sunflower hybrid. Gras. Aceit. 58, 170–178.

Izquierdo, N.G., Aguirrezábal, L.A.N., Andrade, F.H., Geroudet, C., Valentinuz, O., Pereyra Iraola, M., 2009. Intercepted solar radiation affects oil fatty acid composition in crop species. Field Crops Res. 114, 66–74.

Izquierdo, N.G., Nolasco, S., Mateo, C., Santos, D., Aguirrezábal, L.A.N., 2011. Relationship between oil tocopherol concentration and oil weight per grain in several crop species. Crop Past. Sci. 62, 1088–1097.

Izquierdo, N.G., Aguirrezábal, L.A.N., Martínez-Force, E., et al., 2013. Effect of growth temperature on the high stearic and high stearic-high oleic sunflower traits. Crop Past. Sci. 64, 18–25.

Jamieson, P.D., Berntsen, J., Ewert, F., et al., 2000. Modelling CO_2 effects on wheat with varying nitrogen supplies. Agric. Ecosyst. Environ. 82, 27–37.

Jamieson, P.D., Chapman, S.C., Dreccer, M.F., et al., 2008. Modelling wheat production. In: Bonjean, A.P., Angus, W.J., van Ginkel, M. (Eds.), The world wheat book. Lavoisier Publishing, Paris, pp. 40.

Jamieson, P.D., Semenov, M.A., 2000. Modelling nitrogen uptake and redistribution in wheat. Field Crops Res. 68, 21–29.

Jenner, C.F., Ugalde, T.D., Aspinall, D., 1991. The physiology of starch and protein deposition in the endosperm of wheat. Aust. J. Plant Physiol. 18, 211–226.

Jeuffroy, M.-H., Valantin-Morison, M., Champolivier, L., Reau, R., 2006. Azote, rendement et qualité des graines: mise au point et utilisation du modèle Azodyn-colza pour améliorer les performances d. Oléagin. Corps Gras Lipid. 13, 388–392.

Jeuffroy, M.-H., Casadebaig, P., Debaeke, P., Loyce, C., Meynard, J.-M., 2013. Agronomic model uses to predict cultivar performance in various environments and cropping systems. A review. Agron. Sustain. Dev. 34, 121–137.

Kabbaj, A., Vervoort, V., Abbott, A., Tersac, M., Bervillé, A., 1996. Expression of stearate oleate and linoleate desaturase genes in sunflower with normal and high oleic contents. Helia 19, 1–17.

Kade, M., Barneix, A.J., Olmos, S., Dubcovsky, J., 2005. Nitrogen uptake and remobilization in tetraploid 'Langdon' durum wheat and a recombinant substitution line with the high grain protein gene Gpc-B1. Plant Breed. 124, 343–349.

Kamal-Eldin, A., Appelqvist, L., 1996. The chemistry and antioxidant properties of tocopherols and tocotrienols. Lipids 31, 671–701.

Kandil, A., Ibrahim, A., Marquard, R., Taha, R., 1990. Response of some quality traits of sunflower seeds and oil to different environments. J. Agron. Crop Sci. 164, 224–230.

Kane, M.V., Steele, C.C., Grabau, L.J., 1997. Early-maturing soybean cropping system: I. Yield responses to planting date. Agron. J. 89, 454–458.

Karam, F., Lahoud, R., Masaad, R., et al., 2007. Evapotranspiration, seed yield and water use efficiency of drip irrigated sunflower under full and deficit irrigation conditions. Agric. Water Manag. 90, 213–223.

Kelly, M.L., Kolver, E.S., Bauman, D.E., Van Amburgh, M.E., Muller, L.D., 1998. Effect of intake of pasture on concentrations of conjugated linoleic acid in milk of lactating cows. J. Dairy Sci. 81, 1630–1636.

Khatkar, B.S., Fido, R.J., Tatham, A.S., Schofield, J.D., 2002. Functional properties of wheat gliadins. I. Effects on mixing characteristics and bread making quality. J. Cereal Sci. 35, 299–306.

Kibite, S., Evans, L.E., 1984. Causes of negative correlations between grain yield and grain protein concentration in common wheat. Euphytica 33, 801–810.

Kramer, P.J., Kozlowski, T., 1979. Absorption of water, ascent of sap, and water balance. In: Kramer, T. (Ed.), Physiology of woody plants. Academic Press, New York, pp. 445–493.

Kramer, T., 1979. Environmental and genetic variation for protein content in winter wheat (*Triticum aestivum* L.). Euphytica 28, 20–218.

Kris-Etherton, P., Yu, S., 1997. Individual fatty acid effects on plasma lipids and lipoproteins: human studies. Am. J. Clin. Nutr. 65, 1628S–1644S.

Lacombe, S., Bervillé, A., 2000. Analysis of desaturase transcript accumulation in normal and in high oleic oil sunflower development seeds. In: Proceedings 15th International Sunflower Conference. Toulouse, France 1A, F33. .

Lacombe, S., Souyris, I., Bervillé, A.J., 2009. An insertion of oleate desaturase homologous sequence silences via siRNA the functional gene leading to high oleic acid content in sunflower seed oil. Mol. Genet. Genomics 281, 43–54.

Lagravere, T., Kleiber, D., Dayde, J., 1998. Conduites culturales et performances agronomiques du tournesol oléique: réalités et perspectives. Oléagin. Corps Gras Lipid. 5, 477–485.

Landry, J., 2002. A linear model for quantitating the accumulation of zeins and their fractions ($\alpha + \delta$, $\beta + \gamma$) in developing endosperm of wild-type and mutant maizes. Plant Sci. 163, 111–115.

Laperche, A., Brancourt-Hulmel, M., Heumez, E., et al., 2007. Using genotype – nitrogen interaction variables to evaluate the QTL involved in wheat tolerance to nitrogen constraint. Theor. Appl. Genet. 115, 399–415.

Lawlor, D.W., 2002. Carbon and nitrogen assimilation in relation to yield: Mechanisms are the key to understanding production systems. J. Exp. Bot. 53, 773–787.

Lawlor, D.W., Cornic, G., 2002. Photosynthetic carbon assimilation and associated metabolism in relation to water deficits in higher plants. Plant Cell Environ. 25, 275–294.

Lending, C.R., Larkins, B.A., 1989. Changes in the zein composition of protein bodies during maize endosperm development. Plant Cell 1, 1011–1023.

León, A., Andrade, F., Lee, M., 2000. Genetic mapping of factors affecting quantitative variation for flowering in sunflower. Crop Sci. 40, 404–407.

León, A., Lee, M., Rufener, G., Berry, S., Mowers, R., 1996. Genetic mapping of a locus(hyp) affecting seed hypodermis color in sunflower. Crop Sci. 36, 1666–1668.

León, A., Zambelli, A., Reid, R., Morata, M., Kaspar, M. 2013. Nucleotide, sequences mutated by insertion that encode a truncated oleate desaturase protein, proteins, methods and uses. Advanta International Bv. Patent WO2013/004281. WO Patent App. PCT/EP2011/061,165.

Lhuillier-Soundele, A., Munier-Jolain, N., Ney, B., 1999. Influence of nitrogen availability on seed nitrogen accumulation in pea. Crop Sci. 39, 1741–1748.

Lindsey, K., Pullen, M.L., Topping, J.F., 2003. Importance of plant sterols in pattern formation and hormone signaling. Trends Plant Sci. 8, 521–525.

López Pereira, M., Sadras, V.O., Trápani, N., 1999. Genetic improvement of sunflower in Argentina between 1930 and 1995. I. Yield and its components. Field Crops Research 62, 157–166.

López Pereira, M., Trápani, N., Sadras, V.O., 2000. Genetic improvement of sunflower in Argentina between 1930 and 1995. III. Dry matter partitioning and grain composition. Field Crops Res. 67, 215–221.

Mantese, A., Medan, D., Hall, A., 2006. Achene structure, development and lipid accumulation in sunflower cultivars differing in oil content at maturity. Ann. Bot. 97, 999–1010.

Marquard, R., 1990. Untersuchungen tiber dem Einfluss von Sorte und Standort auf den Tocopherogehalt verschiedener Pflanzanole. Fat Sci. Technol. 92, 452–455.

Martínez, R.D., Izquierdo, N.G., Belo, R.G., Aguirrezábal, L.A.N., Andrade, F., Reid, R., 2012. Oil yield components and oil quality of high stearic-high oleic sunflower genotypes as affected by intercepted solar radiation during grain filling. Crop Past. Sci. 63, 330–337.

Martínez de la Cuesta, P., Rus Martínez, E., Galdeano Chaparro, M., 1995. Enranciamiento oxidativo de aceites vegetales en presencia de α-tocoferol. Gras. Aceit. 46, 349–353.

Martínez-Rivas, J.M., Sánchez-García, A., Sicardo, M.D., García-Díaz, M.T., Mancha, M., 2003. Oxygen-independent temperature regulation of the microsomal oleate desaturase (FAD2) activity in developing sunflower (*Helianthus annuus*) seeds. Physiol. Plant. 117, 179–185.

Martre, P., Jamieson, P.D., Semenov, M.A., Zyskowski, R.F., Porter, J.R., Triboi, E., 2006. Modelling protein content and composition in relation to crop nitrogen dynamics for wheat. Eur. J. Agron. 25, 138–154.

Martre, P., Porter, J.R., Jamieson, P.D., Triboi, E., 2003. Modeling grain nitrogen accumulation and protein composition to understand the sink/source regulations of nitrogen remobilization for wheat. Plant Physiol. 133, 1959–1967.

Martre, P., Semenov, M.A., Jamieson, P.D., 2007. Simulation analysis of physiological traits to improve yield, nitrogen use efficiency and grain protein concentration in wheat. In: Spiertz, J.H.J., Struik, P.C., Van Laar, H.H. (Eds.), Scale and complexity in plant systems research, gene-plant-crop relations. Springer, The Netherlands, pp. 181–201.

Martre, P., Bertin, N., Salon, C., Génard, M., 2011. Modelling the size and composition of fruit, grain and seed by process-based simulation models. New Phytol. 191, 601–618.

Mestries, E., Gentzbittel, L., De Labrouhe, D.T., Nicolas, P., Vear, F., 1998. Analyses of quantitative trait loci associated with resistance to *Sclerotinia sclerotiorum* in sunflowers (*Helianthus annuus* L.) using molecular markers. Mol. Breed. 4, 215–226.

Metzger, W.H., 1935. The residual effect of alfalfa cropping periods of various lengths upon the yield and protein content of succeeding wheat crops. J. Am. Soc. Agron. 27, 653–659.

Metzger, J.O., Bornsheuer, U., 2006. Lipids as renewable resources: current state of chemical and biotechnological conversion and diversification. Appl. Microbiol. Biotechnol. 71, 13–22.

Mokrani, L., Gentzbittel, L., Azanza, F., Fitamant, L., Al-Chaarani, G., Sarrafi, A., 2002. Mapping and analysis of quantitative trait loci for grain oil content and agronomic traits using AFLP and SSR in sunflower (*Helianthus annuus* L.). Theor. Appl. Genet. 106, 149–156.

Monaghan, J.M., Snape, J.W., Chojecki, A.J.S., Kettlewell, P.S., 2001. The use of grain protein deviation for identifying wheat cultivars with high grain protein concentration and yield. Euphytica 122, 309–317.

Moreau, D., Allard, V., Gaju, O., Le Gouis, J., Foulkes, M.J., Martre, P., 2012. Acclimation of leaf nitrogen to vertical light gradient at anthesis in wheat is a whole-plant process that scales with the size of the canopy. Plant Physiol. 160, 1479–1490.

Morrison, M.J., Voldeng, H.D., Cober, E.R., 2000. Agronomic changes from 58 years of genetic improvement of short-season soybean cultivars in Canada. Agron. J. 92, 780–784.

Mossé, J., Huet, J.C., Baudet, J., 1985. The amino acid composition of wheat grain as a function of nitrogen content. J. Cereal Sci. 3, 115–130.

Mullor, J., 1968. Improvement of the nutritional value of food oils. Revista Facultad de Ingeniería Química, Universidad Nacional del Litoral 37, 183-210.

Munier-Jolain, N.G., Salon, C., 2005. Are the carbon costs of seed production related to the quantitative and qualitative performance? An appraisal for legumes and other crops. Plant Cell Environ. 28, 1388–1395.

Muratorio, A., Racca, E., González, L., Ketterer, E., Rossi, R., Ferrari, B., 2001. Características de los aceites crudos de soja, extraídos en laboratorio, provenientes de variedades no transgénicas y transgénicas, cultivadas en la Argentina y estudio de sus implicancias agro-industriales. Parte II. Aceit. Gras. 42, 41–53.

Murphy, D.J., 2001. The biogenesis and functions of lipid bodies in animals, plants and microorganisms. Prog. Lipid Res. 40, 325–438.

Neatby, K.W., McCalla, A.G., 1938. Correlation between yield and protein content of wheat and barley in relation to breeding. Can. J. Res. 16, 1–15.

Nolasco, S.M., Aguirrezábal, L.A.N., Crapiste, G.H., 2004. Tocopherol oil concentration in field-grown sunflower is accounted for by oil weight per seed. J. Am. Oil Chem. Soc. 81, 1045–1051.

Nolasco, S.M., Aguirrezábal, L.A.N., Lúquez, J., Mateo, C., 2006. Variability in oil tocopherol concentration and composition of traditional and high oleic sunflower hybrids (*Helianthus annuus* L.) in the Pampean region (Argentina). Gras. Aceit. 57, 260–269.

Olusola Bayo, O., 1997. Genetic and environmental variation for seed yield, protein, lipid and amino acid composition in cowpea (*Vigna unguiculata* (L) Walp. J. Sci. Food Agric. 74, 107–116.

Ohlrogge, J.B., Jaworski, J.G., 1997. Regulation of fatty acid synthesis. Annu. Rev. Plant Biol. 48, 109–136.

Ortiz-Monasterio, J.I., Pena, J.I., Sayre, K.D., Rajaram, S.S., 1997. CIMMYT's genetic progress in wheat grain quality under four nitrogen rates. Crop Sci. 37, 898–904.

Ostlund, Jr., R.E., 2007. Phytosterols, cholesterol absorption and healthy diets. Lipids 42, 41–45.

Oury, F.X., Berard, P., Brancourt-Hulmel, M., et al., 2003. Yield and grain protein concentration in bread wheat: a review and a study of multi-annual data from a French breeding program. J. Genet. Breed. 57, 59–68.

Oury, F.X., Godin, C., 2007. Yield and grain protein concentration in bread wheat: how to use the negative relationship between the two characters to identify favourable genotypes? Euphytica 157, 45–57.

Palta, J.A., Kobata, T., Turner, N.C., Fillery, I.R., 1994. Remobilization of carbon and nitrogen in wheat as influenced by postanthesis water deficits. Crop Sci. 34, 118–124.

Panozzo, J.F., Eagles, H.A., 1999. Rate and duration of grain filling and grain nitrogen accumulation of wheat cultivars grown in different environments. Aust. J. Agric. Res. 50, 1007–1015.

Panozzo, J.F., Eagles, H.A., 2000. Cultivar and environmental effects on quality characters in wheat. II. Protein. Aust. J. Agric. Res. 51, 629–636.

Panozzo, J.F., Eagles, H.A., Wootton, M., 2001. Changes in protein composition during grain development in wheat. Aust. J. Agric. Res. 52, 485–493.

Peak, N.C., Imsande, J., Shoemaker, R.C., Shibles, R., 1997. Nutritional control of soybean storage protein. Crop Sci. 37, 498–503.

Peper, A., Pereyra-Irujo, G., Nolasco, S., Aguirrezábal, L., 2007. Relaciones entre rendimiento y calidad de aceite: investigación experimental y por simulación. 4° Congreso Argentino de Girasol, Buenos Aires. Argentina, 304–305.

Pereyra-Irujo, G.A., Aguirrezábal, L.A.N., 2007. Sunflower yield and oil quality interactions and variability: Analysis through a simple simulation model. Agric. For. Meteorol. 143, 252–265.

Pereyra-Irujo, G.A., Izquierdo, N.G., Covi, M., Nolasco, S.M., Quiroz, F.J., Aguirrezábal, L.A.N., 2009. Variability in sunflower oil quality for biodiesel production: a simulation study. Biomass Bioenerg. 33, 459–468.

Piironen, V., Toivo, J., Lampi, A.-M., 2000. Natural sources of dietary plant sterols. J. Food Comp. Anal. 13, 619–624.

Piper, E., Boote, K., 1999. Temperature and cultivar effects on soybean seed oil and protein concentrations. J. Am. Oil Chem. Soc. 76, 1233–1241.

Pleite, R., Rondanini, D., Garces, R., Martínez-Force, E., 2008. Day/night variation in fatty acids and lipids biosynthesis in sunflower (*Helianthus annuus* L.) seeds. Crop Sci. 48, 1952–1957.

Plessis, A., Ravel, C., Bordes, J., Balfourier, F., Martre, P., 2013. Association study of wheat grain protein composition reveals that gliadin and glutenin composition are trans-regulated by different chromosome regions. J. Exp. Bot. 64, 3627–3644.

Ploschuk, E., Hall, A., 1995. Capitulum position in sunflower affects grain temperature and duration of grain filling. Field Crops Res. 44, 111–117.

Primomo, V., Falk, D., Ablett, G., Tanner, J., Rajcan, I., 2002. Genotype × environment interactions, stability, and agronomic performance of soybean with altered fatty acid profiles. Crop Sci. 42, 37–44.

Pritchard, F., Eagles, H., Norton, R., Salisbury, P., Nicolas, M., 2000. Environmental effects on seed composition of Victorian canola. Aust. J. Exp. Agric. 40, 679–685.

Pushman, F.M., Bingham, J., 1976. The effects of a granular nitrogen fertilizer on a foliar spray of urea on the yield and bread-making quality of ten winter wheats. J. Agric. Sci. 87, 281–292.

Putt, D.E., 1997. Early history of sunflower. In: Schneiter, A.A., (Ed.), Sunflower technology and production. pp. 1–19. Agronomy Series 35. American Society of Agronomy, Madison, WI.

Randall, P.J., Moss, H.J., 1990. Some effects of temperature regime during grain filling on wheat quality. Aust. J. Agric. Res. 41, 603–617.

Ravel, C., Martre, P., Romeuf, I., et al., 2009. Nucleotide polymorphism in the wheat transcriptional activator Spa influences its pattern of expression and has pleiotropic effects on grain protein composition, dough viscoelasticity, and grain hardness. Plant Physiol. 151, 2133–2144.

Rharrabti, Y., Villegas, D., Del Moral, L.F.G., Aparicio, N., Elhani, S., Royo, C., 2001. Environmental and genetic determination of protein content and grain yield in durum wheat under Mediterranean conditions. Plant Breed. 120, 381–388.

Roche, J., Bouniols, A., Mouloungui, Z., Barranco, T., Cerny, M., 2006. Management of environmental crop conditions to produce useful sunflower oil components. Eur. J. Lipid Sci. Technol. 108, 287–297.

Roche, J., Alignan, M., Bouniols, A., et al., 2010. Sterol content in sunflower seeds (*Helianthus annuus* L.) as affected by genotypes and environmental conditions. Food Chem. 121, 990–995.

Rolletschek, H., Borisjuk, L., Sánchez-García, A., et al., 2007. Temperature-dependent endogenous oxygen concentration regulates microsomal oleate desaturase in developing sunflower seeds. J. Exp. Bot. 58, 3171–3181.

Rondanini, D., Mantese, A., Savin, R., Hall, A.J., 2006. Responses of sunflower yield and grain quality to alternating day/night high temperature regimes during grain filling: Effects of timing, duration and intensity of exposure to stress. Field Crops Res. 96, 48–62.

Rondanini, D., Savin, R., Hall, A., 2003. Dynamics of fruit growth and oil quality of sunflower (*Helianthus annuus* L.) exposed to brief intervals of high temperature during grain filling. Field Crops Res. 83, 79–90.

Rondanini, D.P., Vilariño, M.P., Gómez, N.V., Miralles, D.J., 2013. Cambios en el contenido y composición del aceite en colza (*brassica napus* l.) en respuesta a la temperatura y la radiación post-floración. II Workshop Internacional de Ecofisiología de Cultivos. Mar del Plata, Argentina.

Ruiz, R.A., Maddonni, G.A., 2006. Sunflower seed weight and oil concentration under different post-flowering source-sink ratios. Crop Sci. 46, 671–680.

Sadras, V.O., Quiroz, F., Echarte, L., Escande, A., Pereyra, V.R., 2000. Effect of *Verticillium dahliae* on photosynthesis, leaf expansion and senescence of field-grown sunflower. Ann. Bot. 86, 1007–1015.

Salas, J.J., Martínez-Force, E., Garcés, R., 2004. Biochemical characterization of a high-palmitoleic acid *Helianthus annuus* mutant. Plant Physiol. Biochem. 42, 373–381.

Salas, J.J., Youssar, L., Martínez-Force, E., Garcés, R., 2008. The biochemical characterization of a high-stearic acid sunflower mutant reveals the coordinated regulation of stearoyl-acyl carrier protein desaturases. Plant Physiol. Biochem. 46, 109–116.

Salisbury, F.B., Ross, C.W., 1992. Plant physiology. Wadsworth Publishing. Co.: Belmont.

Santalla, E.M., Dosio, G.A.A., Nolasco, S.M., Aguirrezábal, L.A.N., 2002. The effects of intercepted solar radiation on sunflower (*Helianthus annuus* L.). Seed composition from different head positions. J. Am. Oil Chem. Soc. 79, 69–74.

Schaller, H., 2003. The role of sterols in plant growth and development. Progr. Lipid Res. 42, 163–175.

Schnebly, S., Fehr, W., Welke, G., Hammond, E., Duvick, D., 1996. Fatty ester development in reduced- and elevated-palmitate lines of soybean. Crop Sci. 36, 1462–1466.

Schuppert, G.F., Tang, S., Slabaugh, M.B., Knapp, S.J., 2006. The sunflower high-oleic mutant Ol carries variable tandem repeats of FAD2-1, a seed-specific oleoyl-phosphatidyl choline desaturase. Mol. Breed. 17, 241–256.

Seiler, G.J., 1986. Analysis of the relationships of environmental factors with seed oil and fatty acid concentrations of wild annual sunflower. Field Crops Res. 15, 57–72.

Semenov, M.A., Brooks, R.J., 1999. Spatial interpolation of the LARS-WG stochastic weather generator in Great Britain. Clim. Res. 11, 137–148.

Semenov, M.A., Jamieson, P.D., Martre, P., 2007. Deconvoluting nitrogen use efficiency in wheat: A simulation study. Eur. J. Agron. 26, 283–294.

Serrano-Vega, M., Martínez-Force, E., Garcés, R., 2005. Lipid characterization of seed oils from high-palmitic, low-palmitoleic, and very high-stearic acid sunflower lines. Lipids 40, 369–374.

Sexton, P.J., Batchelor, W.D., Boote, K.J., Shibles, R., 1998a. Evaluation of CROPGRO for prediction of soybean nitrogen balance in a Midwestern environment. Trans. ASAE 41, 1543–1548.

Sexton, P.J., Naeve, S.L., Paek, N.C., Shibles, R., 1998b. Sulfur availability, cotyledon nitrogen:sulfur ratio, and relative abundance of seed storage proteins of soybean. Crop Sci. 38, 983–986.

Sezen, S.M., Yazar, A., Tekin, S., 2011. Effects of partial root zone drying and deficit irrigation on yield and oil quality of sunflower in a Mediterranean environment. Irrigat. Drain. 60, 499–508.

Sharma, B.K., Adhvaryu, A., Perez, J.M., Erhan, S.Z., 2005. Soybean oil based greases: influence of composition on thermo-oxidative and tribochemical behavior. J. Agric. Food Chem. 53, 2961–2968.

Shearman, V.J., Sylvester-Bradley, R., Scott, R.K., Foulkes, M.J., 2005. Physiological processes associated with wheat yield progress in the UK. Crop Sci. 45, 175–185.

Shewry, P.R., 2007. Improving the protein content and composition of cereal grain. J. Cereal Sci. 46, 239–250.

Shewry, P.R., Halford, N.G., 2003. Genetics of wheat gluten proteins. Hall, J.C. (Ed.), Advances in Genetics, Vol 49, Academic Press Inc, San Diego, pp. 111–184.

Simmonds, N.W., 1995. The relation between yield and protein in cereal grain. J. Sci. Food Agric. 67, 309–315.

Simmons, S.R., Moss, D.N., 1978. Nitrogen and dry matter accumulation by kernels formed at specific florets in spikelets of spring wheat. Crop Science 1, 139–143.

Sinclair, T.R., Amir, J., 1992. A model to assess nitrogen limitations on the growth and yield of spring wheat. Field Crops Res. 30, 63–78.

Sinclair, T.R., Jamieson, P.D., 2006. Grain number, wheat yield, and bottling beer: An analysis. Field Crops Res. 98, 60–67.

Sinclair, T.R., Muchow, R.C., 1995. Effect of nitrogen supply on maize yield. 1. Modeling physiological-responses. Agron. J. 87, 632–641.

Slafer, G.A., Andrade, F.H., Feingold, S.E., 1990. Genetic improvement of bread wheat (*Triticum aestivum* L.) in Argentina: relationship between nitrogen and dry matter. Euphytica 50, 63–71.

Spano, G., Di Fonzo, N., Perrotta, C., et al., 2003. Physiological characterization of 'stay green' mutants in durum wheat. J. Exp. Bot. 54, 1415–1420.

Spiertz, J.H.J., Hamer, R.J., Xu, H., Primo-Martin, C., Don, C., van der Putten, P.E.L., 2006. Heat stress in wheat (*Triticum aestivum* L.): effects on grain growth and quality traits. Eur. J. Agron. 25, 89–95.

Steer, B., Hocking, P., Kortt, A., Roxburgh, C., 1984. Nitrogen nutrition of sunflower (*Helianthus annuus* L.): yield components, the timing of their establishment and seed characteristics in response to nitrogen supply. Field Crops Res. 9, 219–236.

Steer, B., Seiler, G., 1990. Changes in fatty acid composition of sunflower (*Helianthus annuus*) seeds in response to time of nitrogen application, supply rates and defoliation. J. Sci. Food Agric. 51, 11–26.

Stöckle, C.O., Debaeke, P., 1997. Modeling crop nitrogen requirements: a critical analysis. Eur. J. Agron. 7, 161–169.

Stöckle, C., Donatelli, M., Nelson, R., 2003. CropSyst, a cropping systems simulation model. Eur. J. Agron. 18, 289–307.

Stojšin, D., Luzzi, B., Ablett, G., Tanner, J., 1998. Inheritance of low linolenic acid level in the soybean line RG10. Crop Sci. 38, 1441–1444.

Stone, P.J., Grast, P.W., Nicolas, M.E., 1997. The influence of recovery temperature on the effects of a brief heat shock on wheat. III. Grain protein composition and dough properties. J. Cereal Sci. 25, 129–141.

Stone, P.J., Nicolas, M.E., 1996. Effect of timing of heat stress during grain filling on two wheat varieties differing in heat tolerance. II. Fractional protein accumulation. Aust. J. Plant Physiol. 23, 739–749.

Strecker, L., Bieber, M., Maza, A., Grossberger, T., Doskoczynski, W., 1997. Aceite de maíz. Antecedentes, composición, procesamiento, refinación, utilización y aspectos nutricionales. Aceit. Gras., 507–527.

Tabe, L., Hagan, N., Higgins, T.J.V., 2002. Plasticity of seed protein composition in response to nitrogen and sulfur availability. Curr. Opin. Plant Biol. 5, 212–217.

Tang, S., León, A., Bridges, W.C., Knapp, S.J., 2006. Quantitative trait loci for genetically correlated seed traits are tightly linked to branching and pericarp pigment loci in sunflower. Crop Sci. 46, 721–734.

Tardieu, F., 2003. Virtual plants: Modelling as a tool for the genomics of tolerance to water deficit. Trends Plants Sci. 8, 9–14.

Terman, G.L., 1979. Yields and protein content of wheat grains as affected by cultivar, N and environmental growth factors. Agron. J. 71, 437–440.

Terman, G.L., Ramig, R.E., Dreier, A.F., Olson, R.A., 1969. Yield-protein relationships in wheat grain, as affected by nitrogen and water. Agron. J. 61, 755–759.

Tezara, W., Mitchell, V.J., Driscoll, S.D., Lawlor, D.W., 1999. Water stress inhibits plant photosynthesis by decreasing coupling factor and ATP. Nature 401, 914–917.

Thomas, J.M.G., Boote, K.J., Allen, L.H., Gallo-Meagher, M., Davis, J.M., 2003. Elevated temperature and carbon dioxide effects on soybean seed composition and transcript abundance. Crop Sci. 43, 1548–1557.

Triboi, E., Martre, P., Girousse, C., Ravel, C., Triboi-Blondel, A.M., 2006. Unravelling environmental and genetic relationships between grain yield and nitrogen concentration for wheat. Eur. J. Agron. 25, 108–118.

Triboi, E., Martre, P., Triboi-Blondel, A.M., 2003. Environmentally-induced changes of protein composition for developing grains of wheat are related to changes in total protein content. J. Exp. Bot. 54, 1731–1742.

Triboi, E., Triboi-Blondel, A.M., 2002. Productivity and grain or seed composition: a new approach to an old problem – invited paper. Eur. J. Agron. 16, 163–186.

Uauy, C., Brevis, J.C., Dubcovsky, J., 2006a. The high grain protein content gene Gpc-B1 accelerates senescence and has pleiotropic effects on protein content in wheat. J. Exp. Bot. 57, 2785–2794.

Uauy, C., Distelfeld, A., Fahima, T., Blechl, A., Dubcovsky, J., 2006b. A NAC gene regulating senescence improves grain protein, zinc, and iron content in wheat. Science 314, 1298–1301.

Uribelarrea, M., Below, F.E., Moose, S.P., 2004. Grain composition and productivity of maize hybrids derived from the Illinois protein strains in response to variable nitrogen supply. Crop Sci. 44, 1593–1600.

van Keulen, N., Seligman, N.G. 1987. Simulation of water use, nitrogen nutrition and growth of a spring wheat crop. Pudoc: Wageningen, The Netherlands.

Velasco, L., Fernández-Martínez, J., 2002. Breeding oilseed crops for improved oil quality. J. Crop. Prod. 5, 309–344.

Velasco, L., Fernández-Martínez, J., García-Ruíz, R., Domíínguez, J., 2002. Genetic and environmental variation for tocopherol content and composition in sunflower commercial hybrids. J. Agric. Sci. 139, 425–429.

Velasco, L., Perez-Vich, B., Fernández-Martínez, J., 1999. The role of mutagenesis in the modification of the fatty acid profile of oilseed crops. J. Appl. Genet. 40, 185–209.

Villalobos, F., Hall, A., Ritchie, J., Orgaz, F., 1996. OILCROP-SUN: A development, growth, and yield model of the sunflower crop. Agron. J. 88, 403–415.

Waldron, L.R., 1933. Yield and protein content of hard red spring wheat under conditions of high temperature and low moisture. J. Agric. Res. 47, 129–147.

Wardlaw, I.F., Blumenthal, C., Larroque, O., Wrigley, C.W., 2002. Contrasting effects of chronic heat stress and heat shock on kernel weight and flour quality in wheat. Funct. Plant Biol. 29, 25–34.

White, P., Xing, Y., 1997. Antioxidants from cereals and legumes. In: Shahidi, F. (Ed.), Natural antioxidants: chemistry, health effects, and applications. AOCS Press, Champaign, Illinois, pp. 25–63.

Wieser, H., Zimmermann, G., 2000. Importance of amounts and proportions of high molecular weight subunits of glutenin for wheat quality. Eur. Food Res. Technol. 210, 324–330.

Wilcox, J.R., Cavins, J.F., 1995. Backcrossing high seed protein to a soybean cultivar. Crop Sci. 35, 1036–1041.

Yang, X., Ma, H., Zhang, P., et al., 2012. Characterization of QTL for oil content in maize kernel. Theor. Appl. Genet. 125, 1169–1179.

Zheljazkov, V.D., Vick, B.A., Ebelhar, M.W., et al., 2008. Yield, oil content, and composition of sunflower grown at multiple locations in Mississippi. Agron. J. 100, 635–642.

Zheljazkov, V.D., Vick, B.A., Baldwin, B.S., Buehring, N., Astatkie, T., Johnson, B., 2009. Oil content and saturated fatty acids in sunflower as a function of planting date, nitrogen rate, and hybrid. Agron. J. 101, 1003–1011.

Zheljazkov, V.D., Vick, B., Ebelhar, W., Buehring, N., Astatkie, T., 2013. Effect of N on yield and chemical profile of winter canola in Mississippi. J. Oleo. Sci. 62, 453–458.

Zhu, J., Khan, K., 2001. Effects of genotype and environment on glutenin polymers and breadmaking quality. Cereal Chem. 78, 125–130.

Zimmer, D., Zimmerman, D., 1972. Influence of some diseases on achene and oil quality of sunflower. Crop Sci. 12, 859.

Zuil, S.G., Izquierdo, N.G., Luján, J., Cantarero, M., Aguirrezábal, L.A.N., 2012. Oil quality of maize and soybean genotypes with increased oleic acid percentage as affected by intercepted solar radiation and temperature. Field Crops Res. 127, 203–214.

CHAPTER

18

Integrated views in plant breeding: from the perspective of biotechnology

Grazia M. Borrelli[1], *Luigi Orrù*[2], *Pasquale De Vita*[1], *Delfina Barabaschi*[2], *Anna M. Mastrangelo*[1], *Luigi Cattivelli*[2]

[1]Consiglio per la Ricerca e la sperimentazione in Agricoltura – Cereal Research Centre, Foggia, Italy

[2]Consiglio per la Ricerca e la sperimentazione in Agricoltura – Genomics Research Centre, Fiorenzuola d'Arda, Italy

1 INTRODUCTION

Following their initial domestication, which involved wholesale phenotypic changes in a suite of traits collectively known as the 'domestication syndrome' (Doebley et al., 2006; Paran and van der Knaap, 2007; Sakuma et al., 2011), all major crop lineages have experienced more recent, intensive selection on a variety of agronomic traits. Crop evolution can be divided in three major phases. The initial period of domestication, a subsequent period when crop species have been selected by environmental factors, inter- and intra-specific competition and by empirical farmer selection, and a more recent phase when the application of quantitative genetic principles and, in the last two decades, of DNA-based techniques, has led to significant genetic progress. This last period can be defined as plant breeding.

Crop plants have evolved from their wild ancestors during domestication and selective breeding over the last ≈10 000 years. For most of this time, this has been a 'hit or miss' process. Initially, wild plants carrying promising traits were cultivated, leading eventually to locally adapted landraces. These lost many undesirable alleles and useful alleles became enriched in the cultivated gene pool (Tanksley and McCouch, 1997). Modern breeding has largely continued this process of crossing the 'best with the best' and the successes have been

Crop Physiology. DOI: 10.1016/B978-0-12-417104-6.00018-2

impressive. During the last century, particularly in the second half, the genetic progress has been extremely effective for many crops where improved cultivars significantly out-yield the corresponding landraces (Slafer et al., 1994). Frequently, genetic gain has been quantified as the slope of the regression between yield and year of release of cultivars of different breeding eras grown in the same trial. In general, a genetic gain from 10 to 50 kg ha^{-1} $year^{-1}$ has been recorded for cereals and legumes over the last century in most countries, including those characterized by vast low rainfall regions. These changes were often associated with important variations in morphophysiological traits, for instance, the reduction of plant height and the corresponding increase in harvest index in barley and wheat (e.g. Cattivelli et al., 1994; Sinclair, 1998; De Vita et al., 2007; Giunta et al., 2007, Tian et al., 2011). A similar approach has been used to estimate breeding effects on other traits such as grain nitrogen and phosphorous content, disease tolerance, stomatal conductance, phenology, etc. (e.g. Ortiz-Monasterio et al., 1997; Sinclair, 1998; Sayre et al., 1998; Miadenov et al., 2011; Sadras and Lawson, 2011; Zheng et al., 2011; Xiao et al., 2012). Rates of yield improvement are presented for maize and soybean in cropping systems of the USA (Chapter 2), rice, maize and wheat in China (Chapter 3), cereals and rapeseed in Finland (Chapter 4) and cereals and potato (Chapter 5).

Although important achievements have been made, the accumulation of useful alleles in the cultivated varieties is far from being concluded, rather there is an increasing interest to search for useful alleles and traits in the available germplasm and to introduce them into elite varieties. Many studies have demonstrated the value of alleles originating from wild relatives in plant breeding (e.g. Tanksley and McCouch, 1997; Gur and Zamir, 2004), showing that centuries of selective breeding have thrown away useful alleles in addition to useless ones. The development of plant molecular biology and, more recently, of plant genomics is a promising way for continuing genetic improvement of crops. Three decades after the development of the first genetically modified (GM) plant, GM crops cover more than 10% of arable land worldwide (Clive, 2012), marker-assisted selection (MAS) is routine in many breeding programs and the reference genomes of more than 50 plant species have been released (Michael and Jackson, 2013). This incredible amount of genetic knowledge is changing the way that germplasm characterization and the exploitation of genetic diversity is used which, in turn, will change the traditional breeding strategies. In the future, genomic approaches, such as genome resequencing, allele mining and genomic selection, will integrate the traditional techniques for genotype characterization, germplasm screening and variety selection. This chapter highlights the perspectives in plant breeding that result from the integration of the recent advances in plant genomics into traditional breeding strategy.

2 MODERN VIEWS IN PLANT BREEDING

Plant breeding can be described as the continuous accumulation of superior alleles (i.e. genes encoding useful traits) in the gene pool of the cultivated elite lines that, in turn, represent only a minimal part of the whole germplasm of a given crop species. Crop improvement has two main targets: to search for new useful alleles lost during the past selection process worthy to be introduced into the elite lines (also called 'pre-breeding'), and to promote the recombination within the elite germplasm to find the best combinations among the best alleles (breeding) (Fig. 18.1). Pre-breeding refers to the introgression or to the incorporation of genes and gene combinations from unadapted sources into breeding materials. Introgression

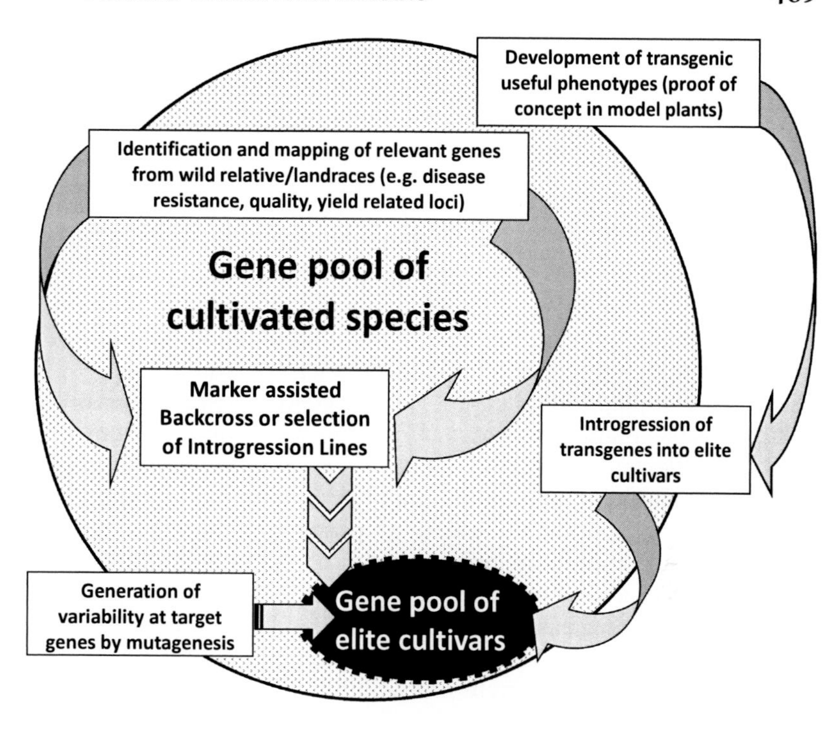

FIG. 18.1 Overview of the modern approaches in plant breeding. Through the breeding process the genes encoding useful traits present in the gene pool of the cultivated species are identified and progressively accumulated in the elite cultivars.

indicates the transfer of one or a few alleles from exotic genotypes to adapted bred cultivars, while incorporation refers to a large-scale effort aiming at developing locally adapted genotypes using exotic germplasm, which broadens the genetic base of new breeding materials (Simmonds, 1993).

Considering available molecular options, different strategies can be used to search for useful traits and transfer them into the cultivated germplasm. New genes and traits identified in genetically distant genotypes and species can be assessed in model plants (e.g, *Arabidopsis*) and then introduced in the modern cultivars via genetic transformation. Genes from wild relatives and landraces can be identified at phenotypic level and then tagged with molecular markers for a fast introgression into the elite germplasm. Alternatively, *in situ* mutagenesis can be used to modify a specific sequence directly into elite cultivars leading to a new allele with a new phenotype (Fig. 18.1).

The knowledge on genome sequence has drastically changed the strategy of modern plant breeding, from the traditional field selection mainly based on phenotype (often yield), to an integrated approach where cultivar development is guided by genomics information linking genetic regions to specific phenotypic traits, and assisted by the deep understanding of physiological and molecular processes involved in plant adaptation and yield. Although the past two decades have shown potentiality as well as limitations of a molecular-based genetic improvement, future breeding will largely depend on the integration between genomics knowledge and classical breeding. This requires two fundamental connections: (1) the relationship between phenotypes and genes; and (2) the relationship between the gene-dependent phenotype and the field performance. While the first one is currently tackled, the second relationship is still largely unexplored. Functional genomics has developed many tools (e.g. transformation, insertional mutagenesis, RNAi, TILLING, etc.) (Varshney et al., 2005; Boutros and Ahringer, 2008; Jung et al., 2008) that allow finding the phenotype associated to a given gene (a strategy known as 'reverse genetics'); while from a phenotype of interest with a simple genetic basis (i.e. disease resistance), it is possible to clone the corresponding gene ('forward genetics'). Nevertheless, most of these works are carried out at single plant level and still few studies have been dedicated to the understanding of the impact of these genes at plant population level. For yield-related traits, plant and crop phenotype are largely unrelated (Pedró et al., 2012).

Using a physiological perspective to link the phenotype and genotype, Chapter 14 presents state-of-the-art modeling approaches, Chapter 15 emphasizes the quantification of yield-related traits in realistic field conditions, and Chapter 19 discusses the complex interactions between traits and the environment, and between relevant traits. The links between genotype and phenotype are further explored with focus on yield potential (Chapter 16) and grain quality (Chapter 17).

3 PRE-BREEDING: A LINK BETWEEN GENETIC RESOURCES AND CROP IMPROVEMENT

Over the past century, plant breeding has significantly contributed to increase crop productivity by systemic genetic improvements (Chapters 2, 3, 4, and 6); nevertheless, these gains have often been accompanied by decreased genetic diversity within elite gene pools (Lee, 1998; Fernie et al., 2006). It has been suggested that the elite germplasm of the major crops, especially self-pollinating cereals, has experienced an overall reduction of its genetic basis as a result of high selection pressures, recurrent use of the adapted elite germplasm and the adoption of breeding schemes not favoring genetic recombination (Hoisington et al., 1999). For example, the majority of hard red winter wheat cultivars in the USA have been mainly derived from two eastern European lines (Harlan, 1987). In rice, molecular analyses reveal that the cultivated gene pool has a limited genetic variation compared to wild relatives (Caicedo et al., 2007; Zhu et al., 2007). Even in maize, which is considered to be a highly polymorphic species, there has been roughly a 30% drop in diversity at the average locus from maize's wild relatives (Buckler et al., 2001). Wheat diversity assessed at nucleotide level showed a general and massive reduction of genetic diversity (Haudry et al., 2007); nevertheless, the loss in diversity was not uniform for all genes in the genome. For genes that do not influence favored phenotypes (neutral genes), the loss in diversity is simply a function of the strength of the bottleneck in terms of the population size and duration (Doebley et al., 2006). On the contrary, genes that influence desirable phenotypes experienced a more drastic loss of diversity because plants carrying favored alleles contributed the most progeny to each subsequent generation, while selection eliminated the other alleles (Wright et al., 2005). Consistent with these results, Laidò et al. (2013) showed the genetic diversity for morphological traits and storage proteins was lower in modern durum wheat cultivars compared to old ones, whereas the reduction of genetic diversity was much less for neutral molecular markers. This suggests that the reduction in diversity is likely associated with selection for a few adaptive or quality traits that strongly correlate with grain yield and technological properties of gluten; Chapter 17 further analyses genetic and environmental influences on wheat proteins. The absence of a parallel reduction in the genetic diversity of neutral markers can be ascribed either to a minimal effect of the genetic drift or to the introduction of new germplasm from different gene pools during durum wheat breeding history.

Wild relatives and landraces carry a number of alleles capable of improving agronomic performance in modern cultivars (Tanksley and McCouch, 1997; Xiao et al., 1998; Gur and Zamir, 2004). In wild species, favorable loci or quantitative trait loci (QTL) often remain cryptic due to several factors including their low frequency, masking effects of deleterious alleles and negative epistatic interactions (Xiao et al., 1998; Gottlieb et al., 2002; Lauter and Doebley, 2002; Peng and Khush, 2003). These alleles have the potential to contribute to crop improvement (Xiao et al., 1998; Poncet et al., 2000, 2002; Peng and Khush, 2003). Thus, although wild and exotic germplasm is perceived to be a poor bet for the improvement of most traits based on phenotypic examination, it is quite possible that some favorable genes (alleles) lie buried amidst the thousands of accessions maintained in gene banks which, if they could be found, might be valuable for crop improvement.

In wheat, the most diffuse example of chromatin introgression from a relative into cultivated varieties is the 1BL/1RS chromosomal translocation (Rabinovich, 1998). Other examples are the introgression in cultivated varieties of almost 30 independent disease resistance genes from wild relatives (reviewed in Hoisington et al., 1999) and of the functional alleles of the *Pin* genes, whose proteins are responsible for endosperm texture in wheat, from the diploid wheat *T. monococcum* (Giroux and Morris, 1998).

Barley improvement has also benefited from genes of wild barleys and Middle Eastern landraces (Ellis et al., 2000), as exemplified by the development of the *mlo*-mediated resistance to powdery mildew (Büschges et al., 1997; Thomas et al., 1998) or the transfer of the scald (*Rhynchosporium secalis*) resistance gene from *Hordeum bulbosum* into *H. vulgare* (Pickering et al., 2006).

Despite some successes, mostly based on single gene traits, traditional breeding has always found difficulties in extracting useful traits from exotic germplasm. The introgression of beneficial alleles from wild species requires repeated backcross to recover most of the desirable agronomic traits and efficient selection to retain the target allele from the exotic donor. Even when these are applicable, linkage drag may compromise the final result. These problems are exacerbated, in the case of complex agronomic traits, by the existence of numerous interacting QTL whose expression is also significantly influenced by the environment. These drawbacks could be partially solved with recent progress in developing large-scale genomic resources and physiological understanding of QTL × environment and QTL × QTL interactions (Chapters 14, 15, 17 and 19). Molecular markers, comprehensive genetic maps, dense consensus maps and identification of marker-trait associations provided a solution to monitor desired and non-desired alien alleles among populations of introgression lines (ILs) developed using advanced backcross methods (Tanksley and Nelson, 1996). In advanced backcross breeding, molecular linkage maps are used to analyze populations obtained by repeatedly backcrossing a wild parent to a recurrent domesticated parent. The outcome of this procedure is a subset of alleles from the wild species that can be mapped and evaluated in a cultivated genetic background (Tanksley and Nelson, 1996). The rationale of this strategy is that beneficial wild alleles can be recovered from transgressive segregations that outperform the cultivated parent (Tanksley and Nelson, 1996; Tanksley and McCouch, 1997).

In rice, QTL introgressed from exotic landraces and/or wild species have improved yield under both irrigated and, more recently, water-limited conditions (Moncada et al., 2001; Bernier et al., 2009a,b; Uga et al., 2011; Venuprasad et al., 2011, 2012). In tomato, yield increase of more than 50% was achieved after the pyramiding of three independent introgressions from a wild tomato in the genetic background of a leading market variety (Gur and Zamir, 2004).

4 DNA TECHNOLOGIES BOOST NEW KNOWLEDGE TO UNDERSTAND PLANT DIVERSITY

Biotechnology offers tools based on molecular markers for the assessment of the genetic variation among individual plants, accessions, populations and species and for monitoring genetic diversity over time and space (Smith and Beavis, 1996; Röder et al., 2004). Molecular markers can reveal polymorphisms in the nucleotide sequence and therefore allow discrimination between different alleles at a given locus. Molecular markers are widely used to confirm identity between parents and progeny, to determine evolutionary relationships and genetic distances, to construct genetic and physical maps, and to localize genes or genomic regions responsible for the expression of a trait of interest. The application of molecular markers can significantly increase the speed and the effectiveness of prebreeding and breeding programs. Introgression and selection of traits with simple genetic bases are routinely carried out with MAS, particularly for disease resistance. Molecular marker characteristics and applications have been extensively reviewed (Röder et al., 2004; Francia et al., 2005; Agarwal et al., 2008) and are summarized in Table 18.1.

Virtually, every kind of molecular marker can be used for MAS; nevertheless, simple sequence repeats (SSRs) and single nucleotide polymorphisms (SNPs) are the markers of choice for most applications due to their easiness, robustness, co-dominant genetic behavior, low cost per sample and possibility of automation (Mammadov et al., 2012).

The recent development of massively parallel sequencing (also called next-generation sequencing – NGS) technologies has reduced the cost of sequencing by more than one thousand times compared to the traditional Sanger method (Perez-de-Castro et al., 2012) making DNA sequencing affordable for many applications. Genome sequencing has greatly improved our understanding of composition, organization, evolution and functional aspects of plant genomes and has enabled efficient and cost-effective genome-wide polymorphism discovery. Although NGS technologies allow the identification of a wide spectrum of molecular markers (SNPs, SSRs, insertion and deletion – InDels), they have

TABLE 18.1 PCR-based molecular markers commonly used in plant breeding

Marker	Description	Suitability for employment in MAS
InDels	Insertion or deletion of nucleotide sequences at a given genomic position	Co-dominant markers, genotyped at low cost based on size separation. Suitable for MAS
SSRs	Simple sequence repeat polymorphisms	Co-dominant markers, also known as microsatellites. Detect a good polymorphism level and are suitable for automation and high-throughput low-cost analysis. Highly recommended for MAS
SNPs	Single nucleotide polymorphisms	Dominant or co-dominant depending on the analysis protocol. Development costs are generally high, but once developed, the marker can be used in high-throughput analysis at extremely low cost per sample, being therefore highly recommended for MAS
DArTs	Diversity array technology markers, based on microarray hybridization of genomic DNA	Dominant markers, their analysis is carried out in a few dedicated laboratories (www.triticarte.com). Due to the dominant behavior, only MAS protocols have been developed for key agronomic traits in wheat at triticarte lab

been used mainly for SNPs discovery, as these are the most abundant class of polymorphism in plant genomes (Mammadov et al., 2012).

The analysis of crop genomes provided surprising findings with important implications to our understanding of plant origin and evolution: genome duplications, ancestral rearrangements and unexpected polyploidization events opened new doors to address fundamental questions related to species proliferation, adaptation and functional modulations. To date, the genome sequence (complete or draft) of more than 50 plant species have been released (Michael and Jackson, 2013). The availability of the complete genome sequence of crop species as well as of a number of accessions within species (resequencing) is providing access to the virtually complete catalog of genes and regulatory elements which, in turn, leads to new keys for biodiversity exploitation and interpretation (Barabaschi et al., 2012). The germplasm of cultivated species made of wild relatives, landraces and modern cultivars shaped over time by natural and human selection, shows a high degree of genetic diversity. Uncovering the forces that have driven the diversification of genomes and the genetic basis of the biodiversity has always been a challenge for crop genetics and breeding. Genome sequence opened the access to an incredible portion of genetic diversity and, in the future, global sequencing will likely replace other tools for evaluating the diversity of the crop gene pool at the finest resolution (Barabaschi et al., 2012).

Growing evidence indicates that in some species, such as maize, a significant part of genetic diversity is explained by presence/absence variants of entire expressed gene (PAVs, sequences that are present in one genome, but absent in the other), and copy number variations (CNVs, sequences that are present in different copy number between compared genomes; see Box 18.1

BOX 18.1

PAVS AND CNVS – A NEW CONCEPT OF GENETIC DIVERSITY

In recent years, whole genome analysis and resequencing studies have documented numerous examples of structural variation in the form of Copy Number Variation (CNV) and Presence/Absence Variants (PAVs) (Schnable and Springer, 2013). Growing evidence suggests that CNV and PAVs contribute to variation in the genome and in the phenotypic diversity of accessions within a species (Springer et al., 2009; Swanson-Wagner et al., 2010). Multiple copies of the *Rhg1* locus in soybean increase the resistance against cyst nematode (Cook et al., 2012). In barley and wheat, variation in copy number at the *CBF* locus is associated with levels of freezing tolerance (Knox et al., 2008). CNV at the *MATE1* locus in maize, resulting in a higher expression of *MATE1*mRNA, increases aluminum tolerance (Maron et al., 2013).

Several works have highlighted that the coding PAV results in a significant enrichment of specific gene categories. In *Cucumis melo*, the most represented functional category among the identified PAV genes was that of stress response proteins (Gonzalez et al., 2013), while an enrichment of nucleotide binding and receptor-like proteins was observed in soybean (McHale et al., 2012). Variability in gene content among maize inbreed lines have been suggested to contribute to heterosis. Inbreed lines show a relatively high rate of CNV and PAV. The cumulative effect of gene loss would result in decreased vigor in the inbreds, while the loss of functions would be complemented in the hybrids resulting in the hybrid's vigor (Swanson-Wagner et al., 2010).

for further details). When PAVs represent a significant part of genetic diversity, the genes of one genotype do not constitute the full set of genes of the species (Schnable and Springer, 2013). Brunner et al. (2005) have compared four allelic chromosomal regions of two different maize inbred lines (B73 and Mo17), revealing that only 50% of the sequences were shared. Results from other resequencing studies (reviewed by Schnable and Springer, 2013) indicate that a reference genome sequence does not describe completely the genetic content of a particular species supporting the concept of 'pan-genome'. The pan-genome is made up of a core genome that consists of genomic features that are in common to all the accessions of the species, and of the dispensable genome, which is composed of the partially shared or not shared genome sequences; therefore a complete description of the genomic composition of a species requires the sequencing of a significant number of different accessions (Barabaschi et al., 2012). Uncovering the composition and the function of the dispensable genome represents a key step towards the understanding of the mechanism generating the phenotypic diversity. Notably, this variation for functional gene content suggests that the loss of functional genes in inbreed lines can be complemented in the hybrid resulting in heterotic effects (Schnable and Springer, 2013).

5 ALLELE MINING: EXPLORE PLANT DIVERSITY BY SEQUENCING

Allelic variants are determined by mutations and, although most mutations are neutral or negative, some have contributed useful agronomic traits such as the reduced plant height in wheat (Peng et al., 1999) or erect growth, greater grain number and higher grain yield in rice (Tan et al., 2008). Recent advances in DNA technologies allow the screening of all mutations in a target gene across a large germplasm collection opening a different dimension in the exploration of plant diversity. 'Allele mining' is the large-scale dissection of the natural variation in the sequence of genes (or candidate loci) associated with specific trait variations, in a representative collection of genotypes. The first technology dedicated to allele mining was Eco-TILLING (targeted-induced local lesions in genomes) (Comai et al., 2004) where SNPs between each tested genotype and a reference genotype are detected through enzymatic cleavage of heteroduplex DNA molecules. The application of NGS technologies to allele mining has made sequence comparison of a potentially unlimited number of alleles very cheap, and allele mining is becoming a common approach for exploration of plant diversity. Furthermore, NGS paves the way for an analysis of haplotype structure, in terms of frequency, type and extension which is essential for association mapping. While early studies of allele mining were based exclusively on the search for mutations in the gene-coding regions and therefore responsible for changes in structure and/or function of proteins, the focus has expanded to non-coding sequences, with important effects on gene expression level, such as introns, 3′UTRs and promoters (Salvi et al., 2007).

Allele mining applied to specific genes in accessions of choice, and associated with clear and obvious phenotypic changes, is playing an increasingly important role in genomics-driven plant breeding. The focused identification of germplasm strategy (FIGS) combines environmental and plant characteristics to facilitate the identification of germplasm with traits of potential interest for particular environments (Chapter 13). Using FIGS, a set of 1320 bread wheat landraces were selected from a virtual collection of 16089 accessions, and subjected to allele mining to search for variations at the powdery mildew resistance locus *Pm3*. Phenotyping the accessions carrying nucleotide variations at the *Pm3* locus revealed seven new resistance alleles (Bhullar et al., 2009). Allele mining in hundreds of barley genotypes has revealed new alleles at the powdery mildew resistance locus *Mla* (Seeholzer et al., 2010), at the leaf stripe resistant locus *Rdg2a* (Biselli et al., 2013) as well as at the *VRN-H1/H4* vernalization locus (Cockram et al., 2007). In rice, allele mining enabled the identification of new alleles at the *Waxy* locus encoding for granule-bound starch synthase and responsible for the amylose content in the kernels (Mikami et al., 2008), and for the *GS3* locus, a gene related to grain size (Takanokai et al., 2009).

While allele mining has mostly focused on single or a few specific loci, the most recent NGS achievements allow the simultaneous targeted resequencing of hundreds of loci (Mamanova et al., 2010; Teer et al., 2010), of the whole exome (Mascher et al., 2013) or, eventually, the resequencing of the whole genome (Huang et al., 2010; Lai et al., 2010). The application of these technologies to a large germplasm collection will reveal in a single run the allele diversity for most loci of a given species.

A large-scale evaluation of genome-wide genetic variation conducted in rice by resequencing more than 500 Chinese landraces identified approximately 3.6 million SNPs that were used to describe the linkage disequilibrium (LD) structure for *indica* and *japonica* landraces and to construct a high-density haplotype map (Huang et al., 2010). The high performance of a genome-wide association study (GWAS) on 14 agronomic traits, demonstrated that this approach is a valuable alternative to genetic mapping, even for complex traits. In soybean, a comprehensive GWAS, performed by resequencing 31 cultivated and wild accessions, detected a high level of LD and allowed identification of about 200000 SNPs, potentially useful in mapping of complex traits (Lam et al., 2010). The greater allelic diversity estimated in wild soybeans, and lost in cultivated ones as a result of human selection, could represent a rich source of novel and interesting alleles for important traits, a conclusion that can probably be extended to most crop species. In crop plants where genome sequencing and resequencing is still difficult, large-scale GWAS can be performed with thousands of SNPs. In a European collection of barley cultivars, the structural and temporal variation in the genetic diversity was characterized with an array carrying 9000 SNPs and used for association mapping of several quantitative traits (Tondelli et al., 2013).

6 GM BREEDING

An important contribution of molecular biology to plant breeding is represented by the understanding of gene actions and functions which, in turn, has suggested the introduction of relevant genes via genetic transformation. Genetic transformation, the non-sexual insertion of alien genes, gene-regulating systems or both, may be applied to endow the new cultivars with useful traits. Plant transformation is the delivery of biologically active and functional foreign genes into plant cells, followed by the integration of the genes into the plant genome, and the recovery of transgenic plants stably expressing the foreign gene. In nature, the soil bacterium *Agrobacteriun tumefaciens* causes tumor formation (so-called crown galls) on many dicotyledonous and some monocotyledonous and gymnosperm species by transferring a specific DNA fragment, the T-DNA (transferred DNA), from its tumor-inducing (Ti) plasmid to the plant cell genome (Gelvin, 1998, 2000). Recombinant *Agrobacterium* strains, in which the native T-DNA has been removed and replaced with genes of interest, are the most efficient vehicles used for the introduction of foreign genes into plant genomes. The use of *Agrobacterium* vectors for genetic transformation is superior to other techniques in terms of stable genome integration and single/low copy number of the intact transgene, though it may lead to variable expression due to genomic position effect and transgene silencing (Lacroix et al., 2005).

Plant transformation provides new variability and novel genes previously inaccessible to breeders, also from unrelated species even in other taxonomic phyla, which can be inserted and functionally expressed into an agronomic background in a single event, without associated deleterious genes. This process also allows changing the level and the spatial and temporal pattern of transgene expression by choosing suitable promoters, constitutive or inducible, or adding introns into the constructs. The engineered genetic enhancement may solve many plant breeding problems particularly when the gene controlling the trait of interest is absent or difficult to access in the gene pool of the major commercial crops. Pest, disease and herbicide resistance provide examples in which GM crops have been obtained and successfully adopted worldwide (Shewry et al., 2008; Ronald, 2011).

Efforts have been made over the past two decades in manipulating genes and inserting them into crop plants to confer desirable attributes, although the success of GM breeding is largely dependent on the trait considered. While GM plants modified for herbicide tolerance and insect resistance have been adopted worldwide, other transgenic events are at a pre-commercial phase, many more are at laboratory level and a few of them are likely to be successfully validated in field trials. Almost all plant features have been targeted with a transgenic approach: (1) increased resistance to biotic and abiotic stress and increased tolerance to herbicides; (2) enhanced nutritional or technological value of crops and improved post-harvest quality; (3) enhanced efficiency of soil phosphorus uptake and nitrogen fixation; (4) improved adaptation to soil salinity and aluminum toxicity; (5) increased photosynthetic rate, sugar and starch production; (6) male-sterility and apomixis to fix hybrid vigor in inbred crops; (7) production of pharmaceuticals and vaccines (Shewry et al., 2008; Ronald, 2011, http://www.isaaa.org). To date, many GM plants have been developed, although still few GM crops and cultivars are used in cropping systems. A number of reasons explain this. The transgenic approach is, by definition, based on a single (or few) gene/s that controls a specific metabolic reaction or cellular process. Therefore, only when the cellular phenotype is positively reflected at plant population and community levels will the transgenic plants have a chance to result in a new GM crop (Chapman et al., 2002). Traits conferring resistance to pathogens have similar effects at cellular and crop levels, while modifications of a single step of a complex metabolic pathway (e.g. photosynthesis, water stress tolerance) can be buffered scaling up from cell to field (see also Chapter 1 for a discussion of trait scalability). Furthermore, the increase in crop yield potential obtained so far have involved, through time, trade-offs between individual fitness and population performance with a stronger effort of the selection on the latter (Denison, 2012; Khush, 2013). For instance, plants with less branching and more erect leaves yield more because crops intercept light more efficiently, due to the non-linear response of leaf photosynthesis, or modern wheat and rice cultivars with short straw reduce lodging and increase both spike weight and grain setting at the expense of stem biomass and competitive ability for light.

GM breeding is nowadays troubled by public constraints related to the commercialization and use of GM crops, particularly in Europe. Risk assessment is a fundamental step for the registration of GM cultivars and for their commercial success (Sparrow, 2010). The development of clearly defined and scientifically-based regulatory frameworks for the introduction of GM crops in the environment and onto market is still a matter of discussion and public concern in many countries.

In 2012, the global area of GM crops had reached 170 million ha, that is about 11% of total arable land (Clive, 2012). More than half (52%) of the global GM crop area is in developing countries and percent growth in GM crop area is higher in the developing (11%) than in industrial countries (3%). More than 92% of commercialized GM crops are concentrated in the USA, Argentina, Brazil, Canada, India, and China (Table 18.2). In spite of the wide diffusion of GM crops, it is noticeable that 98 % of this was represented by four species, soybean (47% of global GM crop area), maize (32%), cotton (14%) and canola (5%). The main traits involved are herbicide tolerance (mainly to glyphosate and gluphosinate, 59%), insect resistance (15%) and the two stacked traits (25%) (Clive, 2012).

TABLE 18.2 Global area of GM crops in 2012

Rank	Country	Area (million hectares)	Transgenic crops
1	USA	69.5	Maize, soybean, cotton, canola, sugar beet, alfalfa, papaya, squash
2	Brazil	36.6	Soybean, maize, cotton
3	Argentina	23.9	Soybean, maize, cotton
4	Canada	11.6	Canola, maize, soybean, sugar beet
5	India	10.8	Cotton
6	China	4.0	Cotton, papaya, poplar, tomato, sweet pepper
7	Paraguay	3.4	Soybean, maize, cotton
8	South Africa	2.9	Maize, soybean, cotton
9	Pakistan	2.8	Cotton
10	Uruguay	1.4	Soybean, maize
11	Bolivia	1.0	Soybean
12	Philippines	0.8	Maize
13	Australia	0.7	Cotton, canola
14	Burkina Faso	0.3	Cotton
15	Myanmar	0.3	Cotton
16	Mexico	0.2	Cotton, soybean
17	Spain	0.1	Maize
18	Chile	<0.1	Soybean, cotton, maize
19	Costa Rica	<0.1	Cotton, soybean
20	Colombia, Sudan	<0.1	Cotton
21	Honduras, Portugal, Czech Republic, Cuba, Egypt, Romania, Slovakia	<0.1	Maize

Source: Clive (2012).

7 BEYOND GM PLANTS: THE NEW BREEDING TECHNIQUES

Molecular and genomic techniques have experienced a remarkable development which has enormously accelerated genetic research. This has made possible the use of recombinant DNA techniques to introduce new traits in early phases of cultivar selection in such a way that, in contrast with standard GM plants, the resulting plant genotype does not contain any exogenous DNA sequence. The techniques that allow these modifications are generally described as new breeding techniques (Lusser et al., 2012; Lusser and Davies, 2013); they aim to improve the efficiency of traditional breeding, even though most of them are at the cutting edge between traditional and GM breeding. The new breeding techniques promote genetic modification events, but their end products are free of foreign genes and the released varieties are similar to those that can be achieved using conventional techniques such as mutagenesis. However, because of the involvement of a genetic modification in early steps of the selection process, these techniques might fall under specific legislation dedicated to biotechnology-derived crops (e.g. the European Directive 2001/18/EC).

The most relevant new breeding techniques are those dedicated to promote site-specific mutagenesis. Since small variations in a gene sequence can have a significant impact on plant phenotype, technology that allows modifying a specific sequence in the plant genome can be used to create new alleles. Techniques such as oligonucleotide-directed mutagenesis (ODM) as well as those based on zinc finger nuclease (ZFN) and transcription activator-like effector nuclease (TALEN) are all capable of specifically modifying a given target sequence leading to genotypes not substantially different from those obtained through traditional mutagenesis (Lusser et al., 2011). Obviously, site-specific mutagenesis is more efficient, faster and cleaner (no additional random mutations in the genome) than traditional mutagenesis. Potentially very relevant also is the reverse breeding technology to derive homozygous parents from any heterozygous genotype. Box 18.2 outlines technical details on new breeding techniques.

The usefulness and impact of new breeding techniques is still to be determined; nevertheless, they open innovative perspectives in plant breeding. For example, ODM has been used to confer resistance to specific herbicides in maize (Zhu et al., 2000), wheat (Dong et al., 2006), rice (Okuzaki and Toriyama, 2004; Iida and Terada, 2005), and tobacco (Kochevenko and Willmitzer, 2003), and antibiotic resistance in canola (Gamper et al., 2000) and banana (Gamper et al., 2000; Rice et al., 2000). ZFN technology has been applied in maize for herbicide resistance (Shukla et al., 2009) and in soybean to modify some genes involved in RNA silencing (Curtin et al., 2011). No practical application has been reported for reverse breeding.

8 GENOMIC SELECTION

Molecular markers are currently used in reconstructing novel genotypes through allele pyramiding of useful loci from many donors, at least for some simple traits. For over two decades, molecular marker technology has been predicted to reshape breeding programs and improve the gain from selection. Nevertheless, molecular breeding has revealed both successful and unsuccessful stories. Most of the traits with a mendelian genetic base were efficiently targeted with molecular breeding technologies, while quantitative traits remain a challenge. In particular, when many alleles with a small effect segregate in a population, substantial and reliable effects could be identified only in rare cases.

The most recent molecular marker technologies offer the possibility of overcoming the limits of traditional MAS. As a matter of fact, the precision of mapping QTL by traditional linkage analysis could be only marginally improved

BOX 18.2

SOME TECHNICAL DETAILS ON 'NEW BREEDING TECHNIQUES'

Oligonucleotide-directed mutagenesis (ODM)

ODM employs chemically synthesized oligonucleotides (single-stranded DNA oligonucleotides or chimeric oligonucleotides) which share homology with the target sequence of the plant genome except for one or a few base pairs, so that they can hybridize with the complementary wild-type DNA in the gene of interest. The hybridized oligonucleotide fragment is then used as a primer by DNA polymerase I which copies the rest of the wild-type gene and induces site-specific nucleotide substitutions, insertions or deletions at the target sequence via the natural repair mechanism of the cell (Edelheit et al., 2009; Lusser and Davies, 2013). Plants regenerated from mutated cells or tissues carry the desired mutation.

Zinc finger nuclease (ZFN) and transcription activator-like effector nuclease (TALEN)

Zinc finger proteins and transcription activator-like effectors are transcription factors that can be converted into site-specific 'DNA scissor' by fusion with an endonuclease (usually *FokI*). ZFNs are a class of engineered DNA-binding restriction enzymes generated by fusing a zinc finger DNA-binding domain to a DNA-cleavage domain (*FokI*) that facilitate targeted editing of the genome by creating double-strand breaks in DNA at specified locations. Zinc finger domains can be engineered to target unique desired DNA sequences within complex genomes. The ZFN technology can be used to introduce site-specific mutations (nucleotide changes or small InDels) (ZFN-1), deletions of specific DNA sequence (ZFN-2), or the insertion of new genes in a specific genome site (ZFN-3) (Carrol, 2011; Isalan, 2012).

Reverse breeding (RB)

RB allows the development of homozygous parental lines from any desired heterozygous hybrid varieties by inserting, via genetic transformation, an RNAi construct in the hybrid line that suppresses recombination during meiosis, giving the haploid gametes of the genetically modified hybrid plant with entirely non-recombined chromosomes. The chromosomes of the microspores are doubled and these gametes are used to produce new homozygous double haploid plants. The plants containing the transgenic sequence are selected out and only double haploid plants that do not contain the RNAi construct are used as parents for the reconstruction and seed production of the original heterozygous genotype (Dirks et al., 2009).

with very dense maps (Darvasi et al., 1993). Therefore, new approaches are needed to use efficiently the marker information generated by the new technologies. An evolution of MAS is represented by genomic selection (GS), an approach which selects favorable individuals based on genomic estimated breeding values (GEBVs) (Nakaya and Isobe, 2012). At the beginning, GS was considered an unrealistic approach due to the need of a very large number of molecular markers. Indeed, GS does not consider the evaluation of a few markers associated with single large effects, rather GS is run through the scoring of a high number of molecular markers covering the whole genome. At such densities, it was assumed that linkage phase between markers or haplotype blocks of markers and casual polymorphisms would be consistent across families so that population-wide estimates of marker effects would be meaningful (Meuwissen et al., 2001). Recent progress in molecular marker technology, making possible many data points at a low price combined with novel statistical methods enabling the simultaneous estimation of all marker effects, allowed GS to be applied in animal breeding (van Raden et al., 2009; Hayes et al., 2009).

MAS for quantitative traits and GS share a common framework composed of two phases: training and breeding (Nakaya and Isobe, 2012). In the training phase, the associations between phenotype and genotype are investigated in a subset of a population (mapping population in MAS and training population in GS). QTL are identified in MAS, while GEBVs are estimated in GS using statistical approaches. A breeding population is then genotyped in the breeding phase, and favorable individuals are selected based on the data of the molecular markers linked to the QTL in MAS, and the GEBVs in GS. Thus, GS analyzes at the same time all the genetic variance of each individual by summing up the effects of GEBV for the markers covering the whole genome (Heffner et al., 2009), and it is expected to address small effect genes that cannot be revealed by MAS (Hayes et al., 2009).

The results in animal breeding (Hayes et al., 2009; van Raden et al., 2009) suggest that GS selection can be valuable for plant improvement, as reinforced by recent studies. Simulation studies and investigations based on empirical data have been carried out to compare the effectiveness of the different statistical methods, and the effect of factors such as LD, population size and structure, and number of molecular markers on the accuracy of GS models. Simulations revealed that the accuracy of GS in estimation of breeding values was higher than pedigree information (Nakaya and Isobe, 2012). Empirical studies were carried out in maize (Lorenzana and Bernardo, 2009; Guo et al., 2011), barley (Piepho, 2009), wheat (Heffner et al., 2011) and *Arabidopsis* (Lorenzana and Bernardo, 2009). In general, the accuracies in plant species were higher than those in animal studies. This is interesting if we consider that fewer molecular markers were used in plant studies with respect to animals (usually few hundreds to less than 2000), probably due to the lower genetic diversity caused by a small number of parental lines and a greater bottleneck in the breeding materials (Nakaya and Isobe 2012).

LD, defined as the non-random association of alleles at different loci, has to be carefully considered in breeding populations as it is an important factor influencing the efficacy of GS. In general, LD is greater in self-crossing than in out-crossing species; moreover, it varies with species, population structure and genome region (Gupta et al., 2005). The number of markers required for GS modeling is determined based on the extension of LD blocks: the wider the LD, the fewer the markers. Furthermore, employing a population that originated from a few parental lines is effective in reducing the number of markers required, especially for species whose LD intensities decay rapidly among unrelated individuals (Nakaya and Isobe, 2012). Dominant markers (e.g. DArT markers) lead to a lower accuracy of GEBV prediction than co-dominant markers (e.g. SNPs, SSRs) due to the lower LD detection power. In this case, the accuracy can be improved by considering haplotypes (Li et al., 2007). Another feature to be taken into account is the number of alleles revealed by the markers. A study in animals showed that SNPs required two to three times greater density compared to SSRs to achieve a similar accuracy of GEBV prediction (Solberg et al., 2008).

Regarding the size of the training and breeding populations, both simulation and empirical GS studies suggest that a larger training population improves the accuracy of GEBV predictions. As an example, Heffner et al. (2011) calculated the average ratio of GS accuracy to phenotypic selection accuracy for grain quality traits in wheat: this ratio was 0.66, 0.54 and 0.42 for training population sizes of 96, 48 and 24, respectively (with breeding populations containing 174 or 209 individuals). More than the absolute size of the training population, the training/breeding population ratio is important for accuracy in GS. In general, the accuracy of the GEBV improves with a higher training/breeding population ratio with greater genetic diversity, smaller-sized breeding populations, lower heritability of traits and larger numbers of existing QTL (Nakaya and Isobe, 2012).

Taken together, these data indicate that GS is a potential method for plant breeding and it can be performed with realistic sizes of populations and markers when the populations are chosen carefully. Nevertheless, GS cannot be considered as a perfect replacement for phenotypic selection. Rather, it should be integrated with classical breeding to make it faster and more effective. GS could also be used together with MAS; for example, Heffner et al. (2010) proposed using traditional MAS for important QTL in the F2 and F3 generations, before GS in the F5 generation, to avoid useless evaluation of lines that do not carry essential QTL alleles.

The application of the genomic tools to plant breeding, besides improving the selection process, offers the possibility of planning and developing novel varieties bearing pre-defined traits (breeding by design). A worldwide genomic-based effort is currently focused to understand the genetic bases of agronomic traits, to analyze the allele variations at the corresponding loci and, ultimately, to enable breeders first to design new ideotypes *in silico*, then construct the new genotypes *in planta* (Chapter 14). Although this approach is currently not an option for breeders, a large amount of public and private resources are committed to reach this goal and a few promising examples have been published. A combination of different approaches has been programmed to develop new rice cultivars referred to as 'Green Super Rice', possessing resistance to multiple insects and diseases, high nutrient efficiency, and drought resistance, promising to reduce the consumption of pesticides, fertilizers, and water (Zhang, 2007). In the first stage, partially achieved, elite lines carrying single genes with a major effect on the expression of key traits are developed and evaluated, which by themselves are useful for cultivar release. In the second stage, a pyramiding of all major genes controlling the target traits is carried out to develop cultivars with multiple resistances and adaptability traits. If fully exploited, the integration of a similar approach with GS would help to design the new plant not only for few selected traits but virtually for all the loci of the genome, fully realizing the concept of breeding by design.

9 CONCLUDING REMARKS

Plant breeding is a continuous accumulation of superior alleles in the gene pool of the cultivated elite lines and recent developments in biotechnology offer new tools for screening and selecting new alleles. Allele mining allows searching for natural existing alleles in a germplasm collection, while new breeding techniques (similar to traditional mutagenesis) or GM breeding allow generating new alleles for traits of interest. Once the alleles of interest have been identified, high-throughput molecular markers can be used to assemble the most favorable combinations in new varieties and to predict their performance. As a result of this progress in genomics, the breeding process is generally moving from phenotypic selection to an integration between phenotypic selection and genotypic data generated with molecular markers either at a few loci (MAS) or at virtually all the loci of interest in the genome (GS). The relevance of the molecular selection in plant breeding is largely dependent on the species of interest, on the trait under selection and on the cost/benefit ratio. In some species, there is no sufficient molecular knowledge to start MAS, some traits can be easily scored and there is no convenience to develop markers for them, other traits are still too complex to be tackled with molecular markers, for some species– trait combinations the molecular selection is too expensive compared to results. Nevertheless, despite all these limitations, the general trend over the last 10 years and all expectations for the future are toward an increasing role for MAS, GS, GM and other new breeding technologies in plant breeding.

Acknowledgments

This work was supported by the Italian Ministry of Agriculture (MiPAAF), project ESPLORA.

References

Agarwal, M., Shrivastava, N., Padh, H., 2008. Advances in molecular marker techniques and their application in plant sciences. Plant Cell Rep. 27, 617–631.

Barabaschi, D., Guerra, D., Lacrima, K., et al., 2012. Emerging knowledge from genome sequencing of crop species. Mol. Biotechnol. 50, 250–266.

Bernier, J., Kumar, A., Venuprasad, R., et al., 2009a. Characterization of the effect of a QTL for drought resistance in rice, qtl12.1, over a range of environments in the Philippines and eastern India. Euphytica. 166, 207–217.

Bernier, J., Serraj, R., Kumar, A., et al., 2009b. The large-effect drought-resistance QTL qtl12.1 increases water uptake in upland rice. Field Crops Res. 110, 139–146.

Bhullar, N.K., Street, K., Mackay, M., Yahiaoui, N., Keller, B., 2009. Unlocking wheat genetic resources for the molecular identification of previously undescribed functional alleles at the Pm3 resistance locus. Proc. Natl. Acad. Sci. USA 106, 9519–9524.

Biselli, C., Urso, S., Tacconi, G., et al., 2013. Haplotype variability and identification of new functional alleles at the Rdg2a leaf stripe resistance gene locus. Theor. Appl. Genet. 126, 1575–1586.

Boutros, M., Ahringer, J., 2008. The art and design of genetic screens: RNA interference. Nat. Rev. Genet. 9, 554–566.

Brunner, S., Fengler, K., Morgante, M., Tingey, S., Rafalski, A., 2005. Evolution of DNA sequence nonhomologies among maize inbreds. Plant Cell 17, 343–360.

Buckler, E.S., Thornsberry, J.M., Kresovich, S., 2001. Molecular diversity, structure and domestication of grasses. Genet. Res. 77, 213–218.

Büschges, R., Hollricher, K., Panstruga, R., et al., 1997. The barley Mlo gene: a novel control element of plant pathogen resistance. Cell 88, 695–705.

Caicedo, A.L., Williamson, S.H., Hernandez, R.D., et al., 2007. Genome-wide patterns of nucleotide polymorphism in domesticated rice. PLOS Genet. 3, 1745–1756.

Carrol, D., 2011. Genome engineering with zinc-finger nucleases. Genetics 188, 773–782.

Cattivelli, L., Delogu, G., Terzi, V., Stanca, A.M., 1994. Progress in barley breeding. In: Slafer, G.A. (Ed.), Genetic improvement of field crops. Marcel Dekker, Inc, New York, pp. 95–181.

Chapman, S.C., Hammer, G.L., Podlich, D.W., Cooper, M., 2002. Linking biophysical, and genetic models to integrate physiology, molecular biology and plant breeding. In: Kang, M.S. (Ed.), Quantitative genetics, genomics and plant breeding. CAB International, Wallingford, pp. 167–187.

Clive, J., 2012. Global status of commercialized biotech/GM crops: ISAAA Briefs N° 44. International Service for the Acquisition of Agri-biotech Applications: Ithaca, New York.

Cockram, J., Chiapparino, E., Taylor, S.A., et al., 2007. Haplotype analysis of vernalization loci in European barley germplasm reveals novel VRN-H1 alleles and a predominant winter VRN-H1/VRN-H2 multi-locus haplotypes. Theor. Appl. Genet. 115, 993–1001.

Comai, L., Young, K., Till, B.J., Reynolds, S.H., Greene, E.A., Codomo, C.A., 2004. Efficient discovery of DNA polymorphisms in natural populations by EcoTILLING. Plant J. 37, 778–786.

Cook, D.E., Lee, T.G., Guo, X., et al., 2012. Copy number variation of multiple genes at Rhg1 mediates nematode resistance in soybean. Science 338, 1206–1209.

Curtin, S.J., 2011. Targeted mutagenesis of duplicated genes in soybean with zinc-finger nucleases. Plant Physiol. 156, 466–473.

Darvasi, A., Weinreb, A., Minke, V., Weller, J.I., Soller, M., 1993. Detecting marker-QTL linkage and estimating QTL gene effect and map location using a saturated genetic map. Genetics 134, 943–951.

De Vita, P., Li Destri Nicosia, O., Nigro, F., et al., 2007. Breeding progress in morpho-physiological, agronomical and qualitative traits of durum wheat cultivars released in Italy during the 20th century. Eur. J. Agron. 26, 39–53.

Denison, R.F., 2012. Darwinian agriculture: how understanding evolution can improve agriculture. Princeton University Press, Princeton, NJ.

Dirks, R., van Dun, K., de Snoo, B., et al., 2009. Reverse breeding: a novel breeding approach based on engineered meiosis. Plant Biotechnol. J. 7, 837–845.

Doebley, J.F., Gaut, B.S., Smith, B.D., 2006. The molecular genetics of crop domestication. Cell 127, 1309–1321.

Dong, C., Beetham, P., Vincent, K., Sharp, P., 2006. Oligonucleotide-directed gene repair in wheat using a transient plasmid gene repair assay system. Plant Cell Rep. 25, 457–465.

Edelheit, O., Hanukoglu, A., Hanukoglu, I., 2009. Simple and efficient site-directed mutagenesis using two single-primer reactions in parallel to generate mutants for protein structure-function studies. BMC Biotechnol. 9, 61.

Ellis, R.P., Forster, B.P., Robinson, D., et al., 2000. Wild barley: a source of genes for crop improvement in the 21st century? J. Exp. Bot. 51, 9–17.

Fernie, A.R., Tadmor, Y., Zamir, D., 2006. Natural genetic variation for improving crop quality. Curr. Opin. Plant Biol. 9, 196–202.

Francia, E., Tacconi, G., Crosatti, C., et al., 2005. Marker assisted selection in crop plants. Plant Cell Tissue Organ Cult. 82, 317–342.

Gamper, H.B., Parekh, H., Rice, M.C., Bruner, M., Youkey, H., Kmiec, E.B., 2000. The strand of chimeric RNA/DNA oligonucleotides can direct gene repair/conversion activity in mammalian and plant cell-free extracts. Nucleic Acids Res. 28, 4332–4339.

Gelvin, S.B., 1998. The introduction and expression of transgenes in plants. Curr. Opin. Biotechnol. 9, 227–232.

Gelvin, S.B., 2000. Agrobacterium and plant genes involved in T-DNA transfer and integration. Annu. Rev. Plant Physiol. Plant Mol. Biol. 51, 223–256.

Giroux, M.J., Morris, C.F., 1998. Wheat grain hardness results from highly conserved mutations in the friabilin components puroindoline a and b. Proc. Natl. Acad. Sci. USA 95, 6262–6626.

Giunta, F., Motzo, R., Pruneddu, G., 2007. Trends since 1900 in the yield potential of Italian-bred durum wheat cultivars. Eur. J. Agron. 27, 12–24.

Gonzalez, V.M., Aventin, N., Centeno, E., Puigdomenech, P., 2013. High presence/absence gene variability in defense-related gene clusters of Cucumis melo. BMC Genomics 14, 782.

Gottlieb, T.M., Wade, M.J., Rutherford, S.L., 2002. Potential genetic variance and the domestication of maize. BioEssays 24, 685–689.

Guo, Z., Tucker, D.M., Lu, J., Kishore, V., Gay, G., 2011. Evaluation of genome wide selection efficiency in maize nested association mapping populations. Theor. Appl. Genet. 124, 261–275.

Gupta, P.K., Pawan, S., Kulwal, P.L., 2005. Linkage disequilibrium and association studies in higher plants: present status and future prospects. Plant Mol. Biol. 57, 461–485.

Gur, A., Zamir, D., 2004. Unused natural variation can lift yield barriers in plant breeding. PLoS Biol. 2, e245.

Harlan, J.R., 1987. Gene centers and gene utilization in American agriculture. In: Yeatman, C.W., Kafton, D., Wilkes, G. (Eds.), Plant genetic resources: A conservation imperative. Westview, Boulder, CO, pp. 111–129.

Haudry, A., Cenci, A., Ravel, C., et al., 2007. Grinding up wheat: a massive loss of nucleotide diversity since domestication. Mol. Biol. Evol. 24, 1506–1517.

Hayes, B.J., Bowman, P.J., Chamberlain, A.J., Goddard, M.E., 2009. Genomic selection in dairy cattle: progress and challenges. J. Dairy Sci. 92, 433–443.

Heffner, E.L., Jannink, J.-L., Iwata, H., Souza, E., Sorrells, M.E., 2011. Genomic selection accuracy for grain quality traits in biparental wheat populations. Crop Sci. 51, 2597–2606.

Heffner, E.L., Lorenz, A.J., Jannink, J.-L., Sorrells, M.E., 2010. Plant breeding with genomic selection: gain per unit time and cost. Crop Sci. 50, 1681–1690.

Heffner, E.L., Sorrells, M.E., Jannink, J.-L., 2009. Genomic selection for crop improvement. Crop Sci. 49, 1–12.

Hoisington, D., Khairallah, M., Reeves, T., et al., 1999. Plant genetic resources: what can they contribute toward increased crop productivity? Proc. Natl. Acad. Sci. USA 96, 5937–5943.

Huang, X., Wei, X., Sang, T., et al., 2010. Genome-wide association studies of 14 agronomic traits in rice landraces. Nat. Genet. 42, 961–967.

Iida, S., Terada, R., 2005. Modification of endogenous natural gene targeting in rice and other higher plants. Plant Mol. Biol. 59, 205–219.

Isalan, M., 2012. Zinc-finger nucleases: how to play two good hands. Nat. Meth. 9, 32–34.

Jung, K.H., An, G., Ronald, P.C., 2008. Towards a better bowl of rice: assigning function to tens of thousands of rice genes. Nat. Rev. Genet. 9, 91–101.

Khush, G.S., 2013. Strategies for increasing the yield potential of cereals: case of rice as an example. Plant Breed. 132, 433–436.

Knox, A.K., Dhillon, T., Cheng, H., Tondelli, A., Pecchioni, N., Stockinger, E.J., 2008. CBF gene copy number variation at Frost Resistance-2 is associated with levels of freezing tolerance in temperate-climate cereals. Theor. Appl. Genet. 121, 21–35.

Kochevenko, A., Willmitzer, L., 2003. Chimeric RNA/DNA oligonucleotide-based site-specific modification of the tobacco acetolactatesyntase gene. Plant Physiol. 132, 174–184.

Lacroix, B., Tzira, T., Vainstein, A., Citovsky, V., 2005. A case of promiscuity: Agrobacterium's endless hunt for new partners. Trends Genet. 22, 29–37.

Lai, J., Li, R., Xu, X., et al., 2010. Genome-wide patterns of genetic variation among elite maize inbred lines. Nat. Genet. 42, 1027–1030.

Laidò, G., Mangini, G., Taranto, F., et al., 2013. Genetic diversity and population structure of tetraploid wheats (*Triticum turgidum* L.) estimated by SSR, DArT and pedigree data. Plos One 8, e67280.

Lam, H.M., Xu, X., Liu, X., et al., 2010. Resequencing of 31 wild and cultivated soybean genomes identifies patterns of genetic diversity and selection. Nat. Genet. 42, 1053–1059.

Lauter, N., Doebley, J., 2002. Genetic variation for phenotypically invariant traits detected in teosinte: implications for the evolution of novel forms. Genetics 160, 333–342.

Lee, M., 1998. Genome projects and gene pools: New germplasm for plant breeding. Proc. Natl. Acad. Sci. USA 95, 2001–2004.

Li, Y., Li, Y., Wu, S., et al., 2007. Estimation of multilocus linkage disequilibria in diploid populations with dominant markers. Genetics 176, 1811–1821.

Lorenzana, R.E., Bernardo, R., 2009. Accuracy of genotypic value predictions for marker-based selection in biparental plant populations. Theor. Appl. Genet. 120, 151–161.

Lusser, M., Davies, H.V., 2013. Comparative regulatory approaches for groups of new plant breeding techniques. New Biotechnol. 30, 437–446.

Lusser, M., Parisi, C., Plan, D., Rodriguez-Cerezo, E., 2011. New plant breeding techniques: state-of-the-art and prospects for commercial development. JRC Technical Report EUR 24760.

Lusser, M., Parisi, C., Plan, D., Rodriguez-Cerezo, E., 2012. Deployment of new biotechnologies in plant breeding. Nat. Biotechnol. 30, 231–239.

Mamanova, L., Coffey, A.J., Scott, C.E., et al., 2010. Target-enrichment strategies for next-generation sequencing. Nat. Meth. 7, 111–118.

Mammadov, J., Aggarwal, R., Buyyarapu, R., Kumpatla, S., 2012. SNP markers and their impact on plant breeding. Int. J. Plant Genomics 2012:728398.

Maron, L.G., Guimarães, C.T., Kirst, M., et al., 2013. Aluminum tolerance in maize is associated with higher MATE1 gene copy number. Proc. Natl. Acad. Sci. USA 110, 5241–5246.

Mascher, M., Richmond, T.A., Gerhardt, D.J., et al., 2013. Barley whole exome capture: a tool for genomic research in the genus Hordeum and beyond. Plant J. 76, 494–505.

McHale, L.K., Haun, W.J., Wayne, W.X., et al., 2012. Structural variants in the soybean genome localize to clusters of biotic stress-response genes. Plant Physiol. 159, 1295–1308.

Meuwissen, T.H.E., Hayes, B.J., Goddard, M.E., 2001. Prediction of total genetic value using genome-wide dense marker maps. Genetics 157, 1819–1829.

Miadenov, N., Hristov, N., Kondic-Spika, A., Djurico, V., Jevtic, R., Mladenov, V., 2011. Breeding progress in grain yield of winter wheat cultivars grown at different nitrogen levels in semiarid conditions. Breed. Sci. 61, 260–268.

Michael, T.P., Jackson, S., 2013. The first 50 plant genomes. Plant Genome 6, 2.

Mikami, I., Uwatoko, N., Ikeda, Y., et al., 2001. Quantitative trait loci for yield and yield components in an Oryza sativa × Oryza rufipogon BC2F2 population evaluated in an upland environment. Theor. Appl. Genet. 102, 41–52.

Moncada, M., Martínez, C., Tohme, J., Guimaraes, E., Chatel, M., Borrero, J., Gauch, H., McCouch, S., 2001. Quantitative trait loci for yield and yield components in an Oryza sativa × Oryza rufipogon BC2F2 population evaluated in an upland environment. Theor. Appl. Genet. 102, 41–52.

Nakaya, A., Isobe, S.N., 2012. Will genomic selection be a practical method for plant breeding? Ann. Bot. 110, 1303–1316.

Okuzaki, A., Toriyama, K., 2004. Chimeric RNA/DNA oligonucleotide-directed gene targeting in rice. Plant Cell Rep. 22, 509–512.

Ortiz-Monasterio, J.I., Sayre, K.D., Rajaram, S., McMahon, M., 1997. Genetic progress in wheat yield and nitrogen use efficiency under four nitrogen rates. Crop Sci. 37, 898–904.

Paran, I., van der Knaap, E., 2007. Genetic and molecular regulation of fruit and plant domestication traits in tomato and pepper. J. Exp. Bot. 58, 3841–3852.

Pedró, A., Savin, R., Slafer, G.A., 2012. Crop productivity as related to single-plant traits at key phenological stages in durum wheat. Field Crops Res. 138, 42–51.

Peng, J.R., Richards, D.E., Hartley, N.M., et al., 1999. Green revolution genes encode mutant gibberellin response modulators. Nature 400, 256–261.

Peng, S.B., Khush, G.S., 2003. Four decades of breeding for varietal improvement of irrigated lowland rice in the international rice research institute. Plant Prod. Sci. 6, 157–164.

Perez-de-Castro, A.M., Vilanova, S., Canizares, J., et al., 2012. Application of genomic tools in plant breeding. Curr. Genomics 13, 179–195.

Pickering, R., Ruge-Wehling, B., Johnston, P.A., Schweizer, G., Ackermann, P., Wehling, P., 2006. The transfer of a gene conferring resistance to scald (*Rhynchosporium secalis*) from *Hordeum bulbosum* into *H-vulgare* chromosome 4HS. Plant Breed. 125, 576–579.

Piepho, H.P., 2009. Ridge regression and extensions for genome wide selection in maize. Crop Sci. 49, 1165–1176.

Poncet, V., Lamy, F., Devos, K.M., Gale, M.D., Sarr, A., Robert, T., 2000. Genetic control of domestication traits in pearl millet (*Pennisetum glaucum* L., Poaceae). Theor. Appl. Genet. 100, 147–159.

Poncet, V., Martel, E., Allouis, S., et al., 2002. Comparative analysis of QTLs affecting domestication traits between two domesticated × wild pearl millet (*Pennisetum glaucum* L., Poaceae) crosses. Theor. Appl. Genet. 104, 965–975.

Rabinovich, S.V., 1998. Importance of wheat-rye translocation for breeding modern cultivars of *Triticum aestivum* L. In: Braun, H.J., Altay, F., Krostand, W.E., Beniwal, S.P.S., McNab, A. (Eds.), Wheat: prospects for global improvement. Kluwer Academic Publishers, pp. 401–418.

Rice, M.C., May, G.D., Kipp, P.B., Parekh, H., Kmiec, E.B., 2000. Genetic repair of mutation in plant cell-free extracts directed by specific chimeric oligonucleotides. Plant Physiol. 123, 427–437.

Röder, M.S., Huang, X.-Q., Ganal, M.W., 2004. Wheat microsatellites: potential and implications. In: Lörz, H., Wenzel, G. (Eds.), Molecular marker systems in plant breeding and crop improvement. Biotechnology in agriculture and forestry, vol 55, Springer, Berlin Heidelberg, pp. 255–266.

Ronald, P., 2011. Plant genetics, sustainable agriculture and global food security. Genetics 188, 11–20.

Sadras, V.O., Lawson, C., 2011. Genetic gain in yield and associated changes in phenotype, trait plasticity and competitive ability of South Australian wheat varieties released between 1958 and 2007. Crop Past. Sci. 62, 533–549.

Sakuma, S., Salomon, B., Komatsuda, T., 2011. The domestication syndrome genes responsible for the major changes in plant form in the Triticeae crops. Plant Cell Physiol. 52, 738–749.

Salvi, S., Sponza, G., Morgante, M., et al., 2007. Conserved noncoding genomic sequences associated with a flowering-time quantitative trait locus in maize. Proc. Natl. Acad. Sci. USA 104, 11376–11381.

Sayre, K.D., Singh, R.P., Huerta-Espino, J., Rajaram, S., 1998. Genetic progress in reducing losses to leaf rust in CIMMYT-derived Mexican spring wheat cultivars. Crop Sci. 38, 654–659.

Schnable, P.S., Springer, N.M., 2013. Progress toward understanding heterosis in crop plants. Annu. Rev. Plant Biol. 64, 71–88.

Seeholzer, S., Tsuchimatsu, T., Jordan, T., et al., 2010. Diversity at the *Mla* powdery mildew resistance locus from cultivated barley reveals sites of positive selection. Mol. Plant Microbe Interact. 23, 497–509.

Shewry, P.R., Jones, H.D., Halford, N.G., 2008. Plant biotechnology: transgenic crops. Adv. Biochem. Engin. Biotechnol. 111, 149–186.

Shukla, V.K., Doyon, Y., Miller, J.C., et al., 2009. Precise genome modification in the crop species *Zea mays* using zinc-finger nucleases. Nature 459, 437–441.

Simmonds, N.W., 1993. Introgression and incorporation: Strategies for the use of plant genetic resources. Biol. Rev. 68, 539–562.

Sinclair, T.R., 1998. Historical changes in harvest index and crop nitrogen accumulation. Crop Sci. 38, 638–643.

Slafer, G.A., Satorre, E.H., Andrade, H., 1994. Increases in grain yield in bread wheat from breeding and associated physiological changes. In: Slafer, G. (Ed.), Genetic improvement of field crops. Marcel Dekker, New York, pp. 1–67.

Smith, S., Beavis, W.D., 1996. Molecular marker assisted breeding in a company environment. In: Sobral, B.W.S. (Ed.), The impact of plant molecular genetics. Birkhäuser Verlag, pp. 260–272.

Solberg, T.R., Sonesson, A.K., Woolliams, J.A., Meuwissen, T.H.E., 2008. Genomic selection using different marker types and densities. J. Anim. Sci. 86, 2447–2454.

Sparrow, P.A., 2010. GM risk assessment. Mol. Biotechnol. 44, 267–275.

Springer, N.M., Ying, K., Fu, Y., et al., 2009. Maize inbreds exhibit high levels of Copy Number Variation (CNV) and Presence/Absence Variation (PAV) in genome content. PLoS Genet. 5, e1000734.

Swanson-Wagner, R.A., Eichten, S.R., Kumari, S., et al., 2010. Pervasive gene content variation and copy number variation in maize and its undomesticated progenitor. Genome Res. 20, 1689–1699.

Takanokai, N., Jiang, H., Kubo, T., et al., 2009. Evolutionary history of *GS3*, a gene conferring grain length in rice. Genetics 182, 1323–1334.

Tan, L., Li, X., Liu, F., et al., 2008. Control of a key transition from prostrate to erect growth in rice domestication. Nat. Genet. 40, 1360–1364.

Tanksley, S., Nelson, J., 1996. Advanced backcross QTL analysis: a method for the simultaneous discovery and transfer of valuable QTLs from unadaptedgermplasm into elite breeding lines. Theor. Appl. Genet. 92, 191–203.

Tanksley, S.D., McCouch, S.R., 1997. Seed banks and molecular maps: Unlocking genetic potential from the wild. Science 277, 1063–1066.

Teer, J.K., Bonnycastle, L.L., Chines, P.S., et al., 2010. Systematic comparison of three genomic enrichment methods for massively parallel DNA sequencing. Genome Res. 20, 1420–1431.

Thomas, C.M., Dixon, M.S., Parniske, M., Goldstein, C., Jones, J.D.G., 1998. Genetics and molecular analysis of tomato *Cf* genes for resistance to *Cladosporium fulvum*. Philosoph. Transact. R. Soc. B 353, 1413–1424.

Tian, Z.W., Jing, Q., Dai, T.B., Jiang, D., Cao, W.X., 2011. Effects of genetic improvements on grain yield and agronomic traits of winter wheat in the Yangtze River Basin of China. Field Crops Res. 124, 417–425.

Tondelli, A., Xu, X., Moragues, M., et al., 2013. Structural and temporal variation in the genetic diversity of a European collection of barley cultivars and utility for association mapping of quantitative traits. Plant Genome 6, 1–14.

Uga, Y., Okuno, K., Yano, M., 2011. *DRO1*, a major QTL involved in deep rooting of rice under upland field conditions. J. Exp. Bot. 62, 2485–2494.

van Raden, P.M., van Tassell, C.P., Wiggans, G.R., et al., 2009. Invited review: reliability of genomic predictions for north American Holstein bulls. J. Dairy Sci. 92, 16–24.

Varshney, R.K., Graner, A., Sorrells, M.E., 2005. Genic microsatellite markers in plants: features and applications. Trends Biotechnol. 23, 48–55.

Venuprasad, R., Bool, M., Quiatchon, L., Atlin, G., 2012. A QTL for rice grain yield in aerobic environments with large effects in three genetic backgrounds. Theor. Appl. Genet. 124, 323–332.

Venuprasad, R., Impa, S.M., Gowda, R.P.V., Atlin, G.N., Serraj, R., 2011. Rice near-isogenic-lines (NILs) contrasting for grain yield under lowland drought stress. Field Crops Res. 123, 38–46.

Wright, S.I., Vroh, B.I., Schroeder, G., et al., 2005. The effects of artificial selection on the maize genome. Science 308, 1310–1314.

Xiao, J., Li, J., Grandillo, S., et al., 1998. Identification of trait-improving quantitative trait loci alleles from a wild rice relative. Oryza rufipogon. Genetics 150, 899–909.

Xiao, Y.G., Qian, Z.G., Wu, K., et al., 2012. Genetic gains in grain yield and physiological traits of winter wheat in Shandong Province, China, from 1969 to 2006. Crop Sci. 52, 44–56.

Zhang, Q., 2007. Strategies for developing Green Super Rice. Proc. Natl. Acad. Sci. USA 104, 16402–16409.

Zheng, T.C., Zhang, X.K., Yin, G.H., et al., 2011. Genetic gains in grain yield, net photosynthesis and stomatal conductance achieved in Henan Province of China between 1981 and 2008. Field Crops Res. 122, 225–233.

Zhu, Q., Zheng, X., Luo, J., Gaut, B., Ge, S., 2007. Multilocus analysis of nucleotide variation of Oryza sativa and its wild relatives: Severe bottleneck during the domestication of rice. Mol. Biol. Evol. 24, 875–888.

Zhu, T., Mettenburg, K., Peterson, D.J., Tagliani, L., Baszczynski, C., 2000. Engineering herbicide-resistant maize using chimeric RNA/DNA oligonucleotides. Nat. Biotechnol. 18, 555–558.

CHAPTER

19

Integration of biotechnology, plant breeding and crop physiology. Dealing with complex interactions from a physiological perspective

Fernando H. Andrade[1], *Rodrigo G. Sala*[2], *Ana C. Pontaroli*[1], *Alberto León*[3], *Sebastián Castro*[4]

[1]INTA-Universidad de Mar del Plata, CONICET, Argentina
[2]Monsanto, Argentina
[3]Advanta Seeds, Argentina
[4]Universidad de Mar del Plata, Argentina

The beauty of nature lies in details;
the message, in generality
Stephen Jay Gould

1 INTRODUCTION

Agriculture faces the biggest challenge since its emergence ten thousand years ago. It has to meet an increasing demand of food and other products and at the same time reduce environmental impact (Andrade, 2011). To achieve this unprecedented goal, higher crop yields must be obtained in all cultivated environments (Hall and Richards, 2013). Crop physiology, the knowledge of factors and mechanisms that determine crop growth and yield in interaction with the environment, can provide conceptual and practical tools to improve crop management and breeding efficiency.

Crop physiology principles are valuable to design knowledge-intensive and sustainable crop management strategies for specific genotype and environment combinations oriented to (1) a high and sustainable production (Andrade

Crop Physiology. DOI: 10.1016/B978-0-12-417104-6.00019-4

et al., 2005, 2010), (2) increase the productivity of multiple cropping systems (Caviglia and Andrade, 2010), and (3) improved grain quality (Izquierdo et al., 2002, 2009; Cirilo et al., 2011; Martínez et al., 2012). This approach constitutes a low-cost technology that can contribute to match crop demands with the particular environmental offer, and to an efficient use of environmental resources and inputs.

Crop physiology contributions to plant breeding have been less significant. The goal of plant breeding is to produce cultivars with high yield potential and stability across environments (Hallauer, 2007). Although production is still increasing, the rate of crop improvement in most of the major cultivated crops is decreasing (Cassman et al., 2010; Grassini et al., 2013). This reduction in crop improvement by breeding is difficult to revert mainly because of the strong dependence of complex traits such as yield on environmental effects (E) and genotype by environment (G × E) interactions (Blum, 2005).

Despite a handful of successful examples (e.g. Bolaños and Edmeades, 1996; Rebetzke et al., 2002; Hall and Sadras, 2009), crop physiology has often failed to guide breeders to improve crop performance (Sinclair et al., 2004; Sinclair and Purcell, 2005). A number of reasons underlie this apparent underperformance including poorly developed links between disciplines (Miflin, 2000) and the limited availability of rapid, accurate and affordable phenotyping methods applicable to large populations (Berger et al., 2010; Chapter 15).

In the last fifteen years, and particularly in the last five, there has been an explosion in the amount of information concerning the structure of plant genomes. The complete genome sequence of ≈50 plant species, including many of the major crops such as rice and maize (http://www.genomevolution.org), and large collections of genomics resources (ESTs, insertion libraries, mutant populations, mapping populations, germplasm resources) are available (Chapter 18). In parallel, there has been a significant development of technologies and resources in proteomics, metabolomics and, importantly, bioinformatics, to pull all of this information together. In a broad sense, along with gene transfer and manipulation, all these technologies can be enclosed in what is known as biotechnology.

Biotechnology has proved valuable to agriculture in three main aspects: (1) improved tolerance to biotic stresses, such as plagues and diseases (Ramesh et al., 2004; Sharma et al., 2004; Creus et al., 2007; Chapter 18) which has led to a reduction in the use of agrochemicals; (2) resistance to herbicides that favored soil conservation practices and opened opportunities for novel farming systems (Elmore et al., 2001); and (3) improved and diversified quality of agricultural products (Sun et al., 2006; Brookes and Barfoot, 2013; Chapter 17). These achievements are a reality now, partly because the relevant traits have a relatively simple inheritance, involve a few genes integrated in linear cascades or small networks, the G × E interaction is small, there are no major scaling-up issues, and trade-offs, if any, are minor (Benedict et al., 1996; Dunwell, 1998; Struik et al., 2007; Salas et al., 2008).

In contrast, the contribution of biotechnology to the increment in crop yield potential and yield stability has been less evident (Edmeades et al., 2004; Passioura, 2007, 2012; Edmeades, 2013) because of neglect or superficial understanding of trade-offs (Denison, 2012), scale issues in the definition of traits (Chapter 1), the complexity of polygenic traits, and G × E interactions (Austin and Lee, 1998; Mackay, 2001; Campos et al., 2004; Holland, 2007; Deikman et al., 2012).

Considering the vast knowledge gained in crop physiology, breeding and biotechnology, a framework integrating these complementary disciplines could be important to achieve steeper gains in plant breeding (Pontaroli et al., 2012). In this context, this chapter focuses on the contribution of crop physiology to plant breeding and biotechnology. Our aim is to highlight the value of a physiological perspective to the understanding and interpretation of complex genetic and physiological mechanisms with strong interaction with the environment. We discuss the contribution of the discipline in (1) analyzing past achievements of plant breeding, (2) identifying relevant traits for yield potential and yield stability across environments, and (3) disentangling complex G × E and G × G interactions. The chapter also discusses the potential contributions of biotechnology to facilitate crop physiology studies as well as the use of physiological traits in plant breeding. Particularly, we focus on how this field could help to bridge the gap and stimulate creative interactions between both disciplines. To illustrate these assertions, we provide some examples mostly derived from our research over the last years. Complementary perspectives to these issues are presented in Chapter 13, emphasizing the need and value of quantitative environmental characterizations; Chapter 14, updating the application of models for the quantitative assessment of G × E and G × G; Chapter 15, highlighting current phenotyping efforts accounting for both physiological principles and breeding relevance, and Chapter 17, combining modeling and experimental methods to untangle the components of the phenotypic variance of grain quality traits.

2 CONTRIBUTIONS OF CROP PHYSIOLOGY TO PLANT BREEDING AND BIOTECHNOLOGY

Many studies have focused on dissecting the after-the-fact genetic progress observed through decades of breeding (Echarte et al., 2004; Sadras and Lawson 2011; Chapter 2). Fewer studies proposed new physiological traits that breeders could use as selection criteria to increase yield in a target population of environments (Morgan and Condon, 1986; Bolaños and Edmeades, 1996; Abbate et al., 2012; Hall and Richards, 2013; see Chapter 13 for methods to characterize target population of environments). Hall and Sadras (2009) have listed past attempts by crop physiologists to identify and evaluate traits related to yield and drought tolerance of interest for plant breeders. Crop physiology would therefore contribute to (1) characterizing target environments (Chapter 13), (2) assessing the potential traits to improve yield potential or crop adaptation in those environments, (3) identifying relevant secondary traits easy to measure, with high heritability and high correlation with crop performance, and (4) understanding complex interactions among traits and with the environment (Schrader, 1985; Bruce et al., 2002).

In the next sections, we concentrate on the contribution of crop physiology to plant breeding related to (1) the analysis of past achievements in increasing yield potential, yield stability, and resource productivity, (2) the identification of traits that could improve efficiency in the selection of genotypes with high yield potential and adaptation to the target environments, and (3) the disentangling of complex interactions among traits and between traits and environment.

2.1 Analysis of past achievements of plant breeding

Crop physiology concepts contribute to analyzing past achievements of plant breeding in increasing yield potential, yield stability, and resource productivity by identifying mechanisms that have been indirectly affected by the selection process. The rationale is that by understanding realized changes in phenotype resulting from breeding for yield and agronomic adaptation, limits and opportunities for future improvement could be identified (Reynolds et al., 2009).

Comparisons of cultivars released in different eras allow the quantification of the contribution of genetic improvement to crop yield, and for the dissection of the traits involved as is shown in the following example. Breeding consistently increased maize yield potential in Argentina over the past decades. The higher yield of modern hybrids was associated with higher biomass production and harvest index (Andrade et al., 1996; Echarte et al., 2004) and mainly resulted from more grains per unit surface area that increased reproductive sink demand during the effective grain-filling period. Compared with their older counterparts, modern hybrids generally produced more stable yield and harvest index in response to increases in plant density and environmental stress, mainly associated with a greater number of grains fixed per unit of plant growth rate at the critical flowering period (Echarte et al., 2004). Modern hybrids, however, could show lower yield stability in response to source reductions during grain filling because of their high reproductive demand (Echarte et al., 2006; Cerrudo et al., 2013). Modern and old hybrids had similar water use and nitrogen absorption during the vegetative stages but former hybrids absorbed more nitrogen after flowering (Nagore et al., 2010; Robles et al., 2011). Finally, modern hybrids presented higher grain yield per unit of nitrogen uptake and per unit evapotranspiration than their older counterparts, associated with improvement in harvest index (Nagore et al., 2010; Robles et al., 2011;). Chapter 15 (section 4) expands on the changes in maize phenotype resulting for selection for yield, and Chapters 2–5 illustrate the synergies between changes in phenotype through breeding and management practices in contrasting cropping systems. This knowledge of the mechanisms underlying the improved performance of modern hybrids could help to identify new avenues for future improvement.

2.2 Identification of traits for high yield potential and adaptation to the target environments

Five main conditions must be met for traits to be successfully employed in crop breeding: (1) relevance for crop growth and productivity; (2) genetic variability; (3) medium to high heritability; (4) easily monitored; and (5) absence of important agronomic trade-offs (Donald, 1968; Schrader, 1985; Bruce et al., 2002). In this section, we emphasize crop physiology contributions to identifying relevant traits related to crop performance in the field following a top-down (phenotype to gene; Chapman et al., 2002) or bottom-up (gene to phenotype; Ishitani et al., 2004) approach (Chapter 14).

2.2.1 The top-down approach

The top-down approach starts with the identification of consistent phenotypic variation for yield or other relevant traits in the field. It continues with the search of the mechanisms underlying those responses at decreasing levels of complexity (crop, plant, tissue, cell, molecule). Then molecular mapping is used to identify quantitative trait loci (QTL) associated with the trait of interest and, finally, genomics-based methods (Chapter 18) allow selection and cloning of genes/QTL controlling those traits of agronomical interest. The top-down approach is illustrated by the following two examples.

The number of grains per unit of spike chaff weight (i.e. spike fertility factor, SF) explained a great proportion of wheat yield variation among cultivars in a wide range of environments (Abbate et al., 1998; Gonzalez et al., 2011). An appropriate method for fast and simple determination of this trait at maturity using a small sample of individual spikes was developed (Abbate et al., 2012). Then, controlled crosses were made between wheat cultivars with contrasting SF under different mating designs to investigate the genetic and molecular mechanisms underlying the trait. These studies showed that SF is controlled by a few genes, with low $G \times E$ interaction and medium to high heritability (Pontaroli et al., 2012). Molecular marker analysis is being currently carried out in two mapping populations. Several markers co-segregating with SF have been detected, which warrant further investigation (Deperi et al., 2012).

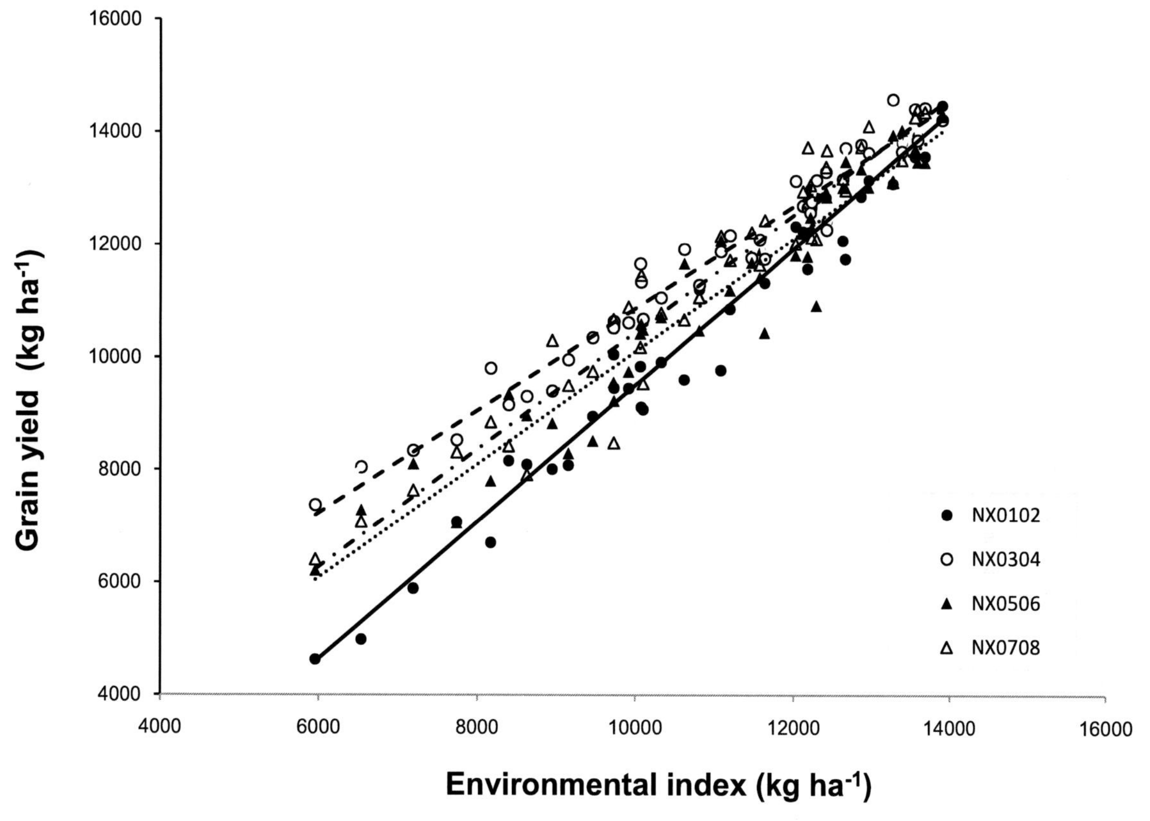

FIG. 19.1 Grain yield of 4 maize hybrids across 42 environments. Each environment was defined as the average yield of 12 hybrids growing under that condition. Fitted linear regressions were $y = -2788.5 + 1.21 \times (R^2 = 0.92)$ for NX0102; and $y = 1786.0 + 0.91 \times (R^2 = 0.96)$ for NX0304. *Source: Castro (2013).*

In a comparison of 12 maize hybrids grown at 42 sites, Castro (2013) identified four maize hybrids of similarly high yield potential and contrasting yield in low-yielding environments leading to contrasting stability (Fig. 19.1). For the two extreme hybrids in this set, he established the underlying physiological mechanisms responsible for the different yield stability in experiments where plant density was used as a surrogate of environmental stress (Andrade et al., 2002). These studies were conducted on the hybrids and their parental lines. At high plant densities, the stable hybrid showed higher grain yield and higher number of grains per unit land area than the unstable hybrid. With respect to secondary traits, the stable hybrid showed greater plant growth rate at flowering and higher number of grains set per unit of ear growth rate during the critical flowering period compared with the unstable hybrid (Fig. 19.2). The same response was observed in their parental lines only for the latter trait. Contrarily, dry matter partitioning to the ear during the flowering period was greater for the unstable hybrid and for one of its inbred lines. As in the previous example, the logical next step would be to map the desired traits and develop fast, effective and easy phenotyping methods to facilitate the selection process. These top-down approaches allowed dissection of a complex trait evaluated

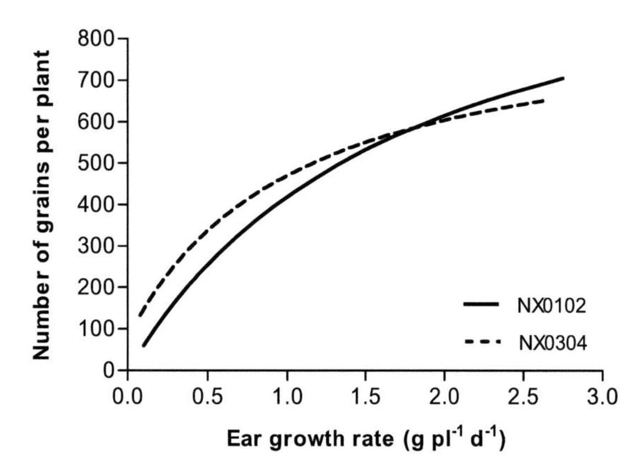

FIG. 19.2 Relationship between number of grains per plant and ear growth rate per plant during the critical period bracketing flowering for two maize hybrids of contrasting yield stability. Fitted hyperbolic equation was $y = \alpha (x - c)/(1 + \beta(x - c))$, where α is the initial slope, β the degree of curvature and c the threshold y value to set grains in the ear. For NX0102, α, β, and c were 662, 0.57 and 0, respectively ($R^2 = 0.85$), for NX0304, α, β, and c were 930, 1.06 and -0.09 respectively ($R^2 = 0.83$). *Source: Castro (2013).*

in a wide range of environments into simpler components under less complex genetic control and with possible complementary effects.

2.2.2 *The bottom-up approach*

In the bottom-up approach, enzymes, metabolites, and transcription factors controlling key biochemical or physiological processes are identified and their genetic control elucidated (Castiglioni et al., 2008; Cattivelli et al., 2008; Gosal et al., 2009; Yang et al., 2010; Grillo et al., 2010). Plant genetic engineering, mutagenesis and marker-assisted selection approaches can be used to develop genotypes with the desired trait (Chapter 18).

The identification of relevant traits or genes associated with crop performance constitutes a huge challenge because of trade-offs, trait attenuation and strong interactions resulting from scaling-up across levels of organization and from complex and redundant regulation of plant systems (Chapman et al., 2002; Sinclair and Purcell, 2005; Passioura, 2012). So, the phenotypic effect of a given trait is generally attenuated when scaling-up across levels of organization because many factors interact and complexity increases. This buffering response may reduce the relevance of many traits to the crop or farming system (Sinclair and Purcell, 2005; Chapter 1).

Molecular genetics has provided, and continues to do so, a large amount of information on QTL; and, to a much lesser extent, on specific genes associated with yield-related traits (Lorenz et al., 2011; Tuberosa et al., 2011; Mastrangelo et al., 2012). Elucidation of gene function could be substantially improved by crop physiology, as it can assist closing the phenotypic gap between the availability of the DNA sequence and gene function related to crop performance in the field (Miflin, 2000; Chapter 15).

2.3 Disentangling complex interactions

Crop physiology also contributes to disentangling the complex interactions between traits and environments and among traits. Relevant examples of the complex interactions that take place at the trait or quantitative trait locus level are presented below to indicate that caution must be taken when interpreting and analyzing the potential of the new available information derived from crop physiology or molecular studies.

2.3.1 *Interactions at the trait level*

Tolerance to water deficits is presented as an example of a complex trait. Effects of water deficit on crop productivity are not easy to predict since they vary with the timing, duration, and intensity of stress, and factors including crop previous history and other environmental aspects. The relevance of a given trait varies with the type of drought. Traits with putative benefits for adaptation to dry environments are not universal and strongly interact with other traits and the environment (Passioura and Angus, 2010; Passioura, 2012; Tardieu, 2012).

Deep roots would be a desirable trait for areas in which substantial amounts of water are left in the soil at physiological maturity and soil stored water can be replenished in the subsequent fallow period (Schwinning and Ehleringer, 2001; Sadras and Rodriguez, 2007). Also, induction of a large resistance to water flow in the plant is a way to save water for later, more critical growth stages, in areas where drought is progressive and becomes more severe towards the end of the growing season (Passioura, 1983; Chapter 11). Similarly, under terminal drought, the most critical component of tolerance in chickpea and pearl millet was the conservative use of water early in the cropping cycle, explained partly by a lower canopy conductance, which resulted in more water available in the soil profile during reproduction (Kholova et al., 2010; Zaman-Allah et al., 2011). Moreover, early flowering would be a favorable trait where crops are exposed to severe terminal water stress, so that the plants can complete their growing cycle without severe stress at the critical stages for grain yield determination (Richards, 1991; Chapman et al., 2003; Debaeke and Aboudrare, 2004).

In environments or seasons with mild or no terminal water stress, however, full season cultivars would be preferred because yield potential is directly associated with maturity group (Capristo et al., 2007). If the crop relies on in-season rainfall, early vigor would increase water use efficiency by reducing soil evaporation (Rebetzke and Richards, 1999; Richards and Lukacs, 2002).

Leaf expansion and stomatal responses to dehydration, traits that show genetic variability (Reymond et al., 2003; Pereyra-Irujo et al., 2008; Sinclair et al., 2010) could have variable value according to the target environment. In environments where crops depend on stored soil water, reduction in leaf area and stomatal conductivity would be desirable to save water for later, more critical reproductive stages. Maintaining high rates of leaf expansion and productivity under stress, however, would be appropriate for short periods of drought during vegetative growth since this strategy would enable the crop to use the incoming radiation and resources more efficiently upon stress relief. Chapter 11 discusses finer aspects of water transport and its implications for growth under favorable or stressful conditions.

2.3.2 Interactions at the QTL or gene level

The following two case studies illustrate physiological interpretations of complex information derived from molecular studies. The first example refers to a genetic linkage analysis of two traits, namely growing degree-days to flowering and photoperiodic response in sunflower families derived from the cross between lines HA89 and ZENB8 (León et al., 2000, 2001). Two QTL, A and B, were highly associated with the photoperiod response that controls thermal time to flowering. Near-isogenic families for those QTL were developed through backcrosses with the aid of molecular markers. This genetic material was used to study the effects of both QTL on sunflower photoperiod response, considering the moment of apex change, i.e. Stage 1.3 in the scale of Mark and Palmer (1981), and its subsequent rate of development (Fonts et al., 2008). The longest time to floral induction and to end of floral differentiation was observed when QTL A was selected homozygous as HA89 parental line and QTL B homozygous as ZENB8 parental line (Fig. 19.3). This previously unnoticed gene × gene interaction is compatible with the web-cascade reaction system proposed for the control of time to flowering and for the transition from vegetative to reproductive stages (Blázquez, 2000; Valverde et al., 2004; Imaizumi and Key, 2006). Additionally, QTL B significantly interacted with photoperiod for both traits (Fonts et al., 2008). Under extended photoperiod, the additive effect of QTL B for phase duration increased, denoting a short-day response. The presence of ZENB8 alleles in QTL B and HA89 alleles in QTL A produced a stronger delay in development up to the end of floral differentiation under long

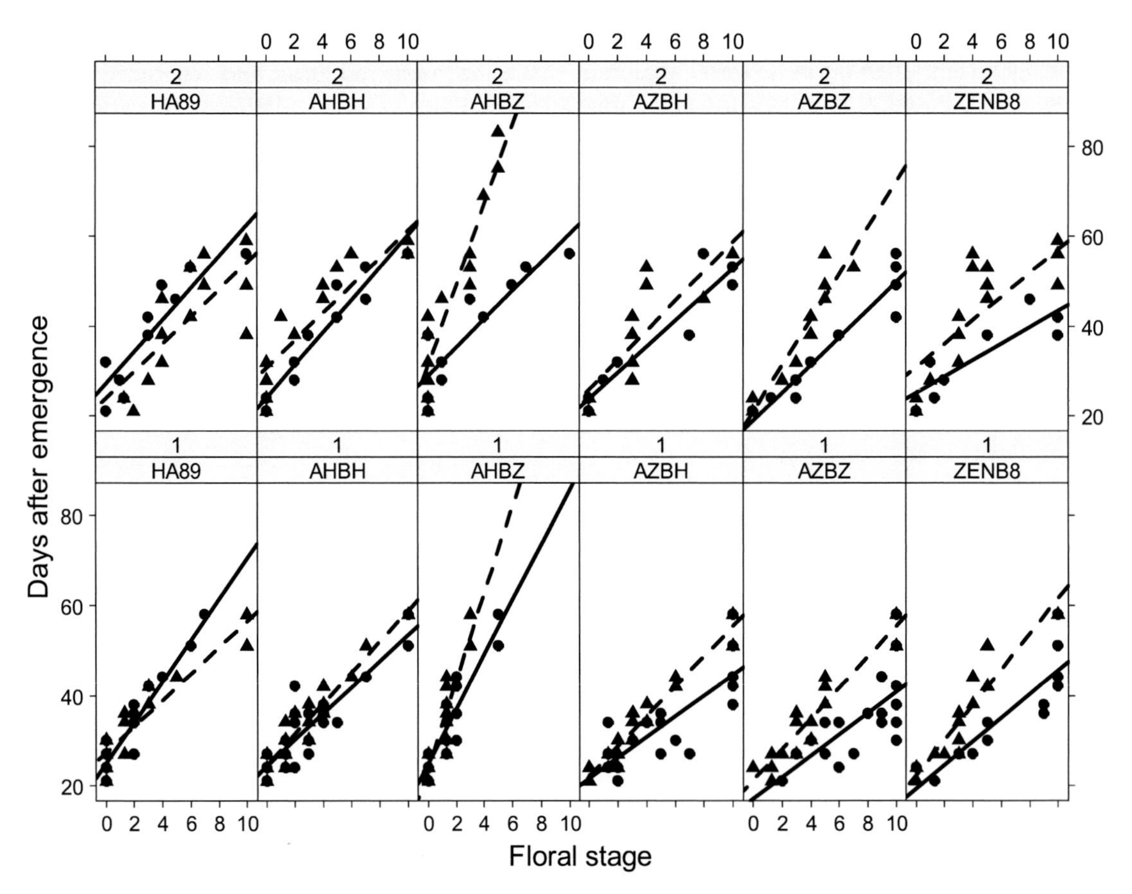

FIG. 19.3 Relationship between floral stage and days after emergence for the near-isogenic families AHBH, AHBZ, AZBH, and AZBZ, and for parental lines HA89 and ZENB8 grown under extended (triangles) and short (circles) photoperiods. Each point indicates an individual observation. Floral stages correspond to the scale of Marc and Palmer (1981). Continuous (short photoperiod) and dashed (extended photoperiod) lines are fitted linear regressions. Upper and lower panels are from replicate experiments (1 and 2). In the near-isogenic families, alleles from HA89 (H) or ZENB8 (Z) in each of two QTLs (A and B) are indicated. *Source: Fonts et al. (2008).*

than under short photoperiods (Fig. 19.3). Thus, a second-degree interaction between QTL A, QTL B and photoperiod was found for rate of development. This higher order interaction is also compatible with the complex web of processes controlling flowering responses to photoperiod, in which a clock-regulated transcription factor is stabilized by light (long days) and activates or suppresses key flowering genes (Turner et al., 2005; Imaizumi and Key, 2006).

A second example deals with the use of molecular markers to locate QTL associated with oil concentration and growing degree days to flowering (DTF) in the same sunflower population grown in a range of latitudes from Fargo (46.8°N) to Venado Tuerto (33.2°S) (León et al., 1995, 2001, 2003). A QTL on linkage group B was associated with both grain oil concentration and DTF. Consistent with the phenotypic correlations, additive effects for higher DTF and lower grain oil concentration in linkage group

TABLE 19.1 Linkage group, position, flanking markers, LOD scores and additive effects of QTL B associated with growing degree-days to flowering and grain-oil concentration in sunflower

Trait	Linkage group	Position (cM)	Left-right locus	Average additive effect	LOD score		LOD score	
					Long photoperiod	Short photoperiod	High latitude	Low latitude
Growing degree-days to flowering	B	64	C1735-C0741	35.2	24.5	1.3		
Grain-oil concentration	B	66	C1735-C0741	−1.0			6.1	0.9

Data obtained from a genetic linkage analysis in families derived from the cross between lines HA89 and ZENB8 (León et al., 2001, 2003). A negative sign for the additive effect means an increase of the mean value of the trait due to HA89 alleles. A positive sign for the additive effect means an increase of the mean value of the trait due to ZEN B8 alleles. HA89 alleles in this QTL decrease the growing degree-days to flowering and increases grain-oil percentage. LOD scores and additive effects of this QTL strongly interact with the environment. Crops were grown in two locations: Fargo at high latitude (46.8°N) and VenadoTuerto at low latitude (33.2°S). Photoperiod was measured at crop emergence.

B were derived from ZENB8 (Table 19.1). Also, these additive effects were largest at high latitudes. The highest LOD scores for this QTL in association with DTF were observed under long photoperiods at crop emergence (Fargo; Table 19.1). The highest LOD score for this QTL in association with grain oil percentage was also observed at Fargo (Table 19.1), the environment with the highest rate of decline in temperature and radiation during grain filling. This environment also showed the most significant phenotypic correlation coefficients between DTF and grain oil concentration ($r = -0.29$; $P < 0.001$). Especially in areas with short growing seasons, such as Fargo, late flowering genotypes (alleles from Zen B8 in QTL on linkage B) were exposed to poor environmental conditions during grain filling and produced less grain oil than early flowering genotypes. This is because low temperature and radiation have negative effects on grain growth rate and grain oil concentration (Andrade and Ferreiro, 1996; Connor and Hall, 1997; Dosio et al., 2000; Chapter 17). Genetic linkage between grain oil concentration and DTF could not be discarded. However, the similarities in QTL position, parental effects, gene action and QTL × E interaction would indicate that the oil concentration effect promoted by QTL B was a consequence of altered phenology (Table 19.1). The modified oil concentration was not a direct effect of the QTL; it was a consequence of trait interaction derived from the phenological changes induced in the plants by that QTL which, in turn, strongly interacted with the environment.

These interactions are magnified and become more difficult to understand when considering more complex traits such as yield potential or yield under restricted water supply (Mackay 2001; Holland, 2007). For example, QTL associated with yield under water deficit are generally cross specific, with strong interaction with the environment and with small individual effects (Campos et al., 2004; Snape et al., 2007; Messmer et al., 2009; Ashraf, 2010; Maccaferri et al., 2011). These QTL or traits could increase, decrease, or be irrelevant to grain yield depending on the environment (Tardieu, 2012; Chapter 13). Accordingly, a considerable difference was observed in gene expression profiles between water-stressed plants in a limited rooting volume compared with those stressed in the field (Habben et al., 2001).

These examples illustrate how confounded conclusions could be reached if interactions between traits and between traits and the environment are omitted. Therefore, applying molecular technologies in breeding such as marker-assisted selection (MAS) and transgenesis may not be completely straightforward and constitutes a continuous challenge (Young, 1999; Xu and Crouch, 2008; Edmeades, 2013). Undoubtedly, conventional breeding, MAS and genetic engineering for higher yield potential and crop adaptation to the target environment can proceed faster if the physiological mechanisms of grain yield determination and their interaction with the environment are better understood.

2.4 Other contributions of crop physiology to plant breeding

Some breeders have been applying useful physiological concepts in their daily work. In maize, breeding programs conducted at high plant densities and evaluation in many sites with different environments rendered some hybrids with high yield stability combined with high yield potential (Fig. 19.1; Castro, 2013; De Santa Eduviges, 2010). This strong G × E interaction for grain yield has a genetic basis and indicates an opportunity for grain yield improvements in medium- and low-yielding environments. This relevant result probably reflects breeders' understanding that high plant density is a surrogate of environmental stress (Andrade et al., 2002).

Weakening of the stem by carbohydrate remobilization induced by a low source–sink ratio during grain filling (Uhart and Andrade, 1991) was also used by breeders as a physiological concept to select by stalk lodging resistance. Moreover, the need to optimize the physiological condition of the crop during the critical periods for grain number determination to obtain high yield (Fischer, 1985; Cantagallo et al., 1997; Andrade et al., 1999; Egli and Bruening, 2005) was another useful concept that assisted breeders in their selection process. Physiological principles can also assist in the development of simple, precise and fast phenotyping techniques which are critical for both top-down and bottom-up approaches (Xu and Crouch, 2008; Berger et al., 2010; Masuka et al., 2012; Pereyra-Irujo et al., 2012). Chapter 15 discusses phenotyping of physiological traits in large breeding populations.

3 CONTRIBUTIONS OF BIOTECHNOLOGY TO PLANT BREEDING AND CROP PHYSIOLOGY

Biotechnology could play an important role in the integration between crop physiologists, breeders and molecular biologists. In the next section, we emphasize how biotechnology could facilitate crop physiology studies as well as the use of physiological traits in plant breeding.

3.1 Biotechnology facilitates the use of key traits in plant breeding

Biotechnology could facilitate two of the five main conditions required for a trait to be successfully employed in crop breeding (section 2.2). First, biotechnology can help generate genetic variability by mutagenesis or transgenesis (Henikoff and Comai, 2003; Castiglioni et al., 2008; Hussain et al., 2011; Edmeades, 2013; Chapter 18). Major crops including wheat, maize, rice, soybean and sunflower are being transformed with techniques such as microprojectile bombardment (or biolistic, Jauhar and Chibbar, 1999), treatments with polyethylene glycol (Funatsuki et al., 1995) and/or manipulation of *Agrobacterium tumefaciens* (Gelvin, 2003). Second, biotechnology can help in trait monitoring by easy screening of the desired traits in large segregant populations, fixing a trait in early generations, or speeding up backcrossed trait introgression by means of molecular markers (Jones et al., 1997; Ashraf, 2010; Xu et al., 2013). Some of the biotechnological tools currently available (genomic libraries, or collections of mutants to name a few) can also assist in the direct transfer of genes from diverse sources in those cases when pre- (Camadro and Peloquin, 1981; Raimondi et al., 2003) and post-fertilization barriers (Johnston et al., 1980; Masuelli and Camadro, 1997) that hinder hybridization and the access to a broader source of genes (Jauhar, 2006), cannot be overcome.

3.2 Contributions of biotechnology to crop physiology

Establishing the association between traits by comparing their respective phenotypic expressions is not always reliable. QTL analysis is a powerful tool for dissecting complex traits, by determining the collocation of QTL related to traits with different levels of organization, or by aiding in the establishment of associations between primary and secondary traits, and between enzymes and their substrates or products. Some examples are the studies of (1) Causse et al. (1995) and Prioul et al. (1997) who used QTL analysis to identify key control factors in carbohydrate metabolism, (2) Reymond et al. (2003) who found that QTL of abscisic acid concentration in the xylem sap co-localized with QTL of leaf growth response to soil water deficit, (3) Sala (2007) who used QTL analysis to study field grain drying rate and its discriminant factors in maize, and (4) Welcker et al. (2007) who found common and independent QTL conferring high leaf elongation rate and short anthesis–silking interval (Chapter 14). More recently, mapping of candidate genes associated with specific traits revealed co-locations between genes with putative functions and QTL. For example, Capelle et al. (2010) co-localized genes and QTL for kernel desiccation and ABA biosynthesis in maize. Uauy et al. (2006) started with a QTL associated with high protein content in wheat grain and arrived at a gene that regulates senescence; the links between crop senescence and grain protein are further discussed in Chapters 8 (section 4.3) and 10 (section 4.8).

Molecular markers and trait dissection could allow identification of key genes and 'design' of superior genotypes with complementation among different traits. Many examples of how QTL and genome analysis could further benefit the integration among disciplines are available in the literature (Simko et al., 1997; Salvi et al., 2002; Cooper et al., 2009; Tuberosa and Salvi, 2009; Landi et al., 2010; Panio et al., 2013).

Technologies from functional genomics (microarrays, ESTs, proteome analyses) to structural genomics (mapping, genome sequencing, synteny and co-linearity with related species) can allow identification of candidate genes that are putatively involved in the expression of diverse traits (Ishitani et al., 2004). The most promising genes could then be assessed for associations with existing genetic information, such as from QTL. After selecting a reasonable number of candidate genes, their expression and phenotypic effect could finally be examined using a number of tools available today, such as TILLING mutants, RNA interference, transgenesis, site-directed mutagenesis, and gene overexpression/silencing (Henikoff and Comai, 2003; Salvi and Tuberosa, 2005; Small, 2007; Voytas, 2013; Chapter 18).

The noise arising from different genetic backgrounds in many traditional crop physiological studies has often been disregarded. The development and use of near-isogenic lines that differ only in the trait of interest (Tanksley and Nelson, 1996; Guo et al., 2007) is useful to study the effects of a particular trait on plant behavior. As an example, Creus et al. (2007) evaluated several pairs of sunflower isohybrids with (R) and without (S) QTL associated with resistance to *Verticillum dahliae* in five locations. They found that (1) disease incidence and severity for all R isohybrids were negligible, (2) grain and oil yield of S isohybrids was nearly 30% less than those of their R counterparts in the most severe disease conditions, and (3) R and S isohybrids presented similar yield under low disease severity. Fonts et al. (2008) used molecular markers and backcrosses to develop and study sunflower near-isogenic lines for the two most important QTL (A and B) associated with growing degree-days to flowering (section 2.4.2). The use of such genetic materials allowed determination of QTL effects for different isohybrids or isolines independently of the genetic background.

Another alternative to reduce the background effect is to pool all the results from QTL analysis into a consensus map. Recently, Sala et al. (2012) used meta-analysis to identify hot-spot regions that were consistently associated with maize grain moisture at harvest across a range of genetic backgrounds for both inbreds and test-cross progenies. This approach reduces the need to map repeatedly individual crosses for a given trait (Tuberosa et al., 2002; Edmeades et al., 2004) and increases the power of QTL detection.

Crop physiology should exploit the numerous tools and pieces of information that are continuously generated by genomics-oriented studies. Molecularly characterized populations and isogenic lines derived from such studies should constitute the main source of genetic material for investigating physiological traits. Along with the use of available information on molecular markers, candidate genes and associated secondary traits in the target crop and related species, this approach constitutes an avenue in the investigation of those mechanisms responsible for crop behavior that remain unknown or poorly understood.

4 CONCLUSIONS

This chapter presents our view on the synergy that can arise at the interface between crop physiology, plant breeding and biotechnology. The most notable examples of connection of transgenic or other biotechnological approaches with field performance come from initiatives which allocate enormous amounts of human and economic resources to multidisciplinary teams (Passioura, 2007). A major challenge and a limiting factor for the success of the interdisciplinary approach is the identification of the main traits that confer yield potential and yield stability in different environments. In this context, crop physiology is crucial to understand and extrapolate processes and mechanisms occurring at different levels of organization, and to explain and predict the complex interactions between relevant traits and/or between traits and the environment. Moreover, the interaction with biotechnology could provide novel and powerful tools that can help to increase the depth and accuracy of crop physiological studies for breeding purposes. Evidence is emerging to show that this type of integrated multidisciplinary framework reinforces and contributes to higher and more stable yield.

References

Abbate, P., Andrade, F., Lazaro, L., et al., 1998. Grain yield increase in modern Argentinean wheat cultivars. Crop Sci. 38, 1203–1209.

Abbate, P., Pontaroli, A., Lázaro, L., Gutheim, F., 2012. A method of screening for spike fertility in wheat. J. Agric. Sci. (Camb.) available on CJO2012 doi:10.1017/S0021859612000068.

Andrade, F., 2011. Latecnología y la producción agrícola. El pasado y los actuales de safíos. Ediciones INTA.

Andrade, F.H., Ferreiro, M., 1996. Reproductive growth of maize, sunflower and soybean at different source levels during grain filling. Field Crops Res. 48, 155–165.

Andrade, F., Vega, C., Uhart, S., Cirilo, A., Cantarero, M., Valentinuz, O., 1999. Kernel number determination in maize. Crop Sci. 39, 453–459.

Andrade, F., Echarte, L., Rizzalli, R., Dellamaggiora, A., Casanovas, M., 2002. Kernel number prediction in maize under nitrogen or water stress. Crop Sci. 42, 1173–1179.

Andrade, F.H., Sadras, V.O., Vega, C.R., Echarte, L., 2005. Physiological determinants of crop growth and yield in maize, sunflower and soybean. Applications to crop management, modeling and breeding. J. Crop Improve. 14, 51–101.

Andrade, F.H., Abbate, P., Otegui, M., Cirilo, A., Cerrudo, A., 2010. Ecophysiological basis for crop management. Am. J. Plant Sci. Biotechnol. 4, 23–34.

Ashraf, M., 2010. Inducing drought tolerance in plants: recent advances. Biotechnol. Adv. 28, 169–183.

Austin, D.F., Lee, M., 1998. Detection of quantitative trait loci for grain yield and yield components in maize across

generations in stress and nonstress environments. Crop Sci. 38, 1296–1308.
Benedict, J.H., Sachs, E.S., Altman, D.W., et al., 1996. Field performance of cotton expressing transgenic CryIA insecticidal proteins for resistance to Heliothis virescens and *Helicover pazea* (Lepidoptera: Noctuidae). J. Econ. Entomol. 89, 230–238.
Berger, B., Parent, B., Tester, M., 2010. High-throughput shoot imaging to study drought responses. J. Exp. Bot. 61, 3519–3528.
Blázquez, M., 2000. Flower development pathways. J. Cell Sci. 113, 3547–3548.
Blum, A., 2005. Drought resistance, water-use efficiency, and yield potential – are they compatible, dissonant, or mutually exclusive? Aust. J. Agric. Res. 56, 1159–1168.
Bolaños, J., Edmeades, G., 1996. The importance of the anthesis-silking interval in breeding for drought tolerance in tropical maize. Field Crops Res. 48, 65–80.
Brookes, G., Barfoot, P., 2013. GM crops: global socio-economic and environmental impacts 1996-2011. PG Economics Ltd, Dorchester, UK.
Bruce, W.B., Edmeades, G.O., Barker, T.C., 2002. Molecular and physiological approaches to maize improvement to drought tolerance. J. Exp. Bot. 53, 13–25.
Camadro, E.L., Peloquin, S.J., 1981. Cross-incompatibility between two sympatric polyploid Solanum species. Theor. Appl. Genet. 60, 65–70.
Campos, H., Cooper, M., Habben, J.E., Edmeades, G.O., Schussler, J.R., 2004. Improving drought tolerance in maize: a view from industry. Field Crops Res. 90, 19–34.
Cantagallo, J., Chimenti, C., Hall, A., 1997. Number of seeds per unit area in sunflower correlates well with a photothermal quotient. Crop Sci. 37, 1780–1786.
Capelle, V., Remoué, C., Moreau, L., et al., 2010. QTLs and candidate genes for desiccation and abscisic acid content in maize kernels. BMC Plant Biol. 10, 2.
Capristo, P., Rizzalli, R., Andrade, F., 2007. Ecophysiological yield components of maize hybrids with contrasting maturity. Agron. J. 99, 1111–1118.
Cassman, K., Grassini, P., van Wart, J., 2010. Crop yield potential, yield trends, and global food security in a changing climate. In: Hillel, D., Rosenzweig, C. (Eds.), Handbook of climate change and agroecosystems. Imperial College Press, London.
Castiglioni, P., Warner, D., Bensen, R., et al., 2008. Bacterial RNA chaperones confer abiotic stress tolerance in plants and improved grain yield in maize under water-limited conditions. Plant Physiol. 147, 446–455.
Castro, S., 2013. Estabilidad de rendimiento y mecanismo seco físiologícos sociados con la inflacion de granos en hibridos de maiz y en suslineas parentales. Tesis de Maestría FCA, UNMP.
Cattivelli, L., Rizza, F., Badeck, F., Mazzucotelli, E., Mastrangelo, A., Francia, E., 2008. Drought tolerance improvement in crop plants: an integrated view from breeding to genomics. Field Crops Res. 105, 1–14.
Causse, M., Rocher, J., Henry, A., Charcosset, A., Prioul, J., de Vienne, D., 1995. Genetic detection of the relationship between carbon metabolism and early growth in maize with emphasis on key-enzyme loci. Mol. Breed. 1, 259–272.
Caviglia, O., Andrade, F., 2010. Sustainable intensification of agriculture in the Argentinean Pampas: Capture and use efficiency of environmental resources. Am. J. Plant Sci. Biotechnol. 3, 1–8.
Cerrudo, A., Fernandez, E., Di Matteo, J., Robles, M., Andrade, F., 2013. Critical period for yield determination in maize. Crop Past. Sci. 64, 580–587.
Chapman, S.C., Cooper, M., Podlich, D.W., Hammer, G.L., 2003. Evaluating plant breeding strategies by simulating gene action and dryland environments effects. Agron. J. 95, 99–113.
Chapman, S.C., Hammer, G.L., Podlich, D.W., Cooper, M., 2002. Linking biophysical and genetic models to integrate physiology, molecular biology and plant breeding. In: Kang, M.S. (Ed.), Quantitative genetics, genomics and plant breeding. CAB International, Wallingford, pp. 167–187.
Cirilo, A.G., Actis, M., Andrade, F., Valentinuz, O., 2011. Crop management affects dry-milling quality of flint maize kernels. Field Crops Res. 122, 140–150.
Connor, D.J., Hall, A.J., 1997. Sunflower physiology. In: Schneiter, A.A. (Ed.), Sunflower technology and production. ASA, CSSA, and SSSA, Madison, WI, pp. 113–182.
Cooper, M., van Eeuwijk, F., Hammer, G., Podlich, D., Messina, C., 2009. Modeling QTL for complex traits: detection and context for plant breeding. Curr. Opin. Plant Biol. 12, 231–240.
Creus, C., Bazzalo, M.E., Grondona, M., Andrade, F.H., León, A.J., 2007. Disease expression and ecophysiological yield components in sunflower isohybrids with and without *Verticillium dahliae* resistance. Crop Sci. 47, 703–710.
De Santa Eduviges, J. M., 2010. Potencial de rendimiento y tolerancia a sequía en híbridos de maíz. Tesis Magister Scientiae, Facultad de CienciasAgrarias, Universidad Nacional de Mar del Plata: Balcarce. Argentina.
Debaeke, P., Aboudrare, A., 2004. Adaptation of crop management to water-limited environments. Eur. J. Agron. 21, 433–446.
Deikman, J., Petracek, M., Heard, J., 2012. Drought tolerance through biotechnology: improving translation from the laboratory to farmers' fields. Curr. Opin. Biotechnol. 23, 243–250.
Denison, R.F., 2012. Darwinian agriculture: how understanding evolution can improve agriculture. Princeton University Press, Princeton.
Deperi, S.I., Alonso, M.P., Woyann, L.G., Pontaroli, A.C., 2012. Detección de marcadores moleculares asociados a

la fertilidad de la espiga de trigo pan. XIV Latin American Genetics Congress (Rosario, Argentina, October 28-31, 2012).

Donald, C.M., 1968. The breeding of crop ideotypes. Euphytica 17, 385–403.

Dosio, G.A., Aguirrezábal, L., Andrade, F.H., Pereyra, V.R., 2000. Solar radiation intercepted during seed filling and oil production in two sunflower hybrids. Crop Sci. 40, 1637–1644.

Dunwell, J.M., 1998. Novel food products from genetically modified crop plants: methods and future prospects. Int. J. Food Sci. Technol. 33, 205–213.

Echarte, L., Andrade, F., Vega, C., Tollenaar, M., 2004. Kernel number determination in Argentinean maize hybrids released between 1965 and 1993. Crop Sci. 44, 1654–1661.

Echarte, L., Andrade, F., Sadras, V., Abbate, P., 2006. Kernel weight and its response to source manipulations during grain filling in Argentinean maize hybrids released in different decades. Field Crops Res. 96, 301–312.

Edmeades, G., 2013. Progress in achieving and delivering drought tolerance in maize. An update. ISAAA, Ithaca, New York.

Edmeades, G.O., McMaster, G.S., White, J.W., Campos, H., 2004. Genomics and the physiologist: bridging the gap between genes and crop response. Field Crops Res. 90, 5–18.

Egli, D., Bruening, W., 2005. Shade and temporal distribution of pod production and pod set in soybean. Crop Sci. 45, 1764–1769.

Elmore, R.W., Roeth, F.W., Klein, R.N., et al., 2001. Glyphosate-resistant soybean cultivar response to glyphosate. Agron. J. 93, 404–407.

Fischer, R.A., 1985. Number of kernels in wheat crops and the influence of solar radiation and temperature. J. Agric. Sci. 105, 447–461.

Fonts, C., Andrade, F.H., Grondona, M., Hall, A.J., León, A.J., 2008. Phenological characterization of near-isogenic sunflower families bearing two QTL for photoperiodic response. Crop Sci. 48, 1579–1585.

Funatsuki, H., Kuroda, M., Lazzeri, P.A., Muller, E., Lorz, H., Kishinami, I., 1995. Fertile transgenic barley generated by direct transfer to protoplasts. Theor. Appl. Genet. 91, 707–712.

Gelvin, S.B., 2003. Agrobacterium-mediated plant transformation: The biology behind the "gene-jockeying" tool. Microbiol. Mol. Biol. Rev. 67, 16–37.

Gonzalez, F., Terrile, I., Falcon, M., 2011. Spike fertility and duration of stem elongation as promising traits to improve potential grain number (and yield): variation in modern Argentinean wheats. Crop Sci. 51, 1693–1702.

Gosal, S., Wani, S., Kang, M., 2009. Biotechnology and drought tolerance. J. Crop. Improve. 23, 19–54.

Grassini, P., Eskridge, K.M., Cassman, K.G., 2013. Distinguishing between yield advances and yield plateaus in historical crop production trends. Nat. Commun. 4 doi:10.1038/ncomms3918..

Grillo, S., Blanco, A., Cattivelli, L., Coraggio, I., Leone, A., Salvi, S., 2010. Plant genetic and molecular responses to water deficit. Ital. J. Agron. 1, 617–638.

Guo, P., Bai, G., Carver, R., Li, A., Bernardo, R., Baum, M., 2007. Transcriptional analysis between two wheat near-isogenic lines contrasting in aluminium tolerance under aluminium stress. Mol. Genet. Genomics 277, 1–12.

Habben, J., Zinselmeier, C., Sun, Y., et al., 2001. Effect of stress on gene expression profiles of corn reproductive tissues. ASA-CSSA-SSSA, Madison, WI, Abstracts of the CSSA. CDROM.

Hall, A.J., Richards, R.A., 2013. Prognosis for genetic improvement of yield potential and water-limited yield of major grain crops. Field Crops Res. 143, 18–33.

Hall, A., Sadras, V., 2009. Whither crop physiology? In: Sadras, V., Calderini, D. (Eds.), Crop physiology: applications for genetic improvement and agronomy. Macmillian Publishing Solutions.

Hallauer, A.R., 2007. History, contribution, and future of quantitative genetics in plant breeding: Lessons from maize. Crop Sci. 47, S4–S19.

Henikoff, S., Comai, L., 2003. Single nucleotide mutations for plant functional genomics. Annu. Rev. Plant Biol. 54, 375–401.

Holland, J.B., 2007. Genetic architecture of complex traits in plants. Curr. Opin. Plant Biol. 10, 156–161.

Hussain, S., Kayani, M., Amjad, M., 2011. Transcription factors as tools to engineer enhanced drought stress tolerance in plants. Biotechnol. Prog. 27, 297–306.

Imaizumi, T., Key, S., 2006. Photoperiodic control of flowering: not only by coincidence. Trends Plant Sci. 11, 550–558.

Ishitani, M., Rao, I., Wenzl, P., Beebe, S., Tohme, J., 2004. Integration of genomics approach with traditional breeding towards improving abiotic stress adaptation: drought and aluminium toxicity as case studies. Field Crops Res. 90, 35–45.

Izquierdo, N., Aguirrezábal, L., Andrade, F., Geroudet, C., Pereyra Iraola, M., Valentinuz, O., 2009. Intercepted solar radiation affects oil fatty acid composition in crop species. Field Crops Res. 114, 66–74.

Izquierdo, N., Aguirrezábal, L., Andrade, F., Pereyra, V., 2002. Night temperature affects fatty acid composition in sunflower oil depending on the hybrid and the phenological stage. Field Crops Res. 77, 115–126.

Jauhar, P.P., 2006. Modern biotechnology as an integral supplement to conventional plant breeding: The prospects and challenges. Crop Sci. 46, 1841–1859.

Jauhar, P.P., Chibbar, R.N., 1999. Chromosome-mediated and direct gene transfers in wheat. Genome 42, 570–583.

Johnston, S.A., den Nijs, T.P.M., Peloquin, S.J., Hanneman, Jr., R.E., 1980. The significance of genic balance to endosperm development in interspecific crosses. Theor. Appl. Genet. 56, 293–297.

Jones, N., Ougham, H., Thomas, H., 1997. Markers and mapping: we are all geneticists now. New Phytol. 137, 165–177.

Kholova, J., Hash, C., Kakkera, A., Kocova, M., Vadez, V., 2010. Constitutive water-conserving mechanisms are correlated with the terminal drought tolerance of pearl millet. J. Exp. Bot. 61, 369–377.

Landi, P., Giuliani, S., Salvi, S., Ferri, M., Tuberosa, R., Sanguineti, M., 2010. Characterization of root-yield-1.06, a major constitutive QTL for root and agronomic traits in maize across water regimes. J. Exp. Bot. 61, 3553–3562.

León, A.J., Andrade, F.H., Lee, M., 2000. Genetic mapping of factors affecting quantitative variation for flowering in sunflower (*Helianthus annuus* L.). Crop Sci. 40, 404–407.

León, A.J., Andrade, F.H., Lee, M., 2003. Genetic analysis of seed-oil concentration across generations and environments in sunflower (*Helianthus annuus* L.). Crop Sci. 43, 135–140.

León, A.J., Lee, M., Andrade, F.H., 2001. Quantitative trait loci for growing degree days to flowering and photoperiod response in sunflower (*Helianthus annuus* L.). Theor. Appl. Genet. 102, 497–503.

León, A.J., Lee, M., Rufener, G.K., Berry, S.T., Mowers, R.P., 1995. Use of RFLP markers for genetic linkage of oil percentage in sunflower seed (*Helianthus annuus*). Crop Sci. 35, 558–564.

Lorenz, A.J., Chao, S., Asoro, F., et al., 2011. Genomic selection in plant breeding: knowledge and prospects. Adv. Agron. 110, 77–123.

Maccaferri, M., Sanguineti, M., Demontis, A., et al., 2011. Association mapping in durum wheat grown across a broad range of water regimes. J. Exp. Bot. 62, 409–438.

Mackay, T., 2001. The genetic architecture of quantitative traits. Annu. Rev. Genet. 35, 303–339.

Marc, J., Palmer, J.H., 1981. Photoperiodic sensitivity of inflorescence initiation and development in sunflower. Field Crops Res. 4, 155–164.

Martínez, R., Izquierdo, N., González Belo, R., Aguirrezábal, L., Andrade, F., Reid, R., 2012. Oil yield components and oil quality of high stearic high oleic sunflower genotypes as affected by intercepted solar radiation during grain filling. Crop Past. Sci. 63, 330–337.

Mastrangelo, A., Mazzucotelli, E., Guerra, D., De Vita, P., Cattivelli, L., 2012. Improvement of drought resistance in crops: from conventional breeding to genomic selection. In: Wenkateswarlu, B., et al. (Eds.), Crop stress and its management: perspectives and strategies. Springer, Netherlands, pp. 225–259.

Masuelli, R.W., Camadro, E.L., 1997. Crossability relationships among wild potato species with different ploidies and endosperm balance numbers (EBN). Euphytica 94, 227–235.

Masuka, B., Araus, J., Das, B., Sonder, K., Cairns, J., 2012. Phenotyping for abiotic stress tolerance in maize. J. Integ. Plant Biol. 54, 238–249.

Messmer, R., Fracheboud, Y., Bänziger, M., Vargas, M., Stamp, P., Ribaut, J., 2009. Drought stress and tropical maize: QTL-by-environment interactions and stability of QTLs across environments for yield components and secondary traits. Theor. Appl. Genet. 119, 913–930.

Miflin, B., 2000. Crop improvement in the 21st century. J. Exp. Bot. 51, 1–8.

Morgan, J., Condon, A., 1986. Water use, grain yield and osmoregulation in wheat. Aust. J. Plant Physiol. 13, 523–532.

Nagore, M., Echarte, L., Della Maggiora, A., Andrade, F., 2010. Rendimiento, consumo y eficiencia de uso del agua del cultivo de maíz bajoes trés hídrico. Actas IX Congreso Nacional de Maíz, Simposio Nacional de Sorgo, 107-109. 17 al 19 de Noviembre de 2010, Rosario, Buenos Aires.

Panio, G., Motzo, R., Mastrangelo, A., et al., 2013. Molecular mapping of stomatal conductance related traits in durum wheat (*Triticumturgidum* ssp. *durum*). Ann. Appl. Biol. 162, 258–270.

Passioura, J.B., 1983. Roots and drought resistance. Agric. Water Manag. 7, 265–280.

Passioura, J.B., 2007. The drought environment: physical, biological and agricultural perspectives. J. Exp. Bot. 58, 113–117.

Passioura, J.B., 2012. Phenotyping for drought tolerance in grain crops: when is it useful to breeders? Funct. Plant Biol. 39, 851–859.

Passioura, J.B., Angus, J., 2010. Improving productivity of crops in water-limited environments. Adv.Agron. 106, 37–75.

Pereyra-Irujo, G., Gasco, E., Peirone, L.,. Aguirrezábal, L., 2012. GlyPh: a low-cost platform for phenotyping plant growth and water use. Funct. Plant Biol. 39, 905–913.

Pereyra-Irujo, G.A., Velázquez, L., Lechner, L., Aguirrezábal, L., 2008. Genetic variability for leaf growth rate and duration under water deficit in sunflower: Analysis of responses at cell, organ, and plant level. J. Exp. Bot. 59, 2221–2232.

Pontaroli, A.C., 2012. How can we foster crop improvement? J. Basic Appl. Genet. 23, 4–6, Available online at http://www.scielo.org.ar/scielo.php?script=sci.

Pontaroli, A.C., Abbate, P.E., Lázaro, L., et al., 2012. Wheat spike fertility: genetics and breeding applications of a key ecophysiological yield component. Workshop 'Applications of physiology in crop management and breeding'; 6th International Crop Science Congress, Bento Gonçalves, Brazil, August 6–10, 2012.

Prioul, J., Quarrie, S., Causse, M., de Vienne, D., 1997. Dissecting complex physiological functions through the use of molecular quantitative genetics. J. Exp. Bot. 48, 1151–1163.

Raimondi, J.P., Sala, R.G., Camadro, E.L., 2003. Crossability relationships among the wild diploid potato species *Solanum kurtzianum, S. chacoense* and *S. ruiz-lealii* from Argentina. Euphytica 132, 287–295.

Ramesh, S., Nagadhara, D., Reddy, V.D., Rao, K.V., 2004. Production of transgenic indica rice resistant to yellow stem borer and sapsucking insects, using super-binary vectors of *Agrobacterium tumefaciens*. Plant Sci. 166, 1077–1085.

Rebetzke, G.J., Condon, A.G., Richards, R.A., Farquhar, G.D., 2002. Selection for reduced carbon isotope discrimination increases aerial biomass and grain yield of rainfed bread wheat. Crop Sci. 42, 39–745.

Rebetzke, G.J., Richards, R.A., 1999. Genetic improvement of early vigour in wheat. Aust. J. Agric. Res. 50, 291–302.

Reymond, M., Muller, B., Leónardi, A., Charcosset, A., Tardieu, F., 2003. Combining quantitative trait loci analysis and an ecophysiological model to analyse the genetic variability of the responses of maize leaf growth to temperature and water deficit. Plant Physiol. 131, 664–675.

Reynolds, M., Foulkes, M., Slafer, G., et al., 2009. Rising yield potential in wheat. J. Exp. Bot. 60, 1899–1918.

Richards, R.A., 1991. Crop improvement for temperate Australia – Future opportunities. Field Crops Res. 26, 141–169.

Richards, R.A., Lukacs, Z., 2002. Seedling vigour in wheat – Sources of variation for genetic and agronomic improvement. Aust. J. Agric. Res. 53, 41–50.

Robles, M., Cerrudo, A., Di Matteo, J., Barbieri, P., Rizzalli, R., Andrade, F., 2011. Nitrogen use efficiency of maize hybrids released in different decades. ASA Congress, USA, 2011.

Sadras, V.O., Lawson, C., 2011. Genetic gain in yield and associated changes in phenotype, trait plasticity and competitive ability of South Australian wheat varieties released between 1958 and 2007. Crop Past. Sci. 62, 533–549.

Sadras, V.O., Rodriguez, D., 2007. The limit to wheat water use efficiency in eastern Australia. II. Influence of rainfall patterns. Aust. J. Agric. Res. 58, 657–669.

Sala, R., Andrade, F., Cerono, J., 2012. Quantitative trait loci associated with grain moisture at harvest for line *per se* and testcross performance in maize: a meta-analysis. Euphytica 185, 429–440.

Sala, R.G., 2007. Morpho-phenological traits associated with grain moisture at harvest in maize (*Zea mays* L.). Genetic linkage analysis. Ph.D. thesis, Universidad Nacional de Mar del Plata, Balcarce, Argentina.

Salas, J., Youssar, L., Martínez-Force, E., Garcés, R., 2008. The biochemical characterization of a high-stearic acid sunflower mutant reveals the coordinated regulation of stearoyl-acyl carrier protein desaturases. Plant Physiol. Biochem. 46, 109–116.

Salvi, S., Tuberosa, R., 2005. To clone or not to clone plant QTLs: present and future challenges. Trends Plant Sci. 10, 297–304.

Salvi, S., Tuberosa, R., Chiapparino, E., et al., 2002. Toward positional cloning of Vgt1, a QTL controlling the transition from the vegetative to the reproductive phase in maize. Plant Mol. Biol. 48, 601–613.

Schrader, L., 1985. Selection for metabolic balance in maize. In: Harper, J., Schraderand, L., Howell, R. (Eds.), Exploitation of physiological and genetic variability to enhance crop productivity. American Society of Plant Physiologists, Rockville, MD, pp. 79–89.

Schwinning, S., Ehleringer, J.R., 2001. Water-use trade-offs and optimal adaptations to pulse-driven arid ecosystems. J. Ecol. 89, 464–480.

Sharma, H.C., Sharma, K., Crouch, J., 2004. Genetic transformation of crops for insect resistance: Potential and limitations. Crit. Rev. Plant Sci. 23, 47–72.

Simko, I., McMurry, S., Yang, H.M., Manschot, A., Davies, P.J., Ewing, E.E., 1997. Evidence from polygene mapping for a causal relationship between potato tuber dormancy and abscisic acid content. Plant Physiol. 115, 1453–1459.

Sinclair, T., Messina, C., Beatty, A., Samples, M., 2010. Assessment across the United States of the benefits of altered soybean drought traits. Agron. J. 102, 475–482.

Sinclair, T.R., Purcell, L.C., 2005. Is a physiological perspective relevant in a 'genocentric' age? J. Exp. Bot. 56, 2777–2782.

Sinclair, T.R., Purcell, L.C., Sneller, C.H., 2004. Crop transformation and the challenge to increase yield potential. Trends Plant Sci. 9, 70–75.

Small, I., 2007. RNAi for revealing and engineering plant gene functions. Curr. Opin. Biotechnol. 18, 148–153.

Snape, J., Foulkes, M., Simmonds, J., et al., 2007. Dissecting gene × environmental effects on wheat yields via QTL and physiological analysis. Euphytica 154, 401–408.

Struik, P.C., Cassman, K.G., Koorneef, M., 2007. A dialogue on interdisciplinary collaboration to bridge the gap between plant genomics and crop science. In: Spiertz, J.H.J., Struik, P.C., Van Laar, H.H. (Eds.), Scale and complexity in plant systems research: gene-plant-crop relations. Springer, pp. 319–328.

Sun, S.S.M., Liu, Q., Chan, R.M.L., 2006. Genetic engineering of crops for improved nutritional quality. In: Xu, Z., Li, J., Xue, Y., Yang, W., (Eds), Biotechnology and sustainable agriculture 2006 and beyond. Proceedings of the 11th IAPTC&B Congress, Beijing, China, pp. 283-287.

Tanksley, D.S., Nelson, C.J., 1996. Advanced backcross QTL analysis: a method for the simultaneous discovery and transfer of valuable QTL from unadapted germplasm into elite breeding lines. Theor. Appl. Genet. 92, 191–203.

Tardieu, F., 2012. Any trait or trait-related allele can confer drought tolerance: just design the right drought scenario. J. Exp. Bot. 63, 25–31.

Tuberosa, R., Salvi, S., 2009. QTL for agronomic traits in maize production. In: Bennetzen, J.L., Hake, S.C. (Eds.), Handbook of maize: its biology. Springer, pp. 501–541.

Tuberosa, R., Salvi, S., Giuliani, S., et al., 2011. Genomics of root architecture and functions in maize. In: Costa de Oliveira, A., Varshney, R. (Eds.), Root genomics. Springer, pp. 179-204.

Tuberosa, R., Salvi, S., Sanguineti, M.C., Landi, P., Maccaferri, M., Conti, S., 2002. Mapping QTLs regulating morpho-physiological traits and yield: case studies, shortcomings and perspectives in drought-stressed maize. Ann. Bot. 89, 941–963.

Turner, A., Beales, J., Faure, S., Dunford, R., Laurie, D., 2005. The pseudo-response regulator Ppd-H1 provides adaptation to photoperiod in barley. Science 310, 1031–1034.

Uauy, C., Distelfeld, A., Fahima, T., Blechl, A., Dubcovsky, J., 2006. A NAC gene regulating senescence improves grain protein, zinc and iron content in wheat. Science 314, 1298–1301.

Uhart, S., Andrade, F., 1991. Source-sink relationship in maize grown in a cool temperate area. Agronomie 11, 863–875.

Valverde, F., Mouradov, A., Soppe, W., Ravenscroft, D., Samach, A., Coupland, G., 2004. Photoreceptor regulation of constans protein in photoperiodic flowering. Science 303, 1003–1006.

Voytas, D., 2013. Plant genome engineering with sequence-specific nucleases. Annu. Rev. Plant Biol. 64, 327–350.

Welcker, C., Boussuge, B., Bencivenni, C., Ribaut, J., Tardieu, F., 2007. Are source and sink strengths genetically linked in maize plants subjected to water deficit? A QTL study of the responses of leaf growth and of anthesis-silking interval. J. Exp. Bot. 58, 339–349.

Xu, Y., Crouch, J.H., 2008. Marker-assisted selection in plant breeding: From publications to practice. Crop Sci. 48, 391–407.

Xu, Y., Xie, C., Wan, J., He, Z., Prasanna, B., 2013. Marker-assisted selection in cereals: platforms, strategies and examples. In: Gupta, P., Varshney, R. (Eds.), Cereal genomics II. Springer, pp. 375–411.

Yang, S., Vanderbeld, B., Wan, J., Huang, Y., 2010. Narrowing down the targets: towards successful genetic engineering of drought-tolerant crops. Mol. Plant 3, 469–490.

Young, N.D., 1999. A cautiously optimistic vision for marker-assisted breeding. Mol. Breed. 3, 505–510.

Zaman-Allah, M., Jenkinson, D., Vadez, V., 2011. A conservative pattern of water use, rather than deep or profuse rooting, is critical for the terminal drought tolerance of chickpea. J. Exp. Bot. 62, 4239–4252.

CHAPTER

20

Crop modeling for climate change impact and adaptation

Senthold Asseng[1], *Yan Zhu*[2], *Enli Wang*[3], *Weijian Zhang*[4]

[1]University of Florida, Gainesville, USA
[2]Nanjing Agricultural University, China
[3]CSIRO, Australia
[4]Institute of Crop Sciences, Chinese Academy of Agricultural Sciences, Beijing, China

1 INTRODUCTION

The global climate is changing and agriculture will have to adapt to ensure sustainability and survival. Due to the complexity of both agricultural systems and climate change, crop models are often used to understand the impact of climate change on agriculture and to assist in the development of adaptation strategies. Crop models integrate the understanding of crop physiology gathered from many years of laboratory and field experimentations and therefore provide an effective means for investigating crop responses to climate change and alternative management scenarios. This chapter describes the application of crop physiology to climate change impact and adaptation research through the use of physiology-based crop models. Agriculture will also have to contribute to the mitigation of climate change, with crop models playing a role. However, mitigation, including aspects of soil carbon sequestration and greenhouse gases emissions (GHAs), is not considered in this chapter. Likewise, other important agricultural issues related to climate change such as land-use patterns, high-level trophic relations (e.g. crop–insect, crop–pathogen), plant-to-plant competition (crop–weed, plant–plant in mixed cropping or pastures) (Chakraborty et al., 1998; Betts et al., 2007; Bhattacharyya et al., 2007; Cerri et al., 2007; Falloon et al., 2007; Kamoni et al., 2007) will receive little or no attention in this chapter.

This chapter focuses on crop models which consider the various aspects of climate change as drivers (including rainfall, atmospheric CO_2, temperature and ozone) and capture the main crop physiological functions and other

Crop Physiology. DOI: 10.1016/B978-0-12-417104-6.00020-0

biophysical aspects of crop–soil–atmosphere systems to address production and natural resource management issues.

2 CLIMATE CHANGE

The global climate has been changing and further changes are inevitable regardless of efforts to reduce global emissions of greenhouse gases (IPCC, 2013).

2.1 Realized trends

The climate has changed significantly over the last century. Atmospheric CO_2 has increased from a pre-industrial concentration of 280 ppm to 379 ppm in 2005 (IPCC, 2013). Global temperatures have increased in the past 100 years by an average of 0.74°C (IPCC, 2013), with minimum temperatures increasing faster than maximum temperatures (Easterling et al., 1997a) and winter temperatures increasing faster than summer temperatures (IPCC, 2013).

Some rainfall changes in the past have shown significant regional differences. One contrasting example of realized change is the rainfall decline by 20% in the wheat-belt of western Australia over the last decades (Smith, 2004; Cai and Cowan, 2006) versus the rainfall increase by 100–200 mm over the last century in the Argentinean Pampas (Viglizzo et al., 1995). Another example is the rainfall change in China. During the past 50 years, the average rainfall increased in northern and eastern China, but decreased in central China (Liu et al., 2005; Jiang et al., 2013; Chapter 3). Over the past four decades, there was no significant difference in the total rainfall during the crop growing seasons in north-east China, however, the number of rainfall days decreased significantly, resulting in an increase in the daily rainfall amounts (Chen et al., 2012b). While some evidence suggests that some of the rainfall trends result from human influences (Cai and Cowan, 2006), this evidence is less convincing than is the case with the increases in temperature (Nicholls and Collins, 2006). In general, rainfall in the high latitudes of the northern hemisphere has increased, while rainfall in eastern Asia, Australia, the Sahel and the Pacific region has declined, with rainfall variability increasing almost everywhere in the world (Dore, 2005).

In addition, many studies have indicated a reduction in the amount of solar radiation reaching the Earth's surface (Liepert and Kukla, 1997; Stanhill, 1998; Chameides et al., 1999; Stanhill and Cohen, 2001; Liepert, 2002; Che et al., 2005; Jiang et al., 2013). This global dimming is probably caused by a combination of increased cloud cover and higher atmospheric concentrations of aerosols (Stanhill and Cohen, 2001). The degree and impact of dimming is also different between the northern and southern hemispheres, usually due to less direct radiation with more cloud cover in the northern hemisphere (Philander et al., 1996). As total radiation has declined in recent decades, there is an accompanying effect of a higher diffused light fraction (Farquhar and Roderick, 2003). While reduction in radiation can reduce crop photosynthesis, the increase in the fraction of diffused radiation can increase the crop radiation-use efficiency, partially compensating the impact of radiation decline.

Other elements that are important for agriculture but often not considered in climate change impact studies are evaporative demand (ETo), vapor pressure deficit (VPD) and wind which, when considered, might overwrite existing assumptions about crop system responses. For instance, ETo is a typical complex variable which will increase or decrease depending on the relative changes of its drivers, i.e. net radiation, vapor pressure, wind speed, and air temperature (Chattopadhyay and Hulme, 1997; Thomas, 2000; Cohen et al., 2002; Roderick and Farquhar, 2002, 2004; Tao et al., 2003; Donohue et al., 2010). To quantify the impact of climate change on ETo and subsequent agricultural processes requires data describing all the drivers (Donohue et al., 2010).

Furthermore, in some parts of the world ozone levels now frequently increase to concentrations harmful to crop growth (IPCC, 2013).

2.2 Future projections

A further increase of CO_2 concentrations in the atmosphere is one of the most certain aspects of global change over the coming decades and, as recent reports indicated, carbon dioxide is accumulating in the atmosphere faster than previously expected (IPCC, 2013). Future atmospheric CO_2 increase will be about 2 ppm $year^{-1}$ (IPCC, 2013). Atmospheric CO_2 is fundamental to the growth and productivity of terrestrial vegetation, as well as critical for the radiative properties of the atmosphere and hence climate. Therefore, the changing atmospheric concentration of CO_2 will have far-reaching effects on agricultural production. Other projected climate changes, such as more severe and frequent droughts in some regions (e.g. Hennessy et al., 2008), increasing temperatures, reduced solar radiation and increased ozone concentrations (IPCC, 2013), will affect cropping systems in particular. Maximum and minimum temperatures are likely to increase at different rates, often with a faster increase in minimum temperature than maximum temperature (Easterling et al., 1997a; Nicholls, 1997; Vose, 2005), reducing the diurnal range (Easterling et al., 1997b). Meanwhile, the increase in air temperature may differ among seasons, and climate warming impacts on crop production can be different for different crops. Nevertheless, projected reduction of rainfall with less cloud cover might increase frost risk frequency in some regions. In addition, an increase in the frequency of extreme high temperatures could potentially result in more severe damage to crops than a small average increase over the year (Stone and Nicolas, 1995; Van Herwaarden et al., 1998; Spiertz et al., 2006; Lobell et al., 2012). Table 20.1 summarizes the main recent and future climate changes and the likely impacts on crop production.

Future climate change scenarios are usually generated using general circulation models (GCMs). It is important to acknowledge that climate change scenarios produced by GCMs have large uncertainties arising from three sources: the modeled processes within the GCMs themselves; initial conditions; and future greenhouse gas emissions (IPCC, 2013). Downscaling GCM scenarios to local scales (e.g. paddock) for use in crop simulation modeling adds another dimension of uncertainty (Wilby et al., 2004). However, the reliability of such downscaling can be critical for the outcomes in terms of crop impact. For example, an increase in average maximum and minimum temperatures during the cool period of a crop season may be positive for growth, while an increase of several degrees in maximum temperature during an already hot period (e.g. during cereal grain filling) for even a few days might be devastating for a crop (Van Herwaarden et al., 1998; Lobell et al., 2012). Recent field observations showed that climate warming might benefit autumn sown crop production (Tian et al., 2012), but be detrimental to summer sown crop production in East China (Dong et al., 2011; Chapter 3). In addition, the change in diurnal temperature range can have a significant impact on crop yields (Lobell, 2007). Such detail in future climate change scenarios is often not available or has significant uncertainty (e.g. Lobell, 2007). These uncertainties and limits in downscaling need to be considered when studying the impact of climate change. One way of dealing with these uncertainties is to use a range of possible future climate change scenarios to assess climate change impact, i.e. using ensemble climate projections (Tao and Zhang, 2012, 2013) rather than a single projection. Nevertheless, to assess the likely impact of these changes requires well validated simulation models which consider these changes as drivers and capture the main crop physiological functions and other

TABLE 20.1 Summary of climate change factors and general impact on crops

Climate variables		Realized trends	Projected trends	General impacts on crop growth
CO_2		1.4 ppm/a (379 ppm in 2005)	1.9 ppm/a (450 ppm by 2050)	Increased net photosynthesis, plant biomass production and transpiration-use efficiency
				Reduced transpiration
				Increased canopy temperature
				Reduced crop nutrient concentration
Temperature	Min	0.56°C (2005) since 1906	0.02°C/a (1.3°C– 1.7°C in 2050)	Reduced frost risk
	Avg	0.74°C (2005) since 1906	1.8–4.0°C in 2100	Increased stomatal conductance, photosynthesis, respiration, and transpiration; faster growth and development, phenological shifts; reduced transpiration efficiency
	Max	0.92°C (2005) since 1906		Increased heat stress
Rainfall		0.11 mm/a	Variable changes across the globe, in general, increase at high latitude and decrease at low latitude	Positive or negative, depending on the direction and other factors
Solar radiation		Reduction in solar radiation and increased diffused light fraction (1365 W m^{-2} in 2005)	Reduction in solar radiation and increased diffused light fraction	Increase photosynthesis and growth due to increase diffused light fraction
Ozone	Troposphere	0.5–2.5%/a (50 ppb in 2000)	0.5–2.5%/a (60– 100 ppb by 2050)	Increased foliar injury, decreased growth and yield
	Stratosphere	0.6 %/a (265 DU in 2000)	0.1–0.2 %/a (275– 286 DU by 2050)	Reduce leaf expansion and biomass accumulation

After IPCC, 2013.

biophysical aspects of soil–crop–atmosphere systems (Jamieson et al., 2010).

3 CROP RESPONSE TO CLIMATE CHANGE

Different aspects of climate change, such as higher atmospheric CO_2 concentration, increasing temperature and changed rainfall (reduced or increased) all have different impacts on plant production and crop yield. In combination, these effects can either increase or reduce plant production, and the net effect of climate change on crop yield depends on the interactions between these different factors.

3.1 Elevated atmospheric CO_2 concentrations

Elevated CO_2 has two main effects on crop growth. It increases the intercellular CO_2

concentration leading to increased net photosynthesis rate and, at the same time, reduces stomatal conductance resulting in reduced transpiration (Farquhar et al., 1978). Many experimental studies have shown that higher CO_2 increases plant biomass production and yield, and increases in crop yield are lower than the photosynthetic response (Morison, 1985; Drake et al., 1997; Garcia et al., 1998; Yang et al., 2006a,b, 2007a,c,e; Tubiello et al., 2007; Ainsworth and McGrath, 2010; Hasegawa et al., 2013). C_3 species (e.g. wheat, soybean, potatoes, sunflower) and C_4 species (e.g. maize, sorghum, millet) have a different degree in response to elevated CO_2. At 500–550 ppm of CO_2 concentration, grain yields of C_3 crops increase by 10–20% while changes in C_4 crop yields are less than 13% (Yang et al., 2006a,b, 2007a,c,d; Tubiello et al., 2007; Ainsworth and McGrath, 2010; Raines, 2011; Davis and Ainsworth, 2012; Lobell et al., 2012). On average, doubling CO_2 concentrations increases photosynthesis by between 30 and 50% in C_3 species, and by 10–25% in C_4 species (Tubiello et al., 2007; Leakey, 2009). Lobell and Field (2007) summarized the CO_2 effect from a number of wheat open top chambers and free air carbon-dioxide enrichment (FACE) experiments with 0.07% grain yield increase in wheat ppm^{-1} CO_2 increase.

The impact of elevated CO_2 on plant production depends on water and nutrient availability. The highest response to elevated CO_2 is found under water-limiting conditions (Kang et al., 2002; Manderscheid and Weigel, 1997) because higher CO_2 concentrations increase leaf and plant level water-use efficiency (WUE) (Wu et al., 2004). Low nutrient availability can reduce the positive impact of elevated CO_2 on yield (Kimball et al., 2001; Yang et al., 2006a,b, 2007a–d). The impact of elevated CO_2 at field and farm-level is probably lower than those estimated in well controlled experimental conditions, due to production limiting factors such as low nutrient availability, pests and weeds (Tubiello et al., 2007). An important indirect effect of higher atmospheric CO_2, is reduced plant nutrient concentrations (Yang et al., 2006a,b, 2007a,c,e) which can result in lower quality in grain and pasture crops (Rogers et al., 1996; Kimball et al., 2001; Wu et al., 2004; Yang et al., 2006a,b, 2007a,c,d; Madan et al., 2012).

Another indirect effect of atmospheric CO_2 increase will be on canopy temperatures via the reduction in stomatal conductance. There is evidence in both wheat and cotton, that selection for improved grain yields in breeding programs has been associated with selection for high stomatal conductance, resulting in 'heat avoidance' through evaporative cooling in hot environments (Lu et al., 1994; Radin et al., 1994; Amani et al., 1996). Reduced stomatal conductance due to atmospheric CO_2 increase might therefore have additional effects on crop growth and development similar to an increase in temperatures.

3.2 Temperature

Temperature affects most plant and crop level processes underlying yield determination, hence the complexity of the final yield response. Where crops are grown near their limits of maximum temperature tolerance, heat spells can be particularly detrimental (Ferris et al., 1998). Conversely, in cooler regions, such as the northeast of China, an increase in annual mean temperature since the 1980s has contributed to the reported increase in agriculture production (Yang et al., 2007e; Chen et al., 2012a). Also, climate warming may benefit crops growing in cooler seasons, such as winter wheat in China (Tian et al., 2012). Warming benefits and associated agronomic adaptation are further illustrated for cropping systems of North America (Chapter 2), China (Chapter 3) and Scandinavia (Chapter 4).

Higher temperature can negatively impact plant production indirectly through accelerated phenology (Menzel et al., 2006; Sadras and Monzon, 2006; Wang et al., 2008a; Lobell et al., 2012) with less time for accumulating biomass (Amthor, 2001; Asseng et al., 2002; Menzel et al., 2006; Liu et al., 2010b, 2012; Wang et al., 2013); Chapter 12 discusses temperature as the key driver of phenological development and its consequences for crop adaptation. Menzel et al. (2006) analyzed phenological data of 125 000 time series of 542 plants in 21 European countries. For the period 1971–2000, they found that 78% of all leafing, flowering and fruiting events advanced. Sadras and Monzon (2006) modeled in detail the effect of realized changes in temperature on phenological development of wheat, and found earlier flowering but, unexpectedly, no changes in the duration from flowering to maturity due to the shift of flowering to 'cooler' parts of the season. Similarly, several studies reported that warming-induced reductions in the length of the wheat growing period mainly occurred during the pre-anthesis phase, while the length of the post-anthesis period stayed unchanged in eastern China (Liu et al., 2010b; Tian et al., 2012; Wang et al., 2013). The non-linearity of warming on plant phenology predicted in these modeling studies has been verified in field experiments with grapevine (Sadras and Moran, 2013).

As a result of climate change, extreme temperature events will occur more often in addition to the increase in average temperature (IPCC, 2013). Extreme temperature events can have large negative impacts on plant growth and yield as shown for wheat by Van Herwaarden et al. (1998). Temperatures below 9°C and above 31°C around anthesis can reduce potential grain weight and therefore grain yield of wheat (Calderini et al., 2001). Temperature increase around anthesis and grain filling can reduce grain yield substantially (Lizana and Calderini, 2013). Higher average temperatures can reduce frost damage due to a reduced frequency of frost (Baethgen et al., 2003). However, indirect effects of temperature may lead to the paradox of increasing frost risk in some systems. Sadras and Monzon (2006) showed that shifting of flowering to 'cooler' parts of the season due to accelerated phenology with higher temperatures could potentially increase the risk of frost at flowering in Australia and Argentina. Belanger et al. (2002) indicated that increased temperature could reduce cold hardening in autumn, and reduced protective snow cover for forage and winter cereal crops during the cold period in continental climates, hence increasing the exposure of plants to killing frosts, soil heaving and ice encasements (Chapter 4). In cool climates and higher altitudes, higher temperatures can increase the length of the potential growing season.

The impact of increasing temperatures can vary widely between crop species. Optimum temperature for leaf photosynthesis and plant growth is higher for C_4 than for C_3 plants (Goudriaan and Van Laar, 1994). Species with a high base temperature for crop emergence like maize, sorghum, millet, sunflower and some of the legumes, e.g. mungbean and cowpea (Angus et al., 1981) could benefit from increasing temperatures in cool regions. Most of the small-grain cereals, legumes like field pea and lentil, linseed and oilseed crops with a low base temperature (Angus et al., 1981) could result in an advanced phenology with increased temperatures. For example, Sadras and Monzon (2006) estimated that wheat flowering in Argentina and Australia could be advanced about 7 days per degree increase in temperature.

When maximum and minimum temperatures change differently with climate change (Nicholls, 1997), the changes in diurnal temperature range can impact differently on various crops. Lobell (2007) analyzed historical yield data of wheat, rice and maize from the leading global producers and showed that an increase in diurnal temperature range was associated with reduced yields of rice and maize in several agricultural regions worldwide. This reflects the non-linear response of yield to temperature, which likely results from greater heat stress during hot days (Lobell, 2007). Peng et al.

(2004a) suggested that rice yields declined by 15% per °C increase in minimum temperature due to wasteful night respiration, but this interpretation was challenged in subsequent studies (Sheehy et al., 2006a,b).

An indirect effect of global warming can be higher plant water demand due to increased stomatal conductance and transpiration at higher temperatures (Donohue et al., 2010; Sadras et al., 2012), which can potentially reduce plant production (Lawlor and Mitchell, 2000; Peng et al., 2004a). In dryland agriculture this can directly limit plant growth, while in irrigated systems increased temperatures could result in higher irrigation water demands in combination with increased losses through evaporation. However, if future temperature changes are similar to the changes in the last 50 years, where global minimum temperatures have generally increased twice as fast as maximum temperatures resulting in a reduced diurnal range (Folland et al., 2001), the impact of increasing temperatures on vapor pressure deficit and therefore on atmospheric evapotranspiration would be very small (Roderick and Farquhar, 2002). In addition, higher atmospheric CO_2 concentrations can partly compensate for the increased water demands due to higher temperatures, through a lower stomatal conductance which reduces transpiration (Kimball et al., 1995; Garcia et al., 1998; Wall, 2001). Reduced leaf transpiration as a consequence of higher CO_2 will also increase leaf temperature with an increased chance of plant damage due to heat stress. Plants grown at higher atmospheric CO_2 tend to have a higher leaf water potential which reduced drought stress (Wall, 2001).

Temperature changes can also affect yield quality as shown for grain protein content (Spiertz, 1977; Triboi et al., 2003; Dias et al., 2008; Zhao et al., 2008; Farooq et al., 2011; Wang et al., 2012) and dough quality of wheat (Randall and Moss, 1990; Wrigley et al., 1994). For some crops, night temperatures are critical for grain quality, as shown for fatty acid composition in sunflower (Izquierdo et al., 2002; Chapter 17).

3.3 Rainfall and rainfall variability

Global warming is likely to change precipitation amounts and patterns differently across the globe. Total annual rainfall tends to increase at the higher latitudes and near the equator, while rainfall in the sub-tropics is likely to decline and become more variable (Giorgi and Bi, 2005). Changing rainfall amounts can have negative or positive impacts on agricultural production. For example, in semi-arid environments higher rainfall could increase growth, where less rainfall could further limit plant production. In contrast, in high rainfall zones, too much rainfall can result in soil waterlogging which damages crop growth (Dracup et al., 1993; Jiang et al., 2008; Robertson et al., 2009; Sharma et al., 2010; Araki et al., 2012; Sadras et al., 2012) or results in nutrient leaching in sandy soils (Anderson et al., 1998). Reduced rainfall on these soils could limit the negative impacts of waterlogging and nutrient leaching.

Not only the total rainfall amount but also the rainfall distribution plays an important role for determining crop yields. Rainfall around anthesis ensuring water supply during grain filling is particularly critical for grain yield in annual crops (Passioura, 1977; Fischer, 1979; Sadras and Connor, 1991; Liu et al., 2010b; Peltonen-Sainio et al., 2011; Araki et al., 2012; Zhang et al., 2012). Changing future in-season rainfall distribution will therefore have an impact on crop growth and yield. Balancing growth and water use before and after anthesis is one of the tools to manage uneven rainfall distribution. Management options to change seasonal water use of crops include sowing time, nutrient management, plant density and cultivar choice (Passioura, 1977; Fischer, 1979; Sadras and Connor, 1991).

Another aspect of future rainfall change is increased rainfall variability through an increase in extreme events. Extreme events could include a higher drought frequency (Hennessy et al., 2008) with long-term effects on farm viability which could reduce crop production and yields below what is expected based on the average climate change (Easterling, 2007; IPCC, 2013). Particularly critical are changes in rainfall intensity and the distribution of small versus large rainfall events, which can have significant consequences for crop production via the impact on soil infiltration depth, water balance, soil mineralization and crop water-use efficiency (Sadras et al., 2003; Rodriguez and Sadras, 2007; Khan et al., 2009; Wang et al., 2009a; Limon-Ortega and Sayre, 2012; Hao et al., 2013).

3.4 Solar radiation

Reduction in solar radiation can potentially have a considerable negative impact on agricultural production (Stanhill and Cohen, 2001; Xiong et al., 2012; Chen et al., 2013), which can be partially or completely compensated by the associated increase in diffused light fraction (Farquhar and Roderick, 2003; Mercado et al., 2009; Zhang et al., 2013). Plant production is primarily driven by sunlight as plants transform solar energy into sugars. Hence, a reduction in solar radiation can potentially reduce photosynthesis and growth. However, photosynthesis is often limited by nutrient and/or water availability. Lower solar radiation also reduces potential evaporation (Roderick and Farquhar, 2002). In water-limited environments, this can increase plant available water and thus increase plant production. Reduced evapotranspiration can also increase drainage and nutrient leaching which can have large negative ecosystem impacts.

A reduction in solar radiation is usually accompanied by an increase in the diffuse light fraction (Farquhar and Roderick, 2003; Liu et al., 2005; Mercado et al., 2009; Zhang et al., 2013). An expected decrease in photosynthesis with less radiation assumes that the maximum photosynthesis rate (A_{max}) is critical for photosynthesis, but this could be overridden by canopy light distribution, that could be favored by dimming (Sinclair et al., 1992; Dang et al., 1997; Dreccer et al., 2000; Farquhar and Roderick, 2003; Gu et al., 2003; Rodriguez and Sadras, 2007; Chen et al., 2013).

3.5 Ozone

Ozone (O_3) is a form of oxygen that is an atmospheric pollutant at ground level. Most of the O_3 in the atmosphere (about 90%) is in the stratosphere, the remaining in the troposphere. The ozone in the troposphere and stratosphere has different effects on life on the Earth depending on its location. Stratospheric ozone plays a beneficial role by absorbing solar ultraviolet radiation (UV-B) from reaching the Earth's surface. Increased levels of UV-B have been measured with a general erosion of the stratospheric ozone layer in the last decades (World Meteorological Organization, 1995). The increased UV-B reduces leaf expansion and biomass accumulation in plants (Ballare et al., 1996; Kakani et al., 2003; Newsham and Robinson, 2009; Yin and Wang, 2012) and could impact on plant–herbivore interactions by increasing plant resistance to insects (Izaguirre et al., 2003; Mazza et al., 2012).

Ozone is a strong oxidizer, and therefore ozone closer to the Earth's surface is potentially destructive to plants. In crops, ozone can create reactive molecules that destroy Rubisco, an enzyme crucial for photosynthesis and accelerates leaf senescence (Giorgi and Bi, 2005; Chapter 10). Increasing concentration of O_3 at ambient CO_2 reduces yield of many crop species (Fuhrer and Booker, 2003). Widespread negative effects of ozone on crops and natural vegetation were found in Europe (Mills et al., 2011), and elsewhere (Emberson et al., 2009; Hollaway et al., 2012). For wheat, high O_3 exposure at grain filling reduced yield in a high-yielding modern cultivar more than in their older counterpart (Xu et al., 2009). The reduction of yield results from a reduction in the activity and amount of Rubisco, reduction in photosynthesis and accelerated leaf senescence (Lehnherr et al., 1987; Grandjean and Fuhrer, 1989). Moreover, ozone can impair the translocation of photosynthates from sources to sinks (Grantz and Yang, 2000). In addition to its effect on crop yield, O_3 may affect yield quality. For example, Pleijel et al. (1999) reported that grain nitrogen concentration generally increased with increasing O_3 leading to a better baking quality of the flour in spring wheat. However, increasing O_3 had a negative impact on tuber quality of potato (Vorne et al., 2002). In general, elevated O_3 has negative effects on crop growth, and limits the magnitude of the yield enhancement by elevated CO_2.

3.6 Combined impact of climate change

Changes in temperature, rainfall and CO_2 concentrations do not act independently but interact with each other. To develop adaptation strategies to maintain crop production in a changing climate, it is important to understand the interactions between different aspects of climate change. This is shown in a few field experiments for CO_2 and water supply interactions with wheat (Kimball et al., 1995) and CO_2 and temperature interactions with a grass–legume pasture (Lilley et al., 2001). Working with wheat in the UK, Wheeler et al. (1996) showed that an increase in mean seasonal temperatures of 1–1.8°C could offset a positive yield effect from 700 ppm elevated CO_2. In experiments in western Australia, Dias de Oliveira et al. (2013) showed that reductions in biomass and grain yield caused by terminal drought induced at 50% anthesis were partially ameliorated by elevated CO_2 and temperature when the temperature was less than 2°C above the ambient. It is important to understand the interactions before developing climate change adaptation strategies because, for example, adaptations to higher temperatures may be different from adapting to reduced rainfall (Tubiello et al., 2007; Ludwig and Asseng, 2010).

4 CROP MODELS FOR CLIMATE CHANGE

Crop modeling is one of the approaches combining the complexity of climate change with the physiological functions and other biophysical aspects of soil–crop–atmosphere systems. The first crop simulation models were developed in the 1980s and were used to simulate wheat growth using conservative crop physiological functions. They include ARCWHEAT1 (Porter, 1984; Weir et al., 1984), five models from the ARS Wheat Yield Project of which CERES-Wheat (Ritchie et al., 1985a), and WINTER WHEAT (Baker et al., 1985), were the most prominent, and the Dutch models SUCROS (Laar et al., 1992) and SWHEAT (Van Keulen and Seligman, 1987). In the 1990s, models for various crops, were merged into crop modeling platforms, including DSSAT (Jones et al., 2003) (http://dssat.net/), APSIM (Keating et al., 2003) (http://www.apsim.info/Wiki/), CropSyst (Stöckle et al., 2003) (http://www.bsyse.wsu.edu/CS_Suite/CropSyst/), Wageningen crop models (Van Ittersum et al., 2003b), STICS (Brisson et al., 2003) (http://www7.avignon.inra.fr/agroclim_stics_eng/) and EPIC (Kiniry et al., 1995). In the 2000s, new models emerged including the SIRIUS model (Martre et al., 2006; Chapters 14 and 17), the WheatGrow model (Pan et al., 2006) which includes grain quality simulations, and the RiceGrow model considering plant architecture (Zhu et al., 2009). Over the last 10 years, researchers developed more crop models that vary in their approach and complexity, including SALUS (Basso et al., 2001), AquaCrop (Steduto et al., 2009) (http://www.fao.org/nr/water/aquacrop.html), HERMES (Kersebaum, 2007), MONICA (Nendel et al., 2011) and LPJmL (Bondeau et al., 2007) (http://www.pik-potsdam.de/research/projects/lpjweb). Other crop specific models include rice (Kropff et al., 1993; Confalonieri and Bocchi, 2005), maize (Jones and Kiniry, 1986), soybean (Sau et al., 1999), velvet bean (Hartkamp et al., 2002), chickpea (Robertson et al., 2002; Soltani et al., 2006), canola (Farre et al., 2002), sugar cane (Keating et al., 1999), potatoes (Peralta and Stockle, 2002), mungbean, peanut (Robertson et al., 2002), lupin (Fernández et al., 1996; Farre et al., 2004), lucerne (Dolling et al., 2005) and cotton (Milroy et al., 2004). Today, at least 30 models exist for wheat, 19 for maize, and 13 for rice (www.agmip.org). One common feature of these models was that all operated on a daily time step, either approximating or aggregating processes that operate on shorter time steps (Jamieson et al., 2010). These models differ in detail in which physiological processes are aggregated, and in which production constraints were addressed. Some of these models are summarized in Asseng et al. (2013) and a special issue of the *European Journal of Agronomy* on crop models (Van Ittersum and Donatelli, 2003).

Most crop models simulate the dynamics of phenological development, biomass growth and partitioning, water and nitrogen cycling in an atmosphere–crop–soil system, driven by daily weather variables of rainfall, maximum and minimum temperatures and solar radiation. Efforts have been made to capture the common physiological processes in a generic modeling framework, thus improvements in understanding can benefit modeling of different crops (Laar et al., 1992; Wang and Engel, 2000; Wang et al., 2002). Phosphorus nutrition is also simulated in some models, e.g. APSIM (Delve et al., 2009) and CERES (Daroub et al., 2003). Pests, diseases, frost and heat damage, biological effects of rotations and lodging that may affect crops are usually not considered in most crop models. These models calculate crop yield for a specific environment, with the maximum yield only limited by temperature, solar radiation, and daylength (Evans and Fischer, 1999), which can be reduced by insufficient water and nitrogen supply. This maximum yield is also called the resource plateau by Sinclair (1997). Thus, the maximum yield will be higher in high-radiation and long-season environments, than in water-limited and short-season growing conditions. Under the latter conditions, high yields are not necessarily linked with higher nitrogen input, due to the interaction between N-induced growth and its effect on water use and water availability for grain filling (Fischer, 1979).

Chapters 2, 5 and 6 use simulation models to investigate yield gaps and agronomic practices, Chapter 13 applies models for quantitative environmental characterizations, Chapter 14 outlines modeling approaches aimed at crop improvement, with emphasis on the links between phenotype and genotype, and Chapter 17 uses models with emphasis on grain quality traits. Comparison of models in those chapters and the models to investigate crop responses to climate change in this chapter highlights common elements, and significant differences depending on the intended applications.

4.1 Modeling CO_2 effect

Elevated CO_2 has two main effects on crop growth. It increases the intercellular CO_2 concentration leading to increased net photosynthesis rates and at the same time reduces stomatal conductance resulting in reduced transpiration (section 3.1). The increased net photosynthesis directly affects radiation-use efficiency (RUE) which, combined with reduced crop transpiration, increases transpiration efficiency (TE). To include these effects in crop models such as APSIM, a relationship as described by Reyenga et al. (1999) has been considered:

$$\Phi p = (Ce - \Gamma)\ (C350 + 2\Gamma)/((Ce + 2\ \Gamma)(C350 - \Gamma)) \tag{20.1}$$

with Φp being the ratio of the light limited photosynthetic response calculated according to Goudriaan et al. (1985) at the enhanced CO_2 concentration, compared with 350 ppm (assumed to be the CO_2 concentration when the model was parameterized) for scaling RUE. C350 is the

CO_2 concentration when most models were initially developed (= 350 ppm), Ce is elevated CO_2 concentration (ppm). The temperature dependent CO_2 compensation point (Γ) is calculated as $\Gamma = (163 - T)/(5 - 0.1T)$, T = temperature (°C), according to Bykov et al. (1981).

Transpiration efficiency (TE = shoot biomass per unit crop water transpiration) is often linearly scaled by a factor that increases linearly from 1 to 1.37 when the CO_2 concentration increases from 350 to 700 ppm (Gifford and Morison, 1993; Duursma et al., 2013). This scaling of TE captures, though not explicitly, the effects of elevated CO_2 on the ratio between internal and external CO_2 concentration.

Other models (e.g. CERES and EPIC) employ a constant multiplier (lower for C_4 than C_3 crops) for daily total crop production under elevated CO_2 (Tubiello et al., 2007). Tubiello and Ewert (2002) and Yin (2013) comprehensively summarize modeling approaches for simulating elevated CO_2 effects and Nendel et al. (2009) tested different CO_2 response algorithms in crop models.

4.2 Modeling temperature effect

Temperature in many crop models causes developmental rates to vary. The concept of thermal time (Cao and Moss, 1997; Tang et al., 2009; Jamieson et al., 2010; Yin and Struik, 2010) or physiological development days (Cao and Moss, 1997; Wang and Engel, 1998) are usually used to predict the progress of development. Thermal time is the time integral of the temperature response function based on either daily (maximum and minimum) or hourly air temperatures. The minimum number of days for development under optimal temperature is defined as the total physiological development days, and a unit number of which is a physiological development day (Wang and Engel, 1998). Temperature response functions used in crop models include segmented linear models with base, optimum and maximum temperatures (Weir et al., 1984) and various curvilinear versions that cover similar temperature ranges (Wang and Engel, 1998; Jame et al., 1999; Streck et al., 2003; Xue et al., 2004). The temperature response function developed by Wang and Engel (1998) has gained wide application due to its simplicity and ability to capture the response to temperature between cardinal temperatures (Streck et al., 2003; Xue et al., 2004). Some crop models (e.g. CERES-Wheat) also simulate the vernalization process (a crop- and cultivar-specific requirement for cold temperature accumulation) and the impact of photoperiod to modify the accumulation of developmental time depending on temperatures affecting the fulfillment of vernalization (Ritchie et al., 1985b; Cao and Moss, 1997; Wang and Engel, 1998). Algorithms to model crop phenology include cultivar-specific parameters but, more recently, attempts have been made to link parameters with genetics, e.g. vernalization and photoperiod responsive genes (Zheng et al., 2013). Chapter 12 discusses the physiological bases of plant development, and the environmental and genetic controls underlying the modeling of crop phenology.

Temperature effect on dry matter production in most crop models is simulated using a temperature response curve to modify either photosynthesis rate or radiation-use efficiency. Thus, temperature changes would have different impact on growth rate and biomass accumulation depending on whether the change is an increase or decrease and whether temperature is above or below the optimal temperature for growth. In a study with wheat in India, Lobell et al. (2012) recently showed that the DSSAT-CERES and APSIM-Wheat models underestimate the impact of high temperature on crop senescence. In contrast, the APSIM-Nwheat model (different to APSIM-Wheat) includes a heat stress routine which accelerates senescence and hence hastens maturity above 34°C (Keating et al., 2001; Asseng et al., 2010); Chapter 10 looks in detail at the physiology of thermal modulation of leaf senescence. Challinor et al. (2005b) included a heat stress impact routine at flowering into the

GLAM-Groundnut model (Challinor et al., 2004) in which temperature above 34°C (moderate cultivar), 36°C (sensitive cultivar) and 37°C (tolerant cultivar) starts to affect pod set; this approach showed good agreement with field observations.

Temperature can affect the vapor pressure deficit, thus affecting the crop water stress status. For example, Lobell et al. (2013) used the APSIM-Maize model to demonstrate how temperatures above 30°C increased vapor pressure deficit, which contributed to water stress and reduction in maize yield by increasing the crop demand for soil water and reducing water supply at later growth stages. However, if minimum temperature increases faster than maximum temperature (Easterling et al., 1997a), the simulated vapor pressure deficit in some crop models (Keating et al., 2001) will result in little changes in evaporation demand, as observed by Roderick and Farquhar (2002). But, if minimum and maximum temperatures increase at a similar rate as reported for a location in Germany (Wessolek and Asseng, 2006), such temperature change would lead to an increase in the evaporative demand and higher water use. In general, most models ignore the impact of diurnal temperature range on grain yield (Lobell, 2007).

Temperature effects on yield quality are considered in some models, for example, for wheat grain protein content (Asseng and Milroy, 2006; Asseng and Turner, 2007) and different wheat grain protein fractions (Martre et al., 2006). For grain legumes and oilseed crops, oil content is an important quality indicator, however, few current models include temperature as a factor affecting oil content (Robertson et al., 2002). However, recent efforts to model thermal effects on concentration and composition of both oil and protein in grain are encouraging (Chapter 17).

4.3 Modeling rainfall and rainfall variability effect

Water from rainfall is absorbed by crop roots from the soil. The state of water in the soil on any particular day depends on the state of the previous day, and any additions (precipitation, irrigation) and subtractions (evaporation, transpiration, drainage, run-off) that occur throughout the day. These processes are simulated usually in a soil water balance sub-module coupled with crop growth modules to determine the water availability for growth. Methods for estimating daily evapotranspiration as combinations of diffusion and energy balance equations have been in existence since the late 1940s (Penman, 1948), with slightly more (Monteith, 1965) or less (Priestley and Taylor, 1972) complex variations (Jamieson et al., 2010). These equations are used to estimate the upper limit of evapotranspiration demand (potential evapotranspiration). Many simulation models use variations of the CERES model (Ritchie, 1972; Baguis et al., 2010) that allows both plant and soil factors to reduce actual evapotranspiration below the potential. Plant factors are mostly related to canopy size and how that limits the interception of energy to drive the transpiration process. Soil factors are considered in crop models when the rate at which the soil can supply water is less than the demand. Evapotranspiration will proceed at a rate limited by either the available energy or water transport through the system. Differences among models are mainly about the calculation of constraints to the uptake rate, and the calculation and application of stress indices to modify canopy expansion, biomass partitioning and growth, and account for temporary reductions in water use such as canopy folding/wilting or stomatal closure (Jamieson et al., 2010).

Water stress is generally simulated by using a stress index that can be the ratio of supply rate to demand rate, constrained to a maximum value of 1 (Ritchie et al., 1985b; Porter, 1993; Gao et al., 2011) or a fraction of available soil water in the rooted soil (Stapper, 1984; Amir and Sinclair, 1991; Keating et al., 2001; Song et al., 2010; Bauwe et al., 2012). The stress index is applied in various ways to reduce biomass production by limiting leaf area expansion or accelerating leaf senescence, and to reduce the photosynthetic rate or radiation-use efficiency (Hu et al., 2004; Liu et al., 2010b; Andarzian et al., 2011). An important application of modeled stress index is the probabilistic mapping of water stress patterns used as a reference for plant breeding (Chapter 13).

4.4 Modeling solar radiation effect

There have been two main approaches in simulating radiation-use efficiency (RUE). One group of models (AFRCWHEAT2, SWHEAT, SUCROS2) calculates the daily growth rate from photosynthesis equations, integrated over a layered canopy at sub-day intervals. The other group uses an aggregated RUE approach (Monteith, 1977). These models include CERES-Wheat and its derivatives such as APSIM-Nwheat (Keating et al., 2001), APSIM-Wheat (Wang et al., 2012; Keating et al., 2003), the models of Amir and Sinclair (1991) and Sinclair and Amir (1992), and Sirius (Jamieson et al., 1998b).

As diffused light is more efficient for canopy photosynthesis (Sinclair et al., 1992), some models consider the impact of diffuse fraction in radiation, e.g. CERES (Ritchie et al., 1985a) and the APSIM-NWheat model (Keating et al., 2001; Yang et al. 2013), while other models ignore it (e.g. Robertson et al., 2002).

4.5 Modeling ozone effect

Ewert et al. (1999) extended the AFRCWHEAT model (Porter, 1993) and the more detailed LINTUL model (Spitters and Schapendonk, 1990), to account for the effects of ozone. In both models, they assumed that ozone damage is caused by ozone uptake. In AFRCWHEAT, ozone reduces the light saturation rate of leaf photosynthesis and induces a short-term response which is reversible dependent on the leaf age. In LINTUL, ozone decreases the Rubisco concentration and increases the costs of detoxification and repair processes. In both models, the plant response to ozone depends on plant development (Ewert et al., 1999).

4.6 Model validation

Simulation models have been tested against measured data under various growing conditions, e.g. wheat models (Ritchie et al., 1985b; Otter-Nacke et al., 1986; Savin et al., 1994; Toure et al., 1994; Keating et al., 1995; Probert et al., 1995, 1998; Asseng et al., 1998, 2001; Jamieson et al., 1998a; O'Leary and Connor, 1998; Wang et al., 2004; Chen et al., 2010; Lv et al., 2013). Figure 20.1 illustrates model validation as simulated grain yield is compared with observed grain yields for a wide range of management options and environments.

Several models have been tested under climate change conditions including elevated atmospheric CO_2 with data from the Maricopa FACE experiment in Arizona, USA (Grant et al., 1995; Kartschall et al., 1995; Tubiello et al., 1999; Jamieson et al., 2000; Grossman-Clarke et al., 2001; Asseng et al., 2004) and experiments with elevated CO_2 in open top chambers (Ewert et al., 1999;

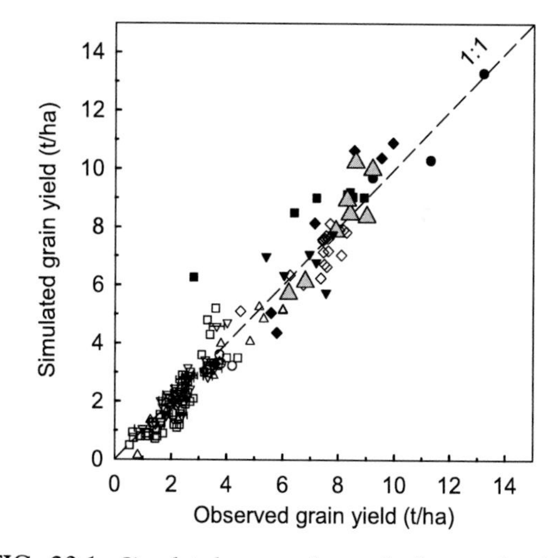

FIG. 20.1 Simulated versus observed wheat grain yield for Western Australia data set I (∇); Western Australia data set II, (□); New South Wales, (○); Gatton, Queensland, (□); Lincoln, New Zealand, (◆); Xiangride, China, (•); FACE experiment, Arizona, USA, (large ▲); Obregon Mexico, (▼); Polder and Wageningen, The Netherlands, (◇) and plant density experiment, The Netherlands (observed after Darwinkel (1978)) (■). *Source: Asseng et al. (2004).*

Rodriguez et al., 2001). While most of this work concentrated on elevated CO_2, a few models have been tested with different intensities and durations of water stress periods (Jamieson et al., 1998a; Asseng et al., 2004) and one model has, in addition, been compared with field data under severe terminal water limitations and rising air temperatures (Asseng et al., 2004).

Some specific aspects of crop models, e.g. the approach of modeling elevated CO_2 impact on crop growth, have been evaluated independently by comparing the calculated RUE and TE using the Reyenga et al. (1999) functions with measured RUE and TE (Asseng et al., 2004). Linear regression of observed vs calculated RUE with intercept set to zero indicated no significant bias ($y = 1.022x$; Mallows C_P statistic = 1). For the FACE experiment at Maricopa, the observed enhancement of ambient concentration of CO_2 by 200 ppm increased TE by 26% for the Dry treatment and 25% for the Wet treatment in 1993–4 and was simulated with 21% across treatments (Asseng et al., 2004).

In addition, the physiological method assuming an average daily reduction of stomatal conductance with elevated CO_2 (Farquhar and Caemmerer, 1982) and the method outlined in section 4.1. (Eq. 20.1) returned similar responses of TE to elevated CO_2, thus confirming the robustness of the approach of Reyenga et al. (1999) used in many crop models.

Ewert et al. (1999) tested the AFRCWHEAT and the LINTUL models with wheat growth data from open top chambers with 2 × ambient CO_2 and 1.5 × ambient ozone. They reported underestimations of the CO_2 effect on biomass with AFRCWHEAT, and overestimations with LINTUL. Both models closely predicted the biomass production with increased ozone, but the LINTUL model was less accurate for leaf area index (LAI) and intercepted radiation. Both models simulated well the observed interaction effects of CO_2 and ozone on biomass (Ewert et al., 1999).

Long et al. (2006) argued that current modeling efforts have overestimated the impact of increasing CO_2 on future crop yields, because crop models used in these studies were parameterized with data obtained from earlier 'enclosure studies' and not from more sophisticated FACE experiments. However, crop models parameterized with data from earlier 'enclosure studies', have been successfully tested with data from FACE experiments (e.g. Grant et al., 1995; Kartschall et al., 1995; Tubiello et al., 1999; Jamieson et al., 2000; Grossman-Clarke et al., 2001; Asseng et al., 2004), as both experimental methods are consistent (Ewert et al., 2007; Tubiello et al., 2007; Ziska and Bunce, 2007). Recent studies suggest that the measured CO_2 effect is underestimated with FACE experiments (Bunce, 2012, 2013).

5 IMPACTS OF CLIMATE CHANGE ON CROP PRODUCTION

Climate change can have a considerable impact on crop production, especially if the amount and distribution of rainfall changes. Rising temperatures, elevated atmospheric CO_2 concentrations (Amthor, 2001; Van Ittersum and Donatelli, 2003), increases in ozone concentration (Ewert et al., 1999), and reduced radiation reaching the Earth (Stanhill and Cohen, 2001) will all affect agricultural production. Elevated atmospheric CO_2 almost always increases plant production if other factors are not limiting (Amthor, 2001; Poorter and Pérez-Soba, 2001). Higher temperatures can potentially increase or decrease grain yields (Van Ittersum and Donatelli, 2003; Peng et al., 2004b). The impact of reduced radiation on crop growth might be compensated for by a parallel increase in the diffuse light fraction (Yang et al., 2013). Increased O_3 will reduce crop yields (Ewert et al., 1999).

5.1 Past trends

Statistical methods have been used to analyze trends in yields driven by climate (Lobell et al., 2011), but interactions between climate and non-climate factors confound results (Lobell and Burke, 2010). This hinders the attribution of causality (Gifford et al., 1998) and development of appropriate adaptation strategies. For example, wheat yields in the past two to three decades increased significantly in many rain-fed environments (Turner and Asseng, 2005; Fischer and Edmeades, 2010; Hall and Richards, 2013) at the same time as temperatures had risen (FAO: http://www.fao.org/). However, attributing these increased yields (Nicholls, 1997; Lobell, 2007) to a single factor such as temperature is difficult due to the confounding effects of other climatic factors such as rainfall (Ciais et al., 2005; Lobell, 2007), radiation (Lobell and Ortiz-Monasterio, 2007), changes in non-climatic factors such as improved cultivars, increased nutrition (Gifford et al., 1998; Godden et al., 1998) and new cropping technologies (Turner and Asseng, 2005). Therefore, glasshouse and field experiments, and long-term yield trends, all suffer from a number of confounding effects and do not often allow climate change effects to be extrapolated to regional and global scales. As an alternative, simulation modeling makes it possible to isolate the effects of climatic and non-climatic factors on yield while others are kept constant (Bell and Fischer, 1994; Lobell et al., 2005). By combining crop modeling and historical yield data analysis, Liu et al. (2010b) separated the impacts of climate and crop varietal changes on wheat and maize yield in the North China Plain, and showed that cultivar selection was able to compensate the negative impact of climate change, leading to increased crop yield since the 1980s (Chapter 3). By using modeling, Asseng et al. (2011) separated the impact of temperature from other factors and showed that the effect of temperature on wheat production has often been underestimated. They showed that observed variations in average growing-season temperatures of ±2°C in the main wheat growing regions of Australia can reduce grain production up to 50%. Most of this has been attributed to increased leaf senescence as a result of temperatures greater than 34°C.

In the north China Plain, a significant decrease in sunshine hours and radiation since 1961 was associated with increased aerosol in the atmosphere (Che et al., 2005; Chen et al., 2010, 2013). The rising air temperature since the early 1980s (Shen and Varis, 2001; Ding, 2006; Tao et al., 2006) was accompanied by decreasing annual precipitation in the north and increasing precipitation in the southeast of the region (Liu et al., 2005; Wang et al., 2008a). Other studies showed that these climatic trends led to changes in potential evapotranspiration (Thomas, 2000) and crop water demand (Tao et al., 2003), and a shortened growing period for wheat and an extended growing season for maize with a tendency to reduce crop yields (Jingyun et al., 2002; Tao et al., 2006). However, the changes in climate were not uniform during various stages of wheat and maize and across the region (Liu et al., 2010b; Wang et al. 2013). Warming mainly occurred during the vegetative growth stage (pre-flowering) of wheat and maize, while there was a cooling trend or no significant change in temperatures during the post-flowering stage of wheat (spring) or maize (autumn). Warming during vegetative stages was predicted to shorten the length of the growing period for both crops, generally leading to a yield reduction. However, adoption of new crop varieties compensated for the warming impact, through stabilizing the length of the pre-flowering period, leading to stabilized or increased wheat and maize yield at most locations.

In east and southern China, Liu et al. (2012, 2013b) modeled the impact of climate trends on rice production from 1981 to 2009. They found that climate change was not uniform across the rice production regions, particularly when changes in rice stages were considered. Past warming led to a reduction in the length of the rice growing season in the single-rice growing region and a reduction in grain yield, if no varietal changes had occurred. However, the adoption of new rice varieties stabilized the growing duration, increased harvest index and grain yield at most studied sites. In the double-rice growing region, a significant increase in temperature occurred before the jointing stage of early planted rice and after jointing stage in late planted rice. Adoption of new cultivars partly mitigated the negative impact of warming. However, the changes of varieties increased the grain yield of both early and late planted rice through an increased harvest index by extending the grain-filling period (Liu et al., 2013a).

In northeast China, the annual air temperature increased by 0.38°C per decade from 1961 to 2007, which caused an expansion of the northern limits of maize (Liu et al., 2013a). According to a regression analysis of the anomalies of maize yield and air temperature (Chun et al., 2011), a 1.0°C increase in daily minimum temperature in May to September led to a corresponding increase of 303 to 284 kg ha^{-1} in maize yields in the region. The existing rice cropping region in the northeast has been extended northwards by about 80 km in 2006 compared to the 1970s (Chen et al., 2012a). Chapter 3 further discusses the impacts of warming on Chinese agriculture, including agronomic and varietal adaptations.

An analysis of trends of on-farm rice and wheat yields in the Indo-Gangetic Plains, starting from the 1980s using the CERES models, revealed that reduced radiation and increased minimum temperatures have caused a decline in the simulated potential yields in several regions (Pathak et al., 2003).

Since the mid-1970s, the Mediterranean and semi-arid region of western Australia experienced a significant decline in winter rainfall (Smith et al., 2000). This decline in rainfall was associated with, and probably caused by, a large scale change in global atmospheric circulation during the mid-1970s with anthropogenic forcing contributing to about 50% of the observed rainfall decline (Cai and Cowan, 2006). The drop in rainfall by up to 20% significantly reduced dam inflow in western Australia (Power et al., 2005), but the impacts on the agricultural sector are less well understood. As most of the agriculture in western Australia is rainfall limited, it has been assumed that a drying trend would also reduce crop production. However, despite the lower rainfall, observed farmers average wheat grain yields have increased from 1.0 t ha^{-1} in 1970 to about 1.7 t ha^{-1} in 2000 (Turner and Asseng, 2005). Ludwig et al. (2009) used the APSIM model in combination with historic climate data to study the impact of past rainfall reductions on wheat yield, deep drainage and nitrate leaching against a constant background of cultivars and management. Unexpectedly, this study revealed that, despite the large decline in rainfall, simulated yields based on the actual weather data did not fall, whereas simulated drainage and nitrogen leaching decreased by up to 95% (Fig. 20.2). These results were due to rainfall reductions mainly occurring in June and July, a period when rainfall often exceeds crop demand, and large amounts of water are lost by deep drainage. These simulated findings have significant implications for estimating future climate change impacts in this region, with rainfall reduction causing non-proportional impacts on production and externalities like deep drainage where proportionality is often presumed (Ludwig et al., 2009).

In Sardinia, Italy, also a Mediterranean environment, a similar decrease in winter rainfall over the last 20 years translated into reduced simulated waterlogging, which improved the yield potential on waterlogged soils (Bassu et al., 2009). This decrease in winter rainfall was also associated with a slight increase in spring rainfall, contributing to additional yield increase in this environment (Bassu et al., 2009).

An opposite rainfall trend has been observed in the Argentinean Pampas where rainfall has increased by 100–200 mm over the last century, with statistically significant increases in December and January rainfall (Viglizzo et al., 1995). Simulations with the CERES-type models showed that summer crops like soybean have mostly

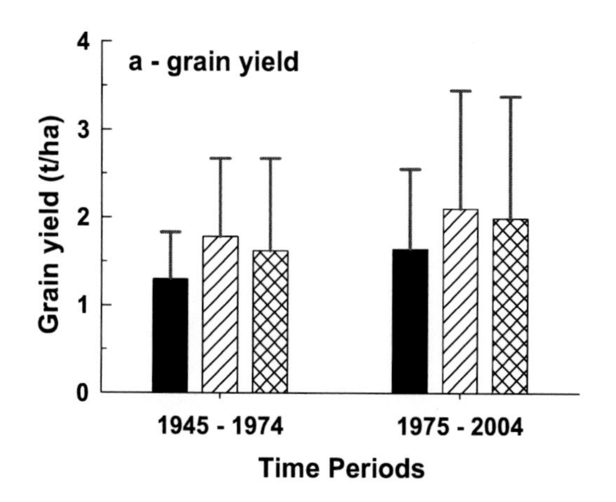

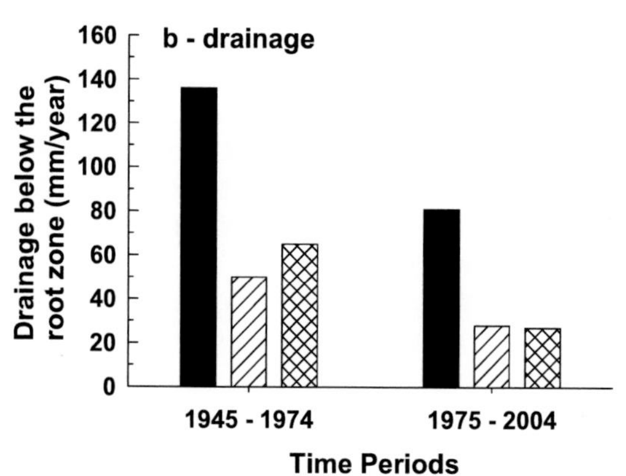

FIG. 20.2 Simulated average (a) wheat grain yield and (b) annual deep drainage below potential root depth for a sand (filled bars), loamy sand (diagonal lines) and a duplex soil (cross bars) at Mingenew, Western Australia. *Source: Ludwig et al. (2009).*

benefited, with a 28% increased yield from the increased summer rainfall in the period 1971–1999, when compared to the period 1950–1970 (Magrin et al., 2005). In another study with CERES-Wheat and CROPGRO-Soybean, an increase in observed average temperature during the same period accelerated wheat maturity, allowing earlier sowing of soybean in double-cropping systems thus increasing simulated soybean yield (Monzon et al., 2007). APSIM simulations indicated that the additional rainfall in the Pampas of Argentina has increased the attainable yield of wheat grown over winter in the currently cropped region, but less than expected from the large amount of additional rainfall. The higher attainable yield from additional rainfall would allow extension of wheat cropping into currently non-cropped areas, where the attainable wheat yield increased on average from 1 t ha^{-1} to currently 2 t ha^{-1}over the last 70 years. However, the poor water-holding capacity of the sandy soils, which dominate the region outside the current cropping area, limits the potential use of most of the increased summer rainfall. Nevertheless, the higher yield potential indicates a suitability of the region for future cropping (Asseng et al., 2012c).

These simulation studies highlighted that the impacts of climate change on yield are often non-linear. These changes are usually complex, leading to not only a shift in average, but also changes in distributions (e.g. rainfall distribution), and resulting in changes in the efficiency of crop to use resources. A crop model that captures the interactions between these changes and their impact on crop growth is therefore a very suitable tool for a realistic assessment of climate change impacts.

5.2 Future scenarios

To generate impact scenarios for agriculture, crop physiology imbedded in crop models is linked with climate projections to predict crop response to future change. However, the climate change information required for such impact studies is of a spatial scale much finer than that provided by GCMs with resolutions of hundreds of kilometers. The most straightforward means of obtaining higher spatial resolution projections is to apply coarse-scale change projections to a high resolution observed baseline – the 'change factor method' or 'delta method' (Wilby et al., 2004). However, this approach does not consider any potential changes in rainfall distribution or intensity. Fine resolution climate information can also be obtained via more sophisticated statistical or dynamic downscaling, assuming that the present day climate is also valid under the different forcing conditions of possible future climate. It is also important to recognize that increased precision of downscaling does not necessarily translate to increased confidence in regional projections. Statistical or dynamic downscaling is usually based on a single GCM output, while comprehensive impact studies must be conducted against multiple GCM outputs to overcome some of the uncertainties associated with each individual GCM. The reliability of future scenarios in terms of rainfall distribution and intensity will be particularly critical for understanding the impact of future rainfall changes on crop production (Mearns et al., 1997; Sadras and Rodriguez, 2007). Other issues with statistical downscaling from GCMs not yet resolved, and which also apply to the change factor method, include the handling of extreme events and potential local feedbacks (e.g. vegetation) (Wilby et al., 2004).

Many impact assessments of future climate change have been carried out across scales, using crop models for specific locations (Semenov et al., 1996), agricultural regions (Ewert et al., 2005; Tao et al., 2009b), and globally (Rosenzweig and Parry, 1994). In such studies, GCM outputs have been directly linked with physiological knowledge (e.g. relative yield response to elevated CO_2) (Olesen and Bindi, 2002; Pielke et al., 2007) or crop models (e.g. CERES, EPIC) which are based on physiology (Wu et al., 2007; Tao et al., 2008). Physiology-based crop models have been extensively used to analyze the impact of climate change on regional or local yield of wheat (Seligman and Sinclair, 1995; Mearns et al., 1996; Wolf et al., 1996; Reyenga et al., 1999, Aggarwal, 2003; Weiss et al., 2003; Asseng et al., 2004; Luo et al., 2005; Richter and Semenov, 2005; Wessolek and Asseng, 2006; Anwar et al., 2007; Kersebaum et al., 2008; Wang et al., 2009a) and other crops (Aggarwal and Mall, 2002; Chipanshi et al., 2003; Holden et al., 2003; Reilly et al., 2003; Tsvetsinskaya et al., 2003; Trnka et al., 2004; Abraha and Savage, 2006; Mall et al., 2006; Olesen et al., 2007, 2012; Elsgaard et al., 2012;). Jamieson and Cloughley (2001) used the Sirius model (Jamieson et al., 1998a; Chapter 17) to assess the impacts of several climate change scenarios (unspecified temperature and CO_2 increase) on both irrigated and dryland wheat production in New Zealand through to 2100. The simulations demonstrated that, although maturity would be advanced by 10 to 20 days, the associated CO_2 fertilization would still raise yield by 15–30% on a base yield of 10 t ha^{-1}.

Wheat crop simulation models were among the first to show the importance of climate variability and extreme events on production (Semenov and Porter, 1995; Porter and Semenov, 1999) and showed that changes in temporal rainfall distribution may be part of the cause of increased risk of crop failure (Porter and Semenov, 1999).

In India, simulations from Aggarwal and Mall (2002) showed a general increase in rice yields (with 400–750 ppm CO_2, 1–5°C temperature increase, irrigated), and decreases or increases, depending on region and scenario in soybean yields (Mall et al., 2004) with future climates (doubling CO_2, various temperature and rainfall scenarios) across the country. In Botswana, assuming doubling CO_2, increased temperatures (2–3°C) and changed rainfall (from increases to declines depending on locations and month), simulations with CERES-wheat and CERES-maize indicated a reduced growing season with yield reductions from 4 to 30% depending on soil type (Chipanshi et al., 2003). For Africa and Latin America, by 2055 (based on IPCC 2001, non-specific atmospheric CO_2, temperature and rainfall changes), maize production was predicted to decline by an average of 10%, based on a simulation study with CERES-maize but with large regional and soil-dependent variations, and therefore large variations in potential negative socioeconomic consequences across these continents (Jones and Thornton, 2003).

Global warming can increase the frequency, intensity and duration of physiologically-relevant extreme temperature events (Porter and Semenov, 2005). However, the risk of extreme high temperature might also be reduced through accelerated phenology and earlier flowering and maturity time due to a simultaneous increase in average temperature. While this may help avoid exposure to high temperature, the early flowering wheat crop may be exposed to higher frost risk (Sadras and Monzon, 2006). Baenziger et al. (2004) studied extreme temperature events by examining late frosts after spike formation, which can be a yield-limiting consequence of global warming.

The apparent yield-enhancing effect of elevated atmospheric CO_2 might come with additional costs. Jamieson et al. (2000), analyzing and extending work from the Maricopa FACE wheat study (Kimball et al., 1999), showed that extra nitrogen was required to realize the potential yield gains from elevated atmospheric CO_2 concentrations (550 ppm), although minimum leaf nitrogen concentrations were slightly less under elevated atmospheric CO_2 (Sinclair et al., 2000). Hungate et al. (2002) have shown that, despite an increase in ecosystem C:N ratio, ecosystem demand for nitrogen will increase with increasing atmospheric CO_2 concentrations. As plant nitrogen nutrition can become yield limiting under elevated CO_2, simulating feedbacks through N cycling will be critical for modeling elevated CO_2 impact assessments on crop production. Nutrients other than N, often not considered in crop models, can also become yield limiting under elevated CO_2. Models lacking these feedbacks are likely to overestimate the yield-enhancing impact of elevated CO_2 (Hungate et al., 2002).

Some modeling studies simulated the individual effects of higher temperatures and CO_2 by artificially constructing climate change scenarios (Wang et al., 1992, 2009a; Van Ittersum and Donatelli, 2003). Ludwig and Asseng (2006) simulated individual effects of increased temperature (2–6°C), increased CO_2 (525, 700 ppm) and reduced rainfall (by 10–60%) on wheat in Australia with specific focus on how different aspects of climate change interact with each other. They found that higher CO_2 increased modeled yield especially at drier sites while higher temperatures had a positive effect in the cooler and wetter region of western Australia. They further showed that heavier clay soils are most vulnerable to reduced rainfall while sandy soils were more vulnerable to higher temperatures, and elevated CO_2 reduced grain-protein concentration and lower rainfall increased grain protein. Wang et al. (2009a) modeled the interaction between temperature increase, raising CO_2 and rainfall reduction on wheat yield, and also investigated the changes in rainfall amount and frequency as they impact yield and soil water balance. They showed that wheat yield reduction caused by 1°C increase in temperature and 10% decrease in rainfall could be compensated by a 266 ppm increase in CO_2 assuming no interactions between the individual effects. An early maturing cultivar (Hartog) was more sensitive in terms of yield response to temperature increase, while a mid-maturing cultivar (Janz) was more sensitive to rainfall reduction. Temperature increase had little impact on long-term average water balance, while CO_2 increase reduced evapotranspiration and increased deep drainage. Rainfall reduction across all rainfall events would have a greater negative impact on wheat yield and WUE than if only smaller rainfall events were reduced in magnitude, even given the same total decrease in annual rainfall. The greater the reduction in rainfall, the larger was the difference.

In a simulation study by Van Ittersum et al. (2003a) in the Mediterranean environment of western Australia, elevated CO_2 concentration (400–700 ppm) increased wheat yields, particularly if nitrogen was sufficient and conditions were relatively dry. This and other simulations showed that higher temperatures can increase plant production (Van Ittersum et al., 2003a; Ludwig and Asseng, 2006). In Mediterranean environments, where crops are grown in winter, plant growth is often limited by low temperatures, and global warming could potentially have a positive effect on crop yields (Van Ittersum et al., 2003a; Ludwig and Asseng, 2006). Higher temperatures had non-linear effects, with initial (up to 3°C) benefits on better water-holding clay soils, and not on poor water-holding sandy soils, but with higher temperature, yields would decline substantially (Van Ittersum et al., 2003a). A similar benefit in yield from a temperature increase of up to 3°C (and 460 ppm CO_2) has been simulated for irrigated and dryland wheat in India (Attri and Rathore, 2003). Differences in crop response to climate change depending on soil type have also been shown in other simulation studies (Chipanshi et al., 2003; Jones and Thornton, 2003). Elevated CO_2 concentrations (400–700 ppm) and increased temperatures (+3°C) both decreased grain protein through an N dilution effect but, in financial terms, this was more than offset by the increase in yield in most cases. If, in addition, precipitation was decreased, financial returns dropped below present levels, particularly in low precipitation regions (Van Ittersum et al., 2003a). Deep drainage as the main cause of dryland salinity in western Australia (George et al., 1997), tended to be slightly higher under elevated CO_2 concentrations (400–700 ppm) but when higher average temperatures (1–7°C) were also simulated this was reversed. Deep drainage was greatly reduced in low precipitation scenarios (Van Ittersum et al., 2003a). Hence, climate change is not only likely to affect productivity, but also deep drainage and dryland salinity in some cropping regions. The impact can vary in direction such that both 'win–win' and 'lose–win' outcomes may occur, particularly depending on the relative change in precipitation (Aggarwal, 2003; Van Ittersum et al., 2003a; Ludwig and Asseng, 2006; Mall et al., 2006).

From a global point of view, crop modeling studies have estimated a positive impact from climate change on agricultural production in the temperate regions up to a temperature increase of about 2–3°C. However, production is likely to be reduced with more severe warming (Easterling, 2007; Tubiello et al., 2007). In tropical and semi-arid regions, with temperature increases of only 1–2°C, climate change has already exhibited a negative impact on food production. Chapter 4 illustrates the benefits of elevated temperature in cropping systems at high latitude.

While crop models are an important tool for understanding the complexity of climate change impact on agriculture, results from such studies still need to be considered as an 'initial estimate'. Most of these models have been tested under a wide range of growing conditions with varying temperatures and rainfall, and some have been compared with elevated CO_2 data from FACE experiments. However, some of the simulated interactions of, for example, CO_2 and temperature have never been tested with measurements.

Studies of climate change impact have recently started to account for simulation uncertainties. Uncertainty, defined as the deviation from the ideal of known deterministic knowledge of a system (Walker et al., 2003), has been addressed through probabilistic projections based on multiple global or regional climate model ensembles (Semenov and Stratonovitch, 2010). Most climate change impact assessments have used only a single-crop model (White et al., 2011), limiting their significance (Muller, 2011). Since crop models differ in the way they simulate dynamic processes, set parameters, and use input variables (White et al., 2011), large differences in their simulation results have been reported (Palosuo et al., 2011). While crop model uncertainty is sometimes assessed by using more than one crop model (Palosuo et al., 2011), or by perturbing crop model parameters (Challinor et al., 2010), comprehensive assessments have proven difficult to coordinate (Rötter et al., 2011b). The recently funded Agricultural Model Intercomparison and Improvement Project (AgMIP; www.agmip.org) is a major international collaborative effort to assess climate impacts on the agricultural sector (Fig. 20.3). AgMIP

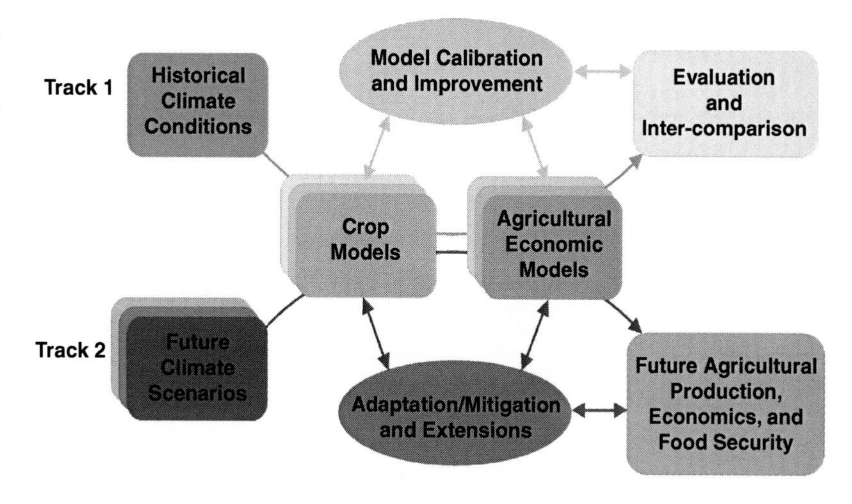

FIG. 20.3 Two-track approach to AgMIP research activities. Track 1: Model intercomparison and improvement; Track 2: Climate change multimodel assessment. *Source: Rosenzweig et al. (2013).*

incorporates state-of-the-art climate products as well as crop and agricultural trade model improvements in coordinated regional and global assessments of future climate impacts. Instead of using single-crop models, the project includes multiple models, scenarios, locations, crops, and participants to standardize climate change impact studies and define the uncertainty of impact assessments (Rosenzweig et al., 2013). As an outcome of AgMIP, the largest standardized model intercomparison for climate change impacts including 27 wheat models was recently published (Asseng et al., 2013). They reported that individual crop models are able to simulate measured wheat grain yields accurately under a range of environments, particularly if the input information is sufficient. But, simulated climate change impact varied across models due to differences in model structure and parameter values. They also showed that a greater proportion of the uncertainty in climate change impact projections was due to variations among crop models than to variations among downscaled GCMs. Uncertainties in simulated impacts increased particularly with increasing warming. They suggested that some of the impact uncertainties can be reduced and quantified through the application of multimodel ensembles (Asseng et al., 2013). Crop model uncertainties can also be due to uncertainties in soil and management input information (van Wart et al., 2013a), initial soil conditions (Moeller et al., 2009) and climate data sources (van Wart et al., 2013b).

6 ADAPTATION TO CLIMATE CHANGE

Due to past emissions of greenhouse gases, the world is already committed to a 0.1°C per decade warming for at least the next 50 years, and because greenhouse gas emissions are unlikely to reduce in the near future, it is likely that the global climate will continue to change even with the most stringent mitigation actions. To reduce the negative impacts of climate change on agriculture, and to benefit from any possible positive impacts, it is necessary to focus on adaptation. For example, relatively small changes in farm-level management and selection of different crop varieties can significantly reduce any negative impact of moderate climate change (Chapters, 2, 3 and 4). However, to adapt to more severe changes in climate, significant changes are needed and farmers might have to switch to new crops and varieties, or even change land use. Crop modeling can play an important role in assisting agriculture to adapt to climate change (Ewert, 2012).

6.1 Management

The beneficial effect of elevated CO_2 concentrations on crop yields depends on soil nutrient availability, as elevated CO_2 reduces grain protein concentration, an important grain quality trait (Amthor, 2001; Van Ittersum et al., 2003a; Ludwig and Asseng, 2006; Fernando et al., 2011; Chapter 17). To benefit fully from the positive impact of higher CO_2 concentrations, and to minimize negative impacts on grain quality, it is necessary to adjust nitrogen fertilization by a relatively small amount under high CO_2 environments (Jamieson et al., 2000; Van Ittersum et al., 2003a; Ludwig and Asseng, 2006; Erbs et al., 2010).

Crop models have been used to manage trade-offs between crop production and externalities under climate change. In the temperate climate of northeast Germany, Wessolek and Asseng (2006) explored the trade-off between grain yield and groundwater recharge management through variable nitrogen inputs, for future climate change scenarios. Groundwater recharge in this region is important for urban water supply, irrigation, forestry and peat protection. The simulations showed that the trade-off between deep drainage and grain yield can be potentially controlled through nitrogen management. However, such control was more effective under current climate conditions than under future climate, and on a better water-holding silt soil compared to a poor water-holding loamy sand. The authors suggested that under future climates areas with poor water-holding soils should be managed extensively for ground-water recharge harvesting, while better water-holding soils should be used for high input grain production (Wessolek and Asseng, 2006).

Increasing average temperatures as a result of future climate change will create new opportunities to increase cropping frequency in some farming systems (Evans, 1993). For example, a simulation study by Monzon et al. (2007) showed recent temperature increases favored wheat–soybean double cropping in the southern Pampas of Argentina. Simulations in the north China Plain by Wang et al. (2012) explored an adaptation strategy in response to recent warming termed the 'Double-Delay' technology, i.e. the delay of both the sowing of wheat and the harvest of maize, which led to 4–6% increase in total grain yield of wheat–maize double cropping systems. The increase in temperature before the over-wintering stage of winter wheat enabled a later harvesting of maize and a delay of sowing of winter wheat.

In many temperate environments, management of winter cropping systems has to be balanced between sowing too early or too late. Higher average temperatures can reduce frost at current flowering windows, but might increase frost risk with earlier flowering (Sadras and Monzon, 2006; Shimono, 2011; Zheng et al., 2012). For example, when wheat is sown too early there is the risk of frost during critical development stages, and when sown too late wheat crops risk high temperature and/or water stress during grain filling. Howden et al. (2003) in north-eastearn Australia showed that with early sowing accounting from recently reduced frost risks, gross margins of wheat cropping systems can be doubled. Another simulation study indicated that future warming in Australia is likely to reduce frost duration, thus allowing earlier planting which could increase yields by up to 36%, assuming 750 ppm CO_2, 4°C warmer and no change in rainfall (Crimp et al., 2008).

Simulations for Bulgaria, suggested early sowing of wheat and maize could reduce yield loss with future warming (Alexandrov and Hoogenboom, 2000). In Italy, 10–40% yield losses were predicted with future climate change for a range of crops using the CropSyst model (Tubiello et al., 2000); sowing earlier summer crops and using slower-maturing winter crops effectively maintained simulated yields at current levels (Tubiello et al., 2000; Moriondo et al., 2010) (Fig. 20.4).

Most of the negative impacts of climate change associated with higher average temperatures in temperate regions are related to

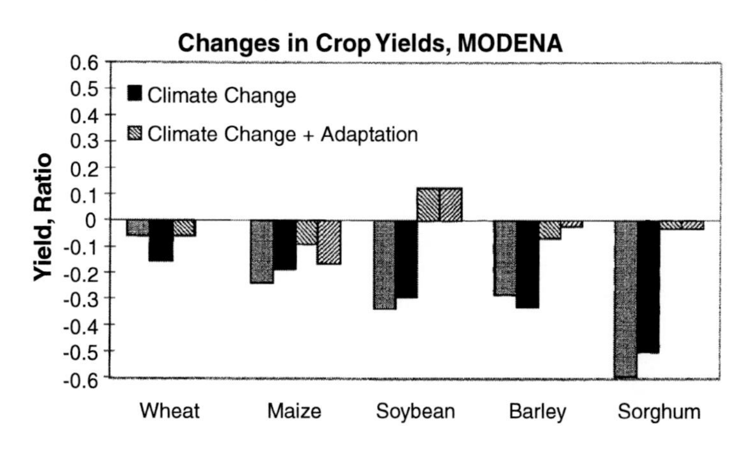

FIG. 20.4 Simulated changes in crop yield in Modena, Italy. Ratio of climate change scenarios with (with pattern) and without adaptation (no pattern) to baseline yields. Within each category (with and without adaptation), results both for general circulation model (GCM) I scenario (gray) and GCM II scenario (black) are shown assuming a doubling of CO_2, a 4°C temperature rise and an increase in rainfall of about 10% in all months with GCM I and up to 30% rainfall decrease in some spring and summer months with GCM II. Note impact of GCM II scenario with adaptation is zero for wheat. *Source: Tubiello et al. (2000).*

accelerated development and less time for biomass accumulation (Rötter et al., 2011a; Turner et al., 2011). Simple adaptations such as using varieties with more thermal time and vernalization requirements, or increased photoperiod sensitivity (Chapter 12) could delay development to counteract these negative impacts (Liu et al., 2010a, 2012). Models which link phenotype with genotype (Hammer et al., 2006; Chapter 14) might be best suited to assist breeding to adapt to these changes (Zheng et al., 2013).

In other regions where rainfall will increase and/or with more extreme rainfall events, measures should be taken to reduce waterlogging, erosion and nutrient leaching (Batisani and Yarnal, 2010; Conway and Schipper, 2011; Arbuckle et al., 2013).

Under the current climate, agricultural production is already significantly affected by climate variability in many semi-arid regions. Changes in climate variability with future climate change are still uncertain (Nicholls and Alexander, 2007), but rainfall variability typically increases as mean annual rainfall decreases (Nicholls and Wong, 1990; Batisani and Yarnal, 2010; Hope et al., 2010). Using CERES-wheat, Mearns et al. (1997) have shown that if the variance of projected temperature and rainfall changes in addition to changes in the mean, this could have further consequences on yield. For example, the positive effect of increased rainfall on yield can be lost through an increased variance of rainfall (Mearns et al., 1997; Patil et al., 2010; Moriondo et al., 2011; Auffhammer et al., 2012).

6.2 Use of seasonal climate forecasting

Improving the management of climate variability will have immediate benefits to improving productivity, and could be a positive step towards adapting to climate change if the future climate becomes more variable. For example, in water-limited environments, seasonal rainfall variability is one of the most important factors for fluctuations in agricultural production and risk. Recent research investigated new approaches to improve managing climate variability using crop models as a key tool (Hammer et al., 1996; Jones et al., 2000; Challinor et al., 2005a; McIntosh et al., 2007; Moeller et al., 2008) (Fig. 20.5). Perceptions about climatic risk and uncertainty about rainfall in the forthcoming season have led in the past to conservative, low input management approaches, which aim to reduce the losses in poor rainfall seasons. However, such low input approaches usually fail to capitalize on the up-sides of climatic variability, i.e. the good rainfall seasons (Meinke and Stone, 2005; Ashok and Sasikala, 2012; Asseng et al., 2012a).

Physiology-based crop simulation models can assist in quantifying the season- and site-specific outcomes of agricultural interventions (Matthews and Stephens, 2002; Luo et al., 2009; Wang et al., 2009b; Challinor et al., 2010; Olesen et al., 2011) and, when integrated with climate data, allow retrospective analysis of the potential value of seasonal climate forecasts for particular decisions (Hansen, 2002; Meinke and Stone, 2005). Seasonal forecasting systems linked with crop simulation models have been employed to establish optimized management strategies for improved risk management, and enabled farmers to tailor better management decisions to the season (Hammer et al., 2001; Hansen, 2005). For example, Hansen et al. (1996) used a wheat simulation model to determine the value of seasonal forecasting to crop management in subtropical northeast Australia, while Moeller et al. (2008) showed simulation-based management strategies linked to specific seasonal rainfall forecasts in the Mediterranean environment of western Australia (Fig. 20.5). Wang et al. (2008b, 2009a,c) and Yu et al. (2008) demonstrated both the economic and environmental value of historical climate knowledge and different seasonal climate forecasts when they are combined with crop modeling to assist in nitrogen management in Australia.

Using yield simulations generated with the DSSAT crop simulation models, Jones et al. (2000) estimated the economic returns of decisions based on perfect predictions of phases of the El-Niño Southern Oscillation (ENSO), which were related to terciles of growing-season rainfall in the Pampas of Argentina. These simulations showed that the mix of rain-fed crops (wheat, maize, peanut and soybean) that maximizes returns differed among ENSO phases. In addition, modifying the maize management (sowing date, plant density, and nitrogen fertilizer) based on rainfall terciles driven by ENSO

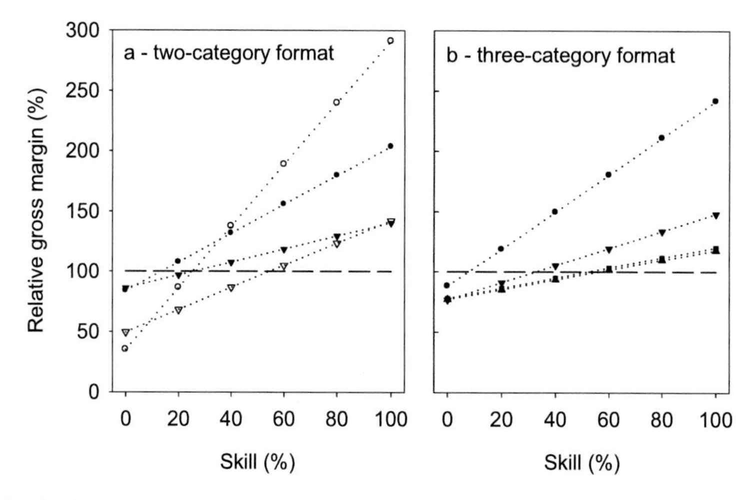

FIG. 20.5 Effect of skill of hypothetical balanced two- and three-category forecasts (a, b) on relative average gross margin (GM) from 105 years of simulated wheat yields for a clay soil at four locations in Western Australia (Mediterranean-type environment in which N fertilizer is often applied conservatively to limit growth and water use before flowering to have enough water from stored soil water and rainfall for the grain-filling period). Results of the conditional nitrogen (N) fertilizer strategies are expressed as percentage of average GM obtained with a fixed N rate and sowing strategy in all years. Fixed management in all years (--); two-category (above and below average rainfall categories) format (a): Merredin, fixed (●) and flexible sowing (○); Buntine fixed (▼) and flexible sowing (∇); three-category (above, average and below average rainfall categories) format (b): Merredin, fixed sowing (●); Buntine, fixed sowing (▼); Wongan Hills, fixed sowing (■); Mingenew, fixed sowing (▲), Western Australia. Only those cases are presented where, at an assumed skill level of 100%, the forecast distribution of GM was significantly different from that associated without forecast knowledge. *Source: Moeller et al. (2008).*

phases returned higher profits than an optimization of management ignoring ENSO phases. Although annual rainfall variability in some regions of the world, including the Pampas of Argentina, the east of South Africa, Australia and parts of India, is influenced by ENSO (5–25%) (Peel et al., 2002), the predictability of ENSO is inherently uncertain (Jones et al., 2000).

Advances in the application of climate prediction in agriculture have been made in regions affected by ENSO, such as northeast Australia (Everingham et al., 2002; McIntosh et al., 2005) or South America (Hansen et al., 1996; Podesta et al., 2002; Monzon et al., 2012). Linking GCMs as seasonal forecasting tools with crop models has been suggested as the next step to manage better increased seasonal rainfall variability in dryland agriculture (Hansen et al., 1996; Tao et al., 2009a; White et al., 2011). Asseng et al. (2012a,b) combined crop modeling with a GCM-based seasonal forecast and showed additional farm economic benefits from such forecast for regions where the forecast skill was high.

Diversifying farming systems can be an adaptive strategy in dryer climates. There are a range of options which can be explored with crop models to diversify systems; for example, using different varieties, varying sowing dates, using different crops or agro-forestry. Also combining livestock and crop production can be a risk-spreading diversifying strategy in some regions, while in some very dry regions the negative impact of livestock on soil cover for erosion control might prohibit livestock in a farming mix.

6.3 Breeding

New crop varieties are often cited as a possible adaptation to climate change (Salinger et al., 2005; Stoddard et al., 2011; Chapman et al., 2012). However, only a few experimental studies have tried to quantify the benefits from cultivar differences under elevated CO_2 conditions (Manderscheid and Weigel, 1997; Slafer and Rawson, 1997; Bunce, 2013). Challinor et al. (2007) used the GLAM model to quantify the impact of different degrees of heat tolerance in groundnut under climate change scenarios for India. They showed that heat stress is currently not a major determinant for groundnut yields but will become so in parts of northern and southern India under warmer climates. Also using a simulation model, Ludwig and Asseng (2010) showed that traits related to early vigor can potentially increase production in future warmer and dryer climates in environments with terminal water shortage (e.g. Mediterranean-type environments; Chapter 7) (Fig. 20.6). However, the rainfall seasonality in terms of water availability before and after flowering, the soil

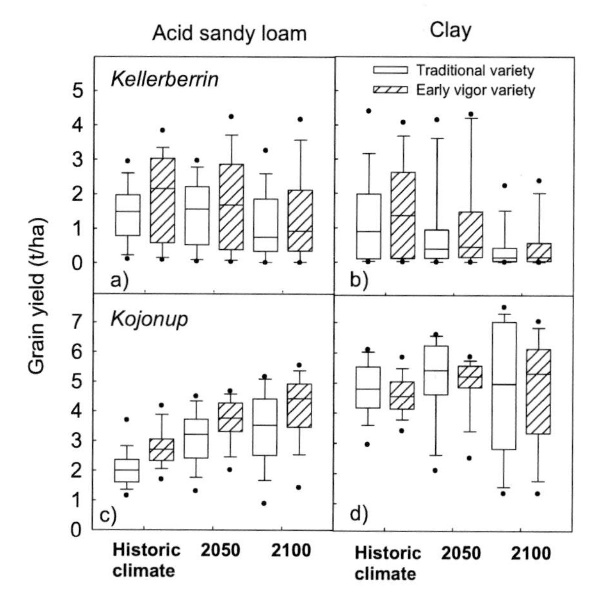

FIG. 20.6 Simulated grain yields for (a, b) Kelleberin (low rainfall location of < 350 mm annual rainfall) and (c, d) Kojonup (high rainfall location (> 450 mm annual rainfall), Western Australia for year 2050 with 525 ppm [CO2], + 2°C and −15% rainfall and for year 2100 with 700 ppm [CO2], + 4°C and −30% rainfall. Standard current cultivar (open bar) and with early vigor trait as an adaptation option (bar with pattern). Box plots represent 5 (dot), 10 (upper error bar), 25 (upper box edge), 50 (line in box), 75 (lower box edge), 90 (lower error bar) and 95 (dot) percentile probability of exceedance. *Source: Ludwig and Asseng (2010).*

water storage capacity and nutrient supply will all influence the benefits from early vigor. For example, on sandy soils in Mediterranean-type environments, early vigor will be most beneficial as yields are often limited by biomass accumulation at flowering (Asseng et al., 2002). In other environments, too vigorous early growth can deplete stored soil water too early, resulting in small grains with low yields (Van Herwaarden et al., 1998). Increased water-use efficiency under elevated CO_2 will reduce some of the latter negative effect. Breeding varieties with slower development in warmer climates, also depending on rainfall pattern changes, could be necessary to reduce the negative impacts of increasing temperatures (Ludwig and Asseng, 2010). Fine-tuning phenotype characteristics of crops for future climate scenarios might require models that link phenotypes to their underlying genetics, with such models currently under development (Hammer et al., 2006; Chapter 14).

Breeding in warm, dry environments should focus on improving agricultural production under even warmer, and more water scarce situations, and how to benefit from higher CO_2 concentrations. For example, a wheat-breeding program has been developed for higher transpiration efficiency by selecting for low carbon isotope discrimination (Condon et al., 1992; Richards, 2006). Breeding needs also to consider that elevated CO_2 will increase both photosynthesis and transpiration efficiency in the future, hence the more restricted benefits of putative improvement in these traits under future climates (Denison, 2009). Improving heat stress tolerance during grain filling of wheat (Spiertz et al., 2006; Semenov and Halford, 2009; Gouache et al., 2012) is another important trait which can help improve agricultural production in future climates with increased hot days.

In high latitude and high altitude environments, where agriculture has not been suitable due to low temperatures and short growing seasons, but become so with global temperature rise, crop simulation models could be used to explore the varietal needs for these potentially new agricultural regions (Chapter 4).

7 CONCLUDING REMARKS AND KNOWLEDGE GAPS

Physiology-based crop simulation models are a key tool in extrapolating the impact of climate change from limited experimental evidence, to broader climatic zones, soil types, management regimens, crops and climate change scenarios. Models also play an important role in assisting agriculture to adapt to these changes (Ewert, 2012). The impact of the individual climate change components, and the combined effect of climate change scenarios on crop production and externalities have been explored with such models. To counteract negative impacts of climate change and to capture some of the new opportunities of these changes, management options have been investigated and, to a lesser extent, trait options for breeding. While these models are a simplification of the reality, they allow a first assessment of the complexity of climate change impact and adaptation options in agriculture. Nevertheless, the simulation outcomes should always be evaluated critically and ground truthing via field experimentations for specific conditions remains essential (Ziska and Bunce, 2007). For example, several of the models used for climate impact studies have been successfully tested with FACE experiments of up to 550 ppm CO_2 (Ewert et al., 2007). However, some of these models, which assume a linear relationship between elevated CO_2 and crop response, have been used for scenarios with CO_2 levels above 550 ppm (Van Ittersum et al., 2003a), but only a few models have been tested with CO_2 up to 700 ppm (Ewert et al., 1999). In an experimental field study with potatoes, the crop response from ambient to 550 ppm atmospheric CO_2 was relatively higher than the response between 550 and 680 ppm CO_2 (Fangmeier et al., 2002).

Other knowledge gaps which require attention include issues related to interactions of climate factors with other environmental factors, water relations, specific crop feedbacks like sink–source relations, and extreme events, which are briefly discussed here.

When comparing many wheat crop models in standardized simulation experiments, model results diverged particularly with increasing temperature, suggesting that better understanding and model improvements of high-temperature interactions are needed for future climate change impact assessments (Asseng et al., 2013). Field experiments with temperature as a treatment factor (Ottman et al., 2012) and across different crop growth stages (Lizana and Calderini, 2013) are needed for model testing and improvements.

Interaction effects of elevated CO_2 with temperature are rarely considered in crop models, but the implications can be important, e.g. causing pollen sterility in rice (Ziska and Bunce, 2007). Whereas elevated CO_2 reduces stomatal conductance and transpiration, and increases TE, elevated temperature can have the opposite effects namely increasing stomata size, stomatal conductance and transpiration at the expense of TE (Soar et al., 2009). The combined effects of high CO_2 and elevated temperature remain largely unknown. As the frequency of extreme high temperatures (>35°C) during the growing season will increase, the interactions with elevated CO_2 need to be understood and considered (Attri and Rathore, 2003). Amthor (2001) indicated a reduction of yield with elevated CO_2 in combination with warming for wheat, compared with elevated CO_2 alone. Alonso et al. (2008) found a higher photosynthetic temperature optimum in wheat under elevated CO_2. No such interactive effects have been reported for soybean (Boote et al., 2005). Crop models need to be tested with such data sets of interactive effects to ensure their validity for climate change scenarios. Similarly, interaction effects of elevated CO_2 and flooded conditions, salinity (Ziska and Bunce, 2007) or soil constraints like soil compaction and sub-soil chemical toxicity are unknown, but need to be considered (Probert and Keating, 2000). Due to the interactive effects and feedbacks that emerge when climate factors are combined, experiments in which only single factors are manipulated are likely to be inadequate to predict fully the impacts of future climate change (Dermody, 2006).

Furthermore, studies with elevated CO_2 and warming at very low water supply, specific rainfall patterns and a range of soil water-holding capacities have been simulated in climate impact scenarios, but these models have never been tested with experimental data with such interactions as treatments. But, the combination effect could be very different to the sum of the single effects. For example, the net primary production response of grassland to interactive global changes (increased atmospheric CO_2, temperature, rainfall and nitrogen deposition) differed greatly from simple combinations of single factor responses (Shaw et al., 2002). In addition, genotype by CO_2 interactions (Manderscheid and Weigel, 1997; Slafer and Rawson, 1997) are usually ignored in simulation studies, mainly due to the lack of experimental data, but different cultivars are often used for various regions and might respond to elevated CO_2 differently. Linking crop models with the genetic underlying structure of crops (Chapter 14) might help to develop better adapted crops. Little is known about CO_2 acclimatization of crops and hence this is ignored in simulation studies.

Nitrogen is considered in most crop models. Other nutrients such as potassium and phosphorus can also become growth-limiting factors under elevated CO_2 (Hungate et al., 2003; Ziska and Bunce, 2007), but are only considered in a very few crop models (e.g. APSIM and DSSAT for P) or in climate impact studies. Nutrients could also become limiting when climate change alters soil factors, e.g. by restricting root growth for nutrient uptake (Brouder and Volenec, 2008).

Quality aspects of yield (e.g. protein or oil content) are often affected by climate change (Rogers et al., 1996; Kimball et al., 2001; Yang et al., 2006a,b, 2007c,d; Wu et al., 2007), but are not, or are less well simulated than yield itself, e.g. protein composition (Martre et al., 2004, 2006); and simulations on the impact of climate change will require a better understanding of the physiology of yield quality and its incorporation into crop models (Chapter 14).

Specific changes in climate could become critical in determining the impact on crops. For example, the importance of minimum temperature, rather than just mean temperature, for grain yield determination in wheat and rice needs more attention and might require modifications on how crop models respond to changes in minimum temperature (Fischer, 2007). Also, if the frequency of dry winds with high vapor pressure deficit does increase in some regions with climate change, their potential damage to crop demands further study as they are ignored in current crop models (Fischer, 2007).

Physiological processes such as sink–source relationships, correctly or incorrectly represented in crop models (Sinclair and Jamieson, 2006; Fischer, 2008), could be differently affected by elevated CO_2 and increased temperature (Jifon and Wolfe, 2005; Triboi et al., 2006; Ziska and Bunce, 2007). For example, wheat grain yield has been reported to be reduced under elevated CO_2 in sink-manipulated shoots, implying that a high source:sink ratio may result in a down-regulation of photosynthesis that more than offsets the direct stimulating effect of elevated CO_2 (Uddling et al., 2008).

In quantifying climate change impact and adaptation with crop models, a challenge remains in dealing with extreme events like heat stress, frost and excess water. While the frequency of extreme events is often predicted to change in the future, predicting the timing of extreme events is still poor (IPCC, 2013). However, the timing of critical events in relation to crop development is crucial in determining the impact on growth and yield (e.g. frost and flowering of cereals).

Despite these many knowledge gaps, models which integrate crop physiological understanding with the dynamics of water, carbon and nutrients have improved our understanding of the impacts of climate change on many aspects of local and world food production. They have also facilitated establishment of a new hypothesis for climate change studies. The application of these models stimulates investigations into climate change adaptation and assists in communicating to the public and policy makers that continued climate change could have a devastating impact on food supply. Continually improving crop models with further physiological understanding will help to improve our understanding of future climate change impact and adaptation options in agriculture.

References

Abraha, M.G., Savage, M.J., 2006. Potential impacts of climate change on the grain yield of maize for the midlands of KwaZulu-Natal, South Africa. Agric. Ecosyst. Environ. 115, 150–160.

Aggarwal, P., 2003. Impact of climate change on Indian agriculture. J. Plant Biol. 30, 189–198.

Aggarwal, P.K., Mall, R.K., 2002. Climate change and rice yields in diverse agro-environments of India. II. Effect of uncertainties in scenarios and crop models on impact assessment. Clim. Change 52, 331–343.

Ainsworth, E., McGrath, J., 2010. Direct effects of rising atmospheric carbon dioxide and ozone on crop yields. In: Lobell, D., Burke, M. (Eds.), Climate change and food security. Springer, The Netherlands, pp. 109–130.

Alexandrov, V.A., Hoogenboom, G., 2000. The impact of climate variability and change on crop yield in Bulgaria. Agric. For. Meteorol. 104, 315–327.

Alonso, A., Perez, P., Morcuende, R., Martinez-Carrasco, R., 2008. Future CO_2 concentrations, though not warmer temperatures, enhance wheat photosynthesis temperature responses. Physiol. Plant. 132, 102–112.

Amani, I., Fischer, R.A., Reynolds, M.P., 1996. Canopy temperature depression association with yield of irrigated spring wheat cultivars in a hot climate. J. Agron. Crop Sci. 76, 119–129.

Amir, J., Sinclair, T.R., 1991. A model of water limitation on spring wheat growth and yield. Field Crops Res. 28, 59–69.

Amthor, J.S., 2001. Effects of atmospheric CO_2 concentration on wheat yield: review of results from experiments using various approaches to control CO_2 concentration. Field Crops Res. 73, 1–34.

Andarzian, B., Bannayan, M., Steduto, P., et al., 2011. Validation and testing of the AquaCrop model under full and deficit irrigated wheat production in Iran. Agric. Water Manag. 100, 1–8.

Anderson, G., Fillery, I., Dunin, F., Dolling, P., Asseng, S., 1998. Nitrogen and water flows under pasture-wheat and lupin-wheat rotations in deep sands in Western Australia – 2. Drainage and nitrate leaching. Aust. J. Agric. Res. 49, 345–361.

Angus, J., Cunningham, R., Moncur, M., Mackenzie, D., 1981. Phasic development in field crops. 1. Thermal response in the seedling phase. Field Crops Res. 3, 365–378.

Anwar, M.R., O'Leary, G., McNeil, D., Hossain, H., Nelson, R., 2007. Climate change impact on rainfed wheat in south-eastern Australia. Field Crops Res. 104, 139–147.

Araki, H., Hamada, A., Hossain, M.A., Takahashi, T., 2012. Waterlogging at jointing and/or after anthesis in wheat induces early leaf senescence and impairs grain filling. Field Crops Res. 137, 27–36.

Arbuckle, Jr., J.G., Morton, L.W., Hobbs, J., 2013. Farmer beliefs and concerns about climate change and attitudes toward adaptation and mitigation: Evidence from Iowa. Clim. Change 118, 1–13.

Ashok, K., Sasikala, C., 2012. Farmers' vulnerability to rainfall variability and technology adoption in rain-fed tank irrigated agriculture. Agric. Econ. Res. Rev. 25, 267–278.

Asseng, S., Bar-Tal, A., Bowden, J.W., et al., 2002. Simulation of grain protein content with APSIM-Nwheat. Eur. J. Agron. 16, 25–42.

Asseng, S., Ewert, F., Rosenzweig, C., et al., 2013. Uncertainty in simulating wheat yields under climate change. Nat. Clim. Change 3, 827–832.

Asseng, S., Foster, I., Turner, N.C., 2011. The impact of temperature variability on wheat yields. Glob. Change Biol. 17, 997–1012.

Asseng, S., Foster, I.A.N., Turner, N.C., 2010. The impact of temperature variability on wheat yields. Glob. Change Biol. 17, 997–1012.

Asseng, S., Jamieson, P.D., Kimball, B., et al., 2004. Simulated wheat growth affected by rising temperature, increased water deficit and elevated atmospheric CO_2. Field Crops Res. 85, 85–102.

Asseng, S., Keating, B.A., Fillery, I.R.P., et al., 1998. Performance of the APSIM-wheat model in Western Australia. Field Crops Res. 57, 163–179.

Asseng, S., McIntosh, P.C., Wang, G.M., Khimashia, N., 2012a. Optimal N fertiliser management based on a seasonal forecast. Eur. J. Agron. 38, 66–73.

Asseng, S., Milroy, S.P., 2006. Simulation of environmental and genetic effects on grain protein concentration in wheat. Eur. J. Agron. 25, 119–128.

Asseng, S., Thomas, D., McIntosh, P., Alves, O., Khimashia, N., 2012b. Managing mixed wheat-sheep farms with a seasonal forecast. Agric. Sys. 113, 50–56.

Asseng, S., Travasso, M., Ludwig, F., Magrin, G., 2012c. Has climate change opened new opportunities for wheat cropping in Argentina? Clim. Change 117, 1–16.

Asseng, S., Turner, N., 2007. Modelling genotype × environment × management interactions to improve yield, water use efficiency and grain protein in wheat. Frontis 21, 91–102.

Asseng, S., Turner, N.C., Keating, B.A., 2001. Analysis of water- and nitrogen-use efficiency of wheat in a Mediterranean climate. Plant Soil 233, 127–143.

Attri, S.D., Rathore, L.S., 2003. Simulation of impact of projected climate change on wheat in India. Internatl. J. Climatol. 23, 693–705.

Auffhammer, M., Ramanathan, V., Vincent, J.R., 2012. Climate change, the monsoon, and rice yield in India. Clim. Change 111, 411–424.

Baenziger, P.S., McMaster, G.S., Wilhelm, W.W., Weiss, A., Hays, C.J., 2004. Putting genes into genetic coefficients. Field Crops Res. 90, 133–143.

Baethgen, W. E., Meinke, H., & Gimenez, A. (2003, November). Adaptation of agricultural production systems to climate variability and climate change: Climate Change Adaptation Conference on Insights and Tools for Adaptation: Learning from Climate Variability. Washington DC, November 18-20, 2013.

Baguis, P., Roulin, E., Willems, P., Ntegeka, V., 2010. Climate change scenarios for precipitation and potential evapotranspiration over central Belgium. Theor. Appl. Climatol. 99, 273–286.

Baker, D.N., Whisler, F.D., Parton, W.J., et al., 1985. The development of winter wheat: a physical physiological process model. ARS United States Department of Agriculture, Agricultural Research Service 38, 176-187.

Ballare, C.L., Scopel, A.L., Stapleton, A.E., Yanovsky, M.J., 1996. Solar ultraviolet-B radiation affects seedling emergence, DNA integrity, plant morphology, growth rate, and attractiveness to herbivore insects in Datura ferox. Plant Physiol. 112, 161–191.

Basso, B., Ritchie, J.T., Pierce, F.J., Braga, R.P., Jones, J.W., 2001. Spatial validation of crop models for precision agriculture. Agric. Syst. 68, 97–112.

Bassu, S., Asseng, S., Motzo, R., Giunta, F., 2009. Optimising sowing date of durum wheat in a variable Mediterranean environment. Field Crops Res. 111, 109–118.

Batisani, N., Yarnal, B., 2010. Rainfall variability and trends in semi-arid Botswana: Implications for climate change adaptation policy. App. Geog. 30, 483–489.

Bauwe, A., Criegee, C., Glatzel, S., Lennartz, B., 2012. Model-based analysis of the spatial variability and long-term trends of soil drought at Scots pine stands in northeastern Germany. Eur. J. For. Res. 131, 1–12.

Belanger, G., Rochette, P., Castonguay, Y., Bootsma, A., Mongrain, D., Ryan, D., 2002. Climate change and winter survival of perennial forage crops in eastern Canada. Agron. J. 94, 1120–1130.

Bell, M.A., Fischer, R.A., 1994. Using yield prediction models to assess yield gains – a case-study for wheat. Field Crops Res. 36, 161–166.

Betts, R.A., Falloon, P.D., Goldewijk, K.K., Ramankutty, N., 2007. Biogeophysical effects of land use on climate: Model simulations of radiative forcing and large-scale temperature change. Agric. For. Meteorol. 142, 216–233.

Bhattacharyya, T., Pal, D.K., Easter, M., et al., 2007. Modelled soil organic carbon stocks and changes in the Indo-Gangetic Plains, India from 1980 to 2030. Agric. Ecosyst. Environ. 122, 84–94.

Bondeau, A., Smith, P., Zaehle, S., et al., 2007. Modelling the role of agriculture for the 20th century global terrestrial carbon balance. Glob. Change Biol. 13, 679–706.

Boote, K.J., Allen, Jr., L.H., Prasad, P.V., et al., 2005. Elevated temperature and CO_2 impact pollination, reproductive growth and yield of globally important crops. J. Agric. Meteorol. Jpn. 60, 469–474.

Brisson, N., Gary, C., Justes, E., et al., 2003. An overview of the crop model STICS. Eur. J. Agron. 18, 309–332.

Brouder, S., Volenec, J., 2008. Impact of climate change on crop nutrient and water use efficiencies. Physiol. Plant. 133, 705–724.

Bunce, J.A., 2012. Responses of cotton and wheat photosynthesis and growth to cyclic variation in carbon dioxide concentration. Photosynthetica 50, 395–400.

Bunce, J.A., 2013. Effects of pulses of elevated carbon dioxide concentration on stomatal conductance and photosynthesis in wheat and rice. Physiol. Plant. 149, 214–221.

Bykov, O.D., Koshkin, V.A., Catsky, J., 1981. Carbon dioxide compensation concentration of C3 and C4 plants: Dependence on temperature [wheat, bean, beet, sugar beet]. Photosynthetica 15, 114–121.

Cai, W., Cowan, T., 2006. SAM and regional rainfall in IPCC AR4 models: Can anthropogenic forcing account for southwest Western Australian winter rainfall reduction? Geophys. Res. Lett. 33. DOI: 10.1029/2006GL028037

Calderini, D.F., Savin, R., Abeledo, L.G., Reynolds, M.P., Slafer, G.A., 2001. The importance of the period immediately preceding anthesis for grain weight determination in wheat. Euphytica 119, 199–204.

Cao, W., Moss, D.N., 1997. Modelling phasic development in wheat: a conceptual integration of physiological components. J. Agric. Sci. 129, 163–172.

Cerri, C.E.P., Easter, M., Paustian, K., et al., 2007. Predicted soil organic carbon stocks and changes in the Brazilian Amazon between 2000 and 2030. Agric. Ecosyst. Environ. 122, 58–72.

Chakraborty, S., Murray, G.M., Magarey, P.A., et al., 1998. Potential impact of climate change on plant diseases of economic significance to Australia. Australas. Plant Pathol. 27, 15–35.

Challinor, A.J., Simelton, E.S., Fraser, E.D.G., Hemming, D., Collins, M., 2010. Increased crop failure due to climate change: assessing adaptation options using models and socio-economic data for wheat in China. Environ. Res. Lett. 5, 034012.

Challinor, A.J., Slingo, J.M., Wheeler, T.R., Doblas-Reyes, F.J., 2005a. Probabilistic simulations of crop yield over western India using the DEMETER seasonal hindcast ensembles. Tellus Ser. a Dynam. Meteorol. Oceanogr. 57, 498–512.

Challinor, A.J., Wheeler, T.R., Craufurd, P.Q., Ferro, C.A.T., Stephenson, D.B., 2007. Adaptation of crops to climate change through genotypic responses to mean and extreme temperatures. Agric. Ecosyst. Environ. 119, 190–204.

Challinor, A.J., Wheeler, T.R., Craufurd, P.Q., Slingo, J.M., 2005b. Simulation of the impact of high temperature stress on annual crop yields. Agric. For. Meteorol. 135, 180–189.

Challinor, A.J., Wheeler, T.R., Craufurd, P.Q., Slingo, J.M., Grimes, D.I.F., 2004. Design and optimisation of a large-area process-based model for annual crops. Agric. For. Meteorol. 124, 99–120.

Chameides, W.L., Yu, H., Liu, S.C., et al., 1999. Case study of the effects of atmospheric aerosols and regional haze on agriculture: An opportunity to enhance crop yields in China through emission controls? Proc. Natl. Acad. Sci. USA 96, 13626–13633.

Chapman, S.C., Chakraborty, S., Dreccer, M.F., Howden, S.M., 2012. Plant adaptation to climate change –opportunities and priorities in breeding. Crop Past. Sci. 63, 251–268.

Chattopadhyay, N., Hulme, M., 1997. Evaporation and potential evapotranspiration in India under conditions of recent and future climate change. Agric. For. Meteorol. 87, 55–73.

Che, H.Z., Shi, G.Y., Zhang, X.Y., et al., 2005. Analysis of 40 years of solar radiation data from China, 1961-2000. Geophys. Res. Lett. 32, L06803.

Chen, C., Baethgen, W., Robertson, A., 2013. Contributions of individual variation in temperature, solar radiation and precipitation to crop yield in the North China Plain, 1961-2003. Clim. Change 116, 767–788.

Chen, C., Qian, C., Deng, A., Zhang, W., 2012a. Progressive and active adaptations of cropping system to climate change in Northeast China. Eur. J. Agron. 38, 94–103.

Chen, C., Wang, E., Yu, Q., Zhang, Y., 2010. Quantifying the effects of climate trends in the past 43 years (1961-2003) on crop growth and water demand in the North China Plain. Clim. Change 100, 559–578.

Chen, G., Liu, H., Zhang, J., Liu, P., Dong, S., 2012b. Factors affecting summer maize yield under climate change in Shandong Province in the Huanghuaihai Region of China. Internatl. J. Biometeorol. 56, 621–629.

Chipanshi, A.C., Chanda, R., Totolo, O., 2003. Vulnerability assessment of the maize and sorghum crops to climate change in Botswana. Clim. Change 61, 339–360.

Chun, J.A., Wang, Q., Timlin, D., Fleisher, D., Reddy, V.R., 2011. Effect of elevated carbon dioxide and water stress on gas exchange and water use efficiency in corn. Agric. For. Meteorol. 151, 378–384.

Ciais, P., Reichstein, M., Viovy, N., et al., 2005. Europe-wide reduction in primary productivity caused by the heat and drought in 2003. Nature 437, 529–533.

Cohen, S., Ianetz, A., Stanhill, G., 2002. Evaporative climate changes at Bet Dagan, Israel, 1964-1998. Agric. For. Meteorol. 111, 83–91.

Condon, A.G., Richards, R.A., Farquhar, G.D., 1992. The effect of variation in soil water availability, vapour pressure deficit and nitrogen nutrition on carbon isotope discrimination in wheat. Aust. J. Agric. Res. 43, 935–947.

Confalonieri, R., Bocchi, S., 2005. Evaluation of CropSyst for simulating the yield of flooded rice in northern Italy. Eur. J. Agron. 23, 315–326.

Conway, D., Schipper, E.L.F., 2011. Adaptation to climate change in Africa: Challenges and opportunities identified from Ethiopia. Glob. Environ. Change 21, 227–237.

Crimp, S., Howden, M., Power, B., Wang, E., De Voil, P., 2008. Global climate change impacts on Australia's wheat crops. Report for the Garnaut Climate Change Review Secretariat, p. 20.

Dang, Q.L., Margolis, H.A., Sy, M., Coyea, M.R., Collatz, G.J., Walthall, C.L., 1997. Profiles of photosynthetically active radiation, nitrogen and photosynthetic capacity in the boreal forest: Implications for scaling from leaf to canopy. J. Geophys. Res. Atmosph. 102, 28845–28859.

Daroub, S.H., Gerakis, A., Ritchie, J.T., Friesen, D.K., Ryan, J., 2003. Development of a soil-plant phosphorus simulation model for calcareous and weathered tropical soils. Agric. Syst. 76, 1157–1181.

Darwinkel, A., 1978. Patterns of tillering and grain production of winter wheat at a wide range of plant densities. Netherl. J. Agric. Sci. 26, 383–398.

Davis, A.S., Ainsworth, E.A., 2012. Weed interference with field-grown soyabean decreases under elevated [CO_2] in a FACE experiment. Weed Res. 52, 277–285.

Delve, R.J., Probert, M.E., Cobo, J.G., et al., 2009. Simulating phosphorus responses in annual crops using APSIM: model evaluation on contrasting soil types. Nutr. Cycl. Agroecosys. 84, 293–306.

Denison, R.F., 2009. Darwinian agriculture: real, imaginary and complex trade-offs as constraints and opportunities. In: Sadras, V.O., Calderini, D.F. (Eds.), Crop physiology: applications for genetic improvement and agronomy. Academic Press, San Diego, pp. 215–234.

Dermody, O., 2006. Mucking through multifactor experiments; design and analysis of multifactor studies in global change research. New Phytol. 172, 598–600.

Dias, A.S., Bagulho, A.S., Lidon, F.C., 2008. Ultrastructure and biochemical traits of bread and durum wheat grains under heat stress. Brazil. J. Plant Physiol. 20, 323–333.

Dias de Oliveira Eduardo, Bramley Helen, Siddique Kadambot H. M., Henty Samuel, Berger Jens, Palta Jairo A., 2013. Can elevated CO2 combined with high temperature ameliorate the effect of terminal drought in wheat?. Functional Plant Biology 40, 160-171. http://dx.doi.org/10.1071/FP12206

Ding, YH., Ren, GY., Shi, GY., Gong, Peng., Zheng, XH., Zhai, PM., et al., 2006. National Assessment report of climate change (I):Climate change in China and its future trend. Adv. Clim. Change Res. 02, 3-08.

Dolling, P.J., Robertson, M.J., Asseng, S., Ward, P.R., Latta, R.A., 2005. Simulating lucerne growth and water use on diverse soil types in a Mediterranean-type environment. Aust. J. Agric. Res. 56, 503–515.

Dong, W., Chen, J., Zhang, B., Tian, Y., Zhang, W., 2011. Responses of biomass growth and grain yield of midseason rice to the anticipated warming with FATI facility in East China. Field Crops Res. 123, 259–265.

Donohue, R.J., McVicar, T.R., Roderick, M.L., 2010. Assessing the ability of potential evaporation formulations to capture the dynamics in evaporative demand within a changing climate. J. Hydrol. 386, 186–197.

Dore, M.H.I., 2005. Climate change and changes in global precipitation patterns: What do we know? Environ. Internatl. 31, 1167–1181.

Dracup, M., Gregory, P., Belford, R., 1993. Restricted growth of lupin and wheat roots in the sandy A horizon of a yellow duplex soil. Aust J. Agric. Res. 44, 1273–1290.

Drake, B.G., Gonzalez-Meler, M.A., Long, S.P., 1997. More efficient plants: A consequence of rising atmospheric CO_2? Annu. Rev. Plant Physiol. Plant Mol. Biol. 48, 609–639.

Dreccer, M.F., Schapendonk, A.H., van Oijen, M., Pot, C.S., Rabbinge, R., 2000. Radiation and nitrogen use at the leaf and canopy level by wheat and oilseed rape during the critical period for grain number definition. Funct. Plant Biol. 27, 899–910.

Duursma, R.A., Payton, P., Bange, M.P., et al., 2013. Near-optimal response of instantaneous transpiration efficiency to vapour pressure deficit, temperature and [CO_2] in cotton (*Gossypium hirsutum* L.). Agric. For. Meteorol. 168, 168–176.

Easterling, D.R., Horton, B., Jones, P.D., et al., 1997a. Maximum and minimum temperature trends for the globe. Science 277, 364–367.

Easterling, W., 2007. Climate change and the adequacy of food and timber in the 21st century. Proc. Natl. Acad. Sci. USA 104, 19679–119679.

Easterling, W.E., Hays, C.J., Easterling, M.M., Brandle, J.R., 1997b. Modelling the effect of shelterbelts on maize productivity under climate change: An application of the EPIC model. Journal: Agriculture, Ecosystems & Environment 61, 163-176.

Elsgaard, L., Borgesen, C.D., Olesen, J.E., et al., 2012. Shifts in comparative advantages for maize, oat and wheat cropping under climate change in Europe. Food Add. Contam. A Chem. Anal. Control Expos. Risk Assess. 29, 1514–1526.
Emberson, L.D., Bueker, P., Ashmore, M.R., et al., 2009. A comparison of North American and Asian exposure-response data for ozone effects on crop yields. Atmosph. Environ. 43, 1945–1953.
Erbs, M., Manderscheid, R., Jansen, G., Seddig, S., Pacholski, A., Weigel, H.-J., 2010. Effects of free-air CO_2 enrichment and nitrogen supply on grain quality parameters and elemental composition of wheat and barley grown in a crop rotation. Agric. Ecosyst. Environ. 136, 59–68.
Evans, L.T., 1993. Crop evolution, adaptation and yield. Cambridge University Press, Cambridge, UK.
Evans, L.T., Fischer, R.A., 1999. Yield potential: Its definition, measurement, and significance. Crop Sci. 39, 1544–1551.
Everingham, Y.L., Muchow, R.C., Stone, R.C., Inman-Bamber, N.G., Singels, A., Bezuidenhout, C.N., 2002. Enhanced risk management and decision-making capability across the sugarcane industry value chain based on seasonal climate forecasts. Agric. Syst. 74, 459–477.
Ewert, F., 2012. ADAPTATION Opportunities in climate change? Nat. Clim. Change 2, 153–154.
Ewert, F., Porter, J.R., Rounsevell, M.D.A., 2007. Crop models, CO_2, and climate change. Science 315, 459–1459.
Ewert, F., Rounsevell, M.D.A., Reginster, I., Metzger, M.J., Leemans, R., 2005. Future scenarios of European agricultural land use I. Estimating changes in crop productivity. Agric. Ecosyst. Environ. 107, 101–116.
Ewert, F., van Oijen, M., Porter, J.R., 1999. Simulation of growth and development processes of spring wheat in response to CO_2 and ozone for different sites and years in Europe using mechanistic crop simulation models. Eur. J. Agron. 10, 231–247.
Falloon, P., Jones, C.D., Cerri, C.E., et al., 2007. Climate change and its impact on soil and vegetation carbon storage in Kenya, Jordan, India and Brazil. Agric. Ecosyst. Environ. 122, 114–124.
Fangmeier, A., De Temmerman, L., Black, C., Persson, K., Vorne, V., 2002. Effects of elevated CO_2 and/or ozone on nutrient concentrations and nutrient uptake of potatoes. Eur. J. Agron. 17, 353–368.
Farooq, M., Bramley, H., Palta, J.A., Siddique, K.H., 2011. Heat stress in wheat during reproductive and grain-filling phases. Crit. Rev. Plant Sci. 30, 491–507.
Farquhar, G.D., Caemmerer, S.V., 1982. Modelling of photosynthetic response to environmental conditions. In : Lange, O.L., Nobel, P.S., Osmond, C.B., Ziegler, H., (eds), Encyclopedia of plant physiology. New series. Volume 12B. Physiological plant ecology. II. Water relations and carbon assimilation. pp. 549-587. Springer Berlin Heidelberg.
Farquhar, G.D., Dubbe, D.R., Raschke, K., 1978. Gain of the feedback loop involving carbon dioxide and stomata. Plant Physiol. 62, 406–412.
Farquhar, G.D., Roderick, M.L., 2003. Atmospheric science: Pinatubo, diffuse light, and the carbon cycle. Science 299, 1997–1998.
Farre, I., Robertson, M.J., Asseng, S., French, R.J., Dracup, M., 2004. Simulating lupin development, growth, and yield in a Mediterranean environment. Aust. J. Agric. Res. 55, 863–877.
Farre, I., Robertson, M.J., Walton, G.H., Asseng, S., 2002. Simulating phenology and yield response of canola to sowing date in Western Australia using the APSIM model. Aust. J. Agric. Res. 53, 1155–1164.
Fernández, E.J., López-Bellido, L., Fuentes, M., Fernández, J., 1996. LUPINMOD: a simulation model for the white lupin crop. Agric. Syst. 52, 57–82.
Fernando, N., Panozzo, J., Tausz, M., Norton, R., Fitzgerald, G., Seneweera, S., 2011. Wheat grain quality at elevated [CO_2] under Mediterranean climate conditions. Universitas 21 Graduate Research Conference on Food, p. 123.
Ferris, R., Ellis, R.H., Wheeler, T.R., Hadley, P., 1998. Effect of high temperature stress at anthesis on grain yield and biomass of field-grown crops of wheat. Ann. Bot. 82, 631–639.
Fischer, R., 1979. Growth and water limitation to dryland wheat yield in Australia – Physiological framework. J. Aust. Inst. Agric. Sci. 45, 83–94.
Fischer, R., 2007. Understanding the physiological basis of yield potential in wheat. J. Agric. Sci. 145, 99–113.
Fischer, R.A., 2008. The importance of grain or kernel number in wheat: A reply to Sinclair and Jamieson. Field Crops Res. 105, 15–21.
Fischer, R.A.T., Edmeades, G.O., 2010. Breeding and cereal yield progress. Crop Sci. 50, 85–98.
Folland, C., Rayner, N., Brown, S., et al., 2001. Global temperature change and its uncertainties since 1861. Geophys. Res. Lett. 28, 2621–2624.
Fuhrer, J., Booker, F., 2003. Ecological issues related to ozone: agricultural issues. Environ. Internatl. 29, 141–154.
Gao, Z., Gao, W., Chang, N.-B., 2011. Integrating temperature vegetation dryness index (TVDI) and regional water stress index (RWSI) for drought assessment with the aid of LANDSAT TM/ETM+ images. Internatl. J. Appl. Earth Observ. Geoinformat. 13, 495–503.
Garcia, R.L., Long, S.P., Wall, G.W., et al., 1998. Photosynthesis and conductance of spring-wheat leaves: field response to continuous free-air atmospheric CO_2 enrichment. Plant Cell Environ. 21, 659–669.
George, R., McFarlane, D., Nulsen, B., 1997. Salinity threatens the viability of agriculture and ecosystems in Western Australia. Hydrogeol. J. 5, 6–21.
Gifford, R., Angus, J., Barrett, D., et al., 1998. Climate change and Australian wheat yield. Nature 391, 448–449.
Gifford, R.M., Morison, J.I.L., 1993. Crop responses to the global increase in atmospheric carbon dioxide concentration. International Crop Science 1. Crop Science Society of America, Inc, Ames, Iowa, pp. 325-331.

Giorgi, F., Bi, X., 2005. Regional changes in surface climate interannual variability for the 21st century from ensembles of global model simulations. Geophys. Res. Lett. 32. DOI: 10.1029/2005GL023002..

Godden, D., Batterham, R., Drynan, R., 1998. Climate change and Australian wheat yield. Nature 391, 447–448.

Gouache, D., Le Bris, X., Bogard, M., Deudon, O., Pagé, C., Gate, P., 2012. Evaluating agronomic adaptation options to increasing heat stress under climate change during wheat grain filling in France. Eur. J. Agron. 39, 62–70.

Goudriaan, J., Laar, H.H.v., Keulen, H.v., Louwerse, W., 1985. Photosynthesis, CO_2 and plant production. NATO advanced study institutes series. Ser. A Life Sci. 86, 107–122.

Goudriaan, J., Van Laar, H.H., 1994. Modelling potential crop growth processes. Kluwer Academic Publishers, The Netherlands.

Grandjean, A., Fuhrer, J., 1989. Growth and leaf senescence in spring wheat (*Triticum aestivum*) grown at different ozone concentrations in open-top field chambers. Physiol. Plant. 77, 389–394.

Grant, R.F., Garcia, R.L., Pinter, P.J., et al., 1995. Interaction between atmospheric CO2 concentration and water deficit on gas exchange and crop growth: Testing of ecosys with data from the Free Air CO_2 Enrichment (FACE) experiment. Glob. Change Biol. 1, 443–454.

Grantz, D.A., Yang, S., 2000. Ozone impacts on allometry and root hydraulic conductance are not mediated by source limitation nor developmental age. J. Exp. Bot. 51, 919–927.

Grossman-Clarke, S., Pinter, Jr., P.J., Kartschall, T., et al., 2001. Modelling a spring wheat crop under elevated CO_2 and drought. New Phytol. 150, 315–335.

Gu, L., Baldocchi, D.D., Wofsy, S.C., et al., 2003. Response of a deciduos forest to the Mount Pinatubo eruption: enhanced photosynthesis. Science 299, 2035–2038.

Hall, A., Richards, R., 2013. Prognosis for genetic improvement of yield potential and water-limited yield of major grain crops. Field Crops Res. 143, 18–33.

Hammer, G., Cooper, M., Tardieu, F., et al., 2006. Models for navigating biological complexity in breeding improved crop plants. Trends Plant Sci. 11, 587–593.

Hammer, G.L., Hansen, J.W., Phillips, J.G., et al., 2001. Advances in application of climate prediction in agriculture. Agric. Syst. 70, 515–553.

Hammer, G.L., Holzworth, D.P., Stone, R., 1996. The value of skill in seasonal climate forecasting to wheat crop management in a region with high climatic variability. Australian Journal of Agricultural Research 47, 717-737.

Hansen, J.M., Ehler, N., Karlsen, P., Høgh-Schmidt, K., Rosenqvist, E., 1996. Decreasing the environmental load by a photosynthetic based system for greenhouse climate control. International symposium on plant production in closed ecosystems. Internatl. Sympos. Plant Product. Closed Ecosyst., 105–110.

Hansen, J.W., 2002. Realizing the potential benefits of climate prediction to agriculture: issues, approaches, challenges. Agric. Syst. 74, 309–330.

Hansen, J.W., 2005. Integrating seasonal climate prediction and agricultural models for insights into agricultural practice. Philosoph. Transact. R. Soc. B Biol. Sci. 360, 2037–2047.

Hao, F., Chen, S., Ouyang, W., Shan, Y., Qi, S., 2013. Temporal rainfall patterns with water partitioning impacts on maize yield in a freeze-thaw zone. J. Hydrol, 486, 412–419.

Hartkamp, A.D., Hoogenboom, G., White, J.W., 2002. Adaptation of the CROPGRO growth model to velvet bean (*Mucunapruriens*): I. Model development. Field Crops Res. 78, 9–25.

Hasegawa, T., Sakai, H., Tokida, T., et al., 2013. Rice cultivar responses to elevated CO_2 at two free-air CO_2 enrichment (FACE) sites in Japan. Funct. Plant Biol. 40, 148–159.

Hennessy, K., Fawcett, R., Kirono, D., et al., 2008. An assessment of the impact of climate change on the nature and frequency of exceptional climatic events. CSIRO - Bureau of Meterology.

Holden, N.M., Brereton, A.J., Fealy, R., Sweeney, J., 2003. Possible change in Irish climate and its impact on barley and potato yields. Agricultural and Forest Meteorology 116, 181-196.

Hollaway, M.J., Arnold, S.R., Challinor, A.J., Emberson, L.D., 2012. Intercontinental trans-boundary contributions to ozone-induced crop yield losses in the Northern Hemisphere. Biogeosciences 9, 271–292.

Hope, P., Timbal, B., Fawcett, R., 2010. Associations between rainfall variability in the southwest and southeast of Australia and their evolution through time. Internatl. J. Climatol. 30, 1360–1371.

Howden, S., Meinke, H., Power, B., McKeon, G., 2003. Risk management of wheat in a non-stationary climate: frost in Central Queensland. International Congress on Modelling and Simulation, pp. 17–22.

Hu, J., Cao, W., Zhang, J., Jiang, D., Feng, J., 2004. Quantifying responses of winter wheat physiological processes to soil water stress for use in growth simulation modeling. Pedosphere 14, 509–518.

Hungate, B., Dukes, J., Shaw, M., Luo, Y., Field, C., 2003. Nitrogen and climate change. Science 302, 1512–1513.

Hungate, B.A., Reichstein, M., Dijkstra, P., et al., 2002. Evapotranspiration and soil water content in a scrub-oak woodland under carbon dioxide enrichment. Glob. Change Biol. 8, 289–298.

IPCC, 2013. Climate Change 2013 The Physical Science Basis. Contribution of Working Group I to the Fifth Assessment Report of the Intergovernmental Panel on Climate Change. Stocker, T.F., Qin, D., Plattner, G.K., Tignor, M., et al. (eds). Cambridge, UK and New York, USA.

Izaguirre, M.M., Scopel, A.L., Baldwin, I.T., Ballaré, C.L., 2003. Convergent responses to stress. Solar ultraviolet-B radiation and Manduca sexta herbivory elicit overlapping transcriptional responses in field-grown plants of Nicotiana longiflora. Plant Physiol. 132, 1755–1767.

Izquierdo, N., Aguirrezábal, L., Andrade, F., Pereyra, V., 2002. Night temperature affects fatty acid composition in sunflower oil depending on the hybrid and the phenological stage. Field Crops Res. 77, 115–126.

Jame, Y.W., Cutforth, H.W., Ritchie, J.T., 1999. Temperature response function for leaf appearance rate in wheat and corn. Can. J. Plant Sci. 79, 1–10.

Jamieson, P., Cloughley, C., 2001. Impacts of Climate Change on Wheat Production. The Effects of Climate Change and Variation in New Zealand, p. 57.

Jamieson, P.D., Asseng, S., Chapman, S.C., et al., 2010. Modelling wheat production. In: The world wheat book, (Eds.). M van Ginkel, A Bonjean, W Angus. Lavoisier Publishing, Paris, p. 40.

Jamieson, P.D., Berntsen, J., Ewert, F., et al., 2000. Modelling CO_2 effects on wheat with varying nitrogen supplies. Agric. Ecosyst. Environ. 82, 27–37.

Jamieson, P.D., Porter, J.R., Goudriaan, J., Ritchie, J.T., van Keulen, H., Stol, W., 1998a. A comparison of the models AFRCWHEAT2, CERES-wheat, Sirius, SUCROS2 and SWHEAT with measurements from wheat grown under drought. Field Crops Res. 55, 23–44.

Jamieson, P.D., Semenov, M.A., Brooking, I.R., Francis, G.S., 1998b. Sirius: a mechanistic model of wheat response to environmental variation. Eur. J. Agron. 8, 161–179.

Jiang, D., Fan, X., Dai, T., Cao, W., 2008. Nitrogen fertiliser rate and post-anthesis waterlogging effects on carbohydrate and nitrogen dynamics in wheat. Plant Soil 304, 301–314.

Jiang, X.-j., Tang, L., Liu, X.-j., Cao, W.-x., Zhu, Y., 2013. Spatial and temporal characteristics of rice potential productivity and potential yield increment in main production regions of China. J. Integ. Agric. 12, 45–56.

Jifon, J.L., Wolfe, D.W., 2005. High temperature-induced sink limitation alters growth and photosynthetic acclimation to elevated CO_2 in bean (*Phaseolus vulgaris* L.). J. Am. Soc. Horticult. Sci. 130, 515–520.

Jingyun, Z., Quansheng, G., Zhixin, H., 2002. Impacts of climate warming on plants phenophases in China for the last 40 years. Chin. Sci. Bull. 47, 1826–1831.

Jones, C.A., Kiniry, J.R., 1986. CERES-Maize: A simulation model of maize growth and development. Texas A&M University Press, College Station, Texas.

Jones, J.W., Hansen, J.W., Royce, F.S., Messina, C.D., 2000. Potential benefits of climate forecasting to agriculture. Agric. Ecosyst. Environ. 82, 169–184.

Jones, J.W., Hoogenboom, G., Porter, C.H., et al., 2003. The DSSAT cropping system model. Eur. J. Agron. 18, 235–265.

Jones, P.G., Thornton, P.K., 2003. The potential impacts of climate change on maize production in Africa and Latin America in 2055. Glob. Environ. Change 13, 51–59.

Kakani, V., Reddy, K., Zhao, D., Sailaja, K., 2003. Field crop responses to ultraviolet-B radiation: a review. Agric. For. Meteorol. 120, 191–218.

Kamoni, P.T., Gicheru, P.T., Wokabi, S.M., et al., 2007. Predicted soil organic carbon stocks and changes in Kenya between 1990 and 2030. Agric. Ecosyst. Environ. 122, 105–113.

Kang, S.Z., Zhang, F.C., Hu, X.T., Zhang, J.H., 2002. Benefits of CO2 enrichment on crop plants are modified by soil water. Plant and Soil. 238 (1), 69–77, doi: 10.1023/A:1014244413067.

Kartschall, T., Grossman, S., Pinter, P.J., et al., 1995. A simulation of phenology, growth, carbon dioxide exchange and yields under ambient atmosphere and free-air carbon dioxide enrichment (FACE) Maricopa, Arizona, for wheat. J. Biogeog. 22, 611–622.

Keating, B.A., Carberry, P.S., Hammer, G.L., et al., 2003. An overview of APSIM, a model designed for farming systems simulation. Eur. J. Agron. 18, 267–288.

Keating, B.A., McCown, R.L., Cresswell, H.P., 1995. Paddock-scale models and catchment-scale problems: The role for APSIM in the Liverpool Plains. In: Proceedings of MODSIM '95. International Congress on Modelling and Simulation. Modelling and Simulation Society of Australia, The University of Newcastle, NSW, Australia, pp. 158–165.

Keating, B.A., Meinke, H., Probert, M.E., Huth, N.I., Hills, I., 2001. NWheat: Documentation and performance of a wheat module for APSIM. CSIRO Tropical Agriculture, Indooroopilly, Queensland, p. 66.

Keating, B.A., Robertson, M.J., Muchow, R.C., Huth, N.I., 1999. Modelling sugarcane production systems I. Development and performance of the sugarcane module. Field Crops Res. 61, 253–271.

Kersebaum, K., 2007. Modelling nitrogen dynamics in soil-crop systems with HERMES. Nutr. Cycl. Agroecosyst. 77, 39–52.

Kersebaum, K.C., Nain, A.S., Nendel, C., Gandorfer, M., Wegehenkel, M., 2008. Simulated effect of climate change on wheat production and nitrogen management at different sites in Germany. J. Agrometeorol. 10, 266–273.

Khan, S., Hanjra, M.A., Mu, J., 2009. Water management and crop production for food security in China: A review. Agric. Water Manag. 96, 349–360.

Kimball, B., Morris, C., Pinter, P., et al., 2001. Elevated CO_2, drought and soil nitrogen effects on wheat grain quality. New Phytol. 150, 295–303.

Kimball, B., Pinter, P., Garcia, R., et al., 1995. Productivity and water use of wheat under free-air CO_2 enrichment. Glob. Change Biol. 1, 429–442.

Kimball, B.A., LaMorte, R.L., Pinter, Jr., P.J., et al., 1999. Free-air CO_2 enrichment and soil nitrogen effects on energy balance and evapotranspiration of wheat. Water Resourc. Res. 35, 1179–1190.

Kiniry, J., Major, D., Izaurralde, R., et al., 1995. EPIC model parameters for cereal, oilseed and forage crops in the northern great plains region. Can. J. Plant Sci. 75, 679–688.

Kropff, M.J., Cassman, K.G., Penning de Vries, F.W.T., Laar, H.H.v., 1993. Increasing the yield plateau in rice and the role of global climate change. J. Agric. Meteorol. 48, 795–798.

Laar, H.H., Goudriaan, J., Keulen, H., 1992. Simulation of crop growth for potential and water-limited production situations (as applied to spring wheat). Simulation reports CABO-TT 27. CABO-DLO, WAU-TPE, Wageningen, The Netherlands, p. 78.

Lawlor, D.W., Mitchell, R.A.C., 2000. Crop ecosystems responses to climatic change: wheat. Climate change and global crop productivity. CAB International, Cambridge, pp. 57–80.

Leakey, A.D.B., 2009. Rising atmospheric carbon dioxide concentration and the future of C4 crops for food and fuel. Proc. R. Soc. B Biol. Sci. 276, 2333–2343.

Lehnherr, B., Grandjean, A., Mächler, F., Fuhrer, J., 1987. The effect of ozone in ambient air on ribulosebisphosphate carboxylase/oxygenase activity decreases photosynthesis and grain yield in wheat. J. Plant Physiol. 130, 189–200.

Liepert, B.G., 2002. Observed reductions of surface solar radiation at sites in the United States and worldwide from 1961 to 1990. Geophys. Res. Lett. 29, 1421.

Liepert, B.G., Kukla, G.J., 1997. Decline in global solar radiation with increased horizontal visibility in Germany between 1964 and 1990. J. Climate 10, 2391–2401.

Lilley, J., Bolger, T., Gifford, R., 2001. Productivity of *Trifolium subterraneum* and *Phalaris aquatica* under warmer, high CO_2 conditions. New Phytol. 150, 371–383.

Limon-Ortega, A., Sayre, K., 2012. Rainfall as a limiting factor for wheat grain yield in permanent raised-beds. Agron. J. 104, 1171–1175.

Liu, B., Xu, M., Henderson, M., Qi, Y., 2005. Observed trends of precipitation amount, frequency, and intensity in China, 1960-2000. J. Geophys. Res. Atmosph. 110, D08103.

Liu, J., Pattey, E., Miller, J.R., McNairn, H., Smith, A., Hu, B., 2010a. Estimating crop stresses, aboveground dry biomass and yield of corn using multi-temporal optical data combined with a radiation use efficiency model. Remote Sens. Environ. 114, 1167–1177.

Liu, L., Wang, E., Zhu, Y., Tang, L., 2012. Contrasting effects of warming and autonomous breeding on single-rice productivity in China. Agric. Ecosyst. Environ. 149, 20–29.

Liu, L., Wang, E., Zhu, Y., Tang, L., Cao, W., 2013a. Effects of warming and autonomous breeding on the phenological development and grain yield of double-rice systems in China. Agric. Ecosyst. Environ. 165, 28–38.

Liu, Y., Wang, E., Yang, X., Wang, J., 2010b. Contributions of climatic and crop varietal changes to crop production in the North China Plain, since 1980s. Glob. Change Biol. 16, 2287–2299.

Liu, Z., Yang, X., Chen, F., Wang, E., 2013b. The effects of past climate change on the northern limits of maize planting in Northeast China. Clim. Change 117, 891–902.

Lizana, X.C., Calderini, D.F., 2013. Yield and grain quality of wheat in response to increased temperatures at key periods for grain number and grain weight determination: considerations for the climatic change scenarios of Chile. J. Agric. Sci. 151, 209–221.

Lobell, D.B., 2007. Changes in diurnal temperature range and national cereal yields. Agric. For. Meteorol. 145, 229–238.

Lobell, D.B., Burke, M.B., 2010. On the use of statistical models to predict crop yield responses to climate change. Agric. For. Meteorol. 150, 1443–1452.

Lobell, D.B., Field, C.B., 2007. Global scale climate – crop yield relationships and the impacts of recent warming. Environ. Res. Lett. 2, 014002.

Lobell, D.B., Hammer, G.L., McLean, G., Messina, C., Roberts, M.J., Schlenker, W., 2013. The critical role of extreme heat for maize production in the United States. Nat. Clim. Change 3, 497–501.

Lobell, D.B., Ortiz-Monasterio, J.I., 2007. Impacts of day versus night temperatures on spring wheat yields: A comparison of empirical and CERES model predictions in three locations. Agron. J. 99, 469–477.

Lobell, D.B., Ortiz-Monasterio, J.I., Asner, G.P., Matson, P.A., Naylor, R.L., Falcon, W.P., 2005. Analysis of wheat yield and climatic trends in Mexico. Field Crops Res. 94, 250–256.

Lobell, D.B., Schlenker, W., Costa-Roberts, J., 2011. Climate trends and global crop production since 1980. Science 333, 616–620.

Lobell, D.B., Sibley, A., Ivan Ortiz-Monasterio, J., 2012. Extreme heat effects on wheat senescence in India. Nat. Clim. Change 2, 186–189.

Long, S.P., Ainsworth, E.A., Leakey, A.D.B., Nosberger, J., Ort, D.R., 2006. Food for thought: Lower-than-expected crop yield stimulation with rising CO_2 concentrations. Science 312, 1918–1921.

Lu, Z., Radin, J.W., Turcotte, E.L., Percy, R., Zeiger, E., 1994. High yields in advanced lines of Pima cotton are associated with higher stomatal conductance, reduced leaf area and lower leaf temperature. Physiol. Plant. 92, 266–272.

Ludwig, F., Asseng, S., 2006. Climate change impacts on wheat production in a Mediterranean environment in Western Australia. Agric. Syst. 90, 159–179.

Ludwig, F., Asseng, S., 2010. Potential benefits of early vigor and changes in phenology in wheat to adapt to warmer and drier climates. Agric. Syst. 103, 127–136.

Ludwig, F., Milroy, S.P., Asseng, S., 2009. Impacts of recent climate change on wheat production systems in Western Australia. Clim. Change 92, 495–517.

Luo, Q., Bellotti, W., Williams, M., Wang, E., 2009. Adaptation to climate change of wheat growing in South Australia: Analysis of management and breeding strategies. Agric. Ecosyst. Environ. 129, 261–267.

Luo, Q.Y., Williams, W., Bryan, B., 2005. Potential impact of climate change on wheat yield in South Australia. Agric. For. Meteorol. 132, 273–285.

Lv, Z., Liu, X., Cao, W., Zhu, Y., 2013. Climate change impacts on regional winter wheat production in main wheat production regions of China. Agric. For. Meteorol. 171, 234–248.

Madan, P., Jagadish, S.V.K., Craufurd, P.Q., Fitzgerald, M., Lafarge, T., Wheeler, T.R., 2012. Effect of elevated CO_2 and high temperature on seed-set and grain quality of rice. J. Exp. Bot. 63, 3843–3852.

Magrin, G.O., Travasso, M.I., Rodriguez, G.R., 2005. Changes in climate and crop production during the 20th century in Argentina. Clim. Change 72, 229–249.

Mall, R.K., Lal, M., Bhatia, V.S., Rathore, L.S., Singh, R., 2004. Mitigating climate change impact on soybean productivity in India: a simulation study. Agric. For. Meteorol. 121, 113–125.

Mall, R.K., Singh, R., Gupta, A., Srinivasan, G., Rathore, L.S., 2006. Impact of climate change on Indian agriculture: A review. Clim. Change 78, 445–478.

Manderscheid, R., Weigel, H., 1997. Photosynthetic and growth responses of old and modern spring wheat cultivars to atmospheric CO_2 enrichment. Agric. Ecosyst. Environ. 64, 65–73.

Martre, P., Jamieson, P.D., Semenov, M.A., Zyskowski, R.F., Porter, J.R., Triboi, E., 2006. Modelling protein content and composition in relation to crop nitrogen dynamics for wheat. Eur. J. Agron. 25, 138–154.

Martre, P., Porter, J.R., Jamieson, P.D., Henton, S.M., Triboi, E., 2004. A dynamic model of the effects of nitrogen fertilization, water deficit, and temperature on grain protein level and composition for bread wheat (*Triticum aestivum* L.). In: Lafiandra, D., Masci, S., Dovidio, R., (eds), Gluten proteins, pp. 196–199. The Royal society of chemistry. Cambridge, UK.

Matthews, R.B., Stephens, W. (Eds.), 2002. Crop-soil simulation models: applications in developing countries. CAB International, Wallingford, UK, p. 277.

Mazza, C.A., Giménez, P.I., Kantolic, A.G., Ballaré, C.L., 2012. Beneficial effects of solar UV-B radiation on soybean yield mediated by reduced insect herbivory under field conditions. Physiol. Plant 147, 307–315.

McIntosh, P.C., Ash, A.J., Stafford Smith, M., 2005. From oceans to farms: The value of a novel statistical climate forecast for agricultural management. J. Clim. 18, 4287–4302.

McIntosh, P.C., Pook, M.J., Risbey, J.S., Lisson, S.N., Rebbeck, M., 2007. Seasonal climate forecasts for agriculture: Towards better understanding and value. Field Crops Res. 104, 130–138.

Mearns, L.O., Rosenzweig, C., Goldberg, R., 1996. The effect of changes in daily and interannual climatic variability on CERES-Wheat: A sensitivity study. Clim. Change 32, 257–292.

Mearns, L.O., Rosenzweig, C., Goldberg, R., 1997. Mean and variance change in climate scenarios: Methods, agricultural applications, and measures of uncertainty. Clim. Change 35, 367–396.

Meinke, H., Stone, R.C., 2005. Seasonal and inter-annual climate forecasting: the new tool for increasing preparedness to climate variability and change in agricultural planning and operations. Clim. Change 70, 1969–1976.

Menzel, A., Sparks, T., Estrella, N., et al., 2006. European phenological response to climate change matches the warming pattern. Glob. Change Biol. 12, 1969–1976.

Mercado, L.M., Bellouin, N., Sitch, S., et al., 2009. Impact of changes in diffuse radiation on the global land carbon sink. Nature 458, 1014–1017.

Milroy, S.P., Bange, M.P., Hearn, A.B., 2004. Row configuration in rain fed cotton systems: modification of the OZCOT simulation model. Agric. Syst. 82, 1–16.

Mills, G., Hayes, F., Simpson, D., et al., 2011. Evidence of widespread effects of ozone on crops and (semi-) natural vegetation in Europe (1990-2006) in relation to AOT40- and flux-based risk maps. Glob. Change Biol. 17, 592–613.

Moeller, C., Smith, I., Asseng, S., Ludwig, F., Telcik, N., 2008. The potential value of seasonal forecasts of rainfall categories – Case studies from the wheatbelt in Western Australia's Mediterranean region. Agric. For. Meteorol. 148, 606–618.

Moeller, C., Asseng, S., Berger, J., Milroy, S.P., 2009. Plant available soil water at sowing in Mediterranean environments – Is it a useful criterion to aid nitrogen fertiliser and sowing decisions? Field Crops Res. 114, 127–136.

Monteith, J.L., 1965. Evaporation and environment. Symp. Soc. Exp. Biol. 19, 205–234.

Monteith, J.L., 1977. Climate and efficiency of crop production in britain. Philosoph. Transact. R. Soc. Lond. Ser. B Biol. Sci. 281, 277–294.

Monzon, J.P., Sadras, V.O., Andrade, F.H., 2012. Modelled yield and water use efficiency of maize in response to crop management and Southern Oscillation Index in a soil-climate transect in Argentina. Field Crops Res. 98, 83–90.

Monzon, J.P., Sadras, V.O., Abbate, P.A., Caviglia, O.P., 2007. Modelling management strategies for wheat-soybean cropping systems in the Southern Pampas. Field Crops Res. 101, 44–52.

Moriondo, M., Bindi, M., Kundzewicz, Z.W., et al., 2010. Impact and adaptation opportunities for European agriculture in response to climatic change and variability. Mitigat. Adapt. Strat. Glob. Change 15, 657–679.

Moriondo, M., Giannakopoulos, C., Bindi, M., 2011. Climate change impact assessment: the role of climate extremes in crop yield simulation. Clim. Change 104, 679–701.

Morison, J., 1985. Sensitivity of stomata and water-use efficiency of high CO_2. Plant Cell Environ. 8, 467–474.

Muller, C., 2011. Agriculture: Harvesting from uncertainties. Nat. Clim. Change 1, 253–254.

Nendel, C., Berg, M., Kersebaum, K., et al., 2011. The MONICA model: Testing predictability for crop growth, soil moisture and nitrogen dynamics. Ecol. Model. 222, 1614–1625.

Nendel, C., Kersebaum, K.C., Mirschel, W., Manderscheid, R., Weigel, H.J., Wenkel, K.O., 2009. Testing different CO_2

response algorithms against a FACE crop rotation experiment. J. Life Sci. 57, 17–25.
Newsham, K.K., Robinson, S.A., 2009. Responses of plants in polar regions to UVB exposure: a meta-analysis. Glob. Change Biol. 15, 2574–2589.
Nicholls, N., 1997. Increased Australian wheat yield due to recent climate trends. Nature 387, 484–485.
Nicholls, N., Alexander, L., 2007. Has the climate become more variable or extreme? Progress 1992-2006. Prog. Phys. Geog. 31, 77–87.
Nicholls, N., Collins, D., 2006. Observed climate change in Australia over the past century. Energ. Environ. 17, 12.
Nicholls, N., Wong, K.K., 1990. Dependence of rainfall variability on mean rainfall, latitude, and the Southern Oscillation. Am. Meteorol. Soc. 3, 8.
O'Leary, G.J., Connor, D.J., 1998. A simulation study of wheat crop response to water supply, nitrogen nutrition, stubble retention, and tillage. Aust. J. Agric. Res. 49, 11–19.
Olesen, J., Carter, T., Diaz-Ambrona, C., et al., 2007. Uncertainties in projected impacts of climate change on European agriculture and terrestrial ecosystems based on scenarios from regional climate models. Clim. Change 81, 123–143.
Olesen, J.E., Bindi, M., 2002. Consequences of climate change for European agricultural productivity, land use and policy. Eur. J. Agron. 16, 239–262.
Olesen, J.E., Borgesen, C.D., Elsgaard, L., et al., 2012. Changes in time of sowing, flowering and maturity of cereals in Europe under climate change. Food Add. Contam. A Chem. Analy. Control Expos. Risk Assess. 29, 1527–1542.
Olesen, J.E., Trnka, M., Kersebaum, K.C., et al., 2011. Impacts and adaptation of European crop production systems to climate change. Eur. J. Agron. 34, 96–112.
Otter-Nacke, S., Godwin, D.C., Richie, J.T., Ag, R.P., United States. Agricultural Research, S., 1986. Testing and validating the ceres-wheat model in diverse environments. Earth Resources Applications Division, Lyndon B. Johnson, Space Center, Houston, Tex.
Ottman, M.J., Kimball, B.A., White, J.W., Wall, G.W., 2012. Wheat growth response to increased temperature from varied planting dates and supplemental infrared heating. Agron. J. 104, 7–16.
Palosuo, T., Kersebaum, K.C., Angulo, C., et al., 2011. Simulation of winter wheat yield and its variability in different climates of Europe: A comparison of eight crop growth models. Eur. J. Agron. 35, 103–114.
Pan, J., Zhu, Y., Jiang, D., Dai, T.B., Li, Y.X., Cao, W.X., 2006. Modeling plant nitrogen uptake and grain nitrogen accumulation in wheat. Field Crops Res. 97, 322–336.
Passioura, J.B., 1977. Grain yield, harvest index, and water use of wheat. J. Aust. Inst. Agric. Sci. 43, 117–120.
Pathak, H., Ladha, J.K., Aggarwal, P.K., et al., 2003. Trends of climatic potential and on-farm yields of rice and wheat in the Indo-Gangetic Plains. Field Crops Res. 80, 223–234.
Patil, R., Laegdsmand, M., Olesen, J.E., Porter, J.R., 2010. Growth and yield response of winter wheat to soil warming and rainfall patterns. J. Agric. Sci. 148, 553–566.
Peel, M.C., McMahon, T.A., Finlayson, B.L., 2002. Variability of annual precipitation and its relationship to the El Niño-Southern Oscillation. J. Clim. 15, 545–551.
Peltonen-Sainio, P., Jauhiainen, L., Hakala, K., 2011. Crop responses to temperature and precipitation according to long-term multi-location trials at high-latitude conditions. J. Agric. Sci. 149, 49.
Peng, S., Huang, J., Sheehy, J., et al., 2004a. Rice yields decline with higher night temperature from global warming. Proc. Natl. Acad. Sci. USA 101, 9971–9975.
Peng, S., Laza, R.C., Visperas, R.M., Khush, G.S., Virk, P., Zhu, D., 2004b. Rice: progress in breaking yield ceiling. Brisbane, Australia.
Penman, H.L., 1948. Natural evaporation from open water, bare soil and grass. Proc. R. Soc. Lond. Ser. A Mathemat. Phys. Sci. 193, 120–145.
Peralta, J.M., Stockle, C.O., 2002. Dynamics of nitrate leaching under irrigated potato rotation in Washington State: a long-term simulation study. Agric. Ecosyst. Environ. 88, 23–34.
Philander, S.G.H., Gu, D., Halpern, D., et al., 1996. Why the ITCZ is mostly north of the equator. J. Clim. 9, 2958–2972.
Pielke, Adegoke, J.O., Chase, T.N., Marshall, C.H., Matsui, T., Niyogi, D., 2007. A new paradigm for assessing the role of agriculture in the climate system and in climate change. Agric. For. Meteorol. 142, 234–254.
Pleijel, H., Mortensen, L., Fuhrer, J., Ojanpera, K., Danielsson, H., 1999. Grain protein accumulation in relation to grain yield of spring wheat (*Triticum aestivum* L.) grown in open-top chambers with different concentrations of ozone, carbon dioxide and water availability. Agric. Ecosyst. Environ. 72, 265–270.
Podesta, G., Letson, D., Messina, C., Royce, F., et al., 2002. Use of ENSO-related climate information in agricultural decision making in Argentina: a pilot experience. Agric. Syst. 74, 371–392.
Poorter, H., Pérez-Soba, M., 2001. The growth response of plants to elevated CO_2 under non-optimal environmental conditions. Oecologia 129, 1–20.
Porter, J.R., 1984. A model of canopy development in winter wheat. J. Agric. Sci. 102, 383–392.
Porter, J.R., 1993. AFRCWHEAT2: A model of the growth and development of wheat incorporating responses to water and nitrogen. Eur. J. Agron. 2, 69–82.
Porter, J.R., Semenov, M.A., 1999. Climate variability and crop yields in Europe. Nature 400, 724–1724.
Porter, J.R., Semenov, M.A., 2005. Crop responses to climatic variation. Philosoph. Transact. R. Soc. B Biol. Sci. 360, 2021–2035.
Power, S., Sadler, B., Nicholls, N., 2005. The influence of climate science on water management in Western

Australia – Lessons for climate scientists. Bull. Am. Meteorol. Soc. 86, 839.

Priestley, C.H.B., Taylor, R.J., 1972. On the assessment of surface heat flux and evaporation using large-scale parameters. Month. Weath. Rev. 100, 81–92.

Probert, M.E., Dimes, J.P., Keating, B.A., Dalal, R.C., Strong, W.M., 1998. APSIM's water and nitrogen modules and simulation of the dynamics of water and nitrogen in fallow systems. Agric. Syst. 56, 1–28.

Probert, M.E., Keating, B.A., 2000. What soil constraints should be included in crop and forest models? Agric. Ecosyst. Environ. 82, 273–281.

Probert, M.E., Keating, B.A., Thompson, J.P., Parton, W.J., 1995. Modelling water, nitrogen, and crop yield for a long-term fallow management experiment. Aust. J. Exp. Agric. 35, 941–950.

Radin, J.W., Lu, Z., Percy, R.G., Zeiger, E., 1994. Genetic variability for stomatal conductance in Pima cotton and its relation to improvements of heat adaptation. Proc. Natl. Acad. Sci. USA 91, 7217–7221.

Raines, C.A., 2011. Increasing photosynthetic carbon assimilation in c3 plants to improve crop yield: current and future strategies. Plant Physiol. 155, 36–42.

Randall, P., Moss, H., 1990. Some effect of temperature regime during grain filling on wheat quality. Aust. J. Agric. Res. 41, 603–617.

Reilly, J., Tubiello, F., McCarl, B., et al., 2003. US agriculture and climate change: New results. Clim. Change 57, 43–69.

Reyenga, P.J., Howden, S.M., Meinke, H., McKeon, G.M., 1999. Modelling global change impacts on wheat cropping in south-east Queensland, Australia. Environ. Model. Software 14, 297–306.

Richards, R.A., 2006. Physiological traits used in the breeding of new cultivars for water-scarce environments. Agric. Water Manag. 80, 197–211.

Richter, G.M., Semenov, M.A., 2005. Modelling impacts of climate change on wheat yields in England and Wales: assessing drought risks. Agric. Syst. 84, 77–97.

Ritchie, J., Godwin, D., Otter-Nacke, S., 1985a. CERES-wheat: A user-oriented wheat yield model. Prelimary documentation. Michigan State University, Michigan, p. 252.

Ritchie, J.T., 1972. Model for predicting evaporation from a row crop with incomplete cover. Water Resour. Res. 8, 1204-&.

Ritchie, J. T., Godwin, D. C., & Otter-Nacke, S., 1985. CERES-Wheat. A simulation model of wheat growth and development. ARS, pp. 159-175. US Department of Agriculture.

Robertson, D., Zhang, H., Palta, J.A., Colmer, T., Turner, N.C., 2009. Waterlogging affects the growth, development of tillers, and yield of wheat through a severe, but transient. N deficiency. Crop Past. Sci. 60, 578–586.

Robertson, M.J., Carberry, P.S., Huth, N.I., et al., 2002. Simulation of growth and development of diverse legume species in APSIM. Aust. J. Agric. Res. 53, 429–446.

Roderick, M.L., Farquhar, G.D., 2002. The cause of decreased pan evaporation over the past 50 years. Science 298, 1410–1411.

Roderick, M.L., Farquhar, G.D., 2004. Changes in Australian pan evaporation from 1970 to 2002. Internatl. J. Climatol. 24, 1077–1090.

Rodriguez, D., Ewert, F., Goudriaan, J., Manderscheid, R., Burkart, S., Weigel, H.J., 2001. Modelling the response of wheat canopy assimilation to atmospheric CO_2 concentrations. New Phytol. 150, 337–346.

Rodriguez, D., Sadras, V.O., 2007. The limit to wheat water use efficiency in eastern Australia. I. Gradients in the radiation environment and atmospheric demand. Aust. J. Agric. Res. 58, 287–302.

Rogers, G., Milham, P., Gillings, M., Conroy, J., 1996. Sink strength may be the key to growth and nitrogen responses in N-deficient wheat at elevated CO_2. Aust. J. Plant Physiol. 23, 253–264.

Rosenzweig, C., Jones, J.W., Hatfield, J.L., et al., 2013. The Agricultural Model Intercomparison and Improvement Project (AgMIP): Protocols and pilot studies. Agric. For. Meteorol. 170, 166–182.

Rosenzweig, C., Parry, M.L., 1994. Potential impact of climate change on world food supply. Nature 367, 133–138.

Rötter, R., Palosuo, T., Pirttioja, N., et al., 2011a. What would happen to barley production in Finland if global warming exceeded 4°C? A model-based assessment. Eur. J. Agron. 35, 205–214.

Rötter, R.P., Carter, T.R., Olesen, J.E., Porter, J.R., 2011b. Crop-climate models need an overhaul. Nat. Clim. Change 1, 175–177.

Sadras, V., Baldock, J., Roget, D., Rodriguez, D., 2003. Measuring and modelling yield and water budget components of wheat crops in coarse-textured soils with chemical constraints. Field Crops Res. 84, 241–260.

Sadras, V.O., Connor, D.J., 1991. Physiological basis of the response of harvest index to the fraction of water transpired after anthesis: A simple model to estimate harvest index for determinate species. Field Crops Res. 26, 227–239.

Sadras, V.O., Lawson, C., Hooper, P., McDonald, G.K., 2012. Contribution of summer rainfall and nitrogen to the yield and water use effi ciency of wheat in Mediterranean-type environments of South Australia. Eur. J. Agron. 36, 41–54.

Sadras, V.O., Moran, M.A., 2013. Nonlinear effects of elevated temperature on grapevine phenology. Agric. For. Meteorol. 173, 107–115.

Sadras, V.O., Montoro, A., Moran, M.A., Aphalo, P.J., 2012. Elevated temperature altered the reaction norms of stomatal conductance in field-grown grapevine. Agric. For. Meteorol. 165, 35–42.

Sadras, V.O., Monzon, J.P., 2006. Modelled wheat phenology captures rising temperature trends: Shortened time to flowering and maturity in Australia and Argentina. Field Crops Res. 99, 136–146.

Sadras, V.O., Rodriguez, D., 2007. The limit to wheat water-use efficiency in eastern Australia. II. Influence of rainfall patterns. Aust. J. Agric. Res. 58, 657–669.

Salinger, M.J., Sivakumar, M.V.K., Motha, R., 2005. Reducing vulnerability of agriculture and forestry to climate variability and change: Workshop summary and recommendations. Clim. Change 70, 341–362.

Sau, F., Boote, K.J., Ruíz-Nogueira, B., 1999. Evaluation and improvement of CROPGRO-soybean model for a cool environment in Galicia, northwest Spain. Field Crops Res. 61, 273–291.

Savin, R., Hall, A.J., Satorre, E.H., 1994. Testing the root growth subroutine of the CERES-Wheat model for two cultivars of different cycle length. Field Crops Res. 38, 125–133.

Seligman, N.A.G., Sinclair, T.R., 1995. Global environment change and simulated forage quality of wheat II. Water and nitrogen stress. Field Crops Res. 40, 29–37.

Semenov, M.A., Halford, N.G., 2009. Identifying target traits and molecular mechanisms for wheat breeding under a changing climate. J. Exp. Bot. 60, 2791–2804.

Semenov, M.A., Porter, J.R., 1995. Climatic variability and the modelling of crop yields. Agric. For. Meteorol. 73, 265–283.

Semenov, M.A., Stratonovitch, P., 2010. Use of multi-model ensembles from global climate models for assessment of climate change impacts. Clim. Res. 41, 1–14.

Semenov, M.A., Wolf, J., Evans, L.G., Eckersten, H., Iglesias, A., 1996. Comparison of wheat simulation models under climate change. 2. Application of climate change scenarios. Clim. Res. 7, 271–281.

Sharma, P.K., Sharma, S., Choi, I., 2010. Individual and combined effects of waterlogging and alkalinity on yield of wheat (*Triticum aestivum* L.) imposed at three critical stages. Physiol. Mol. Biol. Plants 16, 317–320.

Shaw, M., Zavaleta, E., Chiariello, N., Cleland, E., Mooney, H., Field, C., 2002. Grassland responses to global environmental changes suppressed by elevated CO_2. Science 298, 1987–1990.

Sheehy, J.E., Mitchell, P.L., Allen, L.H., Ferrer, A.B., 2006a. Mathematical consequences of using various empirical expressions of crop yield as a function of temperature. Field Crops Res. 98, 216–221.

Sheehy, J.E., Mitchell, P.L., Ferrer, A.B., 2006b. Decline in rice grain yields with temperature: Models and correlations can give different estimates. Field Crops Res. 98, 151–156.

Shen, D., Varis, O., 2001. Climate change in China. AMBIO J. Hum. Environ. 30, 381–383.

Shimono, H., 2011. Earlier rice phenology as a result of climate change can increase the risk of cold damage during reproductive growth in northern Japan. Agric. Ecosyst. Environ. 144, 201–207.

Sinclair, T., Jamieson, P., 2006. Grain number, wheat yield, and bottling beer: An analysis. Field Crops Res. 98, 60–67.

Sinclair, T.R., 1997. Yield 'plateaus' in grain crops: The topography of yield increase. Intensive Sugarcane Production: Meeting the Challenges Beyond 2000. In: Proceedings of the Sugar 2000 Symposium. CAB International, Brisbane, Australia. pp. 87–102.

Sinclair, T.R., Amir, J., 1992. A model to assess nitrogen limitations on the growth and yield of spring wheat. Field Crops Res. 30, 63–78.

Sinclair, T.R., Pinter, P.J., Kimball, B.A., et al., 2000. Leaf nitrogen concentration of wheat subjected to elevated CO_2 and either water or N deficits. Agric. Ecosyst. Environ. 79, 53–60.

Sinclair, T.R., Shiraiwa, T., Hammer, G.L., 1992. Variation in crop radiation-use efficiency with increased diffuse radiation. Crop Sci. 32, 1281–1284.

Slafer, G.A., Rawson, H.M., 1997. CO_2 effects on phasic development, leaf number and rate of leaf appearance in wheat. Ann. Bot. 79, 75–81.

Smith, I., 2004. An assessment of recent trends in Australian rainfall. Aust. Meteorol. Mag. 53, 163–173.

Smith, I.N., McIntosh, P., Ansell, T.J., Reason, C.J.C., McInnes, K., 2000. Southwest Western Australian winter rainfall and its association with Indian Ocean climate variability. Internatl. J. Climatol. 20, 1913–1930.

Soar, C.J., Collins, M.J., Sadras, V.O., 2009. Irrigated Shiraz vines up-regulate gas exchange and maintain berry growth under short spells of high maximum temperature in the field. Funct. Plant Biol. 36, 801–814.

Soltani, A., Robertson, M.J., Rahemi-Karizaki, A., Poorreza, J., Zarei, H., 2006. Modelling biomass accumulation and partitioning in chickpea (*Cicer arietinum* L.). J. Agron. Crop Sci. 192, 379-389.

Song, Y., Birch, C., Qu, S., Doherty, A., Hanan, J., 2010. Analysis and modelling of the effects of water stress on maize growth and yield in dryland conditions. Plant Product. Sci. 13, 199–208.

Spiertz, J.H.J., 1977. Influence of temperature and light-intensity on grain-growth in relation to carbohydrate and nitrogen economy of wheat plant. Netherl. J. Agric. Sci. 25, 182–197.

Spiertz, J.H.J., Hamer, R.J., Xu, H., Primo-Martin, C., Don, C., van der Putten, P.E.L., 2006. Heat stress in wheat (*Triticum aestivum* L.): Effects on grain growth and quality traits. Eur. J. Agron. 25, 89–95.

Spitters, C.J.T., Schapendonk, A.H.C.M., 1990. Evaluation of breeding strategies for drought tolerance in potato by means of crop growth simulation. Plant Soil 123, 193–203.

Stanhill, G., 1998. Long-term trends in, and spatial variation of, solar irradiances in Ireland. Internatl. J. Climatol. 18, 1015–1030.

Stanhill, G., Cohen, S., 2001. Global dimming: a review of the evidence for a widespread and significant reduction in global radiation with discussion of its probable causes and possible agricultural consequences. Agric. For. Meteorol. 107, 255–278.

Stapper, M., 1984. SIMTAG: A simulation model of wheat genotypes. Model documentation. University of New

England and International Center for Agricultural Research in the Dry Areas (ICARDA), Armidale, NSW, Australia.

Steduto, P., Hsiao, T.C., Raes, D., Fereres, E., 2009. AquaCrop – The FAO crop model to simulate yield response to water: I. Concepts and underlying principles. Agron. J. 101, 426–437.

Stöckle, C.O., Donatelli, M., Nelson, R., 2003. CropSyst, a cropping systems simulation model. Eur. J. Agron. 18, 289–307.

Stoddard, F., Mäkelä, P., Puhakainen, T.A., 2011. In: Blanco, J., Kheradmand, H. (Eds.), Adaptation of boreal field crop production to climate change. Climate Change - Research and Technology for Adaptation and Mitigation, InTech., pp. 403–430.

Stone, P.J., Nicolas, M.E., 1995. Comparison of sudden heat stress with gradual exposure to high temperature during grain filling in two wheat varieties differing in heat tolerance. 1. Grain growth. Aust. J. Plant Physiol. 22, 935–944.

Streck, N.A., Weiss, A., Xue, Q., Baenziger, P.S., 2003. Incorporating a chronology response into the prediction of leaf appearance rate in winter wheat. Ann. Bot. 92, 181–190.

Tang, L., Zhu, Y., Hannaway, D., et al., 2009. RiceGrow: A rice growth and productivity model. J. Life Sci. 57, 83–92.

Tao, F., Hayashi, Y., Zhang, Z., Sakamoto, T., Yokozawa, M., 2008. Global warming, rice production, and water use in China: Developing a probabilistic assessment. Agric. For. Meteorol. 148, 94–110.

Tao, F., Yokozawa, M., Hayashi, Y., Lin, E., 2003. Changes in agricultural water demands and soil moisture in China over the last half-century and their effects on agricultural production. Agric. For. Meteorol. 118, 251–261.

Tao, F., Yokozawa, M., Xu, Y., Hayashi, Y., Zhang, Z., 2006. Climate changes and trends in phenology and yields of field crops in China, 1981-2000. Agric. For. Meteorol. 138, 82–92.

Tao, F., Yokozawa, M., Zhang, Z., 2009a. Modelling the impacts of weather and climate variability on crop productivity over a large area: A new process-based model development, optimization, and uncertainties analysis. Agric. For. Meteorol. 149, 831–850.

Tao, F., Zhang, Z., 2012. Climate change, high-temperature stress, rice productivity, and water use in Eastern China: A new superensemble-based probabilistic projection. J. Appl. Meteorol. Climatol. 52, 531–551.

Tao, F., Zhang, Z., 2013. Climate change, wheat productivity and water use in the North China Plain: A new super-ensemble-based probabilistic projection. Agric. For. Meteorol. 170, 146–165.

Tao, F., Zhang, Z., Liu, J., Yokozawa, M., 2009b. Modelling the impacts of weather and climate variability on crop productivity over a large area: A new super-ensemble-based probabilistic projection. Agric. For. Meteorol. 149, 1266–1278.

Thomas, A., 2000. Spatial and temporal characteristics of potential evapotranspiration trends over China. Internatl. J. Climatol. 20, 381–396.

Tian, Y., Chen, J., Chen, C., et al., 2012. Warming impacts on winter wheat phenophase and grain yield under field conditions in Yangtze Delta Plain. China. Field Crops Res. 134, 193–199.

Toure, A.S., Grandtner, M.M., Hiernaux, P.Y., 1994. Relief, soils and vegetation of a Sudano-Saheli savannah in Central Mali. Phytocoenologia 24, 233–256.

Triboi, E., Martre, P., Girousse, C., Ravel, C., Triboi-Blondel, A., 2006. Unravelling environmental and genetic relationships between grain yield and nitrogen concentration for wheat. Eur. J. Agron. 25, 108–118.

Triboi, E., Martre, P., Triboi-Blondel, A., 2003. Environmentally-induced changes in protein composition in developing grains of wheat are related to changes in total protein content. J. Exp. Bot. 54, 1731–1742.

Trnka, M., Dubrovský, M., Semerádová, D., Žalud, Z., 2004. Projections of uncertainties in climate change scenarios into expected winter wheat yields. Theor. Appl. Climatol. 77, 229–249.

Tsvetsinskaya, E.A., Mearns, L.O., Mavromatis, T., Gao, W., McDaniel, L., Downton, M.W., 2003. The effect of spatial scale of climatic change scenarios on simulated maize, winter wheat, and rice production in the southeastern United States. Clim. Change 60, 37–72.

Tubiello, F.N., Donatelli, M., Rosenzweig, C., Stockle, C.O., 2000. Effects of climate change and elevated CO_2 on cropping systems: model predictions at two Italian locations. Eur. J. Agron. 13, 179–189.

Tubiello, F.N., Ewert, F., 2002. Simulating the effects of elevated CO_2 on crops: approaches and applications for climate change. Eur. J. Agron. 18, 57–74.

Tubiello, F.N., Rosenzweig, C., Kimball, B.A., et al., 1999. Testing CERES-Wheat with free-air carbon dioxide enrichment (FACE) experiment data: CO_2 and water interactions. Agron. J. 91, 247–255.

Tubiello, F.N., Soussana, J-F., Howden, S.M., 2007. Climate change and food security special feature: crop and pasture response to climate change. Proc. Natl. Acad. Sci. USA 104, 19686-19690.

Turner, N.C., Asseng, S., 2005. Productivity, sustainability, and rainfall-use efficiency in Australian rainfed Mediterranean agricultural systems. Aust J. Agric. Res. 56, 1123–1136.

Turner, N.C., Molyneux, N., Yang, S., Xiong, Y.-C., Siddique, K.H., 2011. Climate change in south-west Australia and north-west China: challenges and opportunities for crop production. Crop Past. Sci. 62, 445–456.

Uddling, J., Gelang-Alfredsson, J., Karlsson, P., Sellden, G., Pleijel, H., 2008. Source-sink balance of wheat determines responsiveness of grain production to increased [CO_2] and water supply. Agric. Ecosyst. Environ. 127, 215–222.

Van Herwaarden, A., Richards, R., Farquhar, G., Angus, J., 1998. 'Haying-off', the negative grain yield response of dryland wheat to nitrogen fertiliser – III. The influence of water deficit and heat shock. Aust. J. Agric. Res. 49, 1095–1110.

Van Ittersum, M.K., Donatelli, M., 2003. Modelling cropping systems – highlights of the symposium and preface to the special issues. Eur. J. Agron. 18, 187–197.

Van Ittersum, M.K., Howden, S.M., Asseng, S., 2003a. Sensitivity of productivity and deep drainage of wheat cropping systems in a Mediterranean environment to changes in CO_2, temperature and precipitation. Agric. Ecosys. Environ. 97, 255–273.

Van Ittersum, M.K., Leffelaar, P.A., van Keulen, H., Kropff, M.J., Bastiaans, L., Goudriaan, J., 2003b. On approaches and applications of the Wageningen crop models. Eur. J. Agron. 18, 201–234.

Van Keulen, H., Seligman, N.G., 1987. Simulation of water use, nitrogen nutrition and growth of a spring wheat crop. Pudoc, Wageningen, The Netherlands.

van Wart, J., van Bussel, L.G.J., Wolf, J., et al., 2013a. Use of agro-climatic zones to upscale simulated crop yield potential. Field Crops Res. 143, 44–55.

van Wart, J., Grassini, P., Cassman, K.G. 2013b. Estimated impact of weather data source on simulated crop yields. Glob. Change Biol. (in press).

Viglizzo, E.F., Roberto, Z.E., Filippin, M.C., Pordomingo, A.J., 1995. Climate variability and agroecological change in the Central Pampas of Argentina. Agriculture, ecosystems & environment 55, 7–16.

Vorne, V., Ojanperä, K., DeTemmerman, L., et al., 2002. Effects of elevatedcarbondioxide and ozone on potato tuber quality in the European multiple-site experiment 'CHIP-project'. Eur. J. Agron. 17, 369–381.

Vose, R.S., 2005. Reference station networks for monitoring climatic change in the conterminous United States. J. Clim. 18, 5390–5395.

Walker, W.E., Harremoes, P., Rotmans, J., et al., 2003. Defining uncertainty: a conceptual basis for uncertainty management in model-based decision support. Integr. Assess. 4, 5–17.

Wall, G., 2001. Elevated atmospheric CO_2 alleviates drought stress in wheat. Agric. Ecosyst. Environ. 87, 261–271.

Wang, E., Cresswell, H., Xu, J., Jiang, Q., 2009a. Capacity of soils to buffer impact of climate variability and value of seasonal forecasts. Agric. For. Meteorol. 149, 38–50.

Wang, E., Engel, T., 1998. Simulation of phenological development of wheat crops. Agric. Sys. 58, 1–24.

Wang, E., Engel, T., 2000. SPASS: a generic process-oriented crop model with versatile windows interfaces. Environ. Model. Software 15, 179–188.

Wang, E., Robertson, M.J., Hammer, G.L., et al., 2002. Development of a generic crop model template in the cropping system model APSIM. Eur. J. Agron. 18, 121–140.

Wang, E., Smith, C.J., Bond, W.J., Verburg, K., 2004. Estimations of vapour pressure deficit and crop water demand in APSIM and their implications for prediction of crop yield, water use, and deep drainage. Aust. J. Agric. Res. 55, 1227–1240.

Wang, E., Yu, Q., Wu, D., Xia, J., 2008a. Climate, agricultural production and hydrological balance in the North China Plain. Internatl. J. Climatol. 28, 1959–1970.

Wang, H., Gan, Y., Wang, R., et al., 2008b. Phenological trends in winter wheat and spring cotton in response to climate changes in northwest China. Agric. For. Meteorol. 148, 1242–1251.

Wang, J., Wang, E., Feng, L., Yin, H., Yu, W., 2013. Phenological trends of winter wheat in response to varietal and temperature changes in the North China Plain. Field Crops Res. 144, 135–144.

Wang, J., Wang, E., Luo, Q., Kirby, M., 2009b. Modelling the sensitivity of wheat growth and water balance to climate change in Southeast Australia. Clim. Change 96, 79–96.

Wang, S., Wang, E., Wang, F., Tang, L., 2012. Phenological development and grain yield of canola as affected by sowing date and climate variation in the Yangtze River Basin of China. Crop Past. Sci. 63, 478–488.

Wang, Y.-P., Handoko, J., Rimmington, G.M., 1992. Sensitivity of wheat growth to increased air temperature for different scenarios of ambient CO_2 concentration and rainfall in Victoria, Australia – a simulation study. Clim. Res. 2, 131–149.

Wang, Y., Xie, Z., Malhi, S.S., Vera, C.L., Zhang, Y., Wang, J., 2009c. Effects of rainfall harvesting and mulching technologies on water use efficiency and crop yield in the semi-arid Loess Plateau, China. Agric. Water Manag. 96, 374–382.

Weir, A.H., Bragg, P.L., Porter, J.R., Rayner, J.H., 1984. A winter wheat crop simulation model without water or nutrient limitations. J. Agric. Sci. 102, 371–382.

Weiss, A., Hays, C.J., Won, J., 2003. Assessing winter wheat responses to climate change scenarios: a simulation study in the U.S. Great Plains. Clim. Change 58, 119–147.

Wessolek, G., Asseng, S., 2006. Trade-off between wheat yield and drainage under current and climate change conditions in northeast Germany. Eur. J. Agron. 24, 333–342.

Wheeler, T., Batts, G., Ellis, R., Hadley, P., Morison, J., 1996. Growth and yield of winter wheat (*Triticum aestivum*) crops in response to CO_2 and temperature. J. Agric. Sci. 127, 37–48.

White, J.W., Hoogenboom, G., Kimball, B.A., Wall, G.W., 2011. Methodologies for simulating impacts of climate change on crop production. Field Crops Res. 124, 357–368.

Wilby, R., Charles, S., Zorita, E., Timbal, B., Whetton, P., Mearns, L., 2004. Guidelines for use of climate scenarios developed from statistical downscaling methods. IPCC.

WMO 1995. World Meteorological Organization.

Wolf, J., Evans, L.G., Semenov, M.A., Eckersten, H., Iglesias, A., 1996. Comparison of wheat simulation models under climate change. 1. Model calibration and sensitivity analyses. Clim. Res. 7, 253–270.

Wrigley, C., Blumenthal, C., Gras, P., Barlow, E., 1994. Temperature-variation during grain filling and changes in wheat-grain quality. Aust. J. Plant Physiol. 21, 875–885.
Wu, D., Wang, G., Bai, Y., Liao, J., 2004. Effects of elevated CO_2 concentration on growth, water use, yield and grain quality of wheat under two soil water levels. Agric. Ecosys. Environ. 104, 493–507.
Wu, W., Shibasaki, R., Yang, P., Tan, G., Matsumura, K.-i., Sugimoto, K., 2007. Global-scale modelling of future changes in sown areas of major crops. Ecol. Model. 208, 378–390.
Xiong, W., Holman, I., Lin, E., Conway, D., Li, Y., Wu, W., 2012. Untangling relative contributions of recent climate and CO_2 trends to national cereal production in China. Environ. Res. Lett. 7, 044014.
Xu, H., Chen, S.B., Biswas, D.K., Li, Y.G., Jiang, G.M., 2009. Photosynthetic and yield responses of an old and a modern winter wheat cultivar to short-term ozone exposure. Photosynthetica 47, 247–254.
Xue, Q., Weiss, A., Baenziger, P.S., 2004. Predicting phenological development in winter wheat. Clim. Res. 25, 243–252.
Yang, L., Huang, J., Yang, H., et al., 2006a. Seasonal changes in the effects of free-air CO_2 enrichment (FACE) on dry matter production and distribution of rice (*Oryza sativa* L.). Field Crops Res. 98, 12–19.
Yang, L., Huang, J., Yang, H., et al., 2007a. Seasonal changes in the effects of free-air CO_2 enrichment (FACE) on nitrogen (N) uptake and utilization of rice at three levels of N fertilization. Field Crops Res. 100, 189–199.
Yang, L., Huang, J., Yang, H., et al., 2006b. The impact of free-air CO_2 enrichment (FACE) and N supply on yield formation of rice crops with large panicle. Field Crops Res. 98, 141–150.
Yang, L., Wang, Y., Dong, G., et al., 2007b. The impact of free-air CO_2 enrichment (FACE) and nitrogen supply on grain quality of rice. Field Crops Res. 102, 128–140.
Yang, L., Wang, Y., Huang, J., et al., 2007c. Seasonal changes in the effects of free-air CO_2 enrichment (FACE) on phosphorus uptake and utilization of rice at three levels of nitrogen fertilization. Field Crops Res. 102, 141–150.
Yang, L.X., Wang, Y.L., Dong, G.C., et al., 2007d. The impact of free-air CO_2 enrichment (FACE) and nitrogen supply on grain quality of rice. Field Crops Res. 102, 128–140.
Yang, X., Lin, E., Ma, S., et al., 2007e. Adaptation of agriculture to warming in Northeast China. Clim. Change 84, 45–58.
Yang, X., Asseng, S., Wong, M.T.F., Yu, Q., Li, J., Liu, E., 2013. Quantifying the interactive impacts of global dimming and warming on wheat yield and water use in China. Agric. For. Meteorol. 182, 342–351.
Yin, L., Wang, S., 2012. Modulated increased UV-B radiation affects crop growth and grain yield and quality of maize in the field. Photosynthetica 50, 595–601.
Yin, X., 2013. Improving ecophysiological simulation models to predict the impact of elevated atmospheric CO_2 concentration on crop productivity. Ann. Bot. 112, 465–475.
Yin, X., Struik, P.C., 2010. Modelling the crop: from system dynamics to systems biology. J. Exp. Bot. 61, 2171–2183.
Yu, Q., Wang, E., Smith, C.J., 2008. A modelling investigation into the economic and environmental values of 'perfect' climate forecasts for wheat production under contrasting rainfall conditions. Internatl. J. Climatol. 28, 255–266.
Zhang, H., Turner, N.C., Poole, M.L., 2012. Increasing the harvest index of wheat in the high rainfall zones of southern Australia. Field Crops Res. 129, 111–123.
Zhang, H., Yin, Q., Nakajima, T., Makiko, N., Lu, P., He, J., 2013. Influence of changes in solar radiation on changes of surface temperature in China. Acta Meteorol. Sin. 27, 87–97.
Zhao, H., Dai, T., Jiang, D., Cao, W., 2008. Effects of high temperature on key enzymes involved in starch and protein formation in grains of two wheat cultivars. J. Agron. Crop Sci. 194, 47–54.
Zheng, B., Biddulph, B., Li, D., Kuchel, H., Chapman, S., 2013. Quantification of the effects of VRN1 and Ppd-D1 to predict spring wheat (*Triticum aestivum*) heading time across diverse environments. J. Exp. Bot.doi: 10.1093/jxb/ert209.
Zheng, B., Chenu, K., Fernanda Dreccer, M., Chapman, S.C., 2012. Breeding for the future: what are the potential impacts of future frost and heat events on sowing and flowering time requirements for Australian bread wheat (*Triticum aestivium*) varieties? Glob. Change Biol. 18, 2899–2914.
Zhu, Y., Chang, L., Tang, L., Jiang, H., Zhang, W., Cao, W., 2009. Modelling leaf shape dynamics in rice. J. Life Sci. 57, 73–81.
Ziska, L.H., Bunce, J.A., 2007. Predicting the impact of changing CO_2 on crop yields: some thoughts on food. New Phytol. 175, 607–617.

Index

W

X

Y

COLOR PLATES

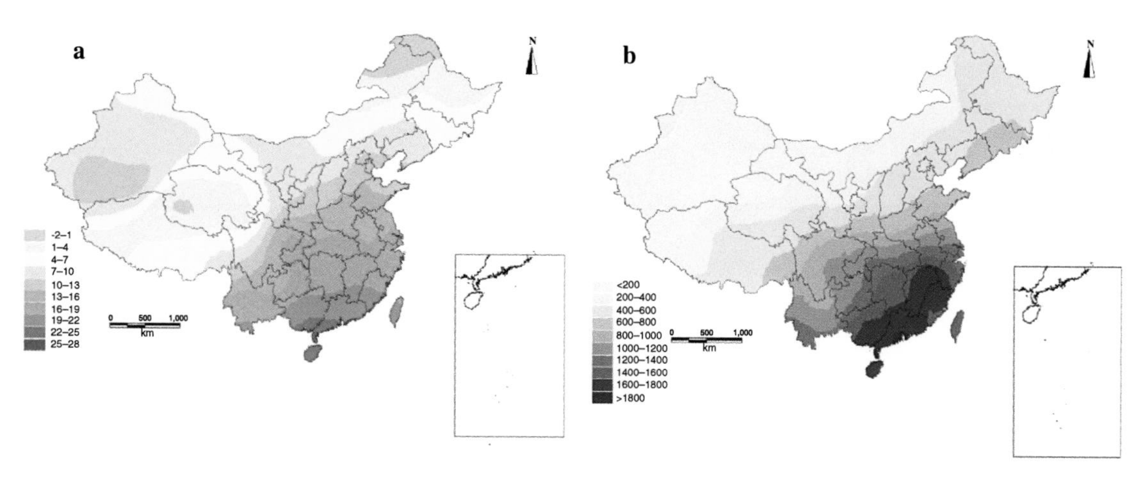

FIG. 3.1 Spatial differences in (a) annual mean air temperature (°C) and (b) the annual precipitation (mm) in China. Data are the mean values during 1990–2010 period.

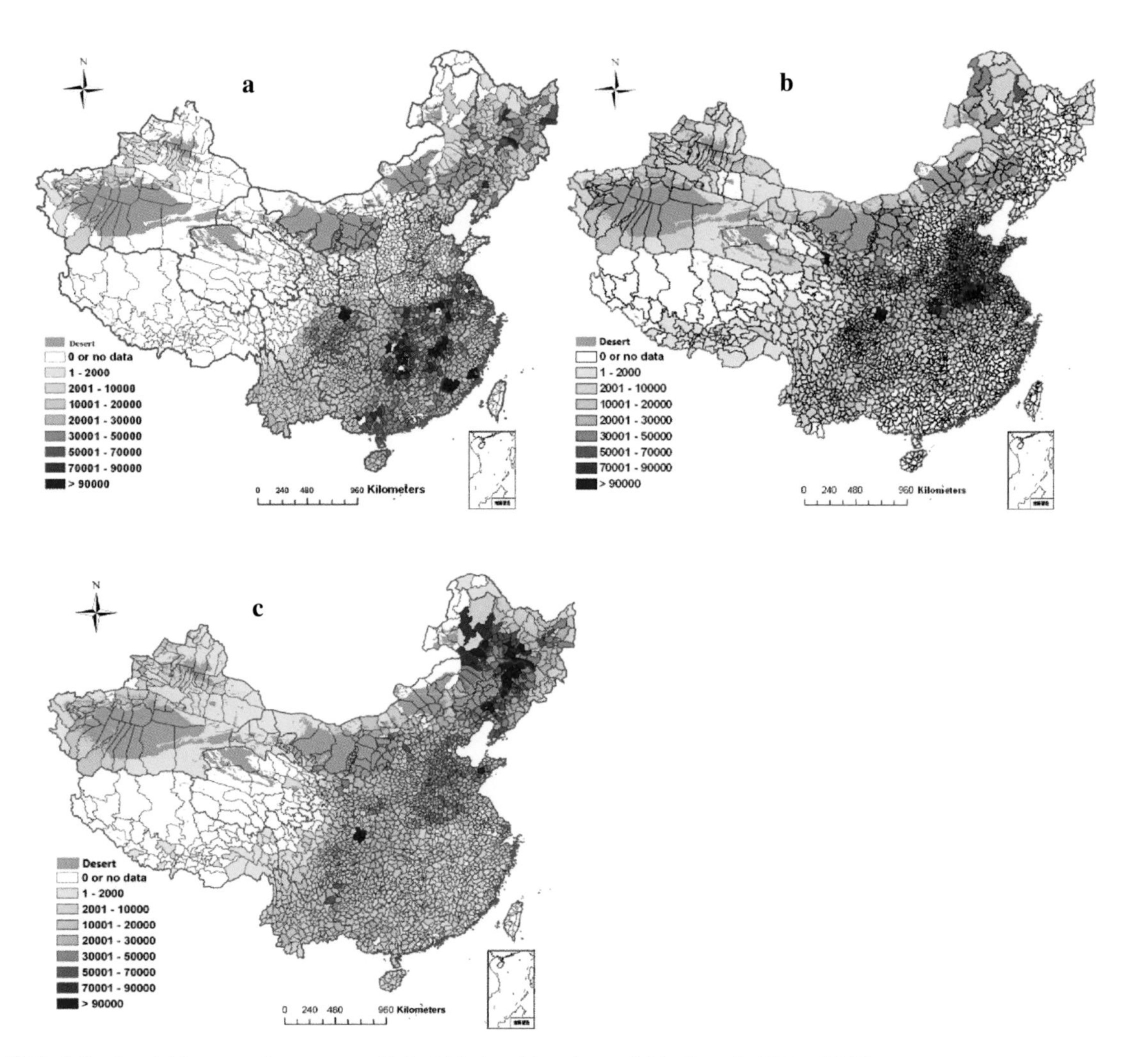

FIG. 3.3 Spatial layouts of sown areas (ha) of (a) rice, (b) maize and (c) wheat in China (2012).

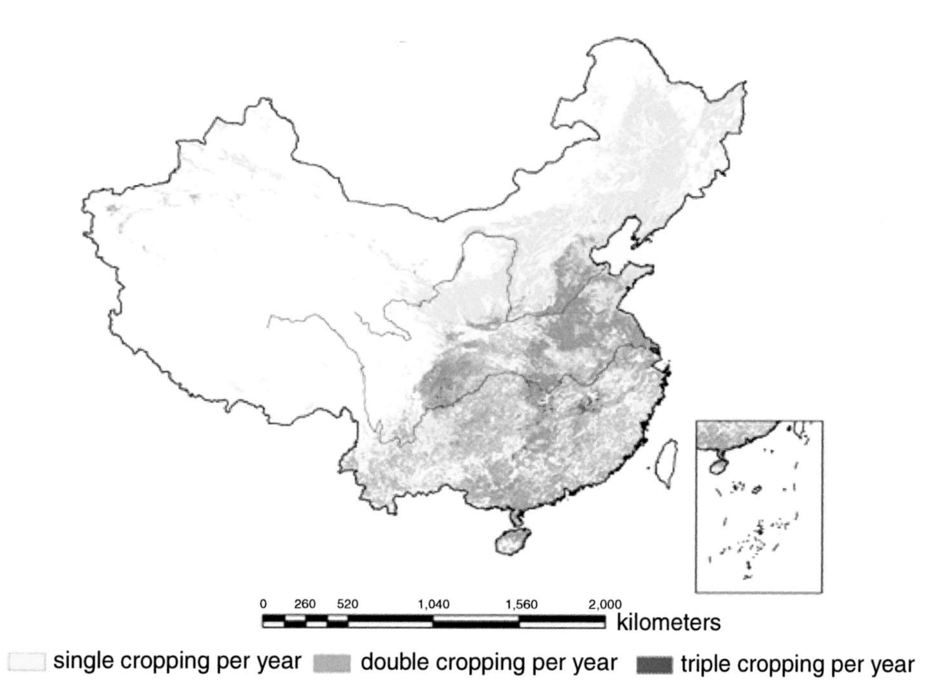

FIG. 3.4 Spatial layout of cropping systems in China, (Liu et al., 2013).

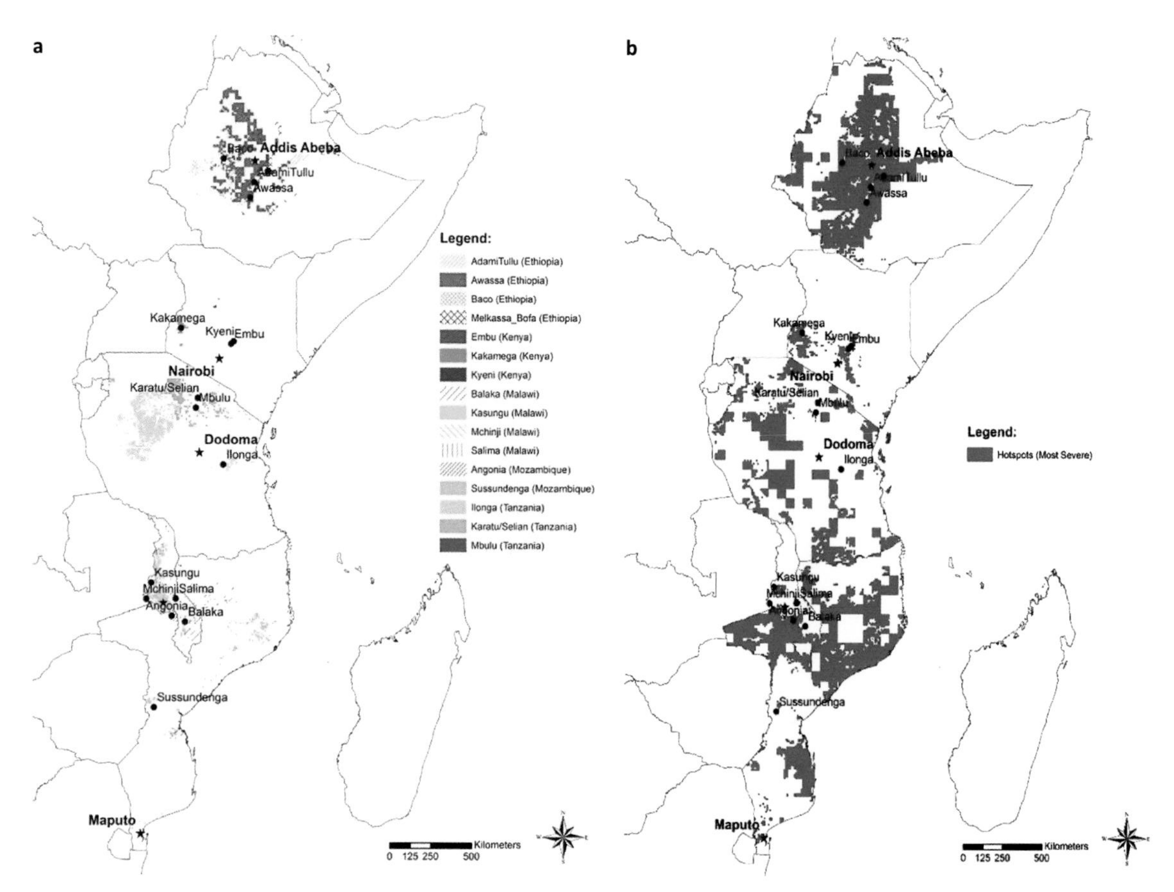

FIG. 5.1 (a) similarity of monthly temperature and rainfall patterns. (b) most severe clustering of food insecurity based on yield gap data (1999–2001) in 5 countries of eastern and southern Africa.

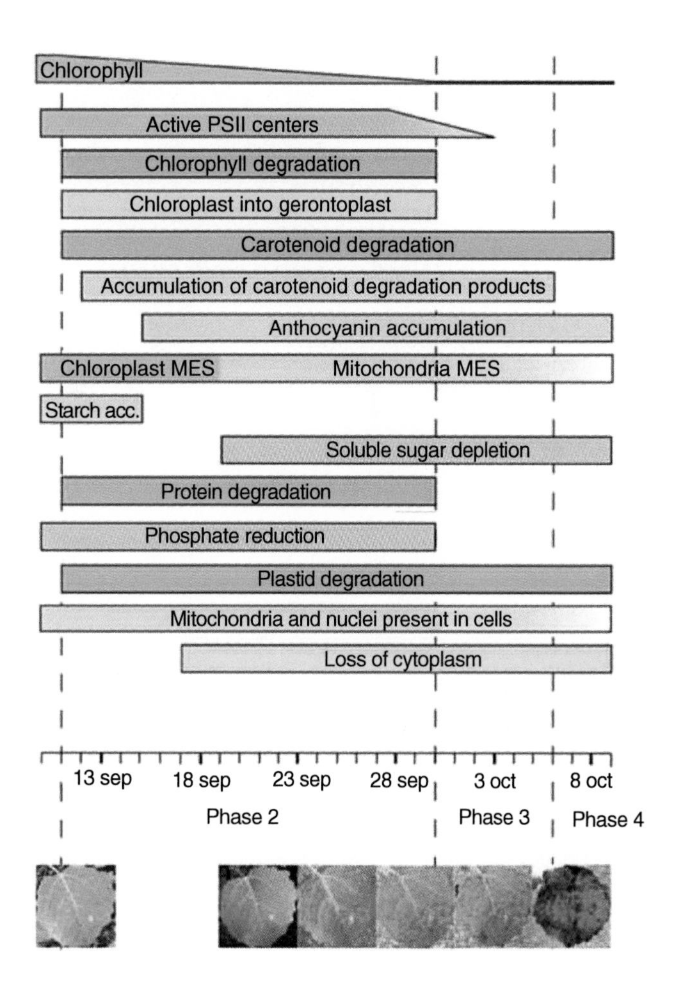

FIG. 10.3 The timetable of events during autumnal senescence in aspen leaves. Senescence is divided into four phases. Phase 1 (the mature, presenescent stage; not shown here) is followed by Phase 2, during which chloroplasts are converted to gerontoplasts, major pigmentation changes occur, N and P are mobilized and sugars metabolized. During Phase 2, the major energy source (MES) switches from chloroplasts to mitochondria. By Phase 3 less than 5% of original chlorophyll remains, cell contents are severely depleted but metabolism continues and viability is sustained in some cells. Phase 4 is the stage at which cell death is complete and few structures are recognizable within residual cell walls. *(From Keskitalo et al., 2005, Figure 10. http://bit.ly/15mqVcH)*

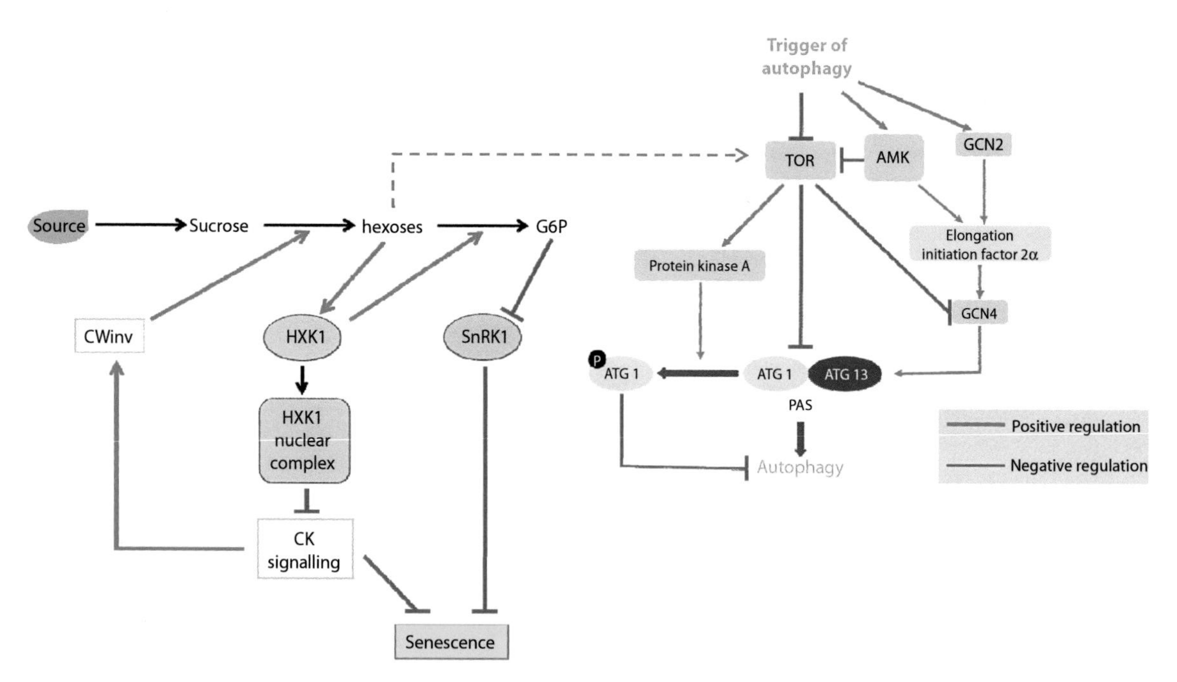

FIG. 10.6 Source–sink regulation of senescence involving global regulators Snf1-related kinase 1 (SnRK1), hexokinase 1 (HXK1) and target of rapamycin (TOR). SnRK1 delays senescence. HXK1 is part of a nuclear complex that promotes senescence by repressing cytokinin (CK) signaling. The products of hydrolysis of sucrose by cell wall invertase (CWinv) are positive regulators of HXK1 and TOR. Glucose-6-phosphate (G6P), the product of HXK1-catalyzed phosphorylation of glucose, is a negative regulator of SnRK1. TOR, protein kinase A, AMK (AMP-activated kinase) and GCN2 (general control non-derepressible 2) are kinases operating in autophagy signaling pathways. Elongation initiation factor 2α and the transcription factor GCN4 regulate expression of ATG1 and ATG13. PAS (pre-autophagosomal structure), the complex containing ATG1 and ATG13, is the precursor of the autophagosome.

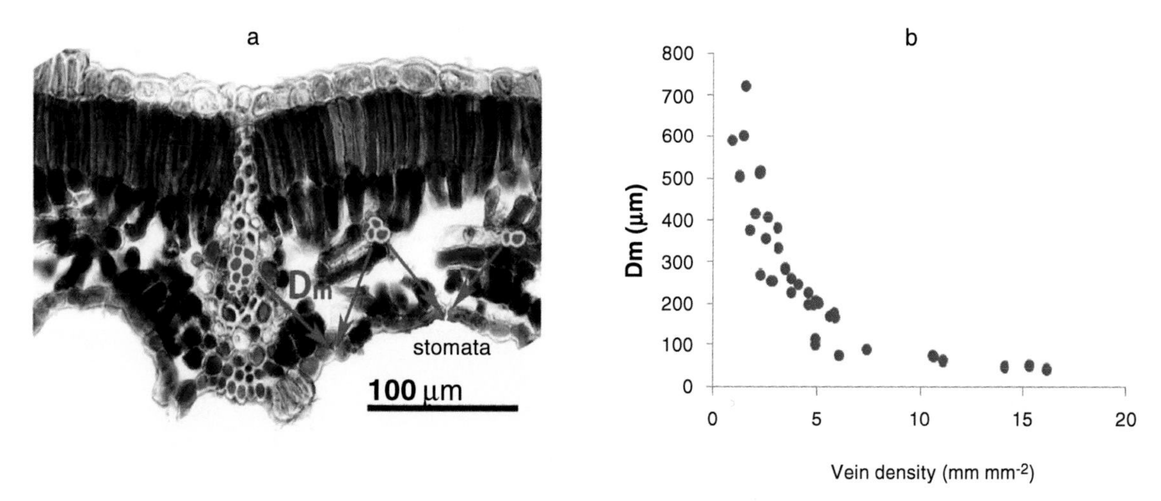

FIG. 11.3 (a) Leaf cross-section from *Curatela americana* showing the pathway for water movement (D_m) from the vein ending (vascular bundle) through the living tissue of the mesophyll to the stomata, where water finally exits as vapor. (b) Mesophyll path length (D_m) decreases as the pathway resistance for water decreases as a function of vein density (D_v) across a range of taxa. *Source: (b) Brodribb et al. (unpublished).*

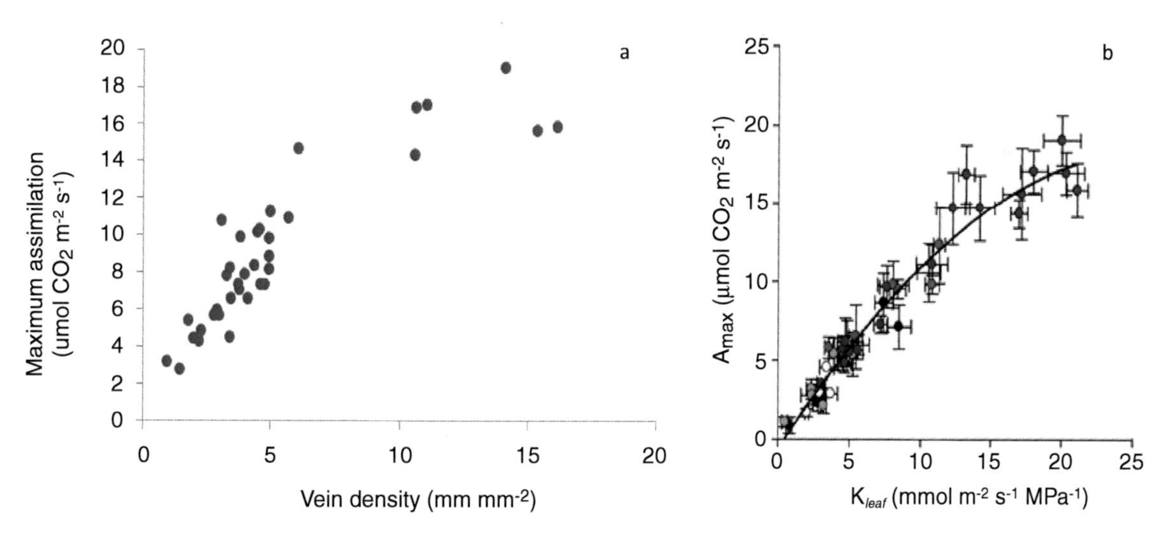

FIG. 11.4 Assimilation rate per unit leaf area can be boosted through increasing the water supply capacity of leaves as represented by (a) vein density and (b) the maximum hydraulic conductivity of the leaves (K_{leaf}) in bryophytes (black), lycopods (white), ferns (green), conifers (red), angiosperms (blue), and gymnosperms (brown) evaluated in the field under conditions of high soil water availability. *Source: Brodribb et al. (2007).*

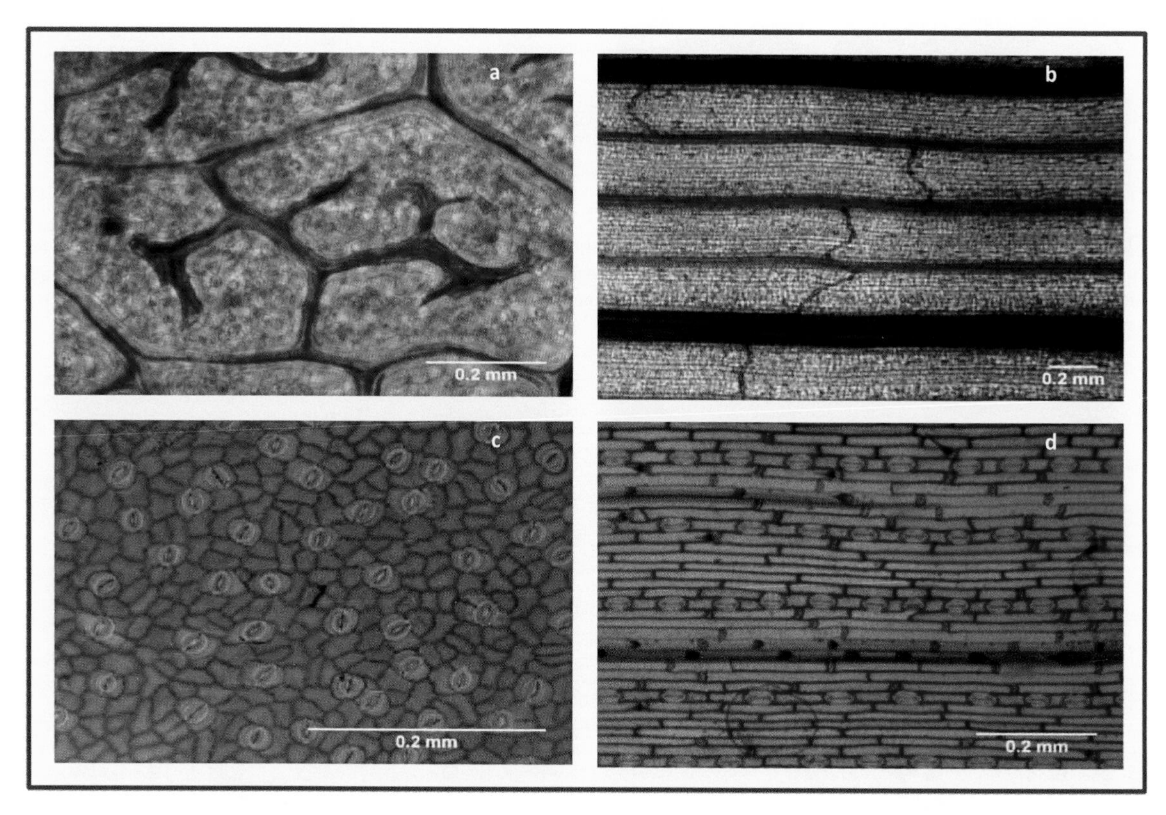

FIG. 11.5 Examples of the reticulated venation of dicots (a) and striate pattern of monocots with longitudinal veins cross-linked by smaller transverse veins (b), and corresponding stomatal patterning (c) dicot and (d) monocot. Source: (a and c) Carins-Murphy (unpublished); (b and d). *Source: Holloway-Phillips et al., unpublished.*

a- Maize mega-environments in southern Africa

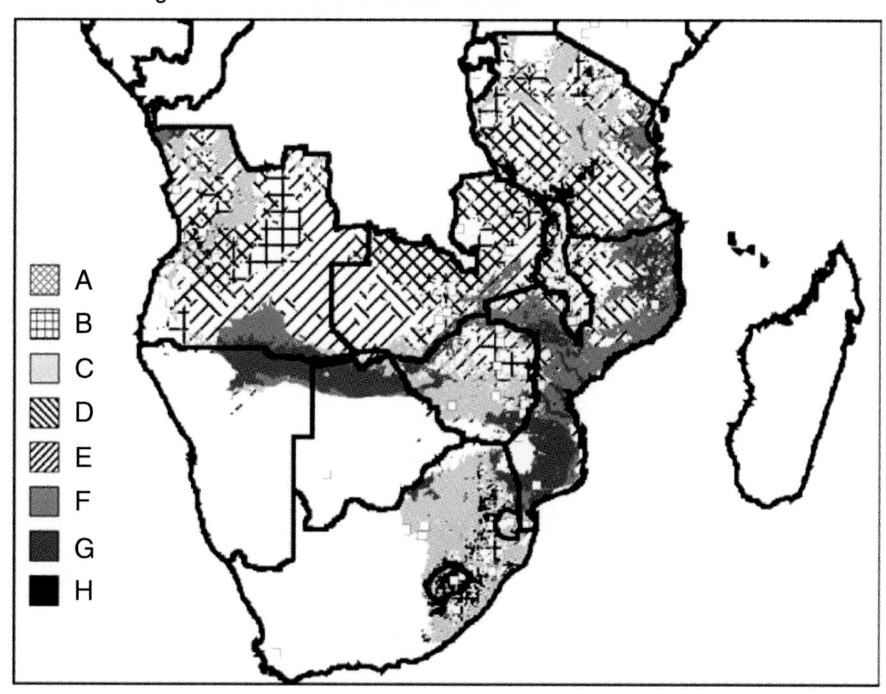

b- Maize environment classes in the U.S.A.

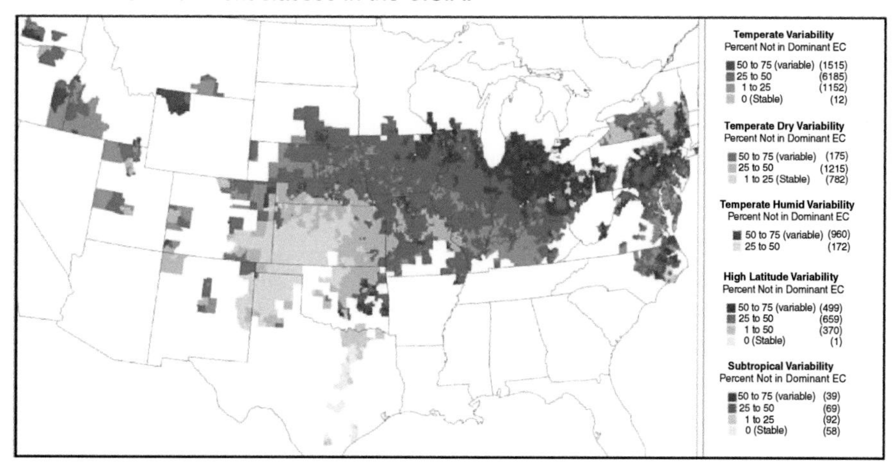

FIG. 13.2 Environment characterization based on pedo-climatic factors. (a) Public breeding organization CIMMYT identified eight maize mega-environments (A–H) in southern Africa that were defined by combination of maximum temperature, seasonal rainfall and subsoil pH. Figure from Bänziger et al. (2006). (b) Private breeding company Pioneer Hi-Bred International Inc. identified five major abiotic environment classes (EC) for the US Corn Belt, for which geographic distribution and variability are presented here. The major environment classes ('Subtropical', 'High Latitude', 'Temperate Dry', 'Temperate Humid' and 'Temperate') were defined in regards to photoperiod, maximum temperature and average solar radiation. *Source: Löffler et al. (2005).*

a- Australian rainfed wheat

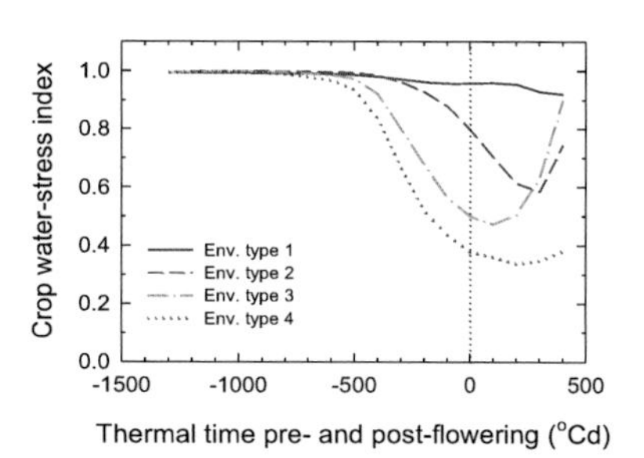

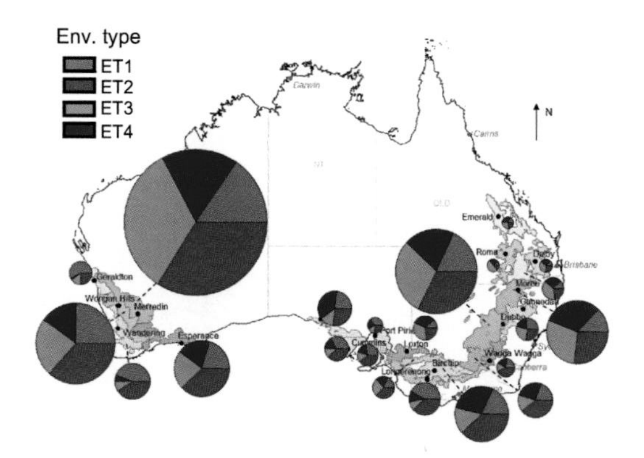

b- European rainfed maize

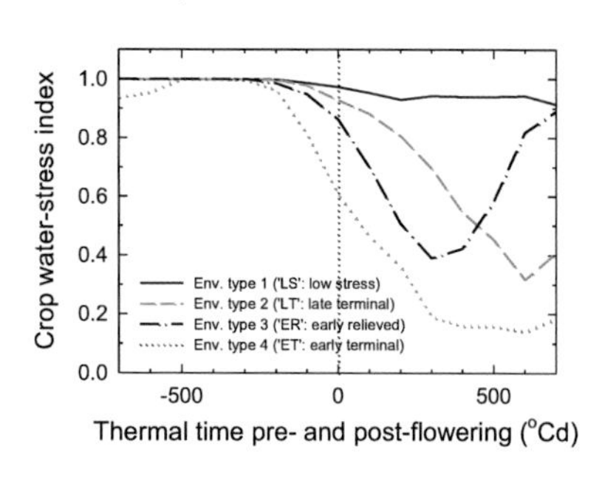

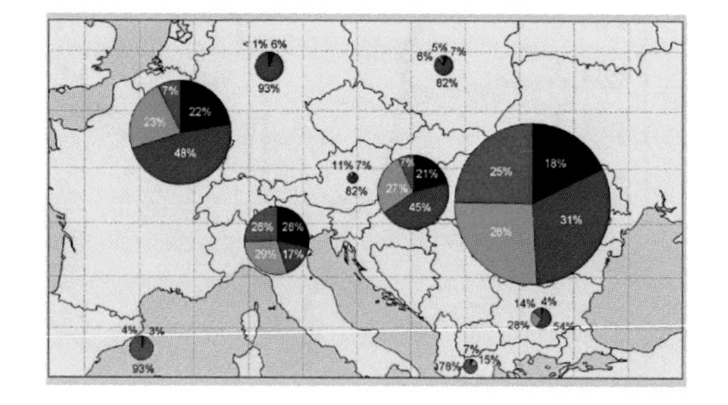

c - Australian rainfed field pea

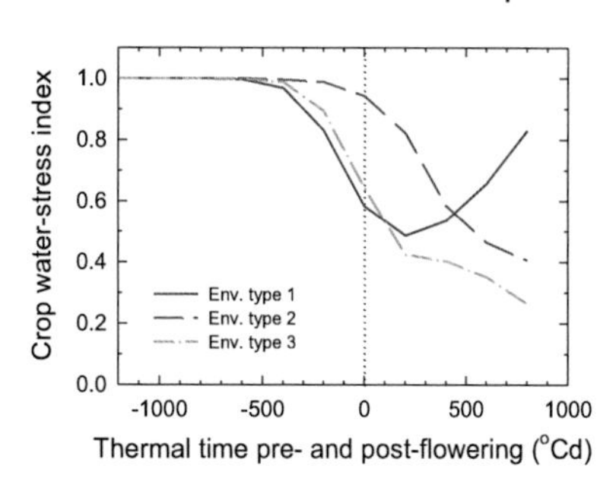

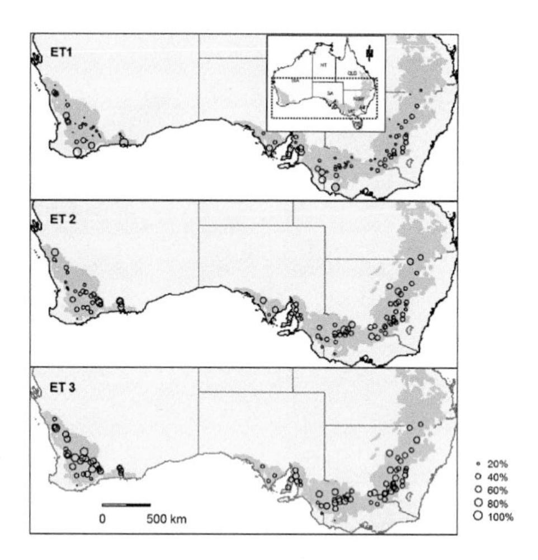

FIG. 13.3 Drought environment characterization for (a) rain-fed wheat in Australia, (b) maize in Europe and (c) field pea in Australia. Dominant water-stress index patterns expressed as thermal time before or after anthesis are presented on the left of the figure, while the distribution of their frequency is displayed on the right. In (a–b), the pie-chart size is proportional to the regional (for wheat) or national (for maize) average cropped area (wheat) or harvest (maize), while in (c), the size of the circle corresponds to the frequency of environment types (ET) at various locations. The crop water-stress index (or 'water supply/demand ratio') indicates the degree to which the potential water supply that depends on the volume and wetness of soil explored by roots ('water supply') is able to match the 'water demand' of the canopy, which is influenced by radiation and temperature and air humidity conditions. See Chapman et al. (1993) or Chenu et al. (2013a) for details.

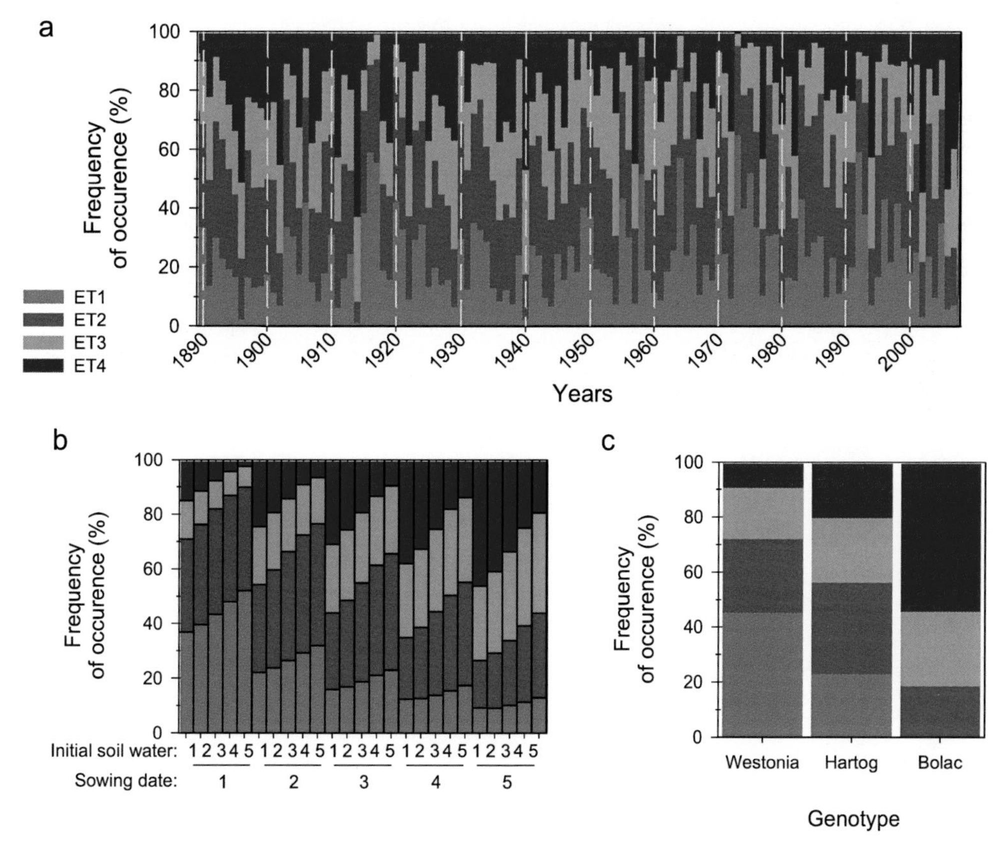

FIG. 13.4 Change in occurrence of drought-environment types for rain-fed Australian wheat (a) over time, (b) with different management practices and (c) for different genotypes. The environment types (ET1–4) correspond to those described in Figure 13.3a. Sowing dates are presented from the earliest (1) to the latest (5), with each sowing date representing 20% of the simulated sowing opportunities. Initial soil water increased from the lowest (1; most severe conditions) to the highest (5; less severe) values, each representing 20% of the simulated initial soil water availability. Simulations were performed for a standard, medium-maturing variety ('Hartog') in (a–b) as well as for an early-maturing ('Westonia') and a late-maturing ('Bolac') variety in (c). For details and information concerning spatial variability, see Chenu et al. (2013a). *Source: Chenu et al. (2013a).*

a- Weather and soil measurements

b- Multiple Environment Trials (MET)

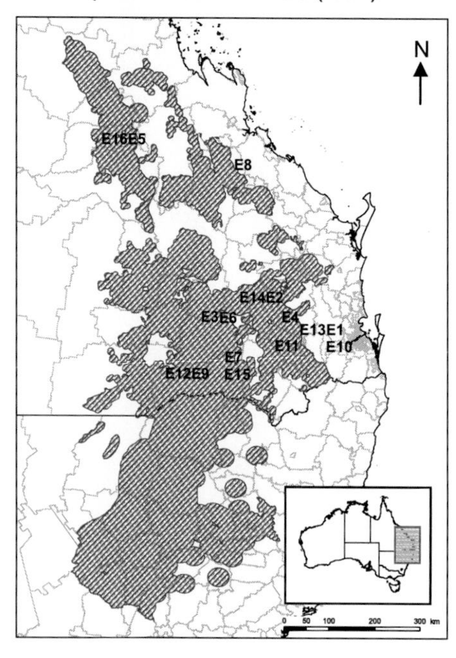

c- Trial simulation

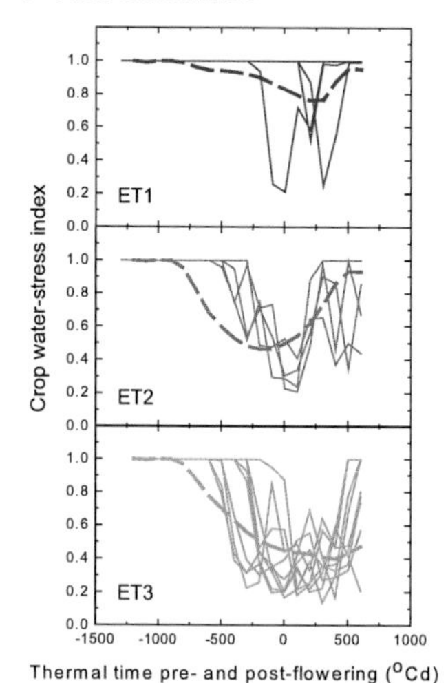

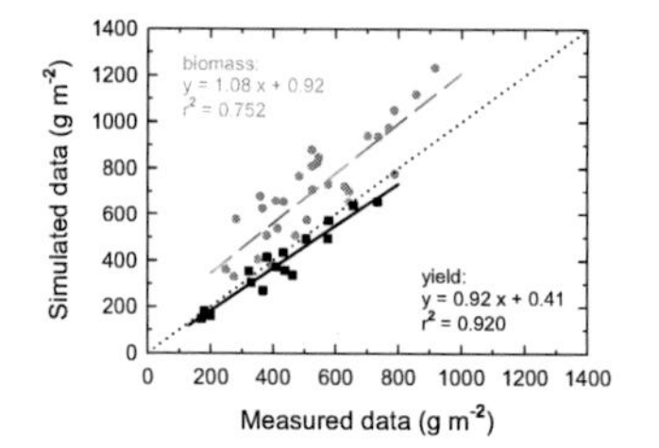

d- Genotype x Environment analysis

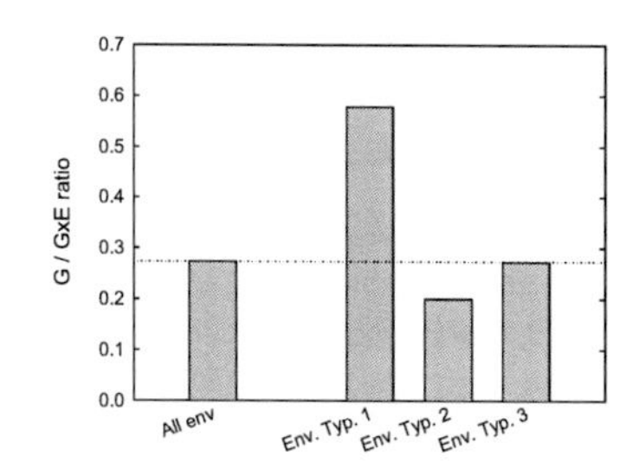

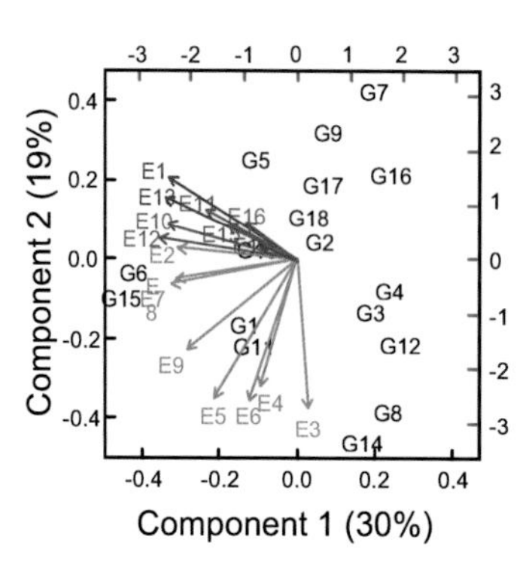

FIG. 13.5 Environment characterization for wheat breeding trials in north-eastern Australia. Measurements of weather and soil data, in particular estimation of the drained upper limit and the crop lower limit (a) were required to simulate crops at specific trials. The example presented here involves 16 wheat trials (E1–E16) of a breeding program in the Australian north-eastern production area (red hashed area) (b), for which drought patterns were characterized using crop modeling (c). Simulations were done for a standard cultivar with the APSIM crop model (Keating et al., 2003). The main drought-environment types (ET1–3) of the studied TPE are presented with dashed lines in (c). Out of the 16 trials, three were classified as environment type 1 (ET1), four trials ET2, and nine trials ET3. Overall, compared to the TPE, this 16-trial MET slightly over-represented ET1 (22% vs 16% in the TPE) and slightly under-represented ET2 (28% vs 34%), while both the TPE and the MET had the same proportion of ET3 (50%). Including the drought characterization in the trial analysis assisted the interpretation of the observed genotype-environment interactions (d). Figure adapted from Chenu *et al.* (2011), with permission from Oxford University Press and the Queensland Government © 2011.

a- Australian managed-environment facilities

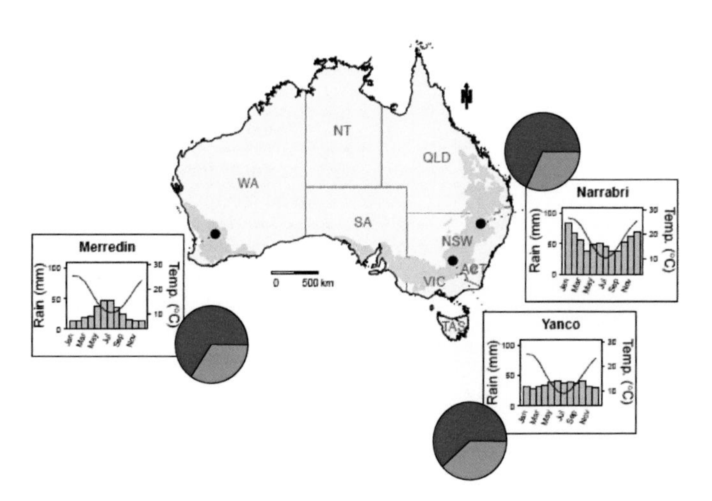

c- Targeted genetic analysis

Traits	No. genetic backgrounds
Awn presence	Five
Canopy stay-green	Two
Canopy temperature	Two
Carbon isotope discrimination	Six
Early vigour	Six
Grain fertility	Three
Leaf glaucousness	Two
Plant development	Two
Reduced-tillering	Six
Root vigour	Two
Stem carbohydrates	Two

b- Modelled-assisted management of drought patterns (StressMaster application)

Input

Past Management
Genotype sown
Sowing date
Sowing depth
Density
Irrigation (amount and timing)
Fertilization (nature, amount and timing)

End-of-season Management
Scenario 1
e.g. no irrigation & no fertilization
Scenario 2
e.g. 1 irrigation & 1 fertilization
Scenario *i*
e.g. 2 irrigations & no fertilization

Soil characteristics
Initial soil water
Initial nitrogen
Plant available water capacity

Climatic data
Min and max air temperature
Rain
Radiation

Output

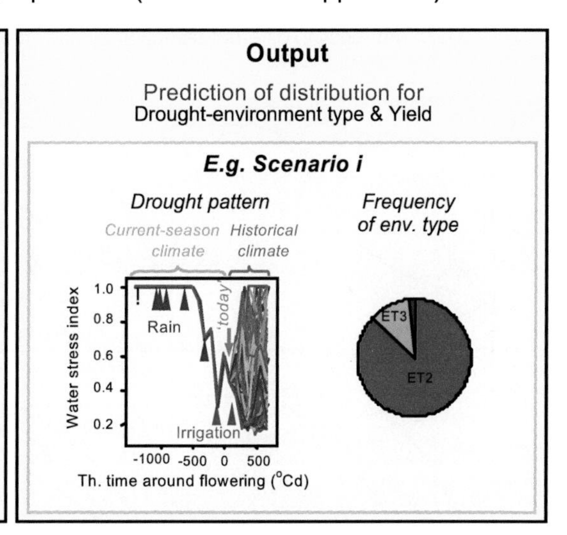

FIG. 13.7 Managed-environment facilities targeting representative drought patterns. Three locations have been chosen across Australia to represent the variability in soil and weather conditions observed across the wheat-belt (a). To adapt for the year-to-year variability that affects the occurrence of drought-environment types (pies in (a); the environment types are defined in Fig. 13.3a), irrigation is applied to target specific drought types. Irrigation scheduling is assisted by a model-based application (StressMaster) that keeps track of the crop water-stress index as the season progresses, and that allows testing of various future end-of-the-season management scenarios (b). The managed-environment facilities are used to test the value of traits with potential for drought adaptation in diverse genetic backgrounds (c). *Sources: Rebetzke et al. (2013) and Chenu et al. (2013b).*

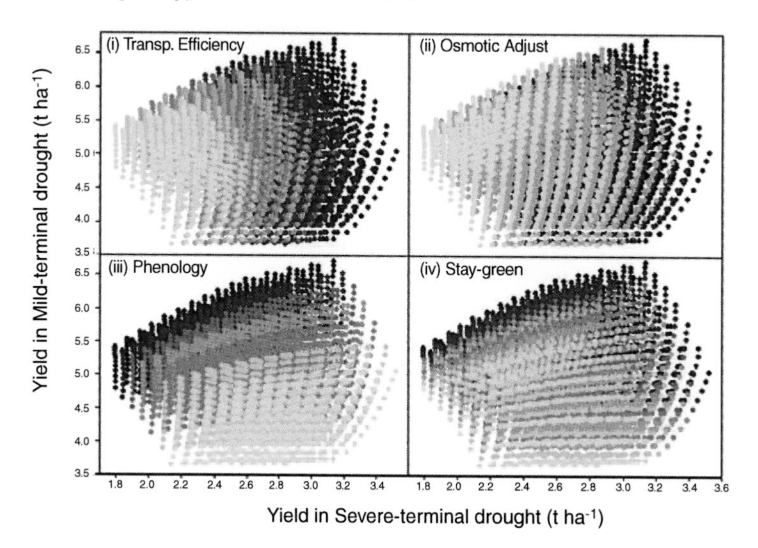

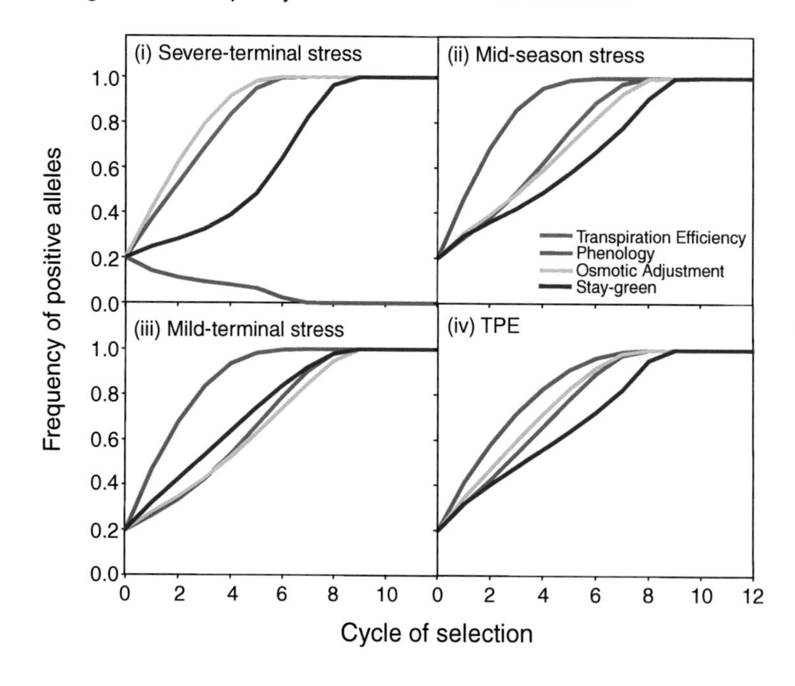

FIG. 13.9 Effect of drought-environment types on (a) simulated yield for trait ideotypes and on (b) the change in allele frequency while selecting for simulated yield. Sorghum simulations generated for 15 genes associated with four adaptive traits identified by genetic and physiological studies as important for drought tolerance: transpiration efficiency (five genes), phenology (three genes), osmotic adjustment (two genes), and stay-green (five genes). In (a), the yield of virtual genotypes in the 'mild terminal stress' and the 'severe terminal stress' environment types is presented with a color code that indicates the genetic distance from the trait ideotype of the TPE (blue: no allele different from the ideotype, yellow: all alleles different). Colors are presented for ideotypes for (i) transpiration efficiency, (ii) osmotic adjustment, (iii) phenology and (iv) stay-green. In (b), change in frequency of alleles with a positive effect on the four traits over cycles of selection, when selection is conducted in the different environment types. Simulations for the sorghum region of north-eastern Australia. *Figures adapted from Cooper et al. (2002) reprinted with permission from IOS Press, and from Chapman et al. (2003).*

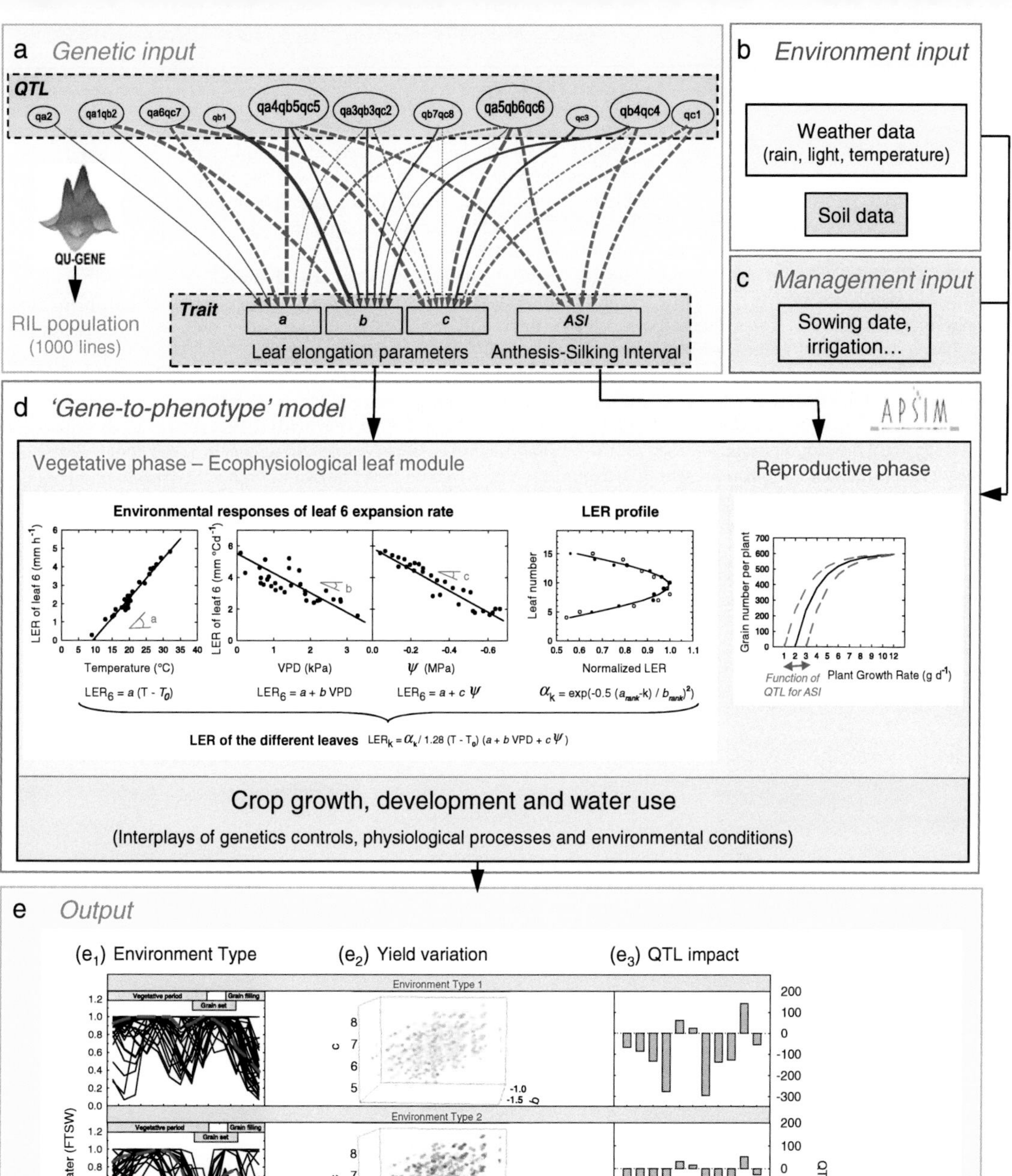

a Genetic input
QTL
qa2
qa1qb2
qa6qc7
qb1
qa4qb5qc5
qa3qb3qc2
qb7qc8
qa5qb6qc6
qc3
qb4qc4
qc1
QU-GENE
RIL population
(1000 lines)
Trait
a
b
c
ASI
Leaf elongation parameters
Anthesis-Silking Interval
b Environment input
Weather data
(rain, light, temperature)
Soil data
c Management input
Sowing date,
irrigation…
d 'Gene-to-phenotype' model
APSIM
Vegetative phase – Ecophysiological leaf module
Reproductive phase
Environmental responses of leaf 6 expansion rate
LER profile
LER of leaf 6 (mm h-1)
Temperature (°C)
LER of leaf 6 (mm °Cd-1)
VPD (kPa)
ψ (MPa)
Leaf number
Normalized LER
LER6 = a (T - T0)
LER6 = a + b VPD
LER6 = a + c ψ
αk = exp(-0.5 (arank-k) / brank)2)
LER of the different leaves
LERk = αk / 1.28 (T - T0) (a + b VPD + c ψ)
Grain number per plant
Plant Growth Rate (g d-1)
Function of QTL for ASI
Crop growth, development and water use
(Interplays of genetics controls, physiological processes and environmental conditions)

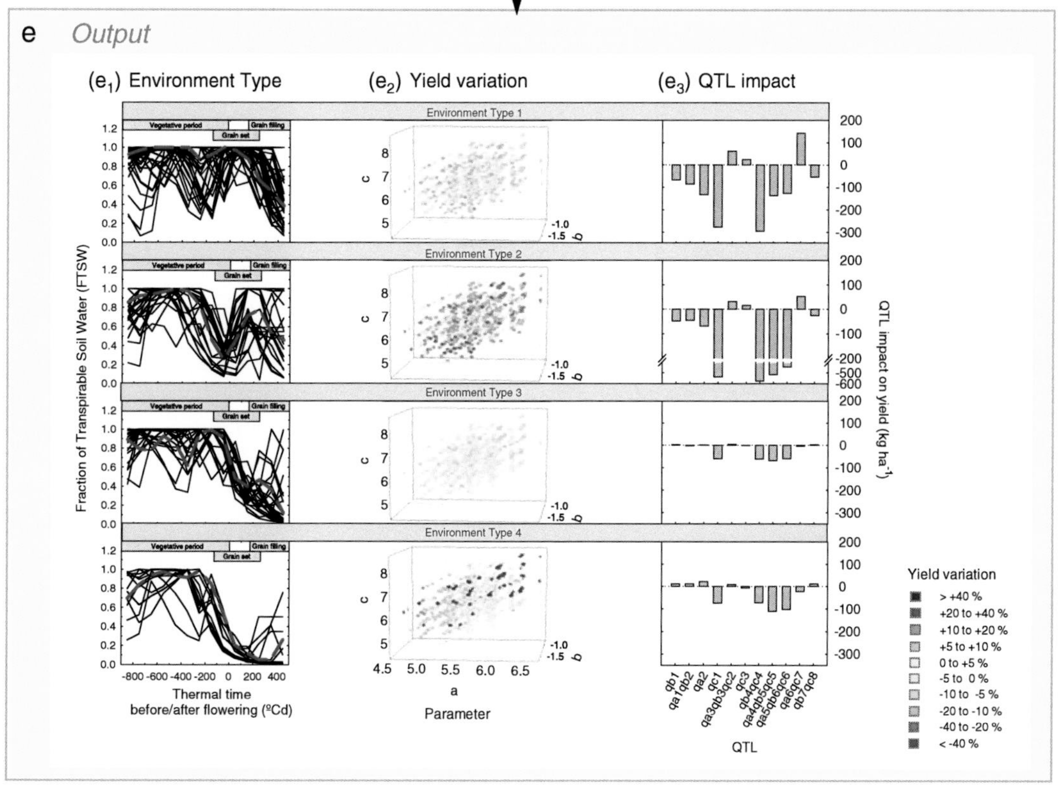

e Output
(e1) Environment Type
(e2) Yield variation
(e3) QTL impact
Environment Type 1
Environment Type 2
Environment Type 3
Environment Type 4
Vegetative period
Grain filling
Grain set
Fraction of Transpirable Soil Water (FTSW)
Thermal time
before/after flowering (°Cd)
Parameter
QTL impact on yield (kg ha-1)
QTL
qb1
qa1qb2
qa2
qc1
qa3qb3qc2
qc3
qb4qc4
qa4qb5qc5
qa5qb6qc6
qa6qc7
qb7qc8
Yield variation
> +40 %
+20 to +40 %
+10 to +20 %
+5 to +10 %
0 to +5 %
-5 to 0 %
-10 to -5 %
-20 to -10 %
-40 to -20 %
< -40 %

◀ **FIG. 14.2** 'Gene-to-phenotype' modeling to capture QTL (quantitative trait loci) effects and gene/QTL × environment interaction from organ to crop levels. Genetic knowledge of 'simple' component traits was used to parameterize the model and to infer the impact of single QTL or QTL combinations on complex traits. In this example, organ-level QTL for leaf and silk elongation of maize were inputs to a modified version of the APSIM ecophysiological model. The impact of the environmentally stable QTL was tested in different environments for 1000 recombinant lines (RILs) simulated with the quantitative genetics model QU-GENE (Podlich and Cooper, 1998) (a). The QTL were associated first with the additive effects (red dashed line, negative; blue solid line, positive) affecting leaf elongation rate (LER) response to temperature (parameter *a*), evaporative demand (parameter *b*) and soil water deficit (parameter *c*) (Reymond et al., 2003), and secondly, with assumed pleiotropic effects on silk elongation and anthesis-silking interval (ASI) under drought (Welcker et al., 2007). These responses were integrated in a leaf module of APSIM, and the response of QTL for ASI was integrated in a reproductive module of APSIM (d). Overall, the model integrated genetic (a), environmental (b) and management (c) information to account for the complex interplay of genetic, physiological and environmental controls throughout the crop cycle (d) (Chenu et al., 2008). After characterizing the drought environment types based on the FTSW (fraction of transpirable soil water) in Sete Laogas, Brazil (e_1), simulations were undertaken for four representative drought patterns (red dashed line). Genotype × environment interaction was generated for simulated yield (e_2) and the impact of the organ-level QTL highly varied depending on the environment considered (e_3). For instance, many positive-effect QTL in low/mild stress environments (Environment Types 1 and 2) had a negative impact in a severe reproductive stress environment (Environment Types 3 and 4) and vice versa. Two QTL (qa6qc7 and qa4qb5qc5) with similar effects on the LER response to temperature (parameter *a*) had contrasting effects on simulated yield (e_3) (Chenu et al., 2009). *Adapted from Chenu et al. (2008, 2009).*